AF839349

QUANTIFICATION IN NONCLASSICAL LOGIC

STUDIES IN LOGIC

AND

THE FOUNDATIONS OF MATHEMATICS

VOLUME 153

Honorary Editor:

P. SUPPES

Editors:

S. ABRAMSKY, *London*
S. ARTEMOV, *Moscow*
D.M. GABBAY, *London*
A. KECHRIS, *Pasadena*
A. PILLAY, *Urbana*
R.A. SHORE, *Ithaca*

ELSEVIER

AMSTERDAM • BOSTON • HEIDELBERG • LONDON • NEW YORK • OXFORD
PARIS • SAN DIEGO • SAN FRANCISCO • SINGAPORE • SYDNEY • TOKYO

QUANTIFICATION IN NONCLASSICAL LOGIC

VOLUME 1

D. M. Gabbay
King's College London, UK
and
Bar-Ilan University, Ramat-Gan, Israel

V. B. Shehtman
Institute for Information Transmission Problems
Russian Academy of Sciences
and
Moscow State University

D. P. Skvortsov
All-Russian Institute of Scientific and Technical Information
Russian Academy of Sciences

ELSEVIER

AMSTERDAM • BOSTON • HEIDELBERG • LONDON • NEW YORK • OXFORD
PARIS • SAN DIEGO • SAN FRANCISCO • SINGAPORE • SYDNEY • TOKYO

Elsevier
30 Corporate Drive, Suite 400, Burlington, MA 01803, USA
Radarweg 29, PO Box 211, 1000 AE Amsterdam, The Netherlands
Linacre House, Jordan Hill, Oxford OX2 8DP, UK

First edition 2009

Copyright © 2009 Elsevier B.V. All rights reserved

No part of this publication may be reproduced, stored in a retrieval system or transmitted in any form or by any means electronic, mechanical, photocopying, recording or otherwise without the prior written permission of the publisher

Permissions may be sought directly from Elsevier's Science & Technology Rights Department in Oxford, UK: phone (+44) (0) 1865 843830; fax (+44) (0) 1865 853333; email: permissions@elsevier.com. Alternatively you can submit your request online by visiting the Elsevier web site at http://elsevier.com/locate/permissions, and selecting *Obtaining permission to use Elsevier material*

Notice
No responsibility is assumed by the publisher for any injury and/or damage to persons or property as a matter of products liability, negligence or otherwise, or from any use or operation of any methods, products, instructions or ideas contained in the material herein. Because of rapid advances in the medical sciences, in particular, independent verification of diagnoses and drug dosages should be made

Library of Congress Cataloging-in-Publication Data
A catalog record for this book is available from the Library of Congress

British Library Cataloguing in Publication Data
A catalogue record for this book is available from the British Library

ISBN: 978-0-444-52012-8
ISSN: 0049-237X

For information on all Elsevier publications
visit our web site at books.elsevier.com

Preface

If some 30 years ago we had been told that we would write a large book on quantification in nonclassical logic, none of us would have taken it seriously: first — because at that time there was no hope of our effective collaboration; second — because in nonclassical logic too much had to be done in the propositional area, and few people could find the energy for active research in predicate logic.

In the new century the situation is completely different. Connections between Moscow and London became easy. The title of the book is not surprising, and we are now late with the first big monograph in this field. Indeed, at first we did not expect we had enough material for two (or more) volumes. But we hope readers will be able to learn the subject from our book and find it quite fascinating.

Let us now give a very brief overview of the existing systematic expositions of nonclassical first-order logic. None of them aims at covering the whole of this large field. The first book on the subject was [Rasiowa and Sikorski, 1963], where the approach used by the authors was purely algebraic. Many important aspects of superintuitionistic first-order logics can be found in the books written in the 1970–80s: [Dragalin, 1988] (proof theory; algebraic, topological, and relational models; realisability semantics); [Gabbay, 1981] (model theory; decision problem); [van Dalen, 1973], [Troelstra and van Dalen, 1988] (realisability and model theory). The book of Novikov [Novikov, 1977] (the major part of which is a lecture course from the 1950s) addresses semantics of superintuitionistic logics and also includes some material on modal logic.

Still, predicate modal logic was partly neglected until the late 1980s. The book [Harel, 1979] and its later extended version [Harel and Tiuryn, 2000] study particular dynamic modal logics. [Goldblatt, 1984] is devoted to topos semantics; its main emphasis is on intuitionistic logic, although modal logic is also considered. The book [Hughes and Cresswell, 1996] makes a thorough study of Kripke semantics for first-order modal logics, but it does not consider other semantics or intermediate logics. Finally, there is a monograph [Gabbay *et al.*, 2002], which, among other topics, investigates first-order modal and intermediate logics from the 'many-dimensional' viewpoint. It contains recent profound results on decidable fragments of predicate logics.

The lack of unifying monographs became crucial in the 1990s, to the extent that in the recent book [Fitting and Mendelsohn, 1998] the area of first-order modal logic was unfairly called 'under-developed'. That original book contains interesting material on the history and philosophy of modal logic, but due to its obvious philosophical flavour, it leaves many fundamental mathematical problems and results unaddressed. Still there the reader can find various approaches to quantification, tableaux systems and corresponding completeness theorems. So there remains the need for a foundational monograph not only addressing areas untouched by all current publications, but also presenting a unifying point of view.

A detailed description of this Volume can be found in the Introduction below. It is worth mentioning that the major part of the material has never been presented in monographs. One of its sources is the paper on completeness and incompleteness, a brief version of which is [Shehtman and Skvortsov, 1990]; the full version (written in 1983) has not been published for technical reasons. The second basic paper incorporated in our book is [Skvortsov and Shehtman, 1993], where so-called metaframe (or simplicial) semantics was introduced and studied. We also include some of the results obtained after 1980 by G. Corsi, S. Ghilardi, H. Ono, T. Shimura, D. Skvortsov, N.-Y. Suzuki, and others.

However, because of lack of space, we had to exclude some interesting material, such as a big chapter on simplicial semantics, completeness theorems for topological semantics, hyperdoctrines, and many other important matters. Other important omissions are the historical and the bibliographical overviews and the discussion of application fields and many open problems. Moreover, the cooperation between the authors was not easy, because of the different viewpoints on the presentation.[1] There may be also other shortcomings, like gaps in proofs, wrong notation, wrong or missing references, misprints etc., that remain uncorrected — but this is all our responsibility.

We would be glad to receive comments and remarks on all the defects from the readers. As we are planning to continue our work in Volume 2, we still hope to make all necessary corrections and additions in the real future.

At present the reader can find the list of corrections on our webpages
http://www.dcs.kcl.ac.uk/staff/dg/
http://lpcs.math.msu.su/~shehtman

Acknowledgements

The second and the third author are grateful to their late teacher and friend Albert Dragalin, one of the pioneers in the field, who stimulated and encouraged their research.

We thank all our colleagues, with whom we discussed the contents of this book at different stages — Sergey Artemov, Lev Beklemishev, Johan van Benthem, Giovanna Corsi, Leo Esakia, Silvio Ghilardi, Dick De Johng, Rosalie Iemhoff, Marcus Kracht, Vladimir Krupski, Grigori Mints, Hiroakira Ono, Nobu-Yuki Suzuki, Albert Visser, Michael Zakharyaschev. Nobu-Yuki and Marcus also kindly sent us Latex-files of their papers; we used them in the process of typing the text.

[1]One of the authors points out that he disagrees with some of the notation and the style of some proofs in the final version.

We would like to thank different institutions for help and support — King's College of London; Institute for Information Transmission Problems, VINITI, Department of Mathematical Logic and Theory of Algorithms at Moscow State University, Poncelet Mathematical Laboratory, Steklov Mathematical Institute in Moscow; IRIT in Toulouse; EPSRC, RFBR, CNRS, and NWO.

We add personal thanks to our teacher Professor Vladimir A. Uspensky, to Academician Sergey I. Adian, and to our friend and colleague Michael Tsfasman who encouraged and supported our work.

We are very grateful to Jane Spurr for her enormous work and patience in preparation of the manuscript — typing the whole text in Latex (several times!), correcting our mistakes, reading multiple pages (with hardly understandable handwriting), making pictures, arranging styles etc. etc. We thank all those people who also essentially helped us in this difficult process — Ilya Shapirovsky, Alexey Romanov, Stanislav Kikot for correcting mistakes and typing, Ilya Vorontsov and Daniel Vorontsov for scanning many hundreds of pages.

We thank all other people for their help and encouragement — our wives Lydia, Marina, Elena, families, friends and colleagues.

Introduction

Quantification and modalities have always been topics of great interest for logicians. These two themes emerged from philosophy and language in ancient times; they were studied by traditional informal methods until the 20th century. Then the tools became highly mathematical, as in the other areas of logic, and modal logic as well as quantification (mainly on the basis of classical first-order logic) found numerous applications in Computer Science.

At the same time many other kinds of nonclassical logics were investigated. In particular, intuitionistic logic was created by L. Brouwer at the beginning of the century as a new basis for mathematical reasoning. This logic, as well as its extensions (superintuitionistic logics), is also very useful for Computer Science and turns out to be closely related to modal logics.

(A) The introduction of quantifier axioms to classical logic is fairly straightforward. We simply add the following obvious postulates to the propositional logic:

1. $\forall x A(x) \supset A(t)$,
   where t is 'properly' substituted for x

2. $\dfrac{A \supset B(x)}{A \supset \forall x B(x)}$,
   where x is not free in A

3. $A(t) \supset \exists x A(x)$,

4. $\dfrac{B(t) \supset A}{\exists x B(x) \supset A}$,
   where x is not free in A.

The passage from the propositional case of a logic L to its quantifier case works for many logics by adding the above axioms to the respective propositional axioms — for example, the intuitionistic logic, standard modal logics **S4**, **S5**, **K** etc. We may need in some cases to make some adjustment to account for constant domains, $\forall x(A \vee B(x)) \supset (A \vee \forall x B(x))$ in case of intuitionistic logic and the Barcan formula, $\forall x \Box A(x) \supset \Box \forall x A(x)$ in the case of modal logics. On the whole the correspondence seems to be working.

The recipe goes on as follows:
take the propositional semantics and put a domain D_u in each world u or take the axiomatic formulation and add the above axioms and you maintain correspondence and completeness.

There were some surprises however. Unexpectedly, this method fails for very simple and well-known modal and intermediate logics: the 'Euclidean logic' $\mathbf{K5} = \mathbf{K} + \Diamond\Box p \supset \Box p$ (see Chapter 6 of this volume), the 'confluence logic' $\mathbf{S4.2} = \mathbf{S4} + \Diamond\Box p \supset \Box\Diamond p$ and for the intermediate logic $\mathbf{KC} = \mathbf{H} + \neg p \vee \neg\neg p$ with constant domains, nonclassical intermediate logics of finite depth [Ono, 1983], etc. All these logics are incomplete in the standard Kripke semantics.

In some other cases, completeness theorems hold, but their proofs require nontrivial extra work — for example, this happens for the logic of linear Kripke frames **S4.3** [Corsi, 1989].

This situation puts at least two difficult questions to us: (1) how should we change semantics in order to restore completeness of 'popular' logics? (2) how should we extend these logics by new axioms to make them complete in the standard Kripke semantics? These questions will be studied in our book, especially in Volume 2, but we are still very far from final answers.

Apparently when we systematically introduce natural axioms and ask for the corresponding semantics, we may not be able to see what are the natural semantical conditions (which may not be expressible in first-order logic) and conversely some natural conditions on the semantics require complex and sometimes non-axiomatisable logics.

The community did not realize all these difficulties. A serious surprise was the case of relevance logic, where the additional axioms were complex and seemed purely technical. See [Mares and Goldblatt, 2006], [Fine, 1988], [Fine, 1989]. For some well-known logics there were no attempts of going first-order, especially for resource logics such as Lambek Calculus.

(B) There are other reasons why we may have difficulties with quantifiers, for example, in the case of superintuitionistic logics. Conditions on the possible worlds such as discrete ordering or finiteness may give the connectives themselves quantificational power of their own (note that the truth condition for $A \supset B$ has a hidden world quantifier), which combined with the power of the explicit quantifiers may yield some pretty complex systems [Skvortsov, 2006].

(C) In fact, a new approach is required to deal with quantifiers in possible world systems. The standard approach associates domains with each possible world and what is in the domain depends only on the nature of the world, i.e. if u is a world, P a predicate, θ a valuation, then $\theta_u(P)$ is not dependent on other $\theta_{u'}(P)$, except for some very simple conditions as in intuitionistic logic.

There are no interactive conditions between existence of elements in the domain and satisfaction in other domains. If we look at some axioms like the Markov principle

$$\neg\neg\exists x A(x) \supset \exists x \neg\neg A(x),$$

we see that we need to pay attention on how the domain is constructed. This is reminiscent of the Herbrand universe in classical logic.

(D) There are other questions which we can ask. Given a classical theory Γ (e.g. a theory of rings or Peano arithmetic), we can investigate what happens if we change the underlying logic to intuitionistic or modal or relevant. Then what kind of theory do we get and what kind of semantics? Note we are not dealing now with a variety of logics (modal or superintuitionistic), but with a fixed nonclassical logic (say intuitionistic logic itself) and a variety of theories.

If intuitionistic predicate semantics is built up from classical models, would the intuitionistic predicate theory of rings have semantics built up from classical rings? How does it depend on the formulation (Γ may be classically equivalent to Γ', but not intuitionistically) and what can happen to different formulations? See [Gabbay, 1981].

(E) One can have questions with quantifiers arising from a completely different angle. E.g. in resource logics we pay attention to which assumptions are used to proving a formula A.

For example in linear or Lambek logic we have that

(1) A

(2) A

(3) $A \to (A \to B)$.

can prove B but (2) and (3) alone cannot prove B; because of resource considerations, we need two copies of A. Such logics are very applicable to the analysis and modelling of natural language [van Benthem, 1991]. So what shall we do with $\forall x A(x)$? Do we divide our resource between all instances $A(t_1), A(t_2), \ldots$ of A? These are design questions which translate into technical axiomatic and semantical questions.

How do we treat systems which contain more than one type of nonclassical connective? Any special problems with regard to adding quantifiers? See, for example, the theory of bunched implications [O'Hearn and Pym, 1999].

(F) The most complex systems with regards to quantifiers are LDS, Labelled Deductive Systems (this is a methodology for logic, cf. [Gabbay, 1996; Gabbay, 1998]). In LDS formulas have labels, so we write $t : A$, where t is a label and A is a formula. Think of t as a world or a context. (This label can be integrated and in itself be a formula, etc.) Elements now have visa rules for migrating between labels and need to be annotated, for example as a_s^t, the element a exists at world s, but was first created (or instantiated) in world t. Surprisingly, this actually helps with the proof theory and semantics for quantifiers, since part of the semantics is brought into the syntax. See [Viganò, 2000]. So it is easier to develop, say, theories of Hilbert ε-symbol using labels. ε-symbols axioms cannot be added simple mindedly to intuitionistic logic, it will collapse [Bell, 2001].

(G) Similarly, we must be careful with modal logic. We have not even begun thinking about ε-symbols in resource logics (consider $\varepsilon x.A(x)$, if there is sensitivity for the number of copies of A, then are we to be sensitive also to copies of elements?).

(H) In classical logic there is another direction to go with quantifiers, namely the so-called generalised quantifiers, for example (many $x)A(x)$ ('there are many

x such that $A(x)$'), or (uncountably many $x)A(x)$ or many others. Some of these can be translated as modalities as van Lambalgen has shown [Alechina, van Lambalgen, 1994], [van Lambalgen, 1996]. Such quantifiers (at least for the finite case) exist in natural language. They are very important and they have not been exported yet to nonclassical logics (only through the modalities e.g. $\Diamond_n A$ ('A is true in n possible worlds'), see [Gabbay, Reynolds and Finger, 2000], [Peters and Westerstahl, 2006]).

Volume 1 of these books concentrates on the landscape described in (A) above, i.e., correspondence between axioms for modal or intuitionistic logic and semantical conditions and vice versa.

Even for such seemingly simple questions we have our hands full. The table of contents for future volumes shows what to be addressed in connection with (B)-(H). It is time for nonclassical logic to pay full attention to quantification. Up to now the focus was mainly propositional. Now the era of the quantifier has begun!

This Volume includes results in nonclassical first-order logic obtained during the past 40 years. The main emphasis is model-theoretic, and we confine ourselves with only two kinds of logics: modal and superintuitionistic. Thus many interesting and important topics are not included, and there remains enough material for future volumes and future authors.

Figure 1. Chapters dependency structure

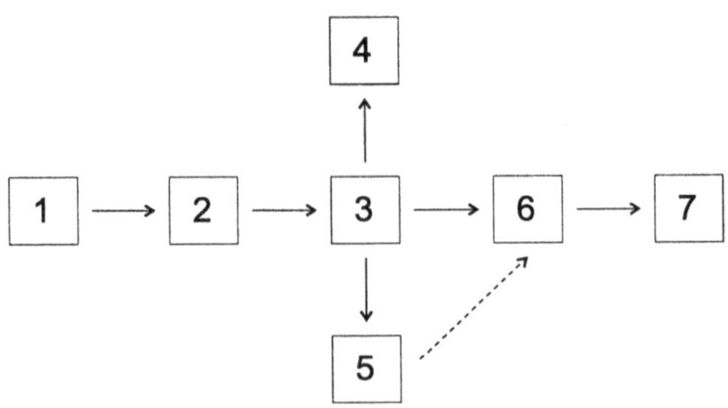

Let us now briefly describe the contents of Volume 1. It consists of three parts. Part I includes basic material on propositional logic and first-order syntax.

Chapter 1 contains definitions and results on syntax and semantics of nonclassical propositional logics. All the material can be found elsewhere, so the proofs are either sketched or skipped.

Chapter 2 contains the necessary syntactic background for the remaining parts of the book. Our main concern is the precise notion of substitution based on re-naming of variables. This classical topic is well known to all students in logic. However none of the existing definitions fits well for our further purposes, because in some semantics soundness proofs may be quite intricate. Our approach is based on the idea that re-naming of bound variables creates different synonymous (or 'congruent') versions of the same predicate formula. These versions are generated by a 'scheme' showing the reference structure of quantifiers. (Schemes are quite similar to formulas in the sense of [Bourbaki, 1968].) Now variable substitutions (acting on schemes or congruence classes) can be easily arranged in an appropriate congruent version. After this preparation we introduce two main types of first-order logics to be studied in the book — modal and superintuitionistic, and prove syntactic results that do not require involved proof theory, such as deduction theorems, Glivenko theorem etc.

In Part II (Chapters 3 – 5) we describe different semantics for our logics and prove soundness results.

Chapter 3 considers the simplest kinds of relational semantics. We begin with the standard Kripke semantics and then introduce two its generalisations, which are equivalent: Kripke frames with equality and Kripke sheaves. The first one (for the intuitionistic case) is due to [Dragalin, 1973], and the second version was first introduced in [Shehtman and Skvortsov, 1990]. Soundness proofs in that chapter are not obvious, but rather easy. We mention simple incompleteness results showing that Kripke semantics is weaker than these generalisations. Further incompleteness theorems are postponed until Volume 2. We also prove results on Löwenheim – Skolem property and recursive axiomatisability using translations to classical logic from [Ono, 1972/73] and [van Benthem, 1983].

Chapter 4 studies algebraic semantics. Here the main objects are Heyting-valued (or modal-valued) sets. In the intuitionistic case this semantics was studied by many authors, see [Dragalin, 1988], [Fourman and Scott, 1979], [Goldblatt, 1984]. Nevertheless, our soundness proof seems to be new. Then we show that algebraic semantics can be also obtained from presheaves over Heyting (or modal) algebras. We also show that for the case of topological spaces the same semantics is given by sheaves and can be defined via so-called 'fibrewise models'. These results were first stated in [Shehtman and Skvortsov, 1990], but the proofs have never been published so far.[2] They resemble the well-known results in topos theory, but do not directly follow from them.

In Chapter 5 we study Kripke metaframes, which are a many-dimensional generalisation of Kripke frames from [Skvortsov and Shehtman, 1993] (where they were called 'Cartesian metaframes'). The crucial difference between frames and metaframes is in treatment of individuals. We begin with two particular cases of Kripke metaframes: Kripke bundles [Shehtman and Skvortsov, 1990] and C-sets (sheaves of sets over (pre)categories) [Ghilardi, 1989]. Their predecessor in philosophical logic is 'counterpart theory' [Lewis, 1968]. In a Kripke bundle individuals may have several 'inheritors' in the same possible world, while in a C-set instead of an inheritance relation there is a family of maps. In

[2]The first author is happy to fulfill his promise given in the preface of [Gabbay, 1981]: "It would require further research to be able to present a general theory [of topological models, second order Beth and Kripke models] possibly using sheaves".

Kripke metaframes there are additional inheritance relations between tuples of individuals.

The proof of soundness for metaframes is rather laborious (especially for the intuitionistic case) and is essentially based on the approach to substitutions from Chapter 2. This proof has never been published in full detail. Then we apply soundness theorem to Kripke bundle and functor semantics. The last section of Chapter 5 gives a brief introduction to an important generalisation of metaframe semantics – so called 'simplicial semantics'. The detailed exposition of this semantics is postponed until Volume 2.

Part III (Chapters 6–7) is devoted to completeness results in Kripke semantics. In Kripke semantics many logics are incomplete, and there is no general powerful method for completeness proofs, but still we describe some approaches.

In Chapter 6 we study Kripke frames with varying domains. First, we introduce different types of canonical models. The simplest kind is rather well-known, cf. [Hughes and Cresswell, 1996], but the others are original (due to D. Skvorstov). We prove completeness for intermediate logics of finite depth [Yokota, 1989], directed frames [Corsi and Ghilardi, 1989], linear frames [Corsi, 1992]. Then we elucidate the methods from [Skvortsov, 1995] for axiomatising some 'tabular' logics (i.e., those with a fixed frame of possible worlds).

Chapter 7 considers logics with constant domains. We again present different canonical models constructions and prove completeness theorems from [Hughes and Cresswell, 1996]. Then we prove general completeness results for subframe and cofinal subframe logics from [Tanaka and Ono, 1999], [Shimura, 1993], [Shimura, 2001], Takano's theorem on logics of linearly ordered frames [Takano, 1987] and other related results.

Here are chapter headings in preparation for later volumes:

Chapter 8. Simplicial semantics

Chapter 9. Hyperdoctrines

Chapter 10. Completeness in algebraic and topological semantics

Chapter 11. Translations

Chapter 12. Definability

Chapter 13. Incompleteness

Chapter 14. Simulation of classical models

Chapter 15. Applications of semantical methods

Chapter 16. Axiomatisable logics

Chapter 17. Further results on Kripke-completeness

Chapter 18. Fragments of first-order logics

Chapter 19. Propositional quantification

Chapter 20. Free logics

Chapter 21. Skolemisation

Chapter 22. Conceptual quantification

Chapter 23. Categorical logic and toposes

Chapter 24. Quantification in resource logic

Chapter 25. Quantification in labelled logics.

Chapter 26. ε-symbols and variable dependency

Chapter 27. Proof theory

Some guidelines for the readers. Reading of this book may be not so easy. Parts II, III are the most important, but they cannot be understood without Part I.

For the readers who only start learning the field, we recommend to begin with sections 1.1–1.5, then move to sections 2.1, 2.2, the beginning parts of sections 2.3, 2.6, and next to 2.16. After that they can read Part II and sometimes go back to Chapters 1, 2 if necessary. We do not recommend them to go to Chapter 5 before they learn about Kripke sheaves. Those who are only interested in Kripke semantics can move directly from Chapter 3 to Part III.

An experienced reader can look through Chapter 1 and go to sections 2.1–2.5 and the basic definitions in 2.6, 2.7. Then he will be able to read later Chapters starting from Chapter 3.

Notation convention

We use logical symbols both in our formal languages and in the meta-language. The notation slightly differs, so the formal symbols $\wedge$, $\supset$, $\equiv$ correspond to the metasymbols &, $\Rightarrow$, $\Leftrightarrow$; and the formal symbols $\vee$, $\exists$, $\forall$ are also used as metasymbols.

In our terminology we distinguish functions and maps. A *function* from A to B is a binary relation $F \subseteq A \times B$ with domain A satisfying the functionality condition $(xFy \;\&\; xFz \Rightarrow x = z)$, and the triple $f = (F, A, B)$ is then called a *map from A to B*. In this case we use the notation $f : A \longrightarrow B$.

Here is some other set-theoretic notation and terminology.

- 2^X denotes the power set of a set X;
- we use $\subseteq$ for inclusion, $\subset$ for proper inclusion;
- $R \circ S$ denotes the composition of binary relations R and S:

$$R \circ S := \{(x, y) \mid \exists z\ (xRz \;\&\; zSy)\};$$

- R^{-1} is the converse of a relation R;
- Id_W is the equality relation in a set W;
- id_W is the identity map on a set W (i.e. the triple (Id_W, W, W));
- for a relation $R \subseteq W \times W$, $R(V)$, or just RV, denotes the *image* of a set $V \subseteq W$ under R, i.e. $\{y \mid \exists x \in V\ xRy\}$; $R(x)$ or Rx abbreviates $R(\{x\})$;
- $\mathrm{dom}(R)$, or $\mathrm{pr}_1(R)$, denotes the *domain* of a relation R, i.e., $\{x \mid \exists y\ xRy\}$;
- $\mathrm{rng}(R)$, or $\mathrm{pr}_2(R)$, denotes the *range* of a relation R, i.e., $\{y \mid \exists x\ xRy\}$;
- for a subset $X \subseteq Y$ there is the *inclusion map* $j_{XY} : X \longrightarrow Y$ (which is usually denoted just by j) sending every $x \in X$ to itself;
- $R \upharpoonright V$ denotes the restriction of a relation R to a subset V, i.e. $R \upharpoonright V = R \cap (V \times V)$, and $f \upharpoonright V$ denotes the restriction of a map f to V;
- for a relation R on a set X $R^- := R - Id_X$ is the 'irreflixivisation' of R;
- $|X|$ denotes the cardinality of a set X;
- I_n denotes the set $\{1, \ldots, n\}$; $I_0 := \varnothing$;
- X^∞ denotes the set of all finite sequences with elements in X;
- $(X_i \mid i \in I)$ (or $(X_i)_{i \in I}$) denotes the family of sets X_i with indices in the set I;
- $\bigsqcup\limits_{i \in I} X_i$ denotes the *disjoint union* of the family $(X_i)_{i \in I}$, i.e. $\bigcup\limits_{i \in I} X_i \times \{i\}$;

- ω is the set of natural numbers, and T_ω denotes ω^∞;
- $\Sigma_{mn} = (I_n)^{I_m}$ denotes the set of all maps $\sigma : I_m \longrightarrow I_n$ (for $m, n \in \omega$);
- Υ_{mn} denotes the set of all injective maps in Σ_{mn};
- Υ_n is the abbreviation for Υ_{nn}, the set of all permutations of I_n.

Note that we use two different notations for composition of maps: the composition of $f : A \longrightarrow B$ and $g : B \longrightarrow C$ is denoted by either $g \cdot f$ or $f \circ g$. So $(f \circ g)(x) = (g \cdot f)(x) = g(f(x))$.

Obviously,
$\Sigma_{mn} \neq \varnothing$ iff $n > 0$ or $m = 0$,
$\Upsilon_{mn} \neq \varnothing$ iff $n \geq m$.

A map $f : I_m \longrightarrow I_n$ (for fixed n) is presented by the table

$$\begin{pmatrix} 1 & \dots & m \\ f(1) & \dots & f(m) \end{pmatrix}$$

We use a special notation for some particular maps.

- *Transpositions* $\sigma^n_{ij} \in \Upsilon_n$ for $n \geq 2$, $1 \leq i < j \leq n$.

$$\sigma^n_{ij} := \begin{pmatrix} 1 & \dots & i & \dots & j & \dots & n \\ 1 & \dots & j & \dots & i & \dots & n \end{pmatrix}$$

  In particular, *simple transpositions* are $\sigma^n_i := \sigma^n_{1i}$ for $1 < i \leq n$;

- *Standard embeddings (inclusion maps).*
  $\sigma^{mn}_+ \in \Upsilon_{mn}$, for $0 \leq m \leq n$ is defined by the table

$$\begin{pmatrix} 1 & \dots & m \\ 1 & \dots & m \end{pmatrix}$$

  In particular, there are *simple embeddings* $\sigma^m_+ := \sigma^{m,m+1}_+$ for $m \geq 0$; $\varnothing_n := \sigma^{0n}_+$ is the empty map $I_0 \longrightarrow I_n$ (and obviously, $\Sigma_{0n} = \{\varnothing_n\}$).

- *Facet embeddings* $\delta^n_i \in \Upsilon_{n-1,n}$ for $n > 0$.

$$\delta^n_i := \begin{pmatrix} 1 & \dots & i-1 & i & \dots & n-1 \\ 1 & \dots & i-1 & i+1 & \dots & n \end{pmatrix}$$

  In particular, $\delta^n_n = \sigma^{n-1}_+$.

- *Standard projections* $\sigma^{mn}_- \in \Sigma_{mn}$ for $m \geq n > 0$.

$$\sigma^{mn}_- := \begin{pmatrix} 1 & \dots & n & \dots & m \\ 1 & \dots & n & \dots & n \end{pmatrix}$$

  In particular, *simple projections* are $\sigma^n_- := \sigma^{n+1,n}_-$ for $n > 0$.

It is well-known that (for $n > 1$) every permutation $\sigma \in \Upsilon_n$ is a composition of (simple) transpositions. One also can easily show that every map from Σ_{mn} is a composition of simple transpositions, simple embeddings, and simple projections. In particular, every injection (from Υ_{mn}) is a composition of simple transpositions and simple embeddings, and every surjection is a composition of simple transpositions and simple projections, cf. [Gabriel and Zisman, 1967].

The *identity map* in Σ_{nn} is $id_n := id_{I_n} = \sigma_+^{nn} = \sigma_-^{nn}$, and it is obvious that $id_n = \sigma_{ji}^n \circ \sigma_{ji}^n$ whenever $n \geq 2$, $j < i$.

Let also $\lambda_i^n \in \Sigma_{1n}$ be the map sending 1 to i; let $\lambda_{ij}^n \in \Sigma_{2n}$ be the map with the table

$$\begin{pmatrix} 1 & 2 \\ i & j \end{pmatrix}$$

For every $\sigma \in \Sigma_{mn}$ we define its *simple extension* $\sigma^+ \in \Sigma_{m+1,n+1}$ such that

$$\sigma^+(i) := \begin{cases} \sigma(i) & \text{for } i \in I_m, \\ n+1 & \text{if } i = m+1. \end{cases}$$

In particular, for any n we have $(\sigma_+^n)^+ = \delta_{n+1}^{n+2} \in \Sigma_{n+1,n+2}$:

$$(\sigma_+^n)^+(i) = \begin{cases} i & \text{for } i \in I_n, \\ n+2 & \text{if } i = n+1. \end{cases}$$

We do not make any difference between words of length n in an alphabet D and n-tuples from D^n. So we write down a tuple $(a_1, \ldots, a_n)$ also as $a_1 \ldots a_n$.

- $\curlywedge$ denotes the void sequence;
- $l(|\alpha|)$ (or $|a|$) denotes the length of a sequence α;
- $\alpha\beta$ denotes the join (the concatenation) of sequences α, β; we often write $x_1 \ldots x_n$ rather than $(x_1, \ldots, x_n)$ (especially if $n = 1$), and also αx or (α, x) rather than the dubious notation $\alpha(x)$;
- For a letter c put
$$c^k := \underbrace{c \ldots c}_{k}.$$

For an arbitrary set S, every tuple $\mathbf{a} = (a_1, \ldots, a_n) \in S^n$ can be regarded as a function $I_n \longrightarrow S$. We usually denote the range of this function, i.e. the set $\{a_1, \ldots, a_n\}$ as $r(\mathbf{a})$. Sometimes we write $b \in \mathbf{a}$ instead of $b \in r(\mathbf{a})$. Every map $\sigma : I_m \longrightarrow I_n$ acts on S^n via composition:

$$\mathbf{a} \cdot \sigma = \sigma \circ \mathbf{a} = (a_{\sigma(1)}, \ldots, a_{\sigma(m)}).$$

Thus every map $\sigma \in \Sigma_{mn}$ gives rise to the map $\pi_\sigma : S^n \longrightarrow S^m$ sending $\mathbf{a}$ to $\mathbf{a} \cdot \sigma$. In the particular case, when $\sigma = \delta_i^n$ is a facet embedding and $\mathbf{a} \in S^n$, we also use the notation $\pi_i^n := \pi_{\delta_i^n}$ and

$$\pi_i^n \mathbf{a} := \mathbf{a} - a_i := \widehat{\mathbf{a}}_i := \mathbf{a} \cdot \delta_i^n = (a_1, \ldots, a_{i-1}, a_{i+1}, \ldots, a_n).$$

Hence we obtain

Lemma 0.0.1 *(1)*

$$\pi_\tau \cdot \pi_\sigma = \pi_{\sigma \cdot \tau},$$

whenever $\mathbf{a} \in S^n$, $\sigma \in \Sigma_{mn}$, $\tau \in \Sigma_{km}$.

(2) If σ *is a permutation* ($\sigma \in \Upsilon_n$)*, then* π_σ *is a permutation of* S^n *and* $\pi_{\sigma^{-1}} = (\pi_\sigma)^{-1}$.

Proof (1) Since composition of maps is associative, we have

$$\mathbf{a} \cdot (\sigma \cdot \tau) = (\mathbf{a} \cdot \sigma) \cdot \tau.$$

(2) $\pi_\sigma \cdot \pi_{\sigma^{-1}} = \pi_{\sigma^{-1}} \cdot \pi_\sigma = \pi_{id_n} = id_{S^n}$. ■

We use the following relations on n-tuples:

$$(\sigma) \qquad \mathbf{a}\, sub\, \mathbf{b} \text{ iff } \forall i, j\, (a_i = a_j \Rightarrow b_i = b_j).$$

Lemma 0.0.2 *Let* $S \neq \varnothing$, $\sigma \in \Sigma_{mn}$. *Then*

$$\pi_\sigma[S^n] = \{\mathbf{a} \in S^m \mid \sigma\, sub\, \mathbf{a}\},$$

where $\sigma\, sub\, \mathbf{a}$ *denotes the property* $\forall i, j\ (\sigma(i) = \sigma(j) \Rightarrow a_i = a_j)$, *cf.* (σ).

Proof In fact, if $\mathbf{a} = \mathbf{b} \cdot \sigma$, then obviously $\sigma(j) = \sigma(k)$ implies $a_j = a_k$. On the other hand, if $\sigma\, sub\, \mathbf{a}$, then $\mathbf{a} = \mathbf{b} \cdot \sigma$ for some $\mathbf{b}$; just put $b_{\sigma(i)} := a_i$ and add arbitrary b_k for $k \notin r(\sigma)$. ■

Lemma 0.0.3 *For* $|S| > 1$, $\sigma \in \Sigma_{mn}$, σ *is injective iff* $\pi_\sigma : S^n \longrightarrow S^m$ *is surjective.*[3]

Proof If σ is injective, then for any $\mathbf{a} \in S^n$, $\sigma(i) = \sigma(j) \Rightarrow i = j \Rightarrow a_i = a_j$, i.e. $\sigma\, sub\, \mathbf{a}$. Hence by Lemma 0.0.2, π_σ is surjective.

The other way round, if $\sigma(i) = \sigma(j)$ for some $i \neq j$, take $\mathbf{a} \in S^m$ such that $a_i \neq a_j$. Then $\sigma\, sub\, \mathbf{a}$ is not true, i.e. $\mathbf{a} \notin \pi_\sigma[S^n]$ by 0.0.2. ■

Lemma 0.0.4 *For* $|S| > 1$, $\sigma \in \Sigma_{mn}$, σ *is surjective iff* π_σ *is injective.*

Proof Suppose $\sigma : I_m \longrightarrow I_n$ is surjective and $\mathbf{a}, \mathbf{b} \in S^n$, $\pi_\sigma \mathbf{a} \neq \pi_\sigma \mathbf{b}$. If $\pi_\sigma \mathbf{a}$ and $\pi_\sigma \mathbf{b}$ differ at the jth component, then $a_i \neq b_i$ for $i \in I_n$ such that $\sigma(i) = j$. On the other hand, let $\sigma \in \Sigma_{mn}$ be non-surjective, $j \in I_m - rng(\sigma)$. Let $c, d \in S$, $c \neq d$. Take $\mathbf{a} = c^n$; $\mathbf{b} = c^{j-1} d c^{n-j}$. Then $\mathbf{a} \neq \mathbf{b}$ and $\pi_\sigma \mathbf{a} = \pi_\sigma \mathbf{b} = c^m$. ■

Hence we obtain

Lemma 0.0.5 *For* $|S| > 1$, $\sigma \in \Sigma_{mn}$, σ *is bijective iff* $\pi_\sigma : S^n \longrightarrow S^m$ *is bijective.*

---

[3] Clearly, if $|S| = 1$, then π_σ is bijective for every $\sigma \in \Sigma_{mn}$.

We further simplify notation in some particular cases. Let $\pi_i^n := \pi_{\delta_i^n}$, so facet embedding δ_i^n eliminates the i th component from an n-tuple $\mathbf{a} \in S^n$. Let also

$$\pi_-^n := \pi_{\sigma_-^n},\ \pi_+^n := \pi_{\sigma_+^n},$$

where $\sigma_-^n \in \Sigma_{n+1,n}$ is a simple projection, $\sigma_+^n \in \Sigma_{n,n+1}$ is a simple embedding. Thus

$$\pi_-^n(a_1, \ldots, a_n) = (a_1, \ldots, a_n, a_n) \text{ for } n > 0,$$
$$\pi_+^n(\mathbf{a}) = \mathbf{a} - a_{n+1} = (a_1, \ldots, a_n) \text{ for } \mathbf{a} = (a_1, \ldots, a_n, a_{n+1}) \in D^{n+1},\ n \geq 0.$$

We say that a sequence $\mathbf{a} \in D^n$ is *distinct*, if all its components a_i are different.

Lemma 0.0.6 *If* $\sigma, \tau : I_m \longrightarrow I_n$, $\sigma \neq \tau$ *and* $|S| \geq n$, *then* $\mathbf{a} \cdot \sigma \neq \mathbf{a} \cdot \tau$ *for any distinct* $\mathbf{a} \in S^n$.

Proof If for some i, $\tau(i) \neq \sigma(i)$, then $a_{\sigma(i)} \neq a_{\tau(i)}$. ■

Lemma 0.0.7 *(1) For* $\tau \in \Sigma_{mn}$, $\sigma \in \Sigma_{km}$,
$(\tau \cdot \sigma)^+ = \tau^+ \cdot \sigma^+$.

(2) For $\sigma \in \Sigma_{mn}$,
$\sigma^+ \cdot \sigma_+^m = \sigma_+^n \cdot \sigma$.

Proof Straightforward. ■

Lemma 0.0.8 *(1) Let* $\mathbf{a} \in S^n$, $\mathbf{b} \in S^m$, $r(\mathbf{b}) \subseteq r(\mathbf{a})$. *Then* $\mathbf{b} = \mathbf{a} \cdot \sigma$ *for some* $\sigma \in \Sigma_{mn}$.

(2) Moreover, if **b** *is distinct,*[4] *then* σ *is an injection.*

Proof Put $\sigma(i) = j$ for some j such that $b_i = a_j$. ■

[4]In other words, **b** is obtained by renumbering a subsequence of **a**.

Contents

Part I

Preliminaries

Chapter 1

Basic propositional logic

This chapter contains necessary information about propositional logics. We give all the definitions and formulate results, but many proofs are sketched or skipped. For more details we address the reader to textbooks and monographs in propositional logic: [Goldblatt, 1987], [Chagrov and Zakharyaschev, 1997], [Blackburn, de Rijke and Venema, 2001], also see [Gabbay, 1981], [Dragalin, 1988], [van Benthem, 1983].

1.1 Propositional syntax

1.1.1 Formulas

We consider *N-modal (propositional) formulas*[1] built from the denumerable set $PL = \{p_1, p_2, \dots\}$ of *proposition letters*, the *classical propositional connectives* $\wedge$, $\vee$, $\supset$, $\bot$ and the unary *modal connectives* $\Box_1, \dots, \Box_N$; the derived connectives are introduced in a standard way as abbreviations:

$$\begin{array}{l} \neg A := (A \supset \bot),\ \top := (\bot \supset \bot), \\ (A \equiv B) := ((A \supset B) \wedge (B \supset A)), \\ \Diamond_i A := \neg\Box_i\neg A \quad \text{for } i = 1, \dots, N. \end{array}$$

To simplify notation, we write p, q, r instead of p_1, p_2, p_3. We also use standard agreements about bracketing: the principal brackets are omitted; $\wedge$ is stronger than $\vee$, which is stronger than $\supset$ and $\equiv$. Sometimes we use dots instead of brackets; so, e.g. $A \supset_{\bullet} B \supset C$ stands for $(A \supset (B \supset C))$.

For a sequence of natural numbers $\alpha = k_1 \dots k_r$ from I_N^∞, $\Box_\alpha$ abbreviates $\Box_{k_1} \dots \Box_{k_r}$. $\Box_\lambda$ denotes the identity operator, i.e. $\Box_\lambda A = A$ for every formula A. If $\alpha = \underbrace{k \dots k}_{r}$, $\Box_\alpha$ is also denoted by $\Box_k^r$ (for $r \geq 0$).

Similarly, we use the notations $\Diamond_\alpha$, $\Diamond_k^r$.

[1] 1-modal formulas are also called *monomodal*, 2-modal formulas are called *bimodal*. Some authors prefer the term 'unimodal' to 'monomodal'.

As usual, for a finite set of formulas Γ, $\bigwedge\Gamma$ denotes its conjunction and $\bigvee\Gamma$ its disjunction; the empty conjunction is $\top$ and the empty disjunction is $\bot$. We also use the notation (for arbitrary Γ)

$$\Box_\alpha\Gamma := \{\Box_\alpha A \mid A \in \Gamma\}.$$

If $n = 1$, we write $\Box$ instead of $\Box_1$ and $\Diamond$ instead of $\Diamond_1$.

The *degree* (or the *depth*) of a modal formula A (denoted by $d(A)$) is defined by induction:

$$\begin{aligned}
&d(p_k) = d(\bot) = 0,\\
&d(A \wedge B) = d(A \vee B) = d(A \supset B) = \max\,(d(A), d(B)),\\
&d(\Box_i A) = d(A) + 1.
\end{aligned}$$

$\mathcal{LP}_N$ denotes the set of all N-modal formulas; $\mathcal{LP}_0$ denotes the set of all formulas without modal connectives; they are called *classical* (or *intuitionistic*[2]).

An *N-modal (propositional) substitution* is a map $S : \mathcal{LP}_N \longrightarrow \mathcal{LP}_N$ preserving $\bot$ and all but finitely many proposition letters and commuting with all connectives, i.e. such that

- $S(\bot) = \bot$;
- $\{k \mid S(p_k) \neq p_k\}$ is finite;
- $S(A \wedge B) = S(A) \wedge S(B)$;
- $S(A \vee B) = S(A) \vee S(B)$;
- $S(A \supset B) = S(A) \supset S(B)$;
- $S(\Box_i A) = \Box_i S(A)$.

Let $q_1, \ldots, q_k$ be different proposition letters. A substitution S such that $S(q_i) = A_i$ for $i \leq k$ and $S(q) = q$ for any other $q \in PL$, is denoted by $[A_1, \ldots, A_k/q_1, \ldots, q_k]$. A substitution of the form $[A/q]$ is called *simple*. It is rather clear that every substitution can be presented as a composition of simple substitutions.

Later on we often write SA instead of $S(A)$; this formula is called the *substitution instance of A under S*, or the *S-instance of A*. For a set of formulas Γ, $Sub(\Gamma)$ (or $Sub_N(\Gamma)$, if we want to specify the language) denotes the set of all substitution instances of formulas from Γ.

An *intuitionistic substitution* is nothing but a 0-modal substitution.

---

[2]In this book intuitionistic and classical formulas are syntactically the same; the only difference between them is in semantics.

1.1.2 Logics

In this book a *logic* (in a formal sense) is a set of formulas. We say that a logic L is *closed under the rule*

$$\frac{A_1, \ldots, A_n}{B}$$

(or that this rule is *admissible* in L) if $B \in L$, whenever $A_1, \ldots, A_n \in L$. A *(normal propositional) N-modal logic* is a subset of $\mathcal{LP}_N$ closed under arbitrary N-modal substitutions, modus ponens $\left(\frac{A,\ A \supset B}{B}\right)$, necessitation $\left(\frac{A}{\Box_i A}\right)$ and containing all classical tautologies and all the formulas

$$AK_i := \Box_i(p \supset q) \supset (\Box_i p \supset \Box_i q),$$

where $1 \leq i \leq N$.

$\mathbf{K}_N$ denotes the minimal N-modal logic, and $\mathbf{K}$ denotes $\mathbf{K}_1$. Sometimes we call N-modal logics (or formulas) just 'modal', if N is clear from the context.

The smallest N-modal logic containing a given N-modal logic $\mathbf{\Lambda}$ and a set of N-modal formulas Γ is denoted by $\mathbf{\Lambda} + \Gamma$; for a formula A, $\mathbf{\Lambda} + A$ is an abbreviation for $\mathbf{\Lambda} + \{A\}$. We say that the logic $\mathbf{K}_N + \Gamma$ *is axiomatised by* the set Γ. A logic is called *finitely axiomatisable* (respectively, *recursively axiomatisable*) if it can be axiomatised by a finite (respectively, recursive) set of formulas. It is well-known that a logic is recursively axiomatisable iff it is recursively enumerable. A logic $\mathbf{\Lambda}$ is *consistent* if $\bot \notin \mathbf{\Lambda}$.

Here is a list of some frequently used modal formulas and modal logics:

$$\begin{array}{ll}
AT & := \Box p \supset p, \\
A4 & := \Box p \supset \Box\Box p, \\
AD & := \Box\Box p \supset \Box p, \\
AM & := \Box\Diamond p \supset \Diamond\Box p \quad \textit{(McKinsey formula)}, \\
A2 & := \Diamond\Box p \supset \Box\Diamond p, \\
A3 & := \Box(p \wedge \Box p \supset q) \vee \Box(q \wedge \Box q \supset p), \\
AGrz & := \Box(\Box(p \supset \Box p) \supset p) \supset p \quad \textit{(Grzegorczyk formula)}, \\
AL & := \Box(\Box p \supset p) \supset \Box p \quad \textit{(Löb formula)}, \\
A5 & := \Diamond\Box p \supset \Box p, \\
AB & := \Diamond\Box p \supset p, \\
At_1 & := \Diamond_1\Box_2 p \supset p, \\
At_2 & := \Diamond_2\Box_1 p \supset p, \\
AW_n & := \bigvee\limits_{i=0}^{n} \Box\left(p_i \wedge \Box p_i \supset \bigvee\limits_{j \neq i} p_j\right), \\
Alt_n & := \bigvee\limits_{i=0}^{n} \Box\left(p_i \supset \bigvee\limits_{j \neq i} p_j\right).
\end{array}$$

D	$:= \mathbf{K} + \Diamond\top$,	**T**	$:= \mathbf{K} + AT$,
K4	$:= \mathbf{K} + A4$,		
S4	$:= \mathbf{K4} + AT$,	**D4**	$:= \mathbf{D} + A4$,
S4.1	$:= \mathbf{S4} + AM$,	**D4.1**	$:= \mathbf{D4} + AM$,
S4.2	$:= \mathbf{S4} + A2$,	**S4.3**	$:= \mathbf{S4} + \Box(\Box p \supset q) \vee \Box(\Box q \supset p)$,
K4.3	$:= \mathbf{K4} + A3$,	**Grz**	$:= \mathbf{S4} + AGrz$ *(Grzegorczyk logic)*,
S5	$:= \mathbf{S4} + AB$,	**GL**	$:= \mathbf{K} + AL$, *(Gödel–Löb logic)*,
K5	$:= \mathbf{K} + A5$,	**K.t**	$:= \mathbf{K}_2 + At_1 + At_2$.

The corresponding N-modal versions are denoted by $\mathbf{D}_N$, $\mathbf{T}_N$ etc.; so for example,

$$\mathbf{D}_N := \mathbf{K}_N + \{\Diamond_i\top \mid 1 \leq i \leq N\},$$

and so on.

A *superintuitionistic logic* is a set of intuitionistic formulas closed under intuitionistic substitutions and modus ponens, and containing the following well-known axioms:

(Ax1) $p \supset (q \supset p)$,

(Ax2) $(p \supset (q \supset r)) \supset ((p \supset q) \supset (p \supset r))$,

(Ax3) $p \wedge q \supset p$,

(Ax4) $p \wedge q \supset q$,

(Ax5) $p \supset (q \supset p \wedge q)$,

(Ax6) $p \supset p \vee q$,

(Ax7) $q \supset p \vee q$,

(Ax8) $(p \supset r) \supset ((q \supset r) \supset (p \vee q \supset r))$,

(Ax9) $\bot \supset p$.

The smallest superintuitionistic logic is exactly the *intuitionistic* (or *Heyting*) *propositional logic*; it is denoted by **H**.

The notation $\mathbf{\Lambda} + \Gamma$ and the notions of finite axiomatisability, etc. are used for superintuitionistic logics as well.

An *m-formula* is a formula without occurrences of letters p_i for $i > m$. $\mathcal{LP}_N\lceil m$ denotes the set of all N-modal m-formulas. If $\mathbf{\Lambda}$ is a modal or a superintuitionistic logic, $\mathbf{\Lambda}\lceil m$ denotes its restriction to m-formulas. The sets $\mathbf{\Lambda}\lceil m$ are called *bounded logics*.

An *extension* of an N-modal logic $\mathbf{\Lambda}$ is an arbitrary N-modal logic containing $\mathbf{\Lambda}$; extensions of a logic $\mathbf{\Lambda}$ are also called $\mathbf{\Lambda}$-*logics*.

Members of a logic are also called its *theorems*; moreover, we use the notation $\mathbf{\Lambda} \vdash A$ as synonymous for $A \in \mathbf{\Lambda}$. A formula A in the language of a logic $\mathbf{\Lambda}$ is $\mathbf{\Lambda}$-*consistent* if $\neg A \notin \mathbf{\Lambda}$. An *N-modal propositional theory* is a set of N-modal propositional formulas. Such a theory is $\mathbf{\Lambda}$-*consistent* if all conjunctions over its finite subsets are $\mathbf{\Lambda}$-consistent and $\mathbf{\Lambda}$-*complete* if it is maximal among $\mathbf{\Lambda}$-consistent theories (in the same language).

In the intuitionistic case we also consider *double theories* that are pairs of sets of intuitionistic formulas. For a superintuitionistic logic $\mathbf{\Lambda}$, a double theory (Γ, Δ) is called $\mathbf{\Lambda}$-*consistent* if for any finite sets $\Gamma_0 \subseteq \Gamma$, $\Delta_0 \subseteq \Delta$, $\mathbf{\Lambda} \nvdash \bigwedge \Gamma_0 \supset \bigvee \Delta_0$. A $\mathbf{\Lambda}$-consistent double theory (Γ, Δ) is called $\mathbf{\Lambda}$-*complete* if $\Gamma \cup \Delta = \mathcal{LP}_0$.

Let us fix names for some particular intuitionistic formulas and superintuitionistic logics:

$$
\begin{array}{lll}
EM & := & p \vee \neg p \quad (\textit{the law of the excluded middle});\\
AJ & := & \neg p \vee \neg\neg p \quad (\textit{the weak law of the excluded middle});\\
AJ^- & := & \neg\neg p \vee (\neg\neg p \supset p);\\
AP_1 & := & EM;\\
AP_n & := & p_n \vee (p_n \supset AP_{n-1}) \quad (\text{for } n > 1);\\
KP & := & (\neg p \supset q \vee r) \supset (\neg p \supset q) \vee (\neg p \supset r) \quad (\textit{Kreisel–Putnam formula});\\
Br_n & := & \left(\bigwedge\limits_{i=0}^{n} \left(p_i \supset \bigvee\limits_{j \neq i} p_j\right) \supset \bigvee\limits_{j \neq i} p_j\right) \supset \bigvee\limits_{i=0}^{n} p_i \ (\textit{Gabbay–De Jongh formulas});\\
AZ & := & (p \supset q) \vee (q \supset p);\\
AIW_n & := & \bigvee\limits_{i=0}^{n} \left(p_i \supset \bigvee\limits_{j \neq i} p_j\right);\\
IG_n & := & \bigvee\limits_{0 \leq i < j \leq n} (p_i \equiv p_j);\\
\mathbf{HJ} & := & \mathbf{KC} := \mathbf{H} + AJ \ (\textit{Jankov's logic});\\
\mathbf{LC} & := & \mathbf{H} + AZ \ (\textit{Dummett's logic});\\
\mathbf{CL} & := & \mathbf{H} + EM \ (\textit{classical, or Boolean logic}).
\end{array}
$$

The following inclusions are well-known:

$$\mathbf{H} \subset \mathbf{HJ} \subset \mathbf{LC} \subset \mathbf{CL},$$

$$\mathbf{H} + AP_{n+1} \subset \mathbf{H} + AP_n.$$

A superintuitionistic logic Σ is called *consistent* if $\bot \notin \Sigma$; Σ is said to be *intermediate* if $\mathbf{H} \subseteq \Sigma \subseteq \mathbf{CL}$. It is well-known that every consistent superintuitionistic propositional logic is intermediate.

Lemma 1.1.1 *(1) Some theorems of* $\mathbf{K}_N$*:*

$$
\begin{array}{l}
\Box_i(p \wedge q) \equiv \Box_i p \wedge \Box_i q;\\
\Diamond_i(p \vee q) \equiv \Diamond_i p \vee \Diamond_i q;\\
\Box_i p \wedge \Diamond_i q \supset \Diamond_i(p \wedge q);\\
\Box_i(p \supset q) \supset (\Diamond_i p \supset \Diamond_i q);\\
\Diamond_i(p \supset q) \supset (\Box_i p \supset \Diamond_i q);\\
\Box_i(p \equiv q) \supset (\Box_i p \equiv \Box_i q).
\end{array}
$$

(2) The following rules are admissible in every modal logic:

Monotonicity rules *Replacement rules*

$$\frac{A \supset B}{\Box_i A \supset \Box_i B} \qquad \frac{A \equiv B}{\Box_i A \equiv \Box_i B}$$

$$\frac{A \supset B}{\Diamond_i A \supset \Diamond_i B} \qquad \frac{A \equiv B}{\Diamond_i A \equiv \Diamond_i B}.$$

(3) Some theorems of **S4***:*

$$\Diamond\Diamond p \equiv \Diamond p;$$
$$\Box(\Box p \vee \Box q) \equiv \Box p \vee \Box q.$$

(4) A theorem of **S4.2***:*

$$\Diamond\Box(\bigwedge_i p_i) \equiv \bigwedge_i \Diamond\Box p_i.$$

Lemma 1.1.2 *Some theorems of* **H***:*

(1) $p \supset \neg\neg p$,

(2) $\neg\neg(p \wedge q) \equiv \neg\neg p \wedge \neg\neg q$,

(3) $\neg\neg(p \equiv q) \supset (\neg\neg p \equiv \neg\neg q)$,

(4) $(p \supset q) \supset (\neg q \supset \neg p)$,

(5) $\bigwedge_{i=1}^{r} ((p_i \supset q) \supset q) \equiv \left(\left(\bigwedge_{i=1}^{r} p_i \supset q\right) \supset q\right)$.

Lemma 1.1.3 (Propositional replacement rule) *The following rule is admissible in every modal or superintuitionistic logic:*

$$\frac{A \equiv A'}{[A/p]B \equiv [A'/p]B}.$$

We can write this rule more loosely as

$$\frac{A \equiv A'}{C(\ldots A \ldots) \equiv C(\ldots A' \ldots)},$$

i.e. in any formula C we can replace some occurrences of a subformula A with its equivalent A'.

To formulate the next theorem, we introduce some notation. For an N-modal formula B, $r \geq 0$, let

$$\Box^{\leq r} B := \bigwedge \{\Box_\alpha B \mid \alpha \in I_N^\infty,\ l(\alpha) \leq r\};$$

for a finite set of N-modal formulas Δ, let

$$\Box^{\leq r} \Delta := \bigwedge \{\Box^{\leq r} B \mid B \in \Delta\}.$$

Theorem 1.1.4 (Deduction theorem)

(I) Let Σ be a superintuitionistic logic, $\Gamma \cup \{A\}$ a set of intuitionistic formulas. Then:

$A \in (\Sigma + \Gamma)$ *iff* $(\bigwedge \Delta \supset A) \in \Sigma$ *for some finite* $\Delta \subseteq Sub(\Sigma)$.

(II) Let $\mathbf{\Lambda}$ be an N-modal logic, $\Gamma \cup \{A\}$ a set of N-modal formulas. Then $A \in (\mathbf{\Lambda} + \Gamma)$ *iff*

$$\left(\bigwedge \Box^{\leq r} \Delta \supset A\right) \in \mathbf{\Lambda} \text{ for some } r \geq 0 \text{ and some finite } \Delta \subseteq Sub(\Gamma).$$

(III) Let $\mathbf{\Lambda}$ be a 1-modal logic, $\Gamma \cup \{A\} \subseteq \mathcal{LP}_1$. Then $A \in (\mathbf{\Lambda} + \Gamma)$ iff

(1) $(\bigwedge\limits_{k=0}^{r} (\bigwedge \Box^k \Delta) \supset A) \in \mathbf{\Lambda}$ *for some* $r \geq 0$ *and some finite* $\Delta \subseteq Sub(\Gamma)$
— *in the general case;*

(2) $(\bigwedge \Box^r \Delta \supset A) \in \mathbf{\Lambda}$ *for some* $r \geq 0$ *and some finite* $\Delta \subseteq Sub(\Gamma)$
— *provided* $\mathbf{T} \subseteq \mathbf{\Lambda}$;

(3) $(\bigwedge \Delta \wedge \bigwedge \Box \Delta \supset A) \in \mathbf{\Lambda}$ *for some finite* $\Delta \subseteq Sub(\Gamma)$
— *provided* $\mathbf{K4} \subseteq \mathbf{\Lambda}$;

(4) $(\bigwedge \Box \Delta \supset A) \in \mathbf{\Lambda}$ *for some finite* $\Delta \subseteq Sub(\Gamma)$
— *provided* $\mathbf{S4} \subseteq \mathbf{\Lambda}$.

Similarly one can simplify the claim (2) for the case when $\mathbf{\Lambda}$ is an N-modal logic containing $\mathbf{T}_N$, $\mathbf{K4}_N$, or $\mathbf{S4}_N$; we leave this as an exercise for the reader. But let us point out that for the case when $\mathbf{S4}_N \subseteq \mathbf{\Lambda}$, $n > 1$, $\Box_\alpha$ is not necessarily an $\mathbf{S4}$-modality, and it may happen that for any Δ, $A \in (\mathbf{\Lambda} + \Gamma)$ is not equivalent to $\left(\bigwedge\limits_{i=1}^{n} \bigwedge \Box_i \Delta \supset A\right) \in \mathbf{\Lambda}$.

Corollary 1.1.5

(1) For superintuitionistic logics:

$$(\Sigma + \Gamma) \cap (\Sigma + \Gamma') = \Sigma + \{A \vee A' \mid A \in \Gamma,\ A' \in \Gamma'\}$$

if formulas from Γ and Γ' do not have common proposition letters.

(2) For N-modal logics:

$$(\mathbf{\Lambda} + \Gamma) \cap (\mathbf{\Lambda} + \Gamma') = \mathbf{\Lambda} + \{\Box_\alpha A \vee \Box_{\alpha'} A' \mid A \in \Gamma,\ A' \in \Gamma';\ \alpha, \alpha' \in I_N^\infty\}$$

if formulas from Γ and Γ' do not have common proposition letters.

(3) For 1-modal logics:

$$(\mathbf{\Lambda} + \Gamma) \cap (\mathbf{\Lambda} + \Gamma') = \mathbf{\Lambda} + \{\Box^r A \vee \Box^s A' \mid A \in \Gamma,\ A' \in \Gamma';\ r, s \geq 0\}$$

if formulas from Γ and Γ' do not have common proposition letters.

In some particular cases this presentation can be further simplified:

(a) for logics above **T**:

$$(\mathbf{\Lambda}+\Gamma)\cap(\mathbf{\Lambda}+\Gamma') = \mathbf{\Lambda}+\{\Box^r A \vee \Box^r A' \mid A \in \Gamma,\ A' \in \Gamma';\ r \geq 0\};$$

(b) for logics above **K4**:

$$(\mathbf{\Lambda}+\Gamma)\cap(\mathbf{\Lambda}+\Gamma') = \mathbf{\Lambda}+\{\Box^r A \vee \Box^s A' \mid A \in \Gamma,\ A' \in \Gamma';\ r,s \in \{0,1\}\};$$

(c) for logics above **S4**:

$$(\mathbf{\Lambda}+\Gamma)\cap(\mathbf{\Lambda}+\Gamma') = \mathbf{\Lambda}+\{\Box A \vee \Box A' \mid A \in \Gamma,\ A' \in \Gamma'\}.$$

Therefore we have:

Proposition 1.1.6

(1) The set of superintuitionistic logics $\mathcal{S}$ is a complete well-distributive lattice:

$$\mathbf{\Lambda}\cap\sum_{i\in I}\mathbf{\Lambda}_i = \sum_{i\in I}(\mathbf{\Lambda}\cap\mathbf{\Lambda}_i).$$

Here the sum of logics $\sum_{i\in I}\mathbf{\Lambda}_i$ is the smallest logic containing their union. The set of finitely axiomatisable and the set of recursively axiomatisable superintuitionistic logics are sublattices of $\mathcal{S}$.

(2) The set of N-modal logics $\mathcal{M}_N$ is a complete well-distributive lattice; the set of recursively axiomatisable N-modal logics is a sublattice of $\mathcal{M}_N$.

Proof In fact, for example, in the intuitionistic case, both parts of the equality are axiomatised by the same set of formulas

$$\left\{A \vee B \mid A \in \mathbf{\Lambda},\ B \in \bigcup_{i\in I}\mathbf{\Lambda}_i\right\}.$$

■

Remark 1.1.7 Although the set of all *finitely axiomatisable* 1-modal logics is not closed under finite intersections [van Benthem, 1983], this is still the case for finitely axiomatisable extensions of **K4**, cf. [Chagrov and Zakharyaschev, 1997].

Theorem 1.1.8 (Glivenko theorem) *For any intermediate logic* $\mathbf{\Sigma}$

$$\neg A \in \mathbf{H} \text{ iff } \neg A \in \mathbf{\Sigma} \text{ iff } \neg A \in \mathbf{CL}.$$

For a syntactic proof see [Kleene, 1952]. For another proof using Kripke models see [Chagrov and Zakharyaschev, 1997], Theorem 2.47.

Corollary 1.1.9 *If* $A \in \mathbf{CL}$, *then* $\neg\neg A \in \mathbf{H}$.

Proof In fact, $A \in \mathbf{CL}$ implies $\neg\neg A \in \mathbf{CL}$, so we can apply the Glivenko theorem. ■

1.2 Algebraic semantics

For modal and intermediate propositional logics several kinds of semantics are known. *Algebraic semantics* is the most general and straightforward; it interprets formulas as operations in an abstract algebra of truth-values. Actually this semantics fits for every propositional logic with the replacement property; completeness follows by the well-known Lindenbaum theorem.

Relational (Kripke) semantics is nowadays widely known; here formulas are interpreted in relational systems, or *Kripke frames.* Kripke frames correspond to a special type of algebras, so Kripke semantics is reducible to algebraic. *Neighbourhood semantics* (see Section 1.17) is in between relational and algebraic.

Let us begin with algebraic semantics.

Definition 1.2.1[3] *A* Heyting algebra *is an implicative lattice with the least element:*

$$\mathbf{\Omega} = (\Omega,\ \wedge,\ \vee,\ \ \to,\ \mathbf{0}).$$

More precisely, $(\Omega,\ \wedge,\ \vee)$ *is a lattice with the least element* $\mathbf{0}$*, and* $\to$ *is the implication in this lattice, i.e. for any* a, b, c

$$c \wedge a \leq b \text{ iff } c \leq (a \to b). \qquad (*)$$

(Here $\leq$ *is the standard ordering in the lattice, i.e.* $a \leq b$ *iff* $a \wedge b = a$*.)*

Recall that *negation* in Heyting algebras is $\neg a := a \to \mathbf{0}$ and $\mathbf{1} = a \to a$ is the greatest element.

Note that $(*)$ can be written as

$$a \to b = \max\{c \mid c \wedge a \leq b\}.$$

In particular,

$$a \to b = \mathbf{1} \text{ iff } a \leq b.$$

Also recall that an implicative lattice is always distributive:

$$(a \vee b) \wedge c = (a \wedge c) \vee (b \wedge c),$$
$$(a \wedge b) \vee c = (a \vee c) \wedge (b \vee c).$$

A lattice is called *complete* if joins and meets exist for every family of its elements:

$$\bigvee_{j \in J} a_j := \min\{b \mid \forall j \in J\ a_j \leq b\}, \quad \bigwedge_{j \in J} a_j := \max\{b \mid \forall j \in J\ b \leq a_j\}.$$

A complete lattice is implicative iff it is well-distributive, i.e., the following holds:

$$a \wedge \left(\bigvee_{j \in J} a_j \right) = \bigvee_{j \in J} (a \wedge a_j).$$

[3] Cf. [Rasiowa and Sikorski, 1963; Borceaux, 1994].

So every complete well-distributive lattice can be turned into a Heyting algebra.

Let us prove two useful properties of Heyting algebras.

Lemma 1.2.2

$$c \wedge \bigwedge_{j \in J} (a_j \to b_j) \leq \bigwedge_{j \in J} (a_j \to b_j \wedge c).$$

Proof We have to prove

$$c \wedge \bigwedge_{j \in J} (a_j \to b_j) \leq a_k \to b_k \wedge c,$$

which is equivalent (by 1.2.1(∗)) to

$$c \wedge a_k \wedge \bigwedge_{j \in J} (a_j \to b_j) \leq b_k \wedge c.$$

But this follows from

$$a_k \wedge \bigwedge_{j \in J} (a_j \to b_j) \leq b_k.$$

The latter holds, since by 1.2.1(∗), it is equivalent to

$$\bigwedge_{j \in J} (a_j \to b_j) \leq a_k \to b_k.$$

■

Lemma 1.2.3

$$u \to \bigwedge_{i \in I} v_i = \bigwedge_{i \in I} (u \to v_i).$$

Proof (≤)

$$\left(u \to \bigwedge_{i \in I} v_i\right) \wedge u \leq \bigwedge_{i \in I} v_i \leq v_i,$$

hence

$$u \to \bigwedge_{i \in I} v_i \leq u \to v_i,$$

and thus

$$u \to \bigwedge_{i \in I} v_i \leq \bigwedge_{i \in I} (u \to v_i).$$

(≥)

$$u \wedge \bigwedge_{i \in I} (u \to v_i) \leq u \wedge (u \to v_i) \leq v_i,$$

hence

$$u \wedge \bigwedge_{i \in I} (u \to v_i) \leq \bigwedge_{i \in I} v_i,$$

and thus

$$\bigwedge_{i\in I}(u \to v_i) \leq u \to \bigwedge_{i\in I} v_i.$$

■

Lemma 1.2.4

$$\bigwedge_{i\in I}(v_i \to u) = (\bigvee_{i\in I} v_i \to u).$$

Proof ($\geq$)
$v_i \leq \bigvee_{i\in I} v_i$ implies

$$\bigvee_{i\in I} v_i \to u \leq v_i \to u;$$

hence

$$\bigvee_{i\in I} v_i \to u \leq \bigwedge_{i\in I}(v_i \to u).$$

($\leq$) Since

$$\bigwedge_{i\in I}(v_i \to u) \leq v_i \to u,$$

it follows that for any $i \in I$

$$v_i \wedge \bigwedge_{i\in I}(v_i \to u) \leq u.$$

Hence

$$(\bigvee_{i\in I} v_i) \wedge \bigwedge_{i\in I}(v_i \to u) = \bigvee_{i\in I}(v_i \wedge \bigwedge_{i\in I}(v_i \to u)) \leq u.$$

Eventually

$$\bigwedge_{i\in I}(v_i \to u) \leq \bigvee_{i\in I} v_i \to u.$$

■

A Boolean algebra is a particular case of a Heyting algebra, where $a \vee \neg a = \mathbf{1}$. In this case $\vee, \wedge, \to, \neg$ are usually denoted by $\cup, \cap, \Supset, -$. Then we can consider $\cup, \cap, -, \mathbf{0}, \mathbf{1}$ (and even $\cup, -, \mathbf{0}$) as basic and define $a \Supset b := -a \cup b$.

We also use the derived operation (*equivalence*)

$$a \leftrightarrow b := (a \to b) \wedge (b \to a)$$

in Heyting algebras and its analogue

$$a \simeq b := (a \Supset b) \cap (b \Supset a)$$

in Boolean algebras.

Definition 1.2.5 *An N-*modal algebra *is a structure*

$$\mathbf{\Omega} = (\Omega,\ \cap,\ \cup,\ -,\ \mathbf{0},\ \mathbf{1}, \Box_1, \ldots, \Box_N),$$

such that its nonmodal part

$$\mathbf{\Omega}^\flat := (\Omega,\ \cap,\ \cup, -,\ \mathbf{0},\ \mathbf{1})$$

is a Boolean algebra, and $\Box_i$ are unary operations in Ω satisfying the identities:

$$\Box_i(a \cap b) = \Box_i a \cap \Box_i b,$$

$$\Box_i \mathbf{1} = \mathbf{1}.$$

$\mathbf{\Omega}$ is called complete *if the Boolean algebra $\mathbf{\Omega}^\flat$ is complete.*

We also use the dual operations

$$\Diamond_i a = -\Box_i - a.$$

For 1-modal algebras we write $\Box$, $\Diamond$ rather than $\Box_1$, $\Diamond_1$ (cf. Section 1.1.1).

Definition 1.2.6 *A* topo-Boolean *(or* interior, *or* **S4**-*)* algebra *is a 1-modal algebra satisfying the inequalities*

$$\Box a \leq a,\ \Box a \leq \Box\Box a.$$

In this case $\Box$ is called the interior operation *and its dual $\Diamond$ the* closure operation. *An element a is said to be* open *if $\Box a = a$ and* closed *if $\Diamond a = a$.*

Proposition 1.2.7 *The open elements of a topo-Boolean algebra $\mathbf{\Omega}$ constitute a Heyting algebra:*

$$\mathbf{\Omega}^0 = (\Omega^0,\ \cap,\ \cup,\ \rightarrow,\ \mathbf{0}),$$

in which $a \rightarrow b = \Box(a \supset b)$. Moreover, if $\mathbf{\Omega}$ is complete then $\mathbf{\Omega}^0$ also is, and

$$\bigvee_{j \in J} a_j = \bigcup_{j \in J} a_j,\ \bigwedge_{j \in J} a_j = \Box\left(\bigcap_{j \in J} a_j\right)$$

Proof Cf. [McKinsey and Tarski, 1944]; [Rasiowa and Sikorski, 1963]. ■

Following [Esakia, 1979], we call $\mathbf{\Omega}^0$ the *pattern* of $\mathbf{\Omega}$. It is known that every Heyting algebra is isomorphic to some algebra $\mathbf{\Omega}^0$ [Rasiowa and Sikorski, 1963]

Definition 1.2.8 *A* valuation *in an N-modal algebra $\mathbf{\Omega}$ is a map $\varphi : PL \longrightarrow \Omega$. The valuation φ has a unique extension to all N-modal formulas such that*

(1) $\varphi(\bot) = \mathbf{0}$;

(2) $\varphi(A \wedge B) = \varphi(A) \cap \varphi(B)$;

(3) $\varphi(A \vee B) = \varphi(A) \cup \varphi(B)$;

(4) $\varphi(A \supset B) = \varphi(A) \Supset \varphi(B)$;

(5) $\varphi(\Box_i A) = \Box_i \varphi(A)$.

The pair $(\mathbf{\Omega}, \varphi)$ *is then called an* (algebraic) model *over* $\mathbf{\Omega}$. *An N-modal formula A is said to be* true in the model $(\mathbf{\Omega}, \varphi)$ *if* $\varphi(A) = \mathbf{1}$ *(notation:* $(\mathbf{\Omega}, \varphi) \vDash A$*); A is called* valid *in the algebra* $\mathbf{\Omega}$ *(notation:* $\mathbf{\Omega} \vDash A$*) if it is true in every model over* $\mathbf{\Omega}$.

Lemma 1.2.9 *Let* $\mathbf{\Omega}$ *be an N-modal algebra, S a propositional substitution. Let* φ, η *be valuations in* $\mathbf{\Omega}$ *such that for any* $B \in PL\lceil k$

(♣) $\eta(B) = \varphi(SB)$.

Then (♣) holds for any N-modal k-formula B.

Proof Easy, by induction on the length of B. ∎

Lemma 1.2.10 (Soundness lemma) *The set*

$$\mathbf{ML}(\mathbf{\Omega}) := \{A \in \mathcal{LP}_N \mid \mathbf{\Omega} \vDash A\}$$

is a modal logic.

Proof First note that $\mathbf{ML}(\mathbf{\Omega})$ is substitution closed. In fact, assume $\mathbf{\Omega} \vDash A$, and let S be a propositional substitution. To show that $\mathbf{\Omega} \vDash SA$, take an arbitrary valuation φ in $\mathbf{\Omega}$, and consider a new valuation η according to (♣) from Lemma 1.2.9. So we obtain

$$\varphi(SA) = \eta(A) = \mathbf{1},$$

i.e. $\mathbf{\Omega} \vDash SA$.

The classical tautologies are valid in $\mathbf{\Omega}$, because they hold in any Boolean algebra. The validity of AK_i follows by a standard argument. In fact, note that in a modal algebra $\Box_i$ is monotonic:
(*) $x \leq y \Rightarrow \Box_i x \leq \Box_i y$,
because $x \leq y$ implies

$$\Box_i x = \Box_i(x \cap y) = \Box_i x \cap \Box_i y.$$

Now since

$$(a \Supset b) \cap a \leq b,$$

by monotonicity (*), we have

$$\Box(a \Supset b) \cap \Box a \leq \Box b,$$

which implies

$$\Box(a \Supset b) \leq (\Box a \Supset \Box b),$$

This yields the validity of AK_i.

Finally, modus ponens and necessitation preserve validity, since in a modal algebra $\mathbf{1} \leq a$ implies $a = \mathbf{1}$, and $\Box_i \mathbf{1} = \mathbf{1}$. ∎

Definition 1.2.11 $\mathbf{ML}(\mathbf{\Omega})$ *is called the* modal logic of the algebra $\mathbf{\Omega}$.
We also define the modal logic of a class $\mathcal{C}$ of N-modal algebras

$$\mathbf{ML}(\mathcal{C}) := \bigcap\{\mathbf{ML}(\mathbf{\Omega}) \mid \mathbf{\Omega} \in \mathcal{C}\}.$$

Note that $\mathbf{ML}(\mathbf{\Omega})$ is consistent iff the algebra $\mathbf{\Omega}$ is nondegenerate, i.e. iff $\mathbf{0} \neq \mathbf{1}$ in $\mathbf{\Omega}$.

Definition 1.2.12 *A* valuation *in a Heyting algebra* $\mathbf{\Omega}$ *is a map* $\varphi : PL \longrightarrow \Omega$. *It has a unique extension* $\varphi^I : \mathcal{LP}_0 \longrightarrow \Omega$ *such that*

(1) $\varphi^I(\bot) = \mathbf{0}$;

(2) $\varphi^I(A \wedge B) = \varphi^I(A) \wedge \varphi^I(B)$;

(3) $\varphi^I(A \vee B) = \varphi^I(A) \vee \varphi^I(B)$;

(4) $\varphi^I(A \supset B) = \varphi^I(A) \rightarrow \varphi^I(B)$.

As in the modal case, the pair $(\mathbf{\Omega}, \varphi)$ *is called an* (algebraic) model *over* $\mathbf{\Omega}$. *An intuitionistic formula A is said to be* true *in* $(\mathbf{\Omega}, \varphi)$ *if* $\varphi^I(A) = \mathbf{1}$ *(notation:* $(\mathbf{\Omega}, \varphi) \vDash A$*); A is called* valid *in the algebra* $\mathbf{\Omega}$ *(notation:* $\mathbf{\Omega} \vDash A$*) if it is true in every model over* $\mathbf{\Omega}$.

We easily obtain an intuitionistic analogue of Lemma 1.2.9:

Lemma 1.2.13 *Let* $\mathbf{\Omega}$ *be a Heyting algebra, S a propositional substitution. Let* φ, η *be valuations in* $\mathbf{\Omega}$ *such that for any* $B \in PL$

(♣) $\eta^I(B) = \varphi^I(SB)$.

Then (♣) holds for any intuitionistic formula B.

Similarly we have

Lemma 1.2.14 (Soundness lemma) *For a Heyting algebra* $\mathbf{\Omega}$, *the set*

$$\mathbf{IL}(\mathbf{\Omega}) := \{A \in \mathcal{LP}_0 \mid \mathbf{\Omega} \vDash A\}$$

is a superintuitionistic logic.

Definition 1.2.15 $\mathbf{IL}(\mathbf{\Omega})$ *is called the* superintuitionistic logic of the algebra $\mathbf{\Omega}$. *Similarly to the modal case, we define the* superintuitionistic logic of a class $\mathcal{C}$ *of Heyting algebras*

$$\mathbf{IL}(\mathcal{C}) := \bigcap\{\mathbf{IL}(\mathbf{\Omega}) \mid \mathbf{\Omega} \in \mathcal{C}\}.$$

Definition 1.2.16 *A valuation* φ *in an* **S4***-algebra* $\mathbf{\Omega}$ *is called* intuitionistic *if it is a valuation in* $\mathbf{\Omega}^0$ *i.e. if its values are open.*

Definition 1.2.17 Gödel–Tarski translation *is the map* $(-)^T$ *from intuitionistic to 1-modal formulas defined by the following clauses:*

$$\begin{aligned}
\bot^T &= \bot; \\
q^T &= \Box q \text{ for every proposition letter } q; \\
(A \wedge B)^T &= A^T \wedge B^T; \\
(A \vee B)^T &= A^T \vee B^T; \\
(A \supset B)^T &= \Box(A^T \supset B^T).
\end{aligned}$$

Lemma 1.2.18 $(\Box A^T \equiv A^T) \in \mathbf{S4}$ *for any intuitionistic formula* A.

Proof Easy by induction; for the cases $A = B \vee C$, $A = B \wedge C$ use Lemma 1.1.1. ■

Lemma 1.2.19 *Let* $\mathbf{\Omega}$ *be an* **S4***-algebra.*

(1) Let φ, ψ *be valuations in* $\mathbf{\Omega}$ *such that for any* $q \in PL$

$$\varphi(q) = \Box\psi(q).$$

Then for any intuitionistic formula A,

$$\varphi^I(A) = \psi(A^T).$$

In particular,

$$\varphi^I(A) = \varphi(A^T),$$

if φ *is intuitionistic.*

(2) For any intuitionistic formula A,

$$\mathbf{\Omega}^0 \vDash A \text{ iff } \mathbf{\Omega} \vDash A^T.$$

Proof
(1) By induction. Consider only the case $A = B \supset C$. Suppose

$$\varphi^I(B) = \psi(B^T),\ \varphi^I(C) = \psi(C^T).$$

Then we have

$$\begin{aligned}
\varphi^I(B \supset C) &= \varphi^I(B) \to \varphi^I(C) = \psi(B^T) \to \psi(C^T) = \Box(\psi(B^T) \Rrightarrow \psi(C^T)) \\
&= \psi(\Box(B^T \supset C^T)) = \psi((B \supset C)^T).
\end{aligned}$$

(2) (Only if.) Assume $\mathbf{\Omega}^0 \vDash A$. Let ψ be an arbitrary valuation in $\mathbf{\Omega}$, and let φ be the valuation in $\mathbf{\Omega}^0$ such that

$$\varphi(q) = \Box\psi(q)$$

for every $q \in PL$. By (1) and our assumption, we have:

$$\psi(A^T) = \varphi^I(A) = \mathbf{1}.$$

Hence $\mathbf{\Omega} \vDash A^T$.

(If.) Assume $\mathbf{\Omega} \vDash A^T$. By (1), for any valuation φ in $\mathbf{\Omega}^0$ we have $\varphi^I(A) = \varphi(A^T) = \mathbf{1}$. Hence $\mathbf{\Omega}^0 \vDash A$. ■

Let us now recall the Lindenbaum algebra construction. For an N-modal or superintuitionistic logic $\mathbf{\Lambda}$, the relation $\sim_{\mathbf{\Lambda}}$ between N-modal (respectively, intuitionistic) formulas such that

$$A \sim_{\mathbf{\Lambda}} B \text{ iff } (A \equiv B) \in \mathbf{\Lambda}$$

is an equivalence.

Let $[A]$ be the equivalence class of a formula A modulo $\sim_{\mathbf{\Lambda}}$.

Definition 1.2.20 *The* Lindenbaum algebra $Lind(\mathbf{\Lambda})$ *of a modal logic* $\mathbf{\Lambda}$ *is the set* $\mathcal{LP}_N / \sim_{\mathbf{\Lambda}}$ *with the operations*

- $[A] \cap [B] := [A \wedge B]$,
- $[A] \cup [B] := [A \vee B]$,
- $-[A] := [\neg A]$,
- $\mathbf{0} := [\bot]$,
- $\mathbf{1} := [\top]$,
- $\Box_i[A] := [\Box_i A]$.

Theorem 1.2.21 *For an* N*-modal logic* $\mathbf{\Lambda}$

(1) $Lind(\mathbf{\Lambda})$ *is an* N*-modal algebra;*

(2) $\mathbf{ML}(Lind(\mathbf{\Lambda})) = \mathbf{\Lambda}$.

Definition 1.2.22 *The* Lindenbaum algebra $Lind(\Sigma)$ *of a superintuitionistic logic* Σ*, is the set* $\mathcal{LP}_0 / \sim_{\Sigma}$ *with the operations*

- $[A] \wedge [B] := [A \wedge B]$,
- $[A] \vee [B] := [A \vee B]$,
- $[A] \to [B] := [A \supset B]$,
- $\mathbf{0} := [\bot]$.

Theorem 1.2.23 *For a superintuitionistic logic* Σ,

(1) $Lind(\Sigma)$ *is a Heyting algebra;*

(2) $\mathbf{IL}(Lind(\Sigma)) = \Sigma$.

Definition 1.2.24 *A set of modal formulas* Γ *is* valid *in a modal algebra* $\mathbf{\Omega}$ *(notation:* $\mathbf{\Omega} \models \Gamma$*) if all these formulas are valid; similarly for intuitionistic formulas and Heyting algebras. In this case* $\mathbf{\Omega}$ *is called a* Γ-algebra. *The set of all* Γ*-algebras is called an* algebraic variety defined by Γ.

Algebraic varieties can be characterised in algebraic terms, due to the well-known Birkhoff theorem [Birkhoff, 1979] (which holds also in a more general context):

Theorem 1.2.25 *A class of modal or Heyting algebras is an algebraic variety iff it is closed under subalgebras, homomorphic images and direct products.*

Since every logic is complete in algebraic semantics, there is the following duality theorem.

Theorem 1.2.26 *The poset $\mathcal{M}_N$ of N-modal propositional logics (ordered by inclusion) is dually isomorphic to the set of all algebraic varieties of N-modal algebras; similarly for superintuitionistic logics and Heyting algebras.*

1.3 Relational semantics (the modal case)

1.3.1 Introduction

First let us briefly recall the underlying philosophical motivation. For more details, we address the reader to [Fitting and Mendelsohn, 2000]. In relational (or Kripke) semantics formulas are evaluated in 'possible worlds' representing different situations. Depending on the application area of the logic, worlds can also be called 'states', 'moments of time', 'pieces of information', etc. Every world w is related to some other worlds called 'accessible from w', and a formula $\Box A$ is true at w iff A is true at all worlds accessible from w; dually, $\Diamond A$ is true at w iff A is true at some world accessible from w.

This corresponds to the ancient principle of Diodorus Cronus saying that

The possible is that which either is or will be true

So from the Diodorean viewpoint, possible worlds are moments of time, with the accessibility relation $\leq$ 'before' (nonstrict).

For polymodal formulas we need several accessibility relations corresponding to different necessity operators.

For the intuitionistic case, Kripke semantics formalises the 'historical approach' to intuitionistic truth by Brouwer. Here worlds represent stages of our knowledge in time. According to Brouwer's truth-preservation principle, the truth of every formula is inherited in all later stages. $\neg A$ is true at w iff the truth of A can never be established afterwards, i.e. iff A is not true at w and always later. Similarly, $A \supset B$ is true at w iff the truth of A implies the truth of B at w and always later. See [Dragalin, 1988], [van Dalen, 1973] for further discussion.

1.3.2 Kripke frames and models

Now let us recall the main definitions in detail.

Definition 1.3.1 *An* N-modal (propositional) Kripke frame *is an* $(N+1)$*-tuple* $F = (W, R_1, \ldots, R_N)$, *such that* $W \neq \varnothing$, $R_i \subseteq W \times W$. *The elements of* W *are called* possible worlds *(or* points*),* R_i *are the* accessibility relations.

Quite often we write $u \in F$ rather than $u \in W$.

For a Kripke frame $F = (W, R_1, \ldots, R_N)$ and a sequence $\alpha \in I_N^\infty$ we define the relation R_α on W:

$$R_{i_1 \ldots i_k} := R_{i_1} \circ \ldots \circ R_{i_k},\ R_\curlywedge := Id_W.$$

(Recall that $\curlywedge$ is a void sequence, Id_W is the equality relation, see the Introduction.)

Every N-modal Kripke frame $F = (W, R_1, \ldots, R_N)$ corresponds to an N-modal algebra

$$MA(F) := (2^W,\ \cap,\ \cup,\ -,\ \varnothing,\ W,\ \Box_1, \ldots, \Box_N),$$

where $\cap$, $\cup$, $-$ are the standard set-theoretic operations on subsets of W, and

$$\Box_i V := \{x \mid R_i(x) \subseteq V\}.$$

$MA(F)$ is called the *modal algebra of the frame* F.

Definition 1.3.2 *A* valuation *in a set* W *(or in a frame with the set of worlds* W*) is a valuation in* $MA(F)$*, i.e. a map* $\theta : PL \longrightarrow 2^W$. *A* Kripke model *over a frame* F *is a pair* $M = (F, \theta)$*, where* θ *is a valuation in* F. θ *is extended to all formulas in the standard way, according to Definition 1.2.8:*

(1) $\theta(\bot) = \varnothing$;

(2) $\theta(A \wedge B) = \theta(A) \cap \theta(B)$;

(3) $\theta(A \vee B) = \theta(A) \cup \theta(B)$;

(4) $\theta(A \supset B) = \theta(A) \Supset \theta(B)$;

(5) $\theta(\Box_i A) = \Box_i \theta(A) = \{u \mid R_i(u) \subseteq \theta(A)\}$.

For a formula A*, we also write:* $M, w \vDash A$ *(or just* $w \vDash A$*) instead of* $w \in \theta(A)$*, and say that* A *is* true at the world w of the model M *(or that* w forces A*).*

The above definition corresponds to the well-known inductive definition of forcing in a Kripke model given by (1)–(6) in the following lemma.

Lemma 1.3.3 *(1)* $M, u \vDash q$ *iff* $u \in \theta(q)$ *(for* $q \in PL$*);*

(2) $M, u \not\vDash \bot$;

(3) $M, u \vDash B \wedge C$ *iff* $(M, u \vDash B$ *and* $M, u \vDash C)$;

(4) $M, u \vDash B \vee C$ *iff* $(M, u \vDash B$ *or* $M, u \vDash C)$;

(5) $M, u \vDash B \supset C$ *iff* ($M, u \vDash B$ *implies* $M, u \vDash C$);

(6) $M, u \vDash \Box_i B$ *iff* $\forall v \in R_i(u)\ M, v \vDash B$;

(7) $M, u \vDash \neg B$ *iff* $M, u \nvDash B$;

(8) $M, u \vDash \Diamond_i B$ *iff* $\exists v \in R_i(u)\ M, v \vDash B$.

(9) $M, u \vDash \Box_\alpha B$ *iff* $\forall v \in R_\alpha(u)\ M, v \vDash B$;

(10) $M, u \vDash \Diamond_\alpha B$ *iff* $\exists v \in R_\alpha(u)\ M, v \vDash B$.

Definition 1.3.4 *An* m-bounded Kripke model *over a Kripke frame* $F = (W, R_1, \ldots, R_N)$ *is a pair* (F, θ), *in which* $\theta : \{p_1, \ldots, p_m\} \longrightarrow 2^W$; θ *is called an* m-valuation. *In this case* θ *is extended only to* m*-formulas, according to Definition 1.3.2.*

Definition 1.3.5 *A modal formula* A *is* true in a model M *(notation:* $M \vDash A$*) if it is true at every world of* M*;* A *is* satisfied in M *if it is true at some world of* M*. A formula is called* refutable *in a model if it is not true.*

Definition 1.3.6 *A modal formula* A *is* valid in a frame F *(notation:* $F \vDash A$*) if it is true in every model over* F*. A set of formulas* Γ *is* valid *in* F *(notation:* $F \vDash \Gamma$*) if every* $A \in \Gamma$ *is valid. In the latter case we also say that* F *is a* Γ-frame. *The* (Kripke frame) variety *of* Γ *(notation:* $\mathbf{V}(\Gamma)$*) is the class of all* Γ*-frames.*

A formula A *is* valid at a world x in a frame F *(notation:* $F, x \vDash A$*) if it is true at* x *in every model over* F*; similarly for a set of formulas.*

A nonvalid formula is called refutable *(in a frame or at a world).*

A formula A *is* satisfiable *at a world* w *of a frame* F *(or briefly,* at F, w*) if there exists a model* M *over* F *such that* $M, w \vDash A$.

Since by Definitions 1.2.8 and 1.3.2, $\theta(A)$ is the same in F and in $MA(F)$, we have

Lemma 1.3.7 *For any modal formula* A *and a Kripke frame* F

$$F \vDash A \text{ iff } MA(F) \vDash A.$$

Thus 1.2.10 implies:

Lemma 1.3.8 (Soundness lemma)

(1) *For a Kripke frame* F*, the set*

$$\mathbf{ML}(F) := \{A \mid F \vDash A\}$$

is a modal logic.

(2) For a class $\mathcal{C}$ of N-modal frames, the set

$$\mathbf{ML}(\mathcal{C}) := \bigcap\{\mathbf{ML}(F) \mid F \in \mathcal{C}\}$$

is an N-modal logic.

Definition 1.3.9 $\mathbf{ML}(F)$ *(respectively,* $\mathbf{ML}(\mathcal{C})$*) is called the* modal logic of F *(respectively, of $\mathcal{C}$), or the* modal logic determined by F *(by $\mathcal{C}$), or* complete w.r.t. F *($\mathcal{C}$).*

For a Kripke model M, the set

$$\mathbf{MT}(M) := \{A \mid M \vDash A\}$$

is called the modal theory *of M.*

$\mathbf{MT}(M)$ is not always a modal logic; it is closed under MP and $\Box$-introduction but not necessarily under substitution.

The following is a trivial consequence of definitions and the soundness lemma.

Lemma 1.3.10 *For an N-modal logic $\mathbf{\Lambda}$ and a set of N-modal formulas Γ, $\mathbf{V}(\mathbf{\Lambda} + \Gamma) = \mathbf{V}(\mathbf{\Lambda}) \cap \mathbf{V}(\Gamma)$. In particular, $\mathbf{V}(\mathbf{K}_N + \Gamma) = \mathbf{V}(\Gamma)$.*

Let us describe varieties of some particular modal logics:

Proposition 1.3.11

- $\mathbf{V}(\mathbf{D})$ *consists of all* serial *frames, i.e. of the frames (W, R) such that $\forall x \exists y\, xRy$;*

- $\mathbf{V}(\mathbf{T})$ *consists of all reflexive frames;*

- $\mathbf{V}(\mathbf{K4})$ *consists of all transitive frames;*

- $\mathbf{V}(\mathbf{S4})$ *consists of all* quasi-ordered *(or* pre-ordered*) sets, i.e. reflexive transitive frames;*

- $\mathbf{V}(\mathbf{S4.1})$ *consists of all* $\mathbf{S4}$*-frames with* McKinsey property*:*

  $$\forall x \exists y\, (xRy \,\&\, R(y) = \{y\});$$

- $\mathbf{V}(\mathbf{S4.2})$ *consists of all* $\mathbf{S4}$*-frames with* Church–Rosser property *(or* confluent, *or* piecewise directed*):*

  $$\forall x, y, z\, (xRy \,\&\, xRz \Rightarrow \exists t\, (yRt \,\&\, zRt)),$$

  or equivalently,

  $$R^{-1} \circ R \subseteq R \circ R^{-1};$$

- **V(K4.3)** *consists of all* piecewise linear *(or* nonbranching*)* **K4**-*frames, i.e. such that*

$$\forall x, y, z\, (xRy \,\&\, xRz \Rightarrow (y = z \vee yRz \vee zRy)),$$

or equivalently

$$R^{-1} \circ R \subseteq I_W \cup R \cup R^{-1};$$

- **V**$(\mathbf{K4} + AW_n)$ *consists of all transitive frames of width* $\leq n$;[4]

- **V(Grz)** *consists of all* Nötherian *posets, i.e. of those without infinite ascending chains* $x_1 R^- x_2 R^- x_3 \ldots$;[5]

- **V(S5)** *consists of all frames, where accessibility is an equivalence relation.*

Due to these characterisations, an N-modal logic is called *reflexive* (respectively, *serial, transitive*) if it contains $\mathbf{T}_N$ (respectively, $\mathbf{D}_N$, $\mathbf{K4}_N$).

Definition 1.3.12 *Let* $\mathbf{\Lambda}$ *be a modal logic.*

- $\mathbf{\Lambda}$ *is called* Kripke-complete *if it is determined by some class of frames;*

- $\mathbf{\Lambda}$ *has the* finite model property *(f.m.p.) if it is determined by some class of finite frames;*

- $\mathbf{\Lambda}$ *has the* countable frame property *(c.f.p.) if it is determined by some class of countable frames.*[6]

The following simple observation readily follows from the definitions.

Lemma 1.3.13

(1) A logic $\mathbf{\Lambda}$ *is Kripke-complete (respectively, has the c.f.p./f.m.p.) iff each of its nontheorems is refutable in some* $\mathbf{\Lambda}$*-frame (respectively, in a countable/finite* $\mathbf{\Lambda}$*-frame).*

(2) $\mathbf{ML}(\mathbf{V}(\mathbf{\Lambda}))$ *is the smallest Kripke-complete extension of* $\mathbf{\Lambda}$*; so* $\mathbf{\Lambda}$ *is Kripke-complete iff* $\mathbf{\Lambda} = \mathbf{ML}(\mathbf{V}(\mathbf{\Lambda}))$.

All particular propositional logics mentioned above (and many others) are known to be Kripke-complete. Kripke-completeness was proved for large families of propositional logics; Section 1.9 gives a brief outline of these results. However not all modal or intermediate propositional logics are complete in Kripke semantics; counterexamples were found by S. Thomason, K. Fine, V. Shehtman, J. Van Benthem, cf. [Chagrov and Zakharyaschev, 1997]. But incomplete propositional logics look rather artificial; in general one can expect that a 'randomly chosen' logic is compete.

Nevertheless every logic is 'complete w.r.t. Kripke models' in the following sense.

---

[4] See Section 1.9.

[5] Recall that xR^-y iff $xRy \,\&\, x \neq y$, see Introduction.

[6] 'countable' means 'of cardinality $\leq \aleph_0$'.

Definition 1.3.14 *An N-modal Kripke model M is* exact *for an N-modal logic $\mathbf{\Lambda}$ if $\mathbf{\Lambda} = \mathbf{MT}(M)$.*

Proposition 1.3.15 *Every propositional modal logic has a countable exact model.*

This follows from the canonical model theorem by applying the standard translation, see below.

1.3.3 Main constructions

Definition 1.3.16 *If $F = (W, R_1, \ldots, R_N)$ is a frame, $V \subseteq W$, then the frame*

$$F \restriction V := (V, R_1 \restriction V, \ldots, R_N \restriction V)$$

is called a subframe *of F (the* restriction of F to V*).*

If $M = (F, \theta)$ is a Kripke model, then

$$M \restriction V := (F \restriction V, \theta \restriction V),$$

where $(\theta \restriction V)(q) := \theta(q) \cap V$ for every $q \in PL$, is called its submodel *(the restriction to V).*

A set $V \subseteq W$ is called stable *(in F) if for every i, $R_i(V) \subseteq V$. In this case the subframe $F \restriction V$ and the submodel $M \restriction V$ are called* generated.

Definition 1.3.17 *$F' = (V, R'_1, \ldots, R'_N)$ is called a* weak subframe *of $F = (W, R_1, \ldots, R_N)$ if $R'_i \subseteq R_i$ for every i and $V \subseteq W$. Then for a Kripke model $M = (F, \theta)$, $M' = (F', \theta \restriction V)$ is called a* weak submodel. *If also $W = V$, F' is called a* full weak subframe *of F.*

We use the signs $\subseteq$, $\lhd$, $\tilde{\subset}$, $\subseteqq$ to denote subframes, generated subframes, weak subframes, and full weak subframes, respectively; the same for submodels.

Definition 1.3.18 *Let F, M be the same as in the previous definition. The smallest stable subset $W{\uparrow}u$ containing a given point $u \in W$ is called the* cone generated by u*; the corresponding subframe $F{\uparrow}u := F \restriction (W{\uparrow}u)$ is also called the* cone (in F) generated by u, *or the* subframe generated by u*; similarly for the submodel $M{\uparrow}u := M \restriction (W{\uparrow}u)$. A frame F (respectively, a Kripke model M) is called* rooted (with the root u) *if $F = F{\uparrow}u$ (respectively, $M = M{\uparrow}u$).*

We skip the simple proof of the following

Lemma 1.3.19 *$W{\uparrow}u = R^*(u)$, where R^* is the reflexive transitive closure of $(R_1 \cup \ldots \cup R_N)$, i.e. $R^* = \bigcup\limits_{\alpha \in I_N^\infty} R_\alpha$.*

Definition 1.3.20 *A* path of length m from u to v *in a frame $F = (W, R_1, \ldots, R_N)$ is a sequence $(u_0, j_0, u_1, \ldots, j_{m-1}, u_m)$ such that $u_0 = u$, $u_m = v$, and $u_i R_{j_i} u_{i+1}$ for $i = 0, \ldots, m-1$.*

For the particular case $N = 1$ we have $j_i = 1$ for any i, so we can denote a path just by $(u_0, u_1, \dots, u_m)$.

Now Lemma 1.3.19 can be reformulated as follows:

Lemma 1.3.21 *$x \in F{\uparrow}u$ iff there exists a path from u to x in F.*

Definition 1.3.22 *The* temporalisation *of a propositional Kripke frame $F = (W, R_1, \dots, R_N)$ is the frame $F^{\pm} := (W, R_1, \dots, R_N, R_1^{-1}, \dots, R_N^{-1})$. A* non-oriented path *in F is a path in $F^{\pm}$.*

Definition 1.3.23 *Let $F = (W, R_1, \dots, R_N)$ be a propositional Kripke frame. A subset $V \subseteq W$ is called* connected *(in F) if it is stable in $F^{\pm}$, i.e. both R_i- and R_i^{-1}-stable for every $i = 1, \dots, N$. F itself is called* connected *if W is connected in F. A cone in $F^{\pm}$ (as a subset) is called a* (connected) component *of F.*

Lemma 1.3.24

(1) The component containing $x \in F$ (i.e. the cone $F^{\pm} \uparrow x$) consists of all $y \in F$ such that there exists a non-oriented path from x to y.

(2) The components of F make a partition.

Proof
(1) Readily follows from Lemma 1.3.21.
(2) Follows from (1) and the observation that

$$\{(x, y) \mid \text{ there exists a nonoriented path from } x \text{ to } y\}$$

is an equivalence relation on W. ∎

The following is well-known:

Lemma 1.3.25 (Generation lemma) *Let V be a stable subset in F, $M = (F, \theta)$ a Kripke model. Then*

(1) For any $u \in V$, for any modal formula A,

$$M \upharpoonright V, u \vDash A \textit{ iff } M, u \vDash A;$$

(2) $\mathbf{ML}(F) \subseteq \mathbf{ML}(F \upharpoonright V)$.

The same holds for bounded models, with obvious changes.

We also have:

Lemma 1.3.26 *(1)* $\mathbf{ML}(F) = \bigcap\limits_{u \in F} \mathbf{ML}(F{\uparrow}u)$.

(2) $\mathbf{MT}(M) = \bigcap\limits_{u \in M} \mathbf{MT}(M \uparrow u)$.

Proof

(1) By Lemma 1.3.25(2), $\mathbf{ML}(F) \subseteq \mathbf{ML}(F{\uparrow}u)$ for any $u \in F$, and thus

$$\mathbf{ML}(F) \subseteq \bigcap_{u\in F} \mathbf{ML}(F{\uparrow}u).$$

Now let $A \not\in \mathbf{ML}(F)$. Then, by definition, there is a Kripke model M over F and a world u such that $M, u \not\vDash A$. By Lemma 1.3.25(1) $M{\uparrow}u, u \not\vDash A$, and so $F{\uparrow}u \not\vDash A$.

(2) By 1.3.25(1), $\mathbf{MT}(M) \subseteq \mathbf{MT}(M \uparrow u)$, so

$$\mathbf{MT}(M) \subseteq \bigcap_{u\in M} \mathbf{MT}(M \uparrow u).$$

On the other hand, if $A \not\in \mathbf{MT}(M)$, then $M, u \not\vDash A$ for some $u \in M$; hence $M \uparrow u, u \not\vDash A$ by 1.3.25 (1), thus $A \not\in \mathbf{MT}(M \uparrow u)$.

■

Definition 1.3.27 *The* disjoint sum *(or* disjoint union*) of a family of frames* $F_j = (W_j, R_{1j}, \dots, R_{nj})$, *for* $j \in J$, *is the frame* $\bigsqcup_{j\in J} F_j := (W, R_1, \dots, R_N)$, *where*

$$W = \bigsqcup_{j\in J} W_j := \bigcup_{j\in J} (W_j \times \{j\}),$$

$$(x, j)R_i(y, j') \text{ iff } j = j' \ \&\ xR_{ij}y.$$

Obviously, in this case $F'_i := \left(\bigsqcup_{j\in J} F_j\right) \restriction (W_i \times \{i\})$ is a generated subframe of $\bigsqcup_{j\in J} F_j$ isomorphic to F_i.

Hence we obtain

Lemma 1.3.28 $\mathbf{ML}\left(\bigsqcup_{j\in J} F_j\right) = \bigcap_{j\in J} \mathbf{ML}(F_j).$

Proof Let $F := \bigsqcup_{j\in J} F_j$. By Lemma 1.3.26,

$$\mathbf{ML}(F) = \bigcap_{v\in F} \mathbf{ML}(F{\uparrow}v) = \bigcap_{j\in J} \bigcap_{u\in F_j} \mathbf{ML}(F{\uparrow}(u, j)).$$

But the embedding of F_j in F yields an isomorphism $F_j{\uparrow}u \cong F{\uparrow}(u, j)$, hence

$$\bigcap_{u\in F_j} \mathbf{ML}(F{\uparrow}(u, j)) = \bigcap_{u\in F_j} \mathbf{ML}(F_j{\uparrow}u) = \bigcap_{j\in J} \mathbf{ML}(F_j)$$

by Lemma 1.3.26, which implies the main statement. ■

Remark 1.3.29 One can show that $MA\left(\bigsqcup_{j\in J} F_j\right) \cong \prod_{j\in J} MA(F_j)$. One can also show that $\mathbf{ML}\left(\prod_{j\in J} \boldsymbol{\Omega}_j\right) = \bigcap_{j\in J} \mathbf{ML}(\boldsymbol{\Omega}_j)$ for a family of modal algebras $(\boldsymbol{\Omega}_j)_{j\in J}$. Together with Lemma 1.3.7, this yields an alternative proof of 1.3.28.

Definition 1.3.30 *Let $F = (W, R_1, \dots, R_N)$, $F' = (W, R'_1, \dots, R'_N)$ be two frames. A map $f : W \to W'$ is called a* morphism *from F to F' (notation: $f : F \longrightarrow F'$) if it satisfies the following conditions:*

(1) $\forall u, v \in W\ \forall i\ (uR_iv \Rightarrow f(u)R'_if(v))$ (monotonicity);

(2) $\forall u \in W\ \forall v' \in W'\ \forall i\ (f(u)R'_iv' \Rightarrow \exists v\ (uR_iv\ \&\ f(v) = v'))$ (lift property).

A surjective morphism is called a p-morphism *(notation: $f : F \twoheadrightarrow F'$).*

f is called an isomorphism *from F onto F' (notation: $f : F \cong F'$) if it is a monotonic bijection and f^{-1} is also monotonic.*

As usual, F and F' are called *isomorphic* (notation: $F \cong F'$) if there exists an isomorphism from F onto F'.

We write $F \twoheadrightarrow F'$ if there exists a p-morphism from F onto F'.

Note that the conjunction (1) & (2) is equivalent to 'cone preservation':

$$\forall u \forall i\ f(R_i(u)) = R'_i(f(u)).$$

Definition 1.3.31 *Let $M = (F, \theta)$, $M' = (F', \theta')$ be two Kripke models. A (p-)morphism (respectively, an isomorphism) of frames $f : F \longrightarrow F'$ is said to be a* (p-)morphism (respectively, an isomorphism) from M (on)to M' *if for every $q \in PL$, $u \in W$*

$$M, u \vDash q \text{ iff } M', f(u) \vDash q.$$

As in the case of frames, morphisms of models are denoted by $\longrightarrow$, p-morphisms by $\twoheadrightarrow$, isomorphisms by $\cong$.

Lemma 1.3.32

(1) The composition of frame morphisms $F \longrightarrow F'$ and $F' \longrightarrow F''$ is a frame morphism $F \longrightarrow F''$; similarly for p-morphisms of frames and for (p-)morphisms of Kripke models.

(2) A subframe $F' \subseteq F$ is generated iff the inclusion map is a morphism $F' \longrightarrow F$; similarly for Kripke models.

(3) A restriction of a frame morphism to a generated subframe is a frame morphism and moreover, a restriction to a cone is a p-morphism onto a cone; similarly for Kripke models.

(4) $\cong$ is an equivalence relation between Kripke frames; the same for Kripke models.

Proof

(1) Easily follows from cone preservation.

(2) In fact, the cone preservation for j is equivalent to $R'(x) = R(x)$ for $x \in F'$.

(3) Follows from (1) and (2); note that the restriction of $f : F \longrightarrow G$ to $F' \subseteq F$ is the composition $f \cdot j$, where $j : F' \longrightarrow F$ is the inclusion map. The image of a cone is a cone by cone preservation.

(4) Trivial, since a composition of isomorphisms is an isomorphism.

■

Lemma 1.3.33 (Morphism lemma)

(1) *Every morphism of Kripke models is* reliable, *i.e. if* $f : M \longrightarrow M'$, *then for any* $u \in M$, *for any modal formula* A,

$$M, u \vDash A \text{ iff } M', f(u) \vDash A.$$

(2) *If* $f : M \twoheadrightarrow M'$, *then for any modal formula* A,

$$M \vDash A \text{ iff } M' \vDash A.$$

(3) *If* $F \twoheadrightarrow F'$ *then* $\mathbf{ML}(F) \subseteq \mathbf{ML}(F')$.

Proposition 1.3.34 *Every variety* $\mathbf{V}(\Gamma)$ *is closed under generated subframes, p-morphic images, and disjoint sums.*

Proof Follows from 1.3.25, 1.3.28 and 1.3.33. ■

Remark 1.3.35 This proposition resembles the 'only if' part of the Birkhoff theorem 1.2.25. The analogy becomes clear if we note the duality between Kripke frames and their modal algebras — generated subframes correspond to homomorphic images of algebras, p-morphic images to subalgebras, and disjoint sums to direct products. However the converse to 1.3.34 is not true. A precise model-theoretic characterisation of Kripke frame varieties (using ultrafilter extensions) was given by van Benthem, cf. [van Benthem, 1983], [Blackburn, de Rijke and Venema, 2001]. We shall return to this topic in Volume 2.

Exercise 1.3.36 Show the following analogue of 1.2.26: the poset of Kripke-complete N-modal propositional logics is dually isomorphic to the poset of all N-modal Kripke frame varieties.

Proposition 1.3.37 *Let* $(F_i \mid i \in I)$ *be a family consisting of all different components of a propositional Kripke frame* F. *Then* $F \cong \bigsqcup_{i \in I} F_i$.

Proof An isomorphism is given by the map f sending x to (x, i) whenever $x \in F_i$. This map is well-defined, since different components are disjoint by 1.3.24.

If xR_jy and $x \in F_i$, then $y \in F_i$ as well, so $f(x) = (x,i)R_{ji}(y,i) = f(y)$. The implication $f(x)R_{ji}f(y) \Rightarrow xR_jy$ is trivial. ■

Proposition 1.3.38 *Let* $(F_i \mid i \in I)$ *be a family of all different components of a propositional Kripke frame* F.

Then

(1) *for any morphism* $f : F \to G$, *every* $f_i := f \upharpoonright F_i$ *is a morphism;*

(2) *for any family of morphisms* $f_i : F_i \longrightarrow G$, *the joined map* $f := \bigcup_{i \in I} f_i$ *is a morphism* $F \longrightarrow G$.

Proof

(1) Since every component is a generated subframe, this follows from Lemma 1.3.32(3).

(2) Let R_j be a relation in F, S_j the corresponding relation in G. For $x \in F_i$ we have $R_j(x) \subseteq F_i$, so

$$f[R_j(x)] = f_i[R_j(x)] = S_j(f_i(x)) = S_j(f(x)),$$

since $f \restriction F_i = f_i$ and f_i is a morphism.

■

In this book we are especially interested in **K4**- and **S4**-frames; by Lemma 1.3.11, **K4**-frames are transitive, and **S4**-frames are quasi-ordered sets.

For a **K4**-frame (W, R) there is the equivalence relation $\approx_R := (R \cap R^{-1}) \cup Id_W$ in W; its equivalence classes are called *(R-)clusters*, and we can consider the quotient set as a frame.

Definition 1.3.39 *Let $F = (W, R)$ be a* **K4***-frame. Let $W^\sim := W/\approx_R$ be the set of all R-clusters, and let $u^\sim$ be the cluster of u. Then the frame $F^\sim := (W^\sim, R^\sim)$, where $R^\sim := \{(u^\sim, v^\sim) \mid uRv\}$, is called the* skeleton *of F.*

A singleton cluster $\{u\}$ is called trivial *(respectively,* degenerate*) if u is reflexive (respectively, irreflexive).*

Lemma 1.3.40

(1) $R^\sim$ is transitive and antisymmetric; if R is reflexive, then $R^\sim$ is a partial order.

(2) The map $u \mapsto u^\sim$ is a p-morphism from F onto $F^\sim$.

Proof Straightforward. ■

Lemma 1.3.41 *If $f : F \twoheadrightarrow G$ and F is connected,then G is connected.*

Proof If $(u_0, j_0, \ldots, j_{m-1}, u_m)$ is a path from u to v in F, then $(f(u_0), j_0, \ldots, f(u_m))$ is a path from $f(u)$ to $f(v)$ in G. ■

1.3.4 Conical expressiveness

Definition 1.3.42 *An N-modal propositional logic L is called* conically expressive *if there exists a propositional N-modal formula $C(p)$ with a single proposition letter p such that for any propositional Kripke model M with $M \vDash L$, for any $u \in M$,*

$$M, u \vDash C(p) \Leftrightarrow M \uparrow u \vDash p.$$

Obviously, an extension of a conically expressive logic (in the same language) is conically expressive. A simple example of a conically expressive logic is **K4**; the corresponding $C(p)$ is $p \wedge \Box p$.

Exercise 1.3.43

(a) Show that $\mathbf{K} + \Box^2 p \supset \Box^4 p$ is conically expressive.

(b) Recall that $\mathbf{K4.t} = \mathbf{K}_2 + \Diamond_1\Box_2 \supset p + \Diamond_1\Box_1 p \supset p + \Box_1 p \supset \Box_1^2 p$. Show that $\mathbf{K4.t} + \Box_1\Box_2 p \supset \Box_2\Box_1 p$ is conically expressive.

Now given C from Definition 1.3.42, for a formula $A \in \mathcal{LP}_N$ put

$$\Box^* A := [A/p]C(p).$$

Let us show that $\Box^*$ behaves like an **S4**-modality in logics containing L.

Lemma 1.3.44 *For any N-modal Kripke model M such that $M \vDash \mathbf{\Lambda}$ for any $u \in M$, for any N-modal formula A*

$$M, u \vDash \Box^* A \Leftrightarrow \forall v \in R^*(u)\, M, v \vDash A,$$

where R^* is the same as in Lemma 1.3.19.

Proof Similar to soundness of the substitution rule. Let M_0 be a Kripke model over the same frame as M, such that for any $v \in M$

$$M_0, v \vDash p \Leftrightarrow M, v \vDash A.$$

Then by induction we obtain for any propositional formula $X(p)$ for any $v \in M$.

$$M_0, v \vDash X(p) \Leftrightarrow M, v \vDash X(A).$$

Hence for any $u \in M$ and for C from 1.3.42

$$M_0, u \vDash C(p) \Leftrightarrow M, u \vDash C(A)\, (= \Box^* A),$$

and so by 1.3.42,

$$M, u \vDash \Box^* A \Leftrightarrow M_0 \uparrow u \vDash p.$$

But by Lemma 1.3.19 and the choice of M_0

$$M_0 \uparrow u \vDash p \Leftrightarrow \forall v \in R^*(u)\ M_0, v \vDash p \Leftrightarrow \forall v \in R^*(u)\ M, v \vDash A.$$

Hence the claim follows. ∎

Lemma 1.3.45 *If a modal logic $\mathbf{\Lambda}$ is conically expressive and $\Box^*$ is defined as above, then the rule $\dfrac{A}{\Box^* A}$ is admissible in $\mathbf{\Lambda}$.*

Proof Since by 1.13.3, $\mathbf{\Lambda}$ has an exact model, it suffices to show that for any Kripke model M, for any A, $M \vDash A$ implies $M \vDash \Box^* A$. But this readily follows from 1.3.44. ∎

Lemma 1.3.46 *If an N-modal propositional logic $\mathbf{\Lambda}$ is conically expressive, then the following formulas are in $\mathbf{\Lambda}$ (for $k \leq N$):*

(1) $\Box^*(p \supset q) \supset_{\bullet} \Box^* p \supset \Box^* q$;

(2) $\Box^* p \supset p$;

(3) $\Box^* p \supset \Box^* \Box^* p$;

(4) $\Box^* p \supset \Box_k \Box^* p$;

(5) $\Box^* p \supset \Box_k p$.

Proof Again we can consider an exact model and show that these formulas are true at every world. This easily follows from 1.3.44, the reflexivity and the transitivity of R^*, and the inclusions $R^* \supseteq R_k \circ R^*$, $R_k \subseteq R^*$. ■

1.4 Relational semantics (the intuitionistic case)

Definition 1.4.1 *Assume that $F = (W, R)$ is an* **S4**-*frame, i.e. $MA(F)$ is a topo-Boolean algebra. Then the algebra of its open elements (or, equivalently, of stable subsets of F)*

$$HA(F) := MA(F)^0$$

is called the Heyting algebra of F.

S4-*frames are also called* intuitionistic propositional.

An intuitionistic valuation *in F is defined as a valuation in $HA(F)$, i.e. as a valuation in W with the following* truth-preservation *(or* monotonicity*) property (for every $q \in PL$):*

$$(TP_0) \qquad \forall u, v \in W \ (uRv \ \& \ u \in \theta(q) \Rightarrow v \in \theta(q)).$$

The corresponding Kripke model $M = (F, \theta)$ is also called intuitionistic. *θ is extended to the map $\theta^I : IF \longrightarrow HA(F)$ according to Definition 1.2.12.*

Definition 1.4.2 *For an intuitionistic Kripke model $M = (W, R, \theta)$ we define the intuitionistic* forcing relation *between worlds and intuitionistic formulas as follows:*

$$M, u \Vdash A \ := \ u \in \theta^I(A).$$

Now we readily obtain an alternative inductive definition of intuitionistic forcing, cf. Lemma 1.3.3.

Lemma 1.4.3 *Intuitionistic forcing has the following properties:*

- $M, u \Vdash q$ *iff* $u \in \theta(q)$ *(for $q \in PL$);*
- $M, u \nVdash \bot$;
- $M, u \Vdash B \wedge C$ *iff* $M, u \Vdash B \ \& \ M, u \Vdash C$;
- $M, u \Vdash B \vee C$ *iff* $M, u \Vdash B \vee M, u \Vdash C$;

- $M, u \Vdash B \supset C$ *iff* $\forall v \in R(u)\ (M, v \Vdash B \Rightarrow M, v \Vdash C)$;
- $M, u \Vdash \neg B$ *iff* $\forall v \in R(u)\ M, v \nVdash B$;
- $M, u \Vdash B \equiv C$ *iff* $\forall v \in R(u)\ (M, v \Vdash B \Leftrightarrow M, v \Vdash C)$.

Lemma 1.4.4 *In intuitionistic Kripke models the truth-preservation holds for any formula A:*

$$(TP) \qquad \forall u, v \in F\ (uRv\ \&\ M, u \Vdash A \Rightarrow M, v \Vdash A).$$

Proof Trivial, since $\theta^I(A)$ is stable. ■

Definition 1.4.5 *An intuitionistic formula A is said to be* valid *in an* **S4**-*frame F (notation: $F \Vdash A$) if $HA(F) \Vdash A$, i.e. if A is true in every intuitionistic model over F.*

Similarly, one can reformulate the definitions of satisfiability, etc. (1.3.5, 1.3.6) for the intuitionistic case.

We also have a relational analogue of Lemma 1.2.19, for which we need the following

Definition 1.4.6 *Let F be an* **S4**-*frame, M a Kripke model over F. The* pattern *of M is the Kripke model M_0 over F such that for any $u \in F$, $q \in PL$*

$$M_0, u \vDash q \text{ iff } M, u \vDash \Box q.$$

Obviously, M_0 is an intuitionistic Kripke model; $M_0 = M$ if M is itself intuitionistic.

Lemma 1.4.7 *Under the conditions of Definition 1.4.6, we have*

(1) for any $u \in F$ and intuitionistic formula A

$$M_0, u \Vdash A \text{ iff } M, u \vDash A^T,$$

(2) $F \Vdash A$ iff $F \vDash A^T$.

Proof Obvious from 1.2.19. ■

Together with Lemma 1.2.14 and the trivial observation that $F \nVdash \bot$, this implies

Lemma 1.4.8 *Let F be an* **S4**-*frame. Then the set of all intuitionistic formulas valid in F is an intermediate logic.*

This logic is called the *intermediate logic of F* and denoted by $\mathbf{IL}(F)$.

For a set of intuitionistic formulas Γ, an *intuitionistic* Γ-*frame* is an intuitionistic propositional frame, in which Γ is intuitionistically valid. The class of all these frames is called the *intuitionistic (Kripke frame) variety* of Γ and denoted by $\mathbf{V}^I(\Gamma)$.

Definition 1.3.9 is obviously transferred to the intuitionistic case. Respectively the notation **ML**, **MT** changes to **IL**, **IT**. Lemmas 1.3.10 and 1.3.13 and Proposition 1.3.15 also have intuitionistic versions; the reader can easily formulate them.

Lemma 1.4.9 *Let F be an* **S4***-frame, M an intuitionistic model over F.*

(1) If M_1 is a generated submodel of M, then M_1 also is intuitionistic, and for any $A \in \mathcal{LP}_0$, $u \in M_1$

$$M_1, u \Vdash A \text{ iff } M, u \Vdash A.$$

(2) If F_1 is a generated subframe of F then

$$\mathbf{IL}(F) \subseteq \mathbf{IL}(F_1).$$

Proof
(1) From the generation lemma 1.3.25 and the definition of intuitionistic forcing (1.4.1).
(2) From Lemma 1.3.25(2) and Lemma 1.4.7(2). ∎

Lemma 1.4.10 *Let F' be a subframe of a Kripke* **S4***-frame F. Then every intuitionistic valuation θ' in F' can be extended to an intuitionistic valuation in F.*

Proof In fact, take a valuation θ in F such that $\theta(q) = R(\theta'(q))$ for any $q \in PL$. ∎

The three subsequent lemmas readily follow from Lemmas 1.4.7, 1.3.26, 1.3.28 and 1.3.33.

Lemma 1.4.11 *For an* **S4***-frame F*

$$\mathbf{IL}(F) = \bigcap_{u \in F} \mathbf{IL}(F{\uparrow}u).$$

Lemma 1.4.12 *For* **S4***-frames F_i, $i \in I$*

$$\mathbf{IL}\left(\bigsqcup_{i \in I} F_i\right) = \bigcap_{i \in I} \mathbf{IL}(F_i).$$

Lemma 1.4.13 (Morphism lemma) *Let M, M' be intuitionistic Kripke models, and let F, F' be* **S4***-frames.*

(1) If $f : M \longrightarrow M'$, then f is reliable: for any $u \in M$, for any intuitionistic formula A,

$$M, u \Vdash A \text{ iff } M', f(u) \Vdash A.$$

(2) If $f : M \longrightarrow M'$, then

$$M \Vdash A \text{ iff } M' \Vdash A.$$

(3) If $F \twoheadrightarrow F'$, then $\mathbf{IL}(F) \subseteq \mathbf{IL}(F')$.

Lemma 1.4.14 *Let F be an* **S4***-frame. Then*

$$\mathbf{IL}(F) = \mathbf{IL}(F^{\sim}).$$

Proof If φ is an intuitionistic valuation in F, consider the valuation $\varphi^{\sim}$ in $F^{\sim}$ defined by

$$\varphi^{\sim}(q) := \{u^{\sim} \mid u \in \varphi(q)\}.$$

It is clear that $\varphi^{\sim}$ is well-defined and intuitionistic. By Lemma 1.3.40, it follows that the map sending u to $u^{\sim}$ is a p-morphism of Kripke models, and thus

$$(F, \varphi) \vDash A \text{ iff } (F^{\sim}, \varphi^{\sim}) \vDash A$$

by Lemma 1.4.13. To complete the proof, note that an arbitrary intuitionistic valuation ψ in $F^{\sim}$ can be presented as $\varphi^{\sim}$, with

$$\varphi(q) = \{u \mid u^{\sim} \in \psi(q)\}.$$

∎

For a class $\mathcal{C}$ of **S4**-frames let $\mathcal{C}^{\sim}$ be the closure of $\{F^{\sim} \mid F \in \mathcal{C}\}$ under isomorphism, and let Posets be the class of all posets.

Lemma 1.4.15

(1) $\mathbf{IL}(\mathcal{C}) = \mathbf{IL}(\mathcal{C}^{\sim})$.

(2) $\mathbf{V}^I(\Gamma)^{\sim} = \mathbf{V}^I(\Gamma) \cap \mathsf{Posets}$ *for any set of intuitionistic formulas* Γ.

(3) Every Kripke-complete intermediate logic is determined by some class of posets: $L = \mathbf{IL}(\mathbf{V}^I(L) \cap \mathsf{Posets})$.

Proof

(1) Follows readily from 1.4.14.

(2) In fact, $F \Vdash \Gamma \Leftrightarrow F^{\sim} \Vdash \Gamma$ by 1.4.14, so $\mathbf{V}^I(\Gamma)^{\sim} \subseteq \mathbf{V}^I(\Gamma)$. The other way round, if a poset $G \in \mathbf{V}^I(\Gamma)$, then $G \cong G^{\sim} \in \mathbf{V}^I(\Gamma)^{\sim}$.

(3) Note that $L = \mathbf{IL}(\mathbf{V}^I(L))$ for a complete L and apply (1), (2).

∎

So instead of $\mathbf{V}^I(\Gamma)$ we can use the *reduced intuitionistic variety*

$$\mathbf{V}^{\sim}(\Gamma) := \mathbf{V}^I(\Gamma) \cap \mathsf{Posets}.$$

Now we obtain an analogue of 1.3.34:

Proposition 1.4.16 *Intuitionistic Kripke frame varieties and reduced intuitionistic Kripke frame varieties are closed under generated subframes, p-morphic images, and disjoint sums.*

Proof For intuitionistic varieties this follows from 1.4.9, 1.4.12 and 1.4.13. For reduced intuitionistic varieties we can also apply 1.4.15(2) and note that the class of posets is closed under the same three operations. ■

Let us describe reduced intuitionistic varieties for some intuitionistic formulas.

Proposition 1.4.17

- $\mathbf{V}^{\sim}(EM) = \mathbf{V}^{\sim}(\mathbf{CL})$ *consists of all* trivial *frames, i.e. frames of the form* (W, Id_W);
- $\mathbf{V}^{\sim}(AJ) = \mathbf{V}^{\sim}(\mathbf{HJ}) = \mathbf{V}(\mathbf{S4.2})^{\sim}$ *consists of all confluent posets;*
- $\mathbf{V}^{\sim}(AZ) = \mathbf{V}^{\sim}(\mathbf{LC}) = \mathbf{V}(\mathbf{S4.3})^{\sim}$ *consists of all nonbranching posets;*
- $\mathbf{V}^{\sim}(AP_n)$ *consists of all posets of depth* $\leq n$;[7]
- $\mathbf{V}^{\sim}(AIW_n)$ *consists of all posets of width* $\leq n$;[8]
- $\mathbf{V}^{\sim}(AG_n)$ *consists of all posets* F, *where for any* $x \in F$, $|HA(F\uparrow x)| \leq n$.

Definition 1.4.18 *A* quasi-morphism *between* **S4**-*frames* $F = (W, R)$ *and* $F' = (W', R')$ *is a monotonic map* $h : W \longrightarrow W'$ *with the* quasi-lift property*:*

$$\forall x \in W\, \forall y' \in W'\, (h(x)R'y' \Rightarrow \exists y\, (xRy\ \&\ h(y) \approx_{R'} y')).$$

For intuitionistic Kripke models $M = (F, \theta)$, $M' = (F', \theta')$ *a quasi-morphism from* M *to* M' *is a quasi-morphism of their frames such that for any* $q \in PL$

$$M, x \Vdash q \text{ iff } M', h(x) \Vdash q.$$

A quasi-p-morphism *is a surjective quasi-morphism.*

The following is clear:

Lemma 1.4.19 *A quasi-(p)-morphism of frames* $h : F \longrightarrow F'$ *gives rise to a quasi-(p)-morphism of their skeletons* $h^{\sim} : F^{\sim} \twoheadrightarrow F'^{\sim}$ *such that* $h^{\sim}(u^{\sim}) = h(u)^{\sim}$.

Quasi p-morphisms are reliable for intuitionistic formulas:

Lemma 1.4.20

(1) If h *is a quasi-morphism from* M *to* M', *then for any* $x \in M$, *for any intuitionistic formula* A,

$$M, x \Vdash A \text{ iff } M', h(x) \Vdash A.$$

(2) If there exists a quasi-p-morphism from F *onto* F' *then* $\mathbf{IL}(F) \subseteq \mathbf{IL}(F')$.

Proof By Lemmas 1.4.20 and 1.4.14. ■

---

[7] See Section 1.15.

[8] See Section 1.9.

1.5 Modal counterparts

The following lemma can be easily proved by induction.

Lemma 1.5.1 *Let* $S = [C/p]$ *be an intuitionistic substitution,* $S^T := [C^T/p]$*. Then* $\mathbf{S4} \vdash (SA)^T \equiv S^T A^T$ *for any intuitionistic formula* A*.*

This lemma implies

Proposition 1.5.2 *For any 1-modal logic* $\mathbf{\Lambda} \supseteq \mathbf{S4}$ *the set*

$$^T\mathbf{\Lambda} := \{A \in \mathcal{LP}_0 \mid A^T \in \mathbf{\Lambda}\}$$

is a superintuitionistic logic.

Lemma 1.5.3 *For any* **S4**-*algebra* $\mathbf{\Omega}$, $^T\mathbf{ML}(\mathbf{\Omega}) = \mathbf{IL}(\mathbf{\Omega}^0)$.

Proof This is a reformulation of 1.2.19(2). ∎

Definition 1.5.4 *The above defined logic* $^T\mathbf{\Lambda}$ *is called the* superintuitionistic fragment *of* $\mathbf{\Lambda}$*; the logic* $\mathbf{\Lambda}$ *is called a* modal counterpart *of* $^T\mathbf{\Lambda}$.

For a set Γ of intuitionistic formulas let $\Gamma^T := \{A^T \mid A \in \Gamma\}$.

Theorem 1.5.5[9] *Every propositional superintuitionistic logic* $L = \mathbf{H} + \Gamma$ *has the smallest modal counterpart:* $\tau(L) := \mathbf{S4} + \Gamma^T$.

Theorem 1.5.6[10] *Every propositional superintuitionistic logic has the greatest modal counterpart. In particular, the greatest modal counterpart of* **H** *is* **Grz**.

The greatest modal counterpart of L is denoted by $\sigma(L)$.

Theorem 1.5.7 (Blok–Esakia)

(1) $\sigma(\mathbf{H} + \Gamma) = \mathbf{Grz} + \Gamma^T$.

(2) The correspondence between superintuitionistic logics and their greatest modal counterparts is an order isomorphism between superintuitionistic logics and modal logics above **Grz**.

See [Chagrov and Zakharyaschev, 1997] for the proof of 1.5.7 (as well as 1.5.5 and 1.5.6).

Proposition 1.5.8 *If a modal logic* $\mathbf{\Lambda} \supseteq \mathbf{S4}$ *is Kripke-complete, then* $^T\mathbf{\Lambda}$ *is also Kripke-complete. More precisely,* $^T\mathbf{ML}(\mathcal{C}) = \mathbf{IL}(\mathcal{C})$.

---

[9] [Dummett and Lemmon, 1959].

[10] [Maksimova and Rybakov, 1974], [Esakia, 1979].

Proof In fact, ${}^{T}\mathbf{ML}(\mathcal{C}) = \bigcap_{F \in \mathcal{C}} {}^{T}\mathbf{ML}(F) = \bigcap_{F \in \mathcal{C}} \mathbf{IL}(F) = \mathbf{IL}(\mathcal{C})$. The second equality follows from 1.4.7. ■

Lemma 1.5.9 *For any intermediate logic L,*

$$\mathbf{V}(\tau(L)) = \mathbf{V}^{I}(L).$$

Proof An exercise. ■

Theorem 1.5.10 (Zakharyaschev) *The map τ preserves Kripke-completeness, so if an intermediate logic L is Kripke-complete, then*

$$\tau(L) = \mathbf{ML}(\mathbf{V}^{I}(L)).$$

For the proof see [Chagrov and Zakharyaschev, 1997]

Remark 1.5.11 Unlike τ, the map σ does not preserve Kripke-completeness; a counterexample can be found in [Shehtman, 1980].

1.6 General Kripke frames

'General Kripke frame semantics' from [Thomason, 1972][11] (cf. also [Chagrov and Zakharyaschev, 1997]) is an extended version of Kripke semantics, which is equivalent to algebraic semantics.

Definition 1.6.1 *A* general modal Kripke frame *is a modal Kripke frame together with a subalgebra of its modal algebra, i.e. $\Phi = (F, \mathcal{W})$, where $\mathcal{W} \subseteq MA(F)$ is a modal subalgebra. $\mathcal{W}$ is called* the modal algebra of Φ *and also denoted by $MA(\Phi)$; its elements are called* interior sets *of Φ.*

So for $F = (W, R_1, \ldots, R_N)$, $\mathcal{W}$ should be a non-empty set of subsets closed under Boolean operations and $\Box_i : V \mapsto \Box_i V$ (Section 1.3). Instead of $\Box_i$ one can use its dual $\Diamond_i V := R_i^{-1} V$.

Definition 1.6.2 *A* valuation *in a general Kripke frame $\Phi = (F, \mathcal{W})$ is a valuation in $MA(\Phi)$. A* Kripke model over Φ *is just a Kripke model $M = (F, \theta)$ over F, in which θ is a valuation in Φ. A modal formula A is* valid *in Φ (notation: $\Phi \vDash A$) if it is true in every Kripke model over Φ; similarly for a set of formulas.*

Analogous definitions are given in the intuitionistic case.

Definition 1.6.3 *A* general intuitionistic Kripke frame *is $\Phi = (F, \mathcal{W})$, where F is an intuitionistic Kripke frame, $\mathcal{W} \subseteq HA(F)$ is a Heyting subalgebra. $\mathcal{W}$ is called* the Heyting algebra of Φ *and denoted by $HA(\Phi)$; its elements are called* interior sets.

[11] In that paper it was called 'first-order semantics'.

Definition 1.6.4 *An* intuitionistic valuation *in a general intuitionistic Kripke frame* $\Phi = (F, \mathcal{W})$ *is a valuation in* $HA(\Phi)$. *An* intuitionistic Kripke model over Φ *is of the form* $M = (F, \theta)$, *where* θ *is an intuitionistic valuation in* Φ. *An intuitionistic formula* A *is* valid *in* Φ *(notation:* $\Phi \Vdash A$*) if it is true in every intuitionistic Kripke model over* Φ*; similarly for a set of formulas.*

Obviously we have analogues of 1.3.7, 1.3.8, 1.4.7 and 1.4.8:

Lemma 1.6.5 *For any modal formula* A *and a general Kripke frame* Φ

$$\Phi \vDash A \ \text{ iff } \ MA(\Phi) \vDash A;$$

analogously, for intuitionistic A *and* Φ

$$\Phi \Vdash A \ \text{ iff } \ HA(\Phi) \vDash A;$$

Lemma 1.6.6

(1) For a general modal Kripke frame Φ *the set*

$$\mathbf{ML}(\Phi) := \{A \mid \Phi \vDash A\}$$

is a modal logic; if Φ *is intuitionistic, then*

$$\mathbf{IL}(\Phi) := \{A \mid \Phi \Vdash A\}$$

is an intermediate logic, and moreover, $\mathbf{IL}(\Phi) = {}^T\mathbf{ML}(\Phi)$.

(2) For a class $\mathcal{C}$ *of* N*-modal general frames the set*

$$\mathbf{ML}(\mathcal{C}) := \bigcap\{\mathbf{ML}(\Phi) \mid \Phi \in \mathcal{C}\}$$

is an N*-modal logic; if the frames are intuitionistic, then*

$$\mathbf{IL}(\mathcal{C}) := \bigcap\{\mathbf{IL}(\Phi) \mid \Phi \in \mathcal{C}\}$$

is an intermediate logic, and $\mathbf{IL}(\mathcal{C}) = {}^T\mathbf{ML}(\mathcal{C})$.

Lemma 1.6.7

(1) For an N*-modal Kripke model* $M = (F, \theta)$ *the set of all* definable sets $\mathcal{W}_M := \{\theta(A) \mid A \in \mathcal{LP}_n\}$ *is a subalgebra of* $MA(F)$.

(2) Similarly, for an intuitionistic Kripke model $M = (F, \theta)$, $\mathcal{W}_M^I = \{\theta^I(A) \mid A \in \mathcal{LP}_0\}$ *is a subalgebra of* $HA(F)$.

Proof (Modal case) In fact, by definition

$$\theta(\neg A) = -\theta(A), \ \theta(A \wedge B) = \theta_L(A) \cap \theta_L(B), \ \theta(\Box_i A) = \Box_i \theta(A).$$

∎

So we define the corresponding general frames

Definition 1.6.8 *For a Kripke model $M = (F, \theta)$, the general frame $GF(M) := (F, \mathcal{W}_M)$ (or $GF^I(M) := (F, \mathcal{W}^I_M)$ in the intuitionistic case) is called* associated.

Lemma 1.6.9 *$GF(M) \vDash A$ iff all modal substitution instances of A are true in M; similarly for the intuitionistic case.*

Proof (If.) For $M = (F, \theta)$, let η be a valuation in F such that $\eta(p_i) = \theta(B_i)$ for every i. An easy inductive argument shows that

$$\eta(A) = \theta([B_1, \dots, B_n/p_1, \dots, p_n]A)$$

for any n-formula A, cf. Lemma 1.2.9.

(Only if.) The same equality shows that for any modal (or intuitionistic) substitution S, $\theta(SA) = \eta(A)$ for an appropriate valuation η in $GF(M)$. ∎

Definition 1.6.10 *If $\Phi = (F, \mathcal{W})$ is a general Kripke frame, $F = (W, R_1, \dots, R_N), V \subseteq W$, then we define the corresponding* general subframe*:*

$$\Phi \restriction V := (F \restriction V, \mathcal{W} \restriction V),$$

where

$$\mathcal{W} \restriction V := \{X \cap V \mid X \in \mathcal{W}\}.$$

If V is stable, $\Phi \restriction V$ is called a generated (general) subframe. *The* cone generated by *u in Φ is the subframe $\Phi{\uparrow}u := \Phi \restriction (W{\uparrow}u)$.*

The definition of a subframe is obviously sound, because $\mathcal{W} \restriction V$ is a modal algebra of subsets of V with modal operations $\Box_{iV} X := \Box_i X \cap V$.

The following is a trivial consequence of 1.3.25 and 1.3.26.

Lemma 1.6.11 *For a general frame Φ,*

(1) if V is a stable subset, then $\mathbf{ML}(\Phi) \subseteq \mathbf{ML}(\Phi \restriction V)$*;*

(2) $\mathbf{ML}(\Phi) = \bigcap_{u \in \Phi} \mathbf{ML}(\Phi{\uparrow}u)$.

Exercise 1.6.12 Define morphisms of general frames and prove their properties.

1.7 Canonical Kripke models

Definition 1.7.1 *The* canonical Kripke frame *for an N-modal propositional logic $\mathbf{\Lambda}$ is $F_\mathbf{\Lambda} := (W_\mathbf{\Lambda}, R_{1,\mathbf{\Lambda}}, \dots, R_{N,\mathbf{\Lambda}})$, where $W_\mathbf{\Lambda}$ is the set of all $\mathbf{\Lambda}$-complete theories, $xR_{i,\mathbf{\Lambda}}y$ iff for any formula A, and $\Box_i A \in x$ implies $A \in y$.*

The canonical model *for $\mathbf{\Lambda}$ is $M_\mathbf{\Lambda} = (F_\mathbf{\Lambda}, \theta_\mathbf{\Lambda})$, where*

$$\theta_\mathbf{\Lambda}(p_i) := \{x \mid p_i \in x\}.$$

Analogously, the canonical frame *for a bounded modal logic* $\mathbf{\Lambda}\lceil m$ *is* $F_{\mathbf{\Lambda}\lceil m} := (W_{\mathbf{\Lambda}\lceil m}, R_{1,\mathbf{\Lambda}\lceil m}, \ldots, R_{N,\mathbf{\Lambda}\lceil m})$, *where* $W_{\mathbf{\Lambda}\lceil m}$ *is the set of all maximal* $\mathbf{\Lambda}$*-consistent sets of* N*-modal* m*-formulas,*

$$xR_{i,\mathbf{\Lambda}\lceil m}y \text{ iff for any } m\text{-formula } A,\ \Box_i A \in x \text{ implies } A \in x.$$

The canonical model *for* $\mathbf{\Lambda}\lceil m$ *is* $M_{\mathbf{\Lambda}\lceil m} := (F_{\mathbf{\Lambda}\lceil m}, \theta_{\mathbf{\Lambda}\lceil m})$, *where* $\theta_{\mathbf{\Lambda}\lceil m}(p_i) = \theta_{\mathbf{\Lambda}}(p_i)$ *for* $i \leq m$.

Definition 1.7.2 *The* canonical frame *for an intermediate propositional logic* $\mathbf{\Sigma}$ *is* $F_{\mathbf{\Sigma}} := (W_{\mathbf{\Sigma}},\ R_{\mathbf{\Sigma}})$, *where* $W_{\mathbf{\Sigma}}$ *is the set of all* $\mathbf{\Sigma}$*-complete intuitionistic (double) theories,* $xR_{\mathbf{\Sigma}}y$ *iff* $x \subseteq y$.

The canonical model *for* $\mathbf{\Sigma}$ *is* $M_{\mathbf{\Sigma}} := (F_{\mathbf{\Sigma}}, \theta_{\mathbf{\Sigma}})$, *where*

$$\theta_{\mathbf{\Sigma}}(p_i) := \{x \mid p_i \in x\}.$$

The corresponding definitions for bounded intermediate logics must be now clear, so we skip them.

The following is well-known, cf. [Chagrov and Zakharyaschev, 1997], [Blackburn, de Rijke and Venema, 2001].

Theorem 1.7.3 (Canonical model theorem) *For any* N*-modal or intermediate logic* $\mathbf{\Lambda}$ *and* m*-formula* A *(of the corresponding kind):*

(1) $M_{\mathbf{\Lambda}}, x \vDash A$ *iff* $A \in x$;

(2) $M_{\mathbf{\Lambda}\lceil m}, y \vDash A$ *iff* $A \in y$;

(3) $M_{\mathbf{\Lambda}\lceil m} \vDash A$ *iff* $M_{\mathbf{\Lambda}} \vDash A$ *iff* $A \in \mathbf{\Lambda}$.

Definition 1.7.4 *The* general canonical frame *of an* N*-modal (respectively, intuitionistic) logic* $\mathbf{\Lambda}$ *is* $\Phi_{\mathbf{\Lambda}} := GF(M_{\mathbf{\Lambda}})$ *(respectively,* $\Phi^I_{\mathbf{\Lambda}} := GF^I(M_{\mathbf{\Lambda}})$*).*

Theorem 1.7.5 (General canonical model theorem) *For a modal (respectively, intermediate) propositional logic* $\mathbf{\Lambda}$

$\mathbf{ML}(\Phi_{\mathbf{\Lambda}}) = \mathbf{\Lambda}$ *(respectively,* $\mathbf{IL}(\Phi_{\mathbf{\Lambda}}) = \mathbf{\Lambda}$*).*

Proof By Lemma 1.6.9, $\mathbf{ML}(\Phi_{\mathbf{\Lambda}})$ consists of all formulas A such that $M_{\mathbf{\Lambda}} \vDash SA$ of any modal substitution S. By 1.7.3, the latter is equivalent to $SA \in \mathbf{\Lambda}$. Since $\mathbf{\Lambda}$ is substitution closed, it follows that $\mathbf{ML}(\Phi_{\mathbf{\Lambda}}) = \mathbf{\Lambda}$. The same argument works for the intuitionistic case. ∎

An alternative proof of 1.7.5 can be obtained from the algebraic completeness theorem 1.2.21 and the following observation:

Proposition 1.7.6 $MA(\Phi_{\mathbf{\Lambda}}) \cong Lind(\mathbf{\Lambda})$ *for a modal logic* $\mathbf{\Lambda}$; $HA(\Phi_{\mathbf{\Sigma}}) \cong Lind(\mathbf{\Sigma})$ *for an intermediate logic* $\mathbf{\Sigma}$.

Proof (Modal case.) The map $\gamma : \ Lind(\mathbf{\Lambda}) \longrightarrow MA(\Phi_{\mathbf{\Lambda}})$ sending $[A]$ to $\theta_{\mathbf{\Lambda}}(A)$ is well-defined, since $M_{\mathbf{\Lambda}} \vDash A \equiv B$ whenever $A \sim_{\mathbf{\Lambda}} B$ by Theorem 1.7.1. γ is obviously surjective and preserves the modal algebra operations. In fact,

$$\gamma(\Box[A]) = \gamma([\Box A]) = \theta_{\mathbf{\Lambda}}(\Box A) = \Box\theta_{\mathbf{\Lambda}}(A)$$

according to Definitions 1.2.20 and 1.3.2, and similarly for the other operations. Finally, note that

$$\gamma([A]) = \gamma([B]) \text{ iff } \theta_{\mathbf{\Lambda}}(A) = \theta_{\mathbf{\Lambda}}(B) \text{ iff } M_{\mathbf{\Lambda}} \vDash A \equiv B \text{ iff } (A \equiv B) \in \mathbf{\Lambda} \text{ iff } [A] = [B]$$

by 1.7.1 and the definitions. So γ is an isomorphism. ■

Definition 1.7.7 *A general frame* $\Phi = ((W, R_1, \ldots, R_N), \mathcal{W})$ *(modal or intuitionistic) is called* descriptive *if it satisfies the following conditions:*

- Φ *is* differentiated (distinguishable)*: for any two different points* x, y *there exists an interior set* $U \in \mathcal{W}$ *separating them, i.e. such that* $x \in U \not\Leftrightarrow y \in U$*;*
- tightness*:*
  $\forall x, y, i\ (\forall U \in \mathcal{W}\,(x \in \Box_i U \Rightarrow y \in U) \Rightarrow xR_iy)$ *(in the modal case),*
  $\forall x, y\ (\forall U \in \mathcal{W}\,(x \in U \Rightarrow y \in U) \Rightarrow xR_1y)$ *(in the intuitionistic case);*
- compactness*:*
  every centered *subset* $\mathcal{X} \subseteq \mathcal{W}$ *(i.e. such that* $\bigcap \mathcal{X}_1 \neq \varnothing$ *for any finite* $\mathcal{X}_1 \subseteq \mathcal{X}$*) has a non-empty intersection (in the modal case);*
  if a pair $(\mathcal{X}, \mathcal{Y})$ *of subsets of* $\mathcal{W}$ *is* centered *(i.e.* $\bigcap \mathcal{X}_1 \not\subseteq \bigcup \mathcal{Y}_1$ *for any finite* $\mathcal{X}_1 \subseteq \mathcal{X}$, $\mathcal{Y}_1 \subseteq \mathcal{Y}$*), then* $\bigcap \mathcal{X} \not\subseteq \bigcup \mathcal{Y}$ *(in the intuitionistic case).*

A differentiated and tight general frame is called refined.

Lemma 1.7.8 *A generated subframe of a refined frame is refined.*

Proof Distinguishability is obviously preserved for subframes.

Let us check tightness in the modal case. We use the same notation as in 1.7.7. Suppose $V \subseteq W$ is stable, $x, y \in V$ and

$$(1) \qquad \forall U \in \mathcal{W} \upharpoonright V\ (x \in \Box_{iV} U \Rightarrow y \in U).$$

This is equivalent to

$$\forall U \in \mathcal{W}\ (x \in \Box_i(U \cap V) \cap V \Rightarrow y \in U \cap V)$$

and thus (since V is stable and $x, y \in V$) to

$$(2) \qquad \forall U \in \mathcal{W}\ (x \in \Box_i U \Rightarrow y \in U).$$

If Φ is tight, (2) implies xR_iy, therefore (1) also implies xR_iy, which means tightness of $\Phi \upharpoonright V$. ■

Remark 1.7.9 Compactness is not always preserved for generated subframes; the reader can try to construct a counterexample.

Descriptive frames can also be characterised as canonical frames of modal algebras. These frames resemble canonical frames of modal logics.

Recall that a *(proper) filter* in a Boolean algebra is a $\leq$-stable (proper) subset closed under meets. A *maximal filter* is maximal among proper filters.

Definition 1.7.10 *For an N-modal algebra $\mathbf{\Omega}$, consider the set Ω_+ of its maximal filters with the accessibility relations:*

$$xR_iy \text{ iff } \forall a\,(\Box_i a \in x \Rightarrow a \in y)$$

and with the interior sets of the form

$$h(a) := \{x \in \Omega_+ \mid a \in x\}$$

for all $a \in \mathbf{\Omega}$. The resulting general frame $\mathbf{\Omega}_+ := (\Omega_+, R_1, \ldots, R_N)$ is called the canonical *or the* dual frame *of $\mathbf{\Omega}$.*

Recall that a *(proper) prime* filter in a Heyting algebra is a $\leq$-stable (proper) subset X closed under $\wedge$ and such that

$$a \vee b \in X \text{ only if } (a \in X \text{ or } b \in X).$$

Definition 1.7.11 *For a Heyting algebra $\mathbf{\Omega}$, consider the set Ω_+ of its prime filters with the accessibility relation xRy iff $x \subseteq y$ and with the interior sets $h(a) = \{x \in \Omega_+ \mid a \in x\}$ for $a \in \mathbf{\Omega}$. The general frame $\mathbf{\Omega}_+ := (\Omega_+, R)$ is called the* canonical *or the* dual frame *of $\mathbf{\Omega}$.*

Theorem 1.7.12 (Tarski–Jonsson)

(1) If $\mathbf{\Omega}$ is a modal algebra, then $MA(\mathbf{\Omega}_+) \cong \mathbf{\Omega}$.

(2) If $\mathbf{\Omega}$ is a Heyting algebra, then $HA(\mathbf{\Omega}_+) \cong \mathbf{\Omega}$.

A required isomorphism is the Stone map h from Definitions 1.7.10 and 1.7.11. In particular, for the Lindenbaum algebra we again obtain the frame isomorphic to $\Phi_{\mathbf{\Lambda}}$ (with an obvious notion of isomorphism):

Theorem 1.7.13 *For a modal or intermediate logic $\mathbf{\Lambda}$*

$$Lind(\mathbf{\Lambda})_+ \cong \Phi_{\mathbf{\Lambda}}.$$

This is because the maximal (prime) filters of $Lind(\mathbf{\Lambda})$ exactly correspond to $\mathbf{\Lambda}$-complete theories.

Theorem 1.7.14 *A general Kripke frame is descriptive iff it is isomorphic to some frame $\mathbf{\Omega}_+$.*

This result was proved in [Goldblatt, 1976] and earlier in [Esakia, 1974] for **S4**-frames. For a more available proof see [Chagrov and Zakharyaschev, 1997].

The subsequent lemma follows from the above, but also has an independent proof based on the properties of $M_{\mathbf{\Lambda}}$.

Lemma 1.7.15 *Every general canonical frame $\Phi_{\mathbf{\Lambda}}$ is descriptive.*

From the definitions it is evident that for any general Kripke frame $(F, \mathcal{W})$, $\mathbf{ML}(F) \subseteq \mathbf{ML}(F, \mathcal{W})$ (or $\mathbf{IL}(F) \subseteq \mathbf{IL}(F, \mathcal{W})$ in the intuitionistic case). Moreover, in many cases these two logics coincide.

Definition 1.7.16 *A set of N-modal formulas Γ is called* d-persistent *(respectively,* r-persistent*) if for any descriptive (respectively, refined) general frame $(F, \mathcal{W})$, $(F, \mathcal{W}) \vDash \Gamma \Rightarrow F \vDash \Gamma$; similarly, for intuitionistic formulas.*

The following observations are trivial consequences of soundness:

Lemma 1.7.17

(1) A modal or intermediate propositional logic is d-persistent (respectively, r-persistent) iff it is axiomatisable by a d-persistent (respectively, r-persistent) set.

(2) A sum of d-persistent (respectively, r-persistent) logics is d-persistent (respectively, r-persistent).

Proposition 1.7.18 *Every d-persistent modal or intermediate logic is canonical, and therefore Kripke complete.*

Proof (Modal case) $\Phi_{\mathbf{\Lambda}} \vDash \mathbf{\Lambda}$ by 1.7.5, and $\Phi_{\mathbf{\Lambda}} = (F_{\mathbf{\Lambda}}, \mathcal{W}_{\mathbf{\Lambda}})$ is descriptive by 1.7.15. So for a d-persistent $\mathbf{\Lambda}$ it follows that $F_{\mathbf{\Lambda}} \vDash \mathbf{\Lambda}$. ■

Thus for propositional modal logics d-persistence implies canonicity and canonicity implies Kripke completeness. The converse implications are not true — however, all three properties are equivalent for elementary logics, by the Fine–van Benthem theorem, see 1.8.3 below.

1.8 First-order translations and first-order definability

Let $\mathcal{L}1_N$ be the classical first-order language with equality and binary predicate letters $R_1, \ldots, R_N$. So classical $\mathcal{L}1_N$-structures are nothing but N-modal frames. The classical truth of an $\mathcal{L}1_N$-formula φ in a frame F is denoted by $F \vDash \varphi$ as usual.

Definition 1.8.1 *An $\mathcal{L}1_N$-formula φ* corresponds *to an N-modal propositional formula A if for any N-modal Kripke frame F,*

$F \models A$ *(modally)* $\Leftrightarrow F \models \varphi$ *(classically).*

Similarly, an $\mathcal{L}1_1$*-formula* φ corresponds *to an intuitionistic propositional formula A if for any intuitionistic Kripke frame F,* $F \Vdash A \Leftrightarrow F \models \varphi$.

Definition 1.8.2 *The class Mod(*Σ*) of all classical models of a first-order theory (i.e. a set of first-order sentences)* Σ *is called* Δ-elementary*; if* Σ *is recursive (respectively, finite), this class is called* R-elementary *(respectively,* elementary*). An* N*-modal or intermediate logic* $\mathbf{\Lambda}$ *is called* Δ-elementary *(respectively,* R-elementary, elementary*) if the class* $\mathbf{V}(\mathbf{\Lambda})$ *(or* $\mathbf{V}^I(\mathbf{\Lambda})$*) is of the corresponding kind in the language* $\mathcal{L}1_N$ *(or* $\mathcal{L}1_1$*).*

A modal or intermediate logic determined by a Δ-elementary *(respectively,* R-elementary, elementary*) class of frames is called* quasi-Δ-elementary *(respectively,* quasi-R-elementary, quasi-elementary*).*[12]

Theorem 1.8.3 (Fine–van Benthem) *Every quasi-*Δ*-elementary modal or intermediate logic (in particular, every Kripke complete* Δ*-elementary logic) is d-persistent.*

For the proof see [van Benthem, 1983], [Chagrov and Zakharyaschev, 1997].

Let us recall the well-known 'standard' translation from modal formulas into first-order formulas, cf. [van Benthem, 1983].[13]

Let $\mathcal{L}1_N^\star$ be the language obtained by adding countably many monadic predicate letters $P_1, P_2, \dots$ to $\mathcal{L}1_N$. Every N-modal formula A is translated into an $\mathcal{L}1_N^\star$-formula $A^\star(t)$ with at most one parameter t, according to the rules:

$p_i^\star = P_i(t)$,

$\bot^\star(t) = \bot$,

$(A \supset B)^\star(t) = A^\star(t) \supset B^\star(t)$,

$(\Box_k A)^\star(t) = \forall x(R_k(t,x) \supset A^\star(x))$, where $A^\star(x)$ is obtained from $A^\star(t)$ by substituting x for t (together with renaming of bound variables, if necessary).[14]

Every Kripke model $M = (F, \varphi)$ over a frame $F = (W, \varrho_1, \dots, \varrho_n)$ clearly corresponds to a classical $\mathcal{L}1_N^\star$-structure $M^\star = (W, \varrho_1, \dots, \varrho_n, \varphi(p_1), \varphi(p_2), \dots)$.

Lemma 1.8.4

(1) Let $M = (F, \varphi)$ *be a Kripke model over a frame* $F = (W, \varrho_1, \dots, \varrho_N)$*. Then for any* $a \in W$*, for any* N*-modal formula* A*,*

$$M, a \models A \text{ iff } M^\star \models A^\star(a).$$

(2) For any N*-modal frame* F*, for any* N*-modal formula* A

$$F \models A \text{ iff } \forall \varphi\, (F, \varphi)^\star \models \forall t A^\star(t).$$

[12] According to another terminology, a quasi-Δ-elementary logic is called Δ-*elementarily determined*, etc.

[13] A similar translation for intuitionistic formulas was first introduced in [Mints, 1967].

[14] Variable substitutions for first-order formulas are considered in detail in Chapter 2.

Proof

(1) Easy by induction. E.g. in the case $A = \Box_k B$ we have:

$$\begin{aligned} M, a \vDash A &\Leftrightarrow \forall b \in \varrho_k(a) \quad M, u_1 \vDash B \\ &\Leftrightarrow \forall b \in \varrho_k(a) \quad M^\star \vDash B^\star(b) \text{ (by the induction hypothesis)} \\ &\Leftrightarrow M^\star \vDash \forall x(R_k(a,x) \supset B^\star(x)) \Leftrightarrow M^\star \vDash (\Box_k B)^\star(a). \end{aligned}$$

$$(2) \quad \begin{aligned} F \vDash A &\Leftrightarrow \forall\varphi\forall a\ (F,\varphi), a \vDash A \Leftrightarrow \forall\varphi\forall a\ (F,\varphi)^\star \vDash A^\star(a) \text{ (by (1))} \\ &\Leftrightarrow \forall\varphi\ (F,\varphi)^\star \vDash \forall t A^\star(t). \end{aligned}$$ ■

In the next proposition we use the following notation. For an $\mathcal{L}1_N$-theory Σ, $[\Sigma[$ denotes the set of its $\mathcal{L}1_N^\star$-theorems (in classical first-order logic).

Proposition 1.8.5

(1) For any $\mathcal{L}1_N$-theory Σ, $\mathbf{ML}(Mod(\Sigma)) \leq_1 [\Sigma[$. In particular, every quasi-R-elementary modal or intermediate logic is recursively axiomatisable.

(2) Every quasi-Δ-elementary modal or intermediate logic has the c.f.p.

(3) Every finitely axiomatisable (quasi-)Δ-elementary modal or intermediate logic is (quasi-)elementary.

Proof

(1) Since all $\mathcal{L}1_N^\star$-models of Σ are exactly the structures of the form $(F,\varphi)^\star$, where $F \vDash \Sigma$, by Lemma 1.8.4 and Gödel's completeness theorem, we obtain

$$(\sharp) \quad \begin{aligned} A \in \mathbf{ML}(Mod(\Sigma)) &\Leftrightarrow \Sigma \vDash \forall t A^\star(t) \text{ (in the classical sense)} \\ &\Leftrightarrow \Sigma \vdash \forall t A^\star(t) \text{ (in classical first-order logic).} \end{aligned}$$

Since A can be uniquely restored from $A^\star$, this proves the 1-reducibility. Next, if Σ is recursive, then $\mathbf{ML}(Mod(\Sigma))$ is RE as it is 1-reducible to the set of theorems of a recursive classical theory. Now (1) follows by the well-known Craig's lemma, cf. [Boolos, Burgess, and Jeffrey, 2002].

To prove (1) in the intuitionistic case, note that by 1.5.8, $\mathbf{IL}(\mathcal{C}) = {}^T\mathbf{ML}(\mathcal{C})$, so $\mathbf{IL}(\mathcal{C})$ is 1-reducible to $\mathbf{ML}(\mathcal{C})$.

(2) Let $\mathcal{C}_0$ be the class of all countable $\mathcal{L}1_N^\star$-models of Σ. Similarly to the above, we obtain that $A \in \mathbf{ML}(\mathcal{C}_0)$ iff for any countable $\mathcal{L}1_N^\star$-structure $\mathcal{M}$, $(\mathcal{M} \vDash \Sigma \Rightarrow \mathcal{M} \vDash \forall t A^\star(t))$. By the Löwenheim–Skolem theorem, the latter is equivalent to $\Sigma \vdash \forall t A^\star(t)$, and thus (by ($\sharp$)) to $A \in \mathbf{ML}(Mod(\Sigma))$. Therefore $\mathbf{ML}(Mod(\Sigma)) = \mathbf{ML}(\mathcal{C}_0)$, which proves (2). The intuitionistic version follows readily.

(3) (Modal case) If $\mathbf{ML}(Mod(\Sigma)) = \mathbf{K}_N + A$ for a formula A, then by ($\sharp$), $\Sigma \vdash \forall t A^\star(t)$ in classical logic, which implies $\Sigma_1 \vdash \forall t A^\star(t)$ for some finite $\Sigma_1 \subseteq \Sigma$. Hence $\varphi \vdash \forall t A^\star(t)$, where $\varphi = \bigwedge \Sigma_1$, so $A \in \mathbf{ML}(Mod(\varphi))$ (by ($\sharp$)), and thus

$$\mathbf{ML}(Mod(\Sigma)) = \mathbf{K}_N + A \subseteq \mathbf{ML}(Mod(\varphi)).$$

On the other hand, $Mod(\Sigma) \subseteq Mod(\varphi)$, hence $\mathbf{ML}(Mod(\varphi)) \subseteq \mathbf{ML}(Mod(\Sigma))$. Therefore $\mathbf{ML}(Mod(\Sigma)) = \mathbf{ML}(Mod(\varphi))$ is quasi-elementary.

The case when $\mathbf{K}_N + A$ is Δ-elementary is left to the reader.

(Intuitionistic case.) Assume that L is finitely axiomatisable and Δ-elementary. Then by definition, $\tau(L)$ is also finitely axiomatisable. Since by 1.5.9, $\mathbf{V}^I(L) = \mathbf{V}(\tau(L))$, it follows that $\tau(L)$ is Δ-elementary. Then (as we have just proved) $\tau(L)$ is elementary. Now the equality $\mathbf{V}^I(L) = \mathbf{V}(\tau(L))$ implies the elementarity of L.

It remains to consider the case when an intermediate L is finitely axiomatisable and quasi-Δ-elementary. Then we can argue as in the modal case. In fact, if $\mathbf{IL}(\mathcal{C}) = \mathbf{H} + A$ for $\mathcal{C} = Mod(\Sigma)$, then $A^T \in \mathbf{ML}(\mathcal{C})$, so by ($\sharp$), $\Sigma \vdash \forall t(A^T)^\star(t)$. Then we can replace Σ with a single φ and obtain that $\mathbf{IL}(\mathcal{C}) = \mathbf{IL}(Mod(\varphi))$. ∎

1.9 Some general completeness theorems

In this section we recall two general results on Kripke-completeness — the Sahlqvist theorem and the Fine theorem.

Definition 1.9.1 *A* Sahlqvist formula *is a modal formula of the form* $\Box_\alpha(A \supset B)$, *where* $\alpha \in I_N^\infty$, B *is a positive modal formula (i.e. built from proposition letters using* $\bot$, $\top$, $\vee$, $\wedge$, $\Diamond_i$, $\Box_i$), *and* A *is built from proposition letters and their negations using the same connectives, so that subformulas of* A *of the form* $C \vee D$ *or* $\Diamond_i C$ *containing negated proposition letters are not within the scope of any* $\Box_j$.

Theorem 1.9.2 (Sahlqvist theorem) *Let* $\mathbf{\Lambda}$ *be a modal logic axiomatised by Sahlqvist formulas. Then* $\mathbf{\Lambda}$ *is d-persistent and* Δ*-elementary.*

For the proof see [Chagrov and Zakharyaschev, 1997], [Blackburn, de Rijke and Venema, 2001].

Definition 1.9.3 *A subset* V *in a transitive frame* $F = (W, R)$ *is called an* antichain *if its different worlds are incomparable:*

$$\forall x, y \in V(xRy \Rightarrow x = y).$$

The width *of* F *is the maximal cardinality of finite antichains in cones of* F *if it exists and* ∞ *otherwise.*

So for a finite n, F is of width $\leq n$ iff

$$\forall y, x_0, \dots, x_n \in W\, (\forall i\; yRx_i \Rightarrow \exists i\; \exists j \neq i\, (x_i R x_j \vee x_j R x_i \vee x_i = x_j)).$$

This property corresponds to the modal formula AW_n (cf. 1.3.11) and to AIW_n in the intuitionistic case (cf. 1.4.17).

Definition 1.9.4 *For a finite* n, a modal logic of width $\leq n$ *is an extension of* $\mathbf{K4} + AW_n$; *an* intermediate logic of width $\leq n$ *is an extension of* $\mathbf{H} + AIW_n$. *All these are called* the logics of finite width.

The *width* of a transitive modal logic $\mathbf{\Lambda}$ can be defined explicitly as the least n such that $AW_n \in \mathbf{\Lambda}$ (and ∞ if $AW_n \notin \mathbf{\Lambda}$ for any n); similarly for the intuitionistic case.

Theorem 1.9.5 (Fine theorem) *Every logic of finite width (modal or intermediate) is Kripke-complete and moreover, has the c.f.p.*

This result (for the modal case) was first proved in [Fine, 1974]; for a shorter proof see [Chagrov and Zakharyaschev, 1997]. The intuitionistic case follows from the observation that $L \vdash AIW_n$ implies $\tau(L) \vdash AW_n$ together with 1.5.8 and 1.5.5.

1.10 Trees and unravelling

Recall that a path in a frame is a sequence of related worlds together with indices of the accessibility relations (Definition 1.3.20).

Definition 1.10.1 *The* depth *of a world u in F (notation: $d_F(u)$ or $d(u)$ if there is no confusion) is the maximal length of a path from u in F if it exists and ∞ otherwise.*

Let us introduce two other kinds of 'paths' and 'depths' in Kripke frames.

Definition 1.10.2 *A* distinct path *in a frame F is a path, in which all the worlds are different. The* distinct depth *of u in F (notation: $dd_F(u)$ or $dd(u)$) is the maximal length of a distinct path from u in F if it exists and ∞ otherwise.*

Definition 1.10.3 *A* strict path *in a transitive frame $F = (W, R)$ is a path $(u_0, u_1, \ldots, u_m)$, in which $u_{i+1} \not\!R\, u_i$ for any $i < m$ (or equivalently, all the worlds are in different clusters). Respectively, the* strict depth *of u in F (notation: $sd_F(u)$ or $sd(u)$) is the maximal length of a strict path from u in F if it exists and ∞ otherwise.*

Obviously, $sd(u) = sd(u) = sd(v)$ if $u \approx_R v$, and $sd_F(u) = sd_{F^\sim}(u^\sim)$.

Exercise 1.10.4 Show that $d(x) \leq n$ iff $x \vDash \Box^n \bot$, where $\Box A := \Box_1 A \wedge \ldots \wedge \Box_N A$.

Definition 1.10.5 *A world u in a frame is called a* dead end *(respectively,* maximal; quasi-maximal*) if $d(u) = 0$ (respectively, $dd(u) = 0$; $sd(u) = 0$). Dead ends in a tree are also called* leaves *(as well as maximal worlds in reflexive or transitive trees).*

A maximal cluster *in a* $\mathbf{K4}$*-frame F is a maximal point in $F^\sim$.*

So quasi-maximal points in a $\mathbf{K4}$-frame are exactly the points of the maximal clusters.

Definition 1.10.6 *A frame* $F = (W, R_1, \ldots, R_N)$ *with a root* u_0 *is called a* tree *(or an N-*tree*) if*

- $\forall i \, \forall y \quad \neg y R_i u_0$,
- *for any* $x \neq u_0$ *there exists a unique pair* (i, y) *such that* $y R_i x$.

Hence we readily have

Lemma 1.10.7 *A frame F is a tree with a root u iff for any* $x \in F$ *there exists a unique path from u to x in F.*

Definition 1.10.8 *The* height *of a world x in a tree F (notation:* $ht_F(x)$*, or* $ht(x)$*) is the length of the (unique) path from the root to x.*

Definition 1.10.9 *A* successor *of a world u in a poset*[15] $F = (W, R)$ *is a minimal element in the set* $R_-(u)$ *of all strictly accessible worlds.* $\beta_F(u)$ *(or* $\beta(u)$*) denotes the set of all successors of u in F. F is called* successive *if for any u,* $R_-(u) = R(\beta_F(u))$*. The* branching at u in F *is* $|\beta_F(u)|$.

Every finite p.o. set is clearly successive. For rooted successive p.o. sets we can also define the height of x as the minimal length of a 'successor-path' from the root to x.

Definition 1.10.10 *Let* $T_\omega = \omega^\infty$ *be the set of all finite sequences of natural numbers. The* 1-modal universal tree *is the frame*

$$F_1 T_\omega := (T_\omega, \sqsubset),$$

where

$$u \sqsubset v := \exists k \in \omega \ v = uk.$$

The N-modal universal tree *is*

$$F_N T_\omega := (T_\omega, \sqsubset_{1N}, \ldots, \sqsubset_{NN}),$$

where

$$u \sqsubset_{iN} v := \exists k \in \omega \, (v = uk \ \& \ k \equiv i - 1 \ (mod \ N)).$$

Sometimes we drop N in $F_N T_\omega$ *and* $\sqsubset_{iN}$ *if it is clear from the context.*

The intuitionistic universal tree *is the p.o. set*[16] $IT_\omega := (T_\omega, \leqslant)$*, where*

$$u \leqslant v := \exists w \, (v = uw).$$

So $\leqslant$ is the reflexive and transitive closure of $\sqsubset$, and the set of immediate successors of u in IT_ω is just $\beta(u) = \sqsubset(u)$. From the definitions it follows easily that $F_N T_\omega$ is really a tree with the root $\curlywedge$.

[15] More generally, we may assume that F is transitive antisymmetric.
[16] This is a p.o. tree according to Definition 1.11.10.

Definition 1.10.11 *A subframe $F \subseteq F_N T_\omega$ or $F \subseteq IT_\omega$, that is stable under $\geqslant$ (or equivalently, under all $\sqsupset_i$, $1 \le i \le N$), is called a* standard tree *(or a* standard subtree *of the corresponding frame).*

Let us fix the following standard trees:

- the *universal n-ary trees*, or the *universal trees of branching n*, for finite $n > 0$:
$$F_N T_n := F_N T_\omega \restriction T_{Nn},\ IT_n := IT_\omega \restriction T_n,$$
where T_n is the set of all $\{0, 1, \ldots, n-1\}$-sequences;
- *universal trees of depth (or height) k*:
$$F_N T_n^k := F_N T_n \restriction T_\omega^k,\ IT_n^k := IT_n \restriction T_\omega^k,$$
where
$$T_\omega^k := \{u \in T_\omega \mid l(u) \le k\},$$
$l(u)$ is the length of sequence u, $n \le \omega$. In particular, T_ω^0 contains only the root $\curlywedge$. Obviously, $l(u) = ht(u)$ in $F_N T_\omega$.

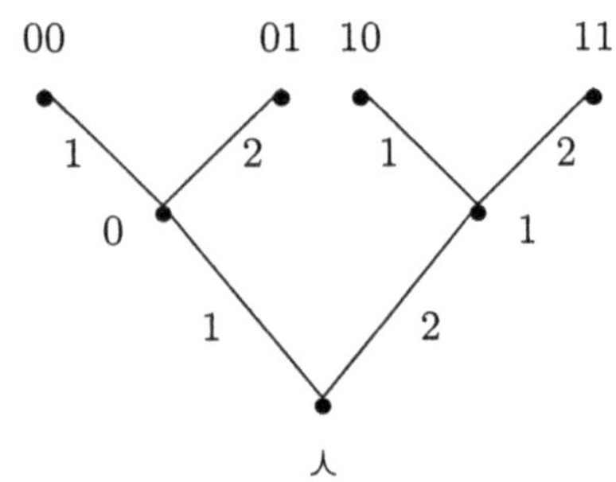

Figure 1.1. $F_2 T_1^1$

Sometimes we also use the notation T_n^ω rather than T_n, for $0 < n \le \omega$.

Definition 1.10.12 *Let F be a successive poset. A subframe $G \subseteq F$ is called* greedy *if for any $u \in G$ either $\beta_F(u) \subseteq G$ or $\beta_F(u) \cap G = \varnothing$. A* greedy standard tree *is a greedy standard subtree of $F_N T_\omega$ or IT_ω.*

Definition 1.10.13 *Let $F = (W, R_1, \ldots, R_N)$ be a frame with a root u_0. The* unravelling *of F is the frame $F^\# = (W^\#, R_1^\#, \ldots, R_N^\#)$, where $W^\#$ is the set of all paths in F starting at u_0 and*

$$\alpha R_j^\# \beta \textit{ iff } \exists v\ \beta = (\alpha, j, v).$$

Unravelling was first introduced probably in [Sahlqvist, 1975]; later it was used by many authors, cf. [Gabbay, 1976], [van Benthem, 1983], [Gabbay and Shehtman, 1998].

The proof of the following lemma is straightforward.

Lemma 1.10.14

(1) $F^{\#}$ is a tree.

(2) The map $\pi : W^{\#} \longrightarrow W$ such that $\pi(u_0, \dots, u_m) = u_m$, is a p-morphism from $F^{\#}$ onto F.

Now let us show how to increase branching in a standard way.

Definition 1.10.15[17] *Let $F = (W, R_1, \dots, R_N)$ be a frame with root u_0. Put*

$$W_1 : \quad = \begin{cases} W & \text{if there exists a path of positive length} \\ & \text{(a 'loop') from } u_0 \text{ to } u_0, \\ W - \{u_0\} & \text{otherwise} \end{cases}$$

$$W^{\blacktriangledown} := (W_1 \times \omega) \cup \{(u_0, -1)\}.$$

Let $R_1^{\blacktriangledown}, \dots, R_N^{\blacktriangledown}$ be the relations on $W^{\blacktriangledown}$ such that

$$(a, n) R_i^{\blacktriangledown} (b, m) \text{ iff } a R_i b \ \& \ m \neq -1.$$

The frame $F^{\blacktriangledown} := (W^{\blacktriangledown}, R_1^{\blacktriangledown}, \dots, R_N^{\blacktriangledown})$ is called the $(\omega-)$ thickening *of F.*

Informally speaking, we make ω copies of every world of W_1 and connect all the copies of two worlds if the original worlds are connected in F; we also add the root $(u_0, -1)$.

From the definition we readily obtain

Lemma 1.10.16

(1) For any $x \in W^{\blacktriangledown}$, $R_i^{\blacktriangledown}(x)$ is either empty or denumerable.

(2) the map $\gamma : \ (a, n) \mapsto a$ is a p-morphism from $F^{\blacktriangledown}$ onto F.

Lemma 1.10.17 *Let $F = (W, R_1, \dots, R_N)$ be a tree, in which every set $R_i(x)$ is either empty or denumerable. Then F is isomorphic to a greedy standard tree.*

Proof Let F^k be the restriction of F to the worlds of height $\leq k$. We define the embeddings $f^k : F^k \rightarrowtail F_N T_\omega^k$ by induction.

If u_0 is the root of F, we put

$$f^0(u_0) := \lambda.$$

[17] [Gabbay and Shehtman, 1998].

If f^k is already defined, we extend it to f^{k+1} as follows. If $ht_F(y) = k$ and $R_i(y) \neq \varnothing$, then $R_i(y)$ is denumerable, by our assumption. So let $e_{y,i} : \omega \to R_i(y)$ be the corresponding bijective enumeration. Then put

$$f^{k+1}(e_{y,i}(j)) := (f^k(y), jN + i - 1).$$

Since F is a tree, every world of height $(k + 1)$ is $e_{y,i}(j)$ for a unique triple (y, i, j), so f^{k+1} is well-defined. Since $e_{y,i}$ is a bijection, it is clear that f^{k+1} is an embedding.

Its image $\mathrm{rng}(f^{k+1})$ is greedy. In fact, if $u = f^k(y)$ and $R_i(y) \neq \varnothing$, then by our construction, f^{k+1} gives rise to a bijection between $R_i(y)$ and

$$\sqsubset_i (u) = \{(u, jN + i - 1) \mid j \geq 0\}.$$

Eventually, the required isomorphism is $f := \bigcup\limits_{k \geq 0} f^k$. ■

Proposition 1.10.18

(1) Every countable rooted frame is a p-morphic image of a greedy standard tree.

(2) Every serial countable rooted N-frame is a p-morphic image of $F_N T_\omega$.

Proof

(1) Let F be the original frame. By 1.10.14 and 1.10.16 we have p-morphisms

$$(F^\blacktriangledown)^\sharp \twoheadrightarrow F^\blacktriangledown \twoheadrightarrow F.$$

By 1.10.17, $(F^\blacktriangledown)^\sharp$ is a tree.

Since by 1.10.16, every $R_i^\blacktriangledown(x)$ is either empty or denumerable, the same property holds for $(F^\blacktriangledown)^\sharp$, and thus it is isomorphic to a greedy standard tree, by 1.10.17 .

(2) If F is serial and $f : G \twoheadrightarrow F$, then G is also serial, due to the lift property. By definition, every serial greedy standard subtree of $F_N T_\omega$ is $F_N T_\omega$ itself.

■

1.11 PTC-logics and Horn closures

Definition 1.11.1 *A modal propositional formula is called* closed *(or* constant*) if it is a 0-formula, i.e. if it does not contain proposition letters.*

Lemma 1.11.2 *If F, F' are N-modal frames and $f : F \twoheadrightarrow F'$, then $F \vDash A \iff F' \vDash A$ for any closed N-modal formula A.*

Proof Let M, M' be arbitrary Kripke models respectively over F, F'. By induction it easily follows that $M, x \vDash A$ iff $M', f(x) \vDash A$ for any $x \in M$ and closed A. Hence $F, x \vDash A$ iff $F', f(x) \vDash A$. This implies $F \vDash A \iff F' \vDash A$ since f is surjective. ■

Proposition 1.11.3 *Every logic axiomatisable by closed formulas is canonical.*

Proof By Theorem 1.7.3, every axiom of the logic is true in the canonical model. For a closed formula, this is equivalent to validity in the canonical frame. ■

Definition 1.11.4 *A* pseudotransitive N-modal formula *has the form* $\Diamond_\alpha \Box_k p \supset \Box_\beta p$, *where* $p \in PL$, $\alpha, \beta \in I_N^\infty$. *This formula is called* one-way *if* $\alpha = \lambda$. *A* PTC-formula *is a formula which is either pseudotransitive or closed. A* PTC-logic *is a modal logic axiomatised by a set of PTC-formulas.* One-way *PTC-formulas and logics are defined similarly.*

For example, the following formulas are pseudotransitive:

$$\Box p \supset p, \quad \Box p \supset \Box\Box p, \quad \Diamond\Box p \supset p, \quad \Diamond_1\Box_2 p \supset p.$$

Thus many well-known logics are PTC, e.g. **D**, **K4**, **D4**, **S4**, **T**, **B**, **K.t**, **K4.t**, **S5**.

Since every pseudotransitive formula is obviously Sahlqvist, from the Sahlqvist theorem and Proposition 1.11.3 we obtain

Proposition 1.11.5

(1) A pseudotransitive N-modal formula $A = \Diamond_\alpha \Box_k p \supset \Box_\beta p$ corresponds to the $\mathcal{L}1_N$-formula[18]

$$A^\sharp := \forall x, v(\exists u(uR_\alpha x \wedge uR_\beta v) \supset xR_k v),$$

(2) Every PTC-logic is canonical and Δ-elementary.

Later to the well-known transitive closure, there exists a 'closure' under pseudotransitivity. This is a rather simple fact from classical logic, but still let us recall its proof briefly.

Further on arbitrary lists of individual variables are denoted by **x**, **y**. etc. and arbitrary lists of individuals by **a**, **b**. etc. As usual, we write $\varphi(x, y, \mathbf{z})$ to indicate that all parameters of the formula φ are among $x, y, \mathbf{z}$.

Definition 1.11.6 *An $\mathcal{L}1_N$-sentence of the form*

$$\psi = \forall x \forall z \forall \bar{z}(\varphi(x, y, \mathbf{z}) \to R_k(x, y))$$

is a called a universal strict Horn clause, *if $\varphi(x, y, \mathbf{z})$ is a (non-empty) conjunction of atomic formulas.*

[18] Strictly speaking, this becomes an $\mathcal{L}1_N$-formula after writing $uR_\alpha x$ and $uR_\beta v$ in terms of the basic predicates $R_1, \ldots, R_N$.

Proposition 1.11.7 *Let $F = (W, \varrho_1, \ldots, \varrho_N)$ be an N-modal frame, $\Gamma = \{\psi_k \mid k \in I\}$ a set of universal (strict) Horn clauses*

$$\psi_k = \forall x \forall y \forall \mathbf{z}_k (\varphi_k(x, y, \mathbf{z}_k) \to R_{i_k}(x, y)).$$

Then there exists a frame F_Γ^+ such that[19]

(1) $F \subseteqq F_\Gamma^+$;

(2) $F_\Gamma^+ \vDash \Gamma$;

(3) If $G \vDash \Gamma$ and $f : F \longrightarrow G$ is monotonic, then $f : F_\Gamma^+ \longrightarrow G$ is also monotonic.

We can say that F_Γ^+ is the 'smallest' full weak extension of F satisfying Γ.

Proof We construct F_Γ^+ as the union of a sequence of frames $F_m = (W, \varrho_{m1}, \ldots, \varrho_{mN})$, beginning with $F_0 = F$. Namely, let:

$$\varrho_{(m+1)j} = \varrho_{mj} \cup \bigcup_{i_k = j} \{(a, b) \mid F_m \vDash \exists \mathbf{z}_k \varphi_k(a, b, \mathbf{z}_k)\},$$

$$\varrho_j^+ = \bigcup_m \varrho_{mj}, \quad F_\Gamma^+ = (W, \varrho_1^+, \ldots, \varrho_N^+).$$

Then $F_0 \subseteqq \ldots F_m \subseteqq F_{m+1} \ldots \subseteqq F_\Gamma^+$, and thus (1) holds. Furthermore, we have:

(4) if $\varphi(\mathbf{z})$ is a conjunction of atomic formulas, then the following conditions are equivalent:

(i) $F_\Gamma^+ \vDash \varphi(\mathbf{a})$;

(ii) $\exists m \forall k \geq m \; F_k \vDash \varphi(\mathbf{a})$;

(iii) $\exists m \; F_m \vDash \varphi(\mathbf{a})$.

In fact, (ii)⇒(iii) is trivial. (iii)⇒(i) holds, since $F_m \subseteqq F_\Gamma^+$, and the truth of positive formulas is preserved by monotonic maps. (i)⇒(ii) easily follows from $\rho_{mj} \subseteq \rho_{(m+1)j}$ and $\rho_j^+ = \bigcup_m \rho_{mj}$.

Now let us prove (2). Assume $F_\Gamma^+ \vDash \varphi_k(a, b, \mathbf{c}_k)$, $i_k = j$. Then by (4), $F_m \vDash \varphi_k(a, b, \mathbf{c}_k)$ for some m, and thus $(a, b) \in \varrho_{(m+1)j}$ by definition. Hence $F_\Gamma^+ \vDash R_j(a, b)$. Therefore $F_\Gamma^+ \vDash \psi_k$.

To check (3), we assume that $G \vDash \Gamma$, $f : F \longrightarrow G$ is monotonic and show by the induction that $f : F_m \longrightarrow G$ is monotonic for any m.

To make the step, assume $a \varrho_{(m+1)j} b$. Then there are two cases.

If $a \varrho_{mj} b$, then $G \vDash R_j(f(a), f(b))$ by induction hypothesis.

If $F_m \vDash \exists \mathbf{z}_k \varphi_k(a, b, \mathbf{z}_k)$, $i_k = j$, then $G \vDash \exists \mathbf{z}_k \varphi_k(f(a), f(b), \mathbf{z}_k)$, since monotonic $f : F_m \longrightarrow G$ preserves the truth of positive formulas. Now $G \vDash \psi_k$ implies $G \vDash R_j(f(a), f(b))$. ∎

[19] Recall that $\subseteqq$ denotes a full weak subframe, cf. 1.3.17.

By Proposition 1.11.5 every pseudotransitive modal formula corresponds to a universal strict Horn formula. So Proposition 1.11.7 yields

Corollary 1.11.8 *Let Γ be a set of pseudotransitive N-modal formulas, F an N-modal frame. Then there exists an N-modal frame F_Γ^+ such that*

(1) $F \subseteqq F_\Gamma^+$;

(2) $F_\Gamma^+ \vDash \Gamma$;

(3) for any N-modal frame G, if $G \vDash \Gamma$ and $f : F \longrightarrow G$ is monotonic, then $f : F_\Gamma^+ \longrightarrow G$ is monotonic.

Definition 1.11.9 *The frame F_Γ^+ described in Corollary 1.11.8 is called the* pseudotransitive closure *of F (under Γ), or the* Γ-closure *of F. If $\mathbf{\Lambda} = K_N + \Gamma + \Delta$ is a PTC-logic with the set of pseudotransitive axioms Γ and a set of closed axioms Δ, then F_Γ^+ is also called the* $\mathbf{\Lambda}$-closure *if it validates Δ.*

Definition 1.11.10 *If $\mathbf{\Lambda}$ is a PTC-logic, then the $\mathbf{\Lambda}$-closure of a tree is called a* $\mathbf{\Lambda}$-tree. **K4**-*trees (respectively,* **S4**-*trees) are also called* transitive trees *(respectively,* p.o. trees*). In the particular case when $\mathbf{\Lambda} = \mathbf{D}_N + \Gamma$, with a set of pseudotransitive axioms Γ, the Γ-closure of $F_N T_\omega$ is called the* full standard $\mathbf{\Lambda}$-tree.

Proposition 1.11.11

(1) Let $\mathbf{\Lambda}$ be a PTC-logic. Then every countable rooted $\mathbf{\Lambda}$-frame is a p-morphic image of a greedy standard $\mathbf{\Lambda}$-tree.

(2) If $\mathbf{\Lambda} = \mathbf{D}_N + \Gamma$, with a set of pseudotransitive axioms Γ, then every serial countable rooted $\mathbf{\Lambda}$-frame is a p-morphic image of the full standard $\mathbf{\Lambda}$-tree.

(3) Every countable rooted **S4**-*frame is a p-morphic image of IT_ω.*

Proof

(1) Let F be such a frame. By Proposition 1.10.18, there exists $f : \Phi \twoheadrightarrow F$ for some greedy standard tree Φ. By Proposition 1.11.7, the map $f : \Phi_\Gamma^+ \longrightarrow F$ is monotonic, where Γ is the set of pseudotransitive axioms of $\mathbf{\Lambda}$. The lift property obviously holds for this map, since it holds for $f : \Phi \longrightarrow F$. So F is a p-morphic image of Φ_Γ^+. Finally note that all closed axioms of $\mathbf{\Lambda}$ are valid in Φ_Γ^+, by Lemma 1.11.2.

(2) By Proposition 1.10.18(2), in this case we can take $\Phi = F_N T_\omega$.

(3) In fact, IT_ω is the full standard **S4**-tree. ∎

Corollary 1.11.12

(1) Every PTC-logic $\mathbf{\Lambda}$ is determined by the class of greedy standard $\mathbf{\Lambda}$-trees.

(2) Every serial PTC-logic $\mathbf{\Lambda}$ is determined by the full standard $\mathbf{\Lambda}$-tree.

(3) $\mathbf{S4} = \mathbf{ML}(IT_\omega)$, $\mathbf{H} = \mathbf{IL}(IT_\omega)$.

Proof
(1) $\mathbf{\Lambda}$ is complete and Δ-elementary, so it has the c.f.p by Proposition 1.8.5. Then by Proposition 1.11.11(1) and the morphism lemma, every $A \notin \mathbf{\Lambda}$ is refuted in a greedy standard $\mathbf{\Lambda}$-tree.
(2) By the same argument using 1.11.11(2).
(3) By (2), since IT_ω is the full standard **S4**-tree. The claim about **H** follows by 1.5.7, since $\mathbf{H} = {}^T\mathbf{S4}$. ■

Proposition 1.11.13 *IT_ω is a p-morphic image of IT_2.*

Proof Let $\tau : \omega \longrightarrow \omega$ be a map such that every set $\tau^{-1}(n)$ is infinite (such a map obviously exists, since there is a bijection between $\omega \times \omega$ and ω). Next, every $\alpha \in T_2$ can be uniquely presented in the form $0^{n_1}1 \dots 0^{n_k}10^{n_{k+1}}$, where $n_1, \dots, n_{k+1} \geq 0$, and 0^n means $\underbrace{0 \dots 0}_{n}$. Now put

$$f(\alpha) := \tau(n_1)\tau(n_2)\dots\tau(n_k).$$

We claim that f is a required p-morphism from IT_ω onto IT_2.

In fact IT_ω is a Horn closure of F, T_ω, so by 1.11.7, it suffices to check monotonicity of f w.r.t. $\sqsubset$. But this is the case, since $f(\alpha 0) = f(\alpha)$, and $f(\alpha 1) = f(\alpha)\tau(n_{k+1})$.

For the lift property, suppose $f(\alpha) \sqsubset u$; then $u = f(\alpha)m$ for some m. By the construction of τ, there exists $n \geq n_{k+1}$ such that $\tau(n) = m$. So we can take

$$\beta = 0^{n_1}1 \dots 0^{n_k}10^{n}1,$$

then $f(\beta) = \tau(n_1)\dots\tau(n_k)\tau(n) = f(\alpha)m = u$, and obviously, $\alpha \leqslant \beta$.

Since $\leqslant$ is the transitive closure of $\sqsubseteq$, the lift property now follows easily.

It remains to note that τ sends the root to the root, which implies the surjectivity of f. ■

Corollary 1.11.14 *Every countable rooted* **S4**-*frame is a p-morphic image of IT_2.*

Proof By Propositions 1.11.13 and 1.11.11. ■

Definition 1.11.15 *A standard 1-tree is called* strongly standard *if together with any sequence αn it contains all the sequences αm for $m < n$.*

Lemma 1.11.16 *Every countable tree is isomorphic to a strongly standard tree.*

Proof Given a tree F, we construct a required isomorphism f. $f(x)$ is defined by induction on $ht(x)$. If x is the root, put $f(x) := \curlywedge$. If $f(x)$ is defined and $\beta(x) = \{y_0, \dots, y_n\}$, put $f(y_i) := f(x)i$; similarly if $\beta(x)$ is countable. (Of course this definition depends on the chosen ordering of $\beta(x)$.) ■

Definition 1.11.17 *An* **S4**-*tree is called* branchy *if for any* u $|\beta(u)| \leq 1$.

Every branchy tree is clearly infinite.

Lemma 1.11.18 *If* F *is a strongly standard branchy tree, then* $F \twoheadrightarrow IT_2$.

Proof We define the p-morphism g as follows:

$$g(n_1 \dots n_k) := \overline{n_1} \dots \overline{n_k},$$

where $\overline{n}$ is the remainder of n modulo 2. g is obviously monotonic. The lift property follows, since $g(x)0 = g(x0)$, $g(x)1 = g(x1)$ and a branchy standard tree always contains $x0$, $x1$ together with x. ∎

Lemma 1.11.19 *If every cone in a countable* **S4**-*tree* F *is nonlinear, then* $F \twoheadrightarrow IT_2$.

Proof By 1.11.18 and 1.11.16, it is sufficient to p-morphically map F onto a branchy tree.

Let $F = (W, R)$. For $x \in W$ let $h(x)$ be the least element in $\{y \in R(x) \mid |\beta(y)| > 1\}$. $h(x)$ clearly exists, since this set has a minimal element, which must be unique.

Now h is a morphism, i.e. a p-morphism onto its image. In fact, the monotonicity is obvious. The lift property follows since

$$h(x)R_{-}h(y) \Rightarrow xR_{-}y.$$

To show the latter, suppose $h(x)R_{-}h(y)$. Since $yRh(y)$ and $xRh(x)Rh(y)$, it follows that x, y are R-comparable (remember that F is an **S4** tree). If yRx, we obtain $yRh(x)R_{-}h(y)$, and $h(x)$ has at least two successors, which contradicts the choice of $h(y)$. Thus $xR_{-}y$.

The image G of h is a branchy tree, since $h(x) = h(h(x))$, and thus $\beta_G(h(x))= h[\beta_F(h(x))]$. ∎

Definition 1.11.20 *A p.o. tree is called* effuse *if it has a cone without linear subcones.*

Proposition 1.11.21 *If* F *is an effuse countable tree, then* $\mathbf{ML}(F) = \mathbf{S4}$ *and thus* $\mathbf{IL}(F) = \mathbf{H}$.

Proof If a cone $F{\uparrow}u$ does not contain linear subcones, then by Lemma 1.11.19 and Proposition 1.11.13, $F{\uparrow}u \twoheadrightarrow IT_2 \twoheadrightarrow IT_\omega$, and thus $\mathbf{ML}(F) \subseteq \mathbf{ML}(IT_\omega) = \mathbf{S4}$ by the generation lemma, the morphism lemma and Corollary 1.11.12. The converse inclusion is trivial by soundness. ∎

1.12 Subframe and cofinal subframe logics

Definition 1.12.1 *A modal or intermediate propositional logic is called* subframe *if its Kripke frame variety is closed under taking subframes.*

Definition 1.12.2 *For a transitive Kripke frame $F = (W, R)$ a subset $V \subseteq W$ and the subframe $F' = F \upharpoonright V$ are called* cofinal *if $R(V) \subseteq R^{-1}(V)$.*

Definition 1.12.3 *A transitive modal or an intermediate propositional logic $\mathbf{\Lambda}$ is called* cofinal subframe *if its Kripke frame variety $\mathbf{V}(\mathbf{\Lambda})$ or $\mathbf{V}^{\sim}(\mathbf{\Lambda})$ is closed under taking cofinal subframes.*

Definition 1.12.4 *Let F, G be Kripke frames of the same kind. A* subreduction *from F to G is a p-morphism from a subframe of F onto G. A* reduction *is a subreduction defined on a cone. A subreduction of transitive frames is called* cofinal *if its domain is cofinal.*

If G is rooted and there exists a reduction (respectively, a subreduction, a cofinal subreduction) from F to G, we say that F is reducible *(respectively,* subreducible, cofinally subreducible*) to G.*

If G is arbitrary we also say that F is reducible to G if it is reducible to every cone in G.

By the morphism and generation lemmas every variety $\mathbf{V}(\mathbf{\Lambda})$ or $\mathbf{V}^{\sim}(\mathbf{\Lambda})$ is closed under reductions. Thus for a subframe (respectively, cofinal subframe) $\mathbf{\Lambda}$ it is closed under subreductions (respectively, cofinal subreductions).

Definition 1.12.5 *A class of first-order structures is called* universal, *if it is a class of models of a set of universal first-order sentences.*

Definition 1.12.6 *A modal or intermediate propositional logic $\mathbf{\Lambda}$ is* universal, *if $\mathbf{V}(\mathbf{\Lambda})$ (or $\mathbf{V}^{\sim}(\mathbf{\Lambda})$) is universal.*

The well known Tarski–Łos theorem states that a class is universal iff it is closed under substructures. So a Δ-elementary propositional logic is universal iff it is subframe.

The following result yields a criterion of universality for subframe logics; for the proof see [Chagrov and Zakharyaschev, 1997], Theorem 11.31, or [Wolter, 1997].

Definition 1.12.7 *A modal or intermediate propositional logic $\mathbf{\Lambda}$ has the* finite embedding property *if for any Kripke frame F, F validates $\mathbf{\Lambda}$ whenever every finite subframe of F validates $\mathbf{\Lambda}$.*

Theorem 1.12.8 (Wolter) *For any subframe modal logic $\mathbf{\Lambda}$ the following properties are equivalent*

(1) $\mathbf{\Lambda}$ is universal and Kripke-complete;

(2) $\mathbf{\Lambda}$ is quasi-Δ-elementary;

(3) $\mathbf{\Lambda}$ *is d-persistent;*

(4) $\mathbf{\Lambda}$ *is r-persistent;*

(5) $\mathbf{\Lambda}$ *has the finite embedding property and is Kripke-complete.*

Later on we will need the implication (1)⇒(4) from this theorem, so let us give some comments on its proof. (1)⇒(2) is obvious, (2)⇒(3) is the Fine–van Benthem theorem. For (3)⇒(4), note that every refined $\mathbf{\Lambda}$-frame $(F, \mathcal{W})$ has a descriptive extension $(F', \mathcal{W}')$ also validating $\mathbf{\Lambda}$ (the so-called 'ultrafilter extension'). By d-persistence it follows that $F' \vDash \mathbf{\Lambda}$, and thus $F \vDash \mathbf{\Lambda}$, since $\mathbf{\Lambda}$ is subframe.

However an example from [Chagrov and Zakharyaschev, 1997] shows that subframe monomodal logics may be Kripke-incomplete.

On the other hand, subframe intermediate and **K4**-logics enjoy better properties. Their main new feature is axiomatisability by special 'subframe formulas' defined below. In this definition we assume that p_a are different proposition letters corresponding to worlds a of a finite Kripke frame F.

Definition 1.12.9 *For a transitive frame* $F = (W, R)$ *with root* 0 *put*

$$SM^-(F) := p_0 \wedge \bigwedge_{aRb} \Box(p_a \supset \Diamond p_b) \wedge \bigwedge_{\neg aRb} \Box(p_a \supset \neg\Diamond p_b) \wedge \bigwedge_{0Ra} \Diamond p_a \wedge$$
$$\bigwedge_{a \neq b} \neg\Diamond(p_a \wedge p_b) \wedge \bigwedge_{a \neq 0} \neg p_a,$$
$$XM^-(F) := SM^-(F) \wedge \Box\left(\bigvee_{a \in W} p_a\right),$$
$$CSM^-(F) := SM^-(F) \wedge \Box\Diamond\left(\bigvee_{a \in W} p_a\right),$$
$$SM(F) := \neg SM^-(F),\ CSM(F) := \neg CSM^-(F),\ XM(F) := \neg XM^-(F).$$

$SM(F)$, $CSM(F)$, $XM(F)$ *are respectively called the* (modal) subframe, the cofinal subframe, and the frame *formula of* F.[20]

Obviously, the conjunct $\bigwedge_{a \neq 0} \neg p_a$ is redundant if the underlying logic is **S4** and all frames are reflexive.

These formulas as well as the next theorem originate from [Fine, 1974], [Fine, 1985], [Zakharyaschev, 1989]. They are particular kinds of *Zakharyaschev canonical formulas*, see [Zakharyaschev, 1989], [Chagrov and Zakharyaschev, 1997].

Theorem 1.12.10 *Let F be a finite rooted transitive Kripke 1-frame. Then for any transitive Kripke 1-frame G*

(1) $G \nvDash XM(F)$ *iff G is reducible to F,*

(2) $G \nvDash SM(F)$ *iff G is subreducible to F,*

[20] $XM(F)$ is also called the *Jankov–Fine*, or the *characteristic* formula.

(3) $G \not\models CSM(F)$ iff G is cofinally subreducible to F.

Let us recall the idea of the proof. For example, if $M = (G, \theta) \not\models CSM(F)$, then we obtain a cofinal subreduction from G to F by putting $f(x) = a$ iff $M, x \models p_a$. The other way round, if f is such a subreduction, we construct a countermodel M for $CSM(F)$ by putting $M, x \models p_a$ iff $f(x) = a$.

The next two theorems from [Zakharyaschev, 1989] are also specific for the transitive case:

Theorem 1.12.11 *A transitive 1-modal logic is subframe (respectively, cofinal subframe) iff it is axiomatisable by subframe (respectively, cofinal subframe) formulas above* **K4**.

Theorem 1.12.12[21] *Every cofinal subframe modal logic has the f.m.p.*

Corollary 1.12.13 *Let $\mathbf{\Lambda}_0$ be a subframe* **K4**-*logic. Then for any 1-modal logic $\mathbf{\Lambda} \supseteq \mathbf{\Lambda}_0$, $\mathbf{\Lambda}$ is subframe (respectively, cofinal subframe) iff it is axiomatisable by subframe (respectively, cofinal subframe) formulas of $\mathbf{\Lambda}_0$-frames above $\mathbf{\Lambda}_0$.*

Proof 'If' easily follows from 1.12.11. To prove 'only if', suppose $\mathbf{\Lambda}$ is subframe; the case of cofinal subframe logics is quite similar. By 1.12.11, we have

$$\mathbf{\Lambda} = \mathbf{K4} + \{SM(F) \mid (F) \in \mathbf{\Lambda}\},\ \mathbf{\Lambda}_0 = \mathbf{K4} + \{SM(F) \mid SM(F) \in \mathbf{\Lambda}_0\}$$

hence

$$\mathbf{\Lambda} = \mathbf{\Lambda}_0 + \{SM(F) \mid SM(F) \in \mathbf{\Lambda} - \mathbf{\Lambda}_0\}.$$

Now $SM(F) \notin \mathbf{\Lambda}_0$ implies $F \models \mathbf{\Lambda}_0$. In fact, since Λ_0 is Kripke-complete by 1.12.12, there exists a $\mathbf{\Lambda}_0$-frame G such that $G \not\models SM(F)$. Then G is subreducible to F by 1.12.10, and since $\mathbf{\Lambda}_0$ is a subframe logic, it follows that $F \models \mathbf{\Lambda}_0$. ∎

We shall use this corollary especially for the case $\mathbf{\Lambda}_0 = \mathbf{S4}$.

Example 1.12.14 $\mathbf{S4.1} = \mathbf{S4} + \Box\Diamond p \supset \Diamond\Box p$ is a cofinal subframe logic (McKinsey property is obviously preserved for cofinal subframes). It can be presented as $\mathbf{S4} + CSM(FC_2)$, where FC_2 is a 2-element cluster — one can check that an **S4**-frame has McKinsey property iff it is not cofinally subreducible to FC_2. Similarly the logic $\mathbf{K4.1}^- := \mathbf{K4} + \Box\bot \vee \Diamond\Box\bot$ is cofinal subframe; it is presented as $\mathbf{K4} + CSM(FC_1)$, where FC_1 is a reflexive singleton and characterised by the (first-order) condition

$$\forall x\, (R(x) = \varnothing \vee \exists y\, (xRy\ \&\ R(y) = \varnothing)).$$

Thanks to completeness stated in 1.12.12, Theorem 1.12.8 has the following transitive version:

[21] For subframe transitive logics this fact was first proved in [Fine, 1985].

Theorem 1.12.15 (Zakharyaschev) *For any subframe modal logic* $\mathbf{\Lambda} \supseteq \mathbf{K4}$ *the following properties are equivalent:*

(1) $\mathbf{\Lambda}$ *is universal;*

(2) $\mathbf{\Lambda}$ *is quasi-*Δ*-elementary;*

(3) $\mathbf{\Lambda}$ *is d-persistent;*

(4) $\mathbf{\Lambda}$ *is r-persistent;*

(5) $\mathbf{\Lambda}$ *has the finite embedding property.*

By applying 1.12.10 to subframe logics $\mathbf{\Lambda} = \mathbf{\Lambda}_0 + \{SM(F_i) \mid i \in I\}$ described in 1.12.13, we can reformulate the finite embedding property as follows:

> *For any* Λ_0*-frame* G*, if* G *is subreducible to some* F_i $(i \in I)$*, then some finite subframe of* G *is subreducible to some* F_j $(j \in I)$.

Hence we obtain a sufficient condition for elementarity of subframe logics above **K4** or **S4**.

Proposition 1.12.16

(1) A subframe **K4***-logic is* Δ*-elementary if above* **K4** *it is axiomatisable by subframe formulas of irreflexive transitive frames.*

(2) A subframe **S4***-logic is* Δ*-elementary if above* **S4** *it is axiomatisable by subframe formulas of posets.*

Proof We prove only (1); the proof of (2) is similar. By 1.12.15, it is sufficient to check the finite embedding property for $\mathbf{K4} + \{SM(F_i) \mid i \in I\}$, where F_i are irreflexive **K4**-frames. So for a **K4**-frame G subreducible to some F_i, we find a finite subframe subreducible to F_i. Given a subreduction $f : G' \twoheadrightarrow F_i$, for $G' \subseteq G$, it is sufficient to construct a finite $G'' \subseteq G'$ such that $f \restriction G'' : G'' \twoheadrightarrow F_i$.

This is done by induction on $|F_i|$. If F_i is an irreflexive singleton, everything is trivial. Otherwise, let u be the root of F_i, and let $f(a) = u$. Obviously, a is irreflexive.

For every $v \in \beta(u)$ there exists a cone $G'_v \subseteq G' \uparrow a$ such that $f \restriction G'_v : G'_v \twoheadrightarrow F_i \uparrow v$ — this follows from 1.3.32(3). Then by the induction hypothesis, there is a finite $G''_v \subseteq G'_v$ such that $f \restriction G''_v : G''_v \twoheadrightarrow F_i \uparrow v$.

Finally put $G'' := \{a\} \cup \bigcup\limits_{v \in \beta(u)} G''_v$ (as a subframe of G'). This G'' is the required one; in fact, monotonicity is preserved by restricted maps, and the lift property easily follows from the construction. ■

For subframe logics axiomatisable by a single subframe formula the converse also holds:

Proposition 1.12.17

(1) A subframe logic $\mathbf{K4} + SM(F)$ *is elementary iff* F *is irreflexive.*

(2) A subframe logic $\mathbf{S4} + SM(F)$ *is elementary iff* F *is a poset.*

Proof

(1) If F contains reflexive points, we can replace each of them by the chain $(\omega, <)$. The resulting frame F' is reducible to F, since a reflexive singleton is a p-morphic image of $(\omega, <)$. However, finite subframes of F' are not reducible to F, since a p-morphic image of a finite irreflexive **K4**-frame is always irreflexive (e.g. because finite irreflexive **K4**-frames are exactly finite **GL**-frames). Thus $\mathbf{K4}+SM(F)$ does not have the finite embedding property in this case.

(2) Similarly, if F contains nontrivial clusters, we can replace each of them by $(\omega, \leq)$. Then we obtain a frame F' reducible to F. Every finite subframe of F' is a poset, so it is not reducible to F (e.g. because it is a **Grz**-frame).

■

Remark 1.12.18 In general, the converse to 1.12.16 is not true. For example, the trivial logic $\mathbf{S4}+p \supset \Box p$ is obviously subframe and elementary, but it cannot be axiomatised by subframe formulas of posets. In fact, every formula $SM(F)$ for a nontrivial F, is valid in every **S5**-frame G (which is a cluster), since G is not subreducible to F. Moreover, there is a conjecture that elementarity of a logic axiomatisable by a finite set of subframe formulas is undecidable.[22]

Theorem 1.12.15 has an analogue for cofinal subframe logics.

Definition 1.12.19 *A world in a transitive frame is called* inner *if its cluster is not maximal. The restriction of a frame* F *to inner worlds is denoted by* F^-.

Definition 1.12.20 *Let* F, G *be* **K4**-*frames and suppose that* G *is finite. A cofinal subreduction* f *from* F *to* G *is called a* cofinal quasi-embedding *if* $f^{-1}(x)$ *is a singleton for any inner* x. *If such a subreduction exists, we say that* G *is a* finite cofinal quasi-subframe *of* F.

Definition 1.12.21 *A modal or intermediate propositional logic* $\mathbf{\Lambda}$ *has the* finite cofinal quasi-embedding property *if for any Kripke frame* F, F *validates* $\mathbf{\Lambda}$ *whenever every its finite cofinal quasi-subframe validates* $\mathbf{\Lambda}$.

Theorem 1.12.22 (Zakharyaschev) *For any cofinal subframe modal logic the following properties are equivalent:*

(1) $\mathbf{\Lambda}$ *is elementary;*

(2) $\mathbf{\Lambda}$ *is quasi-*Δ*-elementary;*

[22] M. Zakharyaschev, personal communication.

(3) $\mathbf{\Lambda}$ *is d-persistent;*

(4) $\mathbf{\Lambda}$ *has the finite cofinal quasi-embedding property.*

Note that unlike the previous theorem, 1.12.22 does not include r-persistence.

Let us now consider the intuitionistic case. Now we assume that q_a are different proposition letters indexed by worlds of a finite poset F.

Definition 1.12.23 *For a poset* $F = (W, \leq)$ *with root* 0 *put*

$$SI^-(F) := \bigwedge_{a<b} (q_b \supset q_a) \wedge \bigwedge_{a} \left(\left(\bigwedge_{a \not\leq b} q_b \bullet \supset q_a \right) \supset q_a \right),$$
$$CSI^-(F) := SI^-(F) \wedge \neg \bigwedge_{a \in W} q_a,$$
$$XI^-(F) := CSI^-(F) \wedge \bigwedge_{a} \left(\bigwedge_{b<a} q_b \supset q_a \vee \bigwedge_{a \not\leq b} q_b \right),$$
$$SI(F) := SI^-(F) \supset q_0,\ CSI(F) := CSI^-(F) \supset q_0,\ XI(F) := XI^-(F) \supset q_0.$$

$SI(F)$, $CSI(F)$, $XI(F)$ *are respectively called the* (intuitionistic) subframe, cofinal subframe, and frame *formula of* F.[23]

Note that $SI(F)$ is an implicative formula and $CSI(F)$ is built from proposition letters and $\supset$, $\perp$.

Then similarly to the modal case, we have (cf. [Zakharyaschev, 1989], [Chagrov and Zakharyaschev, 1997]):

Theorem 1.12.24 *Let* F *be a finite rooted poset. Then for any poset* G

(1) $G \not\models XI(F)$ *iff* G *is reducible to* F*;*

(2) $G \not\models SI(F)$ *iff* G *is subreducible to* F*;*

(3) $G \not\models CSI(F)$ *iff* G *is cofinally subreducible to* F.

The idea of the proof is quite similar to 1.12.10. E.g. if $M = (G, \theta) \not\Vdash XI(F)$, we obtain a reduction f from G to F by putting

$$f(x) = a := (M, x \Vdash \bigwedge_{a \not\leq b} q_b \ \&\ M, x \not\Vdash q_a).$$

The formulas from 1.12.23 can be simplified. For example, in $CSI^-(F)$ we can replace the conjunct $\neg \bigwedge_{a \in W} q_a$ with $\neg \bigwedge_{a \in max(F)} q_a$, where $max(F)$ is the set of maximal points of F; also $\bigwedge_{b<a} q_b$ occurring in the second conjunct of $XI^-(F)$ can be replaced with $\bigwedge_{a \in \beta(b)} q_b$.

Other versions of these formulas are described in [Chagrov and Zakharyaschev, 1997] and [Shimura, 1993].

The next result is also due to [Zakharyaschev, 1989], cf. [Chagrov and Zakharyaschev, 1997].

[23] $XI(F)$ is also called the *Jankov*, or the *characteristic* formula of F.

Theorem 1.12.25 *For an intermediate propositional logic* $\mathbf{\Lambda}$ *the following properties are equivalent:*

(1) $\mathbf{\Lambda}$ *is a subframe (respectively, cofinal subframe) logic;*

(2) $\mathbf{\Lambda}$ *is axiomatisable by subframe (respectively, cofinal subframe) formulas;*

(3) $\mathbf{\Lambda}$ *is axiomatisable by implicative formulas (respectively,* $(\supset, \bot)$*-formulas).*

Let us now formulate analogues of Theorems 1.12.12, 1.12.15 and 1.12.22 for intermediate logics.

Theorem 1.12.26 *Every cofinal subframe intermediate logic has the finite model property.*

This result follows from [McKay, 1968] and 1.12.25; also see [Chagrov and Zakharyaschev, 1997].

Theorem 1.12.27

(1) An intermediate logic is subframe iff it is universal.

(2) Every subframe intermediate logic is r-persistent.

Theorem 1.12.28 *Every cofinal subframe intermediate logic is* Δ*-elementary and d-persistent.*

To formulate a modal analogue of this result, we need

Definition 1.12.29 *A transitive frame is called* almost irreflexive *if all its non-maximal clusters are degenerate. An* **S4***-frame is called* blossom *if all its non-maximal clusters are trivial.*

Theorem 1.12.30

(1) A cofinal subframe **S4***-logic is* Δ*-elementary if above* **S4**.1 *it is axiomatisable by cofinal subframe formulas of blossom frames.*

(2) A cofinal subframe **K4***-logic is* Δ*-elementary if above* $\mathbf{K4}.1^-$ *it is axiomatisable by cofinal subframe formulas of almost irreflexive transitive frames.*

Proof (1) By 1.12.22, it suffices to check the finite cofinal quasi-embedding property for our logic $\mathbf{\Lambda}$. So suppose $G \not\models \mathbf{\Lambda}$. If $G \not\models \mathbf{S4}.1$, the further proof is easy, so let $G \models \mathbf{S4}.1$. Then there is a finite rooted blossom frame F such that $CSM(F) \in \mathbf{\Lambda}$ and a cofinal subreduction f from G to F. Let us show that then there exists a subframe $G_0 \subseteq G$ such that $f \restriction G_0$ is a cofinal subreduction to F and f is injective on $f^{-1}(F^-)$. Moreover, we can choose G_0 so that $f[G_0^-] \subseteq F^-$, and thus G_0^- is finite.

The proof is by induction on the cardinality of F. Let u be an element of G with $f(u) = 0_F$, where 0_F is the root of F. For every $a \in \beta(0_F)$, choose an arbitrary $v \in G$ such that $u \leq_G v$ and $f(v) = a$. Note that if a is maximal in F and G has McKinsey property, we can choose a maximal v, since f is cofinal. Then $f_a := f \restriction (G{\uparrow}v)$, is a cofinal subreduction from $G \uparrow v$ to $F{\uparrow}a$. Hence by the induction hypothesis, there exists a finite subframe G_a of G such that $f_a \restriction G_a$ is a cofinal subreduction from G to $F \uparrow a$. Put

$$G_0 := \{u\} \cup \bigcup_{a \in \beta(0_F)} G_a.$$

Then G_0 obviously has the required property.

(2) is proved similarly. ∎

1.13 Splittings

The following result is a modal version of [Jankov, 1969].

Theorem 1.13.1 *Let F be a finite rooted Kripke frame. Then for any modal algebra $\mathbf{\Omega}$ the following properties are equivalent:*

(1) $\mathbf{\Omega} \not\models XM(F)$;

(2) $MA(F)$ is a subalgebra of a homomorphic image of $\mathbf{\Omega}$.

Lemma 1.13.2 *For any 1-modal logic $\mathbf{\Lambda} \supseteq \mathbf{K4}$ and a finite rooted transitive frame F*

$$XM(F) \in \mathbf{\Lambda} \text{ iff } \mathbf{\Lambda} \not\subseteq \mathbf{ML}(F);$$

in particular, for any 1-modal formula A

$$\mathbf{K4} + A \vdash XM(F) \text{ iff } A \notin \mathbf{ML}(F).$$

If F is reflexive, then the latter is also equivalent to $\mathbf{S4} + A \vdash XM(F)$

Remark This lemma shows that the pair of logics $(\mathbf{K4} + XM(F), \mathbf{ML}(F))$ 'splits' the set of extensions of $\mathbf{K4}$: every $\mathbf{\Lambda} \supseteq \mathbf{K4}$ is either below $\mathbf{ML}(F)$ or above $\mathbf{K4} + XM(F)$.

The original Jankov formula [Jankov, 1969] is defined as follows.

Definition 1.13.3 *Let $F = (W, \leq)$ be a finite p.o. set, $\mathbf{\Omega} = HA(F)$. Then*

$$XI(F) := \left(\bigwedge_{a,b \in \Omega} (p_{a \wedge b} \equiv p_a \wedge p_b) \wedge \bigwedge_{a,b \in \Omega} (p_{a \vee b} \equiv p_a \vee p_b) \wedge \bigwedge_{a,b \in \Omega} (p_{a \to b} \equiv (p_a \supset p_b)) \wedge \neg p_{\mathbf{0}} \right) \supset p_\omega,$$

where $\{p_a \mid a \in \Omega\}$ is a set of distinct propositional variables corresponding to the elements of $\mathbf{\Omega}$ and ω is the subgreatest element of $\mathbf{\Omega}$.

Figure 1.2. F_2

Theorem 1.13.4 *Let F be a finite rooted Kripke frame. Then for any Heyting algebra $\mathbf{\Omega}'$ the following properties are equivalent:*

(1) $\mathbf{\Omega}' \not\models XI(F)$;

(2) $HA(F)$ *is a subalgebra of a homomorphic image of* $\mathbf{\Omega}'$.

Corollary 1.13.5 *Let F be a finite rooted p.o. set. Then for any p.o. set G the following properties are equivalent:*

(1) $G \not\Vdash XI(F)$;

(2) there exists $u \in G$ *such that* $G^u \twoheadrightarrow F$.

Lemma 1.13.6 *For any superintuitionistic logic* $\mathbf{\Lambda}$

$$XI(F) \in \mathbf{\Lambda} \textit{ iff } \mathbf{\Lambda} \not\subseteq \mathbf{IL}(F),$$

in particular, for any intuitionistic formula A

$$XI(F) \in (\mathbf{H} + A) \textit{ iff } A \notin \mathbf{IL}(F).$$

Hence for example, we obtain

Lemma 1.13.7 *Let A be a propositional formula. Then:*

(1) $AJ \in (\mathbf{H} + A)$ *iff* $A \notin \mathbf{IL}(F_2)$ *(Fig. 1.2);*

(2) $AJ^- \in (\mathbf{H} + A)$ *iff* $A \notin \mathbf{IL}(F_3)$ *and* $A \notin \mathbf{IL}(F_4)$ *(Figs. 1.3, 1.4).*

Proof Note that

$$\mathbf{H} + AJ = \mathbf{H} + XI(F_2), \; \mathbf{H} + AJ^- = \mathbf{H} + XI(F_3) \wedge XI(F_4).$$

Then apply Lemma 1.13.6 ∎

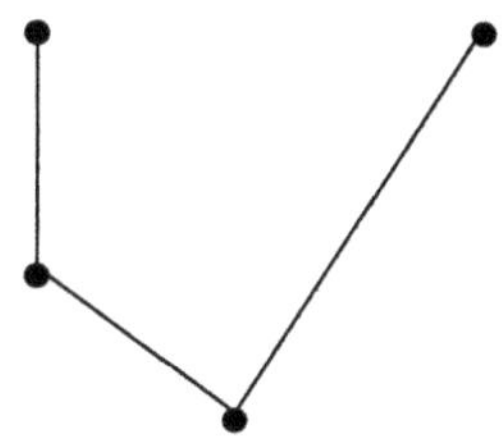

Figure 1.3. F_3

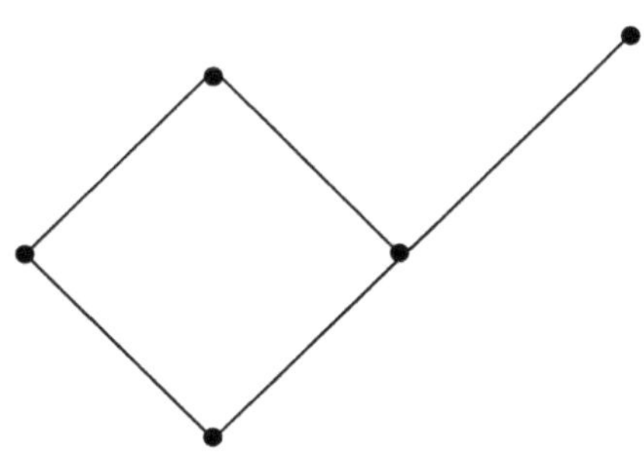

Figure 1.4. F_4

1.14 Tabularity

Definition 1.14.1 *A modal or intermediate propositional logic is called* tabular *if it is determined by a finite modal or Heyting algebra, or equivalently, by a finite Kripke frame.*

Lemma 1.14.2 *In the notation of 1.10.2, if F has a root u_0, $dd(u_0)$ is finite and $R(x)$ is finite for any $x \in F$, then F is finite.*

Proof By induction on $k = dd(u_0)$. The statement is trivial for $k = 0$. For the step note that

$$W = R^*(u_0) = \{u_0\} \cup R(u_0) \cup \bigcup\{W \uparrow u \mid u_0 R u, u_0 \neq u\}.$$

By the induction hypothesis, every $W \uparrow u$ for $u_0 R u$, $u_0 \neq u$ is finite, since $dd(u) < dd(u_0)$.

Hence W is finite. ∎

Consider the following N-modal k-formulas

$$\begin{array}{ll} p_k^i & := \neg p_i \wedge \bigwedge\limits_{i \neq j} p_j; \\ Alt_k & := \neg \bigwedge\limits_{i=1}^{k} \Diamond p_k^i; \\ B_k^1 & := p_k^1; \\ B_k^i & := p_k^i \wedge \Diamond B_k^{i-1} \text{ for } 1 < i \leq k; \\ AH_k & := \neg B_k^k; \end{array}$$

where

$$\Diamond A := \Diamond_1 A \vee \cdots \vee \Diamond_N A.$$

These formulas have the following first-order characterisations.

Lemma 1.14.3 *For a frame $F = (W, R_1, \dots, R_N)$ put $R := R_1 \cup \dots \cup R_N$. Then*

(1) $F, x \vDash Alt_k$ iff $|R(x)| < k$;

(2) $F, x \vDash AH_k$ iff $dd_F(x) < k$.

For intermediate logics we have the same characterisation with different formulas.

Lemma 1.14.4 *For a poset $F = (W, R)$*

$$F, x \vDash IG_k \text{ iff } |R(x)| < k.$$

Hence we obtain a characterisation of tabular modal logics similar to that in [Chagrov and Zakharyaschev, 1997].

Proposition 1.14.5

(1) *An N-modal propositional logic $\mathbf{\Lambda}$ is tabular iff $\mathbf{\Lambda} \vdash Alt_k \wedge AH_k$ for some k.*

(2) *An intermediate propositional logic $\mathbf{\Lambda}$ is tabular iff $\mathbf{\Lambda} \vdash IG_k$ for some k.*

Proof The claim 'only if' follows from 1.14.3 and 1.14.4.

To show 'if', use the canonicity of AH_k, Alt_k, IG_k and the observation that every rooted frame validating $Alt_k \wedge AH_k$ (or IG_k) is finite. ∎

Lemma 1.14.6 *If $\Phi = (F, \mathcal{W})$ is a finite refined frame, then $\mathbf{ML}(\Phi) = \mathbf{ML}(F)$ (or $\mathbf{IL}(\Phi) = \mathbf{IL}(F)$ in the intituionistic case).*

Proof (Modal case) For $x \neq y$ let U_{xy} be an interior set such that $x \in U_{xy}$, $y \notin U_{xy}$. Since F is finite, it follows that

$$\{x\} = \bigcap_{y \neq x} U_{xy} \in \mathcal{W}.$$

Thus all subsets of F are interior, so the valuations in Φ and F are the same. Therefore valid formulas are the same.

(Intuitionistic case.) Suppose $F = (W, R)$ is an intuitionistic frame and for $y \notin R(x)$ let $V_{xy} \in \mathcal{W}$ be such that $x \in V_{xy}, y \notin V_{xy}$. Then

$$R(x) = \bigcap_{y \notin R(x)} V_{xy} \in \mathcal{W}.$$

So for every R-stable U we have

$$U = \bigcup_{x \in U} R(x) \in \mathcal{W},$$

and again it follows that Φ, F have the same valuations. ∎

Proposition 1.14.7 *Every tabular modal or intermediate propositional logic is r-persistent.*

Proof (Modal case) Consider a tabular logic $\mathbf{\Lambda}$ and a refined frame $\Phi = (F, \mathcal{W})$ validating $\mathbf{\Lambda}$. By 1.14.5, $\mathbf{\Lambda} \vdash Alt_k \wedge AH_k$ for some k, hence $\Phi \vDash Alt_k \wedge AH_k$. According to the characterisation given in 1.14.3, the classes $\mathbf{V}(Alt_k)$, $\mathbf{V}(AH_k)$ are universal, so $Alt_k \wedge AH_k$ is r-persistent by Theorem 1.12.8 and thus valid in F and therefore in every cone $F \uparrow u$. So by 1.14.5, $\Phi \uparrow u$ is finite. By 1.7.8, $\Phi \uparrow u$ is refined and by 1.6.11, $\Phi \uparrow u \vDash \mathbf{\Lambda}$. Therefore, $F \uparrow u \vDash \mathbf{\Lambda}$ by 1.14.6. Since u is arbitrary, by Lemma 1.3.26, it follows that $F \vDash \mathbf{\Lambda}$.

In the intuitionistic case use the same argument and 1.14.4 instead of 1.14.5. ∎

Proposition 1.14.8 *For finite modal frames G, F, $\mathbf{ML}(G) \subseteq \mathbf{ML}(F)$ iff G is reducible to F; similarly for the intuitionistic case.*

Theorem 1.14.9 *Every tabular modal or intermediate logic is finitely axiomatisable. Moreover, a transitive modal or intermediate logic is finitely axiomatisable by frame formulas.*

1.15 Transitive logics of finite depth

The results of this section are based on [Hosoi, 1967], [Hosoi, 1969].

Lemma 1.15.1 *For a poset F*

$$F \Vdash AP_n \Leftrightarrow d(F) \leq n.$$

Proposition 1.15.2 *Let Z_n be an n-element chain. Then*

$$\mathbf{IL}(Z_n) = \mathbf{LC} + AP_n,\ \mathbf{LC} = \bigcap_n \mathbf{IL}(Z_n) = \mathbf{IL}(F)$$

for every infinite chain F. In particular, $\mathbf{IL}(Z_1) = \mathbf{H} + AP_1$ is a classical logic and $\mathbf{H} + AP_n \subset \mathbf{IL}(Z_n)$ for $n \geq 2$.

Proposition 1.15.3 $\mathbf{H} + AP_n$ *is determined by the class of finite posets of depth n.*

Proof As we know, AP_n-frames are **S4**-frames of depth $\leq n$, so $\mathbf{H} + AP_n$ is a subframe logic. Thus it is determined by finite posets of depth $\leq n$. It remains to note that every poset of depth $< n$ is a generated subframe in a poset of depth n. ■

Hence by unravelling we obtain

Corollary 1.15.4 $\mathbf{H}+AP_n = \mathbf{IL}(IT^n_\omega)$, *where IT^n_ω is the standard tree of depth n and branching ω.*

Hence by induction on n we obtain

Corollary 1.15.5 *For every $m, n \in \omega$ there exists $k \in \omega$ such that*

$$(\mathbf{H} + AP_n)\lceil m = \mathbf{IL}(IT^n_k)\lceil m,$$

where $\mathbf{IL}(IT^n_k)$ is the tree of depth n and branching k.

So to say, m letters can distinguish only finitely many successors of any point.

Definition 1.15.6 *A logic L is called* locally tabular *if all its finitely generated Lindenbaum algebras $Lind(L\lceil m)$ are finite; in other words, if for every m there exist finitely many non-L-equivalent formulas in m propositional letters.*

Corollary 1.15.7 *Every logic* $\mathbf{H} + AP_n$ *is locally tabular.*

Proposition 1.15.8 *Every locally tabular logic has the f.m.p.*

Lemma 1.15.9 $\mathbf{H} + XI(Z_{n+1}) = \mathbf{H} + AP_n$ *for $n \geq 0$.*

Proof A frame F is of depth $\leq n$ iff F is not reducible to Z_{n+1} iff $F \Vdash XI(Z_{n+1})$. Thus $\mathbf{H}+AP_n$ and $\mathbf{H}+XI(Z_{n+1})$ have the same finite frames, and it remains to show that the latter logic has the f.m.p.

But this logic is subframe, since the class of frames of depth $\leq n$ is closed under subframes. So we can apply Fine's theorem. ■

Hence by Lemma 1.13.6 we obtain

Proposition 1.15.10 *For a superintuitionistic logic L*

$$L \not\subseteq \mathbf{IL}(Z_{n+1}) \text{ iff } \mathbf{H}+AP_n \subseteq L.$$

This proposition shows that the structure of superintuitionistic logics splits into an $(\omega+1)$-sequence of slices.

Definition 1.15.11 *For a finite n, the* nth slice *is the set of superintuitionistic logics*

$$\mathcal{S}_n := \{L \in \mathcal{S} \mid \mathbf{H}+AP_n \subseteq L \subseteq \mathbf{IL}(Z_n)\};$$

the ωth slice *is*

$$\mathcal{S}_\omega := \{L \mid \mathbf{H} \subseteq L \subseteq \mathbf{LC}\}.$$

So

$$\mathcal{S}_0 = \{\mathbf{H}+AP_0\} = \{\mathbf{H}+\bot\},$$

i.e. $\mathcal{S}_0$ contains only the logic of the empty 'chain' Z_0.

Let $\mathcal{S}_n$ be the set of logics of slice n (for $0 \leq n \leq \omega$). Proposition 1.15.10 implies

Proposition 1.15.12

(1) $\mathcal{S} = \bigcup\limits_{n\leq\omega} \mathcal{S}_n$,

(2) $\mathcal{S}_n \bigcap \mathcal{S}_m = \varnothing$ *for* $n \neq m$,

(3) if L is a superintuitionistic logic, then $L \in \mathcal{S}_n$, where

$$n = \begin{cases} \omega & \text{if } \forall n\ L \in \mathbf{IL}(Z_n), \\ \max\{n \in \omega \mid L \subseteq \mathbf{IL}(Z_n)\} & \text{otherwise} \end{cases}$$

(4) the slices $\mathcal{S}_1$ is one-element: $\mathcal{S}_1 = \{\mathbf{H}+AP_1\} = \mathbf{CL}$,

(5) the slice $\mathcal{S}_2$ is a decreasing $(\omega+1)$-chain,

(6) for $n \geq 3$ all slices $\mathcal{S}_n$ are of cardinality $2^{\aleph_0}$.

Proposition 1.15.13 *For a finite n*

$$\mathbf{IL}(F) \in \mathcal{S}_n \Leftrightarrow d(F) = n.$$

Proposition 1.15.14 $\mathcal{S}_n = \{\mathbf{IL}(F) \mid F$ *is a poset of depth* $n\}$. *Every logic of slice* $\mathcal{S}_n$ *is determined by a disjoint union of finite posets of depth* n.

Proof Local tabularity is clearly inherited by extensions, so it holds for all logics of finite slices. Therefore all logics of finite slices have the f.m.p., and hence the proposition follows. ∎

As we mentioned, $\mathbf{LC} = \bigcap_{n\in\omega} \mathbf{IL}(Z_n)$. The set

$$\{L \mid \mathbf{LC} \subset L\} = \{\mathbf{IL}(Z_n) \mid n \in \omega\}$$

is a decreasing ω-chain; this readily follows from Proposition 1.15.14, since $\mathbf{IL}(Z_n) = \mathbf{LC} + AP_n$ is a simple extension of $\mathbf{LC}$ in the nth slice. Also $\mathbf{H} = \bigcap_{n\in\omega} (\mathbf{H} + AP_n)$, since $\mathbf{H}$ has the f.m.p. and every finite poset is of finite depth.

Proposition 1.15.15 $\mathcal{S}_n$ *are sublattices of the lattice of superintuitionistic logics. Every finite slice* $\mathcal{S}_m$ *is embeddable in all slices* $\mathcal{S}_n$ *for* $m \leq n \leq \omega$ *by maps* $\lambda_{mn}\colon L \mapsto L \cap \mathbf{IL}(Z_n)$ *(and* $\mathbf{IL}(Z_\omega) = \mathbf{LC}$*).*

Proof Note that $L_1 \subseteq L_2$ iff $\lambda_{mn}(L_1) \subseteq \lambda_{mn}(L_2)$, since $L = \lambda_{mn}(L) + AP_m = \lambda_{mn}(L) + (H + AP_m)$, for $L \in \mathcal{S}_m$; moreover, $\lambda_{mn}(L) = \lambda_{mk}(L) + AP_n = \lambda_{mn}(L) + (\mathbf{H} + AP_n)$ for $k \geq n$, by distributivity. Also

$$\lambda_{mn}(L \cap L') = \lambda_{mn}(L) \cap \lambda_{mn}(L'),$$

and $\lambda_{mn}(L + L') = \lambda_{mn}(L) + \lambda_{mn}(L')$, again by distributivity. ∎

1.16 Δ-operation

In this section we study an embedding of $\mathcal{S}_n$ in $\mathcal{S}_{n+1}$ (Δ-operation) introduced in [Hosoi, 1969].

Definition 1.16.1 *For* $A \in IF$ *put* $\delta A := p \vee (p \supset A)$, *where* p *is a proposition letter that does not occur in* A. *For a superintuitionistic logic* L *put* $\Delta L := \mathbf{H} + \{\delta A \mid A \in L\}$.[24]

The original definition [Hosoi, 1969] is $\Delta L := \{\delta' A \mid A \in L\}$, where $\delta' A := ((p \supset A) \supset p) \supset p)$ (the 'A-version' of Peirce's law). Let us show that both definitions are equivalent.

Lemma 1.16.2 $\mathbf{H} + \delta A = \mathbf{H} + \delta' A$ *for any formula* A.

[24] Obviously, the definition of ΔL does not depend on the choice of p; but we fix p to make δA unique.

Proof On the one hand, obviously $\mathbf{QH} \vdash \delta A \supset \delta' A$. On the other hand, let us prove

$$\mathbf{H} + \delta' A \vdash \delta A. \tag{0}$$

In fact, obviously

$$p \vdash_{\mathbf{H}} p \vee (p \supset A), \tag{1}$$

hence

$$p,\ p \vee (p \supset A) \supset A \vdash_{\mathbf{H}} A, \tag{2}$$

and thus by the deduction theorem 1.1.4

$$\delta A \supset A \vdash_{\mathbf{H}} p \supset A. \tag{3}$$

This implies

$$\delta A \supset A \vdash_{\mathbf{H}} \delta A, \tag{4}$$

hence

$$\mathbf{H} \vdash (\delta A \supset A) \supset \delta A, \tag{5}$$

by the deduction theorem. Now since

$$((\delta A \supset A) \supset \delta A) \supset \delta A$$

is a substitution instance of $\delta' A$, (5) implies (0). ■

The next lemma shows the semantical meaning of δA.

Lemma 1.16.3 *Let F be a rooted poset with root 0_F. Then*

$$F \Vdash \delta A \textit{ iff } \forall u \neq 0_F\ F \uparrow u \Vdash A.$$

Proof (Only if.) Suppose $F{\uparrow}u \nVdash A$, $u \neq 0_F$, so $M \nVdash A$ for some Kripke model $M = (F{\uparrow}u, \theta)$. By truth preservation it follows that $M, u \nVdash A$. Then consider $M' = (F, \xi)$ such that

$$\xi(q) : \ = \begin{cases} \theta(q) & \text{if } q \neq p, \\ F - \{0_F\} & \text{if } q = p. \end{cases}$$

Obviously M' is intuitionistic. Also $M', u \nVdash A$, since p does not occur in A. Now from $M', u \Vdash p$ and $M', 0_F \nVdash p$ it follows that $M', 0_F \nVdash \delta A$; thus $F \nVdash \delta A$.

(If.) Suppose $F \nVdash \delta A$, i.e. $M \nVdash \delta A$ for some Kripke model M over F. Then by truth preservation $M, 0_F \nVdash \delta A$; hence $M, 0_F \nVdash p$, and $M, u \Vdash p$, $M, u \nVdash A$ for some $u \neq 0_F$. By the generation lemma it follows that $M{\uparrow}u, u \nVdash A$; thus $F{\uparrow}u \nVdash A$. ■

Hence we obtain

Proposition 1.16.4 *If $F \Vdash L$, then $1+F \Vdash \Delta L$, where $1+F$ is obtained by adding a root below F.*

Proof In fact, if $1+F \not\Vdash \delta A$, then by 1.16.3, $F{\uparrow}u \not\Vdash A$ for some $u \in F$; thus $F \not\Vdash A$ by the generation lemma. ■

Lemma 1.16.5 *For any superintuitionistic logic L, $\delta A \in \Delta L$ iff $A \in L$.*

Proof We consider a particular case, when L is Kripke-complete. The claim easily follows from Lemma 1.16.3 (and Proposition 1.16.4). In fact, if $A \notin L = \mathbf{IL}(F)$ for a poset F, then $\Delta L \subseteq \mathbf{IL}(1+F)$ and $\delta A \notin \mathbf{IL}(1+F)$. ■

A predicate analogue of 1.16.5 will be discussed later on in Section 2.13.

Lemma 1.16.6 *(1) $\mathbf{H} \vdash A \supset \delta A$.*

(2) $\mathbf{H} \vdash \delta(A_1 \supset A_2) \supset (\delta A_1 \supset \delta A_2)$.

(3) $\mathbf{H} \vdash \delta\left(\bigwedge\limits_{i=1}^{n} A_i\right) \equiv \bigwedge\limits_{i=1}^{n} \delta A_i$.

Proof An easy exercise; also see Chapter 2. ■

Proposition 1.16.7 *For propositional superintuitionistic logics L_1, L_2.*

(1) $\Delta L \subseteq L$;

(2) $L_1 \subseteq L_2$ iff $\Delta L_1 \subseteq \Delta L_2$;

(3) $L_1 = L_2$ iff $\Delta L_1 = \Delta L_2$.

Proof (1) $\Delta L \subseteq L$ follows from $\mathbf{H} \vdash A \supset \delta A$.

(2) 'Only if' is obvious. To show 'if', suppose $L_1 \not\subseteq L_2$ and $A \in L_1 - L_2$; then $\delta A \in \Delta L_1 - \Delta L_2$ by 1.16.5, and thus $\Delta L_1 \not\subseteq \Delta L_2$.

(3) A trivial consequence of (2). ■

Note that (2) means that Δ is a monotonic embedding $\mathcal{S} \longrightarrow \mathcal{S}$. By (2), $L_1 \subseteq L_2$ also implies $\Delta^n L_1 \subseteq \Delta^n L_2$ for any n.

Now the deduction theorem implies

Lemma 1.16.8 *$\Delta(\mathbf{H}+\Gamma) = \mathbf{H}+\delta\Gamma$, where $\delta\Gamma := \{\delta A \mid A \in \Gamma\}$. Hence $\Delta^n(\mathbf{H}+\Gamma) = \mathbf{H}+\delta^n\Gamma$, where $\delta^n\Gamma := \{\delta^n A \mid A \in \Gamma\}$.*

Proof In fact, if $L = \mathbf{H}+\Gamma \vdash A$ then $\mathbf{H} \vdash \bigwedge\limits_{i=1}^{m} B_i \supset A$ for some $B_1, \dots, B_m \in$ Sub(Γ). Then $\delta B_i \in$ Sub($\delta\Gamma$), and so $\mathbf{H}+\delta\Gamma \vdash \delta A$. Thus $\Delta L = \mathbf{H}+\delta L \subseteq \mathbf{H}+\delta\Gamma$. ■

Obviously $AP_{m+1} = \delta AP_m$, so $\delta^n AP_m = AP_{m+n}$, in particular $AP_m = \delta^m \bot$.

Hence we conclude that $\Delta^n(\mathbf{H} + AP_m) = \mathbf{H} + AP_{m+n}$, and thus for any n $\Delta(\mathbf{H}+AP_n) = \mathbf{H}+AP_{n+1}$; and $\mathbf{H}+AP_n = \Delta^n(\mathbf{H}+\bot) = \Delta^{n-1}(\mathbf{CL})$ for $n > 0$.

Also by Proposition 1.16.7,

$$\Delta(\mathbf{IL}(Z_n)) \subset \mathbf{IL}(Z_{n+1}) \text{ for } n > 0;$$

The inclusion is proper, since $\Delta L \subseteq \Delta\mathbf{CL} = \mathbf{H} + AP_2$ for $L \subseteq \mathbf{CL}$, and thus $AZ = (p \supset q) \vee (q \supset p) \notin \Delta L$.

So we obtain

Proposition 1.16.9 *$\Delta L \in \mathcal{S}_{n+1}$ for $L \in \mathcal{S}_n$, $n \in \omega$. Thus Δ embeds $\mathcal{S}_n$ in $\mathcal{S}_{n+1}$ and $\mathcal{S}_\omega$ in itself.*

Note that Δ is not a lattice embedding; more precisely, it preserves joins not meets.

Lemma 1.16.10 *If $A_1 \in (L_1 - L_2)$ and $A_2 \in (L_2 - L_1)$, then $\delta A_1 \vee \delta A_2 \in \Delta L_1 \cap \Delta L_2 - \Delta(L_1 \cap L_2)$.*

Proof For Kripke-complete L_1 and L_2 (in particular, for all logics of finite slices) this readily follows from Lemma 1.16.3. Namely, if $A_1 \notin L_2 = \mathbf{IL}(F_2)$ and $A_2 \notin L_1 = \mathbf{IL}(F_1)$, then the frame $1 + F_1 \sqcup F_2$ separates $\delta A_1 \vee \delta A_2$ from $\Delta(L_1 \cap L_2)$.

The lemma actually holds for arbitrary logics; its analogue for predicate logics will be discussed in Section 2.13. ∎

Proposition 1.16.11 *For superintuitionistic logics L_1, L_2*

(1) $\Delta(L_1 + L_2) = \Delta L_1 + \Delta L_2$;

(2) $\Delta(L_1 \cap L_2) = \Delta L_1 \cap \Delta L_2$ iff L_1 and L_2 are $\subseteq$-comparable.

Proof (1) follows from Lemma 1.16.5: $L_1+L_2 = \mathbf{H}+L_1\cup L_2$, so $\Delta(L_1+L_2) = \mathbf{H} + \delta L_1 \cup \delta L_2 = (\mathbf{H} + \delta L_1) + (\mathbf{H} + \delta L_2) = \Delta L_1 + \Delta L_2$.

(2) follows from 1.16.10. ∎

Proposition 1.16.12 *$\Delta L = L$ iff $L = \mathbf{H}$. So $\Delta L \subset L$ for any $L \neq \mathbf{H}$.*

Proof If $\Delta L = L$, then $L \subseteq \mathbf{H} + \bot$ implies $L \subseteq \Delta^n(\mathbf{H} + \bot) = \mathbf{H} + AP_n$ for all $n \in \omega$. Hence $L \subseteq \bigcap\limits_n(\mathbf{H} + AP_n) = \mathbf{H}$. ∎

Remark 1.16.13 A similar argument shows that for any superintuitionistic logic L,

$$\bigcap_{n\in\omega} \Delta^n L = \mathbf{H} \qquad (*)$$

in other words, the 'ω-iteration' of Δ is trivial. $(*)$ readily implies Proposition 1.16.12. In Volume 2 we will prove that 1.16.12 and $(*)$ do not transfer to the predicate case.

1.17 Neighbourhood semantics

Neighbourhood semantics is a generalisation of Kripke semantics. In this case 'possible worlds' are regarded as points of an abstract 'space' or a 'neighbourhood frame'. In such a frame every world has a set of 'neighbourhoods', and $\Box A$ is true at w iff A is true in all worlds in some neighbourhood of w, that is, in all the worlds that are 'rather close' to w. So in neighbourhood semantics 'necessary' is interpreted as 'locally true'.

Here is a precise definition.

Definition 1.17.1 *An* n-modal (propositional) neighbourhood frame *is an* $(n+1)$*-tuple* $F = (W, \Box_1, \dots, \Box_N)$*, such that* $W \neq \varnothing$*,* $\Box_i$ *are unary operations in* 2^W *satisfying the identities:*

$$\Box_i(a \cap b) = \Box_i a \cap \Box_i b,$$

$$\Box_i W = W.$$

As in Kripke frames, the elements of W are called *possible worlds*, or *points*; $u \in \Box_i V$ is read as 'V is an *i-neighbourhood* of u'. The basic identities mean that the intersection of two i-neighbourhoods of u is also an i-neighbourhood of u, every extension of an i-neighbourhood is again an i-neighbourhood and that W is an i-neighbourhood of any u. However a neighbourood of u may not contain u, or may even be empty.

Obviously, an N-modal neighbourhood frame F corresponds to the N-modal algebra

$$MA(F) := (2^W,\ \cup,\ \cap,\ -,\ \varnothing,\ W,\ \Box_1, \dots, \Box_N).$$

The following is a trivial consequence of definitions, cf. Section 1.3.

Lemma 1.17.2 *Every Kripke frame* $F = (W, R_1, \dots, R_N)$ *corresponds to a neighbourhood frame* $Nd(F) = (W, \Box_1, \dots, \Box_N)$*, such that* $MA(Nd(F)) = MA(F)$*.*

Definition 1.17.3 *A* neighbourhood model *over a neighbourhood frame* F *is a pair* $M = (F, \theta)$*, in which* $\theta : PL \longrightarrow 2^W$ *is a valuation.* θ *is extended to all formulas, according to Definition 1.2.8.*

We use the same terminology and notation as in Kripke semantics. For a formula A, we write: $M, w \vDash A$ (or $w \vDash A$) instead of $w \in \theta(A)$, and say that A is *true at the world w* of M (or that *w forces A*).

So we have:

$$M, x \vDash \Box_i A \text{ iff } \theta(A) \text{ is an } i\text{-neighbourhood of } x.$$

Definition 1.17.4 *A modal formula* A *is* true in a neighbourhood model M *(notation:* $M \vDash A$*) if it is true at every world of* M*;* A *is* valid *in a neighbourhood frame* F *(notation:* $F \vDash A$*) if it is true in every model over* F*. A set of formulas* Γ *is* valid *in* F *(notation:* $F \vDash \Gamma$*) if every* $A \in \Gamma$ *is valid.*

Similarly to Lemmas 1.3.7 and 1.3.8 we obtain

Lemma 1.17.5 *For any modal formula A and a neighbourhood frame F*

$$F \vDash A \text{ iff } MA(F) \vDash A.$$

Lemma 1.17.6

(1) *For a neighbourhood frame F the set*

$$\mathbf{ML}(F) := \{A \mid F \vDash A\}$$

is a modal logic.

(2) *For a class $\mathcal{C}$ of N-modal neighbourhood frames the set*

$$\mathbf{ML}(\mathcal{C}) := \bigcap \{\mathbf{ML}(F) \mid F \in \mathcal{C}\}$$

is an N-modal logic.

Definition 1.17.7 *The logic* $\mathbf{ML}(F)$ *(respectively,* $\mathbf{ML}(\mathcal{C})$*) is called the* modal logic of F *(respectively, of* $\mathcal{C}$*), or the* modal logic determined by F *(by* $\mathcal{C}$*).*

A modal logic is called neighbourhood complete *if it is determined by some class of neighbourhood frames.*

From Lemma 1.17.2 we have:

Lemma 1.17.8 *Every Kripke-complete propositional logic is neighbourhood complete.*

The converse to the previous Lemma is false [Gabbay, 1975], [Gerson, 1975a], [Shehtman, 1980], [Shehtman, 2005]. There also exist examples of modal logics that are incomplete in neighbourhood semantics [Gerson, 1975b], [Shehtman, 1980], [Shehtman, 2005]. The question of whether all intermediate propositional logics are neighbourhood complete (*Kuznetsov's problem* [Kuznecov, 1974]), is still open.

Chapter 2

Basic predicate logic

2.1 Introduction

The main notion of this chapter is first-order logic. Similarly to the propositional case, we define a logic as a set of formulas that contains some basic axioms and is closed under some basic inference rules.

Here the crucial point is the substitution rule, which is important because we would like to distinguish between logics and theories.

On the one hand, every axiomatic logical calculus (postulated by a set of axioms and inference rules) generates a 'theory' — the set of all theorems. Usually theories are supposed to collect properties of a certain kind of objects. Many well known theories, such as Peano arithmetic, Tarski's elementary geometry, Zermelo–Fraenkel set theory, were developed for that purpose.

On the other hand, we may be interested in theories that do not depend on 'application domains' and express the basic 'logical laws'. For example, the proposition

`Every human has a father and a mother`

expressed by a formula

$$A = \forall x(H(x) \supset \exists y F(y,x) \wedge \exists z M(z,x)),$$

is a specific property of humans that does not hold for all living creatures.

The formula

$$B = \forall x(H(x) \supset \exists y M(y,x))$$

also expresses a true property of humans, which does not hold in other cases.

But the implication

$$A \supset B$$

(allowing us to deduce B from A) is a logical law — its truth does not depend on the meaning of the predicates H, F, M.

So the laws of logic should sustain replacing of predicates by arbitrary formulas. We can regard them as schemata for producing theorems, and define a logic just as a substitution closed theory.[1]

A standard example is classical first-order logic, the set of all theorems of classical predicate calculus. Numerous classical theories contain it as a fixed basic part.

In the nonclassical area there is a great variety of logics deserving special attention. Of course study of nonclassical theories (such as Heyting arithmetic or modal set theories) is also interesting and important, but due to the lack of time, we postpone it until Volume 2.

Study of nonclassical logics in this volume is closely related to study of different semantics. Unlike the classical case, there are many options here. From our viewpoint, a semantics S for a certain class of logics (say, Σ) should include the notions of a 'frame' and 'validity'. A semantics S is 'sound' for Σ if the set of all formulas valid in any S-frame is a logic from Σ.[2] Thus to check soundness, it is necessary to prove that the substitution rule preserves 'validity' in a 'frame'.

In this respect there is a big difference between the classical and nonclassical cases. In classical logic we may not care about formula substitutions, and they are usually not discussed in textbooks and monographs.[3] Classical predicate calculus is traditionally formulated using axiom schemes rather than substitution rule, and soundness is proved without mentioning substitutions. But in our nonclassical studies we deal with rather exotic types of frames and the proof of soundness may be nontrivial.

Therefore let us first take a closer look at the syntactic notion of a logic, and especially at the substitution rule. Its intuitive meaning is clear: given a first-order formula A we can deduce every formula $[C/P(x_1,\ldots,x_n)]A$ obtained by substituting a formula C for an atomic formula $P(x_1,\ldots,x_n)$. More exactly, to obtain $[C/P(x_1,\ldots,x_n)]A$, one should replace every occurrence of $P(x_1,\ldots,x_n)$ with C and every occurrence of $P(y_1,\ldots,y_n)$ (using other variables $y_1,\ldots,y_n$) with the corresponding version of C, $[y_1,\ldots,y_n/x_1,\ldots,x_n]C$. In its turn $[y_1,\ldots,y_n/x_1,\ldots,x_n]C$ is obtained by a 'correct' replacement of parameters $x_1,\ldots,x_n$ with $y_1,\ldots,y_n$.

Thus the definition of a *formula substitution* $[C/P(x_1,\ldots,x_n]$ relies on the definition of a correct *variable substitution* $[y_1,\ldots,y_n/x_1,\ldots,x_n]$.

Our approach to variable substitutions is rather nonstandard and resembles [Bourbaki, 1968]. But it is convenient from the technical viewpoint, see Section 2.3 for further details.

[1]Of course this definition is rather conventional, and there exist examples of 'logics' that are not substitution closed.

[2]Some authors still use 'semantics' that are not sound in this sense [Kracht and Kutz, 2005], [Goldblatt and Maynes, 2006]. This is not so convenient, because it may be difficult to describe all 'frames' characterising a given logic.

[3]With few exceptions, such as [Church, 1996], [Novikov, 1977].

2.2 Formulas

The expansion of a propositional language to a first-order language is defined in a standard way.

Let $Var = \{v_1, v_2, \ldots\}$, $PL^n = \{P_i^n \mid i \geq 0\}$ $(n \geq 0)$ be fixed disjoint countable sets. The elements of Var and PL^n are respectively called *(individual) variables*, and *n-ary predicate letters*.[4] An *atomic formula without equality* is either $\bot$, or P_i^0 (a *proposition letter*), or $P_i^n(x_1, \ldots, x_n)$ for some $n > 0$, $x_1, \ldots, x_n \in Var$. *Atomic formulas with equality* can also be of the form $x = y$, where $x, y \in Var$ and '$=$' is an extra binary predicate letter.[5] Note that our basic language does not include constants or function letters; we will return to these matters in Volume 2. Also note that the language is countable, but we shall consider its uncountable expansions with constants.

Classical (or *intuitionistic*) predicate formulas (with or without equality) are built from atomic formulas using the propositional connectives $\wedge$, $\vee$, $\supset$, and the quantifiers $\forall$, $\exists$; in *N-modal predicate formulas*[6] the unary connectives $\Box_i$, $1 \leq i \leq N$, can also be used. The abbreviations $\neg A$, $\top$, $A \equiv B$, $\Diamond_i A$ have the same meaning as in the propositional case; $x \neq y$ abbreviates $\neg(x = y)$.

For a formula A, a list of variables $\mathbf{x} = x_1 \ldots x_n$ and a quantifier $Q \in \{\forall, \exists\}$, $Q\mathbf{x}A$ denotes $Qx_1 \ldots Qx_n A$.

Definition 2.2.1 *The* (modal) degree $d(A)$ *of a modal predicate formula A is defined by induction:*

$$
\begin{aligned}
&d(A) = 0 \quad \textit{for } A \textit{ atomic};\\
&d(A \wedge B) = d(A \vee B) = d(A \supset B) = \max\,(d(A), d(B));\\
&d(\forall x A) = d(\exists x A) = d(A);\\
&d(\Box_i A) = d(A) + 1.
\end{aligned}
$$

So $d(A) = 0$ iff A is a classical formula.

AF, IF, MF_N denote respectively the sets of atomic, intuitionistic, and N-modal formulas without equality; the corresponding sets of formulas with equality are denoted by $AF^=$, $IF^=$, $MF_N^=$; we omit the subscript N if $N = 1$ or if N is clear from the context. Sometimes we write $IF^{(=)}$, $MF_N^{(=)}$. etc. in order to combine the cases with and without equality in a single statement. The set $MF_N^{(=)}$ is also denoted by $\mathcal{L}_N^{(=)}$ and called the *(basic) N-modal first-order language*; $\mathcal{L}_0^{(=)}$ is *(basic) classical (or intuitionistic) first-order language*.

We skip a routine proof of the following standard statement (cf. slightly different versions in [Shoenfield, 1967], [Bourbaki, 1968]).

Lemma 2.2.2 (Parsing lemma) *Let A, B, A', B' be formulas, $*, *' \in \{\vee, \wedge, \supset\}$ binary connectives such that $A * B = A' *' B'$. Then $A = A'$, $* = *'$, and $B = B'$.*

---

[4] We also use the symbols $P, Q, R, \ldots$ (sometimes with subscripts) as names of predicate letters; $p, q, r, \ldots$ as names of proposition letters and $x, y, \ldots$ as names of variables.

[5] '$=$' is also used as a metasymbol.

[6] To avoid confusion, we use the letter N instead of n, because in many cases n denotes the number of variables in a list.

The definition of *free* and *bound* occurrences of variables in formulas is well-known, but we shall now give it in a more formal way.

Definition 2.2.3 *An* occurrence of a letter c in a word α *is a triple* (α, i, c) *such that* c *is the* i*th letter of* α*;* an occurrence of a word β in α *is a triple* (α, i, β) *such that* β *is a subword of* α *starting from the* i*th letter of* α*.*

Free and bound occurrences of variables in formulas are now defined by induction.

Definition 2.2.4

- *All variable occurrences in atomic formulas are free.*
- *If* (A, i, x) *is free (bound), then* $(\Box_j A, i+1, x)$ *is free (bound).*
- *Let* $C = (A * B)$*, where* $*$ *is a binary connective,* $|A| = l$*.*
  If (A, i, x) *is free (bound), then* $(C, i+1, x)$ *is free (bound).*
  If (B, i, x) *is free (bound), then* $(C, i+l+2, x)$ *is free (bound).*
- *Let* $B = \forall y A$ *or* $\exists y A$*. If* (A, i, x) *is free (bound) and* $x \neq y$*, then* $(B, i+2, x)$ *is free (bound). All occurrences of* y *in* B *are bound.*

The occurrence $(B, 2, y)$ is called *strongly bound.*

$FV(A)$ denotes the set of *parameters (free variables)* of a formula A, i.e. of all variables having free occurrences in A. A *closed formula* (or a *sentence*) is a formula without parameters. $MS_N^{(=)}$ (respectively, $IS^{(=)}$) denotes the set of all N-modal (respectively, intuitionistic) sentences.

$BV(A)$ denotes the set of *bound variables* of A (i.e. variables having bound occurrences in A). We also use the notation

$$BV^-(A) := BV(A) - FV(A), \; V(A) := BV(A) \cup FV(A).$$

A variable is called *new for* A if it does not occur in A.

Recall that a universal closure of a formula is usually understood as the result of the universal quantification over all its parameters. But such a definition is a priori ambiguous. To fix a unique notation for the universal closure, one can take the parameters in a certain order, for example as follows.

Definition 2.2.5 *The* standard list of parameters *of a formula* A *is the set of its parameters* $FV(A)$ *ordered in accordance with their first occurrences in* A*. The* standard universal closure $\overline{\forall} A$ *of a formula* A *is the sentence* $\forall \mathbf{x} A$*, where* $\mathbf{x}$ *is the standard list of parameters of* A*. For any ordering* $\mathbf{y} = y_1 \dots y_n$ *of* $FV(A)$ *the sentence* $\forall \mathbf{y} A$ *is called* a universal closure of A.

The set of all universal closures of formulas from a set Γ is denoted by $\overline{\Gamma}$.

Note that in all the logics considered in this book the universal closures of the same formula are always equivalent, so we can deal only with standard universal closures, i.e. with $\{\overline{\forall} A \mid A \in \Gamma\}$ instead of $\overline{\Gamma}$.

For a subformula QxB of a formula A (where Q is a quantifier), every free occurrence of x in B, as well as the first occurrence of x, is called *referent* to the first occurrence of Q. More precisely:

Definition 2.2.6 *Let (A, i, QxB) be an occurrence of QxB in A. If (B, j, x) is a free occurrence of x in B, then $(A, i + j + 2, x)$ is an occurrence of x* referent *to (A, i, Q); $(A, i + 1, x)$ is also referent to (A, i, Q). All variable occurrences in A referent to the same occurrence of a quantifier are called* coreferent *(or* correlated*). We also say that an occurrence of a quantifier* binds *all referent occurrences of variables.*

Now the *reference structure* of a formula A can be defined as a function sending every bound occurrence of a variable in A to the occurrence of a binding quantifier. But the following definition is more convenient.

Definition 2.2.7 *Let A be a formula of length n. The* reference function *of A is a function rf_A defined on the set*

$$\{i \mid 1 \leq i \leq n \ \&\ (A, i, x) \text{ is a bound variable occurrence for some } x\}$$

such that $rf_A(i) = j$ whenever (A, i, x) is referent to (A, j, Q) (for some variable x and a quantifier Q).

So rf_A sends positions of bound variables to positions of their binding quantifiers.

Definition 2.2.8 *Let $\bullet$ be a new symbol, regarded as an extra variable ('joker'). The* stem *of a formula A is a formula A^- obtained by replacing every bound occurrence of every variable in A with $\bullet$. The* scheme *of A is a pair $\llcorner A \lrcorner := (A^-, rf_A)$.*

Thus $BV(A^-) \subseteq \{\bullet\}$, $FV(A^-) = FV(A)$.

Reference functions can be represented graphically, by connecting occurrences of quantifiers with referent variable occurrences. For example, for

$$A := \forall x(P(x) \supset \exists y Q(y, x))$$

the reference function is pictured as follows:

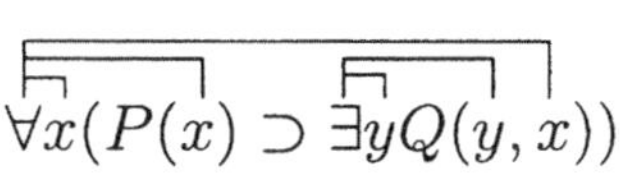

and the scheme as follows:

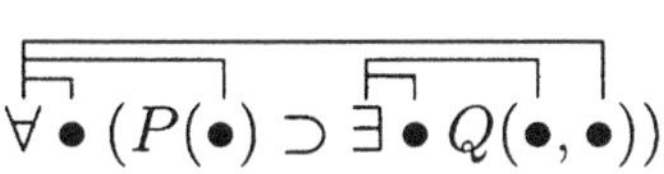

(Note that the stem itself has a different reference function!)

Remark 2.2.9 Instead of this graphic representation, one can number quantifier occurrences and add corresponding superscripts to referent variable occurrences, cf. [Kleene, 1967].

Alternatively, the scheme of a formula can be defined by induction as follows.

Definition 2.2.10

- $\underline{A} := A$ *for* A *atomic;*
- $\underline{(A * B)} := (\underline{A} * \underline{B})$ *for* $* \in \{\vee, \wedge, \supset\}$*;*
- $\underline{\Box_j A} := \Box_j\ \underline{A}$ *;*
- $\underline{QxA}$ *(for* $Q \in \{\forall, \exists\}$*) is obtained from* $Qx\,\underline{A}$ *by replacing every occurrence of* x *with* • *and connecting it with the first occurrence of* Q*.*

Remark 2.2.11 Strictly speaking, this definition shows us how to reconstruct a graphic representation of $\underline{A}$. In terms of functions, it means that $rf_{(A*B)}$ is the union of rf_A and rf_B with some shifts, and so on. Such an explication does not seem useful however.

Schemes can also be defined by induction without appealing to formulas:

Definition 2.2.12

- *Every atomic formula is a scheme.*
- *If* S_1, S_2 *are schemes, then* $(S_1 * S_2)$ *is a scheme.*
- *If* S *is a scheme, then* $\Box_j S$ *is a scheme.*
- *If* S *is a scheme,* $x \in$ *Var, then there is a scheme* $\underline{QxS}$ *obtained from* QxS *by replacing every occurrence of* x *with* • *and connecting it with the first occurrence of* Q*.*

It is obvious (a strict proof is by induction) that for any scheme S in the sense of this definition, there is a formula A such that $S = \underline{A}$. So in our syntax we can deal with schemes rather than formulas. In a systematic way this approach is developed in [Bourbaki, 1968].[7] One can even argue that schemes better correspond to human intuition about first-order logic.[8] But in our book, we prefer to keep to the traditional notion of a formula and use schemes only incidentally, for technical purposes.

---

[7] With minor differences — instead of quantifiers Bourbaki uses the ε-symbol (denoted by τ) and defines $\exists xA$ as an abbreviation for $[\tau_x(A)/x]A$.

[8] Natural language does not use bound variables and hides them in the reference structure, cf. the sentence *Every triangle has at least two acute angles.*

2.3 Variable substitutions

It is well-known that a 'logically correct' variable substitution is not a simple replacement of variables, due to possible variable collisions. For example, if A is $\exists y\,(x \neq y)$, and we want the formula $\forall x\,A \supset [y/x]A$ to be (classically) valid; it is incorrect to define $[y/x]A$ as $\exists y\,(y \neq y)$. Many authors consider such substitutions as 'bad' and simply do not allow them; formally, $[y/x]A$ is a 'good' substitution if free occurrences of x are not within the scope of any quantifier over y; as they say, *y is free for x in A*, cf. [Kleene, 1952], [Kleene, 1967], [Mendelson, 1997].

But this restriction does not help for defining formula substitutions, because e.g.

$$[\exists y Q(x,y)/P(x)]P(y)$$

should be

$$[y/x]\exists y Q(x,y),$$

and the latter substitution $[y/x]$ is 'bad'.

A well-known way to solve the problem is renaming of bound variables, cf. [Kleene, 1963]. For example, in the formula $\exists y\,(x \neq y)$ we can rename the bound y by z and define $[y/x](\exists y\,(x \neq y))$ as $\exists z\,(y \neq z)$. But this variable z can be chosen in many different ways — it can be arbitrary except for the original y. If the result of a substitution should be unique, we have to fix one of these options. For example, we can always take the first variable allowed for renaming (in the list of all variables) [Kolmogorov and Dragalin, 2005]. But this definition is technically inconvenient and rather unnatural , because the alphabetical order of variables is not related to logic at all.

So we propose another approach. A substitution is considered as a relation not function; the result of a substitution is unique *only up to congruence* (correct renaming of bound variables).[9] We regard $[y/x]A$ not as a single formula, but as a member of a certain class of formulas. This resembles the well-known mathematical notation $\int f(x)dx$ of a primitive function, which is unique up to adding a constant. Using schemes instead of formulas simplifies all the details: two formulas are congruent if they have the same scheme.

A formula is called *clean* if all its quantifiers bind different variables and none of its bound variables is free. Every formula A can be transformed into an equivalent clean formula A°, without bound occurrences of x or y. Then we define $[y/x]A$ as $[y/x]A^\circ$. Since there are no variable collisions, the latter substitution is made by a straightforward replacement.

Now we pass to the details. Let $\mathbf{x} = (x_1, \ldots, x_n)$ be a list of variables $(n \geq 0)$.[10] Later on we use the following notation:

[9] Congruence corresponds to α-equivalence in λ-calculi.

[10] If there is no confusion, we also denote the list $(x_1, \ldots, x_n)$ by $x_1 \ldots x_n$.

$r(\mathbf{x})$ for the set $\{x_1, \ldots, x_n\}$;
$z \in \mathbf{x}$ for 'z occurs in $\mathbf{x}$', i.e. for $z \in r(\mathbf{x})$;[11]
$\mathbf{xy}$ for the concatenation of the lists $\mathbf{x}$, $\mathbf{y}$;
$\mathbf{x} \cap S$ for 'the sublist of $\mathbf{x}$ containing the elements of S';
$\mathbf{x} - S$ for 'the sublist of $\mathbf{x}$ obtained by removing the elements of S', etc.

A list is said to be *distinct* if all its members are different.

If $\mathbf{x} = (x_1, \ldots, x_n)$, $\mathbf{y} = (y_1, \ldots, y_n)$ are lists of variables and all $x_1, \ldots, x_n$ are distinct, we define the *variable substitution* $[\mathbf{y}/\mathbf{x}]$ as the following function $Var \longrightarrow Var$:

$$[\mathbf{y}/\mathbf{x}](z) = \begin{cases} y_i \text{ if } z = x_i; \\ z \text{ if } z \notin \mathbf{x}. \end{cases}$$

Note that the same definition can be given in the case when some of the x_i are equal, but $\mathbf{x}$ sub $\mathbf{y}$.

Of course a variable substitution is nothing but a function $Var \longrightarrow Var$ that changes only finitely many variables. So a composition of substitutions is a substitution.

For a list of variables $\mathbf{z} = z_1 \ldots z_m$, put

$$[\mathbf{y}/\mathbf{x}]\mathbf{z} := [\mathbf{y}/\mathbf{x}](z_1) \ldots [\mathbf{y}/\mathbf{x}](z_m).$$

We also use the *dummy substitution* $[/]$, which is the identity function on Var.

Definition 2.3.1 *A* variable transformation *of a (distinct) list* $\mathbf{x}$ *to* $\mathbf{y}$ *is the finite function (the set of pairs)* $\{(x_1, y_1), \ldots, (x_n, y_n)\}$*; this function is denoted by* $[\mathbf{x} \mapsto \mathbf{y}]$.

If a transformation $[\mathbf{x} \mapsto \mathbf{y}]$ *is bijective (i.e.* $\mathbf{y}$ *is distinct), it is called a* variable renaming. *If also* $V(A) \subseteq r(\mathbf{x})$ *for a certain formula* A, $[\mathbf{x} \mapsto \mathbf{y}]$ *is called a* variable renaming in A. *A* bound variable renaming in A *is a variable renaming in* A *fixing all parameters of* A.

Note that $[\mathbf{x} \mapsto \mathbf{y}]$ is a variable renaming iff the corresponding substitution $[\mathbf{y}/\mathbf{x}]$ is a permutation of Var.

For a set of variables S, $[\mathbf{x} \mapsto \mathbf{y}]_S$ denotes the restriction of $[\mathbf{x} \mapsto \mathbf{y}]$ to $r(\mathbf{x}) \cap S$. We also use the abbreviations

$[\mathbf{x} \mapsto \mathbf{y}]_A$ for $[\mathbf{x} \mapsto \mathbf{y}]_{FV(A)}$;
$[\mathbf{x} \mapsto \mathbf{y}]_{-z}$ for $[\mathbf{x} \mapsto \mathbf{y}]_{r(\mathbf{x}) - \{z\}}$.

A transformation $[\mathbf{x} \mapsto \mathbf{y}]$ is called *proper* if $x_i \neq y_i$ for every i. We say that a transformation $[\mathbf{x} \mapsto \mathbf{y}]$ *represents* the substitution $[\mathbf{y}/\mathbf{x}]$.[12] Obviously, every substitution is represented by infinitely many transformations, but only one of them is proper.

It is easy to describe how a transformation $[\mathbf{x} \mapsto \mathbf{y}]$ acts on (modal) predicate formulas. Let $A[\mathbf{x} \mapsto \mathbf{y}]$ be the result of simultaneous replacement of *all* (both

[11] This notation is only occasional.

[12] Sometimes transformations are also called 'substitutions', but we avoid this terminology.

free and bound) occurrences of x_i in A with y_i (for $i = 1, \dots, n$). We can say that $A[\mathbf{x} \mapsto \mathbf{y}]$ is obtained from A by a 'straightforward' renaming of variables.[13]

Lemma 2.3.2 *Let $[\mathbf{x} \mapsto \mathbf{y}']$, $[\mathbf{y} \mapsto \mathbf{z}]$ be variable transformations such that $\mathbf{y}' = \mathbf{y} \cdot \sigma$ for some surjection σ. Then*

(1) $[\mathbf{x} \mapsto \mathbf{y}'] \circ [\mathbf{y} \mapsto \mathbf{z}] = [\mathbf{x} \mapsto \mathbf{z} \cdot \sigma]$;

(2) for any predicate formula A such that $V(A) \subseteq r(\mathbf{x})$
$(A[\mathbf{x} \mapsto \mathbf{y}'])[\mathbf{y} \mapsto \mathbf{z}] = A[\mathbf{x} \mapsto \mathbf{z} \cdot \sigma]$.

In particular, $(A[\mathbf{x} \mapsto \mathbf{y}])[\mathbf{y} \mapsto \mathbf{x}] = A$ if $[\mathbf{x} \mapsto \mathbf{y}]$ is a variable renaming in A.

Proof

(1) Trivial: $[\mathbf{x} \mapsto \mathbf{y}']$ sends x_i to $y_{\sigma(i)}$, and next $[\mathbf{y} \mapsto \mathbf{z}]$ sends $y_{\sigma(i)}$ to $z_{\sigma(i)}$.

(2) Also obvious, formally, one should argue by induction on the length of A.

■

In a more general situation the following holds (the proof is similar).

Lemma 2.3.3 *Let $[\mathbf{x} \mapsto \mathbf{y}']$, $[\mathbf{y} \mapsto \mathbf{z}]$ variable transformations such that $r(\mathbf{y}') \subseteq r(\mathbf{y})$. Then for any predicate formula A,*

$$(A[\mathbf{x} \mapsto \mathbf{y}'])[\mathbf{y} \mapsto \mathbf{z}] = A([\mathbf{x} \mapsto \mathbf{y}'] \circ [\mathbf{y} \mapsto \mathbf{z}]).$$

The formulas A and $A[\mathbf{x} \mapsto \mathbf{y}]$ may be not logically equivalent in classical logic. For example, this is the case for

$$A = \exists x_1 \, (P(x_1) \wedge \neg P(y_1)) \text{ and } A[x_1 \mapsto y_1] = \exists y_1 (P(y_1) \wedge \neg P(y_1)).$$

Later on we will describe 'admissible' transformations that do not affect the truth values of formulas, cf. Lemma 2.3.19.

Lemma 2.3.4 *Let A be a predicate formula, $[\mathbf{x} \mapsto \mathbf{y}]$ a variable renaming in A. Then*

(1) free (respectively, bound) occurrences of variables in A correspond to free (bound) occurrences of variables in $A[\mathbf{x} \mapsto \mathbf{y}]$:
(A, i, x_j) is free (bound) iff $(A[\mathbf{x} \mapsto \mathbf{y}], i, y_j)$ is free (bound);

(2) $rf_A = rf_{A[\mathbf{x} \mapsto \mathbf{y}]}$.

Proof By induction. We denote $A[\mathbf{x} \mapsto \mathbf{y}]$ by A'.

- If A is atomic, then A' is atomic, so all variable occurrences are free, and the claim is trivial.

[13]The notation $[\mathbf{y}/\mathbf{x}]\,A$ is reserved for a 'correct' variable substitution with renaming of bound variables, see below.

- If $A = (B * C)$, then $A' = (B' * C')$. Variable occurrences in A coming from B are of the form $(A, i+1, x_j)$ where (B, i, x_j) is an occurrence in B. Then

  $(A, i+1, x_j)$ is free (bound) iff
  (B, i, x_j) is free (bound) (by Definition 2.2.4) iff
  (B', i, y_j) is free (bound) (by induction hypothesis) iff
  $(A', i+1, y_j)$ is free (bound) (by 2.2.4).

  Moreover, a bound occurrence $(A, i+1, x_j)$ is referent to a quantifier occurrence $(A, k+1, \mathcal{Q})$ iff (B, i, x_j) is referent to $(B, k, \mathcal{Q})$:

  $$A = (\underbrace{\ldots \mathcal{Q}\overset{k+1}{\overset{\downarrow}{x_j}}\ \underbrace{\ldots \overset{i+1}{\overset{\downarrow}{x_j}} \ldots}_{D}\ \ldots}_{B} * C)$$

  In fact, according to 2.2.5, this happens iff either $i = k+1$ or this occurrence of x_j in A comes from a free occurrence in a subformula D starting at position $k+3$ (more precisely, $A, k+1, \mathcal{Q}x_j D)$ is an occurrence in A and $(D, i-k-1, x_j)$ is a free occurrence in D).

  So
  $$rf_A(i+1) = k+1 \Leftrightarrow rf_B(i) = k.$$

  By the same reason,
  $$rf_{A'}(i+1) = k+1 \Leftrightarrow rf_{B'}(i) = k.$$

  Since $rf_B = rf_{B'}$ by induction hypothesis, it follows that
  $$rf_A(i+1) = rf_{A'}(i+1).$$

  A similar argument applies to variable occurrences in C; we leave formal details to the reader.

- The case $A = \Box_k B$ is also left to the reader.

- If $A = \mathcal{Q}x_k B$, then $A' = \mathcal{Q}y_k B'$.

Now if $j \neq k$, then as above, we obtain:
$(A, i+2, x_j)$ is free (bound) iff (B, i, x_j) is free (bound) iff
(B', i, y_j) is free (bound) iff $(A', i+2, y_j)$ is free (bound).

Note that $y_j \neq y_k$, since $[\mathbf{x} \mapsto \mathbf{y}]$ is a bijection.

A bound occurrence $(A, i+2, x_j)$ is referent to $(A, k+2, \mathcal{Q}_1)$ iff (B, i, x_j) is referent $(B, k, \mathcal{Q}_1)$:

$$A = \mathcal{Q}x_j\ \underbrace{\ldots \underbrace{\mathcal{Q}_1\overset{k+2}{\overset{\downarrow}{x_j}} \ldots \overset{i+2}{\overset{\downarrow}{x_j}} \ldots}_{D}\ \ldots}_{B}$$

So

$$rf_A(i+2) = k+2 \Leftrightarrow rf_B(i) = k,$$

and similarly,

$$rf_{A'}(i+2) = k+2 \Leftrightarrow rf_{B'}(i) = k.$$

Hence

$$rf_A(i+2) = rf_{A'}(i+2).$$

And if (A, i, x_k) is an arbitrary occurrence, it is bound, as well as (A', i, y_k). Then

$$rf_A(i) = 1 \Leftrightarrow rf_{A'}(i) = 1.$$

■

Definition 2.3.5 *Two predicate formulas A, B are called* congruent *(notation: $A \doteq B$) if they have the same scheme.*

Definition 2.3.6 *Let A be a predicate formula. A* bound variable renaming *in A is a variable renaming in A fixing all parameters of A.*

So this is a bijective transformation $[\mathbf{x} \mapsto \mathbf{y}]$ such that $V(A) \subseteq r(\mathbf{x})$ and $x_i = y_i$ for $x_i \in FV(A)$.

Definition 2.3.7 *We say that B is strongly congruent to A (notation: $A \stackrel{\circ}{\equiv} B$) if there is a bound variable renaming $[\mathbf{x} \mapsto \mathbf{y}]$ in A such that $B = A[\mathbf{x} \mapsto \mathbf{y}]$.*

Lemma 2.3.2 shows that $\stackrel{\circ}{\equiv}$ is an equivalence relation on formulas.

Lemma 2.3.8 *If $[\mathbf{x} \mapsto \mathbf{y}]$ is a bound variable renaming in A, then $A \doteq A[\mathbf{x} \mapsto \mathbf{y}]$. Thus strong congruence implies congruence.*

Proof By Lemma 2.3.4 the reference functions coincide. As for the stems, A and A^- may differ only in bound variables. Since $x_i = y_i$ for $x_i \in FV(A)$, A and $A' = A[\mathbf{x} \mapsto \mathbf{y}]$ also differ only in bound variables.

Now consider bound variable occurrences. We have:
• occurs in A^- at position j iff some x_i occurs in A at position j iff some y_i occurs in A' at position j (by 2.3.4) iff
• occurs in $(A')^-$ at position j.

Thus $A^- = (A')^-$, and eventually $A \doteq A'$. ■

Lemma 2.3.9

(1) If $A = (B * C) \doteq A'$, then $A' = (B' * C')$ for some $B' \doteq B$, $C' \doteq C$.

(2) If $A = QyB \stackrel{\circ}{\equiv} A'$ and $y \notin BV(B)$, then for some variable $z \notin FV(B)$ and formula B', $A = QzB'$ and $B' \stackrel{\circ}{\equiv} B[y \mapsto z]$.

Proof

(1) Let $A = (B * C) \circeq A'$. Since A' begins with (, it must have the form $(B' *' C')$. Then $\llcorner A \lrcorner = (\llcorner B \lrcorner * \llcorner C \lrcorner)$, and $\llcorner A' \lrcorner = (\llcorner B' \lrcorner *' \llcorner C' \lrcorner)$, hence $\llcorner B \lrcorner = \llcorner B' \lrcorner$, $* = *'$, $\llcorner C \lrcorner = \llcorner C' \lrcorner$ by 2.2.2 (or strictly speaking, 2.2.2 implies the coincidence of the corresponding stems; the reference functions coincide, because they are inherited from A and A').

(2) Assume $A = QyB \circeq A'$ and $y \notin BV(B)$. By definition, $\llcorner A \lrcorner$ is obtained from $Qy \llcorner B \lrcorner$ by replacing y with $\bullet$ and adding connections to the first Q. If $\llcorner A \lrcorner = \llcorner A' \lrcorner$, it follows that A' has the form QzB' (otherwise A' does not begin with Q) and

$$\llcorner B' \lrcorner = \llcorner B \lrcorner [y \mapsto z], \qquad (\sharp)$$

where $\llcorner B \lrcorner [y \mapsto z]$ means $(B^-[y \mapsto z], rf_B)$, and also $z \notin FV(B)$. Since $y \notin BV(B)$, $rf_B = rf_{B[y \mapsto z]}$. It is also clear that $B^-[y \mapsto z] = (B[y \mapsto z])^-$ (a formal proof is routine by induction); therefore

$$\llcorner B \lrcorner [y \mapsto z] = \llcorner B[y \mapsto z] \lrcorner. \qquad (\sharp\sharp)$$

From $(\sharp)$, $(\sharp\sharp)$ we have

$$B' \circeq B[y \mapsto z].$$

■

Definition 2.3.10 *A formula A is called* clean *if there are no variables both free and bound in A and the number of occurrences of quantifiers equals the number of bound variables, i.e. different occurrences of quantifiers in A bind different variables in A.*

This is equivalent to the following inductive definition:

Definition 2.3.11

- *Every atomic formula is clean.*
- *If formulas A, B are clean, $* \in \{\wedge, \vee \supset\}$, and $BV(A) \cap V(B) = BV(B) \cap V(A) = \varnothing$, then $(A * B)$ is clean.*
- *If A is clean, $x \in Var$, and $x \notin BV(A)$, then QxA is clean.*

Lemma 2.3.12 *If $[\mathbf{x} \mapsto \mathbf{y}]$ is variable renaming in a clean formula A, then $A[\mathbf{x} \mapsto \mathbf{y}]$ is clean.*

Proof By induction. Again we denote $A[\mathbf{x} \mapsto \mathbf{y}]$ by A'. If A is atomic, then A' is atomic. In this case, if $[\mathbf{x} \mapsto \mathbf{y}]$ is bound for A, then $A' = A$.

Suppose $A = (B * C)$, B, C are clean,

$$BV(B) \cap V(C) = BV(C) \cap V(B) = \varnothing.$$

B', C' are clean by induction hypothesis. By Lemma 2.3.4,

$$BV(B') = \{u_j \mid x_j \in BV(B)\}, \; V(C') = \{y_j \mid \in V(C)\}.$$

Hence

$$BV(B') \cap V(C') = \varnothing.$$

Similarly we obtain

$$BV(C') \cap V(B') = \varnothing,$$

and thus $A' = (B' * C')$ is clean.

Suppose $A = \mathcal{Q}x_iB$, B is clean, $x_i \notin BV(B)$. Then $A' = \mathcal{Q}y_iB'$, and by induction hypothesis, B' is clean. By Lemma 2.3.4, $y_i \notin BV(B')$. Thus A' is clean by 2.3.11. ■

Lemma 2.3.13 *Let A be a clean formula, $[\mathbf{x} \mapsto \mathbf{y}]$ a variable transformation such that $BV(A) \cap r(\mathbf{xy}) = \varnothing$. Then $A[\mathbf{x} \mapsto \mathbf{y}]$ is clean.*

Proof By induction, similar to the previous lemma. We leave this as an exercise for the reader. ■

Lemma 2.3.14 *If $A \doteq A'$ and A, A' are clean, then $A \overset{\circ}{\equiv} A'$.*

Proof By induction on the complexity of A, we prove the claim for any A'.

- If A is atomic and $A \doteq A'$, then $A' = A$, and the claim is trivial.

- If $A = (B * C)$, then by 2.3.9(1), $A' = (B' * C')$ for $B \doteq B'$, $C \doteq C'$. So $B \overset{\circ}{\equiv} B'$ and $C \overset{\circ}{\equiv} C'$ by the induction hypothesis, i.e.

$$B' = B[\mathbf{x} \mapsto \mathbf{y}], \;\; C' = C[\mathbf{z} \mapsto \mathbf{t}]$$

  for some bound variable renamings $[\mathbf{x} \mapsto \mathbf{y}]$, $[\mathbf{z} \mapsto \mathbf{t}]$. Obviously, we may assume that $r(\mathbf{x}) = V(B)$, $r(\mathbf{z}) = V(C)$. Then $[\mathbf{x} \mapsto \mathbf{y}] \cup [\mathbf{z} \mapsto \mathbf{t}]$ is a function. In fact, the lists $\mathbf{x}, \mathbf{z}$ do not have common bound variables, since A is clean. Both functions $[\mathbf{x} \mapsto \mathbf{y}], [\mathbf{z} \mapsto \mathbf{t}]$ accord on free variables as they fix them.

  The same argument shows that $[\mathbf{y} \mapsto \mathbf{x}] \cup [\mathbf{t} \mapsto \mathbf{z}]$ is a function, and thus $[\mathbf{x} \mapsto \mathbf{y}] \cup [\mathbf{z} \mapsto \mathbf{t}]$ is bijective. So we obtain a bound variable renaming that transforms A into A'.

- If $A = \mathcal{Q}yB$ is clean, y does not have bound occurrences in B. Then by Lemma 2.3.9(2), $A' = \mathcal{Q}zB'$, for $B' \doteq B[y \mapsto z]$. Note that B' is clean as a subformula of A'. B is clean as a subformula of A, $B[y \mapsto z]$ is clean by 2.3.12 ($[y \mapsto z]$ becomes a variable renaming if we prolong it in a trivial way — as the identity function on $V(B) - \{y\}$). So by the induction hypothesis

$$B' \overset{\circ}{\equiv} B[y \mapsto z],$$

i.e.

$$B'[\mathbf{x} \mapsto \mathbf{u}] = B[y \mapsto z]$$

for some bound variable renaming $[\mathbf{x} \mapsto \mathbf{u}]$ in B'. Since A' is clean, $z \notin BV(B')$, so $[\mathbf{x} \mapsto \mathbf{u}]$ fixes z.

Thus

$$(QzB')[\mathbf{x} \mapsto \mathbf{u}] = Qz(B[y \mapsto z]). \tag{1}$$

Hence by applying $[z \mapsto y]$ we obtain

$$(QzB')([\mathbf{x} \mapsto \mathbf{u}] \circ [z \mapsto y]) = QyB. \tag{2}$$

Note that

$$[\mathbf{x} \mapsto \mathbf{u}] \circ [z \mapsto y] = [(\mathbf{x} - z)z \mapsto (\mathbf{u} - z)y]$$

is a bound variable renaming in B', so (2) implies

$$A' = QzB' \overset{\circ}{\equiv} QyB = A.$$

- If $A = \Box_i B \doteq A'$, then $A' = \Box_i B'$, and $B \doteq B'$. By induction hypothesis, $B \overset{\circ}{\equiv} B'$, which easily implies $A \overset{\circ}{\equiv} A'$.

■

Proposition 2.3.15 *For any clean formula A,*

$$\overset{\circ}{\equiv}(A) = \doteq(A) \cap \textit{(clean formulas)}.$$

Proof Immediate from 2.3.12, 2.3.14, 2.3.8. ■

Proposition 2.3.16 *Every predicate formula A is congruent to some clean formula (called a* clean version *of A).*

Proof By induction on the complexity of A.

- If A is atomic, it is already clean.
- If $A = \Box_i B$ and $B \doteq B_0$ for a clean B_0, then obviously $A \doteq \Box_i B_0$ and $\Box_i B_0$ is clean.
- If $A = (B * C)$ and $B \doteq B_0$, $C \doteq C_0$ for clean B_0, C_0, then $A \doteq (B_0 * C_0)$. $(B_0 * C_0)$ may be not clean, but we can make it clean by an appropriate bound variable renaming.

  In fact, let $BV(B_0) = r(\mathbf{x})$, and let $[\mathbf{x} \mapsto \mathbf{y}]$ be a bijection such that $r(\mathbf{y}) \cap V(C_0) = \varnothing$. then $B_1 := B_0[\mathbf{x} \mapsto \mathbf{y}]$ is clean by Lemma 2.3.12, and

  $$BV(B_1) \cap V(C) = \varnothing.$$

  Similarly there exists $C_1 \doteq C_0$ such that

  $$BV(C_1) \cap V(B_1) = \varnothing.$$

Since

$$BV(B_1) \cap FV(C_1) = r(\mathbf{y}) \cap FV(C_0) = \varnothing,$$

by 2.3.11, it follows that $(B_1 * C_1)$ is clean.

$B_1 \stackrel{\circ}{=} B_0$ and $C_1 \stackrel{\circ}{=} C_0$ implies $(B_1 * C_1) \stackrel{\circ}{=} (B_0 * C_0) \stackrel{\circ}{=} A$.

- If $A = \mathcal{Q}xB$ and $B \stackrel{\circ}{=} B_0$ for a clean B_0, there are two cases.

  If $x \in FV(B)$, then also $x \in FV(B_0)$ (since free variable occurrences in a formula are at the same positions as in its stem, and $B^- = B_0^-$). Thus $x \notin BV(B_0)$, so $\mathcal{Q}xB_0$ is clean. Now

$$\llcorner A \lrcorner = \llcorner \mathcal{Q}x \llcorner B \lrcorner \lrcorner = \llcorner \mathcal{Q}x \llcorner B_0 \lrcorner \lrcorner = \llcorner \mathcal{Q}xB_0 \lrcorner,$$

  so $A \stackrel{\circ}{=} \mathcal{Q}xB_0$.

  Finally, if $x \notin FV(B)$, then also $x \notin FV(B_0)$. However, it may be that $x \in BV(B_0)$.

  So we rename x into a new variable $y \notin V(B_0)$. Then $[x \mapsto y]$ can be prolonged to a bound variable renaming in $\mathcal{Q}xB_0$; thus $\mathcal{Q}xB_0 \stackrel{\circ}{=} \mathcal{Q}y(B_0[x \mapsto y])$ by 2.3.8, and $B_0[x \mapsto y]$ is clean by 2.3.12. Hence the formula $\mathcal{Q}y(B_0[x \mapsto y])$ is clean by 2.3.11, and we have proved that it is congruent to A.

■

The next proposition gives us a convenient inductive definition of congruence that does not appeal to schemes.

Proposition 2.3.17 *Congruence is the smallest equivalence relation $\sim$ between N-modal predicate formulas with the following properties:*

(1) $\mathcal{Q}xA \sim \mathcal{Q}y(A[x \mapsto y])$ *for* $x \notin BV(A),\ y \notin V(A),\ \mathcal{Q} \in \{\forall, \exists\}$;

(2) $A \sim B \Rightarrow \mathcal{Q}xA \sim \mathcal{Q}xB$ *for* $\mathcal{Q} \in \{\forall, \exists\}$;

(3) $A \sim A' \ \&\ B \sim B' \Rightarrow (A * B) \sim (A' * B')$ *for* $* \in \{\supset, \wedge, \vee\}$;

(4) $A \sim B \Rightarrow \Box_i A \sim \Box_i B$ *for* $i \in I_N$.

Proof First note that congruence really has properties (1)–(4). (2), (3), (4) follow from the inductive definition of schemes 2.2.12. (1) follows from 2.3.8. In fact, let $\mathbf{z}$ be the list of all other variables occurring in A, then $[x\mathbf{z} \mapsto y\mathbf{z}]$ is a bound variable renaming in A, and it transforms $\mathcal{Q}xA$ into $\mathcal{Q}y(A[x \mapsto y])$. So these formulas are congruent.

Now consider an arbitrary equivalence relation $\sim$ satisfying (1)–(3) and show that congruent formulas are $\sim$-related.

So we prove that for any B

(5) $A \stackrel{\circ}{=} B \Rightarrow A \sim B$

by induction on $|A|$.

If A is atomic, then obviously $A \stackrel{\circ}{=} B$ implies $A = B$, so the claim is trivial.

If $A = (A_1 * A_2)$, then as we saw in the proof of 2.3.14, for some $B_1 \stackrel{\circ}{=} A_1$, $B_1 \stackrel{\circ}{=} A_2$ we have $B = (B_1 * B_2)$.

Then by induction hypothesis, $A_i \sim B_i$, and thus $A \sim B$ by (3).

We skip a similar simple case when $A = \Box_i A'$.

Finally, suppose $A = QxC \stackrel{\circ}{=} B$. Since $\llcorner A \lrcorner = \llcorner B \lrcorner$, it follows that $B = QyD$ for some D, y.

By Proposition 2.3.16, C is congruent to a clean formula C_0. Next, by a bound variable renaming (Lemmas 2.3.8, 2.3.12) we transform C_0 into a congruent clean formula C_1 such that $x, y \notin BV(C_1)$. Since $|C| < |A|$, we have $C \sim C_1$ by the induction hypothesis, and hence $A = QxC \sim QxC_1$ by (2).

We also have

$$QyD \stackrel{\circ}{=} QxC \stackrel{\circ}{=} QxC_1,$$

where the latter readily follows from $C \stackrel{\circ}{=} C_1$.

Since congruent formulas have the same occurrences of parameters, we obtain that $y \notin FV(QxC_1)$, and thus $y \notin V(QxC_1)$ by the choice of C_1. Hence

$$QyD \stackrel{\circ}{=} QxC_1 \stackrel{\circ}{=} Qy(C_1[x \mapsto y]),$$

which implies

$$D \stackrel{\circ}{=} C_1[x \mapsto y].$$

By induction hypothesis,

$$D \sim C_1[x \mapsto y],$$

So

$$B := QyD \sim QyC_1[x \mapsto y]$$

by (2).

On the other hand, by (1),

$$QxC_1 \sim QyC_1[x \mapsto y],$$

and we already know that $A \sim QxC_1$. Therefore $A \sim B$. ■

The characterization given in Proposition 2.3.17 resembles the definition of α-equivalence in λ-calculi, cf. ... It is worth noting that there exists an equivalent description of congruence (or α-equivalence) via variable swaps, cf. [Gabbay and Pitts, 2002]:

Lemma 2.3.18 *Congruence is the smallest equivalence relation $\sim$ between N-modal predicate formulas with the following properties:*

(1′) $A \sim B$ iff B is obtained from A by replacing some of its subformula QxC with $D = Qy(C[xy \mapsto yx])$, where $y \notin FV(C)$, $Q \in \{\forall, \exists\}$,

(2)–(4) from 2.3.17.

Proof Congruence satisfies (1′), since for $y \notin FV(C)$,

$$(\mathcal{Q}xC)[xy\mathbf{z} \mapsto yx\mathbf{z}] = \mathcal{Q}y(C[xy \mapsto yx]),$$

where $\mathbf{z}$ is a list of all other variables from $V(C)$, and thus

$$\mathcal{Q}xC \triangleq \mathcal{Q}y(C[xy \mapsto yx[)$$

by 2.3.12. It also satisfies (2)–(4) as noted in 2.3.17.

The other way round, (1′) obviously implies (1) from 2.3.17, since $A[xy \mapsto yx] = A[x \mapsto y]$ for $y \notin V(A)$. Thus every equivalence relation satisfying (1′), (2)–(4) contains $\triangleq$ by 2.3.17. ∎

To complete the whole picture, let us also give a description of a congruence class of a clean formula in terms of transformations. However this description will not be used in further studies.

Lemma 2.3.19 *Let A be a clean formula. Then every formula congruent to A can be obtained by a transformation $[\mathbf{x} \mapsto \mathbf{y}]$ such that*

- $r(\mathbf{x}) = BV(A)$;
- *for any subformula of A of the form $\mathcal{Q}x_iB$, where $\mathcal{Q}$ is a quantifier, the following holds:*

  (1) if $x_j \in FV(B)$ and $j \neq i$ then $y_j \neq y_i$;

  (2) if $y_i \in FV(B)$ then $y_i \in r(\mathbf{x})$.

Proof [Sketch] If $A \triangleq C$, then all parameters of these formulas are at the same positions.

Since A is clean, all occurrences of every bound variable x_i in A are coreferent and since $rf_A = rf_C$, in C they are replaced with the same bound variable y_i. So $C = A[\mathbf{x} \mapsto \mathbf{y}]$ for a list $\mathbf{x}$ of bound variables and some $\mathbf{y}$. Now suppose $\mathcal{Q}x_iB$ occurs in A. If x_j has a free occurrence in B and $j \neq i$, this occurrence of x_i corresponds to an occurrence of y_j in the formula $\mathcal{Q}y_i(B[\mathbf{x} \mapsto \mathbf{y}])$ occurring in C. So $y_j \neq y_i$, since otherwise $rf_C \neq rf_A$. Thus (1) holds.

If y_i has a free occurrence in B and $y_i \notin r(\mathbf{x})$, this occurrence remains free in $B[\mathbf{x} \mapsto \mathbf{y}]$, and thus referent to the first occurrence of $\mathcal{Q}$ in $\mathcal{Q}y_i(B[\mathbf{x} \mapsto \mathbf{y}])$, so $rf_C \neq rf_A$. Thus (2) holds.

On the other hand, transformations described in the statement of the lemma, preserve the scheme of A.

In fact, any occurrence of x_i free in a subformula B of $\mathcal{Q}x_iB$ occurring in A becomes an occurrence of y_i in $B[\mathbf{x} \mapsto \mathbf{y}]$. This occurrence of y_i is also free, since otherwise it is referent to another occurrence of a quantifier $\mathcal{Q}'$ over y_i in C:

$$C = \ldots \mathcal{Q}y_i \underbrace{\ldots \mathcal{Q}'y_i \ldots y_i \ldots}_{B[\mathbf{x} \mapsto \mathbf{y}]} \ldots$$

But this $Q'y_i$ comes from some occurrence of $Q'x_j$ in A, and $j \neq i$, since A is clean. Then

$$A = \dots Qx_i \; \underbrace{\dots Q'x_j \underbrace{\dots x_i \dots}_{D} \;\dots}_{B} \; \dots$$

Now x_i is free in a subformula D of $Q'x_jD$ occurring in B, while $y_i = y_j$. This contradicts the condition (1).

Also every occurrence of y_i free in a subformula $E[\mathbf{x} \mapsto \mathbf{y}]$ of $Qy_iE[\mathbf{x} \mapsto \mathbf{y}]$ occurring in C comes from a free occurrence of x_i in E from a subformula Qx_iE of A. In fact, otherwise $y_i \in FV(E)$.

$$C = \dots Qy_i \underbrace{\dots y_i \dots}_{E[x \mapsto y]} \; \dots$$

$$A = \dots Qx_i \underbrace{\dots y_i \dots}_{E} \; \dots$$

(since A is clean, $x_i \notin BF(E)$), so $y_i \in r(\mathbf{x})$ by (2).

Then $y_i = x_j$ for $j \neq i$ (since $y_i \in FV(E)$), and $y_i = y_j$, since $[\mathbf{x} \mapsto \mathbf{y}]$ fixes y_i. This again contradicts (1).

Therefore $A \stackrel{\circ}{=} A[\mathbf{x} \mapsto \mathbf{y}]$. ∎

Now we can define how variable substitutions act on formulas.

Definition 2.3.20 *For a variable transformation $[\mathbf{x} \mapsto \mathbf{y}]$ and a scheme S, we define $S[\mathbf{x} \mapsto \mathbf{y}]$ as the result of replacing every occurrence of x_i in S by y_i.*

Thus

$$\llcorner A \lrcorner [\mathbf{x} \mapsto \mathbf{y}] = (A^-[\mathbf{x} \mapsto \mathbf{y}], rf_A).$$

Lemma 2.3.21 *Let A be a formula, $[\mathbf{x} \mapsto \mathbf{y}]$ a variable transformation such that $r(\mathbf{xy}) \cap BV(A) = \varnothing$. Then*

$$\llcorner A \lrcorner [\mathbf{x} \mapsto \mathbf{y}] = \llcorner A[\mathbf{x} \mapsto \mathbf{y}] \lrcorner .$$

Proof Easy by induction on $|A|$. We consider only the case $A = QzB$.

$$\llcorner QzB \lrcorner [\mathbf{x} \mapsto \mathbf{y}] = \llcorner Qz \llcorner B \lrcorner \lrcorner [\mathbf{x} \mapsto \mathbf{y}] = \llcorner Qz \llcorner B \lrcorner [\mathbf{x} \mapsto \mathbf{y}] \lrcorner ,$$

since z does not occur in $\mathbf{x}$.

Now by induction hypothesis

$$\llcorner B \lrcorner [\mathbf{x} \mapsto \mathbf{y}] = \llcorner B[\mathbf{x} \mapsto \mathbf{y}] \lrcorner ,$$

thus

$$\llcorner QzB \lrcorner [\mathbf{x} \mapsto \mathbf{y}] = \llcorner Qz \llcorner B[\mathbf{x} \mapsto \mathbf{y}] \lrcorner \lrcorner = \llcorner Qz(B[\mathbf{x} \mapsto \mathbf{y}]) \lrcorner .$$

But

$$Qz(B[\mathbf{x} \mapsto \mathbf{y}]) = A[\mathbf{x} \mapsto \mathbf{y}],$$

again since z does not occur in $\mathbf{x}$. Hence the claim follows. ∎

Definition 2.3.22 *Let A be a predicate formula, $[\mathbf{y}/\mathbf{x}]$ a variable substitution. Then we define $[\mathbf{y}/\mathbf{x}]A$ as an arbitrary formula B such that $\llcorner A \lrcorner[\mathbf{x} \mapsto \mathbf{y}] = \llcorner B \lrcorner$.*

Let us show soundness of this definition.

Lemma 2.3.23 *Every predicate formula A has a clean version A° such that $r(\mathbf{xy}) \cap BV(A^\circ) = \varnothing$.*

Then $[\mathbf{y}/\mathbf{x}]A \doteq A^\circ[\mathbf{x} \mapsto \mathbf{y}]$.

Proof To obtain A°, take an arbitrary clean version and make an appropriate bound variable renaming. Then

$$\llcorner A \lrcorner[\mathbf{x} \mapsto \mathbf{y}] = \llcorner A^\circ[\mathbf{x} \mapsto \mathbf{y}] \lrcorner$$

by Lemma 2.3.21. ■

From the definition we readily obtain

Lemma 2.3.24 $A \doteq A' \Rightarrow [\mathbf{y}/\mathbf{x}]A \doteq [\mathbf{y}/\mathbf{x}]A'$.

Note that $[\mathbf{y}/\mathbf{x}]$ and $[\mathbf{x} \mapsto \mathbf{y}]$ do not depend on the ordering of $\mathbf{x}$; in precise terms, if $|\mathbf{x}| = n$, $\sigma \in \Upsilon_n$, then $[\mathbf{x} \cdot \sigma \mapsto \mathbf{y} \cdot \sigma] = [\mathbf{x} \mapsto \mathbf{y}]$.

lemma

Therefore congruent formulas have the same substitution instances under every variable substitution.

The next lemma contains some simple properties of variable substitutions.

Lemma 2.3.25 *For any predicate formula A, substitutions $[\mathbf{y}/\mathbf{x}]$, $[\mathbf{y}'/\mathbf{x}']$, $[\mathbf{y}''/\mathbf{x}'']$, $[\mathbf{x}/\mathbf{u}]$, $[\mathbf{z}/\mathbf{u}]$ and quantifier Q*

(1) $FV([\mathbf{y}/\mathbf{x}]A) = (FV(A) - r(\mathbf{x})) \cup \mathrm{rng}[\mathbf{x} \mapsto \mathbf{y}]_A$;

(2) $[\mathbf{x}/\mathbf{x}]A \doteq A$;

(3) $Qy[y/x]A \doteq QxA$ *if* $y \notin FV(A)$ *or* $y = x$;

(4) $[\mathbf{y}''/\mathbf{x}''][\mathbf{y}'/\mathbf{x}']A \doteq [\mathbf{y}/\mathbf{x}]A$, *where* $[\mathbf{y}/\mathbf{x}] = [\mathbf{y}''/\mathbf{x}'']\cdot[\mathbf{y}'/\mathbf{x}']$ *(the composition of functions on* Var *);*

(5) $[\mathbf{y}/\mathbf{x}][\mathbf{z}/\mathbf{u}]A \doteq [[\mathbf{y}/\mathbf{x}]\mathbf{z}/\mathbf{u}]A$ *if* $FV(A) \cap r(\mathbf{x}) \subseteq r(\mathbf{u})$;

(6) $[\mathbf{y}/\mathbf{x}][\mathbf{x}/\mathbf{u}]A \doteq [\mathbf{y}/\mathbf{u}]A$ *if* $FV(A) \cap r(\mathbf{x}) \subseteq r(\mathbf{u})$;

(7) $[v/u][\mathbf{y}/\mathbf{x}]A \doteq [[v/u]\mathbf{y}/\mathbf{x}]A$ *if* $u \notin FV(A) - r(\mathbf{x})$;

(8) $[\mathbf{y}/\mathbf{x}]QzA \doteq Qz[\mathbf{y}/\mathbf{x}]A$ *if* $z \notin r(\mathbf{x})$;

(9) $[\mathbf{y}/\mathbf{x}](A * B) \doteq ([\mathbf{y}/\mathbf{x}]A * [\mathbf{y}/\mathbf{x}]B)$ *for* $* \in \{\vee, \wedge, \supset\}$;

(10) $[\mathbf{y}''/\mathbf{x}''][\mathbf{y}'/\mathbf{x}']A \doteq [\mathbf{y}''\mathbf{y}'/\mathbf{x}''\mathbf{x}']A$ *if* $r(\mathbf{x}'') \cap r(\mathbf{x}'\mathbf{y}') = \varnothing$;

(11) $[\mathbf{y}/\mathbf{x}]A \stackrel{\circ}{=} [y_n/z_n]\ldots[y_1/z_1][z_n/x_n]\ldots[z_1/x_1]A$
for distinct variables $z_1,\ldots,z_n \notin FV(A)\cup r(\mathbf{xy})$*; so every variable substitution in a formula can be presented as a composition of substitutions of the form* $[y/x]$ *(*simple substitutions*);*

(12) $[\mathbf{y}/\mathbf{x}]Q\mathbf{z}A \stackrel{\circ}{=} Q\mathbf{z}[\mathbf{y}/\mathbf{x}]A$ *if* $r(\mathbf{z}\cap r(\mathbf{x}) = \varnothing$*;*

(13) $Q\mathbf{y}[\mathbf{y}/\mathbf{x}[A \stackrel{\circ}{=} Q\mathbf{x}A$ *if* $\mathbf{y}$ *is distinct,* $r(\mathbf{y})\cap FV(A) = r(\mathbf{y})\cap r(\mathbf{x}) = \varnothing$*.*

Proof

(1) Since $\lfloor[\mathbf{y}/\mathbf{x}]A\rfloor = \lfloor A\rfloor[\mathbf{x}\mapsto\mathbf{y}]$, it follows that $FV([\mathbf{y}/x]A) = V(\lfloor A\rfloor[\mathbf{x}\mapsto\mathbf{y}])$ (or, to be more precise, $FV(A^-[\mathbf{x}\mapsto\mathbf{y}]) - \{\bullet\})$. So we should take the set $V(\lfloor A\rfloor) = FV(A)$ and replace every x_i occurring in this set with the corresponding y_i; this gives us exactly

$$(FV(A) - r(\mathbf{x}))\cup \mathrm{rng}[\mathbf{x}\mapsto\mathbf{y}]_A.$$

(2) Trivial.

(3) The case $y = x$ is trivial, so suppose $y \notin FV(A)$. Let A° be a clean version of A such that $BV(A^\circ)\cap r(xy) = \varnothing$. As we know

$$[y/x]A \stackrel{\circ}{=} A^\circ[x\mapsto y],$$

hence by 2.3.17,

(♯1) $Qy[y/x]A \stackrel{\circ}{=} Qy(A^\circ[x\mapsto y])$.

Since $y\notin FV(A)$, by 2.3.17 we also have

(♯2) $Qy(A^\circ[x\mapsto y]) \stackrel{\circ}{=} QxA^\circ$.

Now obviously, $A \stackrel{\circ}{=} A^\circ$ implies

(♯3) $QxA^\circ \stackrel{\circ}{=} QxA$

(by 2.3.17 or just by the definition of a scheme). So from (♯1), (♯2), (♯3) we obtain

$$Qy[y/x]A \stackrel{\circ}{=} QxA.$$

(4) Let A° be a clean version of A such that $BV(A^\circ)\cap r(\mathbf{xx'x''yy'y''}) = \varnothing$. Then

$$[\mathbf{y}'/\mathbf{x}']A \stackrel{\circ}{=} A^\circ[\mathbf{x}'\mapsto\mathbf{y}'].$$

By Lemma 2.3.13, the latter formula is clean, and obviously its bound variables are the same as in A°. So

$$[\mathbf{y}''/\mathbf{x}''][\mathbf{y}'/\mathbf{x}]A \stackrel{\circ}{=} [\mathbf{y}''/\mathbf{x}''](A^\circ[\mathbf{x}'\mapsto\mathbf{y}']) \stackrel{\circ}{=} (A^\circ[\mathbf{x}'\mapsto\mathbf{y}'])[\mathbf{x}''\mapsto\mathbf{y}''].$$

Since $[\mathbf{y}'/\mathbf{x}'] = [\mathbf{y}'\mathbf{t}/\mathbf{x}'\mathbf{t}]$, we can always add variables to both $\mathbf{x}'$ and $\mathbf{y}'$, so we may assume that $r(\mathbf{x}') \supseteq FV(A)(= FV(A^\circ))$,

We can also write

$$A^\circ[\mathbf{x}' \mapsto \mathbf{y}'] = A^\circ[\mathbf{u} \mapsto \mathbf{v}],$$

where

$$[\mathbf{u} \mapsto \mathbf{v}] := [\mathbf{x}' \mapsto \mathbf{y}]_A,$$

since $\mathbf{x}'$ does not contain bound variables of A. Then $r(\mathbf{u}) = FV(A^\circ)$, so $r(\mathbf{v}) = FV(A^\circ[\mathbf{u} \mapsto \mathbf{v}])$.

Similarly we have

$$(A^\circ[\mathbf{x}' \mapsto \mathbf{y}'])[\mathbf{x}'' \mapsto \mathbf{y}''] = (A^\circ[\mathbf{u} \mapsto \mathbf{v}])[\mathbf{x}'' \mapsto \mathbf{y}''] = (A^\circ[\mathbf{u} \mapsto \mathbf{v}])[\mathbf{w} \mapsto \mathbf{z}],$$

where

$$[\mathbf{w} \mapsto \mathbf{z}] := [\mathbf{x}'' \mapsto \mathbf{y}'']_{A^\circ[\mathbf{u}\mapsto\mathbf{v}]}.$$

So $r(\mathbf{w}) = r(\mathbf{v})$, and thus $\mathbf{v} = \mathbf{w}\cdot\sigma$ for some surjective $\sigma \in \Sigma_{|w|,|v|}$. Hence

$$(A^\circ[\mathbf{u} \mapsto \mathbf{v}])[\mathbf{w} \mapsto \mathbf{z}] = A^\circ[\mathbf{u} \mapsto \mathbf{z}\cdot\sigma]$$

by Lemma 2.3.2.

On the other hand, for $[\mathbf{y}/\mathbf{x}] = [\mathbf{y}'/\mathbf{x}'] \circ [\mathbf{y}''/\mathbf{x}'']$ we may assume that $r(\mathbf{x}) \supseteq FV(A)$. Then we have

$$[\mathbf{y}/\mathbf{x}]A \stackrel{\circ}{=} A^\circ[\mathbf{x} \mapsto \mathbf{y}] \stackrel{\circ}{=} A^\circ[\mathbf{x} \mapsto \mathbf{y}]_A,$$

and it remains to show that

$$[\mathbf{u} \mapsto \mathbf{z}\cdot\sigma] = [\mathbf{x} \mapsto \mathbf{y}]_A.$$

In fact, suppose $x_j = u_i \in FV(A)$. Then

$$y_j = [\mathbf{y}''/\mathbf{x}''](y'_j) = [\mathbf{y}''/\mathbf{x}'](v_i) = [\mathbf{y}''/\mathbf{x}''](w_{\sigma(i)}) = z_{\sigma(i)},$$

by the choice of $[\mathbf{u} \mapsto \mathbf{v}]$, $[\mathbf{w} \mapsto \mathbf{z}]$.

(5) By (4), it suffices to check that $[\mathbf{y}/\mathbf{x}] \cdot [\mathbf{z}/\mathbf{u}]$ and $[[\mathbf{y}/\mathbf{x}]\mathbf{z}/\mathbf{u}]$ coincide on parameters of A. In fact, the first substitution sends every u_i to z_i and next to $[\mathbf{y}/\mathbf{x}]z_i$ and every $x_j \notin r(\mathbf{u})$ to y_j. So if $r(\mathbf{x}) \cap FV(A) \subseteq r(\mathbf{u})$, all parameters of A beyond $r(\mathbf{u})$ remain fixed. The second substitution sends u_i directly to $[\mathbf{y}/\mathbf{x}]z_i$ and also fixes other parameters; thus the claim holds.

(6) This is a particular case of (5) when $\mathbf{x} = \mathbf{z}$. Then $[\mathbf{y}/\mathbf{x}]\mathbf{z} = \mathbf{y}$.

(7) Apply (5) to the case $\mathbf{y} := v$, $\mathbf{x} := u$, $\mathbf{z} := \mathbf{u}$, $\mathbf{u} := \mathbf{x}$. Note that

$$FV(A) \cap \{u\} \subseteq r(\mathbf{x}) \text{ iff } \{u\} \subseteq -FV(A) \cup r(\mathbf{x}).$$

(8) Consider a clean version A° of A such that $r(\mathbf{xyz}) \cap BV(A^\circ) = \varnothing$. Then

$$QzA \triangleq QzA^\circ$$

and QzA° is clean by 2.3.11.

So

(8.1) $[\mathbf{y}/\mathbf{x}]QzA \triangleq (QzA^\circ)[\mathbf{x} \mapsto \mathbf{y}] = Qz(A^\circ[\mathbf{x} \mapsto \mathbf{y}])$.

On the other hand,

$$A^\circ[\mathbf{x} \mapsto \mathbf{y}] \triangleq [\mathbf{y}/\mathbf{x}]A,$$

hence

(8.2)

$$Qz(A^\circ[\mathbf{x} \mapsto \mathbf{y}]) \triangleq Qz[\mathbf{y}/\mathbf{x}]A$$

by 2.3.16. Now (8) follows from (8.1) and (8.2).

(9) Let A°, B° be clean versions of A and B such that

$$BV(A^\circ) \cap r(\mathbf{xy}) = BV(B^\circ) \cap r(\mathbf{xy}) = \varnothing,$$

$$BV(A^\circ) \cap FV(A) = BV(B^\circ) \cap FV(B) = \varnothing.$$

Then $(A^\circ * B^\circ)$ is a clean version of $(A * B)$ and by 2.3.21 we have:

$$[\mathbf{y}/\mathbf{x}](A * B) \triangleq (A^\circ * B^\circ)[\mathbf{x} \mapsto \mathbf{y}] = (A^\circ[\mathbf{x} \mapsto \mathbf{y}] * B^\circ[\mathbf{x} \mapsto \mathbf{y}])$$
$$\triangleq ([\mathbf{y}/\mathbf{x}]A * [\mathbf{y}/\mathbf{x}]B).$$

(10) Note that in this case

$$[\mathbf{y}''/\mathbf{x}''] \cdot [\mathbf{y}'/\mathbf{x}'] = [\mathbf{y}''\mathbf{y}'/\mathbf{x}''\mathbf{x}']$$

and apply (4)

(11) Since $z_i \notin FV(A)$, from (6) we obtain

($\sharp$) $$[\mathbf{y}/\mathbf{z}][\mathbf{z}/\mathbf{x}]A \triangleq [\mathbf{y}/\mathbf{x}[A$$

By induction from (10) we also have

($\sharp\sharp$) $$[\mathbf{z}/\mathbf{x}]A \triangleq [z_n/x_n]\ldots[z_1/x_1]A,$$

since $r(\mathbf{x}) \cap r(\mathbf{z}) = \varnothing$ and $\mathbf{z}$ is distinct.

In the same way from (10) we obtain

($\sharp\sharp\sharp$) $$[\mathbf{y}/\mathbf{z}][\mathbf{z}/\mathbf{x}]A \triangleq [y_n/z_n]\ldots[y_1/z_1][\mathbf{z}/\mathbf{x}]A,$$

since $r(\mathbf{z}) \cap r(\mathbf{y}) = \varnothing$ and $\mathbf{z}$ is distinct.

Now (11) follows from ($\sharp$), ($\sharp\sharp$), ($\sharp\sharp\sharp$) and 2.3.24.

(12) Follows from (8) by induction on $|\mathbf{z}|$. For the step, suppose $\mathbf{z} = z_1\mathbf{z}'$ and

$$[\mathbf{y}/\mathbf{x}]\mathcal{Q}\mathbf{z}'A \stackrel{\circ}{=} \mathcal{Q}\mathbf{z}'[\mathbf{y}/\mathbf{x}]A;$$

then by 2.3.17(2),

$$\mathcal{Q}z_1[\mathbf{y}/\mathbf{x}]\mathcal{Q}\mathbf{z}'A \stackrel{\circ}{=} \mathcal{Q}\mathbf{z}[\mathbf{y}/\mathbf{x}]A.$$

On the other hand, by (8)

$$[\mathbf{y}/\mathbf{x}]\mathcal{Q}\mathbf{z}A \stackrel{\circ}{=} \mathcal{Q}z_1[\mathbf{y}/\mathbf{x}]\mathcal{Q}\mathbf{z}'A;$$

hence (12) follows.

(13) Apply induction on $|\mathbf{x}| = |\mathbf{y}|$. The base follows from (3).

For the step, suppose $\mathbf{y} = y_1\mathbf{y}'$, $\mathbf{x} = x_1\mathbf{x}'$ and

$$\mathcal{Q}\mathbf{y}'[\mathbf{y}'/\mathbf{x}']A \stackrel{\circ}{=} \mathcal{Q}\mathbf{x}'A.$$

Then by 2.3.17(2),

$$(*1) \qquad \mathcal{Q}x_1\mathcal{Q}y'[\mathbf{y}'/\mathbf{x}']A \stackrel{\circ}{=} \mathcal{Q}\mathbf{x}A.$$

On the other hand, by (10),

$$[\mathbf{y}/\mathbf{x}]A \stackrel{\circ}{=} [y_1/x_1][\mathbf{y}'/\mathbf{x}']A;$$

hence by 2.3.17(2),

$$(*2) \qquad \mathcal{Q}\mathbf{y}[\mathbf{y}/\mathbf{x}]A \stackrel{\circ}{=} \mathcal{Q}\mathbf{y}[y_1/x_1][\mathbf{y}'/\mathbf{x}']A.$$

By (12),

$$\mathcal{Q}\mathbf{y}'[y_1/x_1][\mathbf{y}'/\mathbf{x}']A \stackrel{\circ}{=} [y_1/x_1]\mathcal{Q}\mathbf{y}'[\mathbf{y}'/\mathbf{x}']A,$$

hence by 2.3.17(2) and (3)

$$(*3) \quad \mathcal{Q}\mathbf{y}[y_1/x_1][\mathbf{y}'/\mathbf{x}']A \stackrel{\circ}{=} \mathcal{Q}y_1[y_1/x_1](\mathcal{Q}\mathbf{y}'[\mathbf{y}'/\mathbf{x}']A) \stackrel{\circ}{=} \mathcal{Q}x_1\mathcal{Q}\mathbf{y}'[\mathbf{y}'/\mathbf{x}']A.$$

Now (13) follows from $(*2), (*3)$, and $(*1)$.

■

Exercise 2.3.26 Describe the composition of substitutions (or of corresponding transformations) explicitly.

2.4 Formulas with constants

Although our basic languages do not contain individual constants, we will need auxiliary languages with constants.

So let D be a non-empty set; we assume that $D \cap Var = \varnothing$. Let $\mathcal{L}_N(D)$ be the language $\mathcal{L}_N$ expanded by individual constants from the set D. Formulas of the language $\mathcal{L}_N(D)$ (respectively, $\mathcal{L}_0(D)$) are called N-modal (respectively, intuitionistic) *D-formulas*; the set of all these formulas is denoted by $MF_N^{(=)}(D)$[14] (respectively, $IF^{(=)}(D)$). Obviously, every predicate formula (in the ordinary sense, i.e. without extra constants) is a D-formula. A *D-sentence* is a D-formula without parameters; $MS_N^{(=)}(D)$ and $IS^{(=)}(D)$ denote the sets of D-sentences of corresponding types.

Definition 2.4.1 *Let* $\mathbf{x} = (x_1, \dots, x_n)$ *be a list of distinct variables,* $\mathbf{a} = (a_1, \dots, a_n)$ *a list of constants (individuals) from D (not necessarily distinct). Then the D-*transformation $[\mathbf{x} \mapsto \mathbf{a}]$ *is a finite function* $\{(x_1, a_1), \dots, (x_n, a_n)\}$ *sending every* x_i *to* a_i, $i = 1, \dots, n$. *The D-*instance $[\mathbf{a}/\mathbf{x}]A$ *of a D-formula A under* $[\mathbf{x} \mapsto \mathbf{a}]$ *is obtained by simultaneous replacement of all free occurrences of* $x_1, \dots, x_n$ *in A respectively with* $a_1, \dots, a_n$.

Strictly speaking, $[\mathbf{a}/\mathbf{x}]A$ is defined by induction on $|A|$:

- $[\mathbf{a}/\mathbf{x}]P(\mathbf{y}) := P([\mathbf{a}/\mathbf{x}]\mathbf{y})$,
  where $[\mathbf{a}/\mathbf{x}]\mathbf{y}$ is a tuple $\mathbf{z}$ such that $|\mathbf{z}| = |\mathbf{y}|$ and for any j,

$$z_j = \begin{cases} a_i \text{ if } y_j = x_i, \\ y_j \text{ if } y_j \notin r(\mathbf{x}). \end{cases}$$

- $[\mathbf{a}/\mathbf{x}]P := P$ if $P \in PL^0$,
- $[\mathbf{a}/\mathbf{x}](B * C) := ([\mathbf{a}/\mathbf{x}[B * [\mathbf{a}/\mathbf{x}]C)$ if $* \in \{\vee, \wedge, \supset\}$,
- $[\mathbf{a}/\mathbf{x}]\bot := \bot$,
- $[\mathbf{a}/\mathbf{x}]\Box_i B := \Box_i [\mathbf{a}/\mathbf{x}]B$,
- $[\mathbf{a}/\mathbf{x}]QzB := Qz[\mathbf{a}/\mathbf{x}]B$ if $z \notin r(\mathbf{x})$,
- $[\mathbf{a}/\mathbf{x}]Qx_iB := Qx_i[\hat{\mathbf{a}}_i/\hat{\mathbf{x}}_i]B$.[15]

So we can also denote $[\mathbf{a}/\mathbf{x}]A$ by $A[\mathbf{x} \mapsto \mathbf{a}]$ if $r(\mathbf{x}) \cap BV(A) = \varnothing$. Normally we use the notation $[\mathbf{a}/\mathbf{x}]A$ in the case when both A is a usual formula and $[\mathbf{a}/\mathbf{x}]A$ is a D-sentence (which is equivalent to $FV(A) \subseteq r(\mathbf{x})$). A formula A is called a *generator* of every D-sentence $[\mathbf{a}/\mathbf{x}]A$.

For D-formulas we define schemes, clean versions and congruence in the natural way.

---

[14] Or briefly, by $MF^{(=)}(D)$.

[15] Recall that $\hat{\mathbf{x}}_i$ is obtained by eliminating x_i from $\mathbf{x}$; similarly for $\hat{\mathbf{a}}_i$.

Lemma 2.4.2 *(1)* $A \stackrel{\circ}{=} B \Rightarrow [\mathbf{a}/\mathbf{x}]A \stackrel{\circ}{=} [\mathbf{a}/\mathbf{x}]B$
for any D-transformation $[\mathbf{x} \mapsto \mathbf{a}]$ *and D-formulas* A, B.

(2) If $\mathbf{x}$ *is a distinct list of variables* $|\mathbf{x}| = n$, $\mathbf{a} \in D^n$, *then for any predicate formula* A, *for any* $\sigma \in \Upsilon_n$

$$[(\mathbf{a} \cdot \sigma)/\mathbf{x}]\, A = [\mathbf{a}/(\mathbf{x} \cdot \sigma^{-1})]\, A.$$

(3) For any predicate formula A, *for any distinct list* $\mathbf{xy}$

$$[\mathbf{a}/\mathbf{y}][\mathbf{y}/\mathbf{x}]\, A \stackrel{\circ}{=} [\mathbf{a}/\mathbf{x}]\, A.$$

(4) Let $\mathbf{x}$, $\mathbf{z}$ *be distinct lists of variables,* $|\mathbf{x}| = n$, $|\mathbf{z}| = m \leq n$, *and let* $\sigma : I_n \longrightarrow I_m$. *Let* A *be a formula such that* $r(\mathbf{z} \cdot \sigma) \cap BV(A) = \varnothing$. *Then*

$$[\mathbf{a}/\mathbf{z}][(\mathbf{z} \cdot \sigma)/\mathbf{x}]\, A \stackrel{\circ}{=} [(\mathbf{a} \cdot \sigma)/\mathbf{x}]\, A.$$

Proof (1) It is clear that

$$\llcorner [\mathbf{a}/\mathbf{x}]A \lrcorner = \llcorner A \lrcorner [\mathbf{x} \mapsto \mathbf{a}]$$

(a strict proof is by induction). So $\llcorner A \lrcorner = \llcorner B \lrcorner$ implies $\llcorner [\mathbf{a}/\mathbf{x}]A \lrcorner = \llcorner [\mathbf{a}/\mathbf{x}]B \lrcorner$.

(2) Note that $[\mathbf{x} \mapsto \mathbf{a} \cdot \sigma] = [\mathbf{x} \cdot \sigma^{-1} \mapsto \mathbf{a}]$ — each of these maps sends x_i to $a_{\sigma(i)}$, and $x_{\sigma^{-1}(j)}$ to a_j. Now the claim follows from 2.4.1.

(3) As noted above, constant substitutions respect congruence. So we can prove the claim for a clean version A° of A, where $BV(A^\circ) \cap r(\mathbf{xy}) = \varnothing$. In this case it is equivalent to

$$(A^\circ[\mathbf{x} \mapsto \mathbf{y}])[\mathbf{y} \mapsto \mathbf{a}] = A^\circ[\mathbf{x} \mapsto \mathbf{a}].$$

This holds, since obviously

$$[\mathbf{x} \mapsto \mathbf{y}] \circ [\mathbf{y} \mapsto \mathbf{a}] = [\mathbf{x} \mapsto \mathbf{a}].$$

But again, a strict proof is by induction.

(4) Similar to (2). Consider a clean version A° of A, with $BV(A^\circ) \cap r(\mathbf{xz}) = \varnothing$. Then the claim reduces to

$$(A^\circ[\mathbf{x} \mapsto \mathbf{z} \cdot \sigma])[\mathbf{z} \mapsto \mathbf{a}] = A^\circ[\mathbf{x} \mapsto \mathbf{a} \cdot \sigma],$$

which follows from

$$[\mathbf{x} \mapsto \mathbf{z} \cdot \sigma] \circ [\mathbf{z} \mapsto \mathbf{a}] = [\mathbf{x} \mapsto \mathbf{a} \cdot \sigma].$$

■

We also use a somewhat ambiguous notation $A(\mathbf{x})$ to indicate that $FV(A) \subseteq r(\mathbf{x})$; in this case $[\mathbf{a}/\mathbf{x}]A$ is abbreviated to $A(\mathbf{a})$. The abbreviation $A(\mathbf{a})$ is convenient and rather common, but it leads to some confusion: it may happen that a D-sentence B can be presented as $[a_1, \ldots, a_n/x_1, \ldots, x_n]A$ for different formulas A. For example, $P(a, a) = [a/x]P(x, x) = [a, a/x, y]P(x, y)$. Such an ambiguity may be undesirable (cf. Section 5.1), so we will mainly use 'maximal' representations described as follows.

Definition 2.4.3 *A formula A is called a* maximal generator *of a D-formula B if $B = [\mathbf{a}/\mathbf{x}]A$ for some bijective D-transformation $[\mathbf{x} \mapsto \mathbf{a}]$.*

Since $[\mathbf{a}/\mathbf{x}]A$ does not depend on the variables x_i that are not parameters of A, in the above definition we may further assume that $r(\mathbf{x}) \subseteq FV(A)$, and thus $\mathbf{a}$ is the list of all constants occurring in B.

Lemma 2.4.4 *Every D-formula has a maximal generator.*

Proof Let $\mathbf{a} = a_1 \ldots a_n$ be a list of all constants occurring in a D-sentence B, $\mathbf{x} = x_1 \ldots x_n$ a list of different new variables for B. The formula $A := B[\mathbf{a} \mapsto \mathbf{x}]$ (obtained by replacing every occurrence of a_i with x_i) is a maximal generator of B, since $x_i \notin BV(B)$, and thus $[\mathbf{a}/\mathbf{x}]A = A[\mathbf{x} \mapsto \mathbf{a}] = (B[\mathbf{a} \mapsto \mathbf{x}])[\mathbf{x} \mapsto \mathbf{a}] = B$. ∎

Lemma 2.4.5

(1) If $B = [\mathbf{a}/\mathbf{x}]A$ for a formula A and a bijection $[\mathbf{x} \mapsto \mathbf{a}]$, then $A = B[\mathbf{a} \mapsto \mathbf{x}]$.

(2) If A_1, A_2 are maximal generators of B, then $A_2 \stackrel{\circ}{=} [\mathbf{y}/\mathbf{x}]A_1$ for some variable renaming $[\mathbf{x} \mapsto \mathbf{y}]$ (and of course, A_1 is obtained from A_2 in the same way).

More precisely, if $B = [\mathbf{a}/\mathbf{x}]A_1 = [\mathbf{a}/\mathbf{y}]A_2$ for bijections $[\mathbf{x} \mapsto \mathbf{a}], [\mathbf{y} \mapsto \mathbf{a}]$, then $A_2 \stackrel{\circ}{=} [\mathbf{y}/\mathbf{x}]A_1$.

(3) A maximal generator of a D-formula B is a substitution instance of any generator of B under some variable substitution.

Proof (1) We check that

$$([\mathbf{a}/\mathbf{x}]A)[\mathbf{a} \mapsto \mathbf{x}] = A$$

by induction on $|A|$.

This is clear for atomic $A = P(\mathbf{y})$, when $[\mathbf{a}/\mathbf{x}]A = A[\mathbf{x} \mapsto \mathbf{a}]$ (cf. Lemma 2.3.2). All induction steps are routine; let us consider only the case $A = \mathcal{Q}x_iB$ for a quantifier $\mathcal{Q}$. By definition,

$$([\mathbf{a}/\mathbf{x}]A)[\mathbf{a} \mapsto \mathbf{x}] = (\mathcal{Q}x_i[\hat{\mathbf{a}}_i/\hat{\mathbf{x}}_i]B)[\mathbf{a} \mapsto \mathbf{x}] = \mathcal{Q}x_i(([\hat{\mathbf{a}}_i/\hat{\mathbf{x}}_i]B)[\mathbf{a} \mapsto \mathbf{x}])$$

Since B is a usual formula and $[\mathbf{x} \mapsto \mathbf{a}]$ is a bijection, a_i does not occur in $[\hat{\mathbf{a}}_i/\hat{\mathbf{x}}_i]B$; thus

$$([\hat{\mathbf{a}}_i/\hat{\mathbf{x}}_i]B)[\mathbf{a} \mapsto \mathbf{x}] = ([\hat{\mathbf{a}}_i/\hat{\mathbf{x}}_i]B)[\hat{\mathbf{a}}_i \mapsto \hat{\mathbf{x}}_i] = B$$

by the induction hypothesis. Therefore (1) holds for A.

(2) If A_1 is a maximal generator of B, then $B = [\mathbf{a}/\mathbf{x}]A_1$ for some bijection $[\mathbf{x} \mapsto \mathbf{a}]$. Similarly, $B = [\mathbf{a}/\mathbf{y}]A_2$ for a bijection $[\mathbf{y} \mapsto \mathbf{a}]$. Now let A_1° be a clean version of A_1 such that $BV(A_1^\circ) \cap r(\mathbf{xy}) = \varnothing$.

By 2.4.2(1), $A_1 \stackrel{\circ}{=} A_1^\circ$ implies

$$B = [\mathbf{a}/\mathbf{x}]A_1 \stackrel{\circ}{=} [\mathbf{a}/\mathbf{x}]A_1^\circ = A_1^\circ[\mathbf{x} \mapsto \mathbf{a}].$$

Hence

$$B[\mathbf{a} \mapsto \mathbf{x}] \stackrel{\circ}{=} (A_1^\circ[\mathbf{x} \mapsto \mathbf{a}])[\mathbf{a} \mapsto \mathbf{x}] = A_1^\circ.$$

Similarly, there exists a clean $A_2^\circ \stackrel{\circ}{=} A_2$ such that

$$B[\mathbf{a} \mapsto \mathbf{y}] \stackrel{\circ}{=} A_2^\circ.$$

Therefore

$$A_2 \stackrel{\circ}{=} B[\mathbf{a} \mapsto \mathbf{y}] \stackrel{\circ}{=} (A_1^\circ[\mathbf{x} \mapsto \mathbf{a}])[\mathbf{a} \mapsto \mathbf{y}] = A_1^\circ[\mathbf{x} \mapsto \mathbf{y}],$$

and the latter formula is $[\mathbf{y}/\mathbf{x}]A_1$ by 2.3.22.

(3) Let C be a generator of B, thus $B = [\mathbf{b}/\mathbf{z}]C$ for some $[\mathbf{z} \mapsto \mathbf{b}]$; and let A be a maximal generator of B, with $B = [\mathbf{a}/\mathbf{x}]A$, for a bijection $[\mathbf{x} \mapsto \mathbf{a}]$. We may assume that $r(\mathbf{x}) \subseteq FV(A)$, $r(\mathbf{z}) \subseteq FV(C)$ and $r(\mathbf{b}) = r(\mathbf{a})$ is the set of all constants occurring in B.

Since $\mathbf{a}$ is distinct, every b_i equals to some a_j, so $\mathbf{b} = \mathbf{a}\cdot\tau$ for some surjective map $\tau : I_n \to I_m$.

Thus

$$B = [\mathbf{a}\cdot\tau/\mathbf{z}]C = [\mathbf{a}/\mathbf{x}]A,$$

and we have by 2.4.2(4)

$$[\mathbf{a}\cdot\tau/\mathbf{z}]C \stackrel{\circ}{=} [\mathbf{a}/\mathbf{x}][\mathbf{x}\cdot\tau/\mathbf{z}]C.$$

Hence by (2), $A \stackrel{\circ}{=} [\mathbf{x}\cdot\tau/\mathbf{z}]C$. ■

2.5 Formula substitutions

Definition 2.5.1 *A (*simple*)* formula substitution *is a pair $(C, P(\mathbf{x}))$, where C is a predicate formula, $P(\mathbf{x})$ is an atomic equality-free formula. The substitution $(C, P(\mathbf{x}))$ is usually denoted by $[C/P(\mathbf{x})]$. More exactly, $[C/P(\mathbf{x})]$ is called an MF_N-, $(MF_N^{=}-$, $IF-$, $IF^{=}-)$* substitution *if the formula C is of the corresponding type.*

Definition 2.5.2 *For a substitution $[C/P(\mathbf{x})]$,*

$$FV[C/P(\mathbf{x})] := FV(C) - r(\mathbf{x})$$

is called the set of parameters,

$$BV[C/P(\mathbf{x})] := r(\mathbf{x})$$

the set of bound variables. *A substitution $[C/P(\mathbf{x})]$ is called* strict *if $FV[C/P(\mathbf{x})] = \varnothing$, i.e. if $FV(C) \subseteq r(\mathbf{x})$.*

Definition 2.5.3 *Let A be a clean predicate formula, $S = [C/P(\mathbf{x})]$ a formula substitution such that $BV(A) \cap FV(S) = \varnothing$. Let B be a result of replacing all subformulas of A of the form $P(\mathbf{y})$ with $[\mathbf{y}/\mathbf{x}]C$.*

Every formula congruent to B is denoted by SA and is called a substitution instance *of A under S.*

More precisely, SA is defined by induction:

$$SP(\mathbf{y}) \stackrel{\circ}{=} [\mathbf{y}/\mathbf{x}]C,$$
$$SA \stackrel{\circ}{=} A \text{ if } A \text{ is atomic and does not contain } P,$$
$$S\Box_i A \stackrel{\circ}{=} \Box_i SA,$$
$$S(A * B) \stackrel{\circ}{=} (SA * SB) \text{ for } * \in \{\vee, \wedge, \supset\},$$
$$SQzA \stackrel{\circ}{=} QzSA \text{ for } Q \in \{\forall, \exists\}.$$

A formula SA is called a substitution instance of A, *or more exactly, an $MF_N^{(=)}$- ($IF^{(=)}$-)* substitution instance *if S is an $MF_N^{(=)}$- ($IF^{(=)}$-) substitution.*

Due to the assumption $BV(A) \cap FV(S) = \varnothing$, in SA the parameters of S do not collide with the existing bound variables from A.

Note that applying S to A does not affect occurrences of equality in A, but may introduce new occurrences if C contains equality.

Lemma 2.5.4 *Let A be a clean formula, $S = [C/P(\mathbf{x})]$ a formula substitution such that $FV(S) \cap BV(A) = \varnothing$. Then*

$$S(A[u \mapsto v]) \stackrel{\circ}{=} [v/u]SA.$$

for any variables u, v such that $v \notin V(A)$, $u \in FV(A)$, and $u, v \notin FV(S)$.

Proof Since $v \notin V(A)$, we can prolong $[u \mapsto v]$ to a variable renaming in A by fixing all variables from $V(A) - \{u\}$. So $A[u \mapsto v]$ is clean by 2.3.11 with the same bound variables as A, and Definition 2.5.3 is applicable to this formula.

Now we argue by induction on $|A|$.

- If $A = P(\mathbf{y})$, then $A[u \mapsto v] = P([v/u]\mathbf{y})$, $SA \stackrel{\circ}{=} [\mathbf{y}/\mathbf{x}]C$, and

$$S(A[u \mapsto v]) \stackrel{\circ}{=} [[v/u]\mathbf{y}/\mathbf{x}]C.$$

By assumption, $u \notin FV(S) = FV(C) - r(\mathbf{x})$, so we obtain

$$S(A[u \mapsto v]) \stackrel{\circ}{=} [v/u]SA$$

by applying 2.3.25 (7).

- Let $A = QzB$, then $SA \stackrel{\circ}{=} QzSB$. Hence

$$(1) \qquad [v/u]SA \stackrel{\circ}{=} [v/u]QzSB \stackrel{\circ}{=} Qz[v/u]SB$$

  by 2.3.25 (8); note that $z \neq u$, since $u \in FV(A)$

By the induction hypothesis,

$$[v/u]SB \mathrel{\mathring{=}} S(B[u \mapsto v]),$$

hence

$$(2) \qquad Qz[v/u]SB \mathrel{\mathring{=}} QzS(B[u \mapsto v])$$

by 2.3.6. Now by 2.5.3

$$(3) \qquad QzS(B[u \mapsto v]) \mathrel{\mathring{=}} SQz(B[u \mapsto v]),$$

so from (1), (2), (3) we have

$$[v/u]SA \mathrel{\mathring{=}} SQz(B[u \mapsto v]).$$

It remains to note that

$$Qz(B[u \mapsto v]) = A[u \mapsto v],$$

since $z \neq u$. Therefore the claim holds for A.

- If $A = (B * C)$, we can use 2.3.25 (9) and the distribution of S and $[u \mapsto v]$ over $*$. Note that if u does not occur in B (or in C), the main statement trivially holds for B (or C), and the argument does not change. The details are left to the reader.

- The case $A = \Box_i B$ is trivial.

■

Lemma 2.5.5 *Let A, B be congruent clean formulas, S a formula substitution such that $BV(A) \cap FV(S) = BV(B) \cap FV(S) = \varnothing$. Then $SA \mathrel{\mathring{=}} SB$.*

Proof By induction on $|A| = |B|$.

- If A is atomic, then $A = B$, and there is nothing to prove.

- If $A = (A_1 * A_2)$, then by Lemma 2.3.9(1), $B = (B_1 * B_2)$ for $A_1 \mathrel{\mathring{=}} B_1$, $A_2 \mathrel{\mathring{=}} B_2$. Hence

$$SA \mathrel{\mathring{=}} (SA_1 * SA_2), \quad SB \mathrel{\mathring{=}} (SB_1 * SB_2),$$

  and $SA_i \mathrel{\mathring{=}} SB_i$ by the induction hypothesis. Eventually $SA \mathrel{\mathring{=}} SB$ by 2.3.17.

- We skip the easy case when $A = \Box_i A_1$.

- Suppose $A = \mathcal{Q}xA_1$ for a quantifier $\mathcal{Q}$. Since A is clean, $x \notin BV(A_1)$, so by Lemma 2.3.9(2), for some $y \notin FV(A_1), B_1$

$$B = \mathcal{Q}yB_1, \ B_1 \stackrel{\circ}{=} A_1[x \mapsto y].$$

We may also assume that $y \notin BV(A_1)$. (Otherwise consider $A_2 \stackrel{\circ}{\equiv} A_1$ such that $y \notin BV(A_2)$, then

$$B_1 \stackrel{\circ}{=} A_1[x \mapsto y] \stackrel{\circ}{=} A_2[x \mapsto y],$$

so A_2 can be used instead of A_1.) Thus

$$SA = \mathcal{Q}xSA_1, \ SB = \mathcal{Q}ySB_1,$$

and

$$SB_1 \stackrel{\circ}{=} S(A_1[x \mapsto y]) \stackrel{\circ}{=} [y/x]SA_1$$

by the induction hypothesis and Lemma 2.5.4 (which is applicable, since $y \notin V(A_1)$ and $x, y \notin FV(S)$ by the assumption of the lemma). Hence by 2.3.24(3)

$$SB = \mathcal{Q}ySB_1 \stackrel{\circ}{=} \mathcal{Q}y[y/x]SA_1 \stackrel{\circ}{=} \mathcal{Q}xSA_1 = SA.$$

■

Now we can define substitution instances of arbitrary formulas.

Definition 2.5.6 *A substitution instance SA of a predicate formula A under a simple substitution S is an arbitrary formula congruent to SA°, where A° is a clean version of A such that $FV(S) \cap BV(A^\circ) = \varnothing$. A* strict *substitution instance is a substitution instance under a strict substitution.*

Lemma 2.5.5 shows soundness of this definition, i.e. that the congruence class of SA° does not depend on the choice of A°.

Note that according to the definition, for a trivial formula substitution $S = [P(\mathbf{x})/P(\mathbf{x})]$ and a formula A, SA denotes an arbitrary formula congruent to A.

Lemma 2.5.7 *Let $[C_1/P(\mathbf{x})]$, $[C_2/P(\mathbf{x})]$ be formula substitutions such that $C_1 \stackrel{\circ}{=} C_2$. Then for any predicate formula A, $[C_1/P(\mathbf{x})]A \stackrel{\circ}{=} [C_2/P(\mathbf{x})]A$.*

Proof We denote $[C_i/P(\mathbf{x})]$ by S_i. Let A° be a clean version of A such that $FV(S_i) \cap BV(A^\circ) = \varnothing$ for $i = 1, 2$. Obviously we can construct such A° by an appropriate bound variable renaming from an arbitrary clean version. Now $S_iA \stackrel{\circ}{=} S_iA^\circ$, so we show $S_1A^\circ \stackrel{\circ}{=} S_2A^\circ$ by induction on $|A^\circ|$. To simplify notation, put $B := A^\circ$.

If $B = P(\mathbf{y})$, then $S_iB = [\mathbf{y}/\mathbf{x}]C_i$, so $S_1B \stackrel{\circ}{=} S_2B$ follows from 2.3.24.

If B is atomic and does not contain P, the claim is trivial.

The induction step easily follows from the distribution of S_i over all connectives and quantifiers. E.g. suppose $B = \mathcal{Q}yB_1$; then $y \notin FV(S_i)$, so

$$S_i\mathcal{Q}yB_1 \stackrel{\circ}{=} \mathcal{Q}yS_iB_1$$

by 2.3.17. By induction hypothesis,

$$S_1B_1 \stackrel{\circ}{=} S_2B_1,$$

hence

$$\mathcal{Q}yS_1B_1 \stackrel{\circ}{=} \mathcal{Q}yS_2B_1$$

by 2.3.17, and therefore

$$S_1B \stackrel{\circ}{=} S_2B.$$

All the remaining cases are left to the reader. ∎

Now let us consider complex substitutions.

Definition 2.5.8 *For atomic equality-free formulas $P_1(\mathbf{x}_1), \dots, P_k(\mathbf{x}_k)$ (with different predicate letters $P_1, \dots, P_k$ and distinct lists $\mathbf{x}_1, \dots, \mathbf{x}_k$) and formulas $C_1, \dots, C_k$ we define the* complex formula substitution

$$[C_1, \dots, C_k/P_1(\mathbf{x}_1), \dots, P_k(\mathbf{x}_k)]$$

as the tuple $(C_1, \dots, C_k, P_1(\mathbf{x}_1), \dots, P_k(\mathbf{x}_k))$.

The set of its parameters *and* bound variables *are respectively*

$$FV[C_1, \dots, C_k/P_1(\mathbf{x}_i), \dots, P_k(\mathbf{x}_k)] := \bigcup_{i=1}^{k} FV[C_i/P_i(\mathbf{x}_i)] = \bigcup_{i=1}^{k} (FV(C_i) - r(\mathbf{x}_i))$$

and

$$BV[C_1, \dots, C_k/P_1(\mathbf{x}_i), \dots, P_k(\mathbf{x}_k)] := r(\mathbf{x}_1 \dots \mathbf{x}_k).$$

A substitution without parameters is called strict.

Now we have an analogue of Definition 2.5.3.

Definition 2.5.9 *For a substitution $S = [C_1, \dots, C_k/_1(x_1), \dots, P_k(x_k)]$ and a clean formula A such that $FV(S) \cap BV(A) = \varnothing$, a* substitution instance SA *is defined up to congruence by induction:*

$$SP_i(\mathbf{y}) \stackrel{\circ}{=} [\mathbf{y}/\mathbf{x}_i]C_i,$$
$$SA \stackrel{\circ}{=} A \text{ if } A \text{ is atomic and does not contain } P_1, \dots, P_k,$$
$$S\Box_i A \stackrel{\circ}{=} \Box_i SA,$$
$$S(A * B) \stackrel{\circ}{=} (SA * SB) \text{ for } * \in \{\vee, \wedge, \supset\},$$
$$S\mathcal{Q}zA \stackrel{\circ}{=} \mathcal{Q}zSA \text{ for } \mathcal{Q} \in \{\forall, \exists\}.$$

We also have an analogue of Lemma 2.5.5.

Lemma 2.5.10 *If A, B are clean formulas, $A \stackrel{\circ}{=} B$, S is a complex formula substitution and*

$$BV(A) \cap FV(S) = BV(B) \cap FV(S) = \varnothing,$$

then $SA \stackrel{\circ}{=} SB$.

Proof The same as in 2.5.5 (including an analogue of 2.5.4). ■

So the following definition is sound.

Definition 2.5.11 *For an arbitrary formula A and a formula substitution S, we define SA as SA°, for a clean version A° of A such that $FV(S) \cap BV(A^\circ) = \varnothing$.*

Hence we readily obtain

Lemma 2.5.12 *For any predicate formulas A, B and a formula substitution S,*

$$A \stackrel{\circ}{=} B \Rightarrow SA \stackrel{\circ}{=} SB.$$

The inductive definition 2.5.9 now extends to arbitrary formulas:

Lemma 2.5.13 *Let S be a formula substitution. Then for any formulas A, B*

(1) $S\Box_i A \stackrel{\circ}{=} \Box_i SA$,

(2) $S(A * B) \stackrel{\circ}{=} (SA * SB)$ for $* \in \{\vee, \wedge, \supset\}$,

(3) $SQzA \stackrel{\circ}{=} QzSA$ for $Q \in \{\forall, \exists\}, z \notin FV(S)$.

Proof

(1) Let A° be a clean version of A such that $FV(S) \cap BV(A^\circ) = \varnothing$. Then $\Box_i A^\circ$ is a clean version of $\Box_i A$, so

$$S\Box_i A \stackrel{\circ}{=} S\Box_i A^\circ \stackrel{\circ}{=} \Box_i SA^\circ.$$

By definition, $SA^\circ \stackrel{\circ}{=} SA$, hence

$$\Box_i SA^\circ \stackrel{\circ}{=} \Box_i SA,$$

therefore (1) holds.

(2) An exercise for the reader.

(3) Let A° be a clean version of A such that $FV(S) \cap BV(A^\circ) = \varnothing$, $z \notin BV(A^\circ)$. Then QzA° is a clean version of QzA and

$$FV(S) \cap BV(QzA^\circ) = \varnothing.$$

By definition,

$$SQzA \stackrel{\circ}{=} SQzA^\circ \stackrel{\circ}{=} QzSA^\circ,\ SA^\circ \stackrel{\circ}{=} SA;$$

hence

$$QzSA^\circ \stackrel{\circ}{=} QzSA.$$

This implies (3). ■

The next lemma shows that the result of applying a substitution does not really depend on the names of its bound variables. We prove this only for simple substitutions, leaving the general case to the reader.

Lemma 2.5.14 *Let $[C/P(\mathbf{x})]$ be a formula substitution, $[\mathbf{x} \mapsto \mathbf{y}]$ a variable renaming such that $r(\mathbf{y}) \cap FV(C) = \varnothing$. Then for any formula A,*

$$[C/P(\mathbf{x})]A \triangleq [[\mathbf{y}/\mathbf{x}]C/P(\mathbf{y})]A.$$

Proof
If $A = P(\mathbf{z})$, we have

$$[C/P(\mathbf{x})]A \triangleq [\mathbf{z}/\mathbf{x}]C, \ \ [[\mathbf{y}/\mathbf{x}]C/P(\mathbf{y})]A \triangleq [\mathbf{z}/\mathbf{y}][\mathbf{y}/\mathbf{x}]C,$$

while

$$[\mathbf{z}/\mathbf{x}]C \triangleq [\mathbf{z}/\mathbf{y}][\mathbf{y}/\mathbf{x}]C$$

by 2.3.25(6).

If A is atomic and does not contain P, the claim is trivial.

Now we can argue by induction on $|A|$. Put

$$S_1 := [C/P(\mathbf{x})], \ \ S_2 := [[\mathbf{y}/\mathbf{x}]C/P(\mathbf{y})].$$

If $A = QuB$, we may assume that $u \notin FV(S_1)(= FV(S_2))$ — otherwise consider $A' \triangleq A$ of the form $Qu'B'$, where $u' \notin FV(S_1)$.

Suppose $S_1B \triangleq S_2B$, then $S_iA \triangleq QuS_iB$ by 2.5.3; hence $S_1A \triangleq S_2A$ by 2.3.17.

Other cases are also based on 2.5.3 and 2.3.17; we leave them to the reader. ■

Lemma 2.5.15 *Let $S = [C/P(\mathbf{u})]$ be a simple formula substitution, $[\mathbf{y}/\mathbf{x}]$ a variable substitution such that $FV(C) \cap r(\mathbf{x}) = \varnothing$. Then for any predicate formula A, $S[\mathbf{y}/\mathbf{x}]A \triangleq [\mathbf{y}/\mathbf{x}]SA$.*

Proof Since S respects congruence by Lemma 2.5.12, $[\mathbf{y}/\mathbf{x}]$ respects congruence by 2.3.24 and both S and $[\mathbf{y}/\mathbf{x}]$ distribute over all connectives and quantifiers in an appropriate clean version of A (Definition 2.5.3, Lemma 2.3.25), it suffices to consider only the case when A is atomic. The nontrivial option is $A = P(\mathbf{z})$. Then

$$S[\mathbf{y}/\mathbf{x}]A \triangleq SP([\mathbf{y}/\mathbf{x}]\mathbf{z}) \triangleq [[\mathbf{y}/\mathbf{x}]\mathbf{z}/\mathbf{u}]C,$$
$$[\mathbf{y}/\mathbf{x}]SA \triangleq [\mathbf{y}/\mathbf{x}][\mathbf{z}/\mathbf{u}]C.$$

Now the claim follows by 2.3.25(5). ■

Lemma 2.5.14 shows that in some cases variable substitutions commute with formula substitutions. The next lemma considers situations where formula substitutions 'absorb' variable substitutions.

Lemma 2.5.16 *Let $[C/P(\mathbf{x})]$ be a simple formula substitution, A a predicate formula, $[\mathbf{y}/\mathbf{z}]$ a variable substitution such that $r(\mathbf{z}) \cap FV(A) = r(\mathbf{z}) \cap r(\mathbf{x}) = r(\mathbf{y}) \cap r(\mathbf{x}) = \varnothing$. Then*

$$[\mathbf{y}/\mathbf{z}][C/P(\mathbf{x})]A \stackrel{\circ}{=} [[\mathbf{y}/\mathbf{z}]C/P(\mathbf{x})]A.$$

Note that $[\mathbf{y}/\mathbf{z}]C$ is defined up to congruence, but the congruence class of $[[\mathbf{y}/\mathbf{z}]C/P(\mathbf{x})]A$ does not depend on the choice of a congruent version of $[\mathbf{y}/\mathbf{z}]C$, thanks to Lemma 2.5.7.

Proof The same idea as in 2.5.14 shows that it is sufficient to consider only the case when $A = P(\mathbf{u})$ is atomic (and by the assumption, $r(\mathbf{z}) \cap r(\mathbf{u}) = \varnothing$).

In this case the claim becomes

$$(*) \qquad [\mathbf{y}/\mathbf{z}][\mathbf{u}/\mathbf{x}]C \stackrel{\circ}{=} [\mathbf{u}/\mathbf{x}][\mathbf{y}/\mathbf{z}]C.$$

The latter congruence follows from 2.3.24. In fact, by 2.3.24,

$$[\mathbf{y}/\mathbf{z}][\mathbf{u}/\mathbf{x}]C \stackrel{\circ}{=} ([\mathbf{y}/\mathbf{z}] \cdot [\mathbf{u}/\mathbf{x}])C.$$

from $r(\mathbf{z}) \cap r(\mathbf{ux}) = \varnothing$, by 2.3.25(10) we have

$$[\mathbf{y}/\mathbf{z}][\mathbf{u}/\mathbf{x}]C \stackrel{\circ}{=} [\mathbf{yu}/\mathbf{zx}]C,$$

and similarly from $r(\mathbf{x}) \cap r(\mathbf{yz}) = \varnothing$,

$$[\mathbf{u}/\mathbf{x}][\mathbf{y}/\mathbf{z}]C \stackrel{\circ}{=} [\mathbf{uy}/\mathbf{xz}]C.$$

Since $[\mathbf{yu}/\mathbf{zx}] = [\mathbf{uy}/\mathbf{xz}]$, this implies $(*)$. ∎

The previous lemma easily transfers to complex sibstitutions:

Lemma 2.5.17 *Let*

$$S = [C_1, \ldots, C_k/P_1(\mathbf{x}_1), \ldots, P_k(\mathbf{x}_k)]$$

be a formula substitution, A a predicate formula, $[\mathbf{y}/\mathbf{x}]$ a variable substitution such that $r(\mathbf{z}) \cap FV(A) = \varnothing$ and $r(\mathbf{yz}) \cap r(\mathbf{x}_1, \ldots, \mathbf{x}_k) = \varnothing$.

Then

$$[\mathbf{y}/\mathbf{z}]SA \stackrel{\circ}{=} S_0A,$$

where $S_0 = [[\mathbf{y}/\mathbf{z}]C_1, \ldots, [\mathbf{y}/\mathbf{z}]C_k/P_1(\mathbf{x}_1), \ldots, P_k(\mathbf{x}_k)]$. Note that $r(\mathbf{z}) \cap FV(S_0) = \varnothing$.

Proof Again everything reduces to the case of atomic A. But in this case S acts as a simple substitution, so we can apply 2.5.16. ∎

Lemma 2.5.18 $[[\mathbf{c}/\mathbf{x}]\,\mathbf{B}/\mathbf{q}]\,A \stackrel{\circ}{=} [\mathbf{c}/\mathbf{x}][\mathbf{B}/\mathbf{q}]\,A$ *for a propositional formula A, a list of proposition letters $\mathbf{q}$, a list of constants $\mathbf{c}$, a distinct list of variables $\mathbf{x}$, a list of predicate formulas $\mathbf{B}$, $r(\mathbf{x}) \cap FV(A) = \varnothing$.*

Proof The same argument as above reduces everything to the case when A is atomic, i.e. a proposition letter. Then the claim is trivial. ■

Lemma 2.5.19 *Every complex substitution acts on formulas as a composition of simple substitutions. More precisely, if $S = [C_1, \dots, C_k / P_1(\mathbf{x}_1), \dots, P_k(\mathbf{x}_k)]$ is a complex substitution, P_i' is of the same arity of P_i and P_i' does not occur in $C_1, \dots, C_k$ for $i = 1, \dots, k$, then for any formula A*

$$SA \mathrel{\dot{=}} [C_1/P_1'(\mathbf{x}_1)] \dots [C_k/P_k'(\mathbf{x}_k)][P_1'(\mathbf{x}_1)/P_1(\mathbf{x}_1)] \dots [P_k'(\mathbf{x}_k)/P_k(\mathbf{x}_k)]A.$$

Proof Since substitutions respect congruence and distribute over all connectives and quantifiers over non-parametric variables (by Lemma 2.5.13), we may prove the claim for a congruent version of A, in which the parameters of S are not bound. In this case it suffices to check the claim for an atomic A. If $P_1, \dots, P_k$ do not occur in A, there is nothing to prove. So let $A = P_i(\mathbf{y})$. Then by definition

$$SA \mathrel{\dot{=}} [\mathbf{y}/\mathbf{x}_i]C_i,$$

while

$$\begin{aligned}&[C_1/P_1'(\mathbf{x}_1)] \dots [P_k'(\mathbf{x}_k)/P_k(\mathbf{x}_k)]A \mathrel{\dot{=}} [C_i/P_i'(\mathbf{x}_i)][P_i'(\mathbf{x}_i)/P_i(\mathbf{x}_i)]P_i(\mathbf{y}) \\ &\mathrel{\dot{=}} [C_i/P_i'(\mathbf{x}_i)]P_i'(\mathbf{y}) \mathrel{\dot{=}} [\mathbf{y}/\mathbf{x}_i]C_i.\end{aligned}$$

So the claim holds. ■

The composition of substitutions reduces to a single (complex) substitution as the following lemma shows.

Lemma 2.5.20 *Let $S_0 = [C_0/P_i(\mathbf{x}_0)]$, $S_1 = [C_1, \dots, C_k/P_1(\mathbf{x}_1), \dots, P_k(\mathbf{x}_k)]$ be formula substitutions. Then for any formula A*

$$S_0 S_1 A \mathrel{\dot{=}} S_2 A,$$

where $S_2 = [S_0C_1, \dots, S_0C_k/P_1(\mathbf{x}_1), \dots, P_k(\mathbf{x}_k)]$.

Proof Similarly to the previous lemma, it suffices to check this only for $A = P_i(\mathbf{y})$. In this case we have

$$S_0 S_1 A \mathrel{\dot{=}} S_0[\mathbf{y}/\mathbf{x}_i]C_i.$$

Lemma 2.5.14 shows that a formula substitution acts in the same way after renaming bound variables. So we may assume that $r(\mathbf{x}_i) \cap FV(C_0) = \varnothing$. Then by Lemma 2.5.15

$$S_0[\mathbf{y}/\mathbf{x}_i]C_i \mathrel{\dot{=}} [\mathbf{y}/\mathbf{x}_i]S_0C_i \mathrel{\dot{=}} S_2P_i(\mathbf{y}).$$

This completes the proof. ■

Lemma 2.5.21 *For any formula substitutions S_0, S_1, there exists a formula substitution S such that for any formula A*

$$S_0 S_1 A \mathrel{\dot{=}} SA.$$

Proof By Lemma 2.5.15

$$S_0 S_1 A \stackrel{\circ}{=} S_2 \dots S_n S_1 A$$

for some simple formula substitutions $S_2, \dots, S_n$. Then we can use induction on n and Lemma 2.5.16. ■

Now let us consider parameters of substitution instances. We begin with a simple remark that a strict substitution instance of a formula A may be not a sentence if A is not a sentence.

Intuitively it is clear that free variable occurrences in a substitution instance $[C/P(\mathbf{x})]A$ may be of three kinds:

(1) those derived from original free occurrences in A if they occur in atoms not containing P (and thus not affected by the substitution);

(2) members of $\mathbf{y}$ in subformulas of the form $[\mathbf{y}/\mathbf{x}]C$ replacing occurrences of $P(\mathbf{y})$ in A;

(3) those produced by parameters of the substitution wherever $P(\mathbf{y})$ is replaced with $[\mathbf{y}/\mathbf{x}]C$.

Parameters of the first two types are called essential. Here is a precise definition for an arbitrary substitution.

Definition 2.5.22 *A parameter $z \in FV(A)$ is called* essential *for a formula substitution $[C/P(\mathbf{x})]$ if one of the following conditions holds:*

(1) there exists a free occurrence of z in A within an atomic subformula that does not contain P;

(2) there exists a free occurrence of z in A as some y_j within an occurrence of $P(\mathbf{y})$, where $\mathbf{y} = y_1 \dots y_n$ and $x_j \in FV(C)$.

The set of all essential parameters of A for S is denoted by $FVe(S, A)$. Now let us prove the above observation on parameters of SA in more detail.

Lemma 2.5.23 *Let A be a formula, $S = [C/P(\mathbf{x})]$ a simple formula substitution. Then*

$$FV(SA) = FV(S) \cup FVe(S, A) \text{ if } P \text{ occurs in } A$$
$$FV(SA) = FVe(S, A) \text{ otherwise.}$$

Proof The second claim obviously follows from 2.5.22(1). To prove the first, we argue by induction. We may assume that A is clean, $BV(A) \cap FV(S) = \varnothing$.

- For atomic A there are two cases.

  (1) $A = P(\mathbf{y})$. Then $SA \stackrel{\circ}{=} [\mathbf{y}/\mathbf{x}]C$ and by Lemma 2.3.25(1), $FV(SA) = FV(S) \cup \text{rng}[\mathbf{x} \mapsto \mathbf{y}]_C$. By definition, $\text{rng}[\mathbf{x} \mapsto \mathbf{y}]_C = FVe(S, A)$ in this case, cf. 2.5.22(2).

(2) A does not contain P. Then $FV(A) = FVe(S, A)$.

- For $A = \Box_i B$ we have $SA = \Box_i SB$ and thus $FV(SA) = FV(SB)$. Since P occurs in A iff it occurs in B and $FVe(S, A) = FVe(S, B)$, the claim follow readily.

- For $A = B * D$, where $*$ is a binary connective, the proof is similar to the previous case; note that $FVe(S, B * D) = FVe(S, B) \cup FVe(S, D)$.

- For $A = QuB$ we have
  $FV(SA) = FV(SB) - \{u\}$. By induction hypothesis,

$$FV(SB) = FV(S) \cup FVe(S, B),$$

since P occurs in B. So it remains to show that

$$FVe(S, A) = FVe(S, B) - \{u\}.$$

In fact,

(1) z has a free occurrence in QuB within an atom that does not contain P iff $z \neq u$ and z has the same kind of occurrence in B;

(2) z has a free occurrence in QuB within $P(\mathbf{y})$ as described in 2.5.22 (2) iff $z \neq u$ and z has the same kind of occurrence in B.

■

From the previous lemma we obtain

Proposition 2.5.24 *Let A be a formula, $S = [C/P(\mathbf{x})]$ a simple formula substitution such that P occurs in A. Then*

(1) $FV(S) \subseteq FV(SA) \subseteq FV(S) \cup FV(A)$,

(2) $FV(SA) = FV(A)$ *if S is strict,*

(3) for any subformula B of A, $FV(SB) \subseteq FV(SA) \cup BV(A)$.

Proof

(1) Note that $FVe(S, A) \subseteq FV(A)$.

(2) Follows from (1).

(3) By 2.5.23,

$$FV(SB) = FV(S) \cup FVe(S, B) \subseteq$$
$$FV(S) \cup FVe(S, A) \cup (FVe(S, B) - FVe(S, A)) =$$
$$FV(SA) \cup (FVe(S, B) - FVe(S, A)).$$

Now note that according to Definition 2.5.22, the set $FVe(S,B) - FVe(S,A)$ contians only variables that are free in B, but not free in A, so this set is contained in $BV(A)$. Hence (3) follows. ∎

Remark 2.5.25 The reader can try to prove this proposition directly without using Lemma 2.5.23. This does not seem easier.

Definition 2.5.26 *For a set of formulas* $\Gamma \subseteq MF_N^{(=)}$ *(respectively,* $IF^{(=)}$*), its* substitution closure *is the set of all their substitution instances of the corresponding kind:*

$$\mathrm{Sub}(\Gamma) := \{SA \mid A \in \Gamma,\ S \textit{ is an } MF_N^{(=)}\text{- } (IF^{(=)}\text{-}) \textit{ formula substitution}\}.$$

The universal substitution closure *of* Γ *is the set* $\overline{\mathrm{Sub}}(\Gamma)$ *of all universal closures*[16] *of formulas from* $\mathrm{Sub}(\Gamma)$.

Since every N-modal formula is also N'-modal for $N' > N$, there is some ambiguity in this definition. But usually it is clear from the context, what kind of formulas we consider.

Lemma 2.5.27

(1) $\mathrm{Sub}(\mathrm{Sub}(\Gamma)) = \mathrm{Sub}(\Gamma)$

(2) $\overline{\mathrm{Sub}}(\overline{\mathrm{Sub}}(\Gamma)) \mathrel{\mathring{=}} \overline{\mathrm{Sub}}(\Gamma)$ *for a set of sentences* Γ *(where* $\mathrel{\mathring{=}}$ *means that these sets are the same up to congruence).*

Proof

(1) Every $B \in \mathrm{Sub}(\Gamma)$ has the form SA for some $A \in \Gamma$ and formula substitution S. Then for any formula substitution S_1, $S_1SA \in \mathrm{Sub}(\Gamma)$ by Lemma 2.5.21.

(2) Let us show that for any $B \in \overline{\mathrm{Sub}}(\Gamma)$ and for any substitution $S_1, \overline{\forall} S_1 B$ is congruent to a formula from $\overline{\mathrm{Sub}}(\Gamma)$. We have

$$B = \forall \mathbf{z} SA$$

for some $A \in \Gamma$, substitution S and $r(\mathbf{z}) = FV(SA)$. We may also assume that $FV(S) \cap BV(A) = \varnothing$ (otherwise we replace A with a congruent formula). Since A is a sentence, we have $FV(SA) = FV(S)$ by 2.5.23.

Now let $\mathbf{y}$ be a distinct list of new variables such that $|\mathbf{y}| = |\mathbf{z}|$ and $r(\mathbf{y}) \cap FV(S_1) = \varnothing$. Then

$$\forall \mathbf{z} SA \mathrel{\mathring{=}} \forall \mathbf{y} [\mathbf{y}/\mathbf{z}] SA$$

by 2.3.25(13), and so by 2.5.12

$$(*) \qquad S_1 B = S_1 \forall \mathbf{z} SA \mathrel{\mathring{=}} S_1 \forall \mathbf{y} [\mathbf{y}/\mathbf{z}] SA.$$

[16] Cf. Definition 2.2.5.

Now by 2.5.16

$$[\mathbf{y}/\mathbf{z}]SA \doteq S_2 A$$

for some formula substitution S_2 (note that the condition $r(\mathbf{z}) \cap BV(S) = \varnothing$ holds, since $r(\mathbf{z}) = FV(S)$).

Hence by 2.3.17(2) and 2.5.12

$$(**) \qquad S_1 \forall \mathbf{y} [\mathbf{y}/\mathbf{z}]SA \doteq S_1 \forall \mathbf{y} S_2 A.$$

Since $r(\mathbf{y}) \cap FV(S_1) = \varnothing$, from 2.5.13(3) it follows that

$$(***) \qquad S_1 \forall \mathbf{y} S_2 A \doteq \forall y S_1 S_2 A.$$

Eventually, by $(*), (**), (***)$ we obtain

$$\overline{\forall} S_1 B \doteq \overline{\forall} \forall \mathbf{y} S_1 S_2 A,$$

and the latter formula is in $\overline{\mathrm{Sub}}(\Gamma)$, by 2.5.21.

■

Now let us define 'minimal' non-strict substitution instances of predicate formulas.

Let $P_1, \ldots, P_k$ be all predicate letters (besides equality) occurring in a formula A, $P_i \in PL^{n_i}$, and put

$$\mathbf{P} := P_1(\mathbf{x}_1) \ldots P_k(\mathbf{x}_k),$$

where every $\mathbf{x}_i$ is a distinct list of variables of length n_i. Next, let $m \geq 0$, and let P_i' be different $(m + n_i)$-ary predicate letters $(i = 1, \ldots, k)$, $\mathbf{z} = z_1 \ldots z_m$ a distinct list of new variables for A. Then we call P_i' the *m-shift* of P_i; an *m-shift* of the formula A is $A^m \doteq [\mathbf{C}/\mathbf{P}]A$, where

$$\mathbf{C} = P_1'(\mathbf{x}_1, \mathbf{z}) \ldots P_k'(\mathbf{x}_k, \mathbf{z}).$$

We also put $A^0 := A$. Obviously $FV(A^m) = FV(A) \cup r(\mathbf{z})$ if A is not purely equational, i.e. it contains some predicate letters other than '='; for purely equational A, $A^m = A$. Note that A^m is a substitution instance of any A^n; this substitution instance is strict iff $m \leq n$.

Lemma 2.5.28 *Let $S = [\mathbf{C}/\mathbf{P}(\mathbf{x})]$ be a formula substitution. Then for any formula A and $m \geq 0$*

$$(SA)^m \doteq [\mathbf{C}^m/\mathbf{P}'(\mathbf{x}, \mathbf{z})]A^m,$$

where we assume that the list of extra parameters $\mathbf{z}$ of A^m, C_i^m is disjoint with $\mathbf{x}$.

Proof By definition, for a certain substitution S_1,

$$(SA)^m \stackrel{\circ}{=} S_1SA,\ A^m \stackrel{\circ}{=} S_1A,$$

and $FV(S_1) = r(\mathbf{z})$. So, as before, we have to check the claim only for atomic A (without equality).

If $A = P_i(\mathbf{y})$, then (for some $\mathbf{z}$)

$$(SA)^m \stackrel{\circ}{=} ([\mathbf{y}/\mathbf{x}]C_i)^m \stackrel{\circ}{=} S_1[\mathbf{y}/\mathbf{x}]C_i,$$

$$[\mathbf{C}^m/\mathbf{P}'(\mathbf{x},\mathbf{z})]A^m \stackrel{\circ}{=} [\mathbf{C}^m/\mathbf{P}'(\mathbf{x},\mathbf{z})]P_i'(\mathbf{y},\mathbf{z}) \stackrel{\circ}{=} [\mathbf{yz}/\mathbf{xz}]C_i^m \stackrel{\circ}{=} [\mathbf{y}/\mathbf{x}]S_1C_i.$$

By our assumption, $FV(S_1) = r(\mathbf{z})$ is disjoint with $\mathbf{x}$. So the claim follows by Lemma 2.5.15. ■

Exercise 2.5.29 Deduce 2.5.28 from 2.5.20.

Lemma 2.5.30 *Every substitution instance SA of a formula A is obtained by a variable renaming from a strict substitution instance of A^m for some $m \geq 0$, e.g. for $m = |FV(S)|$.*

Proof Let us first show this for a simple substitution $S = [C(\mathbf{x},\mathbf{y})/P(\mathbf{x})]$ and a formula A containing P. Let $P_1, \ldots, P_k$ be a list of all other predicate letters occurring in A, and let $P_1', \ldots, P_k'$ be their m-shifts, where $m = |\mathbf{y}|$. Next, let $\mathbf{z}$ be distinct list of new variables. Then

$$[C(\mathbf{x},\mathbf{z})/P(\mathbf{x})]A \stackrel{\circ}{=}$$
$$[P_1(\mathbf{x}_1)\ldots P_k(\mathbf{x}_k)/P_1'(\mathbf{x}_1,\mathbf{z})\ldots P_k'(\mathbf{x}_k,\mathbf{z})][C(\mathbf{x},\mathbf{z})/P'(\mathbf{x},\mathbf{z})]A^m,$$

where every $P_i(\mathbf{x}_i)$ is an atomic formula with distinct $\mathbf{x}_i$.

In fact, by Lemma 2.5.20

$$[C(\mathbf{x},\mathbf{z})/P'(\mathbf{x},\mathbf{z})]A^m \stackrel{\circ}{=}$$
$$[C(\mathbf{x},\mathbf{z})/P'(\mathbf{x},\mathbf{z})][P'(\mathbf{x},\mathbf{z}),P_1'(\mathbf{x}_1,\mathbf{z}),\ldots,P_k'(\mathbf{x}_k,\mathbf{z})/P(\mathbf{x}),P_1(\mathbf{x}_1),\ldots,P_k(\mathbf{x}_k)]A \stackrel{\circ}{=}$$
$$[C(\mathbf{x},\mathbf{z}),P_1'(\mathbf{x}_1,\mathbf{z}),\ldots,P_k'(\mathbf{x}_k,\mathbf{z})/P(\mathbf{x}),P_1(\mathbf{x}_1),\ldots,P_k(\mathbf{x}_k)]A,$$

hence by the same lemma

$$[P_1(\mathbf{x}_1)\ldots P_k(\mathbf{x}_k)/P_1'(\mathbf{x}_1,\mathbf{z})\ldots P_k'(\mathbf{x}_k,\mathbf{z})][C(\mathbf{x},\mathbf{z})/P'(\mathbf{x},\mathbf{z})]A^m$$
$$\stackrel{\circ}{=} [C(\mathbf{x},\mathbf{z}),P_1(\mathbf{x}_1),\ldots,P_k(\mathbf{x}_k)/P(\mathbf{x}),P_1(\mathbf{x}_1),\ldots,P_k(\mathbf{x}_k)]A,$$

and the latter formula is (congruent to) $[C(\mathbf{x},\mathbf{z})/P(\mathbf{x})]A$. So $[C(\mathbf{x},\mathbf{z})/P(\mathbf{x})]A$ is a strict substitution instance of A^m. Since by 2.5.9(2)

$$[C(\mathbf{x},\mathbf{y})/P(\mathbf{x})]A \stackrel{\circ}{=} [\mathbf{y}/\mathbf{z}][C(\mathbf{x},\mathbf{z})/P(\mathbf{x})]A,$$

this proves our claim.

Now we can apply induction. As we know, every complex substitution is a composition of simple substitutions. So it is sufficient to show that applying a simple substitution S to a formula $[\mathbf{y}/\mathbf{z}]S_0A^m$, where S_0 is strict, can also be presented in this form.

Note that we may assume that $r(\mathbf{z}) \cap FV(S) = \varnothing$ — otherwise change the list of extra parameters of A^m. So by 2.5.15 we have

$$S[\mathbf{y}/\mathbf{z}]S_0A^m \stackrel{\circ}{=} [\mathbf{y}/\mathbf{z}]SS_0A^m$$

As we have already proved, for $B \stackrel{\circ}{=} S_0A^m$

$$SB \stackrel{\circ}{=} [\mathbf{u}/\mathbf{t}]S_1B^k$$

for some k, strict substitution S_1 and variable renaming $[\mathbf{t} \mapsto \mathbf{u}]$. By 2.5.28,

$$B^k \stackrel{\circ}{=} (S_0A^m)^k \stackrel{\circ}{=} S_2(A^m)^k \stackrel{\circ}{=} S_2A^{m+k},$$

where S_2 is a strict substitution. Thus

$$[\mathbf{y}/\mathbf{z}]SB \stackrel{\circ}{=} [\mathbf{y}/\mathbf{z}][\mathbf{u}/\mathbf{t}]S_1S_2A^{m+k},$$

and S_1S_2 is strict as required. ∎

Let $\overline{A^m}$ be a universal closure of A^m (for $m \geq 0$); thus $\overline{A^m} = \forall z_1 \dots \forall z_m A^m$ for a sentence A.

For a set of formulas Γ, let $\mathrm{Sub}(\Gamma)$ be the set of all their substitution instances, $\overline{\mathrm{Sub}}(\Gamma)$ the set of all universal closures of formulas from $\mathrm{Sub}(\Gamma)$. Both these sets are closed under congruence if Γ is a set of sentences.

2.6 First-order logics

Definition 2.6.1 *An (N-)*modal predicate logic *(m.p.l.) is a set $L \subseteq MF_N$ such that*

(m0) L contains classical propositional tautologies;

(m1) L contains the propositional axioms

$$AK_i := \Box_i(p \supset q) \supset (\Box_i p \supset \Box_i q).$$

(m2) L contains the predicate axioms (for some fixed P, q and arbitrary x, y):

(Ax12) $\forall x P(x) \supset P(y)$;
(Ax13) $P(y) \supset \exists x P(x)$;
(Ax14) $\forall x(q \supset P(x)) \supset (q \supset \forall x P(x))$;
(Ax15) $\forall x(P(x) \supset q) \supset (\exists x P(x) \supset q)$;

(m3) L is closed under the rules

$$\frac{A,\ (A \supset B)}{B} \quad \text{(Modus Ponens, or MP)};$$

$$\frac{A}{\Box_i A} \quad \textit{(Necessitation, or } \Box\textit{-introduction);}$$

$$\frac{A}{\forall x A} \quad \textit{(Generalisation, or } \forall\textit{-introduction)}$$

(for any $x \in Var$).

(m4) L is closed under MF_N-substitutions.

Definition 2.6.2 *An (N-)*modal predicate logic with equality *(m.p.l.=) is a set $L \subseteq MF_N^=$ satisfying* (m0)–(m3) *from 2.6.1 and also*

(m4$^=$) *L is closed under $MF_N^=$-substitutions;*

(m5$^=$) *L contains the axioms of equality (for arbitrary x, y and fixed P):*

$$\begin{array}{ll} \text{(Ax16)} & x = x, \\ \text{(Ax17)} & x = y \supset (P(x) \supset P(y)). \end{array}$$

Definition 2.6.3 *A* superintuitionistic predicate logic *(s.p.l.) is a set $L \subseteq IF$ such that*

(*s*1) *L contains the axioms of Heyting's propositional calculus* **H** *(cf. Section 1.1.2);*

(*s*2) = (*m*2) *L contains the predicate axioms;*

(*s*3) *L is closed under the rules (MP), $\forall$-introduction, see (m3);*

(*s*4) *L is closed under IF-substitutions.*

Definition 2.6.4 *A* superintuitionistic predicate logic with equality *(s.p.l.=) is a set $L \subseteq IF^=$ satisfying* (*s*1)–(*s*3) *from 2.6.3 and*

(*s*4$^=$) *L is closed under $IF^=$-substitutions;*

(*s*5$^=$) = (*m*5$^=$) *L contains the axioms of equality.*

Further, by a 'first-order logic' we mean an arbitrary logic, modal or superintuitionistic, with or without equality.

Elements of a logic are called *theorems*, and we often write $L \vdash A$ instead of $A \in L$.

Definition 2.6.5 *A logic L (modal or superintuitionistic) is called* consistent *if $\bot \notin L$.*

$\mathcal{M}_N$ (respectively $\mathcal{M}_N^=$, $\mathcal{S}$, $\mathcal{S}^=$) denotes the set of all N-m.p.l. (respectively, N-m.p.l.=; s.p.l.; s.p.l.=). The smallest N-m.p.l. (respectively, N-m.p.l.=, s.p.l., s.p.l.=) is denoted by $\mathbf{QK}_N$ (respectively, by $\mathbf{QK}_N^=$, $\mathbf{QH}$, $\mathbf{QH}^=$). $L + \Gamma$ denotes the smallest m.p.l. containing an m.p.l. L and a set $\Gamma \subseteq MF$. This notation is obviously extended to other cases (m.p.l.=, s.p.l., s.p.l.=).

It is well-known that every theorem of $\mathbf{QH}^{(=)}$ can be obtained by a formal *proof*, which is a sequence of formulas that are either substitution instances of axioms or are obtained from earlier formulas by applying inference rules cited in (*s*3). The same is true for $\mathbf{QK}_N^{(=)}$, but with the rules from (m3). The notion of a formal proof extends to logics of the form $\mathbf{QH}^{(=)} + \Gamma$, $\mathbf{QK}_N^{(=)} + \Gamma$, with the only difference that formulas from Γ can also be used as axioms. By applying deduction theorems, we can reduce the provability in $L + \Gamma$ to provability in L in a more explicit way, see Section 2.8 below.

Definition 2.6.6 *The* quantified version *of a modal (respectively, superintuitionistic) propositional logic* $\mathbf{\Lambda}$ *is*

$$\mathbf{Q\Lambda} := \mathbf{QK}_N + \mathbf{\Lambda} \quad \text{(respectively, } \mathbf{Q\Lambda} := \mathbf{QH} + \mathbf{\Lambda}\text{)}.$$

Definition 2.6.7 *The* propositional part *of a predicate logic L is the set of its propositional formulas:*
$L_\pi := L \cap \mathcal{L}_N$ *(for an N-modal L);*
$L_\pi := L \cap \mathcal{L}_0$ *(for a superintuitionistic L).*

The following is obvious.

Lemma 2.6.8

(1) If L is an N-m.p.l. or an s.p.l., then L_π is a propositional logic of the corresponding kind.

(2) If L is a predicate logic with equality, then $L_\pi = (L^0)_\pi$.

A well-known example of an s.p.l. is the *classical predicate logic*

$$\mathbf{QCL}^{(=)} = \mathbf{Q(CL)}^{(=)} = \mathbf{QH}^{(=)} + EM,$$

where $EM = p \vee \neg p$ (see Section 1.1). An s.p.l.(=) L is called *intermediate* iff $L \subseteq \mathbf{QCL}^{(=)}$. Note that $\mathbf{QCL}^{(=)}$ is included in $\mathbf{QK}_N^{(=)}$ (and thus, in any m.p.l.(=)).

The rule $(m4)$ means that together with a formula A, L contains all its MF_N-substitution instances (and similarly for $(s4)$). In particular, L contains every formula congruent to A, because it is a substitution instance under the dummy substitution. Hence we easily obtain

Lemma 2.6.9 *If $A \doteq B$ then $(A \equiv B) \in L$ (for any m.p.l.(=) or s.p.l.(=) L).*

Proof $A \doteq B$ implies $(A \equiv B) \doteq (A \equiv A)$, and $(A \equiv A) = [A/p](p \equiv p)$, thus $(A \equiv A) \in L$ by $(m0)$, $(m4)$ (or $(s0)$, $(s4)$). Hence $(A \equiv B) \in L$. ∎

Lemma 2.6.10 *Let A, B be formulas in the language of a predicate logic L. Then for a variable $x \notin FV(B)$:*

(1) $L \vdash \forall x(B \supset A) \supset_\bullet B \supset \forall x A$,

(2) $L \vdash \forall x(A \supset B) \supset_\bullet \exists x A \supset B$.

Proof (1) Consider the substitution $S = [A, B/P(x), q]$; note that $x \notin FV(S)$. By Lemma 2.5.13, up to congruence, S distributes over $\supset$ and $\forall x$ (since $x \notin FV(S)$). Congruence also distributes over $\supset$ and $\forall x$, by 2.3.17. Thus

$$S(\text{Ax14}) \mathrel{\mathring{=}} \forall x(B \supset A) \supset_\bullet B \supset \forall x A,$$

and so the latter formula is in L.

The proof of (2) is similar. ■

Definition 2.6.11 *Let Γ be a set of formulas in the language of a predicate logic L. An* L-inference *of a formula B from Γ a sequence $A_1, \ldots, A_n$, in which $A_n = B$ and every A_i is either a theorem of L, or $A_i \in \Gamma$, or A_i is obtained from earlier formulas by applying MP, or A_i is obtained from an earlier formula by $\forall$-introduction over a variable that is not a parameter of any formula from Γ. If such an inference exists, we say that B is* L-derivable from Γ, *notation: $\Gamma \vdash_L B$.*

Note that we distinguish *inferences* from *proofs*; the latter may also use substitution and $\Box$-introduction.

From definitions we easily obtain

Lemma 2.6.12 $\vdash_L A$ *iff* $L \vdash A$.

Proof 'If'. If $L \vdash A$, then A is an L-inference (from $\varnothing$).
'Only if'. By induction on the length of an L-inference of A from $\varnothing$. ■

Recall the simplest first-order analogue of the propositional deduction theorem:

Lemma 2.6.13 *If* $\Gamma \cup \{A\} \vdash_L B$, *then* $\Gamma \vdash_L A \supset B$.

Proof Standard, by induction on the length of an inference of B from $\Gamma \cup \{A\}$.

(i) If $B \in L \cup \Gamma$, then $A \supset B$ follows by MP from B and $B \supset (A \supset B)$, which is a substitution instance of (Ax1).

(ii) If B is obtained by MP from C and $C \supset B$ and by the induction hypothesis $\Gamma \vdash_L A \supset C$, $A \supset_\bullet C \supset B$, note that

$$\Gamma \vdash_L (A \supset_\bullet C \supset B) \supset (A \supset C_\bullet \supset_\bullet A \supset B),$$

from a tautology (or an intuitionistic axiom (Ax2)); hence $\Gamma \vdash_L A \supset B$ by MP.

(iii) Suppose $B = \forall x C$, $\Gamma \vdash_L A \supset C$ by induction hypothesis and x is not a parameter in $\Gamma \cup \{A\}$, then $\Gamma \vdash_L \forall x(A \supset C)$. By Lemma 2.6.10, $L \vdash \forall x(A \supset C) \supset (A \supset B)$, therefore $\Gamma \vdash_L A \supset B$ by MP.

(iv) If $B = A$, then $(A \supset B) = (A \supset A)$, which is L-derivable by a standard argument; see any textbook in mathematical logic. ■

Hence we obtain an equivalent characterisation of L-derivability.

Lemma 2.6.14 *Let Γ be a set of N-modal (or intuitionistic) predicate formulas, L an N-modal (or superintuitionistic) predicate logic (with or without equality). Then for any N-modal (or intuitionistic) formula B, $\Gamma \vdash_L B$ iff there exists a finite $X \subseteq \Gamma$ such that*

$$L \vdash \bigwedge X \supset B.$$

As usual, we also include the case $X = \varnothing$, with $\top$ as the empty conjunction.

Of course the notation $\bigwedge X$ makes sense, due to the commutativity and the associativity of conjunction in intuitionistic logic.

Proof Since every inference from Γ contains a finite number of formulas from Γ, it is clear that $\Gamma \vdash_L B$ iff there exists a finite $X \subseteq \Gamma$ such that $X \vdash_L B$.

So we have to show that

$$(1) \qquad X \vdash_L B \text{ iff } L \vdash \bigwedge X \supset B.$$

The proof is by induction on $|X|$.

If $X = \varnothing$, then $\vdash_L B$ iff $L \vdash B$ by 2.6.12.

But $L \vdash B \supset_\bullet \top \supset B$ (this is an instance of (Ax1)), so by MP, $L \vdash B$ implies $L \vdash \top \supset B$.

The other way round, $L \vdash \top \supset B$ implies $L \vdash B$, since $L \vdash \top$. Therefore $L \vdash B$ iff $L \vdash \top \supset B$.

Suppose (1) holds for X (and any B). Then it also holds for $X \cup \{A\}$.

In fact, by 2.6.13 and our assumption

$$X \cup \{A\} \vdash B \text{ iff } X \vdash_L A \supset B \text{ iff } L \vdash \bigwedge X \supset_\bullet A \supset B.$$

The latter is equivalent to

$$L \vdash (\bigwedge X) \wedge A \supset B,$$

due to

$$(2) \qquad \mathbf{H} \vdash (p \supset_\bullet q \supset r) \equiv (p \wedge q \supset r).$$

(2) follows in a standard way by the deduction theorem from

$$p \supset_\bullet q \supset r,\ p \wedge q \vdash_{\mathbf{H}} r$$

and

$$p \wedge q \supset r,\ p,\ q \vdash_{\mathbf{H}} r.$$

■

The next lemmas collects some useful theorems and admissible rules for different types of logics.

Lemma 2.6.15 *The following theorems (admissible rules) are in every first-order logic L:*

(i) Bernays rules:

$$\frac{B \supset A}{B \supset \forall x A}, \qquad \frac{A \supset B}{\exists x A \supset B}$$

if $x \notin FV(B)$;

(ii)

$$\begin{array}{ll} \forall x A \supset [y/x]A, & \forall \mathbf{x} A \supset A, \\ [y/x]A \supset \exists x A, & A \supset \exists \mathbf{x} A; \end{array}$$

(iii) variable substitution rule:

$$\frac{A}{[\mathbf{y}/\mathbf{x}]A};$$

(iv) $\forall x(A \supset B) \supset (QxA \supset QxB)$;

(v) monotonicity rules for quantifiers

$$\frac{A \supset B}{QxA \supset QxB};$$

(vi) replacement rules for quantifiers

$$\frac{A \equiv B}{QxA \equiv QxB};$$

(vii) $\forall x(A \wedge B) \equiv \forall x A \wedge \forall x B$;

(viii) $\exists x(A \vee B) \equiv \exists x A \vee \exists x B$;

(ix) $\forall x A \equiv A$ *if* $x \notin FV(A)$;

(x) $\exists x A \equiv A$ *if* $x \notin FV(A)$;

(xi) $\forall x\,(C \supset A) \equiv (C \supset \forall x A)$ *if* $x \notin FV(C)$;

(xii) $\forall x\,(A \supset C) \equiv (\exists x A \supset C)$ *if* $x \notin FV(C)$;

(xiii) $\forall x\, \neg A \equiv \neg \exists x\, A$;

(xiv) $\exists x\,(C \supset A) \supset (C \supset \exists x A)$ *if* $x \notin FV(C)$;

(xv) $\exists x\,(A \supset C) \supset (\forall x A \supset C)$ *if* $x \notin FV(C)$;

(xvi) $\exists x\, \neg A \supset \neg \forall x A$;

(xvii) $\exists x(A \vee C) \equiv \exists x A \vee C$ *if* $x \notin FV(C)$;

(xviii) $\mathcal{Q}x(A \wedge C) \equiv \mathcal{Q}xA \wedge C$, *if* $x \notin FV(C)$, $\mathcal{Q} \in \{\forall, \exists\}$;

(xix) $\exists x(A \wedge B) \supset \exists xA \wedge \exists xB$;

(xx) $\forall xA \vee \forall xB \supset \forall x(A \vee B)$;

(xxi) $\forall xA \vee C \supset \forall x(A \vee C)$ *if* $x \notin FV(C)$;

(xxii) $\mathcal{Q}x\mathcal{Q}yA \equiv \mathcal{Q}y\mathcal{Q}xA$ *for* $\mathcal{Q} \in \{\forall, \exists\}$;

(xxiii) $\mathcal{Q}\mathbf{x}A \equiv \mathcal{Q}(\mathbf{x} \cdot \sigma)A$ *for a quantifier* $\mathcal{Q}$, *a distinct list* $\mathbf{x}$ *and a permutation* σ *of* I_n, *where* $n = |\mathbf{x}|$;

(xxiv) $\exists x\forall yA \supset \forall y\exists xA$;

(xxv) $\forall\mathbf{x}A \supset [\mathbf{y}/\mathbf{x}]A$ *for a variable substitution* $[\mathbf{y}/\mathbf{x}]$;

(xxvi) $\forall x(A \equiv B) \supset (\mathcal{Q}xA \equiv \mathcal{Q}xB)$;

(xxvii) $\overline{\forall}(A \equiv A') \supset \overline{\forall}([A/P(\mathbf{x})]B \equiv [A'/P(\mathbf{x})]B)$,
if B *is non-modal (moreover, if* $P(\mathbf{x})$ *is not within the scope of modal operators in* B*)*;

(xxviii) $$\frac{A \equiv A'}{[A/P(\mathbf{x})]B \equiv [A'/P(\mathbf{x})]B} \quad \textit{(replacement rule)}$$

with the same restriction as in (xxvii).

So (xxviii) shows that up to equivalence, the universal closure $\overline{\forall}A$ does not depend on the order of quantifiers.

Similarly to the propositional case (Section 1.1), the replacement rule (xxviii) can be written as follows:

$$\frac{A \equiv A'}{B(\ldots A \ldots) \equiv B(\ldots A' \ldots)}.$$

Proof

(i) Readily follows from 2.6.10.

(ii) By Lemma 2.5.13 we obtain

$$[A/P(x)](\forall xP(x) \supset P(y)) \doteq \forall xA \supset [y/x]A$$

(note that $x \notin FV[A/P(x)]$). So since L contains (Ax12), it also contains (ii).

The particular case of this is $\forall xA \supset A$. Hence $\forall\mathbf{x}A \supset A$ easily follows by induction on $|\mathbf{x}|$ and the trnasitivity of $\supset$.

The dual claims for $\exists$ are proved in a similar way.

(iii) If $L \vdash A$, then $L \vdash \forall x A$. Since $L \vdash \forall x A \supset [y/x]A$ by (ii), we obtain $L \vdash [y/x]A$ by MP. Therefore L is closed under variable substitution, since every variable substitution is a composition of simple substitutions.

(iv) By the deduction theorem, it is sufficient to show

$$\forall x(A \supset B) \vdash_L QxA \supset QxB.$$

First consider the case $Q = \forall$. We have the following 'abridged' L-inference from $\forall x(A \supset B)$:

1. $\forall x(A \supset B) \supset_{\bullet} A \supset B$ by (ii)
2. $\forall x(A \supset B)$ by assumption
3. $A \supset B$ by 1,2, MP
4. $\forall x A \supset A$ by (ii)
5. $\forall x A \supset B$ by 3, 4, transitivity
6. $\forall x A \supset \forall x B$ by 5, (i).

Here we apply the transitivity rule and the Bernays rule to L-derivability from Γ; the reader can easily see that they are really admissible in this situation.

For the case $Q = \exists$ the argument slightly changes in items 4–6.

4. $B \supset \exists x B$ by (ii)
5. $A \supset \exists x B$ by 3, 4, transitivity
6. $\exists x A \supset \exists x B$ by 5, (i).

(v) If $L \vdash A \supset B$, then $L \vdash \forall x(A \supset B)$ by generalisation. Since $L \vdash \forall x(A \supset B) \supset_{\bullet} QxA \supset QxB$ by (iv), we obtain $L \vdash QxA \supset QxB$ by MP.

(vi) If $L \vdash A \equiv B$, then $L \vdash A \supset B,\ B \supset A$ by (Ax3), (Ax4)[17] and MP. Hence $L \vdash QxA \supset QxB,\ QxB \supset QxA$ by (v), and thus $L \vdash QxA \equiv QxB$ by $\wedge$-introduction $\dfrac{C,\ D}{C \wedge D}$, which is admissible in L.

(vii) Since $L \vdash A \wedge B \supset A$ by (Ax3) and substitution, it follows that $L \vdash \forall x(A \wedge B) \supset \forall x A$, by (v), and thus $\forall x(A \wedge B) \vdash_L \forall x A$.

Similarly from (Ax4) we obtain

$$\forall x(A \wedge B) \vdash_L \forall x B;$$

hence

$$\forall x(A \wedge B) \vdash_L \forall x A \wedge \forall x B,$$

[17] In the modal case we may use (Ax3), (Ax4) as classical tautologies.

by $\wedge$-introduction, and therefore

$$L \vdash \forall x(A \wedge B) \supset \forall xA \wedge \forall xB,$$

by the deduction theorem.

To show the converse we may also use the deduction theorem. In fact, we have the following abridged inference from $\forall xA \wedge \forall xB$:

1. $\forall xA \wedge \forall xB$ by assumption
2. $\forall xA \wedge \forall xB \supset \forall xA$ by (Ax3), substitution
3. $\forall xA$ by 1,2, MP
4. $\forall xA \supset A$ by (ii)
5. A by 3,4, MP.

A similar argument shows $\forall xA \wedge \forall xB \vdash_L B$.

Hence $\forall xA \wedge \forall xB \vdash_L A \wedge B$, by $\wedge$-introduction and therefore $\forall xA \wedge \forall xB \vdash_L \forall x(A \wedge B)$.

(viii) It is sufficient to show

$$L \vdash \exists xA \vee \exists xB \supset \exists x(A \vee B)$$

and

$$L \vdash \exists x(A \vee B) \supset \exists xA \vee \exists xB.$$

For the first, we can use the $\vee$-introduction rule:

$$\frac{X \supset Z,\ Y \supset Z}{X \vee Y \supset Z},$$

which is admissible in L, due to (Ax5).

So it remains to show

$$L \vdash \exists xA \supset \exists x(A \vee B),\ \exists xB \supset \exists x(A \vee B).$$

But these follow by (v) from $A \supset A \vee B,\ B \supset A \vee B$, which are substitution instances of (Ax6), (Ax7).

The converse $L \vdash \exists x(A \vee B) \supset \exists xA \vee \exists xB$ follows by the Bernays rule from

$$L \vdash A \vee B \supset \exists xA \vee \exists xB.$$

For the latter we can also use $\vee$-introduction after we show

$$L \vdash A \supset \exists xA \vee \exists xB,\ B \supset \exists xA \vee \exists xB.$$

But $A \supset \exists xA \vee \exists xB$ follows by transitivity from $A \supset \exists xA$ (ii) and $\exists xA \supset \exists xA \vee \exists xB$ (Ax6). The argument for $B \supset \exists xA \vee \exists xB$ is similar.

(ix) $L \vdash \forall xA \supset A$ by (ii). $L \vdash A \supset \forall xA$ follows by the Bernays rule from $L \vdash A \supset A$.

(x) The proof is similar to (ix).

(xi) We have $\forall x(C \supset A) \vdash_L C \supset \forall xA$ by Bernays' rule, and thus $L \vdash \forall x(C \supset A) \supset_\bullet C \supset \forall xA$ by Deduction theorem.

For the converse, first note that

$$C \supset \forall xA,\ C \vdash_L A$$

by the abridged inference

$$C,\ C \supset \forall xA,\ \forall xA,\ \forall xA \supset A,\ A,$$

hence

$$C \supset \forall xA \vdash_L C \supset A$$

by Deduction theorem, and thus

$$C \supset \forall xA \vdash_L \forall x(C \supset A),$$

since $x \notin FV(C)$. Therefore

$$L \vdash C \supset \forall xA_\bullet \supset \forall x(C \supset A).$$

(xii) Along the same lines as in (xi), using the second Bernays' rule and the theorem $A \supset \exists xA$. We leave the details to the reader.

(xiii) Readily follows from (xii), with $C = \bot$.

(xiv) By Deduction theorem, this reduces to $\exists xC \supset A, C \vdash_L \exists xA$. The latter follows by the abridged inference

$$C,\ C \supset \exists xC,\ \exists xC,\ \exists xC \supset A,\ A,\ A \supset \exists xA,\ \exists xA.$$

(xv) By the Bernays rule and deduction theorem from

$$A \supset C \vdash_L \forall xA \supset C.$$

By Deduction theorem, the latter reduces to

$$A \supset C, \forall xA \vdash_L C,$$

which we leave as an easy exercise for the reader.

(xvi) = (xiv) for $C = \bot$.

(xvii) By (x), $L \vdash \exists xC \equiv C$, so the admissible replacement rule

$$\frac{B_1 \equiv B_2}{A \lor B_1 \equiv A \lor B_2}$$

yields

$$L \vdash \exists xA \lor \exists xC \equiv \exists xA \lor C.$$

Since also

$$L \vdash \exists x(A \lor C) \equiv \exists xA \lor \exists xC$$

by (viii), and we obtain (xvi) by transitivity for $\equiv$.

(xviii) If $\mathcal{Q} = \forall$, the argument is similar to (xvi), using (ix), (vii), and the replacement rule

$$\frac{B_1 \equiv B_2}{A \land B_1 \equiv A \land B_2}.$$

Let $\mathcal{Q} = \exists$. Then

$$L \vdash \exists x(A \land C) \supset \exists xA \land C$$

follows from (xvii), (x), and the replacement rule

$$\frac{A_1 \supset A_2 \land B_1, \;\; B_1 \equiv B_2}{A_1 \supset A_2 \land B_2}$$

for $A_1 = \exists x(A \land C)$, $A_2 = \exists xA$, $B_1 = \exists xC$, $B_2 = C$.

Finally, to show

$$L \vdash \exists xA \land C \supset \exists x(A \land C)$$

we argue as follows. First we obtain

$$C \vdash_L A \supset A \land C$$

by the deduction theorem and $\land$-introduction. Hence

$$C \vdash_L \exists xA \supset \exists x(A \land C)$$

by (v); this rule is still admissible in L-inferences from C, since $\forall x$-introduction is admissible.

So by the deduction theorem,

$$\vdash_L C \supset_{\bullet} \exists xA \supset \exists x(A \land C).$$

The latter formula is equivalent to

$$\exists xA \land C \supset \exists x(A \land C).$$

In fact,

$$\exists xA \land C \vdash_L \exists x(A \land C),$$

since $\exists xA \land C \vdash_L C$ and $\exists xA \land C \vdash_L \exists xA$, and we may use $C \supset_{\bullet} \exists xA \supset \exists x(A \land C)$ and MP to obtain $\exists x(A \land C)$.

Therefore

$$\vdash_L \exists xA \land C \supset \exists x(A \land C).$$

(xix) The proof is similar to (vii). From (Ax3), (Ax4) by monotonicity we obtain

$$L \vdash \exists x(A \wedge B) \supset \exists x A,\ \exists x(A \wedge B) \supset \exists x B.$$

Hence $L \vdash \exists x(A \wedge B) \supset \exists x A \wedge \exists x B$ by $\wedge$-introduction and the deduction theorem.

(xx) The proof is similar to (viii).

First we note that

$$L \vdash \forall x A \supset A \vee B$$

by transitivity from $\forall x A \supset A,\ A \supset A \vee B$.

Similarly

$$L \vdash \forall x B \supset A \vee B.$$

Hence by $\vee$-introduction,

$$L \vdash \forall x A \vee \forall x B \supset A \vee B,$$

and (xviii) follows by the Bernays rule.

(xxi) Almost the same as (xx). Apply the Bernays rule to $\forall x A \vee C \supset A \vee C$.

(xxii) By (ii), $L \vdash \forall y A \supset A$; hence

$$L \vdash \forall x \forall y A \supset \forall x A$$

by monotonicity and

$$L \vdash \forall x \forall y A \supset \forall y \forall x A,$$

by the Bernays rule.

The converse is obtained in the same way.

The case of $\exists$ is similar.

(xxiii) Since σ is a composition of elementary transpositions, it is sufficient to consider $\sigma = \sigma^n_{i,i+1}$. So let $\mathbf{x} = \mathbf{y}x_i x_{i+1}\mathbf{z}$, then $\mathbf{x} \cdot \sigma = \mathbf{y}x_{i+1}x_i\mathbf{z}$. We have (in L)

$$\vdash \mathcal{Q}x_i x_{i+1} \mathcal{Q}\mathbf{z}A \equiv \mathcal{Q}x_{i+1}\mathcal{Q}x_i\mathcal{Q}\mathbf{z}A$$

by (xxii), hence

$$\mathcal{Q}\mathbf{y}\mathcal{Q}x_i\mathcal{Q}x_{i+1}\mathcal{Q}\mathbf{z}A \equiv \mathcal{Q}y\mathcal{Q}x_{i+1}\mathcal{Q}x_i\mathcal{Q}\mathbf{z}A$$

by (vi), i.e. we obtain

$$\mathcal{Q}\mathbf{x}A \equiv \mathcal{Q}(\mathbf{x} \cdot \sigma)A.$$

(xxiv) Since $L \vdash \forall y A \supset A$, we obtain $L \vdash \exists x \forall y A \supset \exists x A$ by monotonicity; hence $L \vdash \exists x \forall y A \supset \forall y \exists x A$ by the Bernays rule.

(xxv) First consider the case when $\mathbf{x} \cap \mathbf{y} = \varnothing$. We argue by induction on $n = |\mathbf{x}|$. The base $n = 1$ was proved in (ii).

Next, if $\mathbf{x} = x_1\mathbf{x}'$, $\mathbf{y} = y_1\mathbf{y}'$, and we know that $L \vdash \forall \mathbf{x}' A \supset [\mathbf{y}'/\mathbf{x}']A$, then by (v),

$$L \vdash \forall \mathbf{x}' A \supset \forall x_1 [\mathbf{y}'/\mathbf{x}']A.$$

By (v) again,

$$L \vdash \forall x_1 [\mathbf{y}'/\mathbf{x}']A \supset [y_1/x_1][\mathbf{y}'/\mathbf{x}']A,$$

hence

$$L \vdash \forall \mathbf{x} A \supset [y_1/x_1][\mathbf{y}'/\mathbf{x}']A,$$

by transitivity. Since $x_1 \notin \mathbf{y}'$, the conclusion is $[\mathbf{y}/\mathbf{x}]A$ as we need.

Now in the general case, let $\mathbf{z}$ be a distinct list of new variables, $|\mathbf{z}| = n$. Then as we have proved, $L \vdash \forall \mathbf{x} A \supset [\mathbf{z}/\mathbf{x}]A$, and thus

$$L \vdash \forall \mathbf{x} A \supset \forall \mathbf{z} [\mathbf{z}/\mathbf{x}]A$$

by the Bernays rule. We also have

$$L \vdash \forall \mathbf{z} [\mathbf{z}/\mathbf{x}]A \supset [\mathbf{y}/\mathbf{z}][\mathbf{z}/\mathbf{x}]A$$

from the above, so by transitivity

$$L \vdash \forall \mathbf{x} A \supset [\mathbf{y}/\mathbf{z}][\mathbf{z}/\mathbf{x}]A.$$

Since $[\mathbf{y}/\mathbf{z}][\mathbf{z}/\mathbf{x}]A \stackrel{\circ}{=} [\mathbf{y}/\mathbf{x}]A$, this completes the argument.

(xxvi) We have the following theorems in L:

1. $A \equiv B_\bullet \supset_\bullet A \supset B$ (Ax3)
2. $\forall x(A \equiv B) \supset_\bullet \forall x(A \supset B)$ 1, monotonicity (v)
3. $\forall x(A \supset B) \supset_\bullet QxA \supset QxB$ (iv)
4. $\forall x(A \equiv B) \supset_\bullet QxA \supset QxB$ 2, 3, transitivity.

Hence $\forall x(A \equiv B) \vdash_L QxA \supset QxB$. In the same way (using Ax4) we obtain

$$\forall x(A \equiv B) \vdash_L QxB \supset QxA.$$

Hence by propositional logic

$$\forall x(A \equiv B) \vdash_L QxA \equiv QxB,$$

which implies (xxvi) by the deduction theorem.

(xxvii) To simplify the notation, we write $B(A)$ instead of $[A/P(\mathbf{x})]B$. So we show

$$\overline{\forall}(A \equiv A') \vdash_L \overline{\forall}(B(A) \equiv B(A'))$$

by induction on the length of B and then apply the deduction theorem.

If $B = B_1 * B_2$ for a propositional connective $*$, and by the induction hypothesis we have

$$\overline{\forall}(A \equiv A') \vdash_L \overline{\forall}(B_1(A) \equiv B_1(A')),\ \overline{\forall}(B_2(A) \equiv B_2(A')),$$

hence we deduce (by (xxv))

$$B_1(A) \equiv B_1(A'),\ B_2(A) \equiv B_2(A').$$

Now we can apply the admissible propositional rule

$$\frac{A_1 \equiv A_1',\ A_2 \equiv A_2'}{A_1 * A_2 \equiv A_1' * A_2'}$$

and obtain $B(A) \equiv B(A')$. Since $\overline{\forall}(A \equiv A')$ is closed, $\overline{\forall}$-introduction is also applicable.

If $B = QyB_1$, we change it to a congruent formula so that $y \notin FV[A/P(\mathbf{x})]$. Then $B(A) = QyB_1(A)$, $B(A') = QyB_1(A')$. If by the induction hypothesis

$$\overline{\forall}(A \equiv A') \vdash_L B_1(A) \equiv B_1(A'),$$

then

$$\overline{\forall}(A \equiv A') \vdash_L \forall y(B_1(A) \equiv B_1(A')).$$

Hence we deduce

$$QyB_1(A) \equiv QyB_1(A')$$

by (xxvi) and MP and then apply generalisation.

In the case when $B = \Box_i B_1$ the propositional replacement rule (1.1.1) can be used, the details are left to the reader.

(xxviii) If $L \vdash A \equiv A'$, then by generalisation $L \vdash \overline{\forall}(A \equiv A')$. Hence by (xxvii) and MP, $L \vdash \overline{\forall}(B(A) \equiv B(A'))$. Now we can eliminate $\overline{\forall}$ by (xxv) and MP.

■

Lemma 2.6.16 *Theorems in logics with equality:*

(1) $x = y \supset y = x$;

(2) $x = y \wedge y = z \supset x = z$;

(3) $x = y \supset_\bullet [x/z]A \equiv [y/z]A$;

(4) $x_1 = y_1 \wedge \ldots \wedge x_n = y_n \supset_\bullet P(x_1, \ldots, x_n) \equiv P(y_1, \ldots, y_n)$.

Proof

(1) From (Ax17) by substitution $[x = z/P(x)]$ we obtain

$$L \vdash x = y \supset_{\bullet} x = z \supset y = z.$$

Hence by substitution $[x/z]$ (2.6.15 (iii))

$$L \vdash x = y \supset_{\bullet} x = x \supset y = x.$$

This implies $x = y \vdash_L y = x$ (due to (Ax16)), whence $L \vdash x = y \supset y = x$ by the deduction theorem.

(2) From (Ax17)

$$y = z \supset_{\bullet} P(y) \supset P(z)$$

by substitution $[x = y/P(y)]$ we have

$$y = z \supset_{\bullet} x = y \supset x = z.$$

This is equivalent to (2) by **H**.

(3) By (Ax17)

$$x = y \vdash_L P(x) \supset P(y)$$

and

$$y = x \vdash_L P(y) \supset P(x).$$

Hence by (1), $x = y \vdash_L P(y) \supset P(x)$, so by $\wedge$-introduction $x = y \vdash_L P(x) \equiv P(y)$, and thus

$$L \vdash x = y \supset P(x) \equiv P(y) \qquad (\#1)$$

by the deduction theorem.

Now we apply the substitution

$$S := [B/P(x)],$$

where

$$B := [xx_1/xz]A$$

and $x_1 \notin V(A)$. Then

$$SP(y) \mathrel{\overset{\circ}{=}} [y/x]B \mathrel{\overset{\circ}{=}} [y/x][xx_1/zx]A.$$

Note that

$$[y/x] \cdot [xx_1/zx] = [x_1y/xz],$$

since this substitution sends x to x_1, z to y and does not change other variables. So by 2.3.25(4),

$$SP(y) \mathrel{\overset{\circ}{=}} [x_1y/xz]A,$$

and thus (♯1) implies

$$L \vdash x = y \supset_{\bullet} [x_1 x/xz]A \equiv [x_1 y/sz]A.$$

Hence

$$L \vdash [x/x_1](x = y \supset_{\bullet} [x_1 x/xz]A \equiv [x_1 y/xz]A)$$

by 2.6.15(iii). The latter formula is congruent to

$$x = y \supset_{\bullet} [x/x_1][x_1/xz]A \equiv [x/x_1][x_1 y/xz]A.$$

Now we can again apply 2.3.25(4):

$$[x/x_1][x_1 x/xz]A \stackrel{\circ}{=} [xx/zx_1]A,$$
$$[x/x_1][x_1 y/xz]A \stackrel{\circ}{=} [yx/zx_1]A.$$

It remains to note that

$$[xx/zx_1]A \stackrel{\circ}{=} [x/z]A,$$
$$[yx/zx_1][x_1 y/xz]A \stackrel{\circ}{=} [y/z]A,$$

since x_1 does not occur in A.

(4) By the deduction theorem, it suffices to show

$$x_1 = y_1 \wedge \ldots \wedge x_n = y_n \vdash_L P(x_1, \ldots, x_n) \equiv P(y_1, \ldots, y_n). \qquad (\#2)$$

For this we show by induction that

$$x_1 = y_1 \wedge \ldots \wedge x_m = y_m$$
$$\vdash_L [x_1, \ldots, x_m/z_1, \ldots, z_m]P(\mathbf{z}) \equiv [y_1, \ldots, y_m/z_1, \ldots, z_m]P(\mathbf{z}). \quad (\#3_m)$$

for a list of new variables $\mathbf{z} = (z_1, \ldots, z_n)$.

The case $m = 0$ is trivial.

Suppose $(\#3_m)$ holds; to check $(\#3_{m+1})$, assume

$$x_1 = y_1 \wedge \ldots \wedge x_m = y_m \wedge x_{m+1} = y_{m+1}. \qquad (\#4)$$

Then by $(\forall z_{m+1})$-introduction (since z_{m+1} is new)

$$\forall z_{m+1}(A_m \equiv B_m),$$

where

$$A_m := [x_1, \ldots, x_m/z_1, \ldots, z_m]P(\mathbf{z}),$$
$$B_m := [y_1, \ldots, y_m/z_1, \ldots, z_m]P(\mathbf{z}).$$

Hence by 2.6.15 (ii) and MP

$$[x_{m+1}/z_{m+1}]A_m \equiv [x_{m+1}/z_{m+1}]B_m. \qquad (\#5)$$

The assumption (#4) implies $x_{m+1} = y_{m+1}$, so by (iii) we have

$$[x_{m+1}/z_{m+1}]B_m \equiv [y_{m+1}/z_{m+1}]B_m. \qquad (\#6)$$

From (#5), (#6), by transitivity we obtain $[x_{m+1}/z_{m+1}]A_m \equiv [y_{m+1}/z_{m+1}]B_m$, i.e. $A_{m+1} \equiv B_{m+1}$. Now since (#2) is $(\#3_n)$, the claim is proved.

■

Lemma 2.6.17 *Theorems in* **QCL** *(and thus, in any m.p.l.):*

(1) $\exists x\,(A \supset C) \equiv (\forall x A \supset C)$ *if* $x \notin FV(C)$*;*

(2) $\exists x\,\neg A \equiv \neg\forall x A$*;*

(3) $\exists x\,(C \supset A) \equiv (C \supset \exists x A)$ *if* $x \notin FV(C)$*;*

(4) $\forall x(A \vee C) \equiv \forall x A \vee C$ *if* $x \notin FV(C)$.

Proof

1. We have in **QCL**:

    1. $\neg\exists x(A \supset C) \equiv \forall x\neg(A \supset C)$ (Lemma 2.6.15(xiii))
    2. $\neg(A \supset C) \equiv A \wedge \neg C$ (by a propositional tautology)
    3. $\forall x\neg(A \supset C) \equiv \forall x(A \wedge \neg C)$ (2, replacement)
    4. $\forall x(A \wedge \neg C) \equiv \forall x A \wedge \neg C$ (2.6.15(xix))
    5. $\forall x A \wedge \neg C \equiv \neg(\forall x A \supset C)$ (by a propositional tautology).

    Hence we obtain

    (a) $\neg\exists x(A \supset C) \equiv \neg(\forall x A \supset C)$ (by transitivity from 1, 3, 4, 5).

    This implies (1), due to the admissible rule

$$\frac{\neg A \equiv \neg B}{A \equiv B}\,.$$

(ii) Take $C = \bot$ in (1)

(iii) We have in **QCL**:

(a) $(C \supset A) \equiv (\neg C \vee A)$ (from a propositional tautology)
(b) $\exists x(C \supset A) \equiv \exists x(\neg C \vee A)$ (1, replacement)
(c) $\exists x(\neg C \vee A) \equiv \neg C \vee \exists x A$ (2.6.15(xvii))
(d) $\neg C \vee \exists x A \equiv C \supset \exists x A$ (from a propositional tautology)
(e) $\exists x(C \supset A) \equiv C \supset \exists x A$ (by transitivity from 2, 3, 4).

(iv) We have in **QCL**:

(a) $A \vee C \equiv_{\bullet} \neg C \supset A$ (from a propositional tautology)
(b) $\forall x(A \vee C) \equiv \forall x(\neg C \supset A)$ (1, replacement)
(c) $\forall x(\neg C \supset A) \equiv_{\bullet} \neg C \supset \forall x A$ (2.6.15(xi))
(d) $\neg C \supset \forall x A. \equiv \forall x A \vee C$ (from a propositional tautology)

(e) $\forall x(A \lor C) \equiv \forall x A \lor C$ (by transitivity from 2, 3, 4).

■

Lemma 2.6.18 *Theorems in modal logics (where* $\bigcirc \in \{\Box_i, \Diamond_i\}$, $\mathbf{x}$ *is a list of variables):*

(1) $\bigcirc\forall\mathbf{x}A \supset \forall\mathbf{x}\bigcirc A$;

(2) $\exists\mathbf{x}\bigcirc A \supset \bigcirc\exists\mathbf{x}A$;

(3) $x = y \supset \Box_\alpha(x = y)$ *for* N*-modal logics with equality,* $\alpha \in I_N^\infty$.

Proof (1) $L \vdash \forall\mathbf{x}A \supset A$ (2.6.15 (ii)), hence $L \vdash \bigcirc\forall\mathbf{x}A \supset \bigcirc A$ by monotonicity (1.1.1), which is also admissible in the predicate case. Therefore $L \vdash \bigcirc\forall\mathbf{x}A \supset \forall\mathbf{x}\bigcirc A$, by the Bernays rule.

(2) Similar to (1), using $A \supset \exists\mathbf{x}A$.

(3) Let us first prove $x = y \supset \Box_i(x = y)$. So assuming $x = y$, we prove $\Box_i(x = y)$.

1. $x = y \supset_\bullet \Box_i(z = x) \supset \Box_i(z = y)$ Ax17, substitution $[\Box_i(z = x)/P(x)]$.
2. $\Box_i(z = x) \supset \Box_i(z = y)$ 1, $x = y$, MP.
3. $\forall z(\Box_i(z = x) \supset \Box_i(z = y))$ 2, $\forall z$-introduction (if z is new).
4. $\Box_i(x = x) \supset \Box_i(x = y)$ 3, Ax12, MP.
5. $\Box_i(x = x)$ Ax16, $\Box$-introduction.
6. $\Box_i(x = y)$ 4, 5, MP.

Hence $L \vdash x = y \supset \Box_i(x = y)$.

For arbitrary α apply induction and monotonicity rules, cf. Lemma 1.1.1.

■

We use special notation for some formulas.

Intuitionistic formulas:

$CD := \forall x(P(x) \lor q) \supset \forall x P(x) \lor q$ (*the constant domain principle*);[18]

$CD^- := \forall x(\neg P(x) \lor q) \supset \forall x \neg P(x) \lor q$;

$Ma := \neg\neg\exists x P(x) \supset \exists x \neg\neg P(x)$ (*strong Markov principle*);

$Ma^+ := \neg\exists x P(x) \lor \exists x \neg\neg P(x)$;

$UP := (\neg p \supset \exists x Q(x)) \supset \exists x(\neg p \supset Q(x))$;

$KF := \neg\neg\forall x\,(P(x) \lor \neg P(x))$;

$AP_1^+ := \forall x_1(Q_1(x_1) \lor \neg Q_1(x_1))$;

$AP_n^+ := \forall x_n(Q_n(x_n) \lor (Q_n(x_n) \supset AP_{n-1}^+)) \quad (n > 1)$;

$DE := \forall x \forall y\,(x = y \lor \neg\, x = y)$ (*the decidable equality principle*);

$SE := \forall x \forall y\,(\neg\neg\, x = y \supset x = y)$ (*the stable equality principle*);

$AU_1' := \exists x P(x) \supset \forall x P(x)$;

$$AU_n := \overline{\forall}\left(\bigwedge_{0\leq i\leq n} P_i(x_i) \supset \bigvee_{0\leq i<j\leq n} P_i(x_j)\right) \quad (n > 0);$$

$$AU_n^= := \overline{\forall} \bigvee_{0\leq i<j\leq n} (x_i = x_j) \quad (n > 0).$$

Modal formulas:

$Ba_i := \forall x \Box_i P(x) \supset \Box_i \forall x P(x)$ (Barcan formula for $\Box_i$);
$CE_i := \forall x \forall y (x \neq y \supset \Box_i (x \neq y))$ (the closed equality principle for $\Box_i$).

In particular,

$$AU_1 \doteq \forall x \forall y (P(x) \supset P(y)), \ AU_1^= \doteq \forall x \forall y (x = y).$$

All the above intuitionistic formulas except AU_1', AU_n, and $AU_n^=$ are classical theorems.

Classically both formulas AU_n and $AU_n^=$ state that the individual domain contains at most n elements, so they are logically equivalent. This also holds in intuitionistic logic:

Lemma 2.6.19

(1) $\mathbf{QH}^= \vdash AU_n^= \supset AU_n$ *(and so* $\mathbf{QK}_N^= \vdash AU_n^= \supset AU_n$*).*

(2) $\mathbf{QH}^= + AU_n = \mathbf{QH}^= + AU_n^=$ *(and* $\mathbf{QK}_N^= + AU_n = \mathbf{QK}_N^= + AU_n^=$*).*

(3) $\mathbf{QH} + AU_1' = \mathbf{QH} + AU_1$.

Proof
(1) Since $P_i(x_i) \wedge (x_i = x_j)$ implies $P_i(x_j)$.

(2) Consider the formula

$$AU_n^- := \bigwedge_i P_i(x_i) \supset \bigvee_{i<j} P_i(x_j)$$

(the 'quantifier-free matrix' of AU_n) and the substitution

$$S := [x_1 = z, \ldots, x_n = z / P_1(z), \ldots, P_n(z)].$$

Then

$$S(AU_n^-) = \bigwedge_{i<n} x_i = x_i \supset \bigvee_{i<j} x_i = x_j$$

implies $\bigvee\limits_{i<j} x_i = x_j$; therefore $\mathbf{QH}^= + AU_n \vdash AU_n^=$.

(3) $[\top / P_1(x)] AU_1$ is equivalent to AU_1'. So $\mathbf{QH} + AU_1 \vdash AU_1'$. The converse is trivial. ■

Note that the argument is not valid for $S(AU_n)$, because S renames bound variables x_i in AU_n.

Lemma 2.6.20

(1) For any modal formula A and a list of variables $\mathbf{x}$

$$\mathbf{QS4} \vdash \Box \forall \mathbf{x} \Box A \equiv \Box \forall \mathbf{x} A, \ \Diamond \exists \mathbf{x} \Diamond A \equiv \Diamond \exists \mathbf{x} A.$$

[18] First introduced by A. Grzegorczyk.

(2) For any superintuitionistic formula A

$$\mathbf{QH} + KF \vdash \forall x \neg\neg A \supset \neg\neg\forall x\, A;$$

moreover,

$$\mathbf{QH} + KF = \mathbf{QH} + \forall x \neg\neg P(x) \supset \neg\neg\forall x\, P(x).$$

Proof

(1) $\Box\forall\mathbf{x}\Box A \supset \forall\mathbf{x}\Box A$ follows from $\Box p \supset p$.

The other way round, we obtain $\Box\forall\mathbf{x}A \supset \forall\mathbf{x}\Box A$ by 2.6.16(1), hence $\Box^2\forall\mathbf{x}A \supset \Box\forall\mathbf{x}\Box A$ by monotonicity, and thus $\Box\forall\mathbf{x}A \supset \Box\forall\mathbf{x}\Box A$ (since $\mathbf{S4} \vdash \Box p \supset \Box^2 p$).

For the second formula the proof is similar, using $A \supset \exists\mathbf{x}A$.

(2) For $L := \mathbf{QH} + \forall x\neg\neg P(x) \supset \neg\neg\forall x P(x)$ let us show that $L \vdash KF$.

In fact,

$$L \vdash \forall x\neg\neg(P(x) \vee \neg P(x)) \supset \neg\neg\forall x(P(x) \vee \neg P(x))$$

by substitution.

On the other hand,

$$\mathbf{H} \vdash \neg\neg(p \vee \neg p)$$

by the Glivenko theorem, hence

$$\mathbf{QH} \vdash \neg\neg(P(x) \vee \neg P(x)),$$

and thus

$$\mathbf{QH} \vdash \forall x\neg\neg(P(x) \vee \neg P(x)).$$

So by MP it follows that $L \vdash \neg\neg\forall x(P(x) \vee \neg P(x))(= KF)$.

The other way round, let us show that

$$\mathbf{QHK} := \mathbf{QH} + KF \vdash \forall x\neg\neg P(x) \supset \neg\neg\forall x P(x).$$

By the deduction theorem, this reduces to

$$\forall x\neg\neg P(x) \vdash_{\mathbf{QHK}} \neg\neg\forall x P(x)$$

and next to

$$\forall x\neg\neg P(x), \neg\forall x P(x) \vdash_{\mathbf{QHK}} \bot.$$

It suffices to show that

$$\forall x\neg\neg P(x),\ \neg P(x) \vdash_{\mathbf{QH}} \neg\forall x(P(x) \vee \neg P(x)),$$

which follows from

$$(I) \qquad \forall x\neg\neg P(x), \forall x(P(x) \vee \neg P(x)) \vdash_{\mathbf{QH}} P(x).$$

To show (I), note that

$$(II) \qquad \forall x\neg\neg P(x), \forall x(P(x) \vee \neg P(x)) \vdash_{\mathbf{QH}} \neg\neg P(x), P(x) \vee \neg P(x)$$

and

$$(III) \qquad \neg\neg P(x), P(x) \vee \neg P(x) \vdash_{\mathbf{QH}} P(x),$$

since $P(x) \vdash P(x)$ and $\{\neg\neg P(x), \neg P(x)\}$ is inconsistent. ∎

2.7 First-order theories

In this section we again consider formulas with extra individual constants.

Definition 2.7.1 *Let L be an N-m.p.l.$(=)$ (respectively, an s.p.l.$(=)$). An N-modal (respectively, an intuitionistic) formula with extra constants A is called* L-provable, *or an* L-theorem *(notation: $\vdash_L A$) if it has a maximal generator in L, i.e. if it can be presented in a form $A = [\mathbf{c}/\mathbf{x}]\, B$, with $B \in L$, a distinct list of constants* $\mathbf{c}$, *and a distinct list of variables* $\mathbf{x}$ *(cf. Lemma 2.4.4).*

For A without constants, we also assume that $\vdash_L A$ iff $A \in L$.

Lemma 2.7.2 *$\vdash_L A$ iff A has a generator in L (for A containing constants).*

Proof 'Only if' follows from 2.7.1. The other way round, if A has a generator $B \in L$, then its maximal generator A_1 can be presented as $[\mathbf{y}/\mathbf{x}]B$ (Lemma 2.4.5); so $A_1 \in L$ by 2.6.15 (iii). ∎

L-provability does not depend on the presentation of a formula:

Lemma 2.7.3 *If B, B' are maximal generators of A and $B \in L$, then $B' \in L$.*

Proof By Lemma 2.4.5, $B' \doteq [\mathbf{z}/\mathbf{y}]\, B$ for some variable substitution $[\mathbf{z}/\mathbf{y}]$. So $B' \in L$, by Lemma 2.6.15. ∎

L-provability respects MP:

Lemma 2.7.4 *If $\vdash_L A$ and $\vdash_L A \supset B$, then $\vdash_L B$.*

Proof Suppose $(A \supset B) = [\mathbf{c}/\mathbf{x}](A_1 \supset B_1)$, for distinct $\mathbf{c}$, $\mathbf{x}$, and $(A_1 \supset B_1) \in L$. Then A_1 is a maximal generator of A, so $\vdash_L A$ implies $A_1 \in L$, by Lemma 2.7.3. Hence $B_1 \in L$, by MP. ∎

L-provability respects $\Box$-introduction:

Lemma 2.7.5 *If $\vdash_L A$, then $\vdash_L \Box_i A$.*

Proof Suppose $A = [\mathbf{c}/\mathbf{x}]A_1$ for distinct $\mathbf{c}$, $\mathbf{x}$, and $A_1 \in L$. Then $\Box_i A_1 \in L$ and $\Box_i A = [\mathbf{c}/\mathbf{x}]\Box_i A_1$. ∎

L-provability also respects substitution into propositional formulas:

Lemma 2.7.6 *If L is a predicate logic (of any kind), A is a propositional formula, $A \in L$, $S = [B_1, \dots, B_n/q_1, \dots, q_n]$ is a substitution of formulas with constants for propositional letters, then $\vdash_L SA$.*

Proof Let G_i be a maximal generator of B_i, so $B_i = [\mathbf{c}_i/\mathbf{x}_i]\, G_i$ for an injective $[\mathbf{c}_i/\mathbf{x}_i]$. Take distinct lists $\mathbf{y}_1, \dots, \mathbf{y}_n$ of variables non-occurring in any G_i and put $G'_i := [\mathbf{y}_i/\mathbf{x}_i]\, G_i$, $\mathbf{G}' := G'_1, \dots, G'_n$, $\mathbf{q} := q_1, \dots, q_n$, $\mathbf{c} := \mathbf{c}_1 \dots \mathbf{c}_n$, $\mathbf{y} := \mathbf{y}_1 \dots \mathbf{y}_n$. Then by Lemma 2.4.2(3), $B_i \doteq [\mathbf{c}_i/\mathbf{y}_i]\, G'_i$, and thus

$$SA \doteq [[\mathbf{c}/\mathbf{y}]\, \mathbf{G}'/\mathbf{q}]\, A \doteq [\mathbf{c}/\mathbf{y}][\mathbf{G}'/\mathbf{q}]\, A$$

(Lemma 2.5.18). Since $[\mathbf{G}'/\mathbf{q}]\, A \in L$, it follows that $\vdash_L SA$. ∎

Definition 2.7.7 *An* N-modal theory *(respectively, an* intuitionistic simple theory, *with or without equality) is a set of N-modal (respectively, intuitionistic) sentences with individual constants.*

D_Γ denotes the set of individual constants occurring in a theory Γ. So according to the terminology from Section 2.3, Γ is a set of N-modal or intuitionistic D_Γ-sentences: $\Gamma \subseteq MS_N^{(=)}(D_\Gamma)$ or $\Gamma \subseteq IS^{(=)}(D_\Gamma)$. The set of all D_Γ-sentences (of the corresponding kind) $\mathcal{L}^{(=)}(\Gamma) := MS_N^{(=)}(D_\Gamma)$ (or $IS^{(=)}(D_\Gamma)$) is called the language of Γ.

The next definition is an analogue of 2.6.11.

Definition 2.7.8 *An* L-inference *of a formula with constants B from a theory Γ is a sequence $A_1, \dots, A_n$, in which $A_n = B$ and every A_i is either L-provable, or $A_i \in \Gamma$, or A_i is obtained from earlier formulas by applying MP or $\forall$-introduction.*

If there exists an L-inference of B from Γ, we say that B is L-derivable from Γ *(or an* L-theorem of Γ*); notation: $\Gamma \vdash_L B$.*

Then we easily obtain an analogue of 2.6.14:

Lemma 2.7.9 *Let Γ be an N-modal or intuitionistic first-order theory (with or without equality), L a predicate logic of the corresponding kind. Then for any N-modal or intuitionistic formula B (perhaps, with extra constants)*

$$\Gamma \vdash_L B \text{ iff there exists a finite } X \subseteq \Gamma \text{ such that } \vdash_L \bigwedge X \supset B.$$

Proof The argument is essentially the same as for 2.6.14; we check that

$$(*) \qquad X \vdash_L B \Leftrightarrow \ \vdash_L \bigwedge X \supset B$$

for a finite theory X and a formula with constants B, by induction on $|X|$.

If $X = \varnothing$, $(*)$ means

$$\vdash_L B \Leftrightarrow \ \vdash_L \top \supset B,$$

which follows in the same way as in 2.6.14.

For the induction step we need the equivalence

$$\vdash_L \bigwedge X \supset_\bullet A \supset B \Leftrightarrow \ \vdash_L (\bigwedge X) \wedge A \supset B,$$

which follows by Lemma 2.7.6 from (2) in the proof of 2.6.13. ■

Lemma 2.7.10 *Let Γ be an N-modal theory, $\Box_i\Gamma := \{\Box_i A \mid A \in \Gamma\}$.*

(1) If $\Gamma \vdash_L A$, then $\Box_i\Gamma \vdash_L \Box_i A$.

(2) If $\Gamma \vdash_L A \supset B$, then $\Box_i\Gamma \vdash_L \Box_i A \supset \Box_i B$.

Proof (1) We apply 2.7.9. If $L \vdash \bigwedge X \supset A$ for a finite $X \subseteq \Gamma$, then by 2.7.5, $L \vdash \Box_i(\bigwedge X) \supset \Box_i A$, and thus $L \vdash \bigwedge \Box_i X \supset \Box_i A$. Hence $\Box_i\Gamma \vdash_L \Box_i A$.

(2) By (1), $\Gamma \vdash_L A \supset B$ implies $\Box_i\Gamma \vdash_L \Box_i(A \supset B)$. Then we can apply the axiom AK_i and MP. ■

Another useful fact is the following lemma on new constants.

Lemma 2.7.11 *Let L be an N-modal or superintuitionistic logic, Γ a modal (respectively, intuitionistic) theory, $A(x)$ a formula with constants (resp., modal or intuitionistic), x a variable not bound in $A(x)$, and assume that a constant c does not occur in $\Gamma \cup \{A(x)\}$. Then the following conditions are equivalent:*

(1) $\Gamma \vdash_L A(c)$;

(2) $\Gamma \vdash_L A(x)$;

(3) $\Gamma \vdash_L \forall x A(x)$.

Proof (1) $\Rightarrow$ (2). Assume $\Gamma \vdash_L A(c)$. Then by 2.7.9, $\vdash_L \bigwedge \Gamma_1 \supset A(c)$ for some finite $\Gamma_1 \subseteq \Gamma$. Let $B := \bigwedge \Gamma_1$. Then for some injective $[\mathbf{y}x \mapsto \mathbf{d}c]$,

$$\begin{aligned} &B \supset A(c) = [\mathbf{d}c/\mathbf{y}x](B_0 \supset A_0(x)), \\ &(B_0 \supset A_0(x)) \in L. \end{aligned}$$

Since by assumption c does not occur in Γ, it does not occur in B, so we have

$$\begin{aligned} &B = [\mathbf{d}c/\mathbf{y}x]B_0 = [\mathbf{d}/\mathbf{y}]B_0, \\ &A(c) = [\mathbf{d}c/\mathbf{y}x]A_0(x) = [\mathbf{d}/\mathbf{y}]A_0(c). \end{aligned}$$

Hence

$$A(x) = A(c)[c \mapsto x] = [\mathbf{d}/\mathbf{y}]A_0(x),$$

and thus

$$B \supset A(x) = [\mathbf{d}/\mathbf{y}](B_0 \supset A_0(x)).$$

So $\vdash_L B \supset A(x)$, therefore $\Gamma \vdash_L A(x)$.

(2) $\Rightarrow$ (3). Assume $\Gamma \vdash_L A(x)$, then by 2.7.9, for some finite $\Gamma_1 \subseteq \Gamma$ we have $\vdash_L B \supset A(x)$, where $B = \bigwedge \Gamma_1$. So for some injective $[\mathbf{y} \mapsto \mathbf{d}]$, $B \supset A(x) = [\mathbf{d}/\mathbf{y}](B_0 \supset A_0(x))$ and $(B_0 \supset A_0(x)) \in L$.

Since B is closed, $x \notin FV(B_0)$, thus by the Bernays rule, $(B_0 \supset \forall x A_0(x)) \in L$. We also have

$$[\mathbf{d}/\mathbf{y}]\,(B_0 \supset \forall x A_0(x)) = B \supset \forall x A(x),$$

and so the latter formula is L-provable. Therefore $\Gamma \vdash_L \forall x A(x)$.

(3)$\Rightarrow$(1). Let $A_0(x)$ be a maximal generator of $A(x)$, then $A(x) = [\mathbf{d}/\mathbf{y}]\,A_0(x)$, and also

$$\forall x A(x) \supset A(c) = [\mathbf{d}c/\mathbf{y}x](\forall x A_0(x) \supset A_0(x)).$$

But $(\forall x A_0(x) \supset A_0(x)) \in L$ by 2.6.13 (ii), so

$$\vdash_L \forall x A(x) \supset A(c).$$

Then $\Gamma \vdash_L \forall x A(x)$ implies $\Gamma \vdash_L A(c)$ by MP. ■

In the intuitionistic case it is convenient to use theories of another kind.

Definition 2.7.12 *A* intuitionistic double theory (with or without equality) *is a pair* (Γ, Δ), *in which* Γ, Δ *are intuitionistic sentences (respectively, with or without equality).* $D_{(\Gamma,\Delta)}(= D_{\Gamma\cup\Delta})$ *denotes the set of constants occurring in* $\Gamma \cup \Delta$; *the* language *of* (Γ, Δ) *is* $\mathcal{L}(\Gamma, \Delta) := IF^{(=)}(D_{(\Gamma,\Delta)})$.

Definition 2.7.13 *Let L be an s.p.l.,* (Γ, Δ) *an intuitionistic theory. An intuitionistic formula A (with constants) is L*-provable *in* (Γ, Δ) *if* $\Gamma \vdash_L A \vee \bigvee \Delta_1$ *for some finite* $\Delta_1 \subseteq \Delta$.

So, as we assume $\bigvee \varnothing := \bot$, $\Gamma \vdash_L A$ implies $(\Gamma, \Delta) \vdash_L A$.

This provability respects MP as well:

Lemma 2.7.14 *If* $(\Gamma, \Delta) \vdash_L C$ *and* $(\Gamma, \Delta) \vdash_L C \supset B$, *then* $(\Gamma, \Delta) \vdash_L B$.

Proof First note that $\Gamma \vdash_L A \vee \bigvee \Delta_1$ implies $\Gamma \vdash_L A \vee \bigvee \Delta_2$ for any $\Delta_2 \supseteq \Delta_1$, since $\vdash_{\mathbf{QH}} A \vee \bigvee \Delta_1 \supset A \vee \bigvee \Delta_2$. The latter follows from the intuitionistic tautology

$$p \vee q \supset p \vee (q \vee r).$$

So if $(\Gamma, \Delta) \vdash_L C$ and $(\Gamma, \Delta) \vdash_L C \supset B$, then $\Gamma \vdash_L C \vee \bigvee \Delta_1$ and

$$\Gamma \vdash_L (C \supset B) \vee \bigvee \Delta_1$$

for some finite $\Delta_1 \subseteq \Delta$. But by 2.7.6

$$\vdash_{\mathbf{QH}} (C \vee \bigvee \Delta_1) \wedge ((C \supset B) \vee \bigvee \Delta_1) \supset B \vee \bigvee \Delta_1,$$

since

$$\mathbf{H} \vdash (p \vee r) \wedge ((p \supset q) \vee r) \supset q \vee r$$

(the latter follows from $p \vee r, (p \supset q) \vee r \vdash_{\mathbf{H}} q \vee r$, which we leave to the reader). Hence $\Gamma \vdash_L B \vee \bigvee \Delta_1$, and thus $(\Gamma, \Delta) \vdash_L B$. ■

2.8 Deduction theorems

We begin with an analogue of Lemma 2.6.13.

Lemma 2.8.1 *For a predicate logic L, a first-order theory* Γ *and formulas with constants A, B of the corresponding kind*

$$\Gamma \cup \{A\} \vdash_L B \Rightarrow \Gamma \vdash_L A \supset B.$$

Proof By an easy modification of the proof of 2.6.13, using Lemma 2.7.6. The details are left to the reader. ■

Lemma 2.8.2 *Let A be a predicate (intuitionistic) formula and let S be an* $IF^{(=)}$*-substitution such that* $FV(A) \cap FV(S) = \varnothing$. *Then*

$$\mathbf{QH}^{(=)} \vdash \overline{\forall} SA \supset S\overline{\forall} A.$$

Proof Let $FV(A) = r(\mathbf{x})$, so $\overline{\forall}A = \forall\mathbf{x}A$. Let A° be a clean version of A, such that $BV(A^\circ) \cap FV(S) = \varnothing$. Then $SA \doteq SA^\circ$, and thus

(1) $\overline{\forall}SA \doteq \overline{\forall}SA^\circ$,

(2) $S(\overline{\forall}A) \doteq \forall\mathbf{x}SA^\circ$,

since $r(\mathbf{x}) \cap FV(S) = \varnothing$. But by Lemma 2.6.15 (xx),

$$\mathbf{QH}^{(=)} \vdash \overline{\forall}SA^\circ \equiv \forall\mathbf{y}\forall\mathbf{x}SA^\circ,$$

where $FV(SA^\circ) = r(\mathbf{xy})$. By 2.6.15 (ii),

$$\mathbf{QH}^{(=)} \vdash \forall\mathbf{y}\forall\mathbf{x}SA^\circ \supset \forall\mathbf{x}SA^\circ.$$

Hence

$$\mathbf{QH}^{(=)} \vdash \overline{\forall}SA^\circ \supset \forall\mathbf{x}SA^\circ,$$

and thus by (1), (2) we obtain $\mathbf{QH}^{(=)} \vdash \overline{\forall}SA \supset S(\overline{\forall}A)$. ∎

Here is an example showing that the requirement

$$FV(A) \cap FV(S) = \varnothing$$

is necessary. Let

$$A = P(x),\ \ S = [Q(x,y)/P(y)],$$

so we have $FV(A) \cap FV(S) = \{x\}$. Then

$$\overline{\forall}SA \doteq \forall xQ(x,x), \quad S(\overline{\forall}A) \doteq \forall y\ Q(x,y),$$

but

$$\mathbf{QH} \nvdash \forall xQ(x,x) \supset \forall yQ(x,y).$$

Theorem 2.8.3 (Deduction theorem for superintuitionistic logics) *Let L be an s.p.l.(=), Γ an intuitionistic theory. Then for any $A \in IF^{(=)}$*

$$L + \Gamma \vdash A \ \textit{ iff } \ \overline{Sub}(\Gamma) \vdash_L A.$$

Proof (If.) $Sub(\Gamma) \subseteq L + \Gamma$, hence $\overline{Sub}(\Gamma) \subseteq L + \Gamma$. So $\overline{Sub}(\Gamma) \vdash_L A$ implies $L + \Gamma \vdash_L A$, and thus $L + \Gamma \vdash A$, since $L + \Gamma$ is closed under L-provability.

(Only if.) It is sufficient to show that the set $\{A \mid \overline{Sub}(\Gamma) \vdash_L A\}$ is a superintuitionistic logic. The conditions (s1)–(s3) from Definition 2.6.3 are obvious (e.g. for generalisation we apply the Bernays rule, since $\overline{Sub}(\Gamma)$ consists of sentences).

To check (s4), assume that $\overline{\forall}A_1, \ldots, \overline{\forall}A_k \vdash_L A$ for $A_1, \ldots, A_k \in Sub(\Gamma)$. Consider a substitution $S = [C(\mathbf{xy})/P(\mathbf{x})]$, and let $\mathbf{z}$ be a distinct list such that $r(\mathbf{z}) \cap (r(\mathbf{y}) \cup \bigcup\limits_{i=1}^{k} FV(A_i) \cup FV(A)) = \varnothing$. Let us show that $\overline{Sub}(\Gamma) \vdash_L SA$.

We may also assume that $BV(C) \cap r(\mathbf{xyz}) = \varnothing$. Consider another substitution $S' = [C'/P(\mathbf{x})]$, where $C' := [\mathbf{z}/\mathbf{y}]C$.

Since $\overline{\forall} A_1, \ldots, \overline{\forall} A_k \vdash_L A$, we have $\left(\bigwedge\limits_{i=1}^{k} \overline{\forall} A_i \supset A\right) \in L$, and thus

$$S'\left(\bigwedge_{i=1}^{k} \overline{\forall} A_i \supset A\right) \in L.$$

But

$$S'\left(\bigwedge_{i=1}^{k} \overline{\forall} A_i \supset A\right) \stackrel{\circ}{=} \left(\bigwedge_{i=1}^{k} S'\overline{\forall} A_i \supset S'A\right),$$

and so

$$(1) \quad L \vdash \bigwedge_{i=1}^{k} \overline{\forall} S'(\overline{\forall} A_i) \supset S'A.$$

On the other hand, since $FV(S') \cap FV(A_i) = \varnothing$, by Lemma 2.8.2, we have

$$\mathbf{QH} \vdash \overline{\forall} S'A_i \supset S'(\overline{\forall} A_i),$$

and thus

$$\mathbf{QH} \vdash \bigwedge_{i=1}^{k} \overline{\forall} S'A_i \supset \bigwedge_{i=1}^{k} S'(\overline{\forall} A_i),$$

Hence $\overline{\mathrm{Sub}}(\Gamma) \vdash_L \bigwedge\limits_{i=1}^{k} S'(\overline{\forall} A_i)$, and by (1) we obtain

$$\overline{\mathrm{Sub}}(\Gamma) \vdash_L S'A,$$

which implies

$$\overline{\mathrm{Sub}}(\Gamma) \vdash_L \forall \mathbf{z} S'A,$$

by $\forall$-introduction. But by 2.5.27(ii),

$$[\mathbf{y}/\mathbf{z}]S'A \stackrel{\circ}{=} SA,$$

and thus by 2.6.15(ii),

$$\mathbf{QH} \vdash \forall \mathbf{z} S'A \supset SA.$$

Therefore $\overline{\mathrm{Sub}}(\Gamma) \vdash_L SA$, as stated. ■

Note that the previous proof can be simplified if Γ is a set of sentences: then we choose substitution instances A_s in such a way that $FV(A_s) \cap FV[\mathbf{C}/\mathbf{P}] = \varnothing$, and we can readily apply Lemma 2.8.2.

For an N-modal theory Δ put

$$\Box^\infty \Delta := \{\Box_\alpha B \mid B \in \Delta,\ \alpha \in I_N^\infty\}.$$

Theorem 2.8.4 (Deduction theorem for modal logics) *Let L be a N-m.p.l.(=), Γ an N-modal theory. Then for any N-modal formula A*

$$L + \Gamma \vdash A \ \textit{ iff } \ \Box^\infty \overline{Sub}(\Gamma) \vdash_L A.$$

Proof $\Box^\infty\overline{\mathrm{Sub}}(\Gamma) \vdash_L A$ clearly implies $L + \Gamma \vdash A$. For the converse, note that the set $\{A \mid \Box^\infty\overline{\mathrm{Sub}}(\Gamma) \vdash_L A\}$ is substitution closed; this is proved as in the previous theorem. It is also closed under $\Box$-introduction. In fact,

$$\Box_{\alpha_1}B_1, \ldots, \Box_{\alpha_n}B_n \vdash_L A \text{ iff } \vdash_L \Box_{\alpha_1}B_1 \wedge \ldots \wedge \Box_{\alpha_n}B_n \supset A.$$

By monotonicity and the distribution of $\Box_i$ over $\wedge$ in $\mathbf{K}_N$, this implies

$$\vdash_L \Box_i\Box_{\alpha_1}B_1 \wedge \ldots \wedge \Box_i\Box_{\alpha_n}B_n \supset \Box_i A.$$

Hence $\Box^\infty\overline{\mathrm{Sub}}(\Gamma) \vdash_L \Box_i A$. ∎

Definition 2.8.5 *An m.p.l.(=) is called* conically expressive *if its propositional part is conically expressive.*

Lemma 2.8.6 *For any conically expressive N-m.p.l.(=) L, theory Δ and formula A,*

$$\Box^*\Delta \vdash_L A \textit{ iff } \Box^\infty\Delta \vdash_L A,$$

where

$$\Box^*\Delta := \{\Box^*B \mid B \in \Delta\}.$$

Proof 'If' readily follows from $L \vdash \Box^*p \supset \Box_\alpha p$. For the general proof of the converse we need the strong completeness of $\mathbf{QK}_N$, see below. But in particular cases (like **QS4** or **QK4**), $\Box^*B$ is equivalent to a finite conjunction of formulas from $\Box^\infty\{B\}$, so $\Box^*\Delta \vdash_L A$ obviously implies $\Box^\infty\Delta \vdash_L A$. ∎

Hence we obtain a simplified version of the deduction theorem for conically expressive modal logics:

Theorem 2.8.7 *Let L be a conically expressive N-m.p.l.(=), Γ an N-modal theory. Then for any N-modal formula A*

$$L + \Gamma \vdash A \quad \textit{iff} \quad \Box^*\overline{\mathrm{Sub}}(\Gamma) \vdash_L A,$$

Proof Follows from 2.8.4 and 2.8.6. ∎

Here is a simple application of the deduction theorem.

Lemma 2.8.8 $\mathbf{QH} + W^* \vdash KF$, *where*

$$W^* = \forall x((P(x) \supset \forall y P(y)) \supset \forall y P(y)) \supset \forall x P(x),$$
$$K = \forall x \neg\neg P(x) \supset \neg\neg\forall x P(x).$$

Proof It is sufficient to show that

$$W^*, \forall x\neg\neg P(x), \neg\forall x P(x) \vdash_{\mathbf{QH}} \bot,$$

or, equivalently,

$$W^*, \forall x\neg\neg P(x), \neg\forall x P(x) \vdash_{\mathbf{QH}} \forall x P(x).$$

But this is obvious, since the premise $\forall x((P(x) \supset \forall y P(y)) \supset \forall y P(y))$ of W^* is equivalent to $\forall x((P(x) \supset \bot) \supset \bot)$, i.e. to $\forall x\neg\neg P(x)$, in the presence of $\neg\forall x P(x)$. ∎

2.9 Perfection

Let $A_1 \dots A_m$ be a list of formulas (not necessarily distinct); we define their *disjoint conjunction*

$$A_1 \dot{\wedge} \dots \dot{\wedge} A_m := S_1 A_1 \wedge \dots \wedge S_m A_m,$$

where S_i is a formula substitution transforming A_i in such a way that predicate letters in all conjuncts become disjoint. E.g. we can put

$$S_i P_k^n(\mathbf{x}) := P_{mk+i}^n(\mathbf{x});$$

then P_l^n occurs in $S_i A_i$ only if $l \equiv i \pmod m$. The formula

$$\overline{A_1^k} \dot{\wedge} \dots \dot{\wedge} \overline{A_m^k}$$

is equivalent to

$$\overline{\forall}(A_1^k \dot{\wedge} \dots \dot{\wedge} A_m^k),$$

in all our logics, see below.

Similarly we define a *disjoint disjunction*:

$$A_1 \dot{\vee} \dots \dot{\vee} A_m := S_1 A_1 \vee \dots \vee S_m A_m.$$

For a theory Θ put

$$\Theta^\forall := \{\overline{A^k} \mid A \in \Theta,\ k \geq 0\},$$
$$\Theta^\wedge = \{A_1 \dot{\wedge} \dots \dot{\wedge} A_m \mid m > 0,\ A_1, \dots, A_m \in \Theta\}.$$

Thus $\Theta^\wedge$ contains conjunctions of variants of formulas from Θ in disjoint predicate symbols, and

$$\Theta^{\wedge\forall} = \{\overline{(A_1 \dot{\wedge} \dots \dot{\wedge} A_k)^k} \mid k > 0,\ A_1, \dots, A_k \in \Theta\}.$$

Obviously, $\mathbf{H} + \Theta = \mathbf{H} + \Theta^\forall = \mathbf{H} + \Theta^\wedge = \mathbf{H} + \Theta^{\wedge\forall}$.

For theories $\Theta_1, \dots, \Theta_m$ we also define the *disjoint disjunction*

$$\Theta_1 \dot{\vee} \dots \dot{\vee} \Theta_m := \{A_1 \dot{\vee} \dots \dot{\vee} A_m \mid A_1 \in \Theta_1, \dots, A_m \in \Theta_m\}$$

(the set of disjunctions of variants of formulas from $\Theta_1, \dots, \Theta_m$) and the *extended disjoint disjunction*:

$$(\Theta_1 \dot{\vee} \dots \dot{\vee} \Theta_m)^* = \{\overline{(A_1 \dot{\vee} \dots \dot{\vee} A_m)^k} \mid k > 0,\ A_1 \in \Theta_1, \dots, A_m \in \Theta_m\}.$$

In this section we consider only superintuitionistic logics

Definition 2.9.1 *Let L be an s.p.l.(=), Θ_1, Θ_2 sets of formulas in the language of L. We say that*

- Θ_1 L-implies Θ_2 *(notation:* $\Theta_1 \leq_L \Theta_2$*) if*

$$\forall B \in \Theta_2 \ \exists A \in \Theta_1 \ L \vdash A \supset B.$$

- Θ_1 *and* Θ_2 *are* L-equivalent *(notation:* $\Theta_1 \sim_L \Theta_2$*) if* $\Theta_1 \leq_L \Theta_2$ *and* $\Theta_2 \leq_L \Theta_1$.
- Θ_1 sub-L-implies Θ_2 *(notation:* $\Theta_1 \leq_L^{\text{sub}} \Theta_2$*) if* $\text{Sub}(\Theta_1) \leq_L \text{Sub}(\Theta_2)$;
- Θ_1 *is* sub-L-equivalent to Θ_2 *(notation:* $\Theta_1 \sim_L^{\text{sub}} \Theta_2$*) if* $\text{Sub}(\Theta_1) \sim_L \text{Sub}(\Theta_2)$.

Here are some simple properties of these relations.

Lemma 2.9.2

(1) $\Theta_1 \leq_L \Theta_2 \Longrightarrow \Theta_1 \leq_L^{\text{sub}} \Theta_2$.

(2) $\Theta_1 \sim_L \Theta_2 \Longrightarrow \Theta_1 \sim_L^{\text{sub}} \Theta_2$.

(3) $\Theta_1 \leq_L^{\text{sub}} \Theta_2$ *iff* $\text{Sub}(\Theta_1) \leq_L \Theta_2$.

(4) $\Theta_1 \leq_L^{\text{sub}} \Theta_2 \Longrightarrow L + \Theta_2 \subseteq L + \Theta_1$.

(5) $\Theta_1 \sim_L^{\text{sub}} \Theta_2 \Longrightarrow L + \Theta_1 = L + \Theta_2$.

Proof

(1) Note that $L \vdash A \supset B$ implies $L \vdash SA \supset SB$ for any substitution S.

(2) Follows from (1).

(3) By (1), $\text{Sub}(\Theta_1) \leq_L \Theta_2 \Longrightarrow \text{Sub}(\Theta_1) = \text{Sub}(\text{Sub}(\Theta_1)) \leq_L \text{Sub}(\Theta_2)$. The converse is trivial.

(4) $\Theta_1 \leq_L^{\text{sub}} \Theta_2$ clearly implies $\Theta_2 \subseteq L + \Theta_1$.

(5) Follows from (4).

■

2.9.2(2), (5) allow us sometimes to identify L-(sub)-equivalent sets of formulas.

Definition 2.9.3 *A set of sentences* Θ *is called* $\forall$-perfect *in a logic* L *if the following holds*

($\forall$-p) for any $A \in \text{Sub}(\Theta)$ *there exists* $B \in \text{Sub}(\Theta)$ *such that* $L \vdash B \supset \overline{\forall} A$,

i.e. $\text{Sub}(\Theta) \leq_L \overline{\text{Sub}}(\Theta)$.

Lemma 2.9.4 *($\forall$-p) follows from its weaker version*

($\forall$-p$^-$) for any $A \in \Theta$, $k \geq 0$ there exists $B \in Sub(\Theta)$ such that $L \vdash B \supset \overline{A^k}$.

Proof In fact, assume ($\forall$-p$^-$). If $A' \in Sub(\Theta)$, then $\overline{\forall} A' = S\overline{A^k}$ for some substitution S and $A \in \Theta$, $k \geq 0$. By ($\forall$-p$^-$), $L \vdash B \supset \overline{A^k}$ for some $B \in Sub(\Theta)$; so

$$L \vdash S(B \supset \overline{A^k}) = (SB \supset \overline{\forall} A').$$

Since $SB \in Sub(\Theta)$, ($\forall$-p) holds for A'. ■

Definition 2.9.5 *A set of sentences Θ is called* $\wedge$-perfect *in a logic L if*

($\wedge$-p) for any $A_1, \ldots, A_m \in Sub(\Theta)$ there exists $B \in Sub(\Theta)$ such that $L \vdash B \supset A_1 \wedge \ldots \wedge A_m$.

Note that the condition ($\wedge$-p) for $m = 2$ implies ($\wedge$-p) for arbitrary m, by induction.

Lemma 2.9.6 *Θ is $\wedge$-perfect in L iff $Sub(\Theta) \leq_L \Theta^\wedge$.*

Proof Note that if every A_i is a substitution instance of $A'_i \in \Theta$, then $A_1 \wedge \ldots \wedge A_m$ is a substitution instance of $A'_1 \dot\wedge \ldots \dot\wedge A'_m$. ■

Definition 2.9.7 *A set Θ is called* $\wedge\forall$-perfect *(in L) if it is both $\forall$-perfect and $\wedge$-perfect.*

Obviously, a perfect set (of any kind) in L is also perfect in all every $L' \supseteq L$. The next proposition suggests equivalents to 2.9.7.

Proposition 2.9.8 *Let Θ be a set of sentences. Then the following conditions are equivalent:*

(1) Θ is $\wedge\forall$-perfect in L;

(2) $\mathrm{Sub}(\Theta) \leq_L \Theta^{\wedge\forall}$;

(3) for any $A_1, \ldots, A_m \in \mathrm{Sub}(\Theta)$ there exists $B \in \mathrm{Sub}(\Theta)$ such that $L \vdash B \supset \overline{\forall} A_1 \wedge \ldots \overline{\forall} A_m$ (or equivalently, $L \vdash B \supset \overline{\forall}(A_1 \wedge \ldots \wedge A_m)$);

(4) $L + \Theta \vdash A$ iff $L \vdash B \supset A$ for some $B \in \mathrm{Sub}(\Theta)$.

Informally speaking, (4) means that a $\wedge\forall$-perfect set of axioms allows for the simplest natural form of the deduction theorem.

Proof (1) $\Rightarrow$ (2). If $L \vdash B_i \supset \overline{A_i^m}$ for $i = 1, \ldots, m$ and $L \vdash B \supset B_1 \wedge \ldots \wedge B_m$, then $L \vdash B \supset \overline{A_1^m} \wedge \ldots \wedge \overline{A_m^m}$.

(2) $\Rightarrow$ (3). If $\overline{\forall} A'_i \in \mathrm{Sub}(\overline{A_i^{k_i}})$, for $i = 1, \ldots, m$, then for $k \geq \max(m, k_1, \ldots, k_m)$, a substitution instance of $(A_1 \dot\wedge \ldots \dot\wedge A_k)^k$ implies $\overline{\forall} A'_1 \wedge \ldots \wedge \overline{\forall} A'_m$ in **QH** (where e.g. $A_i = A_m$ for $m < i \leq k$).

(3) $\Rightarrow$ (4). If $L + \Theta \vdash A$, then by the deduction theorem 2.8.3, $L \vdash \overline{\forall} A_1 \wedge \ldots \wedge \overline{\forall} A_m \supset A$ for some $A_1, \ldots, A_m \in \mathrm{Sub}(\Theta)$.

The converse implications (and (4) $\Rightarrow$ (1)) are obvious. ■

Similarly in the case of $\forall$- or $\wedge$-perfection we can somewhat simplify the deduction theorem:

Lemma 2.9.9 *(1) If Θ is $\forall$-perfect in L, then $L + \Theta \vdash A$ iff* $\mathrm{Sub}(\Theta) \vdash_L A$.

(2) If Θ is $\wedge$-perfect in L, then $L + \Theta \vdash A$ iff $L \vdash \overline{\forall} B \supset A$ for some $B \in$ $\mathrm{Sub}(\Theta)$.

Proof (1) By 2.8.3, $L + \Theta \vdash A$ iff $\overline{\mathrm{Sub}}(\Theta) \vdash_L A$, so $\mathrm{Sub}(\Theta) \vdash_L A$ implies $L + \Theta \vdash A$.

The other way round, if $\overline{\mathrm{Sub}}(\Theta) \vdash_L A$, then

$$\overline{\forall} A_1 \wedge \ldots \wedge \overline{\forall} A_m \vdash_L A$$

for some $A_1, \ldots, A_m \in \mathrm{Sub}(\Theta)$. By ($\forall$-p) there exist $B_i \in \mathrm{Sub}(\Theta)$ such that $L \vdash B_i \supset \overline{\forall} A_i$; thus $\mathrm{Sub}(\Theta) \vdash_L A$.

(2) 'If' is again obvious. The other way round, suppose $\overline{\mathrm{Sub}}(\Theta) \vdash_L A$, i.e.

$$\overline{\forall} A_1 \wedge \ldots \wedge \overline{\forall} A_m \vdash_L A$$

for $A_1, \ldots, A_m \in \mathrm{Sub}(\Theta)$. Since

$$\mathbf{QH} \vdash \overline{\forall} A_1 \wedge \ldots \wedge \overline{\forall} A_m \equiv \overline{\forall}(A_1 \wedge \ldots \wedge A_m),$$

by the deduction theorem we obtain

$$L \vdash \overline{\forall}(A_1 \wedge \ldots \wedge A_m) \supset A.$$

By ($\wedge$-p) then

$$L \vdash B \supset A_1 \wedge \ldots \wedge A_m$$

for some $B \in \Theta$; hence by monotonicity

$$L \vdash \overline{\forall} B \supset \overline{\forall}(A_1 \wedge \ldots \wedge A_m)$$

and eventually

$$L \vdash \overline{\forall} B \supset A.$$

∎

Lemma 2.9.10 *(1) For any theory Θ, the theories $\Theta^{\forall}, \Theta^{\wedge}, \Theta^{\wedge\forall}$ are respectively $\forall$-perfect, $\wedge$-perfect, and $\wedge\forall$-perfect.*

(2) For any theories $\Theta_1, \ldots, \Theta_m$, the theory $(\Theta_1 \dot{\vee} \ldots \dot{\vee} \Theta_m)^*$ is $\forall$-perfect

Proof (1) Straightforward.

(2) By 2.6.15(xx)

$$\mathbf{QH} \vdash \overline{\forall} C_1 \vee \ldots \vee \overline{\forall} C_m \supset \overline{\forall}(C_1 \vee \ldots \vee C_m),$$

hence

$$\mathbf{QH} \vdash \overline{(A_1 \dot{\vee} \ldots \dot{\vee} A_m)^{k+l}} \supset \overline{\overline{((A_1 \dot{\vee} \ldots \dot{\vee} A_m)^k)^l}}.$$

Thus ($\forall$-p^-) holds. ∎

So every recursively axiomatisable s.p.l. has a recursive $\wedge\forall$-perfect (in **QH**) axiomatisation.[19]

Sometimes one can construct simpler perfect axiomatisations for $\mathbf{QH} + \Theta$ (or for $L + \Theta$, with a superintuitionistic L). E.g. if a set Θ_0 is $\forall$-perfect, then $\Theta_0^{\wedge}$ (or $(\Theta_0 \cup \Theta^{\forall})^{\wedge}$ for any Θ) is $\wedge\forall$-perfect, etc.

To construct the sets $\Theta^{\forall}, \Theta^{\wedge}, \Theta^{\forall\wedge}$ from Θ we usually need infinitely many additional universal quantifiers or conjuncts, so these sets are infinite for any finite set Θ (unless $\Theta \subseteq \{\bot, \top\}$).[20] But sometimes finite perfect extensions also exist.

Lemma 2.9.11 *If $\Theta_1, \ldots, \Theta_m$ are $\wedge\forall$-perfect, then*

(1) $(\Theta_1 \dot{\vee} \ldots \dot{\vee} \Theta_m)^*$ *is $\wedge\forall$-perfect;*

(2) $\mathbf{QH} + (\Theta_1 \dot{\vee} \ldots \dot{\vee} \Theta_m)^* = \mathbf{QH} + \Theta_1 \dot{\vee} \ldots \dot{\vee} \Theta_m$.

Proof $\forall$-perfection follows from Lemma 2.9.10. To show $\wedge$-perfection, consider $C_1, \ldots, C_k \in (\Theta_1 \dot{\vee} \ldots \dot{\vee} \Theta_m)^*$, with

$$C_i = \overline{A_{i1} \dot{\vee} \ldots \dot{\vee} A_{im}}^{\,l_i},$$

$A_{ij} \in \Theta_j$. As noted above, we may assume that different C_i have no common predicate letters. Then every C_i can be presented as

$$C_i = \overline{\forall}(A'_{i1} \vee \ldots \vee A'_{im}),$$

where $A'_{ij} \in Sub(\Theta_j)$. By $\forall\wedge$-perfection (2.9.8(3)), there exist $B_j \in Sub(\Theta_j)$ such that

$$L \vdash B_j \supset \overline{\forall} \bigwedge_{i=1}^{k} A'_{ij}.$$

Then

$$L \vdash \left(\bigvee_{j=1}^{m} B_j \right) \supset \bigwedge_{i=1}^{k} C_i,$$

with $\left(\bigvee_{j=1}^{m} B_j \right) \in Sub(\Theta_1 \dot{\vee} \ldots \dot{\vee} \Theta_m)$. ∎

Remark 2.9.12 Another kind of perfection was introduced in [Yokota, 1989], cf. also [Skvortsov, 2004]. A theory Θ is called *arity-perfect* in a logic L if for any $A \in$ Sub(Θ) there exists a closed substitution instance B' of some $B \in \Theta$ such that $L \vdash B' \supset A$ (and thus $L \vdash B' \supset \overline{\forall} A$).

[19] Of course every s.p.l. itself is $\wedge\forall$-perfect, but such a trivial axiomatisation is not recursive.

[20] Formally speaking, $\Theta^{\forall}$ and $\Theta^{\wedge}$ are infinite even for $\Theta = \{\bot\}$, but repetitions and dummy quantifiers can be eliminated, so we can say that $\{\bot\}^{\forall} = \{\bot\}^{\forall\wedge} = \{\bot\}$.

Our notion of $\forall$-perfection is weaker, because B in 2.9.3($\forall$-p) is not necessarily closed[21].

Note that the sets $\Theta^{\wedge}$, $\Theta^{\wedge\forall}$ are arity-perfect.

Lemma 2.9.13 *L-sub-equivalence preserves all forms of perfection.*

2.10 Intersections

Proposition 2.10.1 *Let L be a superintuitionistic predicate logic, and let Γ, Γ' be sets of sentences such that formulas from Γ and Γ' have no common predicate letters. Then:*

(1) $(L+\Gamma)\cap(L+\Gamma') = L+\{\overline{A^m}\vee\overline{B^n} \mid A\in\Gamma,\ B\in\Gamma',\ m,n\in\omega\} = L+\{\overline{A^m}\vee\overline{B^m} \mid A\in\Gamma,\ B\in\Gamma',\ m\in\omega\}$;

(2) (cf. [Ono, 1973, Theorem 5.5]) If $L\vdash CD$ then

$$(L+\Gamma)\cap(L+\Gamma') = L+\{A\vee B \mid A\in\Gamma,\ B\in\Gamma'\}.$$

(3) If $L+\Gamma\vdash CD$ and $L+\Gamma'\vdash CD$, then

$$(L+\Gamma)\cap(L+\Gamma') = L+CD+\{A\vee B \mid A\in\Gamma,\ B\in\Gamma'\}.$$

Proof

(1) Let $(L+\Gamma)\cap(L+\Gamma')\vdash A$. Then by Theorem 2.8.3,

$$L\vdash\bigwedge_{i=1}^{k}\overline{\forall}A_i\supset A \quad\text{and}\quad L\vdash\bigwedge_{j=1}^{l}\overline{\forall}B_j\supset A$$

for some $A_1,\dots A_k\in\mathrm{Sub}(\Gamma)$, $B_1,\dots B_l\in\mathrm{Sub}(\Gamma')$. Thus

$$L\vdash\bigwedge_{i,j}(\overline{\forall}A_i\vee\overline{\forall}B_j)\supset A.$$

Assuming that $|FV(A_i)| = m$, $|FV(B_j)| = n$, we obtain that $\overline{\forall}A_i\vee\overline{\forall}B_j$ is a substitution instance of $\overline{A^m}\vee\overline{B^n}$. So we obtain the first equality in (1).

Now note that by substitution and omitting dummy quantifiers,

$$L+\overline{A^{m_1}}\vee\overline{B^{n_1}}\vdash\overline{A^m}\vee\overline{B^n},$$

if $m\leq m_1$, $n\leq n_1$. Hence

$$L+\overline{A^s}\vee\overline{B^s}\vdash\overline{A^m}\vee\overline{B^n},$$

where $s=\max(m,n)$, and the second equality follows.

[21] Recent counterexamples by D. Skvortsov show that $\forall$-perfection is strictly weaker than arity-perfection.

(2) We use (1). Assume that $A \in \Gamma$, $B \in \Gamma'$. Obviously, $L + \overline{A^0} \vee \overline{B^0} \vdash A \vee B$, thus $A \vee B \in (L+\Gamma) \cap (L+\Gamma')$. So it remains to show that for any $m \in \omega$

$$(*) \quad L + A \vee B \vdash \overline{A^m} \vee \overline{B^m}.$$

Let $\overline{A^m} = \forall u_1 \dots \forall u_m\, A^m$, $\overline{B^m} = \forall v_1 \dots \forall v_m\, B^m$, and

$$C := \forall u_1 \forall v_1 \dots \forall u_m \forall v_m (A^m \vee B^m).$$

Then $L + A \vee B \vdash C$ and $L \vdash C \supset \overline{A^m} \vee \overline{B^m}$, by CD. Hence $(*)$ follows.

(3) follows from (2): take $L + CD$ as L.

■

Remark 2.10.2 The assumption $L \vdash CD$ in Proposition 2.10.1(2) is necessary. [Ono, 1972/73] gives an example of predicate logics for which $(**)$ does not hold

Similarly one can describe intersections of modal predicate logics.

Proposition 2.10.3 *Assume that L is a N-modal predicate logic, and Γ, Γ' have no common predicate letters. Then*

(1) $(L+\Gamma) \cap (L+\Gamma') =$
$L + \{\Box_\alpha \overline{A^m} \vee \Box_\beta \overline{B^n} \mid A \in \Gamma,\ B \in \Gamma';\ m, n \in \omega;\ \alpha, \beta \in I_N^\infty\}$.

For particular classes of 1-modal logics this presentation can be simplified, cf. Section 1.1:

(a) For logics above **QT**:

$$\begin{aligned}&(\mathbf{\Lambda} + \Gamma) \cap (\mathbf{\Lambda} + \Gamma') \\ &\quad = L + \{\Box^r \overline{A^m} \vee \Box^s \overline{B^n} \mid A \in \Gamma,\ B \in \Gamma';\ m, n, r, s \in \omega\}.\end{aligned}$$

(b) For logics above **QK4**:

$$(\mathbf{\Lambda} + \Gamma) \cap (\mathbf{\Lambda} + \Gamma') = L + \{\Box^r \overline{A^m} \vee \Box^s \overline{B^n} \mid A \in \Gamma, B \in \Gamma'; m, n, \in \omega; r, s \in \{0,1\}\}.$$

(c) For logics above **QS4**:

$$(\mathbf{\Lambda} + \Gamma) \cap (\mathbf{\Lambda} + \Gamma') = L + \{\Box \overline{A^m} \vee \Box \overline{B^n} \mid A \in \Gamma,\ B \in \Gamma';\ m, n \in \omega\}.$$

(2) If $L \vdash Ba$, then:

$$(L+\Gamma) \cap (L+\Gamma') = L + \{\Box_\alpha A \vee \Box_\beta B \mid A \in \Gamma,\ B \in \Gamma'; \alpha, \beta \in I_N^\infty\}.$$

Similarly, to (i), there exist simpler versions for the cases $L \supseteq$ **QK4** $+ Ba$, **QK4** $+ Ba$ *or* **QS4** $+ Ba$.

Therefore we have

Proposition 2.10.4 *The complete lattices of predicate logics (superintuitionistic or modal) are well-distributive, i.e. they are Heyting algebras. Moreover, the intersection of two recursively axiomatisable logics is recursively axiomatisable. The intersection of finitely axiomatisable logics is finitely axiomatisable for superintuitionistic logics* $\supseteq \mathbf{QH} + CD$ *and for* 1*-modal logics* $\supseteq \mathbf{QK4} + Ba$.

Later in this book we will show that the intersection of finitely axiomatisable superintuitionistic predicate logics may be not finitely axiomatisable, if one of them does not contain CD.

2.11 Gödel–Tarski translation

Definition 2.11.1 Gödel–Tarski translation *for predicate formulas is the map* $(-)^T : IF^= \longrightarrow MF^=$ *defined by the following clauses:*

$$\begin{aligned}
&A^T = \Box A \text{ for } A \text{ atomic};\\
&(A \wedge B)^T = A^T \wedge B^T;\\
&(A \vee B)^T = A^T \vee B^T;\\
&(A \supset B)^T = \Box(A^T \supset B^T);\\
&(\exists x A)^T = \exists x A^T;\\
&(\forall x A)^T = \Box \forall x A^T.
\end{aligned}$$

For a set $\Gamma \subseteq IF^=$ *put* $\Gamma^T := \{A^T \mid A \in \Gamma\}$.

Gödel–Tarski translation is obviously extended to formulas with constants.

Since $\mathbf{QS4}^= \vdash (x = y) \equiv \Box(x = y)$, we may also define $(x = y)^T$ just as $x = y$.

Lemma 2.11.2 $\mathbf{QS4}^{(=)} \vdash \Box A^T \equiv A^T$ *for any* $A \in IF^{(=)}$.

Proof By induction on the length of A. ■

Lemma 2.11.3 *(1) Let* $[x \mapsto y]$ *be a (simple) variable transformation. Then for any* $A \in IF^=$,

$$(A[x \mapsto y])^T = A^T[x \mapsto y].$$

(2) For any formula with constants $A \in IF^=(D)$, *for any* D*-transformation* $[\mathbf{x} \mapsto \mathbf{a}]$, $([\mathbf{a}/\mathbf{x}]A)^T = [\mathbf{a}/\mathbf{x}](A^T)$.

Proof (1) Easy, by induction on $|A|$. We consider only the case $A = \forall z B$. If $z \neq x$, we have

$$\begin{aligned}
&(A[x \mapsto y])^T = (\forall z(B[x \mapsto y]))^T = \Box\forall z(B[x \mapsto y])^T\\
&= \Box\forall z(B^T[x \mapsto y]) \text{ (by the induction hypothesis)}\\
&= (\Box\forall z B^T)[x \mapsto y] = A^T[x \mapsto y].
\end{aligned}$$

If $z = x$, we have

$$\begin{aligned}
&(A[x \mapsto y])^T = (\forall y(B[x \mapsto y]))^T = \Box\forall y(B[x \mapsto y])^T\\
&= \Box\forall y(B^T[x \mapsto y]) \text{ (by the induction hypothesis)}\\
&= (\Box\forall x B^T)[x \mapsto y] = A^T[x \mapsto y].
\end{aligned}$$

(2) An exercise. ■

Lemma 2.11.4 *If $A \stackrel{\circ}{=} B$ for $A, B \in IF^{=}$, then $A^T \stackrel{\circ}{=} B^T$.*

Proof We use Proposition 2.3.17. Consider the equivalence relation

$$A \sim B := (A^T \stackrel{\circ}{=} B^T)$$

on $IF^{=}$. It is sufficient to show that $\sim$ satisfies the conditions (1)–(3) from 2.3.17.

We check only (1) for $\mathcal{Q} = \forall$:

$$(\forall x A)^T \stackrel{\circ}{=} (\forall y (A[x \mapsto y]))^T.$$

The remaining (routine) part is left to the reader. In fact, by 2.11.1 and 2.11.3

$$(\forall y (A[x \mapsto y]))^T = \Box\forall y (A[x \mapsto y])^T = \Box\forall y (A^T[x \mapsto y]),$$

which is congruent to $(\forall x A)^T = \Box\forall x A^T$ by 2.3.17. ■

Lemma 2.11.5 *Let $S = [B(\mathbf{x}, \mathbf{y})/P(\mathbf{x})]$ be an $IF^{(=)}$-substitution, and consider the $MF^{(=)}$-substitution*

$$S^T := [B^T(\mathbf{x}, \mathbf{y})/P(\mathbf{x})].$$

Then for any $A \in IF^{(=)}$

$$(*) \quad \mathbf{QS4}^{(=)} \vdash (SA)^T \equiv S^T A^T.$$

Proof By induction on A.

- If $A = P(\mathbf{z})$, then $(SA)^T = B^T(\mathbf{z}, \mathbf{y})$, and

$$S^T A^T = S^T(\Box P(\mathbf{z})) = \Box B^T(\mathbf{z}, \mathbf{y}),$$

  which is equivalent to $B^T(\mathbf{z}, \mathbf{y})$ by Lemma 2.11.2.

- If A is atomic, $A \neq P(\mathbf{z})$, the claim is trivial.

- If $A = \forall z C$ is clean and also $BV(A) \cap FV(S) = \varnothing$,[22] then

$$(SA)^T = (\forall z SC)^T = \Box\forall z (SC)^T,$$

$$S^T A^T = S^T(\Box\forall z C^T) = \Box\forall z S^T C^T.$$

  But

$$\mathbf{QS4}^{(=)} \vdash (SC)^T \equiv S^T C^T$$

  — by the induction hypothesis, and hence

$$\mathbf{QS4}^{(=)} \vdash \Box\forall z (SC)^T \equiv \Box\forall z S^T C^T$$

  — by replacement. So $(*)$ holds.

[22] Otherwise we can consider a congruent formula with this property.

- If $A = \exists zC$, then

$$\begin{array}{l}(SA)^T = (\exists zSC))^T = \exists z(SC)^T, \\ S^T A^T = S^T(\exists zC^T) = \exists zS^T C^T.\end{array}$$

Now again we have $\mathbf{QS4}^{(=)} \vdash (SA)^T \equiv S^T A^T$ by the induction hypothesis and replacement.

- If $A = C \vee D$, then

$$\begin{array}{l}(SA)^T = (SC \vee SD)^T = (SC)^T \vee (SD)^T, \\ S^T A^T = S^T(C^T \vee D^T) = S^T C^T \vee S^T D^T.\end{array}$$

By the induction hypothesis $\mathbf{QS4} \vdash (SC)^T \equiv S^T C^T$, $(SD)^T \equiv S^T D^T$, so we can apply a replacement.

The case $A = C \wedge D$ is almost the same, and we skip it.

- If $A = C \supset D$, then

$$\begin{array}{l}(SA)^T = (SC \supset SD)^T = \Box((SC)^T \supset (SD)^T), \\ S^T A^T = S^T(\Box(C^T \supset D^T)) = \Box(S^T C^T \supset S^T D^T).\end{array}$$

In this case we can also use **QS4**-theorems

$$(SC)^T \equiv S^T C^T,\ (SD)^T \equiv S^T D^T$$

and replacement.

■

Lemma 2.11.6 $(\mathrm{Sub}(\Gamma))^T \subseteq \mathbf{QS4}^{(=)} + \mathrm{Sub}(\Gamma^T)$ *for any* $\Gamma \subseteq IF^{(=)}$

Proof By Lemma 2.11.5 we have $(SA)^T \in \mathbf{QS4}^{(=)} + S^T A^T$ for any simple substitution S. Now recall that every substitution is reducible to simple ones. ■

Lemma 2.11.7 *For a list of variables* $\mathbf{x}$ *and an* $IF^{(=)}$*-formula* $A(\mathbf{x})$,

$$\mathbf{QS4}^{(=)} \vdash \Box\forall\mathbf{x}A^T \equiv (\forall\mathbf{x}A)^T.$$

Proof By induction on the length of $\mathbf{x}$. The base: $\mathbf{QS4} \vdash A^T \equiv \Box A^T$ (2.11.2).

For the induction step, assume $\Box\forall\mathbf{x}A^T \equiv (\forall\mathbf{x}A)^T$. Then in **QS4** we obtain: $\Box\forall y\forall\mathbf{x}A^T \equiv \Box\forall y\Box\forall\mathbf{x}A^T$ (Lemma 2.6.202)$\equiv \Box\forall y(\forall\mathbf{x}A)^T$ (by assumption and replacement) $\equiv (\forall y\forall\mathbf{x}A)^T$. ■

Proposition 2.11.8 *For any m.p.l.(=) L containing* **QS4** *the set*

$${}^T L := \{A \in IF^{(=)} \mid A^T \in L\}$$

is an s.p.l. (=).

Proof We check that ${}^{T}L$ satisfies the conditions $(s1)$–$(s5)$. By Proposition 1.5.2, $A^T \in \mathbf{QS4}$ for every propositional intuitionistic axiom A. The same holds for predicate axioms and for the axioms of equality, as one can easily see. It is also clear that ${}^{T}L$ satisfies (s3) from Section 1.2. Corollary 2.11.6 shows that ${}^{T}L$ is closed under $IF^{(=)}$- substitutions. ■

Definition 2.11.9 *The above defined s.p.l.(=)* ${}^{T}L$ *is called the* superintuitionistic *fragment of the m.p.l (=)L; an m.p.l.(=) L is called a* modal counterpart *of the s.p.l.(=)* ${}^{T}L$*, cf. Definition 1.5.4.*

Lemma 2.11.10 *For any* $A \in IF^{(=)}$,

$$\mathbf{QH}^{(=)} \vdash A \Rightarrow \mathbf{QS4}^{(=)} \vdash A^T.$$

Proof By induction on the length of the proof of A. An alternative proof makes use of completeness, see below. ■

Lemma 2.11.11 *Let* $L = \mathbf{QH}^{(=)} + \Gamma$ *be an s.p.l.(=). Then for any* $A \in IF^{(=)}$,

$$L \vdash A \Rightarrow \mathbf{QS4}^{(=)} + \Gamma^T \vdash A^T.$$

Proof Suppose $L \vdash A$. Then by Theorem 2.8.3, $\overline{\text{Sub}}(\Gamma) \vdash_{\mathbf{QH}^{(=)}} A$, i.e. there exists formulas $A_1, \ldots, A_n \in \Gamma$ and substitutions $S_1, \ldots, S_n$ such that $\mathbf{QH}^{(=)} \vdash \bigwedge\limits_{i=1}^{n} \overline{\forall} S_i A_i \supset A$. Then by Lemma 2.11.10,

$$\mathbf{QS4}^{(=)} \vdash \left(\bigwedge_{i=1}^{n} \overline{\forall} S_i A_i \bullet \supset A\right)^T \left(= \Box \left(\bigwedge_{i=1}^{n} (\overline{\forall} S_i A_i)^T \bullet \supset A^T\right)\right).$$

Hence

$$\mathbf{QS4}^{(=)} \vdash \bigwedge_{i=1}^{n} (\overline{\forall} S_i A_i)^T \bullet \supset A^T.$$

By definition and Lemma 2.11.5, $\mathbf{QS4}^{(=)} \vdash (\overline{\forall} S_i A_i)^T \equiv \Box\overline{\forall} S_i^T A_i^T$. Therefore $\Box\overline{\text{Sub}}(\Gamma) \vdash_{\mathbf{QS4}^{(=)}} A^T$, i.e. $\mathbf{QS4}^{(=)} + \Gamma^T \vdash A^T$ by Theorem 2.8.4. ■

As we shall see (cf. Proposition 2.16.17), in many cases the converse of 2.11.11 also holds, i.e.

$$\mathbf{QH}^{(=)} + \Gamma = {}^{T}(\mathbf{QS4}^{(=)} + \Gamma^T),$$

or

$$\mathbf{QH} + \Gamma^{(=)} \vdash A \text{ iff } \mathbf{QS4}^{(=)} + \Gamma \vdash A^T. \qquad (*)$$

In particular, it is well known that

$${}^{T}\mathbf{QS4}^{(=)} = \mathbf{QH}^{(=)}$$

[Schütte, 1968], [Rasiowa and Sikorski, 1963].

Let us make another simple observation.

Lemma 2.11.12 *If an s.p.l.(=) $L = \mathbf{QH}^{(=)} + \Gamma$ has modal counterparts, then $\mathbf{QS4}^{(=)} + \Gamma^T$ is the smallest modal counterpart of L.*

Proof Suppose $L = {}^T\mathbf{\Lambda}$. Then obviously, $\mathbf{QS4}^{(=)} + \Gamma^T \subseteq \mathbf{\Lambda}$. By Lemma 2.11.11, $L \vdash A$ only if $\mathbf{QS4}^{(=)} + \Gamma^T \vdash A$. On the other hand, if $L \nvdash A$, then $\mathbf{\Lambda} \nvdash A^T$, and thus $\mathbf{QS4}^{(=)} + \Gamma^T \nvdash A^T$. Therefore $\mathbf{QS4}^{(=)} + \Gamma^T$ is a modal counterpart of L. ∎

Remark 2.11.13 The paper [Pankratyev, 1989] states that every s.p.l. has modal counterparts. However the proof in this paper requires some verification, which we postpone until Volume 2.

We do not know if there exist the greatest modal counterparts in the predicate case, and Theorem 1.5.6 has no direct analogue:

Theorem 2.11.14 *(1) [Pankratyev, 1989]* **QGrz** *is a modal counterpart of* **QH**.

(2) [Naumov, 1991] There exists a proper extension of **QGrz** *which is also a modal counterpart of* **QH**.

These matters will also be discussed in Volume 2.

2.12 The Glivenko theorem

Proposition 2.12.1 (Predicate version of the Glivenko theorem) *For any intuitionistic predicate formula A*

$$\mathbf{QCL} \vdash \neg A \text{ iff } \mathbf{QH} + KF \vdash \neg A.$$

Proof Let $\mathbf{QCL} \vdash \neg A$. Then by the deduction theorem 2.8.3

$$\mathbf{QH} \vdash \bigwedge_s \overline{\forall}(A_s \vee \neg A_s) \supset \neg A$$

for some formulas A_s. Then by Lemma 1.1.2, we have

$$(1) \quad \mathbf{QH} \vdash \bigwedge_s \neg\neg\overline{\forall}(A_s \vee \neg A_s) \supset \neg A.$$

By Lemma 2.6.20(II),

$$(2) \quad \mathbf{QH} + KF \vdash \forall x \neg\neg B \supset \neg\neg \forall x B.$$

By Corollary 1.1.9, $\mathbf{QH} \vdash \overline{\forall}\neg\neg(A_s \vee \neg A_s)$, so by applying (2), we obtain:

$$\mathbf{QH} + KF \vdash \bigwedge_s \neg\neg\overline{\forall}(A_s \vee \neg A_s).$$

Hence by (1), $\mathbf{QH} + KF \vdash \neg A$. ∎

Corollary 2.12.2 *For any* $\neg\wedge\forall$*-formula* A,

$$\mathbf{QCL} \vdash A \text{ iff } \mathbf{QH} + KF \vdash A.$$

Proof By induction on A (exercise). ∎

We say that an intermediate predicate logic L *satisfies the Glivenko property* if for any formula A,

$$L \vdash \neg A \iff \mathbf{QCL} \vdash \neg A$$

and that L *has the classical* $\neg\wedge\forall$*- fragment* if for any $\neg\wedge\forall$-formula A,

$$L \vdash A \iff \mathbf{QCL} \vdash A.$$

Theorem 2.12.3 *The following conditions are equivalent (for a superintuitionistic predicate logic* L*):*

(1) L *satisfies the Glivenko property;*

(2) L *has the classical* $\neg\wedge\forall$*-fragment;*

(3) $\mathbf{QH} + KF \subseteq L \subseteq \mathbf{QCL}$.

In other words, $\mathbf{QH} + KF$ is the smallest intermediate predicate logic with the property (i) or (ii).

Proof By Lemma 2.6.20(II),

$$\mathbf{QH}+KF = \mathbf{QH}+\forall x\neg\neg P(x) \supset \neg\neg\forall x P(x) = \mathbf{QH}+\neg(\forall x\neg\neg P(x) \wedge \neg\forall x P(x)).$$

Thus (i), (ii) imply $\mathbf{QH}+KF \subseteq L$. (i) also implies $L \subseteq \mathbf{QCL}$. In fact, suppose $A \in (L - \mathbf{QCL})$. Since $\mathbf{QH} \vdash A \supset \neg\neg A$ and $\mathbf{QCL} \vdash A \equiv \neg\neg A$, we then have $\neg\neg A \in (L - \mathbf{QCL})$, which contradicts the Glivenko property.

Finally, (ii) implies $L \subseteq \mathbf{QCL}$ as well. In fact, assume (ii) and suppose $A \in (L - \mathbf{QCL})$ and $\mathbf{QCL} \vdash A \equiv B$, where B is a $\neg\wedge\forall$-formula; then $\mathbf{QCL} \vdash \neg\neg(A \equiv B)$, whence $\mathbf{QH}+KF \vdash \neg\neg(A \equiv B)$, by Proposition 2.12.1. As we have proved, $L \vdash KF$. So we obtain $L \vdash \neg\neg(A \equiv B)$, whence $L \vdash \neg\neg A \equiv \neg\neg B$, by Lemma 1.1.2. Since $A \in L$, we also have $\neg\neg A \in L$, and thus $\neg\neg B \in (L - \mathbf{QCL})$, which contradicts (ii). ∎

The same result holds for the logic with equality $\mathbf{QH}^{=} + KF$.

2.13 Δ-operation

Definition 2.13.1 *For a predicate formula* A *put* $\delta A := p \vee (p \supset A)$*, where* p *is a proposition letter that does not occur in* A*. For a superintuitionistic predicate logic* L *put* $\Delta L := \mathbf{QH} + \{\delta A \mid A \in L\}$ *and also* $\Delta^0 L := L$, $\Delta^{n+1} L := \Delta\Delta^n L$ *by induction.*

Lemma 2.13.2 *The following formulas are theorems of* **QH**:

(1) $A \supset \delta A$

(2) $\delta(A \supset B) \supset_{\bullet} \delta A \supset \delta B$,

(3) $\delta\overline{\forall}A \supset \overline{\forall}\delta A$.

Proof (1) $A \vdash_{\mathbf{QH}} p \supset A$ and $p \supset A \vdash_{\mathbf{QH}} \delta A$; hence $A \vdash_{\mathbf{QH}} \delta A$. Then apply the deduction theorem.

(2) First note that by (Ax2),

$$p \supset (A \supset B),\ p \supset A \vdash_{\mathbf{QH}} p \supset B.$$

Hence by ∨-introduction

$$p \vee (p \supset (A \supset B)),\ p \vee (p \supset A) \vdash_{\mathbf{QH}} p \vee (p \supset B).$$

Now (2) follows by the deduction theorem.

(3) We have $\mathbf{QH} \vdash \overline{\forall}A \supset A$ by 2.6.15. Hence $\mathbf{QH} \vdash \delta\overline{\forall}A \supset \delta A$ by (1), and eventually $\mathbf{QH} \vdash \delta\overline{\forall}A \supset \overline{\forall}\delta A$ by the Bernays rule. ∎

So ΔL can be axiomatised as follows.

Lemma 2.13.3 $\Delta L = \mathbf{QH} + \{\delta A \mid A \in \overline{L}\}$.

Proof Let $\overline{\Delta L} := \mathbf{QH} + \{\delta A \mid A \in \overline{L}\}$; then clearly $\overline{\Delta L} \subseteq \Delta L$. The other way round, if $A \in L$, then $\overline{\forall}A \in \overline{L}$, so $\delta\overline{\forall}A \in \overline{\Delta L}$. By Lemma 2.13.2, $\delta\overline{\forall}A$ implies $\overline{\forall}\delta A$; so $\delta A \in \overline{\Delta L}$. Therefore $\Delta L \subseteq \overline{\Delta L}$. ∎

In [Komori, 1983] ΔL was defined as $\mathbf{QH} + \{\delta' A \mid A \in L\}$, where $\delta' A := ((p \supset A) \supset p) \supset p$. As in the propositional case, both definitions are equivalent:

Lemma 2.13.4 $\mathbf{QH} + \delta A = \mathbf{QH} + \delta' A$ *for any formula* A.

Proof Quite similar to 1.16.2 using the deduction theorem. ∎

Lemma 2.13.5 *(1)* $\Delta L \subseteq L$.

(2) $L_1 \subseteq L_2 \Longrightarrow \Delta L_1 \subseteq \Delta L_2$.

(3) $\Delta(\mathbf{QH} + \bot) = \mathbf{QCL}$.

(4) $\Delta L \subset \mathbf{QCL}$ *for every consistent s.p.l.* L.

Proof (1) Since $\mathbf{QH} \vdash A \supset \delta A$ for any A.

(2) Trivial.

(3) $\Delta(\mathbf{QH} + \bot) \subseteq \mathbf{QCL}$, since $\mathbf{QCL} \vdash \delta A$ for every A. The other way round, $\mathbf{QCL} \subseteq \Delta(\mathbf{QH} + \bot)$, since $\mathbf{QCL} = \mathbf{QH} + EM$ and $EM = p \vee \neg p$ is just $\delta\bot = p \vee (p \supset \bot)$.

(4) follows from (3) by Proposition 2.12.5.

∎

In Chapter 6 we will prove the converse to (2), similarly to 1.16.7 for the propositional case.

The logics

$$\mathbf{QHP}_n^+ := \Delta^n(\mathbf{QH} + \bot) = \Delta^{n-1}(\mathbf{QCL})$$

were first introduced in [Komori, 1983] (where they were denoted by $\Delta^k(W)$, for $W := \mathbf{QH} + \bot$).

Let us now turn to a finite axiomatisation of $\mathbf{QHP}_n^+$ presented in [Yokota, 1989].

First we generalise δ-operation as follows [Yokota, 1989].

Definition 2.13.6 *Let*

$$\delta_{k,P}A := \forall \mathbf{y}(P(\mathbf{y}) \vee (P(\mathbf{y}) \supset A)),$$

where $P \in PL^k$, $\mathbf{y}$ is a distinct list of variables, $r(\mathbf{y}) \cap FV(A) = \varnothing$. If P does not occur in A, we use the notation $\delta_k A$ rather than $\delta_{k,P}A$[23] *Also put*

$$\delta_{k_1\ldots k_n}A := \delta_{k_1}\ldots\delta_{k_n}A,\ \ \delta_k^n A := \underbrace{\delta_k\ldots\delta_k}_{n} A^{24}$$

(in particular, $\delta_k^0 A = A$). For a set Θ of formulas and $\gamma = \delta_k,\ \delta_k^n$ etc. put

$$\gamma\Theta := \{\gamma A \mid A \in \Theta\}.$$

We also introduce

$$\delta_\infty^n\Theta := \bigcup_{k\in\omega}\delta_k^n\Theta,$$
$$\delta_*^n\Theta := \underbrace{\delta_\infty^1\ldots\delta_\infty^1}_{n}\Theta = \bigcup_{k_1,\ldots,k_n\in\omega}\delta_{k_1\ldots k_n}\Theta$$

for $n \in \omega$.[25]

Obviously,

$$\delta_\infty^n\Theta \subseteq \delta_*^n\Theta.$$

Also note that $\delta_0 A = \delta A$ and $P_n = \delta_0^n\bot$. Hence

$$P_n^+ = \delta_1^n\bot.$$

Lemma 2.13.7 *(1)* $\mathbf{QH} + \delta A \vdash \delta_k A$.

(2) $\delta_k L \subseteq \Delta L$.

(3) $\delta_{k_1,\ldots,k_n} L \subseteq \Delta^n L$.

---

[23] Cf. Definitions 1.2.1 and 1.2.4 in [Yokota, 1989].

[24] The latter notation is somewhat ambiguous as it means $\delta_{k,P_1}\ldots\delta_{k,P_n}A$ for *different* $P_1,\ldots,P_n$ that do not occur in A.

[25] In [Yokota, 1989] the set $\delta_n^*\{\bot\}$ was denoted by P_n^*.

Proof

(1) $\delta_{k,P}A = \forall\mathbf{y}\,[P(\mathbf{y})/p]\,\delta A$, so we can apply substitution and generalisation.

(2) Follows from (1).

(3) Follows from (2) by induction. ∎

The next lemma shows that δ-type operators behave as $\Box$-modalities.

Lemma 2.13.8 *The following are theorems of* **QH**:

(1) $A \supset \bigcirc A$;

(2) $\bigcirc(A \supset B) \supset\!\bullet\, \bigcirc A \supset \bigcirc B$,

(3) $\bigcirc\left(\bigwedge\limits_{i=1}^{n} A_i\right) \equiv \bigwedge\limits_{i=1}^{n} \bigcirc A_i$;

(4) $\bigcirc\overline{\forall} A \supset \overline{\forall} \bigcirc A$;

where $\bigcirc = \delta_{k,P}$ *or* $\delta_{k_1\ldots k_n}$.

Proof

(1), (2). For $\bigcirc = \delta_{k,P}$ this follows from 2.13.2 (1), (2) by substitution $[P(\mathbf{y})/p]$ and generalisation. For $\bigcirc = \delta_{k_1\ldots k_n}$ apply induction.

(3) Similar to 2.13.2 (3). $\mathbf{QH} \vdash \overline{\forall} A \supset A$ implies $\mathbf{QH} \vdash \bigcirc\overline{\forall} A \supset \bigcirc A$ by (1), (2); hence $\mathbf{QH} \vdash \bigcirc\overline{\forall} A \supset \overline{\forall} \bigcirc A$ by the Bernays rule.

(4) We may suppose $n = 2$ and use induction for the general case. So let us show

$$\mathbf{QH} \vdash \bigcirc(A \wedge B) \equiv \bigcirc A \wedge \bigcirc B.$$

Since $\mathbf{QH} \vdash A \wedge B \supset A$, by (1), (2) and (MP) we obtain

$$\mathbf{QH} \vdash \bigcirc(A \wedge B) \supset \bigcirc A.$$

Similarly

$$\mathbf{QH} \vdash \bigcirc(A \wedge B) \supset \bigcirc B,$$

and thus

$$\mathbf{QH} \vdash \bigcirc(A \wedge B) \supset \bigcirc A \wedge \bigcirc B.$$

For the converse note that

$$A \supset (B \supset A \wedge B)$$

by (Ax5), hence

$$\mathbf{QH} \vdash \bigcirc A \supset \bigcirc(B \supset A \wedge B)$$

by (1), (2) and next

$$\mathbf{QH} \vdash \bigcirc A \supset\!\bullet \bigcirc B \supset \bigcirc(A \wedge B)$$

again by (1), (2) and the transitivity of $\supset$. The latter formula is equivalent to

$$\bigcirc A \wedge \bigcirc B \supset \bigcirc(A \wedge B).$$

∎

Proposition 2.13.9 $\mathbf{QH} \vdash \delta_k^n A \supset \delta_k^m A$ *for* $n \leq m$.

Proof Readily follows from 2.13.8(1), since $\delta_k^m = \delta_k^{m-n}\delta_k^n$. ∎

Lemma 2.13.10 *(1)* $\mathbf{QH} \vdash P_n^+ \supset \delta_1^n A,\ P_n \supset \delta_0^n A$ *for any formula* A.

(2) $\mathbf{QH} + \delta_1^n\Theta \subseteq \mathbf{QH} + P_n^+$ *and* $\mathbf{QH} + \delta_0^n\Theta \subseteq \mathbf{QH} + P_n$ *for any theory* Θ.

Proof
(1) By 2.13.8 (2), since $\mathbf{QH} \vdash \bot \supset A$.
(2) Follows from (1). ∎

Lemma 2.13.11 $\mathbf{QH} + \delta_{k_1\dots k_n} A \vdash \delta_{l_1\dots l_n} A$ *whenever* $l_1 \leq k_1, \dots, l_n \leq k_n$.

Proof By substitution and elimination of dummy quantifiers. ∎

The same argument actually proves a stronger claim. Since elimination of dummy quantifiers is legal in **QH**, we obtain:

Proposition 2.13.12 $\delta_\infty^n\Theta \leq_{\mathbf{QH}}^{\mathrm{Sub}} \delta_*^n\Theta$

Proof ∎

Lemma 2.13.13 *For an s.p.l(=)* L *and sets of formulas* Θ_1, Θ_2

(1) If $L + \Theta_1 \subseteq L + \Theta_2$, *then*

(a) $L + \delta_k^n\Theta_1 \subseteq L + \delta_k^n\Theta_2$ *provided* Θ_2 *is* $\forall$*-perfect in* L;

(b) $L + \delta_k^m(\Theta_1^\forall) \subseteq L + \delta_k^n(\Theta_2^\forall)$.

(2) If $L + \Theta_1 = L + \Theta_2$, *then*

(a) $L + \delta_k^n\Theta_1 = L + \delta_k^n\Theta_2$ *provided* Θ_1, Θ_2 *are* $\forall$*-perfect in* L;

(b) $L + \delta_k^n(\Theta_1^\forall) = L + \delta_k^n(\Theta_2^\forall)$.

Proof

(1) (a) If $A \in \Theta_1 \subseteq L + \Theta_1 \subseteq L + \Theta_2$, then by 2.13.8 $Sub(\Theta_2) \vdash A$, i.e. $L \vdash B_1 \wedge \dots \wedge B_m \supset A$ for some $B_1, \dots, B_m \in Sub(\Theta_2)$. Then by the monotonicity of δ_k^n,

$$L \vdash \delta_k^n(\bigwedge_{i=1}^m B_i) \supset \delta_k^n A;$$

thus

$$L \vdash (\bigwedge_{i=1}^k \delta_k^n B_i) \supset \delta_k^n A.$$

Since $\delta_n^k B_i \in Sub(\delta_n^k\Theta_2)$, it follows that $L + \delta_k^n\Theta_2 \vdash \delta_k^n A$. Therefore $\delta_n^k\Theta_1 \subseteq L + \delta_n^k\Theta_2$.

(b) Readily follows from (a), since $\Theta^\forall$ is $\forall$-perfect in $\mathbf{QH}$ and $\mathbf{QH}+\Theta = \mathbf{QH}+\Theta^\forall$.

(2) Follows from (1).

■

Lemma 2.13.14 [26] *If a set Θ is $\forall$-perfect in an s.p.l.(=) L, then $\delta^1_\infty\Theta$ is also $\forall$-perfect in L.*

Proof For an arbitrary $\delta_k A \in \delta^1_\infty\Theta$, with $A \in \Theta$, let us check the property 2.9.3 ($\forall$-p$^-$). We have

$$\overline{(\delta_k A)^m(\mathbf{z})} = \forall\mathbf{z}\forall\mathbf{x}(P(\mathbf{x},\mathbf{z}) \vee (P(\mathbf{x},\mathbf{z}) \supset A^m(\mathbf{z})))),$$

where $\mathbf{xz}$ is a distinct list of new variables, $|\mathbf{z}| = m$, $|\mathbf{x}| = k$, P does not occur in $A^m(\mathbf{z})$. This formula is clearly $\mathbf{QH}$-equivalent to $\delta_{k+m}A^m(\mathbf{z})$. Since Θ is $\forall$-perfect, there exists $B \in \mathrm{Sub}(\Theta)$ such that

$$L \vdash B \supset \forall\mathbf{z}A^m(\mathbf{z}),$$

and thus

$$L \vdash B \supset A^m(\mathbf{z}),$$

We may also assume that $r(\mathbf{xz}) \cap FV(B) = \varnothing$. Hence

$$L \vdash \delta_{k+m}B \supset \delta_{k+m}A^m(\mathbf{z}),$$

by the monotonicity of δ_{k+m} and

$$\delta_{k+m}B = \forall\mathbf{z}\forall\mathbf{x}(P(\mathbf{x},\mathbf{z}) \vee (P(\mathbf{x},\mathbf{z}) \supset B)) \in \mathrm{Sub}(\delta_{k+m}\Theta),$$

as we can choose P that does not occur in $A^m(\mathbf{z})$ and in B.

$$L \vdash \delta_{k+m}B \supset \overline{(\delta_k A)^m(\mathbf{z})},$$

therefore ($\forall$-p$^-$) holds for $\delta_k A$. ■

Proposition 2.13.15 *If Θ is $\forall$-perfect in L, then $\delta^n_\infty\Theta$ and $\delta^n_*\Theta$ are $\forall$-perfect in L, for all $n \geq 0$.*

Proof For $\delta^n_*\Theta = \underbrace{\delta^1_\infty \ldots \delta^1_\infty}_{n}\Theta$ proceed by induction on n. And for $\delta^n_\infty\Theta$ use sub-equivalence: $\delta^n_\infty\Theta \sim^{\mathrm{sub}}_{\mathbf{QH}} \delta^n_*\Theta$.

Alternatively, for $\delta^n_\infty\Theta$ one can apply a direct argument generalising the proof of Lemma 2.13.14. In fact,

$$L \vdash B \supset \forall\mathbf{z}A^m(\mathbf{z})$$

[26] Cf. [Yokota, 1989, Proposition 1.4.6].

implies

$$L \vdash B \supset A^m(\mathbf{z})$$

and next

$$L \vdash \delta^n_{k+m} B \supset \delta^n_{k+m} A^m(\mathbf{z}),$$

with

$$\mathbf{QH} \vdash \delta^n_{k+m} A^m(\mathbf{z}) \equiv \overline{(\delta^n_k A)^m};$$

again we can present $\delta^n_{k+m}B$, $\overline{(\delta^n_k A)^m}$ as

$$\delta^n_{k+m}B = \forall\mathbf{z}\forall\mathbf{x}[P_n(\mathbf{x},\mathbf{z}) \vee (P_n(\mathbf{x},\mathbf{z}) \supset \forall\mathbf{z}\forall\mathbf{x}(P_{n-1}(\mathbf{x},\mathbf{z}) \vee (P_{n-1}(\mathbf{x},\mathbf{z}) \supset \ldots)))],$$

$$\overline{(\delta^n_k A)^m} = \forall\mathbf{z}\forall\mathbf{x}(P_n(\mathbf{x},\mathbf{z}) \vee P_n(\mathbf{x},\mathbf{z}) \supset \forall\mathbf{x}(P_{n-1}(\mathbf{x},\mathbf{z}) \vee (P_{n-1}(\mathbf{x},\mathbf{z}) \supset \ldots))));$$

with the same set of $(k+m)$-ary predicate letters in both formulas. ■

Lemma 2.13.16 *If $L = \mathbf{QH} + \Theta$ and Θ is $\forall$-perfect in* **QH**, *then*

$$\Delta L = \mathbf{QH} + \delta\Theta = \mathbf{QH} + \delta^1_\infty\Theta.$$

Proof By 2.13.13(2a)

$$\Delta L = \mathbf{QH} + \delta L = \mathbf{QH} + \delta\Theta,$$

since $\mathbf{QH} + L = L = \mathbf{QH} + \Theta$ and both Θ and L are $\forall$-perfect in **QH**. Thus

$$\Delta L = \mathbf{QH} + \delta_0\Theta \subseteq \mathbf{QH} + \delta^1_\infty\Theta.$$

Since by 2.13.7 $\delta^1_\infty\Theta \subseteq \Delta L$, it follows that $\Delta L = \mathbf{QH} + \delta^1_\infty\Theta$. ■

Proposition 2.13.17 *(1) If $L = \mathbf{QH} + \Theta$ and Θ is $\forall$-perfect in* **QH**, *then $\Delta^n L = \mathbf{QH} + \delta^n_*\Theta$. for any $n \geq 0$.*

(2) If $L = \mathbf{QH} + \Theta$, then

$$\Delta^n L = \mathbf{QH} + \delta^n_*(\Theta^\forall).$$

Proof (1) By induction on n. If $n = 0$, then $\Delta L = L$, $\delta^n_*\Theta = \Theta$, so the statement is trivial. For the induction step suppose

$$\Delta^n L = \mathbf{QH} + \delta^n_*\Theta.$$

Then by 2.13.16 and 2.13.15,

$$\Delta^{n+1} L = \mathbf{QH} + \delta^1_\infty\delta^n_*\Theta = \mathbf{QH} + \delta^{n+1}_*\Theta.$$

(2) Follows from (1) and the observation that $\mathbf{QH} + \Theta = \mathbf{QH} + \Theta^\forall$ and $\Theta^\forall$ is $\forall$-perfect in **QH**. ■

In particular, for $\Theta = \{\bot\}$ we have

Corollary 2.13.18 [27] $\mathbf{QHP}_n^+ = \mathbf{QH} + \delta_*^n\{\bot\}$.

Now let us show that every $\delta_k^n A$ for $k > 1$ (and hence every $\delta_{k_1,\ldots,k_n}$), is deducible from $\delta_1^n A$.

Lemma 2.13.19 $\mathbf{QH} + \delta_k^n A \vdash (\delta_k A \supset A) \supset A$ *for any* $k, n \geq 0$.

Proof By 2.13.8(2),

$$B \supset A \vdash_{\mathbf{QH}} \delta_P^k B \supset \delta_P^k A,$$

hence

$$(1) \quad \delta_k A \supset A,\ B \supset A,\ \delta_k B \vdash_{\mathbf{QH}} A.$$

On the other hand, by the deduction theorem

$$(2) \quad \delta_k A \supset A,\ B \supset_\bullet (\delta_k A \supset A) \supset A \vdash_{\mathbf{QH}} B \supset A.$$

From (1), (2) we obtain[28]

$$B \supset_\bullet (\delta_k A \supset A) \supset A,\ \delta_k A \supset A,\ \delta_k B \vdash_{\mathbf{QH}} A.$$

Hence (again by the deduction theorem)

$$(3) \quad B \supset_\bullet (\delta_k A \supset A) \supset A \vdash_{\mathbf{QH}} \delta_k B \supset_\bullet (\delta_k A \supset A) \supset A.$$

Note that $\delta_k A$ is $\delta_{k,P} A$ for some P that does not occur in A, but (3) still holds if P occurs in B.

Now by induction it follows that

$$\mathbf{QH} \vdash \delta_k^n A \supset_\bullet (\delta_k A \supset A) \supset A$$

for any n. In fact, for $n = 0$ this is a substitution instance of an axiom, and for the induction step we can apply (3). ■

Lemma 2.13.20 $\mathbf{QH} + \delta_1^m \delta_j A \vdash \delta_1^m \delta_{j+1} A$ *for* $j > 0,\ m \geq 0$.

Proof Put

$$L := \mathbf{QH} + \delta_1^m \delta_{j,P} A,\ B := \forall z(Q(\mathbf{y}, z) \vee (Q(\mathbf{y}, z) \supset A)),$$

where $|\mathbf{y}| = j,\ z \notin r(\mathbf{y}),\ r(\mathbf{y}z) \cap FV(A) = \varnothing$ and P, Q do not occur in A. Then

$$B = [Q(\mathbf{y}, z)/S(z)]\, \delta_{1,S} A,$$

$$\delta_{j,P} A = \forall \mathbf{y}(P(\mathbf{y}) \vee (P(\mathbf{y}) \supset A)),\ \delta_{j+1,Q} A = \forall \mathbf{y} B.$$

Hence by substitution $[B/P(\mathbf{y})]$

$$(1) \quad L \vdash \delta_1^m \forall \mathbf{y}(B \vee (B \supset A)).$$

[27] Cf. [Yokota, 1989, Theorem 1.4.7].
[28] Cf. Lemma 1.2.3(4) from [Yokota, 1989].

By Lemma 2.13.19 (for $k = 1,\ n = m + 1$) we have

$$(2) \quad \mathbf{QH} + \delta_1^{m+1} A \vdash (\delta_{1,S} A \supset A) \supset A.$$

By 2.13.11, $L \vdash \delta_1^{m+1} A$, and thus

$$(3) \quad L \vdash (\delta_{1,S} A \supset A) \supset A,$$

Hence by substitution $[Q(\mathbf{y}, z)/S(z)]$

$$(4) \quad L \vdash (B \supset A) \supset A.$$

Since $z \notin FV(A)$, we also obtain $\mathbf{QH} \vdash A \supset B$; thus

$$(5) \quad L \vdash (B \supset A) \supset B,$$

which implies

$$(6) \quad L \vdash B \vee (B \supset A) \supset B,$$

and therefore

$$(7) \quad L \vdash \forall \mathbf{y} (B \vee (B \supset A)) \supset \forall \mathbf{y} B,$$

by the monotonicity of $\forall \mathbf{y}$. Eventually from (1), (7) and 2.13.8(2) we obtain

$$L \vdash \delta_1^m \forall \mathbf{y} B = \delta_1^m \delta_{j+1} A$$

as required. ■

Lemma 2.13.21 $\mathbf{QH} + \delta_1^{m+1} A \vdash \delta_1^m \delta_k A$ *for* $k \geq 1,\ m \geq 0$.

Proof By induction on k. The case $k = 1$ is trivial. If the statement holds for k we obtain it for $k + 1$:

$$\mathbf{QH} + \delta_1^m \delta_{k+1} A \subseteq \mathbf{QH} + \delta_1^m \delta_k A \subseteq \mathbf{QH} + \delta_1^m A$$

by Lemma 2.13.20 and the induction hypothesis. ■

Proposition 2.13.22[29] *For any* $A \in IF^{(=)}$

(1) $\mathbf{QH} + \delta_1^n A \vdash \delta_k^n A$ *for* $k, n \geq 0$;

(2) $\mathbf{QH} + \delta_1^n A \vdash \delta_{k_1 \dots k_n} A$ *for* $k_1, \dots, k_n \geq 0$.

Proof (1) implies (2), by 2.13.11. The case $k = 0$ in (1) also follows from 2.13.11, and the case $k = 1$ is trivial. So it remains to prove (1) for $k > 1$.

By induction on $n - m$ let us show

$$(3) \quad \mathbf{QH} + \delta_1^n A \vdash \delta_1^m \delta_k^{n-m} A$$

[29]Theorem 1.3.5 from [Yokota, 1989] is the particular case of this proposition for $A = \bot$. Our proof is similar to that paper.

for $0 \leq m \leq n$. If $m = n$, this is trivial. For the step, suppose (3) for m. By 2.13.21 we have

$$\mathbf{QH} + \delta_1^m \delta_k^{n-m} A \vdash \delta_1^{m-1} \delta_k \delta_k^{n-m} A = \delta_1^{m-1} \delta_k^{n-m+1} A.$$

So by the induction hypothesis we obtain

$$\mathbf{QH} + \delta_1^{m-1} \delta_k^{n-m+1} A \subseteq \mathbf{QH} + \delta_1^m \delta_k^{n-m} A \subseteq \mathbf{QH} + \delta_1^n A,$$

i.e. (3) holds for $m - 1$.

Finally note that (1) is (3) for $m = 0$. ∎

From 2.13.22 it follows that

$$\mathbf{QH} + \delta_1^n A = \mathbf{QH} + \delta_k^n A$$

for any $k > 1$. We also have

Lemma 2.13.23 $\mathbf{QH} + \delta_1 A = \mathbf{QH} + \delta_0 A$.

Proof In fact

$$\mathbf{QH} + \delta_0 A = \mathbf{QH} + p \vee (p \supset A) \vdash \forall y(P(y) \vee (P(y) \supset A)) = \delta_1 A.$$

by substitution $[P(y)/p]$ and generalisation. ∎

However it may happen that $\mathbf{QH} + \delta_0^n A \subset \mathbf{QH} + \delta_1^n A$ for some A and $n > 1$. E.g. $\mathbf{QH} + P_n \not\vdash P_n^+$ for $n > 1$ (cf. [Ono, 1983]); recall that $P_n = \delta_0^n \bot$, $P_n^+ = \delta_1^n \bot$.

Now we can strengthen 2.13.17:

Proposition 2.13.24 *(1) If $L = \mathbf{QH} + \Theta$ and Θ is $\forall$-perfect in $\mathbf{QH}$, then*

$$\Delta^n L = \mathbf{QH} + \delta_1^n \Theta.$$

(2) If $L = \mathbf{QH} + \Theta$, then

$$\Delta^n L = \mathbf{QH} + \delta_1^n (\Theta^\forall).$$

Proof Follows from 2.13.17 and 2.13.22. ∎

By applying 2.13.24 to $\Theta = \{\bot\}$ we obtain

Theorem 2.13.25 $\mathbf{QHP}_n^+ = \mathbf{QH} + P_n^+$ *for $n > 0$.*

Now let us show perfection for finite axiomatisations of $\mathbf{QHP}_n^+$ and obtain an alternative proof of Theorem 2.13.25. This proof is not so straightforward, but does not use the infinite axiomatisation described in 2.13.18. However we still need 2.13.22 for this proof.

Lemma 2.13.26 *$\mathbf{H} \vdash (C \supset q) \supset C$, where $C = \bigwedge_{i=1}^{n} (p_i \vee (p_i \supset q)), n \geq 0$.*

Proof By induction on n. The case $n = 0$ is trivial.

Consider the inductive step from n to $n+1$. By the deduction theorem it suffices to prove

$$C \supset q \vdash_{\mathbf{H}} C,$$

for $C = \bigwedge_{i=1}^{n+1} (p_i \vee (p_i \supset q))$, which obviously follows from

$$(1) \quad C \supset q \vdash_{\mathbf{H}} p_i \supset q$$

for $1 \leq i \leq n+1$. Put

$$B := \bigwedge \{p_j \vee (p_j \supset q) \mid 1 \leq j \leq n+1, \ j \neq i\}.$$

Then

$$\mathbf{H} \vdash C \equiv B \wedge (p_i \vee (p_i \supset q)),$$

and thus

$$C \supset q, \ B, \ p_i \vdash_{\mathbf{H}} q.$$

Hence by the deduction theorem,

$$(2) \quad C \supset q, \ p_i \vdash_{\mathbf{H}} B \supset q.$$

Together with the induction hypothesis

$$\mathbf{H} \vdash (B \supset q) \supset q$$

(2) implies

$$C \supset q, \ p_i \vdash q.$$

Hence (1) follows by the deduction theorem again. ∎

Lemma 2.13.27 *Let L be a predicate superintuitionistic logic, Θ a set of formulas, $k \geq 0$. Then*

(1) If Θ is $\wedge$-perfect in L, then $\delta_k\Theta$ is $\wedge$-perfect in L.

(2) If Θ is $\forall$-perfect in L and $L \supseteq \delta_1^n\Theta$ for some $n > 0$, then $\delta_k\Theta$ is $\forall$-perfect in L.

(3) If Θ is $\wedge\forall$-perfect in L and $L \supseteq \delta_1^n\Theta$ for some $n > 0$, then $\delta_k\Theta$ is $\wedge\forall$-perfect in L.

Proof

(1) Consider arbitrary formulas

$$D_i := \forall \mathbf{y}(A_i(\mathbf{y},\mathbf{x}) \vee (A_i(\mathbf{y},\mathbf{x}) \supset B_i(\mathbf{x}))) \in Sub(\delta_k\Theta)$$

for $i = 1, \ldots, m$, where $\mathbf{y}$ and $\mathbf{x}$ are disjoint lists of variables, $|\mathbf{y}| = k$,

$$B_i(\mathbf{x}) \in Sub(\Theta),\ FV(B_i(\mathbf{x})) \subseteq \mathbf{x},\ FV(A_i(\mathbf{y},\mathbf{x})) \subseteq \mathbf{xy}.$$

By $\wedge$-perfection there exists $B \in Sub(\Theta)$ such that

$$L \vdash B \supset \bigwedge_i B_i(\mathbf{x});$$

we may assume that $r(\mathbf{y}) \cap FV(B) = \varnothing$. Put

$$C := \bigwedge_i (A_i(\mathbf{y},\mathbf{x}) \vee (A_i(\mathbf{y},\mathbf{x}) \supset B)).$$

By Lemma 2.13.26, $\mathbf{QH} \vdash (C \supset B) \supset C$, hence

$$L \vdash C \vee (C \supset B) \supset \bigwedge_i (A_i(\mathbf{y},\mathbf{x}) \vee (A_i(\mathbf{y},\mathbf{x}) \supset B_i(\mathbf{x}))),$$

and so

$$L \vdash \forall \mathbf{y}(C \vee (C \supset B)) \supset \bigwedge_i \forall \mathbf{y}(A_i(\mathbf{y},\mathbf{x}) \vee (A_i(\mathbf{y},\mathbf{x}) \supset B_i(\mathbf{x})))$$

by standard properties of quantifiers (Lemma 2.6.15). Since

$$\forall \mathbf{y}(C \vee (C \supset B)) \in Sub(\delta_k\Theta),$$

we obtain the required property for $\bigwedge_i D_i$.

(2) Suppose $A \in Sub(\Theta)$,

$$\delta_k A = \forall \mathbf{y}(P(\mathbf{y}) \vee (P(\mathbf{y}) \supset A))$$

for $|\mathbf{y}| = k$, where P does not occur in A,

$$\overline{(\delta_k A)^m} = \forall \mathbf{z} \forall \mathbf{y}(Q(\mathbf{z},\mathbf{y}) \vee (Q(\mathbf{z},\mathbf{y}) \supset A^m(\mathbf{z}))),$$

for $|\mathbf{z}| = m,\ r(\mathbf{y}) \cap r(\mathbf{z}) = \varnothing, r(\mathbf{yz}) \cap FV(A) = \varnothing,\ Q \in PL^{k+m}$, where Q does not occur in A^m.

By $\forall$-perfection there is $B \in Sub(\Theta)$ such that $L \vdash B \supset \forall \mathbf{z} A^m(\mathbf{z})$; we may assume that $FV(B) \cap r(\mathbf{yz}) = \varnothing$ and Q does not occur in B. Put

$$C := \delta_{k+m} B = \forall \mathbf{z} \forall \mathbf{y}(Q(\mathbf{z},\mathbf{y}) \vee (Q(\mathbf{z},\mathbf{y}) \supset B)).$$

Since $L \vdash B \supset A^m(\mathbf{z})$, it follows that

$$(3) \quad L \vdash C \supset \overline{(\delta_k A)^m}.$$

Since by the assumption (2) of the lemma $\delta_1^n B \in Sub(\delta_1^n \Theta) \subseteq L$, Proposition 2.13.22 implies $L \vdash \delta_{k+m}^n B$. Thus by Lemma 2.13.19, $L \vdash (C \supset B) \supset B$. Also $L \vdash B \supset C$ by 2.13.8, hence

$$(4) \quad L \vdash (C \supset B) \supset C,$$

and so

$$(5) \quad L \vdash C \vee (C \supset B) \supset \overline{(\delta^k A)^m},$$

by (3) and (4); eventually,

$$(6) \quad L \vdash C' \supset \overline{(\delta^k A)^m},$$

where

$$C' := \forall \mathbf{y}(C \vee (C \supset B)) \in [C/Q(\mathbf{x})]\, \delta_{k,Q} B, \ Q \in PL^k$$

(note that here $\forall \mathbf{y}$ is a dummy quantifier). Since $C' \in Sub(\delta_k \Theta)$, this completes the proof.

■

Similarly to 2.13.27(1) one can show

Proposition 2.13.28

(1) If Θ is $\wedge$-perfect in L, then $\delta_\infty^1 \Theta$ is $\wedge$-perfect in L (and thus $\delta_\infty^n \Theta$, $\delta_*^n \Theta$ are also $\wedge$-perfect in L).

(2) If Θ is $\wedge\forall$-perfect in L, then $\delta_\infty^1 \Theta$ is $\wedge\forall$-perfect in L (and $\delta_\infty^n \Theta$, $\delta_*^n \Theta$ as well).

Lemma 2.13.27(2) allows us to prove 2.13.25 without applying 2.13.17; cf. the proof of 2.13.16.

Since $\{\bot\}$ is $\wedge\forall$-perfect in **QH**, we obtain

Lemma 2.13.29

(1) The sets $\{P_n\}$ and $\{P_n^+\}$ are $\wedge$-perfect in **QH** *for $n > 0$.*

(2) The sets $\{P_n\}$ and $\{P_n^+\}$ are $\wedge\forall$-perfect in **QH**$+P_r^+$ *for all $n, r > 0$ (note that the case $n \geq r$ is trivial).*

Remark 2.13.30 Moreover, these sets are arity-perfect in the sense of [Yokota, 1989].

Proposition 2.13.31 $\mathbf{QHP}_n^+ = \mathbf{QH} + P_r^+ + P_n$ *for $r \geq n \geq 0$.*

So P_n and P_n^+ are deductively equivalent in $\mathbf{QHP}_r^+$ for a sufficiently large r.

Proof By induction on n. Consider the induction step. We have

$$\mathbf{QHP}_{n+1}^+ = \Delta(\mathbf{QHP}_n^+) = \mathbf{QH} + \delta(\mathbf{QHP}_n^+) \subseteq \mathbf{QH} + P_r^+ + P_{n+1},$$

by Lemma 2.13.13 (1a) applied to

$$L = \mathbf{QH} + P_r^+,\ \Theta_1 = \mathbf{QHP}_n^+,\ \Theta_2 := \{P_n\},$$

note that $L + \Theta_1 = \mathbf{QHP}_n^+ = L + \Theta_2$ by the induction hypothesis, and Θ_2 is $\forall$-perfect in L by Lemma 2.13.29 (ii). ∎

Theorem 2.13.25 is clearly a particular case of this statement for $n = r$. Therefore we obtain an alternative proof of Theorem 2.13.25 that does not use 2.13.18.

So we have

Corollary 2.13.32 $\mathbf{QH} + P_n^+ = \mathbf{QH} + P_r^+ + P_n$ *for* $n < r$.

Similarly we obtain

Proposition 2.13.33 *If* $L = L_0 + \Theta$, Θ *is* $\forall$*-perfect in* L_0 *and* $\delta_1^r\Theta \subseteq L_0$ *for some* $r > 0$, *then*

$$L_0 + \Delta^n L = L_0 + \delta_k^n\Theta$$

and $\delta_k^n\Theta$ *is* $\forall$*-perfect in* L_0 *for all* $k, n \geq 0$.

Proof By induction on n. The case $n = 0$ is trivial. Consider the induction step from n to $n + 1$. First note that by 2.9.9,

$$\mathbf{QH} \vdash \delta_1^r A \supset \delta_1^r\delta_k^n A.$$

Hence $\delta_1^r\delta_k^n\Theta \subseteq L$, and so $\delta_k^{n+1}\Theta = \delta_k(\delta_k^n\Theta)$ is $\forall$-perfect in L_0 by the induction hypothesis and Lemma 2.13.27.

Next, by the induction hypothesis and Lemma 2.13.13 (1a)

$$L_0 + \Delta^{n+1}L = L_0 + \delta(\Delta^n L) = L_0 + \delta(L_0 + \delta_k^n\Theta) \subseteq L_0 + \delta(\delta_k^n\Theta) = L_0 + \delta_k^{n+1}\Theta.$$

The converse inclusion

$$L_0 + \delta_k^{n+1}\Theta \subseteq L_0 + \Delta^{n+1}L$$

is trivial. ∎

So $\Delta^n L = L_0 + \delta_k^n\Theta$ in particular, $\Delta^n L = L_0 + \delta_0^n\Theta$, whenever $L_0 \subseteq \Delta^n L$.

Note that $\delta_k^n\Theta \subseteq \Delta^n L$ for any L and $\Theta \subseteq L$ and $n \geq 0$, by Lemma 2.13.7.

Also recall that $\delta_1^r\Theta \subseteq \mathbf{QHP}_r^+$ for any Θ by 2.13.10.

Corollary 2.13.34 *If* $L = L_0 + \Theta$ *and* $\delta_1^r\Theta \subseteq L_0$ *for some* $r > 0$, *then* $L_0 + \Delta^n L = L_0 + \delta_k^n(\Theta^\forall)$ *for all* $k, n \geq 0$.

Actually we need only the case $k = 0$ of this statement, because the case $k = 1$ (and thus, the case $k > 0$) readily follows from 2.13.24, without any restriction on L_0 (recall that $\delta_k^n L_0 \subseteq L_0$ by 2.13.8 (1)). And 2.13.33 shows that if L_0 contains $\delta_r^1\Theta$ for some (sufficiently large) $r > 0$, then one can reduce the case $k = 1$ to $k = 0$ in an axiomatisation of $\Delta^n L$ above L_0.

2.14 Adding equality

In this section we study correlation between a logic without equality L and its minimal extension with equality $L^=$. There exists an obvious translation from formulas with equality to formulas without equality just replacing equality with an ordinary predicate letter. In some cases specific equality axioms can be reduced to finitely many formulas — then we obtain a reduction for the corresponding decision problems.

Definition 2.14.1 *For an N-m.p.l. L, let $L^= := \mathbf{QK}_N^= + L$, and for an s.p.l. L, let $L^= := \mathbf{QH}^= + L$. $L^=$ is called the* equality-expansion of L.

The other way round, for an m.p.l.= (s.p.l.=) L, we define the equality-free fragment *as $L^\circ := L \cap MF_N$ (respectively, $L \cap IF$).*

For any modal formula with equality A and $Q \in PL^2$ that does not occur in A, we define A^Q as the formula obtained from A by replacing all occurrences of '=' with Q. For a set of N-modal formulas Γ put

$$\Gamma^Q := \{A^Q \mid A \in \Gamma\},\ \Box^\infty\Gamma := \{\Box_\alpha B \mid B \in \Gamma,\ \alpha \in I_N^\infty\}.$$

Consider the following sets of formulas (for $N \geq 0$)

$$\begin{aligned}\mathcal{E}_N := &\{\forall x\forall y(x = y \supset y = x),\ \forall x\forall y\forall z(x = y \wedge y = z\ \supset x = z),\ \forall x\ x = x\}\\ &\cup \{\forall x\forall y(x = y \supset \Box_i(x = y)) \mid 1 \leq i \leq N\}\\ &\cup \{\overline{\forall}(\bigwedge_{i=1}^n x_i = y_i \supset_\bullet P_k^n(x_1, \dots, x_n) \equiv P_k^n(y_1, \dots, y_n)) \mid n, k \geq 1,\ P_k^n \neq Q\},\end{aligned}$$

where all x_j, y_j are different.

For a formula A, let $MF_{N,A}$ be the set of all N-modal formulas built from predicate letters occurring in A:

$$\mathcal{E}_{N,A} := \mathcal{E}_N \cap MF_{N,A}.$$

We shall omit N if it is clear from the context and use the notation $\Box^\infty\mathcal{E}$, $\Box^\infty\mathcal{E}^Q$, MF_A, $\mathcal{E}_A$, $\mathcal{E}_A^Q$.

Lemma 2.14.2 *If $A \in \mathrm{Sub}(\mathcal{E}_N)$, then $\Box^\infty\mathcal{E}_N \vdash_{\mathbf{QK}_N} \overline{\forall}A$.*

Proof The only nontrivial case is when

$$A = [B/P(\mathbf{x})]\overline{\forall}(\bigwedge_{i=1}^n x_i = y_i \supset_\bullet P(\mathbf{x}) \equiv P(\mathbf{y}))$$

for $P \in PL^n$, $\mathbf{x} = (x_1, \dots, x_n)$, $\mathbf{y} = (y_1, \dots, y_n)$.

Put

$$\mathbf{x} \equiv \mathbf{y} := \bigwedge_{i=1}^n x_i = y_i.$$

Then

$$\overline{\forall}A = \overline{\forall}(\mathbf{x} \equiv \mathbf{y} \supset_\bullet B \equiv [\mathbf{y}/\mathbf{x}]B).$$

We prove $\Box^\infty\mathcal{E}_N \vdash_{\mathbf{QK}_N} \overline{\forall}A$ by induction on the complexity of B.

- If B is atomic and does not contain P, the claim is trivial.

- If $B = P(\mathbf{z})$ for some list $\mathbf{z}$ (maybe not distinct), then $[\mathbf{y}/\mathbf{x}]B = P(\mathbf{t})$, where $\mathbf{t} = [\mathbf{y}/\mathbf{x}]\mathbf{z}$, i.e. every x_i is replaced with y_i, whenever it occurs in $\mathbf{z}$. Then obviously,

$$\forall x(x = x) \vdash_{\mathbf{QK}_N} \mathbf{x} \equiv \mathbf{y} \supset \mathbf{z} \equiv \mathbf{t}. \tag{1}$$

By Lemma 2.6.10 (xxvii) we also have

$$\overline{\forall}(\mathbf{x} \equiv \mathbf{y} \supset_{\bullet} P(\mathbf{x}) \equiv P(\mathbf{y})) \vdash_{\mathbf{QK}_N} \mathbf{z} \equiv \mathbf{t} \supset_{\bullet} P(\mathbf{z}) \equiv P(\mathbf{t}). \tag{2}$$

Now (1) and (2) imply

$$\Box^{\infty}\mathcal{E}_N \vdash_{\mathbf{QK}_N} \mathbf{x} \equiv \mathbf{y} \supset_{\bullet} P(\mathbf{z}) \equiv P(\mathbf{t}),$$

and it remains to apply generalisation.

- If $B = B_1 \circledast B_2$ for a propositional connective $\circledast$ and

$$\mathbf{QK}_N \vdash \mathbf{x} \equiv \mathbf{y} \supset_{\bullet} B_i \equiv [\mathbf{y}/\mathbf{x}]B_i$$

by the induction hypothesis, then the claim follows by an argument in classical propositional logic – note that the rule

$$\frac{A \equiv A',\ B \equiv B'}{A \circledast B \equiv A' \circledast B'},$$

is admissible in the $\mathbf{QK}_N$-theory $\{\mathbf{x} \equiv \mathbf{y}\}$ (with $\mathbf{x}$, $\mathbf{y}$ considered as constants) and apply the deduction theorem.

- If $B = \Box_i C$ and

$$\Box^{\infty}\mathcal{E}_N \vdash_{\mathbf{QK}_N} \mathbf{x} \equiv \mathbf{y} \supset_{\bullet} C \equiv [\mathbf{y}/\mathbf{x}]C,$$

we obtain by Lemma 2.7.10

$$\Box^{\infty}\mathcal{E}_N \vdash_{\mathbf{QK}_N} \Box_i(\mathbf{x} \equiv \mathbf{y}) \supset \Box_i(C \equiv [\mathbf{y}/\mathbf{x}]C)). \tag{3}$$

Since

$$\mathbf{K}_N \vdash \Box_i(p \equiv q) \supset_{\bullet} \Box_i p \equiv \Box_i q$$

(Lemma 1.1.1), it follows that

$$\Box^{\infty}\mathcal{E}_N \vdash_{\mathbf{QK}_N} \Box_i(\mathbf{x} \equiv \mathbf{y}) \supset_{\bullet} \Box_i C \equiv \Box_i[\mathbf{y}/\mathbf{x}]C. \tag{4}$$

Now

$$\mathcal{E}_N \vdash_{\mathbf{QK}_N} x_j = y_j \supset \Box_i(x_j = y_j).$$

implies

$$\Box^{\infty}\mathcal{E}_N \vdash_{\mathbf{QK}_N} \mathbf{x} \equiv \mathbf{y} \supset \Box_i(\mathbf{x} \equiv \mathbf{y}). \tag{5}$$

From (4) and (5) we have

$$\Box^{\infty}\mathcal{E}_N \vdash_{\mathbf{QK}_N} \mathbf{x} \equiv \mathbf{y} \supset_{\bullet} \Box_i C \equiv \Box_i[\mathbf{y}/\mathbf{x}]C$$

and thus the claim holds for B.

- If $B = \forall zC$, we may assume $z \notin \mathbf{xy}$ (otherwise, after renaming z, B and $[\mathbf{y}/\mathbf{x}]B$ change to equivalent congruent formulas).

  Suppose
  $$\Box^\infty \mathcal{E}_N \vdash_{\mathbf{QK}_N} \mathbf{x} \equiv \mathbf{y} \supset_\bullet C \equiv [\mathbf{y}/\mathbf{x}]C.$$
  Then by $\forall$-introduction we deduce (in the same theory)
  $$\forall z(\mathbf{x} \equiv \mathbf{y} \supset_\bullet C \equiv [\mathbf{y}/\mathbf{x}]C),$$
  and hence
  $$\mathbf{x} \equiv \mathbf{y} \supset_\bullet \forall z(C \equiv [\mathbf{y}/\mathbf{x}]C) \tag{6}$$
  by Lemma 2.6.10(xi).

  Finally
  $$\forall z(C \equiv [\mathbf{y}/\mathbf{x}]C) \supset_\bullet \forall zC \equiv \forall z[\mathbf{y}/\mathbf{x}]C. \tag{7}$$
  by Lemma 2.6.10(xxviii).

  So
  $$\overline{\forall}(\mathbf{x} \equiv \mathbf{y} \supset_\bullet \forall zC \equiv \forall z[\mathbf{y}/\mathbf{x}]C)$$
  follows from (6) and (7) by transitivity and $\overline{\forall}$-introduction.

- If $B = \exists zC$, we also have (6) by the induction hypothesis. Then, instead of (7) we use
  $$\forall z(C \equiv [\mathbf{y}/\mathbf{x}]C) \supset_\bullet \exists zC \equiv \exists z[\mathbf{y}/\mathbf{x}]C, \tag{8}$$
  which follows from 2.6.10(xxvii).

■

Lemma 2.14.2 has an intuitionistic version:

Lemma 2.14.3 *If* $A \in \mathrm{Sub}(\mathcal{E}_0)$, *then* $\mathcal{E}_0 \vdash_{\mathbf{QH}} A$.

Proof Very similar to 2.14.2. Again we prove
$$\mathcal{E}_0 \vdash_{\mathbf{QH}} \overline{\forall}(\mathbf{x} \equiv \mathbf{y} \supset_\bullet B \equiv [\mathbf{y}/\mathbf{x}]B)$$
by induction. The reader can check that in all cases the argument is based only on intuitionistic logic. ■

Proposition 2.14.4 *Let L be an N-m.p.l. Then for any N-modal formula A with equality that does not contain Q*
$$L^= \vdash A \Leftrightarrow \Box^\infty \mathcal{E}^Q \vdash_L A^Q \Leftrightarrow \Box^\infty \mathcal{E}^Q_A \vdash_L A^Q.$$
If L is conically expressive, then
$$L^= \vdash A \Leftrightarrow \Box^* \mathcal{E}^Q_A \vdash_L A^Q.$$

Proof First note that $\mathcal{E}_N$ is a set of axioms for $L^=$ above L (they are all $L^=$-theorems and the usual axioms follow from $\mathcal{E}_N$). So by the deduction theorem 2.8.4,

$$L^= \vdash A \Leftrightarrow \Box^\infty\overline{\mathrm{Sub}}(\mathcal{E}_N) \vdash_L A.$$

All formulas from $\Box^\infty\overline{\mathrm{Sub}}(\mathcal{E}_N)$ are of the form $\Box_\alpha\overline{\forall}B$, where $B \in \mathrm{Sub}(\mathcal{E}_N)$, so they are $\mathbf{QK}_N$-provable in $\Box^\infty\mathcal{E}_N$ by 2.14.2, generalisation and $\Box$-introduction. Thus

$$L^= \vdash A \Leftrightarrow \Box^\infty\mathcal{E}_N \vdash_L A.$$

Replacing '=' with Q does not affect the L-inference (more precisely, the equivalence $\Gamma \vdash_L A$ iff $\Gamma^Q \vdash_L A^Q$ is checked by induction), so

$$L^= \vdash A \Leftrightarrow \Box^\infty\mathcal{E}_N^Q \vdash_L A^Q, \tag{9}$$

and thus

$$\Box^\infty\mathcal{E}_A^Q \vdash_L A^Q \Rightarrow L^= \vdash A.$$

The other way round, suppose $L^= \vdash A$. Then $\Box^\infty\mathcal{E}_N^Q \vdash_L A^Q$. Let S be a formula substitution replacing every atomic formula $P(\mathbf{x})$ with $\top$ for any $P \neq Q$ that does not occur in A. We claim that

$$\Box^\infty\mathcal{E}_N^Q \vdash_L B \Rightarrow \Box^\infty\mathcal{E}_A^Q \vdash_L SB. \tag{10}$$

In fact,

$$S(Q(x,y) \supset_\bullet P(x) \equiv P(y)) = (x = y \supset_\bullet \top \equiv \top)$$

is obviously L-provable, so (10) holds for any $B \in \mathcal{E}_N^Q - \mathcal{E}_A^Q$.

Then we can argue by induction, since S distributes over propositional connectives and quantifiers. Eventually

$$\Box^\infty\mathcal{E}_A^Q \vdash_L SA^Q (= A^Q),$$

which completes the proof.

The observation about conical expressiveness follows from Lemma 2.8.7. ∎

Proposition 2.14.5 *Let L be an s.p.l. Then for any $A \in IF^=$ without occurrences of Q*

$$L^= \vdash A \Leftrightarrow \mathcal{E}_0^Q \vdash_L A^Q \Leftrightarrow \mathcal{E}_A^Q \vdash_L A^Q.$$

Proof Follows the same lines as 2.14.4. By 2.8.3,

$$L^= \vdash A \Leftrightarrow \overline{\mathrm{Sub}}(\mathcal{E}_0) \vdash_L A.$$

Hence by Lemma 2.14.3 and generalisation,

$$L^= \vdash A \Leftrightarrow \mathcal{E}_0 \vdash_L A,$$

which is equivalent to

$$\mathcal{E}_0^Q \vdash_L A^Q.$$

The implication

$$\mathcal{E}_0^Q \vdash_L A^Q \Rightarrow \mathcal{E}_A^Q \vdash_L A^Q$$

is again proved by replacing redundant predicate letters with $\top$. ∎

In many cases the equality-expansion is conservative. To show this, for logics without equality we define a weak analogue of equality – indiscernibility. Namely, for $P \in PL^n$ and variables x, y we put

$$In_P(x,y) := \forall z_1 \dots \forall z_n \bigwedge_{j=1}^{n} ([x/z_j]P(z_1,\dots,z_n) \equiv [y/z_j]P(z_1,\dots,z_n)),$$

where of course, all z_js are different and $x, y \neq z_j$. For a predicate formula A, put

$$In_A(x,y) := \bigwedge\{In_P(x,y) \mid P \text{ occurs in } A\}.$$

Lemma 2.14.6 *For $P \in PL^n$*

$$\mathbf{QH} \vdash \bigwedge_{i=1}^{n} In_P(x_i,y_i) \supset_\bullet P(x_1,\dots,x_n) \equiv P(y_1,\dots,y_n).$$

Proof Almost the same as for 2.6.16(iv).

We show by induction that

$$\bigwedge_{i=1}^{n} In_P(x_i,y_i) \vdash_{\mathbf{QH}} [x_1,\dots,x_m/z_1,\dots,z_m]P(\mathbf{z}) \equiv [y_1,\dots,y_m/z_1,\dots,z_m]P(\mathbf{z})$$

for a list of new variables $\mathbf{z} = (z_1,\dots,z_n)$. For the induction step we again consider

$$A_m := [x_1,\dots,x_m/z_1,\dots,z_m]P(\mathbf{z}),$$
$$B_m := [y_1,\dots,y_m/z_1,\dots,z_m]P(\mathbf{z}).$$

By the induction hypothesis, $\forall z_{m+1}$-introduction and 2.6.10(ii) we obtain

(1) $[x_{m+1}/z_{m+1}]A_m \equiv [x_{m+1}/z_{m+1}]B_m$.

We also have

(2) $[x_{m+1}/z_{m+1}]B_m \equiv [y_{m+1}/z_{m+1}]B_m$.

In fact, $B_m = P(y_1,\dots,y_m,z_{m+1},\dots,z_n)$, so

$$In_P(x_{m+1},y_{m+1}) = \forall \mathbf{z} \bigwedge_{j=1}^{n} ([x_{m+1}/z_j]P(\mathbf{z}) \equiv [y_{m+1}/z_j]P(\mathbf{z}))$$
$$\vdash_{\mathbf{QH}} [y_1,\dots,y_m/z_1,\dots,z_m] \bigwedge_{j=1}^{n} ([x_{m+1}/z_j]P(\mathbf{z}) \equiv [y_{m+1}/z_j]P(\mathbf{z}))$$

by 2.6.10(xxv), and (2) is a conjunct in the latter formula. From (1), (2) by transitivity it follows that

$$A_{m+1} = [x_{m+1}/z_{m+1}]A_m \equiv [y_{m+1}/z_{m+1}]B_{m+1} = B_{m+1}.$$

■

Lemma 2.14.7

(1) *If L is a conically expressive N-m.p.l, then for any N-modal predicate formula A that does not contain Q*

$$[\Box^* In_A(x,y)/Q(x,y)]\Box^*\mathcal{E}_A^Q \subseteq L.$$

(2) *If L is an s.p.l., then for any $A \in IF$*

$$[In_A(x,y)/Q(x,y)]\mathcal{E}_A^Q \subseteq L.$$

Proof We have to prove the corresponding substitution instances of the 'axioms' from $(\Box^*)\mathcal{E}_A^Q$ in L.

(1) Reflexivity of Q.
We have $[x/z_j]P(\mathbf{z}) \equiv [x/z_j]P(\mathbf{z})$ by **H**, hence $In_P(x,x)$ by $\wedge$- and $\forall\mathbf{z}$-introduction. Thus $\forall x In_A(x,x)$ in the intuitionistic case, $\forall x\Box^* In_A(x,x)$ in the modal case — by $\wedge$-, $\Box^*$- and $\forall$-introduction.

(2) Symmetry of Q.
By **H** we have

$$[x/z_j]P(\mathbf{z}) \equiv [y/z_j]P(\mathbf{z}) \vdash_L [y/z_j]P(\mathbf{z}) \equiv [x/z_j]P(\mathbf{z}),$$

hence $In_P(x,y) \vdash_L In_P(y,x)$ and next $In_A(x,y) \vdash_L In_A(y,x)$ by applying $\forall\mathbf{z}B \supset B$, $\wedge$- and $\forall$-introduction. Then $L \vdash In_A(x,y) \supset In_A(y,x)$ by the deduction theorem. In the modal case this implies $L \vdash \Box^* In_A(x,y) \supset \Box^* In_A(x,y)$ by the monotonicity of $\Box^*$. So it remains to generalise over x, y.

(3) Transitivity of Q.
Since $\equiv$ is transitive in **H**, we have

$$[x/z_j]P(\mathbf{z}) \equiv [y/z_j]P(\mathbf{z}),\ [y/z_j]P(\mathbf{z}) \equiv [t/z_j]P(\mathbf{z}) \vdash_L$$
$$[x/z_j]P(\mathbf{z}) \equiv [t/z_j]P(\mathbf{z}),$$

which implies

$$In_P(x,y), In_P(y,t) \vdash_L In_P(x,t)$$

again by standard arguments with $\wedge$ and $\forall$, and hence

$$In_A(x,y), In_A(y,t) \vdash_L In_A(x,t).$$

Since x, y, t are fixed in this proof, in the modal case we may apply Lemma 2.7.10 for $\Box^*$:

$$\Box^* In_A(x,y), \Box^* In_A(y,t) \vdash_L \Box^* In_A(x,t).$$

Thus

$$L \vdash \overline{\forall}(\Box^* In_A(x,y) \wedge \Box^* In_A(y,t) \supset \Box^* In_A(x,t))$$

by the deduction theorem and $\overline{\forall}$-introduction.

(4) $\forall x \forall y(Q(x,y) \supset \Box_i Q(x,y))$.
By 1.3.46 $L \vdash \Box^* p \supset \Box_i \Box^* p$, hence $L \vdash \forall x \forall y(\Box^* In_A(x,y) \supset \Box_i \Box^* In_A(x,y))$ by substitution and $\forall$-introduction.

(5) $\overline{\forall}\left(\bigwedge_{i=1}^{n} (\Box^*) In_A(x_i, y_i) \supset_\bullet P(\mathbf{x}) \equiv P(\mathbf{y})\right)$
for $P \in PL^n \cap MF_{N,A}$, $\mathbf{x} = (x_1, \ldots, x_n)$, $\mathbf{y} = (y_1, \ldots, y_n)$.
By 2.14.6,

$$In_A(x_1, y_1), \ldots, In_A(x_n, y_n) \vdash_L P(\mathbf{x}) \equiv P(\mathbf{y}),$$

hence

$$\Box^* In_A(x_1, y_1), , \ldots, \Box^* In_A(x_n, y_) \vdash_L P(\mathbf{x}) \equiv P(\mathbf{y})$$

by the reflexivity of $\Box^*$. Now we can apply the deduction theorem and $\overline{\forall}$-introduction.

■

Theorem 2.14.8 *$L^=$ is a conservative extension of L, i.e. $(L^=)^\circ = L$ for any superintuitionistic predicate logic L and for any conically expressive m.p.l. L.*

Proof Let A be a formula without equality of the corresponding kind and assume that Q does not occur in A. Then $A^Q = A$. $L^= \vdash A$ implies $\Box^* \mathcal{E}_A^Q \vdash_L A$ in the modal case (by 2.14.4) and $\mathcal{E}_A^Q \vdash_L A$ in the intuitionistic case (by 2.14.5). Hence $L \vdash \bigwedge \Box^* \mathcal{E}_A^Q \supset A$ and respectively $L \vdash \bigwedge \mathcal{E}_A^Q \supset A$ by the deduction theorem.

By applying the substitution $[\Box^* In_A(x,y)/Q(x,y)]$ in the modal case, and $[In_A(x,y)/Q(x,y)]$ in the intuitionistic case, we obtain

$$L \vdash \bigwedge [\Box^* In_A(x,y)/Q(x,y)] \Box^* \mathcal{E}_A^Q \bullet \supset A$$

or respectively,

$$L \vdash \bigwedge [In_A(x,y)/Q(x,y)] \mathcal{E}_A^Q \bullet \supset A.$$

Therefore $L \vdash A$ by Lemma 2.14.7. ■

Problem 2.14.9 Does Theorem 2.14.8 hold for an arbitrary m.p.l. L?

Remark 2.14.10 T. Shimura and N.-Y. Suzuki obtained the following stronger result for the superintuitionistic case. Fix a binary predicate letter P, and let

$$E_P^\bullet(x,y) := \forall z(P(x,z) \equiv P(y,z)).$$

Then for any $L \in \mathcal{S}$, and for any $\Gamma \subseteq IF^=$ consisting of pure equality formulas with only positive occurrences of '=',

$$(L^= + \Gamma)^\circ = L + [E_p^\bullet(x,y)/Q(x,y)]\Gamma^Q$$

(cf. [Shimura and Suzuki, 1993, Theorem 3]).

Now recall a well-known definition from recursion theory [Rogers, 1987].

Definition 2.14.11 *For sets of words (in a finite alphabet $\mathcal{A}$) X, Y we say that X is* m-reducible *to Y (notation $X \leq_m Y$) if there exists a recursive function $f : \mathcal{A}^\infty \to \mathcal{A}^\infty$ such that $X = f^{-1}[Y]$; X is* m-equivalent *to Y (notation: $X \equiv_m Y$) if $X \leq_m Y$ and $Y \leq_m X$.*

Theorem 2.14.12 *If L is an s.p.l. or a conically expressive m.p.l., then $L \equiv_m L^=$.*

Proof $L \leq_m L^=$, since $L^=$ is conservative over L (2.14.8); the corresponding function f sends every formula without equality to itself (and all other words to $\bot$, say).

$L^= \leq_m L$, since for a formula A without Q,

$$L^= \vdash A \text{ iff } L \vdash \bigwedge(\Box^*)\mathcal{E}_A^Q \bullet\supset A^Q$$

by 2.14.4, 2.14.5. So the reducing function sends every formula A without Q to $\bigwedge(\Box^*)\mathcal{E}_A^Q \bullet\supset A^Q$ and every nonformula to $\bot$. If a formula A contains Q, first replace Q with another binary predicate letter that does not occur in A. ∎

Let us now consider some specific axioms of equality.

Definition 2.14.13 *A logic (with equality) L is said to have* stable *(respectively,* decidable, closed) equality *if it contains the corresponding formula (SE, DE, or CE, see Section 2.4). For a superintuitionistic predicate logic L (without equality), let*

$$L^{=d} := L^= + DE, \quad L^{=s} := L^= + SE.$$

Similarly, for an N-m.p.l. L, we define

$$L^{=c} := L^= + \{CE_1, \dots, CE_N\},$$

where CE_i denotes CE for $\Box_i$.

Obviously, $L^= \subseteq L^{=s} \subseteq L^{=d}$.

Soon we will show that $\mathbf{QH}^{=d}$ is conservative over $\mathbf{QH}$. However, $L^{=d}$ is not always conservative over L; the corresponding example will be given later on.

As we do not substitute formulas for '=' in predicate logics with equality, the following lemma is an easy consequence of the deduction theorem.

Lemma 2.14.14

(1) $L + DE \vdash A$ iff $L \vdash DE \supset A$

for a superintuitionistic logic with equality L and a formula A; similarly for SE.

(2) $L + \bigwedge_{i=1}^{N} CE_i \vdash A$ *iff* $L \vdash \Box^{\leq k} \bigwedge_{i=1}^{N} CE_i \supset A$ *for some* $k \in \omega$, *where* L *is an* N*-modal logic with equality.*

For a 1-modal logic $L \supseteq$ **QT** *this can be simplified:*

$$L + CE \vdash A \text{ iff } L \vdash \Box^r CE \supset A \text{ for some } r \in \omega;$$

similarly, for $L \supseteq$ **QK4**, **QS4**, *cf. Section 1.1.*

2.15 Propositional parts

Let us explicitly describe the construction of the propositional part L_π for a predicate logic L without equality, cf. [Ono, 1972/73] for the case of intermediate logics.

Since all the sets PL^n are countable, we can consider a bijection

$$\pi_0 : \left(\bigcup_{n \geq 0} PL^n \right) \longrightarrow PL^0.$$

Then for any formula A, let $\pi(A)$ be the result of replacing each atomic subformula of the form $P(\mathbf{x})$ with $\pi_0(P)$ and erasing all occurrences of quantifiers. For a set of formulas Γ let

$$\pi(\Gamma) = \{\pi(A) \mid A \in \Gamma\}.$$

Obviously $\pi(A)$ is equivalent (in **QH** or in $\mathbf{QK}_N$) to a substitution instance of A (cf. Lemma 2.6.15(I)(iv)). On the other hand, a propositional formula A is a substitution instance of $\pi(A)$. Thus we obtain

Proposition 2.15.1 $L_\pi = \mathbf{K}_N + \pi(L)$ *or* $L_\pi = \mathbf{H} + \pi(L)$ *for a predicate logic* L *without equality (respectively,* N*-modal or superintuitionistic).*

Proposition 2.15.2

(1) $(\mathbf{QK}_N)_\pi = \mathbf{K}_N$;

(2) $(\mathbf{QH})_\pi = \mathbf{H}$.

Proof By induction over a proof of A in $\mathbf{QK}_N$ (or in **QH**) we show that $\pi(A) \in \mathbf{K}_N$ (resp. $\pi(A) \in \mathbf{H}$). It is easily checked that $\pi(A) \in \mathbf{K}_N$ if A is a substitution instance of a $\mathbf{QK}_N$- axiom (for example, $\pi(A) = (B \supset B)$ for predicate axioms). If A is obtained by (MP) or (necessitation), then $\pi(A)$ is also obtained by the same rule. If $A = \forall x B$, then $\pi(A) = \pi(B)$. ∎

Lemma 2.15.3 $\pi(\mathrm{Sub}(\Gamma)) = \mathrm{Sub}_\pi(\pi(\Gamma))$ *for any* $\Gamma \subseteq IF$ *(or* $\Gamma \subseteq MF_N$*), where* Sub_π *denotes closure under propositional substitutions (of the corresponding type).*

Proof Since every substitution is a composition of simple substitutions, we can consider only simple substitutions. It is easily proved that

$$\pi([C/P(x)]B) = [\pi(C)/\pi_0(P)]\pi(B)$$

(by induction on B). On the other hand, every propositional formula can be presented as $\pi(C)$ for some predicate formula C. ■

Proposition 2.15.4 $(L+\Gamma)_\pi = L_\pi + \pi(\Gamma)$ *for a modal (or superintuitionistic) logic* $L+\Gamma$.

In particular, $(\mathbf{QK}_N + \Gamma)_\pi = \mathbf{K}_N + \pi(\Gamma)$, $(\mathbf{QH}+\Gamma)_\pi = \mathbf{H} + \pi(\Gamma)$.

Proof [Modal case.] Let $A \in (L+\Gamma)$. By the deduction theorem,

$$\left(\bigwedge_{s=1}^{k} \Box^{\leq k}\overline{\forall} A_s \supset A\right) \in L$$

for some formulas $A_s \in \mathrm{Sub}(\Gamma)$. Then

$$\left(\bigwedge_{s=1}^{k} \Box^{\leq k}\pi(A_s) \supset \pi(A)\right) \in L_\pi,$$

while $\pi(A_s) \in \mathrm{Sub}_\pi(\pi(\Gamma))$ (by Lemma 2.15.3). Thus $\pi(A) \in (L_\pi + \pi(\Gamma))$.

The converse inclusion is obvious by Proposition 2.15.1. ■

Definition 2.15.5 *The* quantified version *of a modal (respectively, superintuitionistic) propositional logic* Λ *is:*

$$\mathbf{Q}\Lambda := \mathbf{QK}_N + \Lambda \quad \textit{(respectively, } \mathbf{Q}\Lambda := \mathbf{QH} + \Lambda).$$

Lemma 2.15.6

(1) *For any propositional logic* $\mathbf{\Lambda}$ *and a set of propositional formulas* Γ,

$$\mathbf{Q}(\mathbf{\Lambda} + \Gamma) = \mathbf{Q\Lambda} + \Gamma.$$

(2) *For a propositional* **S4***-logic* $\mathbf{\Lambda}$, ${}^T(\mathbf{Q\Lambda}) \supseteq \mathbf{Q}^T\mathbf{\Lambda}$.

Proof

(1) Consider the intuitionistic case. $\mathbf{Q}(\mathbf{\Lambda}+\Gamma)$ is the smallest s.p.l. containing $\mathbf{\Lambda}+\Gamma$, while $\mathbf{Q\Lambda}+\Gamma$ is the smallest s.p.l. containing $\mathbf{Q\Lambda}\cup\Gamma$. These two logics coincide, since every s.p.l. containing $\mathbf{\Lambda}$ also contains $\mathbf{Q\Lambda}$.

(2) By 2.11.11, $\mathbf{Q}^T\mathbf{\Lambda} \vdash A$ implies $\mathbf{QS4} + ({}^T\mathbf{\Lambda})^T \vdash A^T$, hence $\mathbf{Q\Lambda} \vdash A^T$. Thus $\mathbf{Q}^T\mathbf{\Lambda} \subseteq {}^T(\mathbf{Q\Lambda})$.

■

Remark 2.15.7 We do not know if the equality ${}^{T}(\mathbf{Q\Lambda}) = \mathbf{Q}^{T}\mathbf{\Lambda}$ holds for any propositional **S4**-logic. For example, this is unknown for $\mathbf{\Lambda} = \mathbf{S4.2Grz}$.

Definition 2.15.8 [30] *A predicate logic L is called a* predicate extension *of a propositional logic $\mathbf{\Lambda}$ if L is called a conservative extension of $\mathbf{\Lambda}$ (i.e. if $L_\pi = \mathbf{\Lambda}$).*

Proposition 2.15.9 *For any modal or superintuitionistic propositional logic Λ, $\mathbf{Q\Lambda}$ is a predicate extension of $\mathbf{\Lambda}$.*

Proof Follows readily from Proposition 2.15.4 since $\pi(\mathbf{\Lambda}) \subseteq \mathbf{\Lambda}$. ∎

Obviously, $\mathbf{Q\Lambda}$ is the weakest predicate extension of $\mathbf{\Lambda}$.

To describe the greatest predicate extensions of propositional logics, we use the formula

$$AU_1 = \forall x \forall y (P(x) \supset P(y))$$

from Section 2.6; recall that

$$L + AU_1 = L + \exists x P(x) \supset \forall P(x)$$

for any predicate logic L and

$$L + AU_1 = L + \forall x \forall y (x = y)$$

for any predicate logic L with equality.

For a formula A with equality, let $A_\top$ be the result of replacing each occurrence of $(x = y)$ in A with $\top$. Then $(A \equiv A_\top) \in (\mathbf{QK}_N^= + AU_1)$ (respectively, $\mathbf{QH}^= + AU_1$).

Lemma 2.15.10 *Let A be a predicate formula without equality. Then $A \in \mathbf{QK}_N + AU_1 + \pi(A)$ (in the modal case) or $A \in \mathbf{QH} + AU_1 + \pi(A)$ (in the intuitionistic case).*

Proof First, given $\pi(A)$, we can restore all occurrences of quantifiers from A, because they are dummy in $\pi(A)$. Next, every occurrence of $\pi_0(P)$ in $\pi(A)$ coming from some $P(\mathbf{x})$ in A, can be replaced with $\overline{\forall} P(\mathbf{x})$, which is equivalent to $P(\mathbf{x})$ in $\mathbf{QK}_N + AU_1$. ∎

Proposition 2.15.11 *Let L be a predicate logic without equality, modal or superintuitionistic, and let $\mathbf{\Lambda}$ be a propositional logic of the corresponding kind. Then*

$$L_\pi = \mathbf{\Lambda} \text{ iff } \mathbf{Q\Lambda} \subseteq L \subseteq \mathbf{Q\Lambda} + AU_1.$$

Proof $(\mathbf{Q\Lambda} + AU_1')_\pi = \mathbf{\Lambda}$ by Proposition 2.15.4 since

$$\pi(AU_1') = (\pi_0(P) \supset \pi_0(P)) \in \mathbf{\Lambda}.$$

On the other hand, if $L_\pi = \mathbf{\Lambda}$ then $L \subseteq \mathbf{Q\Lambda} + AU_1$. In fact, $A \in L$ implies $\pi(A) \in L_\pi = \mathbf{\Lambda}$ and $A \in \mathbf{Q\Lambda} + AU_1$ by Lemma 2.15.10. ∎

[30] Cf. [Ono, 1973].

Therefore $\mathbf{Q\Lambda} + AU_1$ is the greatest predicate extension of a propositional logic $\mathbf{\Lambda}$.

Corollary 2.15.12 *The greatest intermediate predicate extension of an intermediate propositional logic* $\mathbf{\Lambda}$ *is*

$$(\mathbf{Q\Lambda} + AU_1) \cap \mathbf{QCL} = \mathbf{Q\Lambda} + AU_1 \vee q \vee \neg q.$$

Proof By Proposition 2.10.1 (2); recall that $\mathbf{QCL} = \mathbf{QH} + q \vee \neg q$ and $CD \in \mathbf{QH} + AU_1 \vee q \vee \neg q$. ■

Let us also mention the following result on the number of predicate extensions.

Theorem 2.15.13[31]

(1) Every nonclassical superintuitionistic propositional logic has uncountably many predicate extensions.

(2) Every modal propositional logic which does not contain **S5**, *has uncountably many predicate extensions.*

Now let us give another description of the greatest predicate extensions.

Proposition 2.15.14

(I) Let L be a predicate logic (of any kind). Then the following conditions are equivalent:

(a) $L + A = L + \pi(A)$ *for any predicate formula A without equality;*

(b) $\mathbf{QH}^{(=)} + AU_1 \subseteq L$ *or* $\mathbf{QK}_N^{(=)} + AU_1 \subseteq L$
(for the superintuitionistic or the modal case, respectively);

(c) $L = \mathbf{Q}\Lambda^{(=)} + AU_1$ *for a propositional logic* $\mathbf{\Lambda}$ *(superintuitionistic or modal, respectively).*
Moreover, these conditions imply

(d) for any predicate formula A (in the language of L) there exists a propositional formula A' such that $L + A = L + A'$.

(II) If L is a logic without equality, all conditions (a)–(d) are equivalent.

Proof

(I) *(a)* ⇒ *(b)*. $L + AU_1 = L$ since $\pi(AU_1) = (\pi_0(P) \supset \pi_0(P)) \in \mathbf{H}$ (or $\mathbf{K}_N$).

(b) ⇒ *(a)*. $\pi(A) \in L + A$ by Proposition 2.15.1.

On the other hand, $A \in \mathbf{QH} + AU_1 + \pi(A)$ or $A \in \mathbf{QK}_N + AU_1 + \pi(A)$, by Lemma 2.15.10.

[31] [Suzuki, 1995].

$(a) \Rightarrow (d)$. If L is a logic with equality, we take $A' = \pi(A_\top)$.

$(b) \Rightarrow (c)$. We can replace any A from L with A' because (b) implies (d).

(II) $(d) \Rightarrow (a)$. Let $L + A = L + A'$ for a propositional formula A'. Then

$$A' \in (L + A)_\pi = L_\pi + \pi(A) \subseteq L + \pi(A).$$

On the other hand, $\pi(A) \in (L + A)$.

■

We say that L is a predicate logic *with degenerate predicates* if L satisfies the condition (I)(d) from Proposition 2.15.14. Thus, $\mathbf{QH} + AU_1$ and $\mathbf{QK}_N + AU_1$ are the weakest logics with degenerate predicates. On the other hand, there exist logics with equality incomparable with $\mathbf{QH}^= + AU_1$ and satisfying (I)(d), e.g.

$$L = \mathbf{QH}^= + \exists x \exists y\ (x \neq y \wedge \forall z(z = x \vee z = y)).$$

It is clear how to describe the propositional fragments for these logics with equality.

Lemma 2.15.15 *Let L and L' be predicate logics containing AU_1 such that L is without equality and L' is with equality. Then the following conditions are equivalent:*

(1) $L'_\pi = L_\pi$;

(2) $(L')^0 = L$;

(3) $L' = L^=$.

Proof [32]

(1) $\Rightarrow$ (2). By Proposition 2.15.14, (b) $\Rightarrow$ (a).

(3) $\Rightarrow$ (1). Let A be a propositional formula, $A \in L^=$. Then by the deduction theorem, the following formula is a theorem of L:

$$B := \Box^{\leq k}(\forall x(x = x) \wedge \bigwedge_{s=1}^{k} \overline{\forall}(x = y \supset ([x/z]A_s \supset [y/z]A_s))) \supset A$$

for some $k \geq 0$ and formulas $A_1, \ldots, A_k \in MF_N^=$. Then its substitution instance $B_\top$ belongs to L and $\pi(B_\top) \in L_\pi$. Since $\pi((\forall x(x = x)_\top) = \top$ and $\pi(([x/z]A_s)_\top) = \pi(([y/z]A_s)_\top)$, we obtain $\pi(A) \in L$, i.e. $A \in L_\pi$.

(2) $\Rightarrow$ (3). If $(L')^0 = (L^=)^0 = L$ then $L^= \subseteq L'$, and $L^= = L'$ — because $(A \equiv A_\top) \in L^=$ for any formula with equality A (recall that $A_\top$ is a formula without equality). ■

[32] The implication (3) $\Rightarrow$ (2) is proved in Proposition 2.14.8 for all superintuitionistic logics and for 1-modal logics above **QK4**. Here we give a slightly simpler proof that fits for any modal predicate language.

Corollary 2.15.16 *Let L be a predicate logic with equality, $\mathbf{\Lambda}$ a propositional logic. If*

$$\mathbf{Q\Lambda}^{=} \subseteq L \subseteq \mathbf{Q\Lambda}^{=} + AU_1$$

then $L_\pi = \mathbf{\Lambda}$.

Proof

$$\mathbf{Q\Lambda}^{=} \subseteq L \subseteq \mathbf{Q\Lambda}^{=} + AU_1$$

implies

$$(\mathbf{Q\Lambda}^{=})_\pi \subseteq L_\pi \subseteq (\mathbf{Q\Lambda}^{=} + AU_1)_\pi.$$

Next,

$$(\mathbf{Q\Lambda}^{=} + AU_1)_\pi = (\mathbf{Q\Lambda} + AU_1)^{=}_\pi = (\mathbf{Q\Lambda} + AU_1)_\pi,$$

by Lemma 2.15.15, and $(\mathbf{Q\Lambda} + AU_1)_\pi = \mathbf{\Lambda}$, by Proposition 2.15.11. ∎

Corollary 2.15.17 *Let L be a predicate logic without equality. Then*

(1) $(L^{=})_\pi = (L^{=d})_\pi = (L^{=s})_\pi = L_\pi$ for a superintuitionistic L;

(2) $(L^{=})_\pi = (L^{=c})_\pi = L_\pi$ for modal L.

Proof (Intuitionistic case.) Let $L_\pi = \mathbf{\Lambda}$, $L \subseteq \mathbf{Q\Lambda} + AU_1$. Then

$$\mathbf{Q\Lambda}^{=} \subseteq L^{=} \subseteq L^{=s} \subseteq L^{=d} \subseteq \mathbf{Q\Lambda}^{=} + AU_1,$$

and we can apply Corollary 2.15.16. The modal case is analogous. ∎

Proposition 2.15.14 and Lemma 2.15.15 show that there exists a natural bijection between intermediate propositional logics and extensions of the logics $\mathbf{QH}^{(=)} + AU_1$ (with or without equality), and similarly for the modal case. The logic $\mathbf{Q}\Lambda^{=} + AU_1$ is a *maximal* predicate extension (with equality) of a propositional logic Λ. Nevertheless, for any propositional logic Λ the *greatest* predicate extension with equality does not exist. Indeed, later on we will show that $L_\pi = \Lambda$ for the logic

$$L = \mathbf{Q}\Lambda^{=} + \neg\forall x\forall y(x = y),$$

which is obviously incompatible with $\mathbf{Q}\Lambda^{=} + AU_1$.

2.16 Semantics from an abstract viewpoint

Consider a language with the set of well-formed formulas Φ, and a set of its subsets $\Lambda \subseteq 2^\Phi$ closed under arbitrary intersections. Elements of Λ are called Λ-*logics*. We define a *semantics* for the set Λ as a quadruple $\mathbb{S} = (\Phi, \Lambda, \mathcal{U}, \models)$, in which $\mathcal{U}$ is a class (whose elements are called $\mathcal{U}$-*frames*), and $\models$ is a binary relation between $\mathcal{U}$ and Φ (called the *validity relation*) such that the set

$$L_{\mathbb{S}}(F) = \{A \in \Phi \mid F \models A\}$$

is a Λ-logic for any $F \in \mathcal{U}$. $L_{\mathbb{S}}(F)$ is called the *Λ-logic of the frame F* (in $\mathbb{S}$). Sometimes we write $F \in \mathbb{S}$ instead of $F \in \mathcal{U}$.

Since Λ is intersection closed, we obtain that for any class $\Sigma \subseteq \mathcal{U}$, the set $L_{\mathbb{S}}(\Sigma) = \bigcap\{L_{\mathbb{S}}(F) \mid F \in \Sigma\}$ is also a Λ-logic; it is called the *Λ-logic of Σ*. Logics of this kind are called *$\mathbb{S}$-complete*, and logics of the form $L_{\mathbb{S}}(F)$ are called *simply $\mathbb{S}$-complete*. As stated in [Ono, 1973, Theorem 2.2], completeness does not always imply simple completeness.

A semantics $\mathbb{S}$ gives rise to the *logical consequence relation* (between $\Gamma \subseteq \Phi$ and $A \in \Phi$):

$$\Gamma \models_{\mathbb{S}} A \ \text{ iff } \ \forall F \in \mathcal{U}\, (\Gamma \subseteq L_{\mathbb{S}}(F) \Rightarrow A \in L_{\mathbb{S}}(F)).$$

We say that F is a *Γ-frame (in $\mathbb{S}$)* if $\Gamma \subseteq L_{\mathbb{S}}(F)$. Thus $\Gamma \models_{\mathbb{S}} A$ means that A is valid in every Γ-frame. If F is a Γ-frame such that $F \not\models A$, we say that F *separates* A from a set Γ.

One can recognise a particular case of Galois correspondence here. This correspondence is derived from the relation $\models$ in the standard way; cf. [Chapter 5, Theorem 19][Birkhoff, 1979]. So

$$C_{\mathbb{S}}(\Gamma) = \{A \in \Phi \mid \Gamma \models_{\mathbb{S}} A\}$$

is a closure operation on 2^{Φ}, and the closed sets of formulas are just the S-complete logics.

Hence we have

Lemma 2.16.1 *$C_{\mathbb{S}}(\Gamma)$ is the smallest $\mathbb{S}$-complete logic containing Γ. $C_{\mathbb{S}}(\Gamma)$ is called the* $\mathbb{S}$-completion *of* Γ.

Here is a simple criterion of completeness:

Lemma 2.16.2 *A logic L is complete in a semantics $\mathbb{S}$ iff every formula $A \in (\Phi - L)$ is separated from L by a frame in $\mathbb{S}$.*

Proof In fact, by 2.16.1, L is complete iff $C_{\mathbb{S}}(L) = L$. ∎

We can compare different semantics for the same set of logics. A semantics $\mathbb{S}_1$ is *reducible* to $\mathbb{S}_2$ (notation: $\mathbb{S}_1 \preceq \mathbb{S}_2$) iff $C_{\mathbb{S}_2}(L) \subseteq C_{\mathbb{S}_1}(L)$ for any $L \in \Lambda$.

Lemma 2.16.3

(1) $\mathbb{S}_1 \preceq \mathbb{S}_2$ iff every $\mathbb{S}_1$-complete logic is $\mathbb{S}_2$-complete.

(2) $\mathbb{S}_1 \preceq \mathbb{S}_2$ iff for any $\mathbb{S}_1$-frame F, $L_{\mathbb{S}_1}(F)$ is $\mathbb{S}_2$-complete.

Proof

(1) Suppose $\mathbb{S}_1 \preceq \mathbb{S}_2$. Then for any logic L,

$$L \subseteq C_{\mathbb{S}_2}(L) \subseteq C_{\mathbb{S}_1}(L).$$

If L is $\mathbb{S}_1$-complete then $L = C_{\mathbb{S}_1}(L)$ (by Lemma 2.16.1), and hence $L = C_{\mathbb{S}_2}(L)$ which implies $\mathbb{S}_2$-completeness of Λ (again by Lemma 2.16.1).

Conversely, assume that $\mathbb{S}_1$-completeness implies $\mathbb{S}_2$-completeness. Then $C_{\mathbb{S}_1}(L)$ is $\mathbb{S}_2$-complete, and thus $C_{\mathbb{S}_2}(L) \subseteq C_{\mathbb{S}_1}(L)$ by Lemma 2.16.1.

(2) 'Only if' is an immediate consequence of (i) and Lemma 2.16.1.

To prove 'if' consider an arbitrary $\mathbb{S}_1$-complete logic

$$L = L_{\mathbb{S}_1}(\Sigma) = \bigcap\{L_{\mathbb{S}_1}(F) \mid F \in \Sigma\}.$$

If each $L_{\mathbb{S}_1}(F)$ is $\mathbb{S}_2$-complete, we have that $L_{\mathbb{S}_1}(F) = L_{\mathbb{S}_2}(\Psi_F)$ for some class Ψ_F. Then

$$L = L_{\mathbb{S}_1}(\Sigma) = \bigcap\{L_{\mathbb{S}_2}(\Psi_F) \mid F \in \Sigma\} = L_{\mathbb{S}_2}(\bigcup\{\Psi_F \mid F \in \Sigma\}),$$

showing that Λ is $\mathbb{S}_2$-complete.

■

Definition 2.16.4 *Semantics* $\mathbb{S}_1$ *and* $\mathbb{S}_2$ *are called* equivalent ($S_1 \simeq \mathbb{S}_2$) *if* $\mathbb{S}_1 \preceq \mathbb{S}_2$ *and* $\mathbb{S}_2 \preceq \mathbb{S}_1$*; we say that* $\mathbb{S}_1$ *is* weaker *than* $\mathbb{S}_2$ *(notation:* $\mathbb{S}_1 \prec S_2$*) if* $\mathbb{S}_1 \preceq \mathbb{S}_2$ *but not* $\mathbb{S}_2 \preceq \mathbb{S}_1$.

Lemma 2.16.3 readily implies:

Lemma 2.16.5 $\mathbb{S}_1 \simeq \mathbb{S}_2$ *iff* $\mathbb{S}_1$*-completeness is equivalent to* $\mathbb{S}_2$*-completeness.*

Definition 2.16.6 *A semantics* $\mathbb{S}_1$ *is* simply reducible *to* $\mathbb{S}_2$ *(notation:* $\mathbb{S}_1 \subseteq \mathbb{S}_2$*) if simple* $\mathbb{S}_1$*-completeness implies simple* $\mathbb{S}_2$*-completeness, i.e.*

$$\forall F_1 \in \mathbb{S}_1 \exists F_2 \in \mathbb{S}_2 \ (L_{\mathbb{S}_1}(F_1) = L_{\mathbb{S}_2}(F_2)).$$

Semantics $\mathbb{S}_1$ *and* $\mathbb{S}_2$ *are called* simply equivalent *(notation:* $\mathbb{S}_1 \cong S_2$*) if* $\mathbb{S}_1 \subseteq \mathbb{S}_2$ *and* $\mathbb{S}_2 \subseteq \mathbb{S}_1$.

We do not usually distinguish between simply equivalent semantics; from this viewpoint a semantics $\mathbb{S} = (\Phi, \Lambda, \mathcal{U}, \vDash)$ can be identified with the corresponding set of simply complete logics $\{L_{\mathbb{S}}(F) \mid F \in \mathcal{U}\}$.

Lemma 2.16.7 $\mathbb{S}_1 \subseteq \mathbb{S}_2$ *implies* $\mathbb{S}_1 \preceq \mathbb{S}_2$.

Proof Obvious, since complete logics are intersections of simply complete ones. ■

Example 2.16.8 For any set of logics Λ there exists a 'trivial semantics' $\mathbb{T}$, in which $\mathcal{U} = \Lambda$ (i.e. 'frames' are just logics) and $L \vDash A$ iff $A \in L$, i.e. $L_{\mathbb{T}}(L) = L$).

Definition 2.16.9 *We say that a semantics* $\mathbb{S}$ *has the* collection property *(CP) if every* $\mathbb{S}$*-complete consistent logic is simply* $\mathbb{S}$*-complete.*

By Lemma 2.16.7, for semantics with the (CP), reducibility is equivalent to simple reducibility, so their equivalence implies simple equivalence. But in general equivalent semantics may be not simply equivalent; some counterexamples will be given later on.

Example 2.16.10 In modal propositional logic we have semantics with the (CP):

$$\text{finite} \prec \text{Kripke} \prec \text{topological} \prec \text{algebraic} \cong \text{trivial}.$$

'Finite semantics' consists of all finite Kripke frames, with the usual definition of validity. So in this semantics simple completeness means tabularity and completeness means the f.m.p.

Algebraic semantics is simply equivalent to trivial, since every logic is algebraically complete, by the Lindenbaum theorem.

For superintuitionistic propositional logics we have the following diagram:

$$\text{finite} \prec \text{Kripke} \prec \text{topological} \preceq \text{algebraic} \cong \text{trivial}.$$

The question, whether topological semantics is trivial, is Kuznetsov's problem mentioned in Section 1.17.

Recall that in this book we consider four types of predicate logics — modal or intuitionistic, with or without equality. So we alter the notation $L_{\mathbb{S}}(F)$ respectively. For example, if F is a frame in a semantics $\mathbb{S}$ for N-modal predicate logics with equality, for $\{A \in MF_N^= \mid F \vDash A\}$ we use the notation $\mathbf{ML}^=(F)$ rather than $L_{\mathbb{S}}(F)$; $\mathbf{ML}(F)$ refers to semantics without equality. Similarly in the intuitionistic case we use the notations $\mathbf{IL}^=(F)$, $\mathbf{IL}(F)$.

Every semantics $\mathbb{S}$ for logics with equality generates a semantics $\mathbb{S}^\circ$ for the same kind of logics without equality, with the same frames and validity. So $L_{\mathbb{S}^\circ}(F) = L_{\mathbb{S}}(F)^\circ$. Loosely speaking, we call $\mathbb{S}^\circ$-complete logics '$\mathbb{S}$-complete'.

Remark 2.16.11 However it may happen that there exists a semantics for logics without equality $\mathbb{S}'$, with the same frames as $\mathbb{S}$ (and $\mathbb{S}^\circ$), but with a slightly different notion of validity. In these cases we define $\mathbf{ML}(F)$ (or $\mathbf{IL}(F)$) as $L_{\mathbb{S}'}(F)$ and then prove that $\mathbb{S}' = \mathbb{S}^\circ$. We will encounter such a situation in Chapter 5.

Definition 2.16.12 *Let $\mathbb{S}$ be a semantics for N-m.p.l. We say that $\mathbb{S}$* admits equality *if there exists a semantics $\mathbb{S}_1$ for N-m.p.l.= such that $\mathbb{S} \subseteq \mathbb{S}_1^\circ$, i.e.*

$$\forall F \in \mathbb{S}\, \exists F_1 \in \mathbb{S}_1\; L_{\mathbb{S}}(F) = L_{\mathbb{S}_1}(F_1)^\circ.$$

Proposition 2.16.13 *If $\mathbb{S}$ admits equality and an N-m.p.l. L is $\mathbb{S}$-complete, then $L^=$ is conservative over L.*

Proof Almost obvious. Suppose L is N-modal, $L = \mathbf{ML}(\mathcal{C})$ for a class of $\mathbb{S}$-frames $\mathcal{C}$ and $\mathbf{ML}(F) = \mathbf{ML}^=(F_1)^\circ$ for any $F \in \mathcal{C}$. Then

$$L = \bigcap_{F \in \mathcal{C}} \mathbf{ML}(F) = \bigcap_{F \in \mathcal{C}} \mathbf{ML}^=(F_1)^\circ.$$

Put $\mathcal{C}_1 := \{F_1 \mid F \in \mathcal{C}\}$, then for any $A \in MF_N$

$$A \in L \Leftrightarrow \forall F \in \mathcal{C}\; F_1 \vDash A \Leftrightarrow A \in \mathbf{ML}^=(\mathcal{C}_1).$$

Thus $\mathbf{ML}^=(\mathcal{C}_1)^\circ = L$, i.e. $\mathbf{ML}^=(\mathcal{C}_1)$ is conservative over $L = \mathbf{ML}(\mathcal{C})$. Since $\mathbf{ML}^=(\mathcal{C}_1)$ is a logic with equality containing L, it follows that $L^= \subseteq \mathbf{ML}^=(\mathcal{C}_1)$ and thus $L^=$ is also conservative over L. ■

Now let us prove a simple result on correlation between completeness of L and $L^=$.

Proposition 2.16.14 *Let L be an m.p.l. or an s.p.l., $\mathbb{S}$ a semantics for the corresponding logics with equality. If $L^=$ is conservative over L and $\mathbb{S}$-complete, then L is also $\mathbb{S}$-complete.*

Proof Consider the modal case only. Suppose $L^= = \mathbf{ML}^=(\mathcal{C})$ for a class of $\mathbb{S}$-frames $\mathcal{C}$. Then by conservativity, $L = (L^=)^\circ = \mathbf{ML}^=(\mathcal{C})^\circ$, and obviously, $\mathbf{ML}^=(\mathcal{C})^\circ = \mathbf{ML}(\mathcal{C})$. Thus L is $\mathbb{S}$-complete. ■

Corollary 2.16.15 *If L is an s.p.l. or a conically expressive m.p.l. and $L^=$ is complete in a semantics $\mathbb{S}$, then L is also $\mathbb{S}$-complete.*

Proof By 2.16.14 and 2.14.8. ■

Finally let us show the existence of modal counterparts using completeness.

Definition 2.16.16 *Let $\mathbb{M}$ be a semantics for m.p.l.(=) extending* $\mathbf{QS4}(=)$. *Then we define its* intuitionistic version, *the semantics ${}^T\mathbb{M}$ for s.p.l.(=), with the same frames as $\mathbb{M}$ such that for any $\mathbb{M}$-frame F,*

$$L_{{}^T\mathbb{M}}(F) := {}^T L_{\mathbb{M}}(F).$$

Proposition 2.16.17 *Every ${}^T\mathbb{M}$-complete s.p.l.(=) has a modal counterpart:*

$$\mathbf{QH}^{(=)} + \Gamma = {}^T(\mathbf{QS4}^{(=)} + \Gamma^T).$$

Proof The inclusion $\subseteq$ is proved in Lemma 2.11.11. For the converse suppose $\mathbf{QH} + \Gamma \nvdash A$. By assumption $\mathbf{QH} + \Gamma$ is ${}^T\mathbb{M}$-complete, so there exists an $\mathbb{M}$-frame F such that $\Gamma \subseteq L_{{}^T\mathbb{M}}(F)$, but $A \notin L_{{}^T\mathbb{M}}(F)$. By Definition 2.16.16, $L_{{}^T\mathbb{M}}(F) = {}^T L_{\mathbb{M}}(F)$, hence $\Gamma^T \subseteq L_{\mathbb{M}}(F)$, $A^T \notin L_{\mathbb{M}}(F)$. Since $\mathbb{M}$ is a semantics for modal logics above $\mathbf{QS4}$, we have $L_{\mathbb{M}}(F) \supseteq \mathbf{QS4} + \Gamma^T$. Consequently $\mathbf{QS4} + \Gamma^T \nvdash A^T$, i.e. ${}^T(\mathbf{QS4} + \Gamma^T) \nvdash A$. ■

As we shall see in Volume 2, the smallest modal counterpart of an ${}^T\mathbb{M}$-complete logic may be $\mathbb{M}$-incomplete. But completeness always transfers in the other direction:

Proposition 2.16.18 *If $L \supseteq \mathbf{QS4}^{(=)}$ is $\mathbb{M}$-complete, then ${}^T L$ is ${}^T\mathbb{M}$-complete.*

Proof Suppose $L = L_{\mathbb{M}}(\mathcal{C})$ for a class of $\mathbb{M}$-frames $\mathcal{C}$. Then

$$L_{{}^T\mathbb{M}}(\mathcal{C}) = \bigcap_{F\in\mathcal{C}} L_{{}^T\mathbb{M}}(F) = \bigcap_{F\in\mathcal{C}} {}^T L_{\mathbb{M}}(F) = \{A \mid \forall F \in \mathcal{C}\ A^T \in L_{\mathbb{M}}(F)\} = {}^T L_{\mathbb{M}}(\mathcal{C}),$$

so ${}^T L$ is $\mathbb{M}$-complete. ■

Part II

Semantics

Introduction: What is semantics?

In Chapter 2 we proposed a general approach to semantics, which is very formal and does not help understand the meaning of language expressions.

For example, this approach allows for a degenerate semantics, where 'frames' are just logics and $L \models A$ iff $L \vdash A$. In this case the completeness theorem is a triviality, but of course such a semantics does not explain anything.[1]

To describe more plausible semantics we need to define notions of a *model* and of the *truth* in a model that could tell us 'how the logic works'. This is not a serious problem for classical first-order logic: in this case model theory can be developed within the standard semantics based on the well-known Tarski truth definition. Due to Gödel's completeness theorem (GCT), the standard semantics works properly, and thus alternative types of semantics (such as sheaves, forcing, polyadic algebras) are of less importance in the classical case.

The situation in nonclassical first-order logic is quite different. GCT has no direct analogues, and incompleteness phenomena enable us to consider various semantics, without any obvious preference between them. Nonclassical model theory is still rather miscellaneous, and this book is aimed at organising some part of it.

The book does not cover all the semantics in equal proportion. So we begin this part with a brief survey of important results in the area and some references for further reading.

Gödel's completeness theorem: discussion

First let us explain why GCT is not always transferred to nonclassical logics. Actually there exist two forms of Gödel's theorem:

(GCT1) A formula A is a theorem of classical predicate calculus iff A is valid in any domain.

(GCT2) Every consistent classical theory has a model.

These two statements are more or less equivalent: (GCT1) is equivalent to (GCT2) for finitely axiomatisable theories.

However in the nonclassical area it is essential to distinguish between logics and theories. A *theory* is an extension of some basic calculus by additional axioms (and perhaps inference rules) postulating specific properties of objects being studied; a typical example is Heyting arithmetic, which formalises the intuitionistic viewpoint on natural numbers. A *logic* is a theory whose theorems may be considered as 'logical laws' common for a certain class of theories and not depending on specific 'application domains'. We can treat logical laws as

[1] Of course in general we may have a new logic and several syntactical translations into well-known logics. Such translations can be viewed as semantics and can be very illuminating. See [Gabbay, 1996, Chapter 1].

schemata for producing theorems, and define a logic just as a substitution-closed theory.[2] So we can say that (GCT2) is a property of classical theories, and (GCT1) is a property of classical logic.

Some results similar to (GCT2) can be proved for nonclassical theories. In fact, it is well known that every consistent first-order normal modal theory S is satisfied in some Kripke model, i.e. there exists a Kripke model M and a possible world $w \in M$ such that $(M, w) \models A$ for any $A \in S$. Moreover, M can be chosen uniformly for all consistent theories [Gabbay, 1976]. Analogous claims are true for superintuitionistic theories considered as pairs of sets of formulas, cf. [Dragalin, 1988].

However finding nonclassical analogues of (GCT1) is more problematic. These are completeness theorems of the following form:

(GCT1′) a formula A is a theorem of a logic L iff A is L-valid.

The main problem is in finding 'reasonable' notions of validity, for which (GCT1′) may be true. In classical logic this is validity in a domain. In a general case, because we require logics to be substitution-closed, we also need 'substitution-invariant' notions of validity. So for a nonclassical logic L, analogues of domains are *frames* (or *model structures*), from which we can obtain *models* of theories based on $\mathbf{\Lambda}$ if we specify interpretations of basic predicates; a formula is valid in a frame iff it is true in every model over this frame.

It is usually required that the set of all formulas true in a model is a theory, and thus the set $\mathbf{L}(\mathbb{F})$ of all formulas valid in a frame $\mathbb{F}$ is also a theory.[3] The difference between models and frames is the following *substitution property* implying that $\mathbf{L}(\mathbb{F})$ is a logic:

(SP) The set of all formulas valid in a frame $\mathbb{F}$ is substitution closed.

Generally speaking, there may be many kinds of 'frames' and 'validity'. We say that a class of frames C *generates a semantics*

$$S(C) = \{\mathbf{L}(\mathbb{F}) \mid \mathbb{F} \in C\}.$$

A logic L is called *complete* in $S(C)$ iff for every formula $A \notin L$ there exists a frame $\mathbb{F} \in C$, such that formulas from L are valid in $\mathbb{F}$, whereas A is not. Equivalently, a logic is complete iff it can be presented as an intersection of logics of the form $\mathbf{L}(\mathbb{F})$ for some frames $\mathbb{F}$.

For example, in the *standard classical semantics* 'frames' are just sets, and the notion of validity is well known. So (GCT1) means that the classical predicate logic is complete in this semantics.

[2] Of course this definition is rather conventional, and there exist examples of 'logics' which are not substitution-closed.

[3] This set depends also on the language, where formulas are constructed.

Examples of incompleteness in first-order logic

In Chapter 1 we gave a picture of semantics in nonclassical propositional logic and pointed out some rather strong completeness results. But in first-order logic the situation becomes worse: there are very few completeness theorems known, and incompleteness is very frequent.

Incompleteness already appears for logics extending the classical predicate logic **QCL**. In fact, consider any formula A valid in all finite domains, but refutable in some infinite domain (and thus refutable in any infinite domain, by the Löwenheim–Skolem theorem). The logic $\mathbf{QCL}+A$ is incomplete, because the set of all finitely valid formulas is not recursively axiomatisable [Trachtenbrot, 1950].

This argument seems to be just a trick with definitions. But in nonclassical logic there exist more natural examples of incompleteness.

The first example of this kind was discovered by H.Ono [Ono, 1973], who proved that the intermediate logic of the strong Markov principle

$$\mathbf{QHE} = \mathbf{QH} + \neg\neg\exists x P(x) \supset \exists x \neg\neg P(x)$$

is incomplete in the standard Kripke semantics ($\mathcal{K}$).

Recall that $\mathcal{K}$ is generated by *predicate Kripke frames*; such a frame is a triple $\mathbb{F} = (W, R, D)$, in which (W, R) is a propositional Kripke frame; D is a system of 'expanding individual domains', i.e. a family of non-empty sets indexed by possible worlds $(D_w)_{w \in W}$, such that

$$\forall u, v\ (uRv \Rightarrow D_u \subseteq D_v).$$

The logics **QS4** and **QH** are known to be $\mathcal{K}$-complete. But their minimal equality extensions ($\mathbf{QS4}^=, \mathbf{QH}^=$) are incomplete if the symbol '=' is interpreted in every domain just as the identity. This is due to the formulas

$$DE = \forall x \forall y\ (x = y \vee \neg(x = y)) \qquad \text{(decidable equality principle)}$$

and

$$CE = \forall x \forall y\ (\Diamond(x = y) \supset x = y) \qquad \text{(closed equality principle)}$$

which are valid in all Kripke frames, but nonprovable respectively in $\mathbf{QH}^=$ and in $\mathbf{QS4}^=$.

But we can also interpret equality in a world w in another natural way, as an equivalence relation in D_w. Then we come to the semantics $\mathcal{KE}$ of *Kripke frames with equality* (KFEs), in which $\mathbf{QS4}^=, \mathbf{QH}^=$ become complete; so $\mathcal{KE}$ is stronger than $\mathcal{K}$. The latter observation and the definition of $\mathcal{KE}$ for the intuitionistic case first appeared in [Dragalin, 1973] and then in [Dragalin, 1988].

Note that in the classical case these two approaches are equivalent: every 'non-normal' model, in which equality is an arbitrary equivalence relation can be 'normalised' by identifying equal elements, so that its elementary theory does not change.

For a Kripke frame with equality a similar construction is possible: one can factorise every individual domain through the corresponding equivalence relation. Thus we obtain a *Kripke sheaf*; it can be defined as a propositional Kripke frame (W, R) together with a system of individual domains $(D_w)_{w \in W}$ and *transition maps* $\rho_{uv} : D_u \longrightarrow D_v$ for every R-related pair u, v, satisfying the natural transitivity condition:

$$uRvRw \;\Rightarrow\; \rho_{uw} = \rho_{uv} \circ \rho_{vw}$$

and such that ρ_{uu} is the identity map for every R-reflexive u. So every KFE corresponds to a Kripke sheaf with the same modal logic. Note that predicate Kripke frames correspond to those Kripke sheaves in which all transition maps are inclusions.

Although the semantics $\mathcal{KE}$ was introduced for dealing with equality, it happens to be stronger than $\mathcal{K}$ also for logics *without equality*. This can be seen again by analysing Ono's counterexample **QHE**.

Algebraic and Kripke-type semantics

The semantics $\mathcal{KE}$ also is inadequate in many cases, in particular for quantified versions of all intermediate propositional logics of finite depths. $\mathcal{K}$-incompleteness of these logics was proved by H. Ono in [Ono, 1973], and the proof is easily transferred to $\mathcal{KE}$. Another counterexample (found independently by S. Ghilardi and V. Shehtman & D. Skvortsov) is presented by the intermediate logic of 'the weak excluded middle' with constant domains:

$$\mathbf{QHJD} = \mathbf{QH} + \neg p \vee \neg\neg p + \forall x(P(x) \vee q) \supset (\forall x P(x) \vee q).$$

So one can try to generalise $\mathcal{KE}$ in a 'reasonable way' . Generalisations can be done at least in two directions.

The first direction leads to the *algebraic semantics* $\mathcal{AE}$. In the intuitionistic case a frame in $\mathcal{AE}$ is a *Heyting-valued set* [Fourman and Scott, 1979]. This is a set of 'individuals', in which every individual has a 'measure of existence' (an element of some Heyting algebra) and every pair of individuals has a 'measure of equality'. For the modal case Heyting algebras are replaced by modal algebras. The *neighbourhood* (or *topological*) *semantics* $\mathcal{TE}$ is a particular case of $\mathcal{AE}$, involving only algebras of topological spaces (or algebras of propositional neighbourhood frames for the modal case).

Both semantics $\mathcal{AE}$, $\mathcal{TE}$ are not much investigated and seem to be rather strong. We do not know if there exist $\mathcal{AE}$-incomplete logics. On the other hand, we do not know simple and natural examples of logics that are $\mathcal{AE}$-complete, but $\mathcal{KE}$-incomplete.

Another direction leads from $\mathcal{KE}$ to the semantics of *Kripke quasi-sheaves* ($\mathcal{KQ}$), and further to the semantics of *Kripke bundles* ($\mathcal{KB}$). In Kripke bundles transition maps $\rho_{uv} : D_u \longrightarrow D_v$ are replaced with transition relations $\rho_{uv} \subseteq D_u \times D_v$ collected in the unified accessibility (or *inheritance*, or *counterpart*) relation between individuals. Every individual has an inheritor (not necessarily unique) in any accessible world.

The idea of counterparts first appeared in [Lewis, 1968], and in a formal setting — in [Shehtman and Skvortsov, 1990]. A Kripke bundle can be defined as a p-morphism from a (propositional) frame of individuals (D^+, ρ) onto a frame of possible worlds (W, R). Then Kripke sheaves correspond exactly to etale maps, similarly to the well-known fact in sheaf theory [Godement, 1958].

In Kripke bundles an individual may have several inheritors even in its own possible world. *Kripke quasi-sheaves* are a subclass of Kripke bundles, in which this is not allowed. Further generalisations of $\mathcal{KB}$ are the *functor semantics* ($\mathcal{FS}$), in which 'frames' are set-valued functors (or presheaves over categories); the *metaframe semantics*, and finally the *hyperdoctrine semantics* generalising both metaframe and algebraic semantics. In hyperdoctrine semantics every logic is complete, but this semantics is very abstract, and thus not suitable for large-scale applications.

There are several other options, which will be considered later in this book (Volume 3), such as Kripke–Joyal–Reyes semantics [Goldblatt, 1984; Makkai and Reyes, 1995] or 'abstract realisability' [Dragalin, 1988]. Also we will discuss the approaches where the basic language is modified, e.g. by adding the existence predicate [Fine, 1978], various kinds of quantifiers [Garson, 1978] or nonstandard substitutions [Ghilardi and Meloni, 1988]. This may help the reader to find his way through the landscape of first-order logic.

Substitution property: discussion

General Kripke-type semantics have some peculiarities: it turns out that in Kripke bundles (and even in Kripke quasi-sheaves) the above mentioned substitution property fails. In fact, a very simple counterexample shows that the (intuitionistic) validity of $p \vee \neg p$ does not necessarily imply the validity of $P(x) \vee \neg P(x)$ (cf. Chapter 5). To avoid this, the 'straightforward' notion of validity has to be changed. So we introduce *strong validity* in a frame: a formula is said to be strongly valid iff all its substitution instances are valid (in the original sense). Then the set $\mathbf{L}(\mathbb{F})$ of all formulas strongly valid in a frame $\mathbb{F}$ is a logic (called the *logic of* $\mathbb{F}$).

One may argue that after such a modification the notion of frame becomes almost useless and can be replaced by the notion of model. For, in the same manner we can define a 'logic' $\mathbf{L}(M)$ of an arbitrary Kripke model M as the set of all the formulas 'strongly verifiable' in M, i.e. $\mathbf{L}(M)$ contains a formula A iff all substitution instances of A are true in M. Then we get a first-order version of general propositional frames mentioned above. In this semantics (GCT2) follows from (GCT1), and thus every modal or superintuitionistic first-order logic is complete.

None the less, there is some advantage in dealing with Kripke bundles (or metaframes) rather than Kripke models. For, let us recall the two kinds of substitutions considered in Chapter 2: a *strict* substitution does not add new parameters to atomic formulas; a *shift* adds a certain fixed list of new parameters to every atomic formula. Recall that a simple substitution instance A^k of a formula A is obtained by replacing every atom $P_i^n(x_1, \ldots, x_n)$ by $P_i^{n+k}(x_1, \ldots, x_n, y_1, \ldots, y_k)$, where $y_1, \ldots y_k$ are variables not occurring in A. As we know, every substitution is a composition of some strict and some simple substitution and a variable renaming. One can prove that the set $\mathbf{L}^{-}(\mathbb{F})$ of all formulas valid in a Kripke bundle $\mathbb{F}$ is closed under strict substitutions and variable renaming. Therefore the logic of $\mathbb{F}$ can be derived from $\mathbf{L}^{-}(\mathbb{F})$ as follows:

$$\mathbf{L}(\mathbb{F}) = \{A \mid \forall k \geq 0 \ A^k \in \mathbf{L}^{-}(\mathbb{F})\}.$$

So to check strong validity of a formula A there is no need to verify *all* its substitution instances, but only the A^k are sufficient. In other words, strong validity is nothing but validity with arbitrarily many extra parameters.

Unlike that, the set of formulas true in a Kripke model M is usually not closed under strict substitutions. So in general there is no way to describe the logic of M other than checking the truth of all possible substitution instances of formulas. This may be very difficult even in the propositional case; in algebraic terms, this means describing equations in a subalgebra of the modal algebra of a Kripke frame with a given set of generators. So the semantics of Kripke bundles is in some sense 'more constructive' than the semantics of Kripke models (not frames!). Anyway the algebraic semantics $\mathcal{AE}$ is the strongest semantics with (SP), that we know of. The Kripke-type semantics stronger than the Kripke bundle semantics $\mathcal{KB}$ (such as the functor semantics $\mathcal{FS}$ or the metaframe semantics $\mathcal{MF}$) do not have the (SP) either, but here again only simple substitutions are essential for strong validity.

Chapter 3

Kripke semantics

3.1 Preliminary discussion

There is no remembrance in former things; neither shall there be any remembrance of things that are to come with those that shall come after.
(Ecclesiastes, 1.11.)

In propositional modal logic Kripke semantics is widely used and is very helpful. As pointed out in Chapter 1, natural logics turn out to be Kripke-complete, and moreover, many of them have finite model property. This makes Kripke semantics an efficient model-theoretic instrument in the propositional case.

In the first-order case one can try to generalise the propositional Kripke semantics in a straightforward way.

Consider the first-order language $\mathcal{L}_1$ with a single modal operator $\Box$ (Section 2.1). Let (W, R) be a 1-modal propositional Kripke frame (Definition 1.3.1). We can define a 'first-order Kripke model' over (W, R) as a collection of classical models parametrised by possible worlds:

$$M = (M_u)_{u \in W},$$

where every M_u is a classical $\mathcal{L}$-structure, i.e. at the world u the basic predicates are interpreted as in M_u. This is not yet enough for a formal definition, because we have to answer the following two questions:

- What are the individuals in M?
- How are the quantifiers interpreted?

The simplest way is to assume that every individual is an element of some M_u; more exactly, if D_u is the domain of M_u, then M has the set of individuals

$$D^+ = \bigcup_{u \in W} D_u.$$

We can say that D_u consists of the individuals *existing in the world* u. Thus some individuals may exist in one world, but may not exist in another world. A more realistic example is given by a moving lift. Here W is the set of moments of time, R is the earlier–later relation, D_u is the set of people inside the lift at the moment u.

From this viewpoint, it is natural to quantify only over existing individuals; thus $\forall x\varphi(x)$ must be true at the world u iff $\varphi(a)$ is true for every $a \in D_u$.

In our example, if $P(x)$ is interpreted as

`x is a child,`

then $u \models \forall x P(x)$ means that at the moment u (it is true that)

`only children are in the lift.`

Eventually, we can define the forcing relation $u \models \varphi$ between a world u and a D_u-sentence (cf. 2.2) φ by induction on the complexity of φ, so that

$$\begin{array}{lll}
u \models P_k^n(a_1, \dots, a_n) & \text{iff} & M_u \models P_k^n(a_1, \dots, a_n) \text{ (classically)} \\
u \models \forall x\varphi & \text{iff} & \forall a \in D_u \;\; u \models [a/x]\varphi \\
u \models \exists x\varphi & \text{iff} & \exists a \in D_u \;\; u \models [a/x]\varphi \\
u \models \Box\varphi & \text{iff} & \forall v \, (uRv \,\&\, \varphi \text{ is a } D_v\text{-sentence} \Rightarrow v \models \varphi)
\end{array}$$

and with the standard clauses for the classical propositional connectives.

Returning again to our example, for a certain individual a (say, Robert Smith) $u \models \Box P(a)$ means: at the moment u it is true that

`always in the future Robert Smith will use the lift only`
`while he is a child.`

This is an essential point: we cannot state $u \models P(a)$ for an individual a that is not present in D_u; and thus to check $u \models \Box P(a)$, we have to consider only those v in $R(u)$, which have a in their domain.

Once the forcing relation is defined, we can say that a formula φ is *true* in M if $u \models \bar{\forall}\varphi$ for any $u \in W$. If $FV(\varphi) = \{x_1, \dots, x_n\}$, the latter is equivalent to

$$u \models [a_1, \dots, a_n/x_1, \dots, x_n]\,\varphi \text{ for any } a_1, \dots, a_n \in D_u.$$

Now what formulas are true in every Kripke model? This set can be axiomatised, but unfortunately, it may *not* be a modal predicate logic in the sense of Chapter 2. To see what happens, consider the formula

$$\alpha = \Box(p \wedge q) \supset \Box p$$

which is a theorem of **K** (propositional) and is obviously true in all models defined above. However take the substitution instance of α:

$$\beta = \Box(P(x) \wedge P(y)) \supset \Box P(x).$$

This formula is not always true. For, consider the model with two possible worlds u, v and two individuals a, b, such that

$$R = \{(u,u),(u,v)\}, D_u = \{a,b\}, D_v = \{a\},$$
$$u \models P(a) \wedge P(b), v \not\models P(a), \text{ see Fig. 3.1.}$$

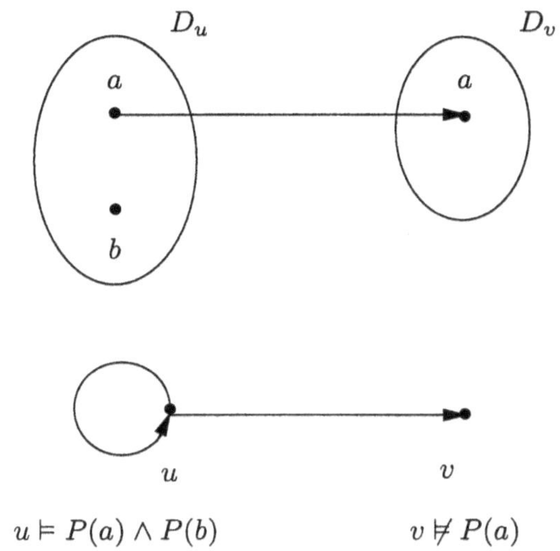

Figure 3.1.

Then we have $u \models \Box(P(a) \wedge P(b))$, since $P(a) \wedge P(b)$ is not a D_v-sentence, but $u \not\models \Box P(a)$.

Here is a more natural analogue of this example: if

`John and Mary always talk when they are in the lift together,`

it may be not the case that

`John always talks in the lift.`

Now there are several options to choose:

A. If we need semantics for the logics described in the previous chapter, we can try to amend the above definition.

B. Alternatively, we can change (actually, extend) the notion of a logic and try to axiomatise the 'logic' which is complete with respect to the above interpretation.

C. We can accept other definitions of semantics, using different kinds of individuals and different interpretations of quantifiers.

In this Volume we accept OPTION A and keep the same notion of logic. There is not so much to change in the original definition: it is enough to assume that 'individuals are immortal', which in precise terms, means the *expanding domain condition*:

$$(ED) \quad uRv \Rightarrow D_u \subseteq D_v.$$

Of course this requirement contradicts the viewpoint expressed in the epigraph to this section.[1]

But it saves the logic, and as we will see later on, this semantics is sound and complete for a few well-known logics, such as **QK**, **QS4**, etc.

Furthermore, (ED) is essential in semantics of intuitionistic logic. In this case (W, R) is an **S4**-frame, and we would like to keep the intuitionistic truth-preservation principle (TP) from Lemma 1.4.4 for atomic D_u-sentences:

$$u \Vdash P_k^n(a_1, \ldots, a_n) \,\&\, uRv \,\&\, a_1, \ldots, a_n \in D_v \Rightarrow v \Vdash P_k^n(a_1, \ldots, a_n).$$

The inductive definition of forcing from Section 1.3 can be extended to the first-order case so that

$$u \Vdash \exists x\varphi \text{ iff } \exists a \in D_u \; u \Vdash [a/x]\varphi.$$

Now, without (ED), we cannot guarantee the truth-preservation for all D_u-sentences. For, take a model based on a two-element chain:

$$W = \{u, v\}, R = \{(u, u), (u, v), (v, v)\}, \; D_u = \{a\}, \; D_v = \{b\},$$
$$u \Vdash P(a), \; v \nVdash P(b).$$

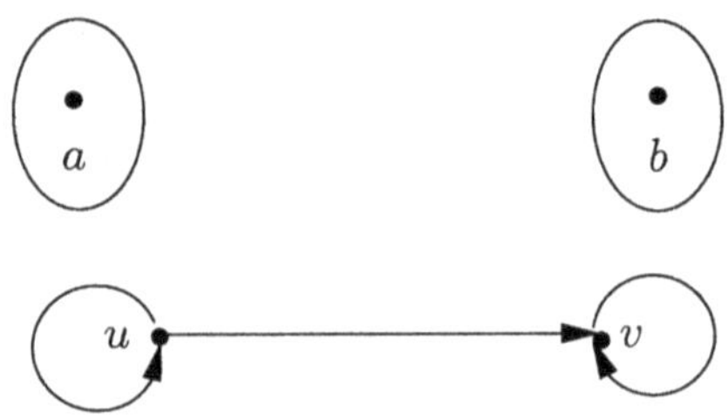

Figure 3.2.

In this case the truth-preservation holds for $P(a)$ (because a exists only in u), but it fails for $\exists x P(x)$, because obviously,

$$u \vDash \exists x P(x), \; v \nvDash \exists x P(x).$$

To restore the truth-preservation, we can change the definition, e.g. as follows:

$$u \Vdash \exists x\varphi \text{ iff } \forall v \in R(u) \; \exists a \in D_v \; u \Vdash [a/x]\varphi,$$

[1]Commonsense understanding of individuals is not easy to formalise, and there were many philosophical debates on that subject. 'What is an individual? — This is a good question!' — Dana Scott writes in [Scott, 1970]. Note that in labelled deductive systems, individuals are labelled and may be deemed to have internal structure. Wait for a later volume on this.

cf. the clause for the implication in the intuitionistic case. Another alternative is:

$$u \Vdash \exists x\varphi \text{ iff } \exists a \in D_u\ \forall v \in R(u)\ u \Vdash [a/x]\varphi,$$

which means that only 'immortal' individuals are considered as existing. But nevertheless without (ED), the semantics is not sound, and some intuitionistic theorems may be false. For example, consider the formula $P(x) \supset \exists x P(x)$. In the above model we have $u \Vdash P(a)$, but $u \nVdash \exists x P(x)$, since uRv, and $v \nVdash P(b)$, b being the only individual in v. Thus $u \nVdash P(a) \supset \exists x P(x)$.

We will return to this subject in Volume 2 of our book. Note that anyway, the principle (ED) is quite natural for intuitionistic logic. It normalises the situation, and the logic **QH** becomes sound and complete in an appropriate semantics.

OPTION B is close to Kripke's treatment in [Kripke, 1963]; it will be also considered later on (in Volume 2).

OPTION C includes many different approaches; we mention some of them.

OPTION C_1 (modal case). As before, a Kripke model is a collection of classical structures, but quantifiers now range over *all* possible individuals i.e. over the whole set D^+. This is so-called *possibilist quantification*.

As individuals may not exist in some worlds, we have to add the unary *existence predicate* E to our language. Now instead of D_u-sentences we evaluate D^+-sentences (where arbitrary individuals are used as extra constants). The truth condition for $\forall$ becomes

$$u \models \forall x\varphi \quad \text{iff} \quad \forall a \in D^+\ u \models [a/x]\varphi.$$

Note that $u \models [a/x]\varphi$ may also hold if $a \notin D_u$.[2]

Option B can be realised within this approach as well, if we interpret $\forall x\varphi(x)$ as $\forall x(E(x) \supset \varphi(x))$.

It turns out that the minimal logic **QK** is sound, but incomplete in this semantics. This is due to the formula

$$\forall x \Box P(x) \supset \Box \forall x P(x)$$

found by Ruth Barcan and named after her. It is easily checked that the Barcan formula is true in every model, but is not a theorem of **QK**, as the following countermodel (of the type A) shows (Fig. 3.3).

Here we have $u \models \Box P(a)$, and so $u \models \forall x \Box P(x)$. On the other hand, $v \not\models \forall x P(x)$, and thus $u \not\models \Box \forall x P(x)$ since uRv.

Although axiomatising the minimal complete logic for the C_1-semantics is simple, this approach is controversial, because quantification over all individuals is rather ambiguous, at least in natural language.[3]

Still option C_1 is possible and useful by formal reasons.

---

[2] In natural language a non-existing individual can still have a name in the world u. We can say: `Sir Isaac Newton discovered the laws of motion`, using the name of a person given after the event had happened.

[3] Quantification in natural language is a complicated subject, not to be discussed here. We remark only that in most cases quantification refers to the actual world, as in `All students`

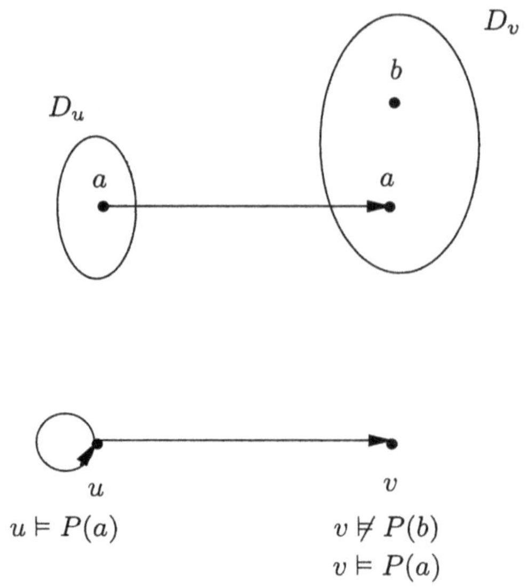

Figure 3.3.

OPTION C_2. This is *actualist quantification* (see [Fitting and Mendelsohn, 1998]), a combination of B and C_1.

Now $u \models \varphi$ is defined when φ is a D^+-sentence, but $u \models P_k^n(a_1, \ldots, a_n)$ is put to be false when some of $a_1, \ldots, a_n$ do not exist in the world u: only actual individuals are allowed to have atomic properties.[4]

The inductive definition of forcing is the same as in the cases A, B. However, there is a difference in the definition of the truth in a model:

> $M \models \varphi(x_1, \ldots, x_n)$ iff $u \models \varphi(a_1, \ldots, a_n)$ for any u, for any $a_1, \ldots, a_n$ in D^+ (not only in D_u).

This approach again breaks the logic: the formula $\forall x P(x) \supset P(y)$ may be false, while $\forall y(\forall x P(x) \supset P(y))$ is always true.

FURTHER OPTIONS. One may argue that one and the same individual cannot exist in different worlds. In fact, nobody would identify a newly born baby and an old man or woman. This suggests we consider disjoint individual domains, together with transition maps (or relations) between them, as mentioned in the Introduction to Part II.

Another idea is to treat individuals as changing entities and consider *individual concepts*, that are partial functions from W to the set D^+ (of 'individual images').

`have supervisors`, but there also exists possibilist quantification, as in `Any student has a supervisor`. In the latter case it is not quite clear whether the quantifier ranges over the whole D^+ or not. Cf. [Krongauz, 1998] studying these matters in Russian.

[4]Note that in the case C_1 this is not required, and basic predicates may be true for non-existing individuals. This happens in natural languages as well. For example, Mike likes Socrates may be evaluated as true in the actual world.

Now $u \models P(a_1, \ldots, a_n)$ is defined if $a_1(u), \ldots, a_n(u)$ exist, and

$$u \models \forall x\varphi(x) \text{ iff } \forall a\, (a(u) \text{ exists } \Rightarrow u \models \varphi(a)).$$

In this case the logic grows much larger than **QK** and even becomes not recursively axiomatisable. We shall return to this topic in Volume 2.

3.2 Predicate Kripke frames

Now let us turn to precise definitions and statements.

Definition 3.2.1 *A* system of domains *over a set $W \neq \varnothing$ is a family of non-empty sets $D = (D_u)_{u \in W}$.*

Definition 3.2.2 *Let $F = (W, R_1, \ldots, R_N)$ be a propositional Kripke frame. An* expanding system of domains over *F is a system of domains D over W such that*

$$\forall i \in I_N\ \forall u, v \in W\ (uR_iv \Rightarrow D_u \subseteq D_v).$$

A predicate Kripke frame over *F is a pair $\mathbf{F} = (F, D)$, in which D is an expanding system of domains over F.*

The set D_u (sometimes denoted by $D(u)$) is called the individual domain *of the world u; the set*

$$D^+ := \bigcup_{u \in W} D_u$$

is the total domain *of $\mathbf{F}$; the frame F (also denoted by $\mathbf{F}_\pi$) is called the* propositional base *of $\mathbf{F}$.*

The following observation is an easy consequence of the definition.

Lemma 3.2.3 *Let $\mathbf{F} = (F, D)$ be a predicate Kripke frame, $v \in F{\uparrow}u$. Then $D_u \subseteq D_v$.*

Proof By Lemma 1.3.19, $v \in R_\alpha(u)$ for some $\alpha \in I_N^\infty$. Then we can apply induction on $|\alpha|$. In fact, for $\alpha = \lambda$ we have $v = u$, so $D_u = D_v$; if $\alpha = \beta i$ and $uR_\alpha v$, then $uR_\beta wR_iv$ for some w, so $D_u \subseteq D_w \subseteq D_v$ by the induction hypothesis and 3.2.2. ∎

By default, we denote an arbitrary propositional Kripke frame by $F = (W, R_1, \ldots, R_N)$ and an arbitrary predicate Kripke frame by $\mathbf{F} = (F, D)$.

For an individual $a \in D^+$ the set

$$E(a) := \{u \in W \mid a \in D_u\}$$

is called the *measure of existence* (or the *extent*). Since the system of domain D is expanding, $E(a)$ is a stable subset of F.

For a tuple $\mathbf{a} \in (D^*)^n$ we also introduce the *measure of existence*

$$E(\mathbf{a}) := \bigcap_{i=1}^{n} E(a_i) = \{u \in W \mid r(\mathbf{a}) \subseteq D_u\}.$$

The set is also stable in F. In particular, for an empty $\mathbf{a}$ we define $E(\curlywedge)$ as W (since $r(\curlywedge) = \varnothing$).

Definition 3.2.4 *A* (modal) *valuation ξ in a predicate Kripke frame* $\mathbf{F}$ *is a function associating with every predicate letter P_k^m a member of the set* $\prod_{u \in W} 2^{D_u^m}$, *i.e. a family of m-ary relations on individual domains:*

$$\xi(P_k^m) = (\xi_u(P_k^m))_{u \in W},$$

where $\xi_u(P_k^m) \subseteq D_u^m$. To include the case $m = 0$, we assume that $D_u^0 = \{u\}$; so $\xi_u(P_k^0)$ is either $\{u\}$ or $\varnothing$. The pair $M = (\mathbf{F}, \xi)$ is called a (predicate) Kripke model *over* $\mathbf{F}$.

We may call the function ξ_u sending every n-ary predicate letter to an n-ary relation on D_u, a *local valuation in* $\mathbf{F}$ *at* u; this is nothing but a classical valuation in D_u. So we can say that ξ is a family of local valuations $(\xi_u)_{u \in W}$.

Definition 3.2.5 *Let $M = (\mathbf{F}, \xi)$ be a Kripke model. For $u \in M$, the classical model structure $M_u := (D_u, \xi_u)$ is called the* stalk *(or the* fibre*) of M at u.*

On the other hand, we may call a function sending every n-ary predicate letter to an n-ary relation on D^+, a *global valuation* in $\mathbf{F}$.

Let us also define another kind of valuation in predicate Kripke frames.

Definition 3.2.6 *A* global valuation *in a predicate Kripke frame* $\mathbf{F}$ *is a function γ sending every n-ary predicate letter $P(n > 0)$ to an 'n-ary 2^W-valued predicate on D^+', i.e. to a function $\gamma(P) : (D^+)^n \to 2^W$ such that for any $\mathbf{a} \in (D^+)^n$, $\gamma(P)(\mathbf{a}) \subseteq E(\mathbf{a})$, and every $q \in PL^0$ to a subset of W.*

So $\gamma \restriction PL^0$ is a propositional valuation in the propositional frame F. We can also regard it as an 0-ary predicate, i.e. as a function $(D^+)^0 \to 2^W$, where $(D^+)^0 = \{\curlywedge\}$. In this case the condition

$$\gamma(q)(\mathbf{a}) \subseteq E(\mathbf{a})$$

trivially holds for any $\mathbf{a} \in (D^+)^0$ (i.e. for $\curlywedge$), since $E(\curlywedge) = W$.

Lemma 3.2.7 *Let $\mathbf{F} = (F, D)$ be a predicate Kripke frame.*

(1) For any valuation θ in F there exists a global valuation θ^+ such that for any $P \in PL^m$, $m > 0$, for any $a \in (D^+)^m$

$$\theta^+(P)(\mathbf{a}) = \{u \in W \mid \mathbf{a} \in \theta_u(P)\}$$

and for $q \in PL^0$

$$\theta^+(q) = \{u \mid \theta_u(P) = \{u\}\},$$

(2) Every global valuation γ equals ξ^+ for a unique valuation ξ.

Proof

(1) From the definition it is clear that

$$\theta^*(P)(\mathbf{a}) \subseteq E(\mathbf{a})$$

for any $P \in PL^n,\ n > 0,\ \mathbf{a} \in (D^+)^n$.

(2) Put

$$\xi_u(P_k^m) := \{\mathbf{a} \mid u \in \gamma(P_k^m)(\mathbf{a})\}$$

for $m > 0$ and

$$\theta_u(P_k^0) := \begin{cases} \{u\} & \text{if } u \in \gamma(P_k^0), \\ \varnothing & \text{otherwise.} \end{cases}$$

Then for $P \in PL^m,\ m > 0$

$$u \in \gamma(P)(\mathbf{a}) \text{ iff } \mathbf{a} \in \xi_u(P) \text{ iff } u \in \xi^+(P)(\mathbf{a}),$$

and similarly for $m = 0$.

The uniqueness of ξ is also clear; $\gamma = \xi^+$ means that for $P \in PL^m$ ($m > 0$), $\mathbf{a} \in (D^+)^m,\ u \in W$

$$u \in \gamma(P)(\mathbf{a}) \Leftrightarrow \mathbf{a} \in \xi_u(P),$$

and for $q \in PL^0$

$$u \in \gamma(q) \Leftrightarrow u \in \xi_u(q).$$

■

Definition 3.2.8 *A 1-modal predicate Kripke frame* **F** *is called* **S4**-based, *or* intuitionistic *if its propositional base* $\mathbf{F}_\pi$ *is an* **S4**-*frame. In this case a modal valuation* ξ *in* **F** *(and the corresponding Kripke model) is called* intuitionistic *if it has the* truth preservation *(or* monotonicity*)* property:[5]

(TP) $uRv \Rightarrow \xi_u(P_k^m) \subseteq \xi_v(P_k^m)$ (*if* $m > 0$);
$uRv \,\&\, u \in \xi_u(P_k^0) \Rightarrow v \in \xi_v(P_k^0)$.

To define forcing, we use the notion of a D-sentence introduced in Section 2.4.

Definition 3.2.9 *For a Kripke model* $M = (\mathbf{F}, \xi)$, $\mathbf{F} = (F, D)$, $F = (W, R_1, \ldots, R_N)$ *we define the* (modal) forcing relation $M, u \vDash A$ *(in another notation:* $\xi, u \vDash A$*, or briefly:* $u \vDash A$*) between worlds* $u \in W$[6] *and* D_u*-sentences* A. *The definition is inductive:*

[5] Recall that we omit the subscript '1' in the 1-modal case.

[6] Sometimes we write $u \in \mathbf{F}$ or $u \in M$ instead of $u \in W$.

- $M, u \models P_k^0$ *iff* $u \in \xi_u(P_k^0)$;
- $M, u \models P_k^m(\mathbf{a})$ *iff* $\mathbf{a} \in \xi_u(P_k^m)$ *(for* $m > 0$*)*;
- $M, u \models a = b$ *iff* a *equals* b;
- $M, u \not\models \bot$;
- $M, u \models B \lor C$ *iff* ($M, u \models B$ *or* $M, u \models C$);
- $M, u \models B \land C$ *iff* ($M, u \models B$ *and* $M, u \models C$);
- $M, u \models B \supset C$ *iff* ($M, u \not\models B$ *or* $M, u \models C$);
- $M, u \models \Box_i B$ *iff* $\forall v \in R_i(u)\ M, v \models B$;[7]
- $M, u \models \exists x B$ *iff* $\exists a \in D_u\ M, u \models [a/x]B$;
- $M, u \models \forall x B$ *iff* $\forall a \in D_u\ M, u \models [a/x]B$.

Remark 3.2.10 The above truth definition for propositional letters seems peculiar. It is more natural to use truth values 1 and 0 rather than $\{u\}$ and $\varnothing$. But our definition will be convenient later on, in Chapter 5.

Now we readily obtain an analogue of Lemma 1.3.3:

Lemma 3.2.11 *For any Kripke model:*
$u \models \Diamond_i B$ *iff* $\exists v \in R_i(u)\ v \models B$;
$u \models \neg B$ *iff* $u \not\models B$;
$u \models \Box_\alpha B$ *iff* $\forall v \in R_\alpha(u)\ v \models B$;
$u \models \Diamond_\alpha B$ *iff* $\exists v \in R_\alpha(u)\ v \models B$.

Similarly to the propositional case, we give the following inductive definition of intuitionistic forcing.

Definition 3.2.12 *Let* $M = (\mathbf{F}, \xi)$ *be an intuitionistic Kripke model. We define the intuitionistic forcing relation* $M, u \Vdash A$ *between a world* $u \in \mathbf{F}$ *and an intuitionistic* D_u*-sentence* A *(also denoted by* $\xi, u \Vdash A$ *or just by* $u \Vdash A$*) by induction:*

- $M, u \Vdash A$ *iff* $M, u \models A$ *for* A *atomic;*
- $M, u \Vdash B \land C$ *iff* $M, u \Vdash B \,\&\, M, u \Vdash C$;[8]
- $M, u \Vdash B \lor C$ *iff* ($M, u \Vdash B$ *or* $M, u \Vdash C$);
- $M, u \Vdash B \supset C$ *iff* $\forall v \in R(u)\ (M, v \Vdash B \Rightarrow M, v \Vdash C)$;

[7] Note that B is a D_v-sentence, since D is expanding.

[8] Here & is an abbreviation of 'and', and further on in this definition $\forall v \in R(u)$ abbreviates 'for any v in $R(u)$', etc.

- $M, u \Vdash \exists x B$ *iff* $\exists a \in D_u\ M, u \Vdash [a/x]B$;
- $M, u \Vdash \forall x B$ *iff* $\forall v \in R(u)\ \forall a \in D_v\ M, v \Vdash [a/x]B$.

Hence we easily obtain

Lemma 3.2.13 *The intuitionistic forcing has the following properties*

- $M, u \Vdash \neg B$ *iff* $\forall v \in R(u)\ M, v \nVdash B$;
- $M, u \Vdash a \neq b$ *iff* $a, b \in D_u$ *are not equal.*

Definition 3.2.14 *Let* $\mathbf{F}$ *be an* **S4**-*based Kripke frame;* M *a Kripke model over* $\mathbf{F}$. *The* pattern *of* M *is the Kripke model* M_0 *over* $\mathbf{F}$ *such that for any* $u \in \mathbf{F}$ *and any atomic* D_u-*sentence without equality* A

$$M_0, u \models A \text{ iff } M, u \models \Box A.$$

Obviously, every Kripke model M over F has a unique pattern; M_0 is an intuitionistic Kripke model and $M_0 = M$ if M is itself intuitionistic.

Lemma 3.2.15 *If* M_0 *is a pattern of* M, *then for any* $u \in M$, *for any intuitionistic* D_u-*sentence* A

$$M_0, u \Vdash A \text{ iff } M, u \models A^T;$$

Proof Easy, by induction on the length of A. The atomic case follows from Definition 3.2.14 (and is trivial for A of the form $a = b$). Let us check only the case $A = \forall x B$; other cases are left to the reader.

$M_0, u \Vdash \forall x B$ iff $\forall v \in R(u)\ \forall a \in D_v\ M_0, u \Vdash [a/x]B$ (by Definition 3.2.12)
iff $\forall v \in R(u)\ \forall a \in D_v\ M, u \models ([a/x]B)^T$ (by the induction hypothesis).

On the other hand, by Definition 3.2.9

$$M, u \Vdash A^T (= \Box\forall x B^T) \text{ iff } \forall v \in R(u)\ \forall a \in D_v\ M, u \models [a/x](B^T),$$

and it remains to note that $([a/x]B)^T = [a/x](B^T)$, by 2.11.3. ∎

Lemma 3.2.16 *Let* u, v *be worlds in an intuitionistic Kripke model* M. *Then for any intuitionistic* D_u-*sentence* A

$$M, u \Vdash A \ \&\ uRv \ \Rightarrow\ M, v \Vdash A.$$

Proof This easily follows by induction. The case when $A = P(\mathbf{a})$ is atomic follows from the monotonicity property (TP). The cases $A = B \supset C, A = \forall x B$ follow from the transitivity of R. For $A = \exists x B$ note that $M, u \Vdash A$ implies $M, u \Vdash [a/x]B$ for some $a \in D_u$, and hence $M, v \Vdash [a/x]B$ by the induction hypothesis, which yields $M, v \Vdash B$.

Other cases are trivial. ∎

Lemma 3.2.17 *Let M be an N-modal predicate Kripke model, $A(\mathbf{x})$ an N-modal formula with all its parameters in the list $\mathbf{x}$, $|\mathbf{x}| = n$. Then for any $u \in M$*

$$M, u \vDash \forall \mathbf{x} A(\mathbf{x}) \textit{ iff } \forall \mathbf{a} \in D_u^n \ M, u \vDash A(\mathbf{a}).$$

Proof By induction on n. The base is trivial and the step is almost trivial:

$$u \vDash \forall y \forall \mathbf{x} A(y, \mathbf{x}) \text{ iff } \forall b \in D_u \ u \vDash \forall \mathbf{x} A(b, \mathbf{x})$$

iff $\forall b \in D_u \ \forall \mathbf{c} \in D_u^n \ u \vDash A(b, \mathbf{c})$ (by the induction hypothesis)
iff $\forall \mathbf{a} \in D_u^{n+1} \ u \vDash A(\mathbf{a})$. ■

In the intuitionistic case we have the following

Lemma 3.2.18 *Let M be an intuitionistic Kripke model with the accessibility relation R, $u \in M$, $A(\mathbf{x})$ an intuitionistic D_u-formula with all its parameters in the list $\mathbf{x}$, $|\mathbf{x}| = n$. Then*

$$M, u \Vdash \forall \mathbf{x} A(\mathbf{x}) \textit{ iff } \forall v \in R(u) \forall \mathbf{a} \in D_v^n \ \ M, v \Vdash A(\mathbf{a}).$$

Proof By induction, similar to the previous lemma. The base (with $n = 0$) follows from Lemma 3.2.16. For the induction step we have:

$$u \Vdash \forall y \, \forall \mathbf{x} \, A(y, \mathbf{x}) \text{ iff } \forall v \in R(u) \, \forall b \in D_v \ \ v \Vdash \forall \mathbf{x} \, A(b, \mathbf{x}) \ \text{(by 3.2.13)}$$

iff $\forall v \in R(u) \, \forall b \in D_v \, \forall w \in R(v) \, \forall \mathbf{c} \in D_w^n \, w \Vdash A(b, \mathbf{c})$ (by the induction hypothesis) iff $\forall w \in R(u) \, \forall \mathbf{a} \in D_w^{n+1} \, w \Vdash A(\mathbf{a})$.

In the latter equivalence, 'if' follows from the transitivity of R and the inclusion $D_v \subseteq D_w$ for $w \in R(v)$; to prove 'only if', for given $u, w, \mathbf{a}$, take $v = w$ and $b = a_1$, $\mathbf{c} = (a_2, \ldots, a_n)$. ■

Definition 3.2.19 *A modal (respectively, intuitionistic) predicate formula A is said to be* true *in a Kripke model (respectively, intuitionistic Kripke model) M if its universal closure $\bar{\forall} A$ is true at every world of M.*

This is denoted by $M \vDash A$ in the modal case and by $M \Vdash A$ in the intuitionistic case.

The set of all modal (respectively, intuitionistic) sentences that are true in M is denoted by $\mathbf{MT}^{(=)}(M)$ (respectively, $\mathbf{IT}^{(=)}(M)$) and called the modal *(respectively, the* intuitionistic*)* theory *of M (with or without equality).*

Lemma 3.2.20 *Let M be an N-modal or intuitionistic predicate Kripke model with a system of domains D, $A(x_1, \ldots, x_n)$ a predicate formula of the corresponding type. Then*

$$M \vDash (\Vdash) \ A(x_1, \ldots, x_n) \ \textit{ iff } \ \forall u \in M \, \forall a_1, \ldots, a_n \in D_u \ \ M, u \vDash (\Vdash) \ A(a_1, \ldots, a_n).$$

Proof Follows easily from Lemma 3.2.17, 3.2.18. E.g. in the intuitionistic case we have

$$M \Vdash \forall \mathbf{x} \, A(\mathbf{x}) \text{ iff } \forall u \, \forall v \in R(u) \, \forall \mathbf{a} \in D_v^n \, M, v \Vdash A(\mathbf{a})$$

iff $\forall v \, \forall \mathbf{a} \in D_v^n \, M, v \Vdash A(\mathbf{a})$.

In the latter equivalence, 'if' is trivial; to show 'only if', take $u = v$. ■

Definition 3.2.21 *A modal (respectively, intuitionistic) predicate formula A is said to be* valid *in a Kripke frame (respectively,* **S4***-based Kripke frame)* $\mathbf{F}$ *if it is true in all Kripke models (respectively, intuitionistic Kripke models) over* $\mathbf{F}$.

Validity is denoted by $\mathbf{F} \vDash A$ in the modal case and by $\mathbf{F} \Vdash A$ in the intuitionistic case.

Definition 3.2.22 *Let Σ be a set of modal (respectively, intuitionistic) sentences. We say that Σ is* valid *in a Kripke frame* $\mathbf{F}$ *(of the corresponding type) if every formula from Σ is valid in* $\mathbf{F}$*; or equivalently, we say that* $\mathbf{F}$ *is a* Σ-frame *(or* $\Sigma \subseteq \mathbf{ML}^{(=)}(\mathbf{F})$, $\Sigma \subseteq \mathbf{IL}^{(=)}(\mathbf{F})$*).*

$\mathbf{F} \vDash \Sigma$ *(or* $\mathbf{F} \Vdash \Sigma$*) denotes that Σ is valid in* $\mathbf{F}$*. The class of all Σ-frames is denoted by* $\mathbf{V}(\Sigma)$ *and called* modally *(respectively,* intuitionistically*)* definable *(by Σ).*

Lemma 3.2.23 *Let* $\mathbf{F}$ *be an N-modal (respectively, intuitionistic) Kripke frame, A an N-modal (respectively, intuitionistic) propositional formula. Then* $\mathbf{F} \vDash A$ *iff* $\mathbf{F}_\pi \vDash A$ *(respectively,* $\mathbf{F} \Vdash A$ *iff* $\mathbf{F}_\pi \Vdash A$*), i.e.* $\mathbf{L}(\mathbf{F})_\pi = \mathbf{L}(\mathbf{F}_\pi)$*, where* $\mathbf{L}$ *is* $\mathbf{ML}^{(=)}$ *or* $\mathbf{IL}^{(=)}$.

Proof Easy from definitions. Valuations in $\mathbf{F}_\pi$ are in principle the same as valuations of proposition letters in $\mathbf{F}$; more precisely, a valuation ξ in $\mathbf{F}$ corresponds to ξ' in $\mathbf{F}_\pi$ such that

$$\xi'(q) = \xi^+(q) = \{u \mid u \in \xi_u(q)\},$$

for any $q \in PL^0$, and both ξ, ξ' are extended to propositional formulas in the same way. So $(\mathbf{F}, \xi), u \vDash A$ iff $(\mathbf{F}_\pi, \xi'), u \vDash A$ and similarly for the intuitionistic case. Finally note that every propositional valuation in $\mathbf{F}_\pi$ is ξ' for some valuation ξ in $\mathbf{F}$. ∎

Lemma 3.2.24 *Let M be an* **S4***-based predicate Kripke model, M_0 its pattern. Then for any $A \in IF^=$, $M_0 \Vdash A$ iff $M \vDash A^T$, and thus*

$$A \in \mathbf{IT}(M_0) \textit{ iff } A^T \in \mathbf{MT}(M).$$

Proof From 3.2.13, 3.2.18. ∎

Proposition 3.2.25 *Let* $\mathbf{F}$ *be an* **S4***-based Kripke frame, A an intuitionistic formula with equality. Then*

$$\mathbf{F} \Vdash A \textit{ iff } \mathbf{F} \vDash A^T.$$

Proof Let $A = A(\mathbf{x})$, $|\mathbf{x}| = n$.

(Only if.) Assume $\mathbf{F} \Vdash A$. Let M be a Kripke model over $\mathbf{F}$. By Lemma 3.2.20, we have to show $M, u \vDash A^T(\mathbf{a})$ $(= A(\mathbf{a})^T)$ for any $u \in \mathbf{F}$, $\mathbf{a} \in D_u^n$. By Lemma 3.2.15 we have:

$$M_0, u \Vdash A(\mathbf{a}) \text{ iff } M, u \vDash A(\mathbf{a})^T.$$

By Lemma 3.2.20, $\mathbf{F} \Vdash A$ implies $M_0, u \Vdash A(\mathbf{a})$, and thus we obtain $\mathbf{F} \models A^T$.

(If.) Assume $\mathbf{F} \models A^T$. Then for any intuitionistic model M and $u \in F$, $\mathbf{a} \in D_u^n$ we have $M, u \models A(\mathbf{a})^T (= A^T(\mathbf{a}))$ by Lemma 3.2.20, i.e. $M, u \Vdash A(\mathbf{a})$ by Definition 3.2.12. Hence $\mathbf{F} \Vdash A$, by Lemma 3.2.20. ∎

Lemma 3.2.26 *Let $A(\mathbf{x})$, $B(\mathbf{x})$ be congruent N-modal (respectively, intuitionistic) formulas, $|\mathbf{x}| = n$, and let M be an N-modal (respectively, intuitionistic) Kripke model. Then for any $u \in M$, $\mathbf{a} \in D_u^n$*

$$M, u \models (\Vdash)\, A(\mathbf{a}) \text{ iff } M, u \models (\Vdash)\, B(\mathbf{a}).$$

Proof We begin with the modal case. Consider the following equivalence relation between N-modal formulas:

$A \sim B := FV(A) = FV(B)$ and for any distinct list of variables $\mathbf{x}$ such that $FV(A) = r(\mathbf{x})$, for any $u \in M$, $\mathbf{a} \in D_u^{|\mathbf{x}|}$

$$M, u \models [\mathbf{a}/\mathbf{x}]A \Leftrightarrow M, u \models [\mathbf{a}/\mathbf{x}[B. \qquad (\sharp)$$

Our aim is to show that $A \doteq B$ implies $A \sim B$.

By Proposition 2.3.17 this implication follows from the properties (1)–(4). So let us check these properties for $\sim$

(1) $QyA \sim Qz(A[y \mapsto z])$ for $y \notin BV(A)$, $z \notin V(A)$.

We consider only the case $Q = \exists$. Suppose $FV(\exists yA) = r(\mathbf{x})$ for a distinct $\mathbf{x}$. Then there are two subcases.

(i) $y \notin FV(A)$ (and thus $y \notin V(A)$).
In this case $A[y \mapsto z] = A$, so

$$\exists z(A[y \mapsto z]) = \exists zA,\ [\mathbf{a}/\mathbf{x}]\exists yA = \exists y[\mathbf{a}/\mathbf{x}]A,$$
$$[\mathbf{a}/\mathbf{x}]\exists zA = \exists z[\mathbf{a}/\mathbf{x}]A.$$

By 3.2.9 we have $M, u \models \exists y[\mathbf{a}/\mathbf{x}]A \Leftrightarrow M, u \models [\mathbf{a}/\mathbf{x}]A$, since y does not occur in $[\mathbf{a}/\mathbf{x}]A$. Similarly,

$$M, u \models \exists z[\mathbf{a}/\mathbf{x}]A \Leftrightarrow M, u \models [\mathbf{a}/\mathbf{x}]A,$$

so (1) holds.

(ii) $y \in FV(A)$. Then again (since $y \notin r(\mathbf{x})$)

$$[\mathbf{a}/\mathbf{x}]\exists yA = \exists y[\mathbf{a}/\mathbf{x}[A,\ [\mathbf{a}/\mathbf{x}]\exists z(A[y \mapsto z]) = \exists z[\mathbf{a}/\mathbf{x}[(A[y \mapsto z]).$$

Hence

$$M, u \models [\mathbf{a}/\mathbf{x}]\exists yA \Leftrightarrow \exists d \in D_u M, u \models [d/y][\mathbf{a}/\mathbf{x}]A (= [\mathbf{a}d/\mathbf{x}y]A = [\mathbf{a}/\mathbf{x}][d/y]A)$$

Similarly

$$M, u \models [\mathbf{a}/\mathbf{x}]\exists z(A[y \mapsto z]) \Leftrightarrow \exists d \in D_u\ M, u \models [\mathbf{a}/\mathbf{x}][d/z](A[y \mapsto z]).$$

But since $y \notin BV(A)$, $z \notin V(A)$, we have

$$[d/y]A = A[y \mapsto d] = (A[y \mapsto z])[z \mapsto d] = [d/z](A[y \mapsto z]).$$

Thus (1) holds in this case too.

(2) Supposing $A \sim B$, let us show that

$$\mathcal{Q}yA \sim \mathcal{Q}yB \text{ for } \mathcal{Q} = \forall;$$

the case $\mathcal{Q} = \exists$ is quite similar.

Obviously, $FV(A) = FV(B)$ implies $FV(\forall yA) = FV(\forall yB)$; let $r(\mathbf{x}) = FV(\forall yA)$. Then

$$M, u \vDash [\mathbf{a}/\mathbf{x}]\forall yA (= \forall y[\mathbf{a}/\mathbf{x}]A) \Leftrightarrow \forall d \in D_u\, M, u \vDash [d/y][\mathbf{a}/\mathbf{x}]A (= [\mathbf{a}d/\mathbf{x}y]A),$$

and similarly

$$M, u \vDash [\mathbf{a}/\mathbf{x}]\forall yB \Leftrightarrow \forall d \in D_u M, u \vDash [\mathbf{a}d/\mathbf{x}y]B.$$

Now if $y \in FV(A) = FV(B)$, then we use the hypothesis ($\sharp$) for $A, B, \mathbf{x}y, \mathbf{a}d$. If $y \notin FV(A)$, then $FV(A) = FV(B) = r(\mathbf{x})$,

$$[\mathbf{a}d/\mathbf{x}y]A = [\mathbf{a}/\mathbf{x}]A, \ [\mathbf{a}d/\mathbf{x}y]B = [\mathbf{a}/\mathbf{x}]B,$$

so we can use ($\sharp$) for $A, B, \mathbf{x}, \mathbf{a}$. Anyway

$$M, u \vDash [\mathbf{a}/\mathbf{x}]\forall yA \Leftrightarrow M, u \vDash [\mathbf{a}/\mathbf{x}]\forall yB.$$

(3) Supposing $A \sim A'$, $B \sim B'$, we prove that $(A * B) \sim (A' * B')$. Let us consider the case $* = \wedge$.

Obviously

$$FV(A \wedge B) = FV(A) \cup FV(B) = FV(A') \cup FV(B') = FV(A' \wedge B').$$

Let $r(\mathbf{x}) = FV(A \wedge B)$. Then its subset $FV(A)$ is $r(\mathbf{x} \cdot \sigma)$ for some injection σ and $FV(B) = r(\mathbf{x} \cdot \tau)$ for some injection τ. Obviously

$$[\mathbf{a}/\mathbf{x}]A = [\mathbf{a} \cdot \sigma/\mathbf{x} \cdot \sigma]A, \ [\mathbf{a}/\mathbf{x}]B = [\mathbf{a} \cdot \tau/\mathbf{x} \cdot \tau]B,$$

and similarly for A', B';

hence

$$\begin{array}{l} M, u \vDash [\mathbf{a}/\mathbf{x}](A \wedge B) \Leftrightarrow M, u \vDash [\mathbf{a}/\mathbf{x}]A \ \& \ M, u \vDash [\mathbf{a}/\mathbf{x}]B \Leftrightarrow \\ M, u \vDash [\mathbf{a} \cdot \sigma/\mathbf{x} \cdot \sigma]A \ \& \ M, u \vDash [\mathbf{a} \cdot \tau/\mathbf{x} \cdot \tau]B. \end{array}$$

Similarly

$$M, u \vDash [\mathbf{a}/\mathbf{x}](A' \wedge B') \Leftrightarrow M, u \vDash [\mathbf{a} \cdot \sigma/\mathbf{x} \cdot \sigma]A' \ \& \ M, u \vDash [\mathbf{a} \cdot \tau/\mathbf{x} \cdot \tau]B'.$$

Since $A \sim A'$, $B \sim B'$, this implies $(A \wedge B) \sim (A' \wedge B')$.

(4) Supposing $A \sim B$, let us show $\Box_i A \sim \Box_i B$.

Obviously, $FV(\Box_i A) = FV(A) = FV(B) = FV(\Box_i B)$. Next,

$$M, u \vDash [\mathbf{a}/\mathbf{x}]\Box_i A (= \Box_i [\mathbf{a}/\mathbf{x}]A) \Leftrightarrow \forall v \in R_i(u)\ M, v \vDash [\mathbf{a}/\mathbf{x}[A,$$

and similarly for B.

Since $A \sim B$, these two conditions are equivalent.

In the intuitionistic case note that $A \doteq B$ implies $A^T \doteq B^T$, by 2.11.3. So by applying the modal case and Lemma 3.2.15 we obtain:

$$M, u \Vdash A(\mathbf{a}) \text{ iff } M, u \vDash A^T(\mathbf{a}) \text{ iff } M, u \vDash B^T(\mathbf{a}) \text{ iff } M, u \Vdash B(\mathbf{a}).$$

■

Lemma 3.2.27 *Let A be a modal (respectively, intuitionistic) formula valid in a Kripke frame (respectively, an* **S4***-based Kripke frame)* **F**. *Then every formula congruent to A is valid in* **F**.

In other words, the sets $\mathbf{ML}^{(=)}(\mathbf{F})$, $\mathbf{IL}^{(=)}(\mathbf{F})$ *are closed under congruence.*

Proof By Lemmas 3.2.26, 3.2.17. ■

Lemma 3.2.28 *For any classical D_u-formula A*

$$M, u \vDash A \textit{ iff } M_u \vDash A \textit{ (in the classical sense).}$$

Proof By induction. The classical and modal truth definitions coincide in this case. ■

Theorem 3.2.29 (Soundness theorem)

(I) The set $\mathbf{ML}^{(=)}(\mathbf{F})$ *of all modal predicate formulas (with equality) valid in a predicate Kripke frame* **F** *is a modal predicate logic (with equality).*

(II) The set $\mathbf{IL}^{(=)}(\mathbf{F})$ *of all intuitionistic predicate formulas (with equality) valid in an* **S4***-based Kripke frame* **F** *is a superintuitionistic predicate logic (with equality). Moreover,* $\mathbf{IL}^{(=)}(\mathbf{F}) = {}^T\mathbf{ML}^{(=)}(\mathbf{F})$.

Proof

(I) Let **F** be a predicate Kripke frame. The axioms of $\mathbf{K}_n$ are obviously valid, since they are valid in every propositional Kripke frame and we can apply 3.2.23.

Let M be a Kripke model over **F**. The classical first-order axioms and the axioms of equality are true in every M_u; so by Lemma 3.2.28, they are true in M.

It is also easy to check that modus ponens preserves the truth in a Kripke model. In fact, suppose $M \vDash A,\ A \supset B$, and let $\mathbf{x}$ be a list of parameters of $(A \supset B)$, $|\mathbf{x}| = n$. Then by 3.2.20(1), for any $u \in M$, $\mathbf{a} \in D_u^n$

$$M, u \vDash [\mathbf{a}/\mathbf{x}]A,\ [\mathbf{a}/\mathbf{x}]A \supset [\mathbf{a}/\mathbf{x}]B,$$

hence $M, u \vDash [\mathbf{a}/\mathbf{x}]B$. Thus $M \vDash B$ by 3.2.20(1).

To verify $\Box$-introduction, note that

$$M, u \vDash [\mathbf{a}/\mathbf{x}]\Box A \text{ iff } \forall v \in R(u)\ M, v \vDash [\mathbf{a}/\mathbf{x}]A,$$

and the right hand side of the equivalence follows from $M \vDash A$ by 3.2.20.

Next, consider the $\forall$-introduction rule. If $M \vDash A(y, x_1, \dots, x_n)$, then by Lemma 3.2.20,

$$\forall u\, \forall b, a_1, \dots, a_n \in D_u\ M, u \vDash A(b, a_1, \dots, a_n),$$

and thus

$$\forall u\, \forall a_1, \dots, a_n \in D_u\ M, u \vDash \forall y A(y, a_1, \dots, a_n);$$

hence $M \vDash \forall y A(y, x_1, \dots, x_n)$ again by Lemma 3.2.20. Thus $\forall$-introduction preserves the truth in M.

The only nontrivial part of the proof is to show that the substitution rule preserves validity.

So suppose $\mathbf{F} \vDash A$, and consider a simple formula substitution $S = [C(\mathbf{x}, \mathbf{y})/P(\mathbf{x})]$, where P occurs[9] in A, $P \in PL^n$, $n > 0$[10] and $\mathbf{x}$, $\mathbf{y}$ are disjoint lists of different variables such that

$$r(\mathbf{y}) \subseteq FV(C) \subseteq r(\mathbf{xy}).$$

Recall that SA is obtained by appropriate replacements from a clean version A° of A such that $BV(A^\circ) \cap r(\mathbf{y}) = \varnothing$. By Lemma 3.2.27, A and A° are equivalent with respect to validity, so we may assume that A is clean and $BV(A) \cap r(\mathbf{y}) = \varnothing$. As we know from 2.5.24,

$$r(\mathbf{y}) = FV(S) \subseteq FV(SA) \subseteq FV(A) \cup FV(S),$$

so $BV(A) \cap FV(SA) = \varnothing$.

Let us fix a distinct list $\mathbf{z}$ such that $r(\mathbf{z}) = FV(A) \cup FV(S)$ (for example, put $\mathbf{z} = \mathbf{yt}$, where $r(\mathbf{t}) = FV(A)$); so $r(\mathbf{y}) \subseteq FV(SA) \subseteq r(\mathbf{z})$. Due to our assumption, $r(\mathbf{z}) \cap BV(A) = \varnothing$.

Now for an arbitrary Kripke model $M = (\mathbf{F}, \xi)$, let us show that $M \vDash SA$, i.e. for any $u \in \mathbf{F}$ and $\mathbf{c} \in D_u^m$

$$M, u \vDash [\mathbf{c}/\mathbf{z}]SA, \tag{1}$$

where $m = |\mathbf{z}|$.

To verify this (for certain fixed u and $\mathbf{c}$), consider another model $M_1 = (\mathbf{F}, \eta)$ such that

- for any $v \in F{\uparrow}u$, $\mathbf{a} \in D_v^n$

  $$M_1, v \vDash P(\mathbf{a}) \text{ iff } M, v \vDash C(\mathbf{a}, \mathbf{c}'),$$

  where $\mathbf{c}'$ is the part of $\mathbf{c}$ corresponding to $\mathbf{y}$ (i.e. if $\mathbf{y} = z_1 \dots z_k$, then $\mathbf{c}' = c_1 \dots c_k$);

---

[9]The case when P does not occur in A is trivial.

[10]There is a little difference in the case when $n = 0$ (then $\mathbf{x}$ is empty), but we leave this to the reader.

- for any other atomic D_v-sentence Q

$$M_1, v \vDash Q \text{ iff } M, v \vDash Q.$$

Thus $\eta_v(P) = \{\mathbf{a} \in D_v^n \mid M, v \vDash C(\mathbf{a}, \mathbf{c}')\}$.

Now consider a subformula B of A. Since $FV(B) \subseteq FV(A) \cup BV(A)$, we can present B as $B(\mathbf{z}, \mathbf{q})$[11], where $\mathbf{q}$ is distinct, $r(\mathbf{q}) = BV(A)$.

Then by 2.5.24

$$FV(SB) \subseteq FV(S) \cup FV(B) \subseteq r(\mathbf{zq}),$$

So we can also present SB as $(SB)(\mathbf{z}, \mathbf{q})$.

Certainly we can use the same presentation in the trivial case when P does not occur in B (and $SB = B$). Now let us prove the following

Claim. For any $v \in F \uparrow u$ and tuple $\mathbf{a}$ in D_v such that $|\mathbf{a}| = |\mathbf{q}|$

$$M_1, v \vDash B(\mathbf{c}, \mathbf{a}) \text{ iff } M, v \vDash (SB)(\mathbf{c}, \mathbf{a}). \tag{2}$$

This is proved by induction. It is essential that B (as a subformula of A) is clean, otherwise the argument is inapplicable, because SB can be constructed by induction only for clean formulas.

- The case when B is atomic and does not contain P is trivial, then $SB = B$ and

  $$M_1, v \vDash B(\mathbf{c}, \mathbf{a}) \Leftrightarrow M, v \vDash B(\mathbf{c}, \mathbf{a})$$

  by the definition of M_1.

- If B is atomic and contains P, it has the form $P(t_1, \ldots, t_n)$, where $t_1, \ldots, t_n$ are variables from the list $\mathbf{zq}$, $\mathbf{q}$ consists of parameters of B that are not in $\mathbf{z}$ (i.e. $r(\mathbf{q}) = FV(B) \cap BV(A)$). So we have $B = P((\mathbf{zq}) \cdot \sigma)$ for some $\sigma : I_n \longrightarrow I_k,\ k = |\mathbf{zq}|$. Then

  $B(\mathbf{c}, \mathbf{a}) = [\mathbf{ca}/\mathbf{zq}]P((\mathbf{zq}) \cdot \sigma) = P((\mathbf{ca}) \cdot \sigma)$,
  $(SB)(\mathbf{c}, \mathbf{a}) = [\mathbf{ca}/\mathbf{zq}]SP((\mathbf{zq}) \cdot \sigma) = [\mathbf{ca}/\mathbf{zq}]C((\mathbf{zq}) \cdot \sigma, \mathbf{y}) = C((\mathbf{ca}) \cdot \sigma, \mathbf{c}')$.
  Recall that $\mathbf{y}$ transforms to $\mathbf{c}'$ when $\mathbf{z}$ transforms to $\mathbf{c}$. So in this case

  $$M_1, v \vDash B(\mathbf{c}, \mathbf{a}) \text{ iff } M, v \vDash (SB)(\mathbf{c}, \mathbf{a})$$

  by the definition of M_1.

- If $B = B_1 \vee B_2$, then
  $M_1, v \vDash B(\mathbf{c}, \mathbf{a})$ iff $(M_1, v \vDash B_1(\mathbf{c}, \mathbf{a}) \vee M_1, v \vDash B_2(\mathbf{c}, \mathbf{a}))$ iff
  $(M, v \vDash (SB_1)(\mathbf{c}, \mathbf{a})$ or $M, v \vDash (SB_2)(\mathbf{c}, \mathbf{a}))$ iff $M, v \vDash (SB)(\mathbf{c}, \mathbf{a})$.

[11]Of course, this only means that $FV(B) \subseteq r(\mathbf{zq})$.

- If $B = \exists t B_1$, then

  $M_1, v \vDash B(\mathbf{c}, \mathbf{a})$ iff $\exists d \in D_v\ M_1, v \vDash B_1(\mathbf{c}, \mathbf{a}')$ iff $\exists d \in D_v\ M, v \vDash (SB_1)(\mathbf{c}, \mathbf{a}')$

  (by the induction hypothesis) iff $M, v \vDash (\exists t\, SB_1)(\mathbf{c}, \mathbf{a})(= (SB)(\mathbf{c}, \mathbf{a}))$.

  Here $\mathbf{a}'$ denotes the tuple obtained from $\mathbf{a}$ by putting d in the position corresponding to t, i.e. if $t = q_i$ (in $\mathbf{q}$), then $b_i = d$ and $b_j = a_j$ for $j \neq i$.

  Let us explain the first equivalence in more detail; the third equivalence is checked in the same way. In fact, we have

  $$B(\mathbf{c}, \mathbf{a}) = [\mathbf{ca}/\mathbf{zq}]B = [\mathbf{c}\widehat{\mathbf{a}}_i/\mathbf{z}\widehat{\mathbf{q}}_i]B,$$

  where $\widehat{\mathbf{q}}_i$ is obtained by eliminating $q_i = t$ from $\mathbf{q}$ and respectively $\widehat{\mathbf{a}}_i$ is obtained by eliminating a_i from $\mathbf{a}$, since t is bound in B. So by definition,

  $$M, v \vDash B(\mathbf{c}, \mathbf{a}) \text{ iff } \exists d \in D_v\ M, v \vDash [d/t][\mathbf{c}\widehat{\mathbf{a}}_i/\mathbf{z}\widehat{\mathbf{q}}_i]B_1,$$

  and the latter D_v-sentence is $B_1(\mathbf{c}, \mathbf{a}')$.

- If $B = \Box_i B_1$, then $M_1, v \vDash B(\mathbf{c}, \mathbf{a})$ iff

  $\forall w \in R_i(v)\ M_1, w \vDash B_1(\mathbf{c}, \mathbf{a})$ iff $\forall w \in R_i(v)\ M, w \vDash (SB_1)(\mathbf{c}, \mathbf{a})$ (by the induction hypothesis) iff $M, v \vDash \Box_i(SB_1)(\mathbf{c}, \mathbf{a})(= (SB)(\mathbf{c}, \mathbf{a}))$.

- The remaining cases: $B = C \supset D$, $C \wedge D$, $\forall x\, C(x, \mathbf{z})$ are quite similar.

Now (1) easily follows from (2). In fact, (1) is equivalent to

$$M, u \vDash [\mathbf{ac}/\mathbf{qz}]\, SA$$

for any $\mathbf{a} \in D_u^n$ since $r(\mathbf{q}) \cap FV(SA) = \varnothing$. By (2), this is equivalent to

$$M_1, u \vDash [\mathbf{ac}/\mathbf{qz}]A\, (= [\mathbf{c}/\mathbf{z}]A),$$

which holds since $\mathbf{F} \vDash A$.

(II) By Proposition 3.2.25, $\mathbf{F} \Vdash A$ iff $\mathbf{F} \vDash A^T$. This is exactly the same as $A \in \mathbf{IL}^{(=)}(\mathbf{F})$ iff $A^T \in \mathbf{ML}^{(=)}(\mathbf{F})$. So we obtain: $\mathbf{IL}^{(=)}(\mathbf{F}) = {}^T\mathbf{ML}^{(=)}(\mathbf{F})$, and thus $\mathbf{IL}^{(=)}(\mathbf{F})$ is an s.p.l. $(=)$, by Proposition 2.11.8. ∎

In accordance with Section 2.16, the set $\mathbf{ML}^{(=)}(\mathbf{F})$ is called the *modal logic of* $\mathbf{F}$ (respectively, with or without equality). Similarly, for an intuitionistic $\mathbf{F}$, its *superintuitionistic logic* is $\mathbf{IL}^{(=)}(\mathbf{F})$.

For a class of N-modal frames $\mathcal{C}$ we define the modal logic *determined by* $\mathcal{C}$ (or *complete w.r.t.* $\mathcal{C}$)

$$\mathbf{ML}^{(=)}(\mathcal{C}) := \bigcap\{\mathbf{ML}^{(=)}(\mathbf{F}) \mid \mathbf{F} \in \mathcal{C}\},$$

and similarly for superintuitionistic logics.

Definition 3.2.30 (N-modal) Kripke (frame) semantics $\mathcal{K}_N$ *(or* $\mathcal{K}_N^=$*, for logics with equality) is generated by the class of all N-modal predicate Kripke frames. Similarly intuitionistic Kripke (frame) semantics* $\mathcal{K}_{int}^{(=)}$ *is generated by the class of all intuitionistic predicate Kripke frames.* $\mathcal{K}_N^{(=)}$*-complete or* $\mathcal{K}_{int}^{(=)}$*-complete logics are called* Kripke (frame) complete *(or* $\mathcal{K}$-complete *if there is no confusion).*

Here is another version of soundness.

Lemma 3.2.31 *Let M be a modal (respectively, intuitionistic) Kripke model, L an m.p.l.(=) (respectively, s.p.l.(=)), Γ a modal (respectively, intuitionistic) theory without constants, such that $M \vDash (\Vdash)\ \overline{L} \cup \Gamma$. Then for any $A \in IF^{(=)}$,*

$$\Gamma \vdash_L A \Rightarrow M \vDash (\Vdash)\ \overline{\forall} A.$$

Proof By induction on the length of an L-inference of A from Γ.

If $A \in \Gamma$ the claim is trivial.

If $A \in L$, then $\overline{\forall} A \in \overline{L}$, so the claim is also trivial.

It is clear that the truth in M respects MP, cf. the proof of 3.2.29.

Finally, if $A = \forall x B$ and $M \vDash (\Vdash)\ \overline{\forall} B$, then also $M \vDash (\Vdash)\ \overline{\forall} A$, since $\mathbf{QH} \vdash \overline{\forall} A \equiv \overline{\forall} B$ and we can apply soundness. ∎

Definition 3.2.32 *If F is a propositional Kripke frame, $\mathcal{K}(F)$ (or $\mathcal{K}F$) denotes the class of all predicate Kripke frames based on F.* $\mathbf{ML}^{(=)}(\mathcal{K}(F))$ *is called the* modal logic (with equality) determined over F. *Similarly we define $\mathcal{K}(\mathcal{C})$ (or $\mathcal{K}\mathcal{C}$) for a class of N-modal propositional frames $\mathcal{C}$ and the* modal logic determined over $\mathcal{C}$.

We also define the superintuitionistic logic determined over *a propositional* **S4***-frame F as* $\mathbf{IL}^{(=)}(\mathcal{K}F)$ *and similarly for a class of* **S4***-frames $\mathcal{C}$.*

We may regard $\mathcal{K}\mathcal{C}$ as a semantics for a certain class of logics. In particular, for a propositional modal logic $\mathbf{\Lambda}$, the class $\mathcal{K}\mathbf{V}(\Lambda)$ generates the semantics of all Kripke-complete logics containing $\mathbf{\Lambda}$.

Proposition 3.2.33 $\mathcal{K}_{\text{int}}^{(=)}$ *is the intuitionistic version of* $\mathcal{K}\mathbf{V}(\mathbf{S4})$. *Thus* $\mathbf{IL}^{(=)}(\mathcal{C}) = {}^T\mathbf{ML}^{(=)}(\mathcal{C})$ *for any class of* **S4***-based predicate frames $\mathcal{C}$.*

Proof According to Definition 2.16.16, the first assertion is equivalent to 3.2.25. The second one is checked in the proof of 2.16.18. ∎

Lemma 3.2.34 *Let* $\mathbf{F}$ *be a predicate Kripke frame over a propositional Kripke frame F. Then*

$$\mathbf{ML}(\mathbf{F})_\pi = \mathbf{ML}(F).$$

Proof Rather trivial. The definition of validity for propositional formulas in $\mathbf{F}$ is the same as in F. ∎

Proposition 3.2.35 *For a class of propositional Kripke frames $\mathcal{C}$,*

$$(\mathbf{ML}^{(=)}(\mathcal{KC}))_\pi = \mathbf{ML}(\mathcal{C})$$

and

$$(\mathbf{IL}^{(=)}(\mathcal{KC}))_\pi = \mathbf{IL}(\mathcal{C})$$

in the intuitionistic case.

Proof We begin with the modal case. By Lemma 3.2.34, we obtain

$$(\mathbf{ML}^{(=)}(\mathcal{KC}))_\pi = (\bigcap\{\mathbf{ML}^{(=)}(\mathbf{F}) \mid \mathbf{F} \in \mathcal{KC}\})_\pi = \bigcap\{\mathbf{ML}^{(=)}(\mathbf{F})_\pi \mid \mathbf{F} \in \mathcal{KC}\} = \bigcap\{\mathbf{ML}^{(=)}(F) \mid F \in \mathcal{C}\} = \mathbf{ML}^{(=)}(\mathcal{C}).$$

■

3.3 Morphisms of Kripke frames

In this section we extend the notions of a frame morphism, a generated subframe and a p-morphism to the predicate case.

Definition 3.3.1 *Let $\mathbf{F} = (F, D)$, $\mathbf{F}' = (F', D')$ be predicate Kripke frames based on $F = (W, R_1, \dots, R_N)$, $F' = (W', R'_1, \dots, R'_N)$ respectively. A* morphism *from $\mathbf{F}$ to $\mathbf{F}'$ is a pair $\mathbf{f} = (f_0, f_1)$ such that*

(1) $f_0 : F \longrightarrow F'$ is a morphism of propositional frames;

(2) $f_1 = (f_{1u})_{u \in W}$;

(3) every $f_{1u} : D_u \longrightarrow D'_{f_0(u)}$ is a surjective map;

(4) $uR_i v \Rightarrow f_{1u} = f_{1v} \upharpoonright D_u$.

f_0 is called the world component*, f_1 the* individual component *of (f_0, f_1).*

Definition 3.3.2 *Let $\mathbf{f} = (f_0, f_1)$ be a morphism from $\mathbf{F}$ to $\mathbf{F}'$.*

- *$\mathbf{f}$ is called an* equality-morphism *(briefly,* =-morphism*) if every f_{1u} is a bijection.*
- *$\mathbf{f}$ is called a* p-morphism *if f_0 is surjective (i.e. f_0 is a p-morphism of propositional frames).*
- *$\mathbf{f}$ is called a* p$^=$-morphism *if it is a p-morphism, and an =-morphism.*
- *$\mathbf{f}$ is called an* isomorphism *if it is an =-morphism, and f_0 is an isomorphism of propositional frames.*

Definition 3.3.3 *A* morphism of predicate Kripke models $M = (\mathbf{F}, \xi)$ *and* $M' = (\mathbf{F}', \xi')$ *is a morphism* (f_0, f_1) *of their frames* $\mathbf{F}$ *and* $\mathbf{F}'$ *preserving the truth values of atomic formulas, i.e. such that for any* $P \in PL^m$, $m \geq 0$, $u \in F$; $b_1, \ldots, b_m \in D_u$

$$M, u \vDash P(b_1, \ldots, b_m) \quad \textit{iff} \quad M', f_0(u) \vDash P(f_{1u}(b_1), \ldots, f_{1u}(b_m)).$$

A p-morphism *of Kripke models is a morphism of Kripke models, which is a p-morphism of their frames. Similarly for* $=$-, $p^=$*-morphisms and isomorphisms.*

The notation $(f_0, f_1) : M \longrightarrow M'$ means that (f_0, f_1) is a morphism from M to M', and similarly for frames. As in the propositional case, p-morphisms of predicate Kripke frames (models) are denoted by $\twoheadrightarrow$. We also use the notation $\longrightarrow^=$ for $=$-morphisms, $\twoheadrightarrow^=$ for p$^=$-morphisms, $\cong$ for isomorphisms.

Lemma 3.3.4

(1) For a predicate Kripke frame $\mathbf{F} = (F, D)$, *the* identity morphism $id_{\mathbf{F}} := (id_W, f_1)$, *where* $f_{1u} := id_{D_u}$ *for any* u, *is an isomorphism.*

(2) For morphisms of predicate Kripke frames $(f_0, f_1) : \mathbf{F} \longrightarrow \mathbf{F}'$ *and* $(g_0, g_1) : \mathbf{F}' \longrightarrow \mathbf{F}''$, *consider the* composition

$$(f_0, f_1) \circ (g_0, g_1) := (g_0, g_1) \cdot (f_0, f_1) := (f_0 \circ g_0, (f_{1u} \circ g_{1u})_{u \in F}).$$

This is a morphism $\mathbf{F} \longrightarrow \mathbf{F}''$.

(3) The composition of $=$*-morphisms is an* $=$*-morphism; similarly for p(=)-morphisms and isomorphisms.*

(4) The statements (1)–(3) also hold for Kripke models, with obvious changes.

Proof An easy exercise; use Lemma 1.3.32. ∎

Thus we obtain the categories $\mathsf{PKF}_N^{(=)}$ of N-modal predicate Kripke frames and $(=)$-morphisms, with the composition $\circ$ and the identity morphism $id_{\mathbf{F}}$. One can easily check the following

Lemma 3.3.5 *Isomorphisms in* $\mathsf{PKF}_N^{(=)}$ *are exactly isomorphisms in the sense of Definition 3.3.2.*

In a similar way we can define the categories $\mathsf{PKF}_{int}^{(=)}$ of intuitionistic Kripke frames, $\mathsf{PKM}_N^{(=)}$ of N-modal Kripke models, $\mathsf{PKM}_{int}^{(=)}$ of intuitionistic Kripke models; the details are left to the reader.

Definition 3.3.6 *Let* $F = (W, R_1, \ldots, R_N)$ *be a propositional Kripke frame. A* predicate Kripke frame morphism over F *is a morphism of the form* $(id_W, f_1) : (F, D) \rightarrow (F, D')$.

The following is obvious

Lemma 3.3.7 *The composition of (=)-morphisms over F is an (=)-morphism over F.*

Thus we can also consider the categories of frames over F and (=)-morphisms. But the case with equality is not so interesting, because all =-morphisms over F are isomorphisms.

Definition 3.3.8 *Let F, F' be propositional Kripke frames, $h : F' \longrightarrow F$, and let $\mathbf{F} = (F, D)$ be a predicate Kripke frame. Then we say that a Kripke frame $\mathbf{F}' = (F', D')$ is obtained from $\mathbf{F}$ by* changing the base along h *if $D'_u = D_{h(u)}$ for all $u \in F$. This frame $\mathbf{F}'$ is denoted by $h_*\mathbf{F}$.*

Lemma 3.3.9 *Under the conditions of Definition 3.3.8, there exists a 'canonical' =-morphism $(h, g) : \ h_*\mathbf{F} \longrightarrow^= \mathbf{F}$; if h is a p-morphism, then (h, g) is a p=-morphism.*

Proof Put

$$g_u := \mathrm{id}_{D'_u} : D'_u \longrightarrow D_{h(u)}.$$

Then $(h, g) : \ (F', D') \longrightarrow (F, D)$ by Definition 3.3.1. ■

Now let us show that every morphism is represented as a specific composition:

Proposition 3.3.10 *Every (=)-morphism of predicate Kripke frames $(f_0, f_1) : \mathbf{F}' \longrightarrow \mathbf{F}$ can be presented as a composition of an (=)-morphism over the propositional base F' of $\mathbf{F}'$ and the canonical morphism:*

$$\mathbf{F}' \xrightarrow{(id_{W'}, f_1)} (f_0)_*\mathbf{F} \xrightarrow{(f_0, g)} \mathbf{F}.$$

Moreover, $(id_{W'}, f_1)$ is a unique morphism $(id_{W'}, h_1)$ over F' such that $(f_0, f_1) = (id_{W'}, h_1) \circ (f_0, g)$.

Proof In fact,

$$(id_{W'}, h_1) \circ (f_0, g) = (f_0, f_1)$$

iff $id_{W'} \circ f_0 = f_0$ (which is true) and for any u

$$f_{1u} = h_{1u} \circ g_u = h_{1u} \circ id_{D'_u} = h_{1u},$$

i.e. iff $f_1 = h_1$. ■

Proposition 3.3.11 *If $(f_0, f_1) : \ M \longrightarrow^{(=)} M'$, then for any $u \in M$, $B \in MS_N^{(=)}(D_u)$*

$$M, u \vDash B \ \textit{ iff } \ M', f_0(u) \vDash f_{1u} \cdot B,$$

where $f_{1u} \cdot B$ is obtained from B by replacing occurrences of every $a \in D_u$ with $f_{1u}(a)$.

If the models are intuitionistic, the same holds for the intuitionistic forcing and $B \in IS^{(=)}(D_u)$.

Proof By induction on the complexity of B we prove the claim for any u. Let us check only two cases

(1) $B = \Box_j C$. Then

$$M, u \vDash B \text{ iff } \forall v \in R_j(u)\ M, v \vDash C \text{ iff } \forall v \in R_j(u)\ M', f_0(v) \vDash f_{1v} \cdot C$$

by the induction hypothesis.

Since f_0 is a morphism, we have $R'_j(f_0(u)) = f_0[R_j(u)]$. By Definition 3.3.1, $f_{1u}(a) = f_{1v}(a)$ for any $a \in D_u$, $v \in R_j(u)$; hence $f_{1v} \cdot C = f_{1u} \cdot C$ for any D_u-sentence C.

So

$M, u \vDash B$ iff $\forall v' \in R'_j(f_0(u))\ M', v' \vDash f_{1u} \cdot C$ iff $M', f_0(u) \vDash \Box_j(f_{1u} \cdot C)$ ($= f_{1u} \cdot B$).

(2) $B = \exists x C(x, \mathbf{a})$, where $\exists x C(x, \mathbf{y})$ is a generator of B, $\mathbf{a}$ is a tuple from D_u. Then

$M, u \vDash B$ iff $\exists b \in D_u\ M, u \vDash C(b, \mathbf{a})$ iff

$\exists b \in D_u\ M', f_0(u) \vDash C(f_{1u}(b), f_{1u} \cdot \mathbf{a})$ (by induction hypothesis) iff

$\exists c \in D'_{f_0(u)}\ M', f_0(u) \vDash C(c, f_{1u} \cdot \mathbf{a})$ (since $f_{1u}[D_u] = D'_{f_0(u)}$) iff

$M', f_0(u) \vDash \exists x C(x, f_{1u} \cdot \mathbf{a})$.

The intuitionistic case now follows easily by Gödel–Tarski translation. ∎

Proposition 3.3.12 *Let $(f_0, f_1) : \mathbf{F} \longrightarrow \mathbf{F}'$, and let M' be a Kripke model over $\mathbf{F}'$. Then there exists a unique model M over $\mathbf{F}$ such that $(f_0, f_1) : M \longrightarrow M'$, and similarly for $\twoheadrightarrow$, $\longrightarrow^=$, $\twoheadrightarrow^=$. If M' is intuitionistic, then M is also intuitionistic.*

Proof In fact, if $M' = (\mathbf{F}', \xi')$, then $M = (\mathbf{F}, \xi)$ is uniquely determined by the equalities

$$\xi_v(P_k^m) := \{\mathbf{b} \in D_v^m \mid f_{1v} \cdot \mathbf{b} \in \xi'_{f_0(v)}(P_k^m)\}$$

for any $v \in \mathbf{F}$, $m > 0$, where

$$f_{1v} \cdot (b_1, \ldots, b_m) := (f_{1v}(b_1), \ldots, f_{1v}(b_m)),$$

and

$$\xi_v(P_k^0) := \begin{cases} \{v\} & \text{if } \xi'_{f_0(v)}(P_k^0) = \{f_0(v)\}, \\ \varnothing & \text{otherwise.} \end{cases}$$

Now suppose ξ' is intutionistic. Then vRu implies $f_0(v)R'f_0(u)$ and next $\xi'_{f_0(v)}(P_k^m) \subseteq \xi'_{f_0(u)}(P_k^m)$ (for $m > 0$), whence by definition

$$\xi_v(P_k^m) \subseteq \xi_u(P_k^m).$$

In the same way, we obtain

$$vRu \ \&\ v \in \xi_v(P_k^0) \Rightarrow u \in \xi_u(P_k^0).$$

Thus ξ is intutionistic. ∎

Now let M be a Kripke model over $\mathbf{F}$ and consider changing the base of $\mathbf{F}$. By applying 3.3.12 to the canonical morphism $(h, g) : h_*\mathbf{F} \to \mathbf{F}$, we obtain a unique Kripke model M' over $h_*\mathbf{F}$ such that $(h, g) : M' \to M$. We also say that M' is obtained from M by *changing the base along* h and denote M' by h_*M.

Proposition 3.3.13[12] *If there exists a* $p^{(=)}$*-morphism* $\mathbf{F} \twoheadrightarrow^{(=)} \mathbf{F}'$, *then* $\mathbf{ML}^{(=)}(\mathbf{F}) \subseteq \mathbf{ML}^{(=)}(\mathbf{F}')$, *and similarly,* $\mathbf{IL}^{(=)}(\mathbf{F}) \subseteq \mathbf{IL}^{(=)}(\mathbf{F}')$ *for intuitionistic Kripke frames.*

Proof Consider the modal case only. Let $(f_0, f_1) : \mathbf{F} \twoheadrightarrow \mathbf{F}'$, and assume that $\mathbf{F}' \not\models A(x_1, \ldots, x_n)$. So by Lemma 3.2.20, for some $u' \in \mathbf{F}'$, $a_1, \ldots, a_n \in D'_{u'}$, for some model M' over $\mathbf{F}'$ we have

$$M', u' \not\models A(a'_1, \ldots, a'_n).$$

By Proposition 3.3.12, there exists a model M over $\mathbf{F}$ such that

$$(f_0, f_1) : M \twoheadrightarrow M'.$$

Now since f_0 and f_{1u} are surjective, there exist $u, a_1, \ldots, a_n$, such that

$$u' = f_0(u),\ a'_1 = f_{1u}(a_1), \ldots,\ a'_n = f_{1u}(a_n),$$

and thus by Lemma 3.3.11, we obtain $M, u \not\models A(a_1, \ldots, a_n)$.

Therefore, $\mathbf{F}' \not\models A$ implies $\mathbf{F} \not\models A$. ∎

Proposition 3.3.14 *If there exists a p-morphism* $F' \twoheadrightarrow F$, *then* $\mathbf{ML}^{(=)}(\mathcal{K}F') \subseteq \mathbf{ML}^{(=)}(\mathcal{K}F)$
(and $\mathbf{IL}^{(=)}(\mathcal{K}F') \subseteq \mathbf{IL}^{(=)}(\mathcal{K}F)$ *in the intuitionistic case).*

Proof Consider the modal case. Let $h : F' \twoheadrightarrow F$. For any $\mathbf{F} \in \mathcal{K}F$ we have

$$\mathbf{ML}^{(=)}(\mathcal{K}F') \subseteq \mathbf{ML}^{(=)}(h_*\mathbf{F}) \subseteq \mathbf{ML}^{(=)}(\mathbf{F})$$

by Lemma 3.3.9 and Proposition 3.3.13. Hence

$$\mathbf{ML}^{(=)}(\mathcal{K}F') \subseteq \mathbf{ML}^{(=)}(\mathcal{K}F).$$

∎

Definition 3.3.15 *Let* $F = (W, R_1, \ldots, R_N)$ *be a propositional Kripke frame,* $\mathbf{F} = (F, D)$ *a predicate Kripke frame,* $V \subseteq W$. *A* subframe *of* F *obtained by restriction to* V *is defined as*

$$\mathbf{F} \restriction V := (F \restriction V, D \restriction V),$$

where $D \restriction V := (D_u)_{u \in V}$. *If* $M = (\mathbf{F}, \xi)$ *is a Kripke model, we define the* submodel

$$M \restriction V := (\mathbf{F} \restriction V, \xi \restriction V),$$

where

$$(\xi \restriction V)(P_k^m) := (\xi_u(P_k^m))_{u \in V}.$$

If V *is stable, the subframe* $\mathbf{F} \restriction V$ *and the submodel* $M \restriction V$ *are called* generated.

[12]Cf. [Ono, 1972/73], Theorem 3.4.

The notation $M_1 \subseteq M$ means that M_1 is a submodel of M.

Definition 3.3.16 *A submodel $M_1 \subseteq M$ is called* reliable *if for any $u \in M_1$, for any D_u-sentence A*

$$M_1, u \vDash A \text{ iff } M, u \vDash A.$$

In this case obviously, $\mathbf{MT}(M) \subseteq \mathbf{MT}(M_1)$.

Definition 3.3.17 *Similarly to the propositional case (Definition 1.3.14), we define* cones, rooted frames *and* rooted models*:*

$$\mathbf{F}{\uparrow}u := \mathbf{F} \restriction (W{\uparrow}u), \;\; M{\uparrow}u := M \restriction (W{\uparrow}u).$$

Lemma 3.3.18 (Generation lemma) *Let $\mathbf{F}$ be a predicate Kripke frame, M a Kripke model over $\mathbf{F}$, V a stable set of worlds in $\mathbf{F}$. Then*

(1) there is a morphism $(j, i) : \mathbf{F} \restriction V \longrightarrow \mathbf{F}$ (and $M \restriction V \longrightarrow M$), in which $j : V \longrightarrow W$ is the inclusion map and every i_u is the identity map on the corresponding domain;

(2) $M \restriction V$ is a reliable submodel of M;

(3) $\mathbf{ML}^{(=)}(\mathbf{F}) \subseteq \mathbf{ML}^{(=)}(\mathbf{F} \restriction V)$; similarly, for the intuitionistic case.

Proof

(1) Obvious.

(2) Apply Lemma 3.3.11 and (1).

(3) Every valuation ξ' in $\mathbf{F} \restriction V$ equals $\xi \restriction V$ for a valuation ξ in $\mathbf{F}$ such that $\xi_u = \xi'_u$ whenever $u \in V$. Such a ξ obviously exists, e.g. put

$$\xi_u(P) := \begin{cases} \xi'_u(P) & \text{if } u \in V, \\ \varnothing & \text{otherwise.} \end{cases}$$

If ξ' is intuitionistic, then ξ is also intuitionistic, since V is stable. Now the claim follows from (2).

■

Definition 3.3.19 *If $\mathbf{f} : \mathbf{F} \longrightarrow \mathbf{F}'$ is a morphism of Kripke frames and V is stable in $\mathbf{F}$, then $\mathbf{f} \restriction V := (j, i) \circ \mathbf{f}$, where (j, i) is a morphism from 3.3.18(1), is called the* restriction of $\mathbf{f}$ to V. *Restrictions of Kripke model morphisms are defined in the same way.*

Lemma 3.3.4 readily implies

Lemma 3.3.20 *A restriction of an $(=)$-morphism (of Kripke frames or models) to a generated subframe (or submodel) is an $(=)$-morphism (respectively, of frames or models).*

If $\mathbf{G} = \mathbf{F} \restriction V$, we also denote $\mathbf{f} \restriction V$ by $\mathbf{f} \restriction \mathbf{G}$.

Lemma 3.3.21 *(1)* $\mathbf{ML}^{(=)}(\mathbf{F}) = \bigcap_{u \in F} \mathbf{ML}^{(=)}(\mathbf{F}{\uparrow}u)$
and analogously in the intuitionistic case.

(2) Every Kripke complete modal or superintuitionistic logic is determined by a class of rooted predicated Kripke frames:

$$\mathbf{L}^{(=)}(\mathcal{C}) = \mathbf{L}^{(=)}(\mathcal{C}\uparrow),$$

where $\mathbf{L}$ *is* $\mathbf{ML}$ *or* $\mathbf{IL}$*,*

$$\mathcal{C}\uparrow := \{\mathbf{F}\uparrow u \mid \mathbf{F} \in \mathcal{C},\ u \in \mathbf{F}\}.$$

Proof

(1) Similar to Lemma 1.3.26 (an exercise).

(2) Follows from (1).

■

Definition 3.3.22 *The notions 'path', 'connectedness', 'component', 'non-oriented path' are obviously extended to the predicate case. Viz., a predicate Kripke frame is called* connected *if its propositional base is connected; a* non-oriented path *in (F, D) is the same as in F; a* component *in (F, D) is its restriction to a component in F.*

Now we have a predicate version of 1.3.38.

Proposition 3.3.23 *Let $(\mathbf{F}_i \mid i \in I)$ be a family of all different components of a predicate Kripke frame* $\mathbf{F}$*.*
Then

(1) for any morphism $\mathbf{f} : \mathbf{F} \longrightarrow \mathbf{G}$, every $\mathbf{f} \restriction \mathbf{F}_i$ is a morphism;

(2) for any family of morphisms $\mathbf{f}_i = (g_i, h_i) : \mathbf{F}_i \longrightarrow \mathbf{G}$ there is a joined morphism $\bigcup_{i \in I} \mathbf{f}_i : F \longrightarrow G$ defined as (g, h), with
$g := \bigcup_{i \in I} g_i$, $h := (h_u)_{u \in F}, h_u := (h_i)_u$ for $u \in \mathbf{F}_i$;

(3) every morphism $\mathbf{f} : \mathbf{F} \longrightarrow \mathbf{G}$ is presented as $\bigcup_{i \in I} (\mathbf{f} \restriction \mathbf{F}_i)$.

Proof

(1) Follows from Lemma 3.3.20.

(2) g is a morphism of propositional bases, by Propositon 1.3.38. Every h_u is surjective, since it coincides with some $(h_i)_u$. Finally, if $u \in \mathbf{F}_i$, uR_jv, then $v \in \mathbf{F}_i$, thus for any $a \in D_u$

$$h_u(a) = (h_i)_u(a) = (h_i)_v(a) = h_v(a),$$

i.e. $h_u = h_v \restriction D_u$.

(3) Trivial, by definition.

■

We write $\mathbf{F}_1 \twoheadrightarrow \mathbf{F}_2$ to denote that there is a p-morphism from $\mathbf{F}_1$ onto $\mathbf{F}_2$. Similarly to the propositional case, we give

Definition 3.3.24 *A predicate Kripke frame* $\mathbf{F}_1$ *is called* (=)-reducible *to a rooted predicate Kripke frame* $\mathbf{F}_2$ *if there exists* $u \in \mathbf{F}_1$ *such that* $\mathbf{F}_1{\uparrow}u \twoheadrightarrow^{(=)} \mathbf{F}_2$.

Definition 3.3.25 *Let* $\mathcal{C}_1$, $\mathcal{C}_2$ *be classes of predicate Kripke frames. We say that* $\mathcal{C}_1$ *is* (=)-reducible *to* $\mathcal{C}_2$ *if for any* $\mathbf{F}_2 \in \mathcal{C}_2$ *for any* $v \in \mathbf{F}_2$ *there exists* $\mathbf{F}_1 \in \mathcal{C}_1$ *that is (=)-reducible to* $\mathbf{F}_2 \uparrow v$.

In the similar way reducibility is defined for classes of propositional Kripke frames.

$\mathcal{C}_1\, red^{(=)}\, \mathcal{C}_2$ denotes that $\mathcal{C}_1$ is (=)-reducible to $\mathcal{C}_2$. It is clear that =-reducibility implies reducibility.

Proposition 3.3.26 *If* $\mathcal{C}_1\, red^{(=)}\, \mathcal{C}_2$ *for classes* $\mathcal{C}_1, \mathcal{C}_2$ *of predicate Kripke frames, then* $\mathbf{L}^{(=)}(\mathcal{C}_1) \subseteq \mathbf{L}^{(=)}(\mathcal{C}_2)$ *(where* $\mathbf{L}$ *denotes* **ML** *in the modal case and* **IL** *in the intuitionistic case).*

Proof Suppose $\mathcal{C}_1\, red^{(=)}\, \mathcal{C}_2$. Then for any $\mathbf{F}_2 \in \mathcal{C}_2$, $v \in \mathbf{F}_2$ there is $\mathbf{F}_1 \in \mathcal{C}_1$, $u \in \mathbf{F}_1$ such that $\mathbf{F}_1{\uparrow}u \twoheadrightarrow^{(=)} \mathbf{F}_2{\uparrow}v$. Hence by 3.3.18 and 3.3.13

$$\mathbf{L}^{(=)}(\mathbf{F}_1) \subseteq \mathbf{L}^{(=)}(\mathbf{F}_1{\uparrow}u) \subseteq \mathbf{L}^{(=)}(\mathbf{F}_2{\uparrow}v),$$

and thus $\mathbf{L}^{(=)}(\mathbf{F}_1) \subseteq \mathbf{L}^{(=)}(\mathbf{F}_2)$ by 3.3.21. Therefore $\mathbf{L}^{(=)}(\mathcal{C}_1) \subseteq \mathbf{L}^{(=)}(\mathbf{F}_2)$ for any $\mathbf{F}_2 \in \mathcal{C}_2$, which implies $\mathbf{L}^{(=)}(\mathcal{C}_1) \subseteq \mathbf{L}^{(=)}(\mathcal{C}_2)$. ■

Lemma 3.3.27 *Let* F *be a propositional Kripke frame,* $u \in F$. *Then*

$$\mathcal{K}(F{\uparrow}u) = \{\mathbf{F}{\uparrow}u \mid \mathbf{F} \in \mathcal{K}F\}.$$

We shall denote the latter class by $(\mathcal{K}F) \uparrow u$.

Proof The inclusion $(\mathcal{K}F) \uparrow u \subseteq \mathcal{K}(F \uparrow u)$ is trivial. The other way round, if $\mathbf{F}' = (F \uparrow u, D')$, then $\mathbf{F}' = \mathbf{F} \uparrow u$ for the frame $\mathbf{F} = (F, D)$, where

$$D_v := \begin{cases} D'_v \text{ if } v \in F{\uparrow}u, \\ D'_u \text{ otherwise.} \end{cases}$$

The system of domains D is expanding, since by 3.2.3, $D'_u \subseteq D'_v$ for any $v \in F{\uparrow}u$. Thus $\mathbf{F}' \in (\mathcal{K}F) \uparrow u$. ■

Proposition 3.3.28 *Let $\mathcal{C}$ be a class of N-modal propositional Kripke frames,*

$$\mathcal{C}\uparrow := \{F\uparrow u \mid u \in F,\ F \in \mathcal{C}\}.$$

Then

(1) $\mathcal{K}(\mathcal{C}\uparrow) = (\mathcal{K}\mathcal{C})\uparrow$ $(= \{\mathbf{F}\uparrow u \mid u \in \mathbf{F},\ \mathbf{F} \in \mathcal{K}\mathcal{C}\})$

(2) $\mathbf{ML}^{(=)}(\mathcal{K}\mathcal{C}) = \mathbf{ML}^{(=)}(\mathcal{K}(\mathcal{C}\uparrow))$,

(3) $\mathbf{IL}^{(=)}(\mathcal{K}\mathcal{C}) = \mathbf{IL}^{(=)}(\mathcal{K}(\mathcal{C}\uparrow))$ *if $\mathcal{C}$ is a class of* **S4**-*frames.*

Proof (1) In fact,

$$\mathcal{K}(\mathcal{C}\uparrow) = \bigcup\{\mathcal{K}(F\uparrow) \mid u \in F,\ F \in \mathcal{C}\},$$
$$(\mathcal{K}\mathcal{C})\uparrow = \bigcup\{(\mathcal{K}F)\uparrow u \mid u \in F,\ F \in \mathcal{C}\},$$

and we can apply 3.3.27.

(2), (3) follow from (1) and 3.3.21(2). ∎

Lemma 3.3.29 *Let F_1, F_2 be propositional Kripke frames such that $F_1\, red\, F_2$. Then $(\mathcal{K}F_1)\, red^{=}\, (\mathcal{K}F_2)$.*

Proof By assumption, there exists $h : F_1\uparrow u \twoheadrightarrow F_2$ for some u. So by 3.3.9, for any $\mathbf{F}_2 \in \mathcal{K}F_2$ there is a canonical

$$\gamma : h_*\mathbf{F}_2 \twoheadrightarrow^{=} \mathbf{F}_2,$$

and $h_*\mathbf{F}_2 \in \mathcal{K}(F_1\uparrow u) = (\mathcal{K}F_1)\uparrow u$ (by 3.3.27). Since h is surjective, for any $v \in F_2$, there is $w \in F_1\uparrow u$ such that $h(w) = v$. Then by 1.3.32, the restriction of h is a p-morphism

$$F_1\uparrow w = (F_1\uparrow u)\uparrow w \twoheadrightarrow F_2\uparrow v,$$

and it follows that

$$(h_*\mathbf{F}_2)\uparrow w \twoheadrightarrow \mathbf{F}_2\uparrow v$$

by the restriction of γ. Since $(h_*\mathbf{F}_2)\uparrow w \in (\mathcal{K}F_1)\uparrow w$, we obtain $(\mathcal{K}F_1)\, red^{=}\, (\mathcal{K}F_2)$. ∎

Proposition 3.3.30 *If $\mathcal{C}_1\, red\, \mathcal{C}_2$ for classes $\mathcal{C}_1$, $\mathcal{C}_2$ of propositional Kripke frames, then $(\mathcal{K}\mathcal{C}_1)\, red^{=}\, (\mathcal{K}\mathcal{C}_2)$.*

Proof Let $\mathbf{F}_2 \in \mathcal{K}\mathcal{C}_2$, i.e. $\mathbf{F}_2 \in \mathcal{K}F_2$ for some $F_2 \in \mathcal{C}_2$; then $\mathbf{F}_2\uparrow v \in (\mathcal{K}F_2)\uparrow v = \mathcal{K}(F_2\uparrow v)$. Since $\mathcal{C}_1\, red\, \mathcal{C}_2$, there exists $F_1 \in \mathcal{C}_1$ such that $F_1\, red\, (F_2\uparrow v)$.

By Lemma 3.3.29, $(\mathcal{K}F_1)\, red^{=}\, \mathcal{K}(F_2\uparrow v)$; thus $\mathbf{F}_1\, red^{=} (\mathbf{F}_2\uparrow v)$ for some $\mathbf{F}_1 \in \mathcal{K}F_1 \subseteq \mathcal{K}\mathcal{C}_1$. Therefore $(\mathcal{K}\mathcal{C}_1)\, red^{=}\, (\mathcal{K}\mathcal{C}_2)$. ∎

Corollary 3.3.31 *If $\mathcal{C}_1\, red\, \mathcal{C}_2$ for classes $\mathcal{C}_1, \mathcal{C}_2$ of propositional Kripke frames, then $\mathbf{ML}^{(=)}(\mathcal{K}\mathcal{C}_1) \subseteq \mathbf{ML}^{(=)}(\mathcal{K}\mathcal{C}_2)$ (and $\mathbf{IL}^{(=)}(\mathcal{K}\mathcal{C}_1) \subseteq \mathbf{IL}^{(=)}(\mathcal{K}\mathcal{C}_2)$ for* **S4**-*frames).*

Proof By 3.3.30 and 3.3.26. ∎

Definition 3.3.32 *We say that a Kripke frame (F, D') is obtained by* domain restriction *from (F, D) if for some $V \subseteq D^+$, $D'_u = D_u \cap V$.*

In this case we denote (F, D') by $(F, D) \cap V$.

Obviously this definition is sound iff $V \cap D_u \neq \varnothing$ for any $u \in F$.

Proposition 3.3.33 *(F, D) red $(F, D) \cap V$.*

Proof If F has a root v, there exists a p-morphism over F

$$(id_F, g) : (F, D) \twoheadrightarrow (F, D') = (F, D) \cap V.$$

In fact, we can define g_u as an identity map on D'_u sending every $a \notin V$ to some fixed element of V. Then obviously, $g_v \upharpoonright D_u = g_u$ whenever uR_iv. ∎

Definition 3.3.34 *Let $(\mathbf{F}_i)_{i\in I}$ be a family of predicate Kripke frames, $\mathbf{F}_i = (F_i, D_i)$. The* disjoint sum *(or* disjoint union*) of the family $(\mathbf{F}_i)_{i\in I}$ is the frame*

$$\bigsqcup_{i\in I} \mathbf{F}_i := \left(\bigsqcup_{i\in I} F_i, D\right),$$

where

$$D_{(u,i)} := (D_i)_u \times \{i\}$$

for $i \in I$, $u \in F_i$.

If for each $i \in I$, $M_i = (\mathbf{F}_i, \theta_i)$ is a Kripke model over $\mathbf{F}_i$, the disjoint sum (union) *of $(M_i)_{i\in I}$ is the Kripke model*

$$\bigsqcup_{i\in I} M_i := \left(\bigsqcup_{i\in I} \mathbf{F}_i, \theta\right),$$

where

$$\theta_{(u,i)}(P) := \{((a_1, i), \dots, (a_n, i)) \mid (a_1, \dots, a_n) \in (\theta_i)_u(P)\}$$

for $P \in PL^n$, $n > 0$ and

$$\theta_{(u,i)}(P) := \begin{cases} \{(u,i)\} & \text{if } (\theta_i)_u(P) = \{u\}, \\ \varnothing & \text{otherwise} \end{cases}$$

for $P \in PL^0$.

Obviously, there exists an isomorphism from $\mathbf{F}_k$ (respectively, M_k) onto a generated subframe in $\bigsqcup_{i\in I} \mathbf{F}_i$ (respectively, $\bigsqcup_{i\in I} M_i$), given by the pair (f_0, f_1) such that

$$f_0(u) := (u, k), \ f_{1u}(a) := (a, k).$$

In particular, the definition of $\theta_{(u,i)}$ yields

$$\bigsqcup_{i\in I} M_i, (u,k) \models P(f_{1u}(a_1), \ldots, f_{1u}(a_n)) \text{ iff } M_k, u \models P(a_1, \ldots, a_n).$$

Now we have an analogue of 1.3.37.

Proposition 3.3.35 *Let $(\mathbf{F}_i \mid i \in I)$ be a family consisting of all different components of a predicate Kripke frame $\mathbf{F}$. Then $\mathbf{F} \cong \bigsqcup_{i\in I} \mathbf{F}_i$.*

Proof A required isomorphism is the map f from the proof of 1.3.37 together with the family $(id_{D_u})_{u\in F}$ (as usual, we assume that $\mathbf{F} = (F, D)$). ∎

Proposition 3.3.36

(1) $\mathbf{ML}^{(=)}\left(\bigsqcup_{i\in I} \mathbf{F}_i\right) = \bigcap_{i\in I} \mathbf{ML}^{(=)}(\mathbf{F}_i)$, $\mathbf{MT}^{(=)}\left(\bigsqcup_{i\in I} M_i\right) = \bigcap_{i\in I} \mathbf{MT}^{(=)}(M_i)$ *for the modal case.*

(2) $\mathbf{IL}^{(=)}\left(\bigsqcup_{i\in I} \mathbf{F}_i\right) = \bigcap_{i\in I} \mathbf{IL}^{(=)}(\mathbf{F}_i)$, $\mathbf{IT}^{(=)}\left(\bigsqcup_{i\in I} M_i\right) = \bigcap_{i\in I} \mathbf{IT}^{(=)}(M_i)$ *for the intuitionistic case.*

Proof Similar to Lemma 1.3.28 using Lemma 3.3.21; a simple exercise for the reader. ∎

Corollary 3.3.37 *Kripke frame semantics $\mathcal{K}_N^{(=)}$, $\mathcal{K}_{int}^{(=)}$ have the Collection Property (cf. Definition 2.16.9).*

Proposition 3.3.38 *Every modally or intuitionistically definable class of predicate Kripke frames is closed under generated subframes, disjoint sums and $\mathrm{p}^{(=)}$-morphic images.*[13]

Proof Validity is preserved for generated subframes by 3.3.18, for $\mathrm{p}^{(=)}$-morphic images by 3.3.13, for disjoint sums by 3.3.36. ∎

Finally let us prove a predicate analogue of Lemma 1.3.44.

Lemma 3.3.39 *Let L be a conically expressive N-m.p.l.$(=)$. Then for any N-modal Kripke model M such that $M \models L$ for any $u \in M$, for any D_u-sentence A*

$$M, u \models \Box^* A \Leftrightarrow \forall v \in R^*(u)\, M, v \models A,$$

where R^* is the same as in Lemma 1.3.19.

[13] $\mathrm{p}^{=}$-morphic images — if the class is definable by a set of formulas with equality.

Proof Similar to 1.3.44 and the soundness of the substitution rule in 3.2.29. Let M_0 be a propositional model over the same propositional frame as M, such that for any $v \in M$

$$M_0, v \vDash p \Leftrightarrow M, v \vDash A.$$

Then by induction we obtain for any propositional formula $X(p)$ for any $v \in M$.

$$M_0, v \vDash X(p) \Leftrightarrow M, v \vDash X(A).$$

Hence for any $u \in M$ and C from 1.3.42

$$M_0, u \vDash C(p) \Leftrightarrow M, u \vDash C(A)\,(= \Box^* A)$$

and so by 1.3.42,

$$M, u \vDash \Box^* A \Leftrightarrow M_0 \uparrow u \vDash p.$$

But by Lemma 1.3.19 and the choice of M_0

$$M_0 \uparrow u \vDash p \Leftrightarrow \forall v \in R^*(u)\ M_0, v \vDash p \Leftrightarrow \forall v \in R^*(u)\ M, v \vDash A.$$

Hence the claim follows. ■

3.4 Constant domains

Definition 3.4.1 *A predicate Kripke frame* **F** has a constant domain *(or* **F** is a CD-frame *) iff $D_u = D_v$ for any $u, v \in$* **F**.

A CD-frame (F, D), in which $V = D_u$ for all $u \in F$, is denoted by $F \odot V$.

Proposition 3.4.2

(1) A rooted N-modal predicate Kripke frame **F** *has a constant domain iff $\forall i \in I_N$* **F** $\vDash Ba_i$.

(2) A rooted intuitionistic Kripke frame **F** *has a constant domain iff* **F** $\Vdash CD$.

Proof (1) (Only if.) Suppose M is a Kripke model over **F** and $M, u \models \Diamond_i \exists x P(x)$. Then $M, v \models P(a)$ for some $v \in R_i(u)$, $a \in D_v = D_u$. Hence $M, u \models \Diamond_i P(a)$ and thus $M, u \vDash \exists x\ \Diamond_i P(x)$.

(If.) Let u_0 be a root of **F**. If **F** does not have a constant domain, then $D_u \neq D_v$ for some $u, v \in$ **F**, and thus $D_u \neq D_{u_0}$ or $D_v \neq D_{u_0}$. Consider the first option. Then there exists a path $u_0 R_{i_1} u_1, \ldots, u_{k-1} R_{i_k} u_k = u$, so $D_{u_0} \subseteq D_{u_1} \subseteq \ldots \subseteq D_u$. It follows that for some k $D_{u_k} \subset D_{u_{k+1}}$.

Thus for some i, u, v we have $D_u \subset D_v$, $v \in R_i(u)$. Let $a_0 \in (D_v - D_u)$. Consider a model $M = (\mathbf{F}, \xi)$ such that (for a certain $P \in PL^1$)

$$\xi_w(P) := \begin{cases} \{a_0\} & \text{if } a_0 \in D_w, \\ \varnothing & \text{otherwise.} \end{cases}$$

Then we have $M, v \models \exists x P(x)$, and thus $M, u \models \Diamond_i \exists x P(x)$; but $M, u \not\models \exists x \Diamond P(x)$, since $a_0 \notin D_u$.

(2) Let R be the relation in $\mathbf{F}$.

(Only if.) If M is an intuitionistic Kripke model over $\mathbf{F}$ and $M, u \Vdash \forall x(P(x) \vee q)$, but $M, u \not\Vdash q$ and $M, u \not\Vdash \forall x P(x)$, then $M, v \not\Vdash P(a)$ for some $v \in R(u)$, $a \in D_v = D_u$. Hence $M, u \not\Vdash P(a) \vee q$, which contradicts $M, u \Vdash \forall x(P(x) \vee q)$ and uRu.

(If.) Let u be root of $\mathbf{F}$, and suppose $D_u \neq D_v$ for some $v \in W = R(u)$. Since uRv, we have $D_u \subset D_v$, so there exists $a_0 \in (D_v - D_u)$. Consider the valuation ξ in $\mathbf{F}$ such that for any w

$$\xi_w(P) := D_u,$$

$$\xi_w(q) := \begin{cases} \varnothing & \text{if } wRu, \\ \{w\} & \text{otherwise,} \end{cases}$$

for a certain $P \in PL^1$, $q \in PL^0$ and $\xi_w(Q) = \varnothing$ for all other predicate letters Q. It is clear that ξ is intuitionistic. Under this valuation we have $u \not\Vdash q$, $v \not\Vdash P(a_0)$, and thus $u \not\Vdash \forall x P(x) \vee q$; but $u \Vdash \forall x(P(x) \vee q)$.

In fact, suppose uRw. If wRu, then obviously $D_w = D_u$, and thus $w \Vdash P(a)$ for any $a \in D_w$; on the other hand, $w \not R u$ implies $w \Vdash q$.

Therefore $u \not\Vdash CD$. ■

From Definition 3.4.1 it is clear that the class of CD-frames is closed under generated subframes.

Definition 3.4.3 *Modal and intuitionistic* Kripke semantics with constant domains *are generated by CD-frames:*

$$\mathcal{CK}_N^{(=)} := \{\mathbf{ML}^{(=)}(\mathcal{X}) \mid \mathcal{X} \textit{ is a class of N-modal CD-frames}\};$$
$$\mathcal{CK}_{int}^{(=)} := \{\mathbf{IL}^{(=)}(\mathcal{X}) \mid \mathcal{X} \textit{ is a class of intuitionistic CD-frames}\}.$$

The class of CD-frames is not closed under disjoint sums (the reader can easily construct a counterexample), but there is an equivalent semantics generated by a class with this property.

Definition 3.4.4 *A predicate Kripke frame* $\mathbf{F} = ((W, R_1, \ldots, R_N), D)$ *is called a* local CD-frame *if it satisfies*

(*) $\forall i \forall u \forall v\ (uR_iv \Rightarrow D_u = D_v)$.

Lemma 3.4.5 *(1) For an N-modal frame $\mathbf{F}$, $\mathbf{F} \models \bigwedge_{i \in I_N} Ba_i$ iff every cone $\mathbf{F}{\uparrow}u$ has a constant domain iff $\mathbf{F}$ is a local CD-frame.*
(2) Similarly, for an intuitionistic $\mathbf{F}$, $\mathbf{F}$ is a local CD-frame iff $\mathbf{F} \Vdash CD$ iff all cones in $\mathbf{F}$ have constant domains.

Proof Easily follows from 3.4.2. ■

Proposition 3.4.6

(1) *The class of local CD-frames is closed under disjoint sums, generated subframes and p-morphic images.*

(2) *CD-frames and local CD-frames generate equivalent semantics.*

Proof
(1) By 3.4.5 and 3.3.38 this class is modally (or intuitionistically) definable, so we can apply 3.3.38.
(2) Note that every CD-frame is a local CD-frame; on the other hand, if $\mathcal{X}$ is a class of (N-modal) local CD-frames, then by Lemma 3.3.21,

$$\begin{aligned}\mathbf{ML}^{(=)}(\mathcal{X}) &= \bigcap\{\mathbf{ML}^{(=)}(\mathbf{F}) \mid \mathbf{F} \in \mathcal{X}\} = \bigcap\{\mathbf{ML}^{(=)}(\mathbf{F}{\uparrow}u) \mid \mathbf{F} \in \mathcal{X},\ u \in \mathbf{F}\} \\ &= \mathbf{ML}^{(=)}(\{\mathbf{F}{\uparrow}u \mid \mathbf{F} \in \mathcal{X}, u \in \mathbf{F}\}),\end{aligned}$$

and all the $\mathbf{F}{\uparrow}u$ are CD-frames. ■

Exercise 3.4.7 Show that the class of CD-frames is not closed under p-morphic images, but closed under p$^=$-morphic images.

Let us now describe morphisms of local CD-frames.

Lemma 3.4.8 *Every connected local CD-frame is a CD-frame.*

Proof In a local CD-frame, if u, v are in the same component, then $D_u = D_v$. This easily follows by induction on the length of a non-oriented path from u to v. ■

Proposition 3.4.9

(1) *If $(f_0, f_1) : (F, D) \twoheadrightarrow (F', D')$ for a connected CD-frame (F, D), then all the maps f_{1u} (and certainly their targets $D'_{f_0(u)}$) for $u \in F$ coincide and (F', D') is a connected CD-frame.*

(2) *Conversely, assume that (F, D), (F', D') are connected CD-frames, $f : F \twoheadrightarrow F'$ and $g : D_u \longrightarrow D'_u$ is a surjective map. Then $(f, f_1) : (F, D) \twoheadrightarrow (F', D')$ for $f_1 = (g)_{u \in F}$.*

Proof

(1) Again we show $f_{1u} = f_{1v}$ by induction on the length of a non-oriented path from u to v. It suffices to consider the case when uR_iv. Then $f_{1u} = f_{1v} \upharpoonright D_u$, so $f_{1u} = f_{1v}$ as functions, since $D_u = D_v$. The targets coincide, due to the surjectivity.

We know from Lemma 1.3.39 that F' is connected. By 3.4.6, (F', D') is local CD, so it is CD by 3.4.8.

(2) Trivial by definition.

∎

The p-morphism described in 3.4.9 is briefly denoted by $f \odot g$.

Corollary 3.4.10 *Let* $\mathbf{F}$ *be a local CD-frame,* $(\mathbf{F}_i \mid i \in I)$ *a family of all its different components. Then the p-morphisms* $\mathbf{F} \twoheadrightarrow \mathbf{G}$ *are exactly the maps of the form* $\bigcup_{i \in I} \mathbf{f}_i$*, where every* $\mathbf{f}_i : \mathbf{F}_i \twoheadrightarrow \mathbf{G}$ *(as a p-morphism onto its image) has the form described in 3.4.9.*

Proof Note that every $\mathbf{F}_i$ is a CD-frame by 3.4.8 and apply Proposition 3.3.23. ∎

Every Kripke frame over a propositional frame F is reducible to some Kripke frame $F \odot V$; for example, with a singleton V. More precisely, the following proposition holds (its intuitionstic version was proved in [Ono, 1972/73]).

Proposition 3.4.11 *Let* $\mathbf{F} = (F, D)$ *and* $\mathbf{F}' = F' \odot V$ *be predicate Kripke frames such that* F *is reducible to* F' *and* $|D_u| \geq |V|$ *for any* $u \in F$.

Then $\mathbf{F}$ *is reducible to* $\mathbf{F}'$ *and thus*[14] $\mathbf{ML}(\mathbf{F}) \subseteq \mathbf{ML}(\mathbf{F}')$ *(and* $\mathbf{IL}(\mathbf{F}) \subseteq \mathbf{IL}(\mathbf{F}')$ *for the intuitionistic case).*

Proof Assuming $h : F \uparrow v \twoheadrightarrow F' \uparrow u$, let us show

$$\mathbf{F} \uparrow v \twoheadrightarrow \mathbf{F}' \uparrow u.$$

To simplify notation, we assume that $\mathbf{F} = \mathbf{F}{\uparrow}v$ and $\mathbf{F}' = \mathbf{F}'{\uparrow}u$. Let a_0 be a fixed element of V, $g_v : D_v \longrightarrow V$ a surjective map. For every $u \in F$ consider the following surjective function $g_u : D_u \longrightarrow V$.

$$g_u(a) := \begin{cases} g_v(a) & \text{if } a \in D_v, \\ a_0 & \text{otherwise.} \end{cases}$$

It is clear that uR_iu' implies $g_u = g_{u'} \upharpoonright D_u$, so we obtain $(h, g) : \mathbf{F} \twoheadrightarrow \mathbf{F}'$, and thus $\mathbf{ML}(\mathbf{F}) \subseteq \mathbf{ML}(\mathbf{F}')$ by Proposition 3.3.13. ∎

Remark 3.4.12 Proposition 3.4.11 is not transferred to the case with equality. In fact, 3.4.11 implies that

$$\mathbf{IL}(F \odot V) \subseteq \mathbf{IL}(F \odot V') \text{ if } |V| \geq |V'|.$$

But a similar assertion does not hold for logics with equality (even for *infinite* V'). In fact, consider the following formula [Skvortsov, 1989]:

$$A_0 := \exists x \exists y\, (x \neq y \wedge (P(x) \equiv P(y))).$$

[14] By Proposition 3.3.26.

Lemma 3.4.13

$$F \odot V \Vdash A_0 \textit{ iff } \forall u \in F\ |HA(F{\uparrow}u)| \not\geq |V|.$$[15]

Proof Let M be an intuitionistic model over $F \odot V$ and let $u \in F$. Consider the sets

$$\Xi_a := \{v \in F{\uparrow}u \mid M, v \Vdash P(a)\}$$

for $a \in V$. It is clear that $\Xi_a \in HA(F{\uparrow}u)$.

(If.) If $|HA(F{\uparrow}u)| \not\geq |V|$ for every $u \in F$, then $\Xi_a = \Xi_b$ for some $a \neq b$ (from V) — otherwise the map $a \mapsto \Xi_a$ embeds V is $HA(F \uparrow u)$. Then[16] $M, u \Vdash a \neq b \wedge (P(a) \equiv P(b))$, and thus $M, u \Vdash A_0$.

(Only if.) Suppose $|V| \leq |HA(F{\uparrow}u)|$ for some $u \in F$. Let $h : V \longrightarrow HA(F{\uparrow}u)$ be an injection, and consider a model M over $\mathbf{F}$ such that for any w

$$M, w \Vdash P(a) \text{ iff } w \in h(a).$$

Then $\Xi_a = h(a)$, and thus $\Xi_a \neq \Xi_b$, whenever $a \neq b$. Hence $M, u \nVdash P(a) \equiv P(b)$ whenever $a, b \in V$, $a \neq b$, and thus $M, u \nVdash A_0$. ∎

Therefore, if a p.o. set F contains an infinite cone $F{\uparrow}u$ and V' is countable, but $|V| > 2^{|F|}$, we have $A_0 \in (\mathbf{IL}^=(F \odot V) - \mathbf{IL}^=(F \odot V'))$.

3.5 Kripke frames with equality

3.5.1 Introduction

Let us first make some informal comments about interpreting equality. In Kripke frame semantics this interpretation is the simplest (Definition 3.2.9):

$$u \vDash a = b \ \text{ iff } \ a \text{ equals } b.$$

But does this properly correspond to our intuitive understanding of equality?

In fact, we should evaluate $a = b$ *with respect to a certain world*, so it might happen that a and b are the same in a world u, but different in another world. Examples of this kind are quite popular in literature.

But as mentioned in the Introduction to Part II, there are more formal reasons for modifying the notion of predicate Kripke frame. For example, Lemma 3.10.5 (see below) shows that the principle of decidable equality DE is valid in any intuitionistic Kripke frame. But from the intuitionistic point of view, equality may be undecidable. In particular, this happens in intuitionistic analysis — for real numbers or functions. So we should choose the interpretations of equality appropriately.

In classical model theory there exist two ways of dealing with equality. The standard way is to interpret equality as coincidence of individuals ('normal

[15] Of course, we can replace $\not\geq$ with $<$ if we accept Axiom of Choice.

[16] Recall that $M, u \Vdash a \neq b$ iff a, b are different.

models'). The second way is to interpret equality as an equivalence relation preserving values of basic predicates. It is well-known that in classical logic these two approaches are equivalent, because we can always take the quotient domain modulo the equivalence relation corresponding to equality, and obtain a logically equivalent 'normal' interpretation.

Kripke semantics for intuitionistic and modal logics corresponds to classical 'normal models', but the second approach also makes sense. So we can interpret equality in a predicate Kripke frame as an equivalence relation on every individual domain D_u. Thus we obtain the notion of a *predicate Kripke frame with equality* (KFE) and the corresponding semantics $\mathcal{KE}$. This semantics is stronger than the semantics of Kripke frames $\mathcal{K}$ for formulas with equality — the crucial formula DE can be refuted in a KFE. Moreover, as we shall see, $\mathcal{KE}$ is stronger than $\mathcal{K}$ for logics without equality.

We will also describe an equivalent semantics of 'Kripke sheaves'. Kripke sheaves are obtained from KFEs by taking quotients of individual domains through the corresponding equivalence relation.

3.5.2 Kripke frames with equality

Definition 3.5.1 *A* predicate Kripke frame with equality *(KFE) is a triple $(F, D, \asymp)$, in which (F, D) is a predicate Kripke frame and $\asymp$ is a valuation for the binary predicate symbol '$=$' satisfying the standard equality axioms. In more detail, if $F = (W, R_1, \ldots, R_N)$, then $\asymp = (\asymp_u)_{u \in W}$ is a family of equivalence relations ($\asymp_u \subseteq D_u \times D_u$), which is R_i-stable for every $i \in I_N$:*

$$uR_iv \ \&\ a \asymp_u b \Rightarrow a \asymp_v b.$$

The following lemma gives an alternative definition of KFEs presenting them as a kind of Ω-sets (see Chapter 4).

Lemma 3.5.2 *(1) Let $(F, D, \asymp)$ be a KFE, $F = (W, R_1, \ldots, R_N)$. Consider the function $E : D^+ \times D^+ \longrightarrow 2^W$, such that for any $a, b \in D^+$*

$$E(a, b) = \{u \mid a \asymp_u b\}.$$

Then the following holds:

(E1) $E(a, b) = E(b, a)$*;*

(E2) $E(a, b) \cap E(b, c) \subseteq E(a, c)$*;*

(E3) $\bigcup_{a \in D^+} E(a, a) = W$*;*

(E4) $R_i(E(a, b)) \subseteq E(a, b)$ *for* $i \in I_N$*.*

(2) Given a propositional Kripke frame F, a non-empty set D^+, and a function E satisfying (E1)–(E4), we can uniquely restore the corresponding KFE as $(F, D, \asymp)$, with

$$D_u := \{a \in D^+ \mid u \in E(a, a)\},\ \asymp_u := \{(a, b) \mid u \in E(a, b)\}.$$

Proof

(1) (E1), (E2) follow respectively from the symmetry and the transitivity of $\asymp_u$. The reflexivity of $\asymp_u$ implies (E3) — it follows that $u \in E(a, a)$ for $a \in D_u$, and such an a exists since $D_u \neq \varnothing$. (E4) follows from the R_i-stability of $\asymp$.

(2) Note that

$$E(a, b) = E(a, b) \cap E(b, a) \subseteq E(a, a)$$

by (E1), (E2) and similarly, $E(a, b) \subseteq E(b, b)$. Thus $u \in E(a, b)$ implies $u \in E(a, a) \cap E(b, b)$, i.e. $\asymp_u \subseteq D_u \times D_u$.

(E1) implies the symmetry of $\asymp_u$, and (E2) implies the transitivity.

(E3) implies the non-emptiness of every D_u. The reflexivity of $\asymp_u$ is obvious.

The R_i-stability of $\asymp$ follows from (E4).

■

$E(a, b)$ is called the *measure of identity* of a and b. $E(a, a)$ is called the *measure of existence* (or the *extent*) of a, and is also denoted by $E(a)$.

Definition 3.5.3 *A* KFE-model *over a Kripke frame with equality* $\mathbf{F} = (F, D, \asymp)$ *is a pair* $M = (\mathbf{F}, \xi)$, *where* ξ *is a valuation in* (F, D) *respecting the relations* $\asymp_u$, *i.e. such that for every* $P \in PL^n$, $n \geq 1$ *and* $a_1, \ldots, a_n, b_1, \ldots, b_n \in D^+$

$$(a_1, \ldots, a_n) \in \xi_u(P) \ \&\ a_1 \asymp_u b_1 \ \&\ \ldots\ \&\ a_n \asymp_u b_n \Rightarrow (b_1, \ldots, b_n) \in \xi_u(P).$$

Then we define forcing $M, u \vDash A$ *(for* $u \in F$ *and a* D_u*-sentence* A*) by the same conditions as in Definition 3.2.9, with the following difference:*

- $M, u \vDash a = b$ *iff* $a \asymp_u b$.

A KFE-model M *is called* intuitionistic *if the corresponding model without equality* (F, ξ) *is intuitionistic. In this case we define forcing* $M, u \Vdash A$ *according to Definition 3.2.12 with the only difference:*

- $M, u \Vdash a = b$ *iff* $a \asymp_u b$.

A modal predicate formula A *is called* true *in* M *if* $\overline{\forall} A$ *is true at every world of* M*; similarly for the intuitionistic case.*

Definition 3.5.4 *A modal (respectively, intuitionistic) predicate formula* A *is called* valid *in a KFE (respectively,* **S4***-based KFE)* $\mathbf{F}$ *iff it is true in all KFE-models (respectively, intuitionistic KFE-models) over* $\mathbf{F}$.

Again we use the notation $M \vDash A$, $\mathbf{F} \vDash A$ for the modal case; $M \Vdash A$, $\mathbf{F} \Vdash A$ for the intuitionistic case.

Lemma 3.2.11 obviously transfers to KFE-models. Lemma 3.2.13 has the following analogue.

Lemma 3.5.5 • $M, u \Vdash \neg B$ *iff* $\forall v \in R(u)\ M, v \nVdash B$;

• $M, u \Vdash a \neq b$ *iff* $\forall v \in R(u)\ a \not\asymp_v b$.

We also have an analogue of 3.2.16:

Lemma 3.5.6 *For an intuitionistic model M and D_u-sentence A*

$$M, u \Vdash A \ \&\ uRv \ \Rightarrow\ M, v \Vdash A$$

Next we obtain analogues of 3.2.17, 3.2.18, 3.2.20 by the same arguments.

Lemma 3.5.7

(1) Let M be an N-modal KFE-model, $A(\mathbf{x})$ an N-modal formula with $FV(A) = r(\mathbf{x})$, $|\mathbf{x}| = n$. Then

(i) for any $u \in M$

$$M, u \vDash \forall \mathbf{x} A(\mathbf{x}) \text{ iff } \forall \mathbf{a} \in D_u^n\ M, u \vDash A(\mathbf{a}),$$

(ii) $M \vDash A(\mathbf{x})$ iff $\forall u \in M\ \forall \mathbf{a} \in D_u^n\ M, u \vDash A(\mathbf{a})$.

(2) Let M be an intuitionistic KFE-model with the accessibility relation R, $A(\mathbf{x})$ an intuitionistic formula with $FV(A) = r(\mathbf{x})$, $|\mathbf{x}| = n$. Then

(i) for any $u \in M$

$$M, u \Vdash \forall \mathbf{x} A(\mathbf{x}) \text{ iff } \forall v \in R(u)\ \forall \mathbf{a} \in D_v^n\ M, v \Vdash A(\mathbf{a}),$$

(ii) $M \Vdash A(\mathbf{x})$ iff $\forall u \in M\ \forall \mathbf{a} \in D_u^n\ M, u \Vdash A(\mathbf{a})$.

Definition 3.5.8 *Let* **F** *be an* **S4***-based KFE, M a KFE-model over* **F***. The* pattern *of M is the intuitionistic KFE-model M_0 over* **F** *such that for any $u \in$* **F** *and any atomic D_u-sentence A without equality*

$$M_0, u \Vdash A \text{ iff } M, u \vDash \Box A.$$

Let us check soundness of this definition:

Lemma 3.5.9 *The pattern exists for every* **S4***-based KFE-model.*

Proof According to 3.5.3, we have to show that for any $P \in PL^n$, $\mathbf{a}, \mathbf{b} \in D_u^n$ such that $\forall i\ a_i \asymp_u b_i$,

$$M, u \vDash \Box P(\mathbf{a}) \text{ iff } M, u \vDash \Box P(\mathbf{b}). \qquad (*)$$

In fact, by 3.5.1 for any $v \in R(u)$, $a_i \asymp_v b_i$; hence

$$M, v \vDash P(\mathbf{a}) \text{ iff } M, v \vDash P(\mathbf{b})$$

by 3.5.3. Then

$$\forall v \in R(u)\ M, v \vDash P(\mathbf{a}) \text{ iff } \forall v \in R(u)\ M, v \vDash P(\mathbf{b}),$$

which implies (*).

Thus M_0 always exists; it is intuitionistic by definition. ∎

Now there is an analogue of 3.2.15.

Lemma 3.5.10 *Let M_0 be the pattern of M; then for any $u \in M$ and for any intuitionistic D_u-sentence A*

$$M_0, u \Vdash A \text{ iff } M, u \vDash A^T,$$

and for any intuitionistic sentence A

$$M_0 \Vdash A \text{ iff } M \vDash A^T$$

Proof The same as for 3.2.15, with a difference in the case when A is atomic of the form $a = b$. Then we have

$$M_0, u \Vdash a = b \Leftrightarrow a \asymp_u b,$$
$$M, u \vDash A^T (= \Box(a = b)) \Leftrightarrow \forall v \in R(u)\ a \asymp_v b,$$

which is equivalent to $a \asymp_u b$ by Definition 3.5.1. ∎

Proposition 3.5.11 *Let **F** be an **S4**-based KFE, $A \in IF^=$. Then*

$$\mathbf{F} \Vdash A \text{ iff } \mathbf{F} \vDash A^T.$$

Proof The same as for 3.2.25 based on 3.5.7 and 3.5.10.

∎

Corollary 3.5.12 *For a class $\mathcal{C}$ of **S4**-based KFEs*

$$\mathbf{IL}^{(=)}(\mathcal{C}) = {}^T\mathbf{ML}^{(=)}(\mathcal{C}).$$

Proof Follows from 3.5.11 as in the proof of 2.16.18. ∎

The set of formulas valid in a KFE **F** is denoted by $\mathbf{ML}^{(=)}(\mathbf{F})$ (or $\mathbf{IL}^{(=)}(\mathbf{F})$). This notation is not quite legal before we show soundness; the proof of soundness is postponed until Section 3.6.

Lemma 3.5.13 *For any KFE $(F, D, \asymp)$, $\mathbf{ML}(F, D) \subseteq \mathbf{ML}(F, D, \asymp)$.*

Proof For formulas without equality the definitions of forcing in $(F, D, \asymp)$ and (F, D) are the same. Every valuation in $(F, D, \asymp)$ is a valuation in (F, D), so refutability of a formula in $(F, D, \asymp)$ implies its refutability in (F, D). ∎

Remark 3.5.14 Not all valuations in (F, D) are admissible in $(F, D, \asymp)$, so it may happen that $\mathbf{ML}(F, D, \asymp) \neq \mathbf{ML}(F, D)$. A trivial counterexample is $(F, D, \asymp)$ where F is a reflexive singleton, D is two-element and $\asymp$ is universal. Then obviously $(F, D) \nvDash \exists x P(x) \supset \forall x P(x)$, while $(F, D, \asymp) \vDash \exists x P(x) \supset \forall x P(x)$.

On the other hand, predicate Kripke frames can be regarded as a particular kind of KFEs.

Lemma 3.5.15

(1) *Every Kripke frame* $\mathbf{F} = (F, D)$ *is associated with a* simple *KFE* $\mathbf{F}_= = (F, D, \asymp)$, *in which* $\asymp_u = id_{D_u}$ *for any* $u \in F$. *Then* $\mathbf{ML}^{(=)}(\mathbf{F}) = \mathbf{ML}^{(=)}(\mathbf{F}_=)$, *and* $\mathbf{IL}^{(=)}(\mathbf{F}) = \mathbf{IL}^{(=)}(\mathbf{F}_=)$ *in the intutionistic case.*

(2) *A KFE* $(F, D, \asymp)$ *is simple iff*

$$\forall a, b \in D^+(a \neq b \Rightarrow E(a,b) = \varnothing),$$

where E *is the measure of identity (3.5.2).*

Proof

(1) Valuations and the corresponding forcing relations in $\mathbf{F}$ and $\mathbf{F}_=$ are just the same.

(2) In fact, in a simple KFE, $u \in E(a,b)$ implies $a = b$. The other way round, if $E(a,b) = \varnothing$ whenever $a \neq b$, then $a \asymp_u b$ holds only for $a = b$.

∎

Proposition 3.5.16 *Let* $\mathcal{C}$ *be a class of* N*-modal propositional frames. Then* $\mathbf{ML}(\mathcal{KEC}) = \mathbf{ML}(\mathcal{KC})$. *Similarly, for a class* $\mathcal{C}$ *of intuitionistic propositional frames,* $\mathbf{IL}(\mathcal{KEC}) = \mathbf{IL}(\mathcal{KC})$.

Proof Consider the modal case. If $A \in \mathbf{ML}(\mathcal{KEC})$, then A is valid in every KFE over $\mathcal{C}$. In particular, for any $\mathbf{F} \in \mathcal{KC}$, A is valid in the associated simple frame $\mathbf{F}_=$. So by Lemma 3.5.15, $A \in \bigcap\{\mathbf{ML}(\mathbf{F}) \mid \mathbf{F} \in \mathcal{KC}\} = \mathbf{ML}(\mathcal{KC})$.

The other way round, suppose $A \in \mathbf{ML}(\mathcal{KC})$. Then for any $(F, D, \asymp)$ with $F \in \mathcal{C}$, we have $A \in \mathbf{ML}(F, D) \subseteq \mathbf{ML}(F, D, \asymp)$ by Lemma 3.5.13. Hence $A \in \mathbf{ML}(\mathcal{KEC})$. ∎

3.5.3 Strong morphisms

In this section we consider strong morphisms defined in an obvious way — as morphisms of Kripke frames preserving equality. A larger class of morphisms will be considered in Section 3.7.

Definition 3.5.17 *Let* $\mathbf{F} = (F, D, \asymp)$, $\mathbf{F}' = (F', D', \asymp')$ *be Kripke frames with equality. A* strong (p-)morphism *from* $\mathbf{F}$ *to* $\mathbf{F}'$ *is a (p-)morphism of frames without equality* $\mathbf{f} = (f_0, f_1) : (F, D) \longrightarrow (F', D')$ *(Definition 3.3.1) such that for any* $u \in F$, $a, b \in D^+$

$$a \asymp_u b \text{ iff } f_{1u}(a) \asymp'_{f_0(u)} f_{1u}(b).$$

A strong isomorphism *is a strong morphism which is an isomorphism of frames without equality.*[17] Strong morphisms of KFE-models *are defined as morphisms of their frames satisfying the reliability condition from Definition 3.3.2; similarly for strong p-morphisms and strong isomorphisms.*

---

[17]Definition 3.3.2.

Obviously, if the KFEs $\mathbf{F} = (F, D, \asymp)$, $\mathbf{F}' = (F', D', \asymp)$ are simple, then a strong KFE-morphism from $\mathbf{F}$ to $\mathbf{F}'$ is nothing but an $=$-morphism from (F, D) to (F', D').

Strong morphisms of KFE-frames and models are denoted by $\longrightarrow^=$, strong p-morphisms by $\twoheadrightarrow^=$, strong isomorphisms by $\cong$.

Lemma 3.3.4 easily transfers to KFEs:

Lemma 3.5.18

(1) For a KFE $\mathbf{F} = (F, D, \asymp)$ the identity morphism $id_F := id_{(F,D)}$ is a strong isomorphism.

(2) The composition of strong morphisms (in the sense of 3.3.4) is a strong morphism; similarly for p-morphisms and isomorphisms.

This yields the categories of N-modal KFEs and strong morphisms, intuitionistic KFEs and strong morphisms, and similarly, of KFE-models.

Lemma 3.5.19 *Strong isomorphisms are exactly isomorphisms in the category of N-modal KFEs and strong morphisms.*

Proof An exercise. ∎

Now we have an analogue of Proposition 3.3.11

Lemma 3.5.20 *If $(f_0, f_1) : M \longrightarrow^= M'$, for KFE-models M, M', then for any $u \in F$ and for any D_u-sentence B*

$$M, u \vDash B \text{ iff } M', f_0(u) \vDash f_{1u} \cdot B,$$

where $f_{1u} \cdot B$ is obtained from B by replacing occurrences of every $c \in D_u$ with $f_{1u}(c)$.

If the models are intuitionistic, the same holds for any intuitionistic D_u-sentence (and the intuitionistic forcing).

Lemma 3.5.21 *Let $(f_0, f_1) : \mathbf{F} \longrightarrow^= \mathbf{F}'$ be a KFE-morphism, and let M' be a KFE-model over $\mathbf{F}'$. Then there exists a unique model M over $\mathbf{F}$ such that $(f_0, f_1) : M \longrightarrow^= M'$, and similarly for $\twoheadrightarrow^=$.*

Proof As in the proof of 3.3.12, we put

$$\xi_u(P) := \{\mathbf{b} \mid M', f_0(u) \vDash P(f_{1u} \cdot \mathbf{b})\}$$

for any $u \in \mathbf{F}$, $P \in PL^m$. Then $M = (\mathbf{F}, \xi)$ is a KFE-model. In fact, we have to check that

$$\mathbf{a} \in \xi_u(P) \ \&\ \forall i\, a_i \asymp_u b_i \Rightarrow \mathbf{b} \in \xi_u(P),$$

i.e.

$$M', f_0(u) \vDash P(f_{1u} \cdot \mathbf{a}) \ \&\ \mathbf{a} \asymp_u \mathbf{b} \Rightarrow M', f_0(u) \vDash P(f_{1u} \cdot \mathbf{b}), \qquad (1)$$

where

$$\mathbf{a} \asymp_u \mathbf{b} := \forall i\, a_i \asymp_u b_i.$$

But (1) holds, since

$$\mathbf{a} \asymp_u \mathbf{b} \Rightarrow (f_{1u} \cdot \mathbf{a}) \asymp_{f_0(u)} (f_{1u} \cdot \mathbf{b})$$

by the definition of a strong KFE-morphism, and

$$M', f_0(u) \vDash P(f_{1u} \cdot \mathbf{a}) \ \&\ (f_{1u} \cdot \mathbf{a}) \asymp_{f_0(u)} (f_{1u} \cdot \mathbf{b}) \Rightarrow M', f_0(u) \vDash P(f_{1u} \cdot \mathbf{b})$$

by the definition of a KFE-model. The claim $(f_0, f_1) : M \longrightarrow M'$ and the uniqueness of M' follow easily. ■

Hence we have an analogue of 3.3.13:

Proposition 3.5.22 *If* $\mathbf{F} \twoheadrightarrow^{=} \mathbf{F}'$ *for KFEs* $\mathbf{F}$, $\mathbf{F}'$, *then* $\mathbf{ML}^{=}(\mathbf{F}) \subseteq \mathbf{ML}^{=}(\mathbf{F}')$, *and similarly,* $\mathbf{IL}^{=}(\mathbf{F}) \subseteq \mathbf{IL}^{=}(\mathbf{F}')$ *in the intuitionistic case.*

Proof Along the same lines as 3.3.13, now using 3.5.21, 3.5.7, 3.5.20. ■

3.5.4 Main constructions

Now let us extend the definition of subframes and submodels 3.3.15 to the case with equality.

Definition 3.5.23 *Let* $F = (W, R_1, \ldots, R_N)$ *be a propositional Kripke frame,* $\mathbf{F} = (F, D, \asymp)$ *a KFE,* $V \subseteq W$. *A* subframe *of* $\mathbf{F}$ *obtained by restriction to* V *is*

$$\mathbf{F} \restriction V := (F \restriction V, D \restriction V, \asymp\restriction V),$$

where $D \restriction V$ *is the same as in 3.3.15,* $\asymp\restriction V := (\asymp_u)_{u \in V}$.

If $M = (\mathbf{F}, \xi)$ *is a KFE-model, we define the* submodel

$$M \restriction V := (\mathbf{F} \restriction V, \xi \restriction V),$$

where $\xi \restriction V$ *is the same as in 3.3.15. If* V *is stable,* $\mathbf{F} \restriction V$, $M \restriction V$ *are called* generated.

It is obvious that $M \restriction V$ is a KFE-model, since $\xi \restriction V$ coincides with ξ on V.

The definitions of *reliability, rooted frames (models)* and *cones* are trivially extended to frames and models with equality.

Lemma 3.5.24 (Generation lemma) *Let* $\mathbf{F}$ *be a KFE,* M *a KFE-model over* $\mathbf{F}$, V *a stable set of worlds in* $\mathbf{F}$. *Then*

(1) $M \restriction V$ *is a reliable submodel of* M*;*

(2) $\mathbf{ML}^{(=)}(\mathbf{F}) \subseteq \mathbf{ML}^{(=)}(\mathbf{F} \restriction V)$*; similarly, for the intuitionistic case.*

Proof Almost the same as for 3.3.18. We use the same map (j, i) and apply Lemma 3.5.20. To prove (2), we need a valuation ξ in $\mathbf{F}$ such that $\xi_u = \xi'_u$ for $u \in V$, viz.

$$\xi_u(P) := \begin{cases} \xi'_u(P) & \text{if } u \in V, \\ \varnothing & \text{otherwise.} \end{cases}$$

This is really a valuation; ξ respects $\asymp_u$ for $u \in V$, since it coincides with the valuation ξ'; and for $u \notin V$ there is nothing to prove. ■

We define *cones* exactly as in 3.3.17:

Definition 3.5.25

$$\mathbf{F}{\uparrow}u := \mathbf{F} \restriction (W{\uparrow}u), \;\; M{\uparrow}u := M \restriction (W{\uparrow}u).$$

Then we obtain an analogue of 3.3.21:

Lemma 3.5.26

(1) $\mathbf{ML}^{(=)}(\mathbf{F}) = \bigcap\limits_{u \in F} \mathbf{ML}^{(=)}(\mathbf{F}{\uparrow}u)$

and similarly for the intuitionistic case.

(2) For a class $\mathcal{C}$ of N-modal or intuitionistic KFEs,

$$\mathbf{L}^{(=)}(\mathcal{C}) = \mathbf{L}^{(=)}(\mathcal{C}\uparrow),$$

where $\mathbf{L}$ *is* $\mathbf{ML}$ *or* $\mathbf{IL}$*,* $\mathcal{C}\uparrow$ *is the class of all cones of frames from* $\mathcal{C}$*.*

Lemma 3.5.27 *Let F be a propositional Kripke, $u \in F$, then*

$$\mathcal{KE}(F \uparrow u) = (\mathcal{KE}F) \uparrow u,$$

where $(\mathcal{KE}F) \uparrow u := \{\mathbf{F} \uparrow u \mid \mathbf{F} \in \mathcal{KE}F\}$.

Proof Similar to 3.3.27. Given a KFE $\mathbf{G} = (F \uparrow u, D, \asymp)$, we have to show that $\mathbf{G}$ is $\mathbf{F} \uparrow u$ for some KFE $\mathbf{F}$ over F. It suffices to extend the domain function D to the whole F; $\asymp$ is extended in the trivial way (as identity at every world). We do this by putting

$$D'_w := \begin{cases} D_u \text{ if } w \notin F{\uparrow}u, \\ D_w \text{ if } w \in F{\uparrow}u. \end{cases}$$

Then D' is obviously expanding, since $F{\uparrow}u$ is a generated subframe and $D_u \subseteq D'_w$ for any $w \in F$. ■

This implies an analogue of 3.3.28:

Proposition 3.5.28 *Let C be a class of N-modal or intuitionistic propositional Kripke frames. Then*

(1) $\mathcal{KE}(\mathcal{C}\uparrow) = (\mathcal{KEC})\uparrow$

(2) $\mathbf{L}^{(=)}(\mathcal{KEC}) = \mathbf{L}^{(=)}(\mathcal{KE}(\mathcal{C}\uparrow))$, *where* $\mathbf{L}$ *is* **ML** *or* **IL** *respectively.*

Proof Similar to 3.3.28; apply 3.5.27 and 3.5.26(2). ∎

Let us also define disjoint sums:

Definition 3.5.29 *Let* $\mathbf{F}_i = (F_i, D_i, \asymp_i)$ *be predicate Kripke frames with equality. Then*

$$\bigsqcup_{i\in I} \mathbf{F}_i := \left(\bigsqcup_{i\in I} F_i, D, \asymp\right),$$

where

$$D(u,i) := D_i(u) \times \{i\},\ \ (a,i) \asymp_{u,i)} (b,i) := a \asymp_i b$$

for $i \in I$, $u \in F_i$, $a, b \in D_u$.

Since $\mathbf{F}_i$ is isomorphic to a generated subframe of $\bigsqcup_{i\in I} \mathbf{F}_i$, we obtain an analogue of 3.3.36:

Lemma 3.5.30 $\mathbf{ML}^{(=)}\left(\bigsqcup_{i\in I} \mathbf{F}_i\right) = \bigcap_{i\in I} \mathbf{ML}^{(=)}(\mathbf{F}_i)$, *and similarly* $\mathbf{IL}^{(=)}\left(\bigsqcup_{i\in I} \mathbf{F}_i\right) = \bigcap_{i\in I} \mathbf{IL}^{(=)}(\mathbf{F}_i)$ *for the intuitionistic case.*

3.6 Kripke sheaves

Now let us consider semantical equivalents of KFEs called 'Kripke sheaves'. To define them, let us begin with the **S4**-case. In this case a Kripke sheaf is a set-valued functor defined on a certain category.

Recall that a *category* consists of *objects* ('points') and *morphisms* ('arrows') between some of these points. In precise terms, it can be defined as a tuple.

$$\mathcal{C} = (\mathbf{X}, \mathbf{Y}, \mathbf{o}, \mathbf{t}, \mathbf{I}, \circ),$$

where $\mathbf{X}$, $\mathbf{Y}$ are non-empty classes, $\mathbf{o}$, $\mathbf{t}$, $\mathbf{I}$ are functions:

$$\mathbf{o}, \mathbf{t} : \mathbf{Y} \longrightarrow \mathbf{X};$$

$$\mathbf{I} : \mathbf{X} \longrightarrow \mathbf{Y};$$

$\circ$ is a partial function $\mathbf{Y} \times \mathbf{Y} \longrightarrow \mathbf{Y}$.

$\mathbf{X}$ is called the *class of objects* of $\mathcal{C}$ and is also denoted by $Ob\mathcal{C}$. $\mathbf{Y}$ is called the *class of morphisms* of $\mathcal{C}$ and is also denoted by $Mor\mathcal{C}$. For a morphism f, the object $\mathbf{o}(f)$ is called the *origin* of f, and $\mathbf{t}(f)$ the *target* of f.

The notation $f : a \longrightarrow b$ or $a \xrightarrow{f} b$ means that $\mathbf{o}(f) = a$ and $\mathbf{t}(f) = b$; this is read as 'f is a morphism *from* a *to* b' (or *between* a *and* b).

$\mathcal{C}(a, b)$ denotes the class of all morphisms in $\mathcal{C}$ between a and b. $\mathbf{I}(a)$ is called the *identity morphism* of a and is also denoted by 1_a.

There are also the following conditions ('axioms'):

(1) $1_a : a \longrightarrow a$;

(2) $\alpha \circ \beta$ is defined iff $o(\beta) = t(\alpha)$ (i.e. iff arrows α, β are consecutive);

(3) if $x \xrightarrow{\alpha} y \xrightarrow{\beta} z$, then $x \xrightarrow{\alpha\circ\beta} z$;

(4) if $x \xrightarrow{\alpha} y \xrightarrow{\beta} z \xrightarrow{\gamma} u$, then $(\alpha \circ \beta) \circ \gamma = \alpha \circ (\beta \circ \gamma)$;

(5) if $x \xrightarrow{\alpha} y$, then $1_x \circ \alpha = \alpha \circ 1_y = \alpha$.

A standard example of a category is SET, the category of sets, in which objects are arbitrary sets, morphisms are maps.

There is a well-known canonical way of associating a category $\mathcal{C} =$ Cat$F$ with an **S4**-frame $F = (W, R)$ [Goldblatt, 1984]. Viz., we put

$$Ob\,\mathcal{C} := W,\ \ Mor\,\mathcal{C} := R,\ \ \mathcal{C}(u,v) := \{(u,v)\},\ \ 1_u := (u,u).$$

So arrows of $\mathcal{C}$ just represent the relation R. The composition is defined according to the 'triangle rule':

$$(u,v) \circ (v,w) := (u,w).$$

Definition 3.6.1 *An (***S4***-based, or intuitionistic) Kripke sheaf over an* **S4***-frame F is a SET-valued co-functor defined on Cat$F$. This means that a Kripke sheaf is a triple $\Phi = (F, D, \rho)$ where (F, D) is a system of domains, $\rho = (\rho_{uv} \mid uRv)$ is a family of functions $\rho_{uv} : D_u \longrightarrow D_v$ (transition maps) satisfying the following functoriality conditions:*

(1) for every $u \in F$, $\rho_{uu} = id_{D_u}$ (the identity function on D_u);

(2) $uRvRw$ (in F) implies $\rho_{uv} \circ \rho_{vw} = \rho_{uw}$.

We call $\rho_{uv}(a)$ the inheritor *of the individual a (from D_u) in the world v. The domains D_u are also called the* fibres *or the* stalks *of Φ.*

To extend this definition to arbitrary frames, recall that an arbitrary propositional Kripke frame $F = (W, R_1, \ldots, R_N)$ is associated with an **S4**-frame $F^* := (W, R^*)$, where R^* is the reflexive transitive closure of $R_1 \cup \ldots \cup R_N$.

Definition 3.6.2 *A* Kripke sheaf *over a propositional Kripke frame F is a triple $\Phi = (F, D, \rho)$ such that $\Phi^* := (F^*, D, \rho)$ is an* **S4***-based Kripke sheaf over F^*.*

The frame F is called the *(propositional) base* of $\Phi = (F, D, \rho)$ and denoted by Φ_π. (F, D, ρ) is called *N-modal* if F is N-modal.

Note that in Definition 3.6.20 ρ is a system of functions parameterised by R^*, i.e. $\rho = (\rho_{uv} \mid uR^*v)$, satisfying the conditions 3.7.1 (1), (2) for R^*.

Lemma 3.6.3 *Let $F = (W, R_1, \ldots, R_N)$ be a propositional Kripke frame, (F, D) a predicate Kripke frame, $\rho = (\rho_{iuv} \mid uR_iv,\ 1 \leq i \leq N)$ a family of functions $\rho_{iuv} : D_u \longrightarrow D_v$ satisfying the following 'coherence' conditions:*

(1) *if* $uR_{i_0}u_1R_{i_1}u_2\dots u_kR_{i_k}u$ *for some* $k \geq 0$,
then $\rho_{i_0uu_1}\circ\rho_{i_1u_1u_2}\circ\dots\circ\rho_{i_ku_ku} = id_{D_u}$;

(2) *if* $uR_{i_0}u_1R_{i_1}u_2\dots u_kR_{i_k}v$ *and* $uR_{j_0}v_1R_{j_1}v_2\dots v_mR_{j_m}v$, *then*
$\rho_{i_0uu_1}\circ\rho_{i_1u_1u_2}\circ\dots\circ\rho_{i_ku_kv} = \rho_{i_0uv_1}\circ\rho_{j_1v_1v_2}\circ\dots\circ\rho_{j_mv_mv}$.

Then there exists a unique Kripke sheaf (F, D, ρ^*) *such that* $\rho_{iuv} = \rho^*_{uv}$ *whenever* uR_iv.[18]

Proof For any pair $(u, v) \in R^*$ we can define a function $\rho^*_{uv} : D_u \longrightarrow D_v$ such that $\rho^*_{uu} := id_{D_u}$ and $\rho^*_{uv} := \rho_{i_0uu_1}\circ\rho_{i_1u_1u_2}\circ\dots\circ\rho_{i_ku_kv}$ for any path $(u, i_0, u_1, i_1, \dots, u_k, i_k, v)$. The conditions (1), (2) show that ρ^*_{uv} is well defined. Then it follows that ρ^* determines a Kripke sheaf over F^*. ■

Of course for **S4**-based Kripke sheaves the above conditions (1), (2) follow from Definition 3.6.1. So every Kripke sheaf can be presented in the form described in 3.6.3. Viz. for $\Phi = (F, D, \rho)$ put $\rho_{iuv} := \rho_{uv}$ whenever uR_iv. Then the conditions 3.6.3 (1), (2) hold and the corresponding Kripke sheaf (F, D, ρ^*) is just Φ. So 3.6.3 yields an alternative definition of a Kripke sheaf.

If (F, D, ρ) is a Kripke sheaf, $a \in D_u$, uR^*v, we sometimes use the notation $a|v := \rho_{uv}(a)$ and $\mathbf{a}|v := (a_1|v, \dots, a_n|v)$ for $\mathbf{a} = (a_1, \dots, a_n) \in D_u^n$. But this notation should be used carefully, because it is ambiguous if the fibres are not disjoint (then it may happen that $\rho_{u_1v}(a) \neq \rho_{u_2v}(a)$ for $a \in D_{u_1} \cap D_{u_2}$).

Definition 3.6.4 *A* valuation *in a Kripke sheaf* $\Phi = (F, D, \rho)$ *is just a valuation in the frame* (F, D). *If* ξ *is a valuation,* $M = (\Phi, \xi)$ *is called a* Kripke sheaf model *over* D.

Forcing relation $M, u \models A$ *between a world* u *in an* N*-modal Kripke sheaf model and an* N*-modal* D_u*-sentence.* A *is defined by the same clauses as in predicate Kripke frames (Definition 3.2.9), with the only difference:*

- $M, u \models \Box_i B(a_1, \dots, a_n)$ *iff* $\forall v \in R_i(u)\ M, v \models B(\rho_{uv}(a_1), \dots, \rho_{uv}(a_n))$.

Definition 3.6.5 *A valuation* ξ *in an* **S4**-*based Kripke sheaf (and the corresponding Kripke sheaf model) is called* intuitionistic *if it satisfies the following conditions:*

- $uRv\ \&\ (a_1, \dots, a_n) \in \xi_u(P_k^n) \Rightarrow (\rho_{uv}(a_1), \dots, \rho_{uv}(a_n)) \in \xi_v(P_k^n)$,
- $uRv\ \&\ u \in \xi_u(P_k^0) \Rightarrow v \in \xi_v(P_k^0)$.

Then the intuitionistic forcing in M is defined by the following clauses (cf. Definition 3.2.10):

- $u \Vdash P_k^n(a_1, \dots, a_n)$ iff $(a_1, \dots, a_n) \in \xi_u(P_k^n)$ (for $n > 0$);
- $u \Vdash P_k^0$ iff $u \in \xi_u(P_k^0)$;

[18]So in particular, $\rho_{iuu} = id_{D_u}$ if uR_iu; $\rho_{iuv} = \rho_{juv}$ if uR_iv and uR_jv.

- $u \nVdash \bot$;
- $u \Vdash B \wedge C$ iff $u \Vdash B \,\&\, u \Vdash C$;
- $u \Vdash B \vee C$ iff $u \Vdash B \vee u \Vdash C$;
- $u \Vdash (B \supset C)(a_1, \ldots, a_n)$ iff

  $\forall v \in R(u)\ (v \Vdash B(\rho_{uv}(a_1), \ldots, \rho_{uv}(a_n)) \Rightarrow v \Vdash C(\rho_{uv}(a_1), \ldots, \rho_{uv}(a_n)))$;
- $u \Vdash \exists x A(x)$ iff $\exists a \in D_u\ u \Vdash A(a)$;
- $u \Vdash \forall x B(x, a_1, \ldots, a_n)$ iff $\forall v \in R(u)\, \forall c \in D_v\, v \Vdash B(c, \rho_{uv}(a_1), \ldots, \rho_{uv}(a_n))$.

Lemma 3.6.6 *Let u, v be worlds in an intuitionistic Kripke sheaf model M. Then for any intuitionistic D_u-sentence $A(\mathbf{a})$*

$$M, u \Vdash A(\mathbf{a}) \ \&\ uRv \ \Rightarrow\ M, v \Vdash A(\rho_{uv} \cdot \mathbf{a}).$$

Proof Similar to 3.2.16. ■

Lemma 3.6.7 *The intuitionistic forcing relation has the following properties:*

$$M, u \Vdash \neg B(\mathbf{a}) \textit{ iff } \forall v \in R(u)\ M, v \nVdash B(\rho_{uv} \cdot \mathbf{a}),$$

$$M, u \Vdash a \neq b \textit{ iff } \forall v \in R(u)\ \rho_{uv}(a) \neq \rho_{uv}(b).$$

Proof Obvious from 3.6.5. ■

Definition 3.6.8 *An N-modal predicate formula A is called* true *in an N-modal Kripke sheaf model M if $\overline{\forall} A$ is true at every world of M. A is called* valid *in a Kripke sheaf Φ if it is true under any valuation in Φ.*

The definitions for the intuitionistic case are similar.

The notation for the truth and validity is the same as in the case of Kripke frames.

Lemma 3.6.9 *Let M be an N-modal Kripke sheaf model, $A(\mathbf{x})$ an N-modal formula with $FV(A(\mathbf{x})) = r(\mathbf{x})$, $|\mathbf{x}| = n$. Then for any $u \in M$*

$$M, u \vDash \forall \mathbf{x} A(\mathbf{x}) \textit{ iff } \forall \mathbf{a} \in D_u^n\ M, u \vDash A(\mathbf{a}).$$

Proof Similar to 3.2.17. ■

Lemma 3.6.10 *Let M be an intuitionistic Kripke sheaf model with the accessibility relation R, $u \in M$, $A(\mathbf{x})$ an intuitionistic D_u-formula with $FV(A(\mathbf{x})) = r(\mathbf{x})$, $|\mathbf{x}| = n$. Then*

$$M, u \Vdash \forall \mathbf{x} A(\mathbf{x}) \textit{ iff } \forall v \in R(u)\, \forall \mathbf{a} \in D_v^n\, M, v \Vdash A(\rho_{uv} \cdot \mathbf{a}).$$

Proof Similar to 3.2.18. ■

Definition 3.6.11 *Let Φ be an* **S4**-*based Kripke sheaf; M a model over* **F**. *The* pattern *of M is the model M_0 over* **F** *such that for any $u \in \mathbf{F}$ and any atomic D_u-sentence without equality A*

$$M_0, u \vDash A \text{ iff } M, u \vDash \Box A.$$

As in the case of Kripke frames, the pattern is an intuitionistic model and it always exists.

Lemma 3.6.12 *If M_0 is a pattern of a Kripke sheaf model M, then*

(1) for any $u \in M$, for any intuitionistic D_u-sentence A

$$M_0, u \Vdash A \text{ iff } M, u \vDash A^T;$$

(2) for any $A \in IF^=$

$$M_0 \Vdash A \text{ iff } M \vDash A^T.$$

Proof Similar to 3.2.15 and 3.2.24. ■

Definition 3.6.13 *For a set Σ of modal (or intuitionistic) sentences, a Σ-*sheaf *is a Kripke sheaf (of the corresponding type) validating every formula from Σ. $\Phi \vDash \Sigma$ (or $\Phi \Vdash \Sigma$) denotes that Σ is valid in Φ.*

The class of all Σ-sheaves is denoted by $\mathbf{V}_{\mathcal{KS}}(\Sigma)$ and called modally *(respectively,* intuitionistically*)* definable *(by Σ).*

Lemma 3.6.14 *Let Φ be an N-modal (respectively, intuitionistic) Kripke sheaf over a propositional frame F, A an N-modal (respectively, intuitionistic) propositional formula. Then $\Phi \vDash (\Vdash)A$ iff $\Phi_\pi \vDash (\Vdash)A$.*

Proof Similarly to the case of Kripke frames (Lemma 3.3.32), validity for propositional formulas in Φ is the same as in F. ■

Proposition 3.6.15 *Let Φ be an* **S4**-*based Kripke sheaf, $A \in IF^=$. Then*

$$\Phi \Vdash A \text{ iff } \Phi \vDash A^T.$$

Proof Along the same lines as 3.2.25, now using 3.6.11. ■

Now we have an analogue of 3.2.26:

Lemma 3.6.16 *Let $A(\mathbf{x}), B(\mathbf{x})$ be congruent modal (or intuitionistic) formulas, $|\mathbf{x}| = n$, and let M be a modal (respectively, intuitionistic) Kripke sheaf model. Then for any $u \in M$, $\mathbf{a} \in D_u^n$*

$$M, u \vDash (\Vdash)\ A(\mathbf{a}) \text{ iff } M, u \vDash (\Vdash)\ B(\mathbf{a}).$$

Thus the set of formulas valid in a Kripke sheaf is closed under congruence.

Proof Along the same lines as 3.2.26. Again we consider the equivalence relation on modal formulas

$$A \sim B \text{ iff } FV(A) = FV(B)$$

and for any distinct $\mathbf{x}$ with $FV(A) = r(\mathbf{x})$, for any $u \in M, \mathbf{a} \in D_u^{|\mathbf{x}|}$

$$M, u \vDash [\mathbf{a}/\mathbf{x}]A \Leftrightarrow M, u \vDash [\mathbf{a}/\mathbf{x}]B.$$

We have to check the properties 2.3.17(1)–(4) for this relation. For (1)–(3) the proof is the same as in 3.2.26. For (4) there is a slight difference: now

$$M, u \vDash [\mathbf{a}/\mathbf{x}]\Box_i A \Leftrightarrow \forall v \in R_i(u)\ M, v \vDash [\rho_{uv} \cdot \mathbf{a}/\mathbf{x}]A$$

and similarly for B. So $A \sim B$ implies $\Box_i A \sim \Box_i B$. ■

Theorem 3.6.17 (Soundness theorem)

(I) The set of all modal predicate formulas (with equality) valid in a Kripke sheaf is a modal predicate logic (with equality).

(II) The set of all intuitionistic predicate formulas (with equality) valid in an **S4***-based Kripke sheaf is a superintuitionistic predicate logic (with equality).*

Proof Along the same lines as 3.2.29. The main thing is to check that formula substitutions preserve validity.

So we assume that $\Phi \vDash A$ for a formula A and a Kripke sheaf Φ and show that $\Phi \vDash SA$ for $S = [C(\mathbf{x}, \mathbf{y})/P(\mathbf{x})]$, where $P \in PL^n$ occurs in A, the list $\mathbf{xy}$ is distinct, and $r(\mathbf{y}) \subseteq FV(C) \subseteq r(\mathbf{xy})$.

By Lemma 3.6.16 we can replace A with a congruent formula, so we assume that A is clean, $BV(A) \cap r(\mathbf{y}) = \varnothing$. Again we choose a distinct list $\mathbf{z}$ such that $FV(A) \cup r(\mathbf{y}) = r(\mathbf{z})$; then

$$r(\mathbf{y}) \subseteq FV(A) \subseteq r(\mathbf{z}), \quad r(\mathbf{z}) \cap BV(A) = \varnothing.$$

Let $m = |\mathbf{z}|$. Given a model $M = (\Phi, \xi)$, we show that for any $u \in \Phi$, $\mathbf{c} \in D_u^m$

$$M, u \vDash [\mathbf{c}/\mathbf{z}]SA.$$

Let $\mathbf{c}'$ be the part of $\mathbf{c}$ corresponding to $\mathbf{y}$. We define $M_1 = (\Phi, \eta)$ similarly to 3.2.29:

- for any $v \in \Phi{\uparrow}u$, $\mathbf{a} \in D_v^n$

$$M_1, v \vDash P(\mathbf{a}) \text{ iff } M, v \vDash C(\mathbf{a}, \mathbf{c}'|v);$$

- for any other atomic D_v-sentence Q

$$M_1, v \vDash Q \text{ iff } M, v \vDash Q.$$

Then every subformula of A has the form $B(\mathbf{z}, \mathbf{q})$, where $\mathbf{q}$ is distinct, $r(\mathbf{q}) = BV(A)$; so by 2.5.24 we present SB as $(SB)(\mathbf{z}, \mathbf{q})$.

Then we prove the claim:

$$\forall v \in \Phi \uparrow u \; \forall \mathbf{a} \in D_v^{|q|} \, (M_1, v \vDash B(\mathbf{c}|v, \mathbf{a}) \Leftrightarrow M, v \vDash (SB)(\mathbf{c}|v, \mathbf{a}))$$

by induction. The only difference with 3.2.29 is in the case $B = \Box_i B_1$:
$M_1, v \vDash B(\mathbf{c}|v, \mathbf{a})$ iff $\forall w \in R_i(v) \; M_1, w \vDash B_1(\mathbf{c}|w, \mathbf{a}|w)$
iff $\forall w \in R_i(v) \; M, w \vDash (SB_1)(\mathbf{c}|w, \mathbf{a}|w)$ (by the induction hypothesis)
iff $M, v \vDash \Box_i(SB_1)(\mathbf{c}|v, \mathbf{a})(= (SB)(\mathbf{c}|v, \mathbf{a}))$.

Here we use the equality $(\mathbf{a}|v)|w = \mathbf{a}|w$, which follows from Definition 3.6.1.

■

Remark 3.6.18 One can similarly define forcing and validity for the case when ρ does not satisfy the coherence conditions (1), (2) from 3.6.3. But then the set of validities is not necessarily substitution-closed. These 'frames' are a special kind of Kripke bundle considered in Chapter 5; note that validity in Kripke bundles is not substitution closed either, cf. Exercise 5.2.13.

Every predicate Kripke frame corresponds to a Kripke sheaf, in which uR_iv always implies $D_u \subseteq D_v$ and ρ_{uv} is the inclusion map, i.e. $\rho_{uv}(a) = a$ for any $a \in D_u$.

More generally, every KFE $\mathbf{F} = (W, R_1, \ldots, R_N, D, \asymp)$ corresponds to a Kripke sheaf $\Theta(\mathbf{F})$ constructed as follows. Let a_u be the class of $a \in D_u$ modulo $\asymp_u$. We define $\Theta(\mathbf{F})$ as the Kripke sheaf with the fibres $D'_u := \{a_u \mid a \in D_u\}$ and the transition maps $\rho_{uv}(a_u) := a_v$ for uR_iv; $\Theta(\mathbf{F})$ is well defined, due to Lemma 3.6.3 and since uR_iv implies $D_u \subseteq D_v$ and $\asymp_u \subseteq \asymp_v$.

The following is almost obvious:

Lemma 3.6.19 *Valuations (both modal and intuitionistic) in a KFE* $\mathbf{F}$ *and in the Kripke sheaf* $\Theta(\mathbf{F})$ *are associated. Namely, a valuation* ξ *in* $\mathbf{F}$ *corresponds to the valuation* $\Theta(\xi)$ *in* $\Theta(\mathbf{F})$ *such that*

$$(\Theta(\xi))_u(P) = \{((a_1)_u, \ldots, (a_n)_u) \in (D'_u)^n \mid (a_1, \ldots, a_n) \in \xi_u(P)\}$$

for $P \in PL^n, n > 0$, *and* $\Theta(\xi)_u$, ξ_u *coincide on* PL^0. *The other way round, every valuation in* $\Theta(\mathbf{F})$ *has the form* $\Theta(\xi)$ *for some valuation* ξ *in* $\mathbf{F}$. ξ *is intuitionistic iff* $\Theta(\xi)$ *is intuitionistic.*

Proof The above definition of $\Theta(\xi)$ is sound; recall that $(a_1, \ldots, a_n) \in \xi_u(P)$ depends only on classes of $a_1, \ldots, a_n$ modulo $\asymp_u$.

Now if η is a valuation in $\Theta(\mathbf{F})$, we define ξ by

$$\xi_u(P) := \{\mathbf{a} \mid \mathbf{a}_u \in \eta_u(P)\}$$

where

$$\mathbf{a}_u := ((a_1)_u, \ldots, (a_n)_u) \text{ for } \mathbf{a} = (a_1, \ldots, a_n), \; n > 0.$$

This definition is sound, because $\mathbf{a}_u = \mathbf{b}_u$ iff $\forall i \; a_i \asymp_u b_i$. Then obviously $\eta = \Theta(\xi)$.

The argument for the intuitionistic case is left to the reader. ■

Lemma 3.6.20

(1) For any N-modal formula $A(x_1, \ldots, x_n)$, a valuation ξ in $\mathbf{F}$, for any $u \in F$; $a_1, \ldots, a_n \in D_u$:

$$\xi, u \models A(a_1, \ldots, a_n) \text{ (in } \mathbf{F}) \text{ iff } \Theta(\xi), u \models A((a_1)_u, \ldots, (a_n)_u) \text{ (in } \Theta(\mathbf{F}));$$

and similarly for the intuitionistic case.

(2) If $\mathbf{F}$ is an N-modal KFE, A is an N-modal formula, then $\mathbf{F} \vDash A$ iff $\Theta(\mathbf{F}) \vDash A$.

(3) If $\mathbf{F}$ is an intuitionistic KFE, A is an intuitionistic formula, then $\mathbf{F} \Vdash A$ iff $\Theta(\mathbf{F}) \Vdash A$.

Proof (1) By induction on the complexity of A. For example,

$$\xi, u \models a = b \text{ iff } a \asymp_u b \text{ iff } a_u = b_u \text{ iff } \Theta(\xi), u \models a = b;$$

$\xi, u \models \Box_i B(\mathbf{a})$ iff $\forall v \in R_i(u)\ \xi, v \models B(\mathbf{a})$ iff $\forall v \in R_i(u)\ \Theta(\xi), v \models B(\mathbf{a}_v)$ iff $\Theta(\xi), u \models \Box_i B(\mathbf{a}_u)$, since $\mathbf{a}_v = \rho_{uv} \cdot \mathbf{a}_u$.

The remaining cases are left to the reader.

The claims (2), (3) now follow from 3.6.19, 3.6.9, 3.6.10. ■

Due to Theorem 3.6.17 and Lemma 3.6.20, we obtain

Theorem 3.6.21 (Soundness theorem)

(I) The set of all modal predicate formulas (with equality) valid in a KFE is a modal predicate logic (with equality).

(II) The set of all intuitionistic predicate formulas (with equality) valid in an **S4***-based KFE is a superintuitionistic predicate logic (with equality).*

These logics are denoted by $\mathbf{ML}^{(=)}(\mathbf{F})$, $\mathbf{IL}^{(=)}(\mathbf{F})$ as usual.

According to the general definitions from Section 2.16, the *modal predicate logic* of a class of Kripke sheaves $\mathcal{F}$ is

$$\mathbf{ML}^{(=)}(\mathcal{F}) := \bigcap \{\mathbf{ML}^{(=)}(\Phi) \mid \Phi \in \mathcal{F}\}.$$

The superintuitionistic logic $\mathbf{IL}^{(=)}(\mathcal{F})$ is defined analogously.

The other way round, every Kripke sheaf is equivalent to one of the form $\Theta(\mathbf{F})$. To show this, let us introduce a convenient subclass of Kripke sheaves.

Definition 3.6.22 *A Kripke sheaf is said to be* disjoint *if all its fibres are disjoint.*

Definition 3.6.23 *An* isomorphism *between Kripke sheaves (F, D, ρ) and (F, D', ρ') is a family of bijections $f_u : D_u \longrightarrow D'_u$ such that $f_u \cdot \rho'_{uv} = f_v \cdot \rho_{uv}$ whenever uR^*v.*

A more general notion of morphism will be discussed in the next section. It is almost obvious that isomorphic Kripke sheaves have the same modal (or superintuitionistic) logics; for a precise proof, one should check the equivalence

$$M, u \vDash B(\mathbf{a}) \text{ iff } M', u \vDash B(f_u \cdot \mathbf{a})$$

for any D_u-sentence $B(\mathbf{a})$ if it holds for any atomic D_u-sentence. We leave this as an exercise for the reader.

Lemma 3.6.24 *Every Kripke sheaf is isomorphic to a disjoint Kripke sheaf over the same propositional frame.*

Proof In fact, we can replace each D_u with $D'_u = \{(a,u) \mid a \in D_u\}$ and change the functions ρ_{uv} appropriately, viz., put $\rho'_{uv}(a,u) := (\rho_{uv}(a), v)$. ∎

For a disjoint Kripke sheaf (F, D, ρ) and $a \in D_u$, uR_iv, we sometimes write aR_iv and say that a is *R_i-related* to v.

Now let $\Phi = (F, D, \rho)$ be a disjoint Kripke sheaf. Consider the KFE $\mathbf{G}(\Phi) := (F, D', \asymp')$, where

$$D'_u := \bigcup\{D_w \mid w \in F,\ u \in F{\uparrow}w\},$$

$$\asymp'_u := \{(a,b) \in (D'_u)^2 \mid (a|u) = (b|u)\}.$$

Here is an equivalent presentation of $\mathbf{G}(\Phi)$ in the form described in Lemma 3.5.2:

$$D'^{+} := \bigcup_{u \in F} D_u,$$

$$E'(a,b) := \{w \in (F{\uparrow}u \cap F{\uparrow}v) \mid (a|w) = (b|w)\}$$

for $a \in D_u$, $b \in D_v$; $u, v \in F$.

It follows easily that $D'_u \subseteq D'_v$ and $\asymp'_u \subseteq \asymp'_v$, whenever uR_iv; thus $\mathbf{G}(\Phi)$ is really a KFE.

Speaking informally, D'_u absorbs D_u and the domains of all R^*-predecessors of u; $\asymp'_u$ makes every individual from D_u equivalent to all its predecessors. So there is a natural bijection between D_u and $D'_u/\asymp'_u$. This observation is used in the proof of the following

Lemma 3.6.25 *The Kripke sheaves Φ and $\Theta(\mathbf{G}(\Phi))$ are isomorphic.*

Proof We have $\Theta(\mathbf{G}(\Phi)) = (F, D'', \rho'')$, where $D'' = (D'_u/\asymp'_u)_{u \in F}$, and the transition map ρ''_{uv} sends every equivalence class $a/\asymp'_u$ to $a/\asymp'_v$ (for $a \in D'_u$).

Now, every class $a'' = (b/\asymp'_u) \in D''_u$ contains a single element a from D_u, namely, $a = (b|u)$. So there exists a well-defined bijection $\theta_u : D''_u \to D_u$ such that $\theta_u(b/\asymp'_u) = b|u$. (For the surjectivity, note that $(a|u) = a$ for $a \in D_u$.) Finally, uR_iv implies

$$\rho_{uv}(\theta_u(b/\asymp'_u)) = \rho_{uv}(b|u) = b|v,$$

and

$$\theta_v(\rho''_{uv}(b/\asymp'_u)) = \theta_v(b/\asymp'_v) = b|v.$$

Thus $\rho_{uv} \cdot \theta_u = \theta_v \cdot \rho''_{uv}$, which means that the family of functions $(\theta_u \mid u \in F)$ is an isomorphism between the Kripke sheaves $\Theta(\mathbf{G}(\Phi))$ and Φ. ■

So we can introduce semantics generated by predicate Kripke frames with equality. Due to Lemmas 3.6.20 and 3.6.25, the same semantics are generated by Kripke sheaves.

Definition 3.6.26 *The* N-modal Kripke sheaf semantics $\mathcal{KE}_N^{(=)}$ *is generated by the class of all N-modal predicate Kripke frames with equality (or Kripke sheaves). Similarly the* intuitionistic Kripke sheaf semantics $\mathcal{KE}_{int}^{(=)}$ *is generated by the class of all intuitionistic KFEs (or Kripke sheaves). Logics complete in these semantics are called* Kripke sheaf complete, *or just* $\mathcal{KE}$-complete.

As we pointed out at the end of Subsection 3.4.2, every predicate Kripke frame is equivalent a simple KFE. Thus $\mathcal{K}_N^{(=)} \preceq \mathcal{KE}_N^{(=)}$ and actually $\mathcal{K}_N^{(=)} \prec \mathcal{KE}_N^{(=)}$, as we shall see later on; the same is true for the intuitionistic semantics.

Similarly to Corollary 3.3.37 we obtain

Proposition 3.6.27 *Kripke sheaf semantics has the collection property.*

Proof Obvious, by Lemma 3.5.30. ■

For Kripke and Kripke sheaf semantics there also exists a stronger version of completeness.

Definition 3.6.28 *An* N*-modal theory* Γ *is called* satisfiable *in an* N*-modal Kripke frame (respectively, KFE)* F *if there exists a model (respectively, KFE-model)* M *over* F *and a world* $u \in M$ *such that* $M, u \vDash \Gamma$, *i.e.* $M, u \vDash A$ *for any* $A \in \Gamma$.

An intuitionistic theory (Γ, Δ) *is called* satisfiable *in an intuitionistic Kripke frame (or KFE)* F *if there exists an intuitionistic model (or KFE-model)* M *over* F *and a world* $u \in M$ *such that* $M, u \Vdash (\Gamma, \Delta)$, *i.e.* $M, u \Vdash A$ *for any* $A \in \Gamma$ *and* $M, u \nVdash B$ *for any* $B \in \Delta$.

Lemma 3.6.29 *A theory (modal or intuitionistic) is satisfiable in an* L*-KFE iff it is satisfiable in a Kripke sheaf validating* L.

Proof In fact, by Lemma 3.6.20, $\mathbf{F}$ and $\Theta(\mathbf{F})$ have the same logic (of the corresponding kind); a sentence A is satisfiable at $\mathbf{F}, u$ iff it is satisfiable at $\Theta(\mathbf{F}), u$. On the other hand, by Lemma 3.6.25, every Kripke sheaf Φ is isomorphic to $\Theta(\mathbf{G}(\Phi))$, so satisfiability in Φ and $\mathbf{G}(\Phi)$ is the same. ■

Definition 3.6.30 *An N-modal predicate logic L (with or without equality) is called* strongly Kripke complete *(respectively,* strongly Kripke sheaf complete*) if every L-consistent N-modal theory (respectively, with or without equality) is satisfiable in some Kripke L-frame (respectively, L-KFE). Similarly, an intuitionistic predicate logic L (with/without equality) is called* strongly Kripke (Kripke sheaf) complete *if every L-consistent intuitionistic theory (with/ without equality) is satisfiable in some Kripke L-frame (L-KFE).*

Thanks to Lemma 3.6.29, the notions of strong completeness are the same for Kripke sheaves and KFEs.

Lemma 3.6.31 *Every strongly complete m.p.l.(=) or s.p.l. (=) is complete (in the corresponding semantics of Kripke frames or Kripke sheaves).*

Proof In fact, in the modal case, if $A \notin L$, then $\{\neg A\}$ is L-consistent. So if L is strongly complete, then $\neg A$ is satisfiable in an L-frame F, and thus F separates A from L. Therefore L is complete by Lemma 2.16.2.

In the intuitionistic case, if $A \notin L$, then $(\varnothing, \{A\})$ is L-consistent; the remaining argument is the same as in the modal case. ∎

3.7 Morphisms of Kripke sheaves

Kripke sheaf morphisms are a natural generalisation of predicate Kripke frame morphisms defined in 3.3.1:

Definition 3.7.1 *Let $\Phi = (F, D, \rho)$, $\Phi' = (F', D', \rho')$ be Kripke sheaves over the frames $F = (W, R_1, \dots, R_N)$, $F' = (W', R'_1, \dots, R'_N)$. A* morphism *from Φ to Φ is a pair $\mathbf{f} = (f_0, f_1)$ satisfying the conditions (1)–(3) from Definition 3.3.1 and also*

(5) the following diagram commutes whenever uR_iv:

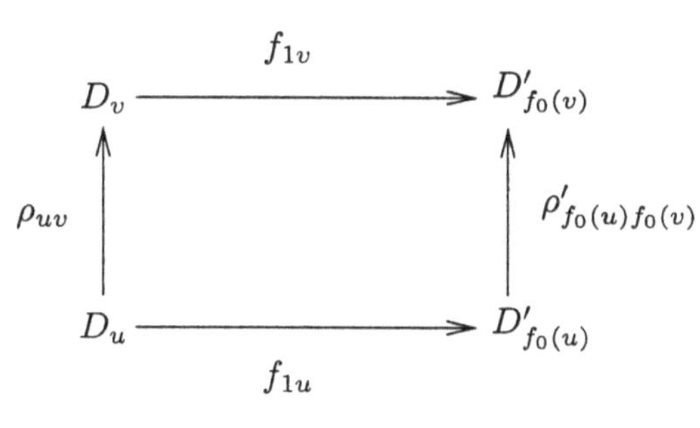

The latter condition can be briefly written as

$$f_{1v}(a|v) = f_{1u}(a)|f_0(v)$$

(for $a \in D_u$, uR_iv), but strictly speaking, this makes sense only for disjoint sheaves.

Exercise 3.7.2 Show that under the conditions of 3.7.1 the diagram (5) commutes whenever uR^*v.

Definition 3.7.3 *A* morphism *of Kripke sheaf models* $M = (\Phi, \xi)$ *and* $M' = (\Phi', \xi')$ *is a morphism* $(f_0, f_1) : \Phi \longrightarrow \Phi'$ *such that for any* $P \in PL^m$, $m \geq 0$, $u \in \Phi$, $\mathbf{a} \in D_u^m$

$$M, u \vDash P(\mathbf{a}) \text{ iff } M', f_0(u) \vDash P(f_{1u} \cdot \mathbf{a}).$$

The notions '*=-morphism*', '$p^{(=)}$*-morphism*', '*isomorphism*' are transferred to Kripke sheaves in an obvious way.

Now we easily obtain an analogue to Lemma 3.3.4:

Lemma 3.7.4

(1) The identity morphism $id_\Phi := (id_W, (id_{D_u})_{u \in W})$ *is an isomorphism of Kripke sheaves.*

(2) The composition *of morphisms* $(f_0, f_1) : \Phi \longrightarrow \Phi'$ *and* $(g_0, g_1) : \Phi' \longrightarrow \Phi''$ *defined as* $(f_0 \circ g_0, f_{1u} \circ g_{1u})_{u \in W})$ *is a morphism* $\Phi \longrightarrow \Phi'$, *similarly for all other kinds of morphism.*

This yields us different categories of Kripke sheaves as in the case of Kripke frames.

Next, we have analogues to 3.3.6, 3.3.8–3.3.10:

Definition 3.7.5 *A Kripke sheaf morphism over a propositional Kripke frame* F *is a morphism of Kripke sheaves over* F, *in which the world component is the identity map.*

Definition 3.7.6 *Let* $\Phi = (F, D, \rho)$, $\Phi' = (F', D', \rho')$ *be Kripke sheaves,* $h : F' \longrightarrow F$ *a morphism of propositional frames. We say that* Φ' *is obtained by* changing the base along h *if* $D'_u = D_{h(u)}$ *for any* $u \in F$ *and* $\rho'_{uv} = \rho_{h(u)h(v)}$ *for any pair* $(u, v) \in R_i$. *We use the notation* $h_*\Phi$ *for* Φ'.

Remark 3.7.7 This definition is sound in the case when h is only monotonic, i.e., for a 'more traditional category of sheaves'.

Proposition 3.7.8 *Under the conditions of the Definition 3.7.6, there exists a 'canonical' =-morphism* $(h, g) : h_*\Phi \stackrel{=}{\longrightarrow} \Phi$. *Every morphism* $(h, f_1) : \Phi'' \longrightarrow \Phi$, *where* $\Phi''_\pi = F'$, *can be uniquely presented as a composition* $\Phi'' \longrightarrow h_*\Phi \longrightarrow \Phi$ *of a morphism over* F' *and the canonical morphism.*

The next claim is analogous to 3.3.11:

Proposition 3.7.9 *If* $(f_0, f_1) : M \stackrel{=}{\longrightarrow} M'$ *for* N*-modal Kripke sheaf models* M, M', *then for any* $u \in M$, $A \in MS_N^{(=)}(D_u)$

$$M, u \vDash A \text{ iff } M', f_0(u) \vDash f_{1u} \cdot A,$$

where $f_{1u} \cdot A$ *is obtained by replacing every* $a \in D_u$ *with* $f_{1u}(a)$.

The same holds in the intuitionistic case.

Proof We check only the $\Box$-case. Let R_i, R'_i be the accessibility relations in M, M'. If $A = \Box_i B(\mathbf{a})$ (for a formula $B(\mathbf{x})$), then

$$M, u \vDash A \text{ iff } \forall v \in R_i(u)\ M, v \vDash B(\rho_{uv}\cdot\mathbf{a}) \text{ iff } \forall v \in R_i(u)\ M', f_0(v) \vDash B(f_{1v}\cdot(\rho\cdot\mathbf{a}))$$

by the induction hypothesis. By 3.7.1 (5), the latter is equivalent to

$$\forall v \in R_i(u)\ M', f_0(v) \vDash B(\rho'_{f_0(u)f_0(v)} \cdot (f_{1u} \cdot \mathbf{a})),$$

which is the same as

$$\forall w \in f_0[R_i(u)]\ M', w \vDash B(\rho'_{f_0(u)w} \cdot (f_{1u} \cdot \mathbf{a})).$$

Since $\mathbf{f}$ satisfies the conditions 3.3.1 (1)–(3), we have $f_0[R_i(u)] = R'_i(f_0(u))$. Eventually

$$M, u \vDash A \text{ iff } M', f_0(u) \vDash \Box_i B(f_{1u} \cdot \mathbf{a})$$

as required. ∎

Lemma 3.7.10 *Let $\mathbf{f} : \Phi \longrightarrow \Phi'$ be a morphism of Kripke sheaves, M' a Kripke sheaf model over Φ'. Then there exists a unique model M over Φ such that $\mathbf{f} : M \longrightarrow M'$. The same holds for all kinds of morphism. If M' is intuitionistic, then M is also intuitionistic.*

Proof Similar to 3.3.12; an exercise for the reader. ∎

Proposition 3.7.11 *Let Φ_1 and Φ_2 be Kripke sheaves. If there exists a $\mathrm{p}^{(=)}$-morphism from Φ_1 onto Φ_2, then* $\mathbf{ML}^{=}(\Phi_1) \subseteq \mathbf{ML}^{=}(\Phi_2)$ *(or* $\mathbf{IL}^{=}(\Phi_1) \subseteq \mathbf{IL}^{=}(\Phi_2)$ *in the intuitionistic case).*

Proof Similar to 3.3.13. Use Lemmas 3.6.9, 3.6.10 and 3.7.10. ∎

In the case of disjoint Kripke sheaves there exists en equivalent definition of a morphism. Let us first present disjoint Kripke sheaves in an equivalent form.

Definition 3.7.12 *A morphism of propositional Kripke frames*

$$f : F = (W, R_1, \ldots, R_N) \longrightarrow F' = (W', R'_1, \ldots, R'_N)$$

is called etale *if it has the* unique lift property

$$\forall u \in W\ \forall v' \in W'\ (f(u)R'^{*}v' \Rightarrow \exists! v\ (f(v) = v' \ \&\ uR^{*}v)).$$

In this case every restriction $f \restriction (F \uparrow u)$ is an isomorphism to $F' \uparrow f(u)$. In fact, this is a bijection, due to the monotonicity and the unique lift property. Its converse is also monotonic, due to the lift property.

Proposition 3.7.13

(1) *Let $\Phi = (F, D, \rho)$ be a disjoint Kripke sheaf over a frame $F = (W, R_1, \ldots, R_N)$. Consider the* frame of individuals $F^+ := (D^+, \rho_1^+, \ldots, \rho_N^+)$, *where*

$$D^+ = \bigcup_{u \in W} D_u, \quad \rho_i^+ := \bigcup_{uR_iv} \rho_{uv}.$$

Let $\tau : D^+ \longrightarrow W$ be the map sending every individual to its world (i.e. $\tau(a) = u \iff a \in D_u$). Then $\tau : F^+ \twoheadrightarrow F$ is etale.

(2) *Conversely, for any etale p-morphism of propositional frames $\tau : F' \twoheadrightarrow F$ there exists a disjoint Kripke sheaf $\Phi = (F, D, \rho)$, in which F' is the frame of individuals and $D_u = \tau^{-1}(u)$.*

Proof

(1) The monotonicity and the lift property of τ hold, since ρ_{uv} is a function from D_u to D_v whenever uR_iv. The unique lift property

$$uR^*v \ \&\ a \in D_u \Rightarrow \exists! b \in D_v\ a(\rho^+)^*b$$

holds, since $(\rho^+)^*$ induces a function $\rho_{uv} : D_u \longrightarrow D_v$ for any pair $(u, v) \in R^*$, according to 3.6.1, 3.6.2.

(2) In fact, for uR^*v we can define the function $\rho_{uv} : D_u \longrightarrow D_v$ by the condition $\rho_{uv}(a) = b \iff \tau(b) = v \ \&\ a(\rho')^*b$. Since τ is etale, this b is always unique. The unique lift property also implies the properties 3.6.1 (1), (2) for (F^*, D, ρ).

■

Proposition 3.7.14

(1) *Let $\Phi = (F, D, \rho)$, $\Phi' = (F', D', \rho')$ be disjoint Kripke sheaves over N-modal frames, and let $\tau : F^+ \twoheadrightarrow F$, $\tau' : F'^+ \twoheadrightarrow F'$ be the corresponding etale morphisms. Also let $(f_0, f_1) : \Phi \longrightarrow \Phi'$ be a morphism (in the sense of 3.7.1). Consider the map $f_1^+ : D^+ \longrightarrow D'^+$ of total domains such that $f_1^+(a) = f_{1u}(a)$ for $a \in D_u$ (i.e., $f_1^+ = \bigcup_{u \in F} f_{1u}$). Then $f_1^+ : F^+ \twoheadrightarrow F'^+$ and the following diagram commutes:*

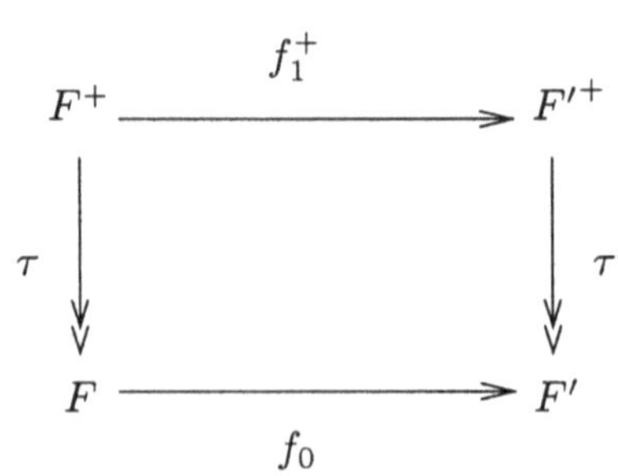

f_1^+ is surjective on all fibres; it is a p-morphism whenever f_0 is a p-morphism.

(2) *Conversely, every morphism $g : F^+ \longrightarrow F'^+$, for which every $g \upharpoonright D_u$ is a surjective map onto $D_{f_0(u)}$, equals f_1^+ for some morphism $(f_0, f_1) : \Phi \longrightarrow \Phi'$.*

Proof (1) Let us show that f_1^+ is monotonic. Suppose $a\rho_i^+ b$, $a \in D_u$, $b \in D_v$. Then by definition, $b = a|v$, $f_1^+(a) = f_{1u}(a)$, $f_1^+(b) = f_{1v}(b)$, uR_iv. By 3.7.1,

$$f_{1v}(b) = f_{1u}(a)|f_0(v), \quad f_0(u)R_i'f_0(v),$$

and thus $f_1^+(a)\rho_i'^+ f_1^+(b)$.

The commutativity of the diagram is almost obvious: for $a \in D_u$,

$$\tau'(f_1^+(a)) = \tau'(f_{1u}(a)) = f_0(u),$$

since $f_{1u} : D_u \longrightarrow D'_{f_0(u)}$. But $f_0(u) = f_0(\tau'(a))$.

Now we can check the lift property for f_1^+. In fact, suppose $a \in D_u$, $f_1^+(a) = f_{1u}(a)\rho_i'^+ b'$. Then $b' = f_{1u}(a)|v'$ for some $v' \in R_i'(u')$, where $f_{1u}(a) \in D_{u'}$, i.e.,

$$u' = \tau'(f_1^+(a)) = f_0(\tau(a)) = f_0(u).$$

Since f_0 has the lift property, we obtain $v \in R_i(u)$ such that $f_0(v) = v'$. Then by definition, $a\rho_i^+(a|v)$. On the other hand, we obtain

$$f_1^+(a|v) = f_{1v}(a|v) = f_{1u}(a)|f_0(v) = b',$$

by 3.7.1. Therefore f_1^+ is a morphism. It is surjective, since every f_{1u} is surjective, by 3.3.1.

(2) Given g, we define $f_1 := (f_{1u})_{u \in F}$, with $f_{1u} = g \upharpoonright D_u$. Then for any $a \in D_u$, uR_iv

$$f_{1v}(a|v) = g(a|v) \in D_{f_0(v)}.$$

Since $a\rho_i^+(a|v)$, it follows that $f_{1u}(a) = g(a)\rho_i'^+ g(a|v)$, i.e. $g(a|v) = g(a)|f_0(v)$. Thus (f_0, f_1) satisfies 3.7.1 (5), and so it is a morphism, $g = f_1^+$. ■

The definitions of a subframe, etc. from Section 3.3 can also be transferred to Kripke sheaves.

Definition 3.7.15 *Let $\Phi = (F, D, \rho)$ be a Kripke sheaf over a propositional frame $F = (W, R_1, \dots, R_N)$, and let $V \subseteq W$. Then we define the corresponding* subsheaf *as follows:*

$$\Phi \upharpoonright V := (F \upharpoonright V, D \upharpoonright V, \rho \upharpoonright V),$$

where $F \upharpoonright V$ is the same as in Definition 1.3.13.

$$D \upharpoonright V := (D_u)_{u \in V}, \ \rho \upharpoonright V := (\rho_{uv})_{(u,v) \in R^* \upharpoonright V}.$$

For $u \in F$, the subsheaf generated by u *or the* cone $\Phi{\uparrow}u$ *is defined as $\Phi \upharpoonright (W{\uparrow}u)$.*

Exercise 3.7.16 (1) Prove an analogue to Generation lemma 3.3.18 for Kripke sheaves.

(2) Define disjoint sums of Kripke sheaves and prove an analogue to Proposition 3.3.36.

Lemma 3.7.17 *For a Kripke sheaf Φ over a frame F*

$$\mathbf{ML}^{(=)}(\Phi) = \bigcap\{\mathbf{ML}^{(=)}(\Phi{\uparrow}u) \mid u \in F\},$$

and similarly for the intuitionistic case.

Proof Cf. Lemma 3.3.21. ■

Note that if $u \approx_R v$ (i.e. if u, v are in the same cluster) in a Kripke **S4**-frame F, then in every predicate Kripke frame $\mathbf{F} = (F, D)$ we have $D_u = D_v$. So we can define the *skeleton of* $\mathbf{F}$:

$$\mathbf{F}^{\sim} := (F^{\sim}, D^{\sim}),$$

where $F^{\sim}$ is the skeleton of F (Definition 1.3.40) and for a cluster $u^{\sim}$, $D^{\sim}_{u^{\sim}} := D_u$.

Analogously, one can easily see that for $u \approx_R v$ in an **S4**-based Kripke sheaf $\Phi = (F, D, \rho)$, the function ρ_{uv} is a bijection between D_u and D_v with the converse ρ_{vu}. Thus for any model over Φ,

$$u \vDash A(a_1, \ldots, a_n) \text{ iff } v \vDash A(a_1|v, \ldots, a_n|v).$$

So we can properly define the *skeleton* $\Phi^{\sim}$ of an **S4**-based Kripke sheaf Φ. Then we obtain an analogue to 1.4.14:

Lemma 3.7.18 *For an intuitionistic Kripke frame* $\mathbf{F}$, $\mathbf{IL}(\mathbf{F}) = \mathbf{IL}(\mathbf{F}^{\sim})$*; similarly, for intuitionistic Kripke sheaves.*

Proof Consider the case of Kripke sheaves. For an intuitionistic valuation ξ in Φ there exists an intuitionistic valuation $\xi^{\sim}$ in $\Phi^{\sim}$ such that

$$\xi^{\sim}_{u^{\sim}}(P) = \xi_u(P)$$

for any P, u. Then the map $u \mapsto u^{\sim}$ is a p-morphism $(\Phi, \xi) \twoheadrightarrow (\Phi^{\sim}, \xi^{\sim})$. It remains to note that every intuitionistic valuation in $\Phi^{\sim}$ has the form $\xi^{\sim}$. ■

Therefore, in the semantics $\mathcal{KE}_{int}$ is generated by Kripke sheaves over posets (and similarly, for the semantics $\mathcal{K}_{int}$).

Finally note that Kripke sheaf morphisms are appropriate also for KFEs. Viz., we can define a morphism $\mathbf{F} \longrightarrow \mathbf{F}'$ of KFEs just as an arbitrary morphism of associated Kripke sheaves $\Theta(\mathbf{F}) \longrightarrow \Theta(\mathbf{F}')$.

Every strong morphism of KFEs corresponds to a morphism in this sense, but not the other way round. E.g. consider KFEs $\mathbf{F} = (F, D, E)$ and $\mathbf{F}' = (F, D', E')$, where F is a reflexive singleton u, $|D_u| = 1$, $|D'_u| = 2$ and E, E' are universal. Then obviously $\Theta(\mathbf{F}) \cong \Theta(\mathbf{F}')$, but there is no strong isomorphism from $\mathbf{F}$ to $\mathbf{F}'$.

3.8 Transfer of completeness

In this section we show that in some simple cases Kripke and Kripke sheaf completeness transfer to extensions.

We begin with a lemma analogous to 1.3.45. Its proof is based on the existence of exact KFE-models for logics with equality, which will be proved in Chapter 6 (cf. Proposition 6.1.27).

Lemma 3.8.1 *If an m.p.l.(=) L is conically expressive, then the rule* $\dfrac{A}{\Box^* A}$ *is admissible in L.*

Proof Since L has a characteristic model, it suffices to show that for any KFE-model, for any A, $M \vDash A$ implies $M \vDash \Box^* A$. We can argue similarly to 3.2.29, if we consider the Kripke model with the accessibility relation R^*.

If $M \vDash A(x_1, \dots, x_n)$, then for any $u \in M$ for any $a_1, \dots, a_n \in D_u$, $M, u \vDash A(a_1, \dots, a_n)$, by 3.5.7.

Hence, $M, u \vDash \Box^* A(a_1, \dots, a_n)$ by 3.3.39.

But obviously, $\Box^* A(a_1, \dots, a_n) = (\Box^* A)(a_1, \dots, a_n)$. Thus $M \vDash \Box^* A(x_1, \dots, x_n)$, by 3.5.7. ∎

Definition 3.8.2 *A* pure equality sentence *is a predicate sentence with equality that does not contain predicate letters other than '='.*

In particular, every closed propositional formula is a pure equality sentence.

Theorem 3.8.3 *If a conically expressive m.p.l.= or an s.p.l.= L is Kripke frame (respectively, Kripke sheaf) complete and C is a pure equality sentence in the language of L, then $L + C$ is also Kripke frame (respectively, Kripke sheaf) complete.*

Proof (I) Modal case. If $L+C \nvdash A$, then $L \nvdash \Box^* C \supset A$, since by Lemma 3.8.1, $L + C \vdash \Box^* C$. By completeness, there exists a Kripke frame (or a KFE) $\mathbf{F} \vDash L$ and a Kripke (respectively, KFE) model M over F such that $M, u \vDash \Box^* C \wedge \neg A$ for some u. Hence $M{\uparrow}u, u \vDash \Box^* C \wedge \neg A$ by Lemma 3.3.18 (1), and thus $M{\uparrow}u \vDash C$, by Lemmas 3.3.39 and 1.3.19. But C is a simple equality sentence, so its truth value at every world does not depend on the valuation. Thus $\mathbf{F}{\uparrow}u \vDash C$. We also have $\mathbf{F}{\uparrow}u \vDash L$ by Lemma 3.3.18 (2), and therefore A is refuted in an $(L + C)$-frame.

(II) Intuitionistic case. The proof is very similar. If $L + C \nvdash A$, then $L \nvdash C \supset A$. By completeness, there exists an intuitionistic Kripke frame $\mathbf{F} \Vdash L$ and a Kripke model M over $\mathbf{F}$ such that $M, u \nVdash C \supset A$ for some u. Hence $M, v \Vdash C$ and $M, v \nVdash A$ for some v accessible from u. Then $M{\uparrow}v, v \Vdash C$ and $M{\uparrow}v, v \nVdash A$, by Lemma 3.3.18 (1), and thus $M{\uparrow}v \Vdash C$ (by 3.2.16). Since C is a pure equality sentence, it follows that $\mathbf{F}{\uparrow}v \vDash C$. $\mathbf{F}{\uparrow}v \vDash L$ by Lemma 3.3.18 (2), so A is refuted in an $(L + C)$-frame. ∎

An analogue of this theorem holds for strong completeness:

Theorem 3.8.4 *Let L be an s.p.l.(=) or an m.p.l.(=), Γ a set of pure equality sentences in the language of L. If L is strongly Kripke (respectively, Kripke sheaf) complete, then $L+\Gamma$ is also strongly Kripke (Kripke sheaf) complete.*

Proof (I) Modal case. If a theory Δ is $(L+\Gamma)$-consistent, then it is $(L+\Box^\infty\Gamma)$-consistent, since obviously $L+\Gamma \vdash \Box^\infty\Gamma$. So the theory $\Box^\infty\Gamma\cup\Delta$ is L-consistent. By strong completeness, there exists a Kripke frame (or a KFE) $\mathbf{F} \models L$ such that $M, u \models \Box^\infty\Gamma \cup \Delta$ for some world u in a Kripke model M over $\mathbf{F}$. Then $M, u \models \Box_\alpha B$ for any $B \in \Gamma$, $\alpha \in I_N^\infty$, so by 3.2.11 it follows that $M, v \models \Gamma$ for any $v \in R^*(u)$. Then as in the proof of 3.8.3, it follows that $\mathbf{F}{\uparrow}u$ is an $(L+\Gamma)$-frame satisfying Δ at u.

(II) Intuitionistic case. If a theory (Δ, Ξ) is $L+\Gamma$-consistent, then the theory $(\Gamma\cup\Delta, \Xi)$ is L-consistent. By strong completeness, it is satisfiable in an L-frame Φ at some world u. Then similarly to the modal case, we obtain that (Δ, Ξ) is satisfiable in the $(L+\Gamma)$-frame $\Phi{\uparrow}u$. ∎

Now let us consider equality expansions and prove some results generalising [Shimura and Suzuki, 1993].

Lemma 3.8.5 *Let $M = (\Phi, \xi)$ be an L-modal (respectively intuitionistic) Kripke model over N-modal (respectively, an* **S4**-*)KFE $\Phi = (F, D^+, E)$ such that $M \models (\Vdash)\ \mathcal{E}_A^Q$ and Q does not occur in A.*

Let $\Phi' := (F, D^+, E')$, where

$$E'(a,b) := \{u \mid M, u \models Q(a,b)\},$$

$M' := (\Phi', \xi')$, where for any $u \in F$, $P \in PL$

$$\xi'_u(P) := \begin{cases} \xi_u(P) & \text{if } P \text{ occurs in } A \text{ or } P = Q, \\ \varnothing & \text{otherwise.} \end{cases}$$

Then

(1) Φ' is an N-modal (respectively, **S4**-*) KFE;*

(2) M' is a modal (respectively, intuitionistic) KFE-model;

(3) for any $u \in M$, for any modal (respectively, intuitionistic) D_u-sentence B in the same predicate letters as A

$$M, u \models (\Vdash)\ B^Q \text{ iff } M', u \models (\Vdash)\ B.$$

The same statements are true under the assumption $M \models (\Vdash)\ \mathcal{E}^Q$, with the following changes: $\xi' = \xi$; (3) holds for any D_u-sentence B without occurrences of Q.

Proof (1) In fact,

$$M \vDash \forall x \forall y (Q(x,y) \supset Q(y,x))$$

implies

$$E'(a,b) \subseteq E'(b,a)$$

for any $a, b \in D^+$; hence obviously $E'(a,b) = E'(b,a)$.

Similarly the other members of $\mathcal{E}$ are responsible for the properties (2)–(4) from Lemma 3.5.2.

For example,

$$M, u \vDash \forall x Q(x,x)$$

means that

$$M, u \vDash Q(a,a)$$

for any $a \in D_u$, hence

$$u \in \bigcup_{a \in D^+} E'(a,a).$$

(2) Next, let us show that M' is a KFE-model. In fact, according to 3.5.2 and 3.5.3, we have to check

$$(a_1, \ldots, a_n) \in \xi'_u(P^n_k) \ \&\ u \in \bigcap_{i=1}^{n} E'(a_i, b_i) \Rightarrow (b_1, \ldots, b_n) \in \xi'_u(P^n_k). \qquad (*)$$

If P^n_k occurs in A, this means

$$M, u \vDash P^n_k(a_1, \ldots, a_n) \wedge Q(a_1, b_1) \wedge \ldots \wedge Q(a_n, b_n) \supset P^n_k(b_1, \ldots, b_n),$$

which follows from $M \vDash \mathcal{E}^Q_A$.

If $P^n_k = Q$, (*) is equivalent to

$$M, u \vDash Q(a_1, a_2) \wedge Q(a_1, b_1) \wedge Q(a_2, b_2) \supset Q(b_1, b_2),$$

which also follows from $\mathcal{E}^Q_A$.

If P^n_k does not occur in A, $P^n_k \neq Q$, (*) holds trivially.

(3) Let $MF^{(=)}_A(D_u)$ be the set of all D_u sentences constructed from $MF^{(=)}_A$. By definition of M',

$$M', u \vDash B \text{ iff } M, u \vDash B$$

for any $u \in M$, for any atomic $B \in MF_A(D_u)$, and thus the equivalence

$$M, u \vDash B^Q \text{ iff } M', u \vDash B$$

holds for any atomic $B \in MF^{=}_A(D_u)$.

By induction we obtain that it holds for any $B \in MF^{=}_A(D_u)$, since Φ, Φ' are based on the same (F, D). ∎

Theorem 3.8.6

(1) Let $\mathcal{C}$ be a class of N-modal propositional Kripke frames such that $\mathbf{ML}(\mathcal{C})$ *is conically expressive. Then* $\mathbf{ML}^{=}(\mathcal{KEC}) = \mathbf{ML}(\mathcal{KEC})^{=}$.

(2) If $\mathcal{C}$ is a class of propositional Kripke **S4**-*frames, then* $\mathbf{IL}^{=}(\mathcal{KEC}) = \mathbf{IL}(\mathcal{KEC})^{=}$.

Proof (1) Since $\mathbf{ML}(\mathcal{C})$ is conically expressive, $\mathbf{ML}(\mathcal{KEC})$ is also conically expressive. So let us show that if a modal sentence A does not contain Q, then

$$(\sharp) \quad A \in \mathbf{ML}^{=}(\mathcal{KEC}) \Leftrightarrow \Box^* \left(\bigwedge \mathcal{E}^Q_A\right) \supset A^Q \in \mathbf{ML}(K\mathcal{EC}).$$

($\Rightarrow$) Suppose

$$\Box^* \left(\bigwedge \mathcal{E}^Q_A\right) \supset A^Q \notin \mathbf{ML}(K\mathcal{EC}).$$

Then

$$M, u \vDash \Box^* \left(\bigwedge \mathcal{E}^Q_A\right) \wedge \neg A^Q$$

for a KFE-model M over $F \in \mathcal{C}$ and for some $u \in M$. By the generation lemma

$$M \uparrow u, u \vDash \Box^* \left(\bigwedge \mathcal{E}^Q_A\right) \wedge \neg A^Q.$$

Hence by a KFE-analogue of Lemma 3.3.39, $M \uparrow u \vDash \mathcal{E}^Q_A$.

Then by Lemma 3.8.5, $M \uparrow u, u \vDash \neg A^Q$ implies

$$(M \uparrow u)', u \vDash \neg A.$$

Since $(M \uparrow u)'$ is a KFE-model over $F \uparrow u$, we have $A \notin \mathbf{ML}^{=}(\mathcal{KE}(\mathcal{C} \uparrow u))$, whence $A \notin \mathbf{ML}^{=}(\mathcal{KEC})$ by Proposition 3.5.28.

($\Leftarrow$) Suppose for some $F \in \mathcal{C}$, $A \notin \mathbf{ML}^{=}(\mathcal{KE}F)$. Then there exists a KFE $\Phi = (F, D, \asymp)$ and a model $N = (\Phi, \xi)$ such that $N, u \nvDash A$ at some world u.

Consider a model M over Φ with the trivial equality such that for any $v \in M, a, b \in D_v$

$$M, v \vDash Q(a, b) \text{ iff } a \asymp_v b,$$

and

$$M, v \vDash B \text{ iff } N, v \vDash B$$

for any D_v-sentence B without Q and equality. Then $M \vDash \mathcal{E}^Q_A$, as one can easily check, and $N = M'$, so we can apply Lemma 3.8.5. Thus $N, u \nvDash A$ implies $M, u \nvDash A^Q$, and therefore $\Box^*(\bigwedge \mathcal{E}^Q_A) \supset A^Q \notin \mathbf{ML}(\mathcal{KE}F) \subseteq \mathbf{ML}(\mathcal{KEC})$.

The equivalence ($\sharp$) together with Proposition 2.14.4 show that sentences without Q are the same in $\mathbf{ML}^{=}(\mathcal{KEC})$ and $\mathbf{ML}(\mathcal{KE}F)^{=}$. Now if A contains Q, it can be replaced with another letter Q' that does not occur in A. If $A' := [Q'(x,y)/Q(x,y)]\, A$, then $A = [Q(x,y)/Q'(x,y)]\, A'$; so for any logic L, $A \in L$ iff $A' \in L$. Therefore all sentences in these logics are the same. Since for any formula A and a logic L, $A \in L$ iff $\overline{\forall} A \in L$, the logics are really equal.

(2) Similarly, in the intuitionistic case due to Proposition 2.14.5, it is sufficient to show that for a sentence A and an extra predicate letter Q

$$A \in \mathbf{IL}^{=}(\mathcal{KEC}) \Leftrightarrow \bigwedge \mathcal{E}_A^Q \supset A^Q \in \mathbf{IL}(\mathcal{KEC}). \qquad (\sharp\sharp)$$

The proof is quite similar to the modal case.

($\Rightarrow$) Suppose $\mathcal{E}_A^Q \supset A^Q \notin \mathbf{IL}(\mathcal{KEC})$. Then for some intuitionistic KFE-model M over $F \in \mathcal{C}$ and for some $u \in M$

$$M, u \Vdash \mathcal{E}_A^Q \text{ and } M, u \not\Vdash A^Q,$$

so by the generation lemma

$$M{\uparrow}u \Vdash \mathcal{E}_A^Q \text{ and } M{\uparrow}u, u \not\Vdash A^Q.$$

Then by Lemma 3.8.5,

$$(M{\uparrow}u)', u \not\Vdash A,$$

which implies $A \notin \mathbf{IL}^{=}(\mathcal{KE}(\mathcal{C}{\uparrow}))$, and so $A \notin \mathbf{IL}^{=}(\mathcal{KEC})$ by Proposition 3.5.28.

($\Leftarrow$) Given $\Phi = (F, D, \asymp)$ and an intuitionistic $N = (\Phi, \xi)$ such that $N, u \not\Vdash A$, the same construction as in the modal case yields an intuitionistic M such that $N = M'$. We leave the details to the reader. ■

Theorem 3.8.7 *Let L be a strongly $\mathcal{KE}$-complete m.p.l. or s.p.l. Then $L^{=}$ is also strongly $\mathcal{KE}$-complete.*

Proof

(I) Consider the modal case first. Let Γ be an $L^{=}$-consistent N- modal theory. We may assume that Q does not occur in Γ — to avoid Q, we can appropriately rename all letters in Γ. Then the theory $\Box^\infty \mathcal{E}_N^Q \cup \Gamma^Q$ is L-consistent.

In fact, otherwise by 2.8.1

$$\Box^\infty \mathcal{E}_N^Q \vdash_L \neg(A_1^Q \wedge \ldots \wedge A_k^Q)$$

for some $A_1, \ldots, A_k \in \Gamma$. Hence $L^{=} \vdash \neg(A_1 \wedge \ldots \wedge A_k)$ by 2.14.4, which means that Γ is $L^{=}$-inconsistent.

Since L is strongly $\mathcal{KE}$-complete, there exists a KFE $\Phi = (F, D^+, E)$[19] such that $\Phi \vDash L$ and for some model $M = (\Phi, \xi)$ and some world u_0 we have $M, u_0 \vDash \Gamma^Q \cup \Box^\infty \mathcal{E}_N^Q$.

We may also assume that u_0 is the root of M; in fact, otherwise M and Φ can be replaced with $M{\uparrow}u_0$, and $\Phi{\uparrow}u_0$, respectively, since by Lemma 3.3.18,

$$M{\uparrow}u_0, u_0 \vDash \Gamma^Q \cup \Box^\infty \mathcal{E}_N^Q \text{ and } \Phi{\uparrow}u_0 \vDash L.$$

Thus $M, u_0 \vDash \Box_\alpha B^Q$ for any $B \in \mathcal{E}_N$, and so by 3.2.11, $M \vDash \mathcal{E}_N^Q$, since every $u \in M$ is covered by some $R_\alpha(u_0)$ (Lemma 1.3.19).

[19] We use an equivalent definition of a KFE from Lemma 3.5.2.

Now consider $\Phi' = (F, D^+, E')$, where

$$E'(a,b) := \{u \mid M, u \vDash Q(a,b)\},$$

and $M' = (\Phi', \xi')$ as in Lemma 3.8.5. Then Φ' is a KFE and $M', u_0 \vDash \Gamma$ by Lemma 3.8.5.

Since all equality axioms are valid in Φ', it remains to show that $\Phi' \vDash L$. It is again sufficient to check that $\Phi' \vDash A$ for any $A \in \overline{L}$ without Q, since otherwise Q can be renamed. So for an arbitrary KFE-model $M_2 = (\Phi', \theta)$, let us show that $M_2 \vDash A$.

Since Q does not occur in A the truth values of A do not depend on $\theta^+(Q)$. So we may further assume that for any u,

$$M_1, u \vDash Q(a,b) \Leftrightarrow u \in E'(a,b).$$

M_2 remains a KFE-model under this assumption, because

$$E'(a_1,a_2) \cap E'(a_1,b_1) \cap E'(a_2,b_2) \subseteq E'(b_1,b_2).$$

Now put $M_1 := (\Phi, \theta)$. Then M_1 is also a KFE-model.

In fact, the latter means

(1) $(a_1, \ldots, a_n) \in \theta_u(P_k^n)$ & $u \in \bigcap_{i=1}^{n} E(a_i,b_i) \Rightarrow (b_1, \ldots, b_n) \in \theta_u(P_k^n)$.

Since M_1' is a KFE-model, we have

(2) $(a_1, \ldots, a_n) \in \theta_u(P_k^n)$ & $u \in \bigcap_{i=1}^{n} E'(a_i,b_i) \Rightarrow (b_1, \ldots, b_n) \in \theta_u(P_k^n)$.

Now (1) follows from (2), and the observation that

(3) $E(a,b) \subseteq E'(a,b)$ holds for any $a, b \in D^+$.

To check (3), note that it is equivalent to

(4) $M, u \vDash a = b \supset Q(a,b)$
provided $a, b \in D_u$.

But this holds, since

$$M, u \vDash a = b \wedge Q(a,a) \supset Q(a,b)$$

as soon as M is a KFE-model, and $M, u \vDash Q(a,a)$ as we already know.

By definition, it follows that $M_2 = M_1'$ as in Lemma 3.8.5. Now $\Phi \vDash L$ implies $M_1 \vDash A$. $A = A^Q$, since A is without equality; hence $M_2 \vDash A$ by Lemma 3.8.5.

(II) The intuitionistic case is considered in a very similar way. If (Γ, Δ) is an $L^=$-consistent intuitionistic theory, then $(\Gamma^Q \cup \mathcal{E}_0^Q, \Delta^Q)$ is also L-consistent.

In fact, otherwise by 2.7.13,

$$\Gamma^Q \cup \mathcal{E}_0^Q \vdash_L \bigvee \Delta_1^Q$$

for some finite $\Delta_1 \subseteq \Delta$, which implies $\bigwedge \Gamma_1^Q \cup \mathcal{E}_0^Q \vdash \bigvee \Delta_1^Q$ for some finite $\Gamma_1 \subseteq \Gamma$ by 2.8.1. Hence $L^= \vdash \bigwedge \Gamma_1 \supset \bigvee \Delta_1$ by 2.14.5, so (Γ, Δ) is $L^=$-inconsistent.

By strong completeness, there exists a KFE $\Phi = (F, D^+, E)$ such that $\Phi \models L$ and for some intuitionistic model $M = (\Phi, \xi)$ and some world u_0, $M, u_0 \Vdash (\Gamma^Q \cup \mathcal{E}_0^Q, \Delta^Q)$. As in the modal case, we may assume that u_0 is the root of M, and thus $M \Vdash \mathcal{E}_0^Q$. So we can define a KFE $\Phi' := (F, D^+, E')$ such that

$$E'(a, b) := \{u \mid M, u \Vdash Q(a, b)\}$$

and $M' = (\Phi', \xi')$ as in 3.8.5. Then by 3.8.5, M' is a KFE-model such that

$$M, u \Vdash A^Q \text{ iff } M', u \Vdash A$$

for any $u \in M$ and D_u-sentence A. Hence $M', u_0 \Vdash (\Gamma, \Delta)$.

$\Phi' \Vdash L$ is proved as in the modal case — every intuitionistic model over Φ' corresponds to an intuitionistic model over Φ (for the basic language $\mathcal{L}_0$). ■

Theorem 3.8.8 *Let L be a $\mathcal{KE}$-complete s.p.l. or a conically expressive m.p.l. Then $L^=$ is also $\mathcal{KE}$-complete.*

Proof Essentially the same as for 3.8.7, but now we should take care of finiteness.

(I) Let us begin with the modal case. Consider an arbitrary formula A in the language of L such that $L^= \nvdash A$.

Then by 2.14.4

$$\Box^* \mathcal{E}_A^Q \nvdash_L A^Q,$$

i.e.

$$L \nvdash (\bigwedge \Box^* \mathcal{E}_A^Q) \supset A^Q,$$

and thus by completeness, there exists a KFE $\Phi = (F, D^+, E)$ such that $\Phi \models L$ and for some world u_0 in some KFE-model $M = (\Phi, \xi)$ we have $M, u_0 \models \Box^* \mathcal{E}_A^Q \cup \{\neg A^Q\}$.

As in 3.8.7, we may assume that u_0 is the root of M and define

$$\Phi' := (F, D^+, E'),$$

where

$$E'(a, b) := \{u \mid M, u \models Q(a, b)\}.$$

and $M' = (\Phi', \xi')$ according to Lemma 3.8.5. Then M' is a KFE-model.

By 3.8.5 we obtain

$$M, u \models B^Q \text{ iff } M', u \models B$$

for any $u \in M$, for any D_u-sentence B using predicate letters only from A, and so $M', u_0 \not\models A$. The proof of $\Phi' \models L$ is exactly the same as in 3.8.7.

(II) In the intuitionistic case suppose $L^= \nvdash A$. Then by 2.14.5, $\mathcal{E}_A^Q \nvdash_L A^Q$, i.e. $L \nvdash \bigwedge \mathcal{E}_A^Q \supset A^Q$, and thus the latter formula is refuted in an intuitionistic

KFE-model $M = (\Phi, \xi)$ over a KFE Φ validating L. Again we may assume that M is rooted and

$$M, u_0 \Vdash (\mathcal{E}_A^Q, \{A^Q\})$$

for the root u_0 of M. Now we define Φ', M' as in 3.8.5. Then M' is intuitionistic and

$$M, u \Vdash B^Q \text{ iff } M', u \Vdash B$$

for any u, whenever B is a D_u-sentence in predicate letters from A. Eventually we obtain $\Phi' \Vdash L$ and $M', u_0 \not\Vdash A$. i.e. Φ' separates A from $L^=$. ■

Remark 3.8.9 The analogue of 3.8.8 does not hold for $\mathcal{K}$-completeness. In fact, **QH** and **QS4** are $\mathcal{K}$-complete, as we shall see in Chapter 7, but the equality-expansions **QH**$^=$, **QS4**$^=$ are $\mathcal{K}$-incomplete, by 3.10.6 below.

Corollary 3.8.10 *If L is an s.p.l. or a conically expressive m.p.l., then L is $\mathcal{KE}$-complete iff $L^=$ is $\mathcal{KE}$-complete.*

Proof 'If' follows from 2.16.15, 'only if' – from 3.8.8. ■

Now let us prove an analogue of 2.16.14 for strong completeness.

Proposition 3.8.11 *Let L be an m.p.l. or an s.p.l., $\mathcal{S}$ a semantics of Kripke frames or Kripke sheaves for the corresponding logics with equality. If $L^=$ is conservative over L and strongly $\mathcal{S}$-complete, then L is also strongly $\mathcal{S}$-complete.*

Proof Consider the modal case. Suppose Γ is an L-consistent theory. Then Γ is $L^=$-consistent, due to the conservativity. So Γ is satisfiable in an $L^=$-frame (in the corresponding semantics), which is obviously an L-frame. ■

Corollary 3.8.12 *If L is an s.p.l. or a conically expressive m.p.l., then L is strongly $\mathcal{KE}$-complete iff $L^=$ is strongly $\mathcal{KE}$-complete.*

Proof 'If' follows from 2.14.8 and 3.8.11, 'only if' – from 3.8.7. ■

3.9 Simulation of varying domains

Consider a formula A in a language with or without equality, and let U be a new unary predicate letter. Let A_U be the formula obtained from A by relativising all quantifiers under U; formally:

$$\begin{aligned}
&A_U := A \text{ for } A \text{ atomic;}\\
&(\exists x B)_U = \exists x (U(x) \wedge B_U);\\
&(\forall x B)_U = \forall x (U(x) \supset B_U);\\
&(B * C)_U = B_U * C_U \quad \text{for } * \in \{\supset, \vee, \wedge\}.
\end{aligned}$$

Next, put

$$A_U^* := \exists x U(x) \supset A_U.$$

Let $M = (\mathbf{F}, \theta)$ be a Kripke model over a rooted frame $\mathbf{F} = (W, R, D)$ with root v_0, such that $M, v_0 \Vdash \exists x U(x)$. Consider the frame $\mathbf{F}^U := (W, R, D^U)$ with

$$D_v^U := \{a \in D_v \mid M, v \Vdash U(a)\}$$

and a Kripke model $M^U := (\mathbf{F}^U, \theta_U)$ such that

$$(\theta^U)^+(P(a_1, \ldots, a_n)) := \{v \mid a_1, \ldots, a_n \in D_v^U\} \cap \theta^+(P(a_1, \ldots, a_n)) =$$
$$\{v \mid M, v \Vdash P(a_1, \ldots, a_n) \wedge \bigwedge_{i=1}^{n} U(a_i)\}$$

for $P \neq U$,

$$(\theta^U)^+(U(a)) := \{v \mid a \in D_v\}.$$

$\mathbf{F}^U$ is well defined. In fact, $D_{v_0}^U \neq \varnothing$ since $M, v_0 \Vdash \exists x U(x)$; by truth preservation $D_w^U \subseteq D_v^U$ whenever wRv.

The model M^U is intuitionistic, since M is intuitionistic.

We can obviously extend the relativisation to formulas with constants. Then the following holds:

Lemma 3.9.1 *For any world v in M, for any D_v-sentence B that does not contain U*

$$M, v \Vdash B_U \Leftrightarrow M^U, v \Vdash B.$$

Proof Easy by in induction. By definition,

$$M, v \Vdash B \Leftrightarrow M^U, v \Vdash B.$$

for any atomic D_v-sentence B. ■

Lemma 3.9.2 *For every Kripke model $M' = (F, D', \theta')$ over a poset F with root v_0 there exists a Kripke model $M = (F \odot V, \theta)$ over F with a constant domain such that $M, v_0 \Vdash \exists x U(x)$ and $M^U = M'$.*

Proof In fact, put

$$V := (D')^+;$$
$$\theta^+(U(a)) := \{v \mid a \in D'_v\};$$
$$\theta^+(P(\mathbf{a})) := \{v \mid M', v \Vdash P(\mathbf{a}) \wedge \bigwedge_{i=1}^{n} U(a_i)\}$$

for $P \in PL^n$, $\mathbf{a} = (a_1, \ldots, a_n)$.

Then obviously $M, u_0 \Vdash \exists x U(x)$ and

$$a \in D_v^U \Leftrightarrow M, v \Vdash U(a) \Leftrightarrow a \in D'_v,$$

and for any $v \in M$, $P \in PL^n$

$$a \in (D_v)^n \text{ iff } M, v \Vdash P(\mathbf{a}) \text{ iff } M', v \Vdash P(\mathbf{a}).$$

Therefore $M' = M^U$. ■

Theorem 3.9.3

(1) Let F be a rooted poset, A a sentence without equality and without occurrences of U. Then

$$A \in \mathbf{IL}(\mathcal{K}F) \Leftrightarrow A_U \in \mathbf{IL}(\mathcal{CK}F).$$

(2) Similarly, if $A \in IF^=$, then

$$A \in \mathbf{IL}^=(\mathcal{KE}F) \Leftrightarrow A_U \in \mathbf{IL}^=(\mathcal{CKE}F).$$

Proof We prove only (1); the proof of (2) is similar. First, there exist a model M_0 over $F \odot V$ and a world $v_0 \in F$ such that $M_0, v_0 \Vdash \exists x U(x)$ and $M_0, v_0 \nVdash A_U$. Put $M := M_0 \uparrow v_0$. Then by the generation lemma and Lemma 3.9.1, $M^U, v_0 \nVdash A$. Thus $A \notin \mathbf{IL}(\mathcal{K}(F \uparrow v_0))$, and so $A \notin \mathbf{IL}(\mathcal{K}F)$.

The other way round, suppose $A \notin \mathbf{IL}(\mathcal{K}F)$, i.e. there exists a frame $\mathbf{F}' = (F, D')$ with a model M' such that $M', v_0 \nVdash A$; now v_0 is root of F. Consider a model M according to Lemma 3.9.2. Then $M' = M^U$, so $M, v_0 \nVdash A_U$ by 3.9.1. Hence $A_U \in \mathbf{IL}(\mathcal{CK}F)$. ∎

Corollary 3.9.4

(1) $\mathbf{IL}(\mathcal{K}F) \leq_m \mathbf{IL}(\mathcal{CK}F)$.

(2) $\mathbf{IL}^=(\mathcal{KE}F) \leq_m \mathbf{IL}^=(\mathcal{CKE}F)$.

3.10 Examples of Kripke semantics

According to the general definition from Chapter 2, the *Kripke semantics generated by* a class $\mathcal{C}$ of Kripke sheaves (or KFEs) is

$$S(\mathcal{C}) = \{\mathbf{ML}(\Phi) \mid \Phi \in \mathcal{C}\}.$$

Let us consider three examples of Kripke semantics.

Definition 3.10.1 *A Kripke frame with equality $\mathbf{F} = (F, D, \asymp)$ is called* monic *if*

$$(*) \quad \forall u, v \in F\ \forall a, b \in (D_u \cap D_v)\ \ (a \asymp_u b \Leftrightarrow a \asymp_v b);$$

$\mathbf{F}$ *is a* KFE with closed equality *(or* CE-KFE*) if for any i for any $a, b \in D^+$ for any $u, v \in F$*

$$(**) \quad \forall i\ \forall u, v \in F\ (uR_iv \Rightarrow \forall a, b \in D_u\ (a \asymp_u b \Leftrightarrow a \asymp_v b));$$

or equivalently, the premise uR_iv can be replaced with uR^*v. An intuitionistic KFE with closed equality is also called a KFE with decidable equality, *or a* DE-KFE.

If $\mathbf{F}$ is presented in the form (F, D^+, E) as in Lemma 3.3.2, then the above conditions are rewritten as follows:

$$(*') \qquad \forall a, b \in D^+ \ (E(a,b) = \varnothing \vee E(a,b) = E(a) \cap E(b));$$

$$(**') \qquad \forall a, b \in D^+ \ \forall i \ E(a) \cap E(b) \cap R_i^{-1}E(a,b) \subseteq E(a,b).$$

Definition 3.10.2 *A disjoint Kripke sheaf* $\Phi = (F, D, \rho)$ *is called* monic *if for any* $a, b \in D^+, \ u, v \in F$

$$(1) \quad aR^*u \ \& \ bR^*u \ \& \ aR^*v \ \& \ bR^*v \ \& \ (a|u) = (b|u) \Rightarrow (a|v) = (b|v);$$

Φ *is a* CE-sheaf *(or a* DE-sheaf *in intuitionistic case) if*

$$(2) \quad \forall u, v \in F \ \forall a, b \in D_u \ \forall i \ (uR_iv \ \& \ (a|v) = (b|v) \Rightarrow a = b).$$

The necessary modification for the non-disjoint case is quite obvious.

These three types of KFEs and Kripke sheaves fully correspond to each other and generate equal semantics:

Lemma 3.10.3

(1) A disjoint Kripke sheaf Φ *is monic iff the KFE* $\mathbf{G}(\Phi)$ *is monic. The other way round, a KFE* $\mathbf{F}$ *is monic iff the Kripke sheaf* $\Theta(\mathbf{F})$ *is monic.*

(2) $\mathbf{F}$ *is a CE-KFE iff* $\Theta(\mathbf{F})$ *is a CE-sheaf.*

Thus a Kripke sheaf Φ is CE iff the KFE $\mathbf{G}(\Phi)$ is CE. Recall that $\Theta(\mathbf{G}(\Phi))$ is isomorphic to Φ.

Proof

Consider the monic case (the CE-case is quite obvious).

Let $\mathbf{F} = (F, D, \asymp)$ be a monic KFE,

$$\Theta(\mathbf{F}) = (F, D', \rho), \ D'_u = (D_u / \asymp_u), \ \rho_{uv}(a_u) = a_v \text{ if } v \in F^u, a \in D_u \subseteq D_v.$$

Assume wR^*y, wR^*v, $w'R^*u$, $w'R^*v$, $a_w \in D'_w$, $b_{w'} \in D_u \cap D_v$ and $a \asymp_u b$. Then $a \asymp_v b$ since $\mathbf{F}$ is monic, i.e. $\rho_{wv}(a_w) = \rho_{w'v}(b_{w'})$.

Now, let $\Phi = (F, D, \rho)$ be a monic disjoint Kripke sheaf, $\mathbf{G}(\Phi) = (F, D', \asymp')$. Let $u, v \in F$, $a, b \in D'_u \cap D'_v, a \asymp'_u b$, i.e. $a \in D_w, b \in D_{w'}$ for some $w, w' \in (R^*)^{-1}(u) \cap (R^*)^{-1}(v)$ and $\rho_{wu}(a) = \rho_{w'u}(b)$. Then $\rho_{wv}(a) = \rho_{w'v}(b)$ since Φ is monic, i.e. $a \asymp'_v b$. ∎

Example 3.10.4 (Warning) In general if $\Theta(\mathbf{F})$ is monic, then $\mathbf{F}$ is not necessarily monic (even in the intuitionistic case). In fact, let us consider the KFE $\mathbf{F} = (F, D, \asymp)$ based on the poset $F = \{u, v\}$, in which u and v are incomparable, $D_u = D_v = \{a, b\}$, $\asymp_u = (D_u)^2$, $\asymp_v = id_{D_v}$. Then $\mathbf{F}$ is not monic, since $a \asymp_u b$, but not $a \asymp_v b$. On the other hand,

$$\Theta(\mathbf{F}) = (F, D', \rho), D'_u = \{a_u\}, D'_v = \{a_v, b_v\}, \rho_{uu} = id_{D'_u}, \rho_{vv} = id_{D'_v}.$$

This Kripke sheaf is monic, since

$$aR^*w \ \& \ aR^*w' \Rightarrow w = w'$$

in $\Theta(\mathbf{F})$.

Speaking informally, the individual $a_u = b_u$ in $\Theta(\mathbf{F})$ does not know that it corresponds to $a_v \neq b_v$.

One can construct a similar example based on a rooted **S4**-frame.

Lemma 3.10.5

(1) For any N-modal predicate Kripke frame $\mathbf{F}$, $i \leq N$

$$\mathbf{F} \vDash \forall x \forall y (x \neq y \supset \Box_i (x \neq y)).$$[20]

(2) For any intuitionistic Kripke frame $\mathbf{F}$, $\mathbf{F} \vDash DE$.

Proof

(1) If $u \vDash a \neq b$ for $a, b \in D_u$, then $a \neq b$, and thus $v \vDash a \neq b$ for any $v \in R_i(u)$; hence $u \vDash \Box_i(a \neq b)$.

(2) We have to show that $u \Vdash a = b \vee a \neq b$ for any $a, b \in D_u$. In fact, if $a = b$, then $u \Vdash a = b$, by definition. If $a \neq b$, then $v \nVdash a = b$ for any $v \in R(u)$; thus $u \Vdash a \neq b$, by definition.

∎

Proposition 3.10.6 *The logics* $\mathbf{QK}^=, \mathbf{QH}^=$ *are $\mathcal{K}$-incomplete.*

Moreover, let $\mathbf{\Lambda}$ *be a modal propositional logic such that* $\mathbf{K} \subseteq \mathbf{\Lambda} \subseteq \mathbf{S4}$ *or* $\mathbf{S4} \subseteq \mathbf{\Lambda} \subset \mathbf{S5}$, *or an intermediate propositional logic* $\neq \mathbf{CL}$. *Then* $\mathbf{Q\Lambda}^=$ *is $\mathcal{K}$-incomplete.*

Proof In this case a two-element reflexive chain Z_2 validates $\mathbf{\Lambda}$. Thus $\mathbf{\Lambda}$ (and $\mathbf{Q\Lambda}^=$) is valid in the Kripke sheaf $\Phi = (Z_2, D, \rho)$, in which $D = \{a, b\}$, $D_1 = \{c\}$, $a|1 = b|1 = c$. Obviously, $\Phi \nvDash CE$ and $\Phi \nVdash DE$. Therefore we obtain $\mathbf{Q\Lambda}^= \nvdash CE$ in the modal case and $\mathbf{Q\Lambda}^= \nvdash DE$ in the intuitionistic case. But by Lemma 3.10.5, $\mathbf{QK}^= \vDash_{\mathcal{K}} CE$, $\mathbf{QH}^= \vDash_{\mathcal{K}} DE$, which yields incompleteness.

∎

The semantics generated by monic KFEs (or Kripke sheaves) is denoted by $\mathcal{MK}_N^{(=)}$, $\mathcal{MK}_{int}^{(=)}$, and the semantics generated by CE-KFEs (or Kripke sheaves) is denoted by $\mathcal{KCE}_N^{(=)}$, $\mathcal{KCE}_{int}^{(=)}$. Obviously every monic KFE is CE (and similarly for Kripke sheaves). Lemma 3.10.5 shows that every simple KFE (or a *simple Kripke sheaf* — with inclusions ρ_{uv}) is also monic. So we have:

$$\mathcal{K}_N^{(=)} \subseteq \mathcal{MK}_N^{(=)} \subseteq \mathcal{KCE}_N^{(=)},$$

and similarly for superintuitionistic logics. All these inclusions are actually strict, as we will see later on.

[20]Recall that this formula is denoted by CE_i (Section 2.6).

Lemma 3.10.7

(1) A KFE (or a Kripke sheaf) is CE iff it validates $CE : \forall x \forall y\ (\Diamond(x = y) \supset (x = y))$.

(2) An intuitionistic KFE (or a Kripke sheaf) is DE iff it validates $DE : \forall x \forall y\ ((x = y) \lor (x \neq y))$.

Proof

(1) immediate

(2) (for KFEs). Let $a, b \in D_u$. Clearly $u \not\models (a = b) \lor (a \neq b)$ iff $\neg(a \asymp_u b)\ \&\ \exists v\ (uRv\ \&\ a \asymp_v b)$.

■

Example 3.10.4 shows that monic KFEs do not have an adequate logical characterisation. In fact, KFEs $\mathbf{F}$ and $\mathbf{G}(\Theta(\mathbf{F}))$ have the same modal logic, but only one of them is monic. A similar example can be constructed for monic Kripke sheaves.

In some special cases the three semantics introduced are equivalent. Let us give two examples.

Recall that *simple* Kripke sheaves correspond to Kripke frames with trivial equality.

Lemma 3.10.8 *Every CE-Kripke sheaf over an* **S4***-tree* F *is isomorphic to a simple Kripke sheaf over* F.

Proof Let $\Phi = (F, D, \rho)$ be the original Kripke sheaf (which we suppose disjoint) and let $F = (W, R)$. Consider the following relation between individuals:

$$a \sim b := \exists d\ (d\rho a\ \&\ d\rho b).$$

Since F is a tree, $\sim$ is an equivalence relation. In fact, if $a \sim b$, $b \sim c$ and $d\rho a$, $d\rho b$, $e\rho c$, $e\rho b$, $d \in D_u$, $e \in D_v$, then both u, v see the world of b, and thus they are comparable. We may assume that uRv. So $d\rho d'$ for some $d' \in D_v$, and also $d'\rho b$. Hence $d\rho d'$, so by (CE) it follows that $d' = e$, and thus $d\rho c$, which implies $a \sim c$.

Then we factorise the domains: $\widetilde{D_u} := D_u/\sim$. It follows that $\widetilde{D_u} \subseteq \widetilde{D_v}$ whenever uRv, hence we obtain a PKF $(F, \widetilde{D})$. One can easily show that Φ is isomorphic to $\Theta(F)$. ■

Definition 3.10.9 *A propositional Kripke frame* (W, R) *is called*

- directed *iff* $\forall x, y \in W\ R(x) \cap R(y) \neq \varnothing$*;*
- weakly directed *iff* $\forall x, y \in W\ (R(x) \cap R(y)$ *is directed (if it is non-empty) as the frame with the relation* R *or* R^{-1} *restricted to it).*

Lemma 3.10.10 *Every CE-sheaf over a weakly directed frame is monic.*

Proof Assume that individuals a, b in a given CE-sheaf Φ are both related to worlds u, v. Also assume that $(a|u) = (b|u)$, uRw, vRw. Then

$$(a|w) = (a|u)|w = (b|u)|w = (b|w),$$

i.e.

$$(a|v)|w = (b|v)|w,$$

and thus $(a|v) = (b|v)$ by (2). Similarly we obtain $(a|v) = (b|v)$ in the case, when a, b are related to some world in $R^{-1}(u) \cap R^{-1}(v)$. ∎

Lemma 3.10.11 *Let* $\Phi = (W, R, D, \rho)$ *be a monic Kripke sheaf over a directed frame. Then* $\mathbf{ML}(\Phi) = \mathbf{ML}(\mathbf{F})$ *for some PKF* $\mathbf{F}$ *over* (W, R).

Proof Since (W, R) is directed and the inheritance relation ρ is transitive, we can construct the direct limit

$$D_0^+ = \varinjlim(D_u : u \in W).$$

Namely, we take $D_0^+ = D^+ / \sim$, where

$$a \sim b \text{ iff } a,\ b \text{ have a common inheritor.}$$

There exist natural embeddings $i_u : D_u \longrightarrow D_0^+$ (the injectivity follows from (1)), such that the diagram

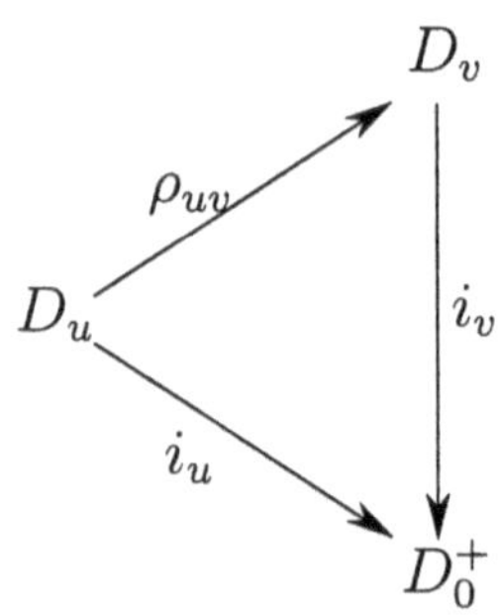

commutes. Consider the PKF $\mathbf{F} = (W, R, D_0)$ with $(D_0)_u = i_u(D_u)$. It can be easily proved that Φ and $\Theta(\mathbf{F})$ (see Section 3.3) are isomorphic. Thus, $\mathbf{ML}(\Phi) = \mathbf{ML}(\mathbf{F})$ by Lemma 3.3.4. ∎

Corollary 3.10.12 $\mathcal{K}_{\mathrm{dir}} = \mathcal{MK}_{\mathrm{dir}} = \mathcal{KCE}_{\mathrm{dir}}$, *where* dir *means the restriction to directed frames.*

Definition 3.10.13 *A KFE is called a* KFE with a constant domain *if all its individual domains are equal.*

We use the notation $\mathbf{F} = (F \odot V, \asymp)$ for KFEs with a constant domain (where $V = D_u$ for all $u \in F$).

Definition 3.10.14 *A KFE with a set of possible worlds W is called* flabby *if the measure of existence of every its individual is W.*

So, a KFE is flabby iff it has a constant domain.

Now let us describe flabby Kripke sheaves corresponding to flabby KFEs.

Definition 3.10.15 *A Kripke sheaf $\Phi = (F, D, \rho)$ is called* flabby *if there exists a domain D_0 and a family of surjections $\rho_{0u} : D_0 \longrightarrow D_u$ for $u \in F$ such that $\rho_{0v} = \rho_{0u} \circ \rho_{uv}$, whenever uR^*v in F,*

Lemma 3.10.16

(1) If $\mathbf{F} = (F, D, \asymp)$ is a flabby KFE, then $\Theta(\mathbf{F})$ is a flabby Kripke sheaf.

(2) For a flabby Kripke sheaf $\Phi = (F, D, \rho)$ there exists a flabby KFE $\mathbf{F}$ such that $\Theta(\mathbf{F})$ is isomorphic to Φ.

Proof

(1) Let D_0 be the domain of any world in $\mathbf{F}$ and $\rho_0(a) = a_u (= a/\asymp_u)$ for $u \in F$.

(2) Let $\mathbf{F} = (F \odot D_0, \asymp)$, where $a \asymp_u b$ iff $\rho_{0u}(a) = \rho_{0u}(b)$.

■

Definition 3.10.17 *A Kripke sheaf (F, D, ρ) is called* meek *if all the maps ρ_{uv} (for uR^*v) are surjective. A KFE $(F, D, \asymp)$ is called* meek *if*

$$\forall i \forall u\, \forall v \in R_i(u)\, \forall b \in D_v\, \exists a \in D_u\, a \asymp_u b,$$

i.e. if any individual in any accessible world is locally equal to some individual from the present world.

Obviously, every flabby Kripke sheaf (or KFE) is meek — if $\rho_{0u} \circ \rho_{uv}$ is surjective, then ρ_{uv} is surjective. Every flabby KFE is meek, since it has a constant domain.

Lemma 3.10.18 *Meek Kripke sheaves, meek KFEs, flabby Kripke sheaves, and flabby KFEs generate the same semantics.*

Proof Note that every meek Kripke sheaf over a rooted frame is flabby. ■

Lemma 3.10.19 $CD^T \in (\mathbf{QS4} + Ba)$.

Proof This follows from Kripke completeness of $\mathbf{QS4} + Ba$, see Chapter 7. A syntactic proof is an exercise for the reader. ■

Remark 3.10.20 In Chapter 7 we will show that

$$\mathbf{IL}(\mathbf{S4}\text{-PKFs with constant domains}) = \mathbf{QH} + CD,$$

$$\mathbf{ML}(N\text{-modal PKFs with constant domains}) = \mathbf{QS4} + \{Ba_1, \ldots, Ba_N\}.$$

Remark 3.10.21 In a meek Kripke sheaf the following holds:

$$u \vDash \forall x B(x, a_1, \ldots, a_n) \text{ iff } \forall c \in D_u\ u \vDash B(c, a_1, \ldots, a_n).$$

Lemma 3.10.22 *(i) Let Φ be an* **S4***-based Kripke sheaf. Then:*

$$CD \in \mathbf{IL}(\Phi) \text{ iff } Ba \in \mathbf{ML}(\Phi) \text{ iff } \Phi \text{ is meek.}$$

(ii) For an **S4***-based KFE* **F***,*

$$\mathbf{F} \Vdash CD \text{ iff } \mathbf{F} \text{ is meek.}$$

(iii) For an **S4***-based PKF* **F***,*

$$\mathbf{F} \Vdash CD \text{ iff every its cone } \mathbf{F}{\uparrow}u \text{ has a constant domain.}$$

(1m) For a N-modal Kripke sheaf Φ,

$$\Phi \vDash \bigwedge_{i=1}^{N} Ba_i \text{ iff } \Phi \text{ is meek.}$$

(2m) For an N-modal KFE **F**

$$\mathbf{F} \vDash \bigwedge_{i=1}^{N} Ba_i \text{ if } \mathbf{F} \text{ is meek.}$$

(3m) For an N-modal PKF **F***,*

$$\mathbf{F} \vDash \bigwedge_{i=1}^{N} Ba_i \text{ iff every } \mathbf{F}{\uparrow}u \text{ has a constant domain.}$$

Proof
(1) Let us consider only the 1-modal case; the intuitionistic case is similar.

(Only if.) Let $v, w \in F, vRw$ and suppose there is $b_0 \in D_w - \rho_{vw}(D_v)$. Take the valuation ξ in Φ such that $\xi, u \vDash P(a)$ iff vRu and $a \in \rho_{vu}(D_v)$.

Then $v \vDash \forall x \Box P(x)$, since $u \vDash P(\rho_{vu}(a))$ for any u such that vRu and for any $a \in D_v$. On the other hand, $w \nvDash P(b_0)$. It follows that $w \nvDash \forall x P(x)$ and $v \nvDash \Box \forall x P(x)$.

(If.) Suppose that $\xi, u \vDash \forall x \Box P(x)$ and $\xi, u \nvDash \Box \forall P(x)$ for some valuation ξ in Φ and $u \in F$. Then $\xi, v \nvDash P(b)$ for some v such that uRv and some $u \in D_v$. Since Φ is meek, we can take $a \in D_u$ such that $b = \rho_{uv}(a)$. Then $u \nvDash \Box P(a)$ and $u \nvDash \forall x \Box P(x)$. This is a contradiction.

(2) (Only if.) Let $v, w \in F, vRw$, and suppose there is $b_0 \in D_w$ such that $\forall b \in D_v \; b \not\asymp_w b_0$. Take the valuation ξ in $\mathbf{F}$ such that:

$$\xi, u \Vdash Q \text{ iff } vRu \,\&\, u \neq v,$$

$$\xi, u \Vdash P(a) \text{ iff } vRu \,\&\, \exists a' \in D_v \; a \asymp_u a'.$$

Then $v \Vdash \forall x(Q \vee P(x))$, since $v \Vdash P(a)$ for any $a \in D_v$ and $u \Vdash Q$ for any $u \in R(v)$ with $u \neq v$. On the other hand, $v \nVdash Q$ and $v \nVdash \forall x P(x)$ since $w \nVdash P(b_0)$.

(If.) Suppose that $\xi, u \Vdash \forall x(Q \vee P(x))$, $\xi, u \nVdash Q, u \nVdash \forall x P(x)$ for some valuation ξ in $\mathbf{F}$ and $u \in F$. Then $v \nVdash P(b)$ for some $v \in R(u)$, $b \in D_v$. Since $\mathbf{F}$ is meek, we can take $a \in D_u$ such that $b \asymp_v a$. Then $u \nVdash P(a)$ and $u \nVdash Q \vee P(a)$. This is a contradiction.

(3) This directly follows from (2) and from the obvious fact that a PKF $\mathbf{F}$ has a constant domain iff its corresponding KFE with trivial equality is meek. ∎

As we noticed, every flabby Kripke sheaf is meek. On the other hand, a rooted meek Kripke sheaf is flabby.

Therefore, flabby Kripke sheaves generate the semantics equivalent to meek Kripke sheaves. Moreover, this semantics equals the semantics of flabby KFEs.

Let $\mathcal{CK}$ be the semantics generated by all PKFs with constant domains.

Definition 3.10.23 *A Kripke sheaf is called* bijective *if all its transition maps are bijections.*

Let $\mathcal{BK}$ be the semantics generated by all bijective Kripke sheaves.

Proposition 3.10.24 $\mathcal{CK} \subset \mathcal{BK}$.

Proof $\mathcal{CK} \subseteq \mathcal{BK}$ since every PKF with a constant domain $\mathbf{F}$ corresponds to the bijective Kripke sheaf $\Theta(\mathbf{F})$. To prove that the converse inclusion is not true, consider the bijective Kripke sheaf $\Phi_0 := (W, R, D, \rho)$, where

$$W = \{u, v, w\},\; D = \{a, b, c, d\},\; D_u = \{a\},\; D_v = \{b\},\; D_w = \{c, d\},$$

$$R = id_W \cup \{(u, v)\},\; \rho = id_D \cup \{(a, b)\},$$

see Fig. 3.4.

Let

$$C := \forall x \forall y (P(x) \supset P(y)),$$

$$K' := q \supset \Box q.$$

Then obviously,

$$\Phi_0 \models C' \vee K',$$

but

$$\Phi_0 \not\models C'.$$

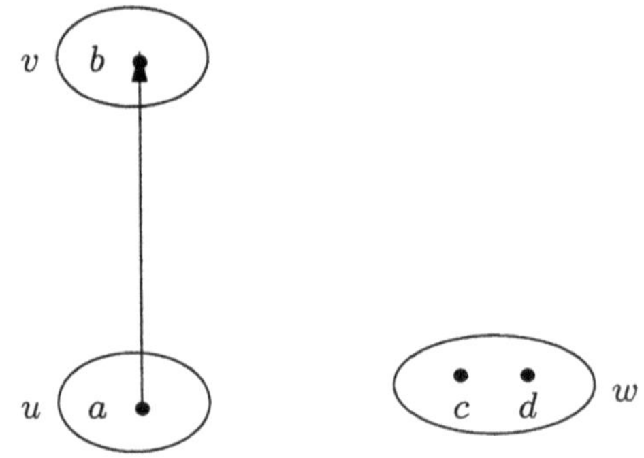

Figure 3.4.

However, for every PKF $\mathbf{F} = F \odot V$, we have

$$|V| = 1 \Rightarrow \mathbf{F} \models C,$$

and

$$|V| > 1 \,\&\, \mathbf{F} \not\models K' \Rightarrow \mathbf{F} \not\models C' \vee K';$$

so

$$\mathbf{F} \models C' \vee K' \text{ only if } (\mathbf{F} \models C' \text{ or } \mathbf{F} \models K').$$

Thus $\mathbf{ML}(\Phi_0) \notin \mathcal{CK}$. ∎

The above example also shows:

Lemma 3.10.25 *$\mathcal{CK}$ lacks the collection property (CP) from Section 2.16.*

Proof In fact,

$$\mathbf{ML}(\Phi_0) = \mathbf{ML}(\Phi_1) \cap \mathbf{ML}(\Phi_2),$$

where Φ_1 are Φ_2 the restrictions of Φ_0 respectively to $\{u, v\}$ and $\{w\}$; on the other hand,

$$\mathbf{ML}(\Phi_1), \mathbf{ML}(\Phi_2) \in \mathcal{CK}.$$

∎

Nevertheless, $\mathcal{CK}$ and $\mathcal{BK}$ are equivalent, as we shall see below.

Lemma 3.10.26 *Let $\Phi = (F, D, \rho)$ be a bijective Kripke sheaf over a rooted frame F. Then* $\mathbf{ML}(\Phi) \in \mathcal{CK}$.

Proof Let u_0 be the root of F, $D_0 = D_{u_0}$, and consider the PKF $\mathbf{F} = F \odot D_0$. Since ρ gives rise to a family of bijections between D_0 and each D_u, the Kripke sheaves $\Theta(\mathbf{F})$ and Φ are isomorphic. So, $\mathbf{ML}(\Phi) = \mathbf{ML}(\mathbf{F})$ by Lemma 3.3.4. ∎

Corollary 3.10.27 *$\mathcal{CK}_r = \mathcal{BK}_r$ (r means the restriction to rooted frames).*

Lemma 3.10.28 *Every logic from $\mathcal{BK}$ is $\mathcal{CK}$-complete.*

Proof Let Φ be a bijective Kripke sheaf over F. Then

$$\mathbf{ML}(\Phi) = \bigcap_{u \in F} \mathbf{ML}(\Phi{\uparrow}u)$$

by Lemma 3.7.17. Thus $\mathbf{ML}(\Phi)$ is $\mathcal{CK}$-complete by Lemmas 3.10.26 and 2.16.3. ∎

Lemmas 3.10.24, 3.10.28 and Corollary 2.12.4 yield

Corollary 3.10.29 $\mathcal{CK} \simeq \mathcal{BK}$.

Here is the sequence of all Kripke semantics considered above:

$$\mathcal{BK} \simeq \mathcal{CK} \prec \mathcal{K} \prec \mathcal{MK} \prec \mathcal{KCE} \prec \mathcal{KE}.$$

3.11 On logics with closed or decidable equality

In this section we show that unlike equality-expansions, extensions with closed or decidable equality may be nonconservative.

3.11.1 Modal case

Consider the following modal formula:

$$Ba^1 := \Diamond \exists x P(x) \supset \exists x \Diamond \exists y (P(y) \wedge \Diamond (x = y)).$$

Note that

$$\mathbf{QS4}^{=} + Ba^1 \subseteq \mathbf{QS4}^{=} + Ba.$$

We will see that these inclusions are actually strict (Lemma 3.11.4). Nevertheless:

Lemma 3.11.1 $\mathbf{QS4}^{=c} + Ba = \mathbf{QS4}^{=c} + Ba^1$.

Proof We shall use 'naive reasoning' in $\mathbf{QS4}^{=c} + Ba^1$. Assume that $\Diamond \exists x P(x)$. Then (from Ba^1)

$$\exists x \Diamond \exists y (P(y) \wedge \Diamond (x = y)).$$

Using $\Diamond(x = y) \supset x = y$, we obtain

$$\exists x \Diamond \exists y (P(y) \wedge (x = y)),$$

i.e. $\exists x \Diamond P(x)$. ∎

Corollary 3.11.2 $\mathbf{QS4}^{=} + Ba^1 \vDash_{\mathcal{K}} Ba$.

Lemma 3.11.3 *Let Φ be an* **S4***-based Kripke sheaf. Then $\Phi \vDash Ba_1$ iff Φ satisfies:*

(1) $\forall u \forall v\ (uRv \Rightarrow \forall b \in D_v\ \exists a \in D_u\ \exists w\ (vRw \& \rho_{uw}(a) = \rho_{vw}(b)))$.

Proof (Only if.) Let $u, v \in F, uRv$ and suppose there is $b \in D_v$ such that $\forall a \in D_u\ \forall w\ (uRw \Rightarrow \rho_{uw}(a) \neq \rho_{uw}(b))$. Take the valuation ξ in Φ such that

$$\xi, s \vDash P(c) \text{ iff } s = v \text{ and } c = b.$$

Then clearly $\xi, u \vDash \Diamond\exists x P(x)$. On the other hand, suppose $\xi, u \vDash \exists x \Diamond \exists y(P(y) \wedge \Diamond(x = y))$. Then there are $a \in D_u, w \in F$ and $c \in D_w$ such that $\xi, w \vDash P(c) \wedge \Diamond(\rho_{uv}(a) = c)$. Since $\xi, w \vDash P(a)$, we have $w = v$ and $c = b$. We have $\xi, v \vDash \Diamond(\rho_{uw}(a) = b)$, which leads to a contradiction.

(If.) Suppose there is a valuation ξ in Φ and $u \in F$ such that $\xi, u \vDash \Diamond\exists x P(x)$ and $\xi, u \nvDash \exists x \Diamond(\exists y(P(y) \wedge \Diamond(x = y))$. Then there are $v \in F$ and $b \in D_v$ such that uRv and $\xi, v \vDash P(b)$. Since Φ satisfies (1), there are $a \in D_u$ and $w \in F$ such that vRw and $\rho_{uw}(a) = \rho_{vw}(b)$. It follows that $\xi, v \vDash \Diamond(\rho_{uv}(a) = b)$ and thus $\xi, v \vDash \exists y(P(y) \wedge \Diamond(\rho_{uv}(a) = y))$. We have a contradiction. ∎

Lemma 3.11.4 **QS4**$^=$ $+ Ba_1 \nvDash_{\mathcal{KE}} Ba$.

Proof Consider the Kripke sheaf Φ depicted at Figure 3.5. It is easily seen that Φ satisfies (2), but it is not meek since b_1 does not have a predecessor in u_0; thus $\Phi \vDash Ba^1$, $\Phi \nvDash Ba$. ∎

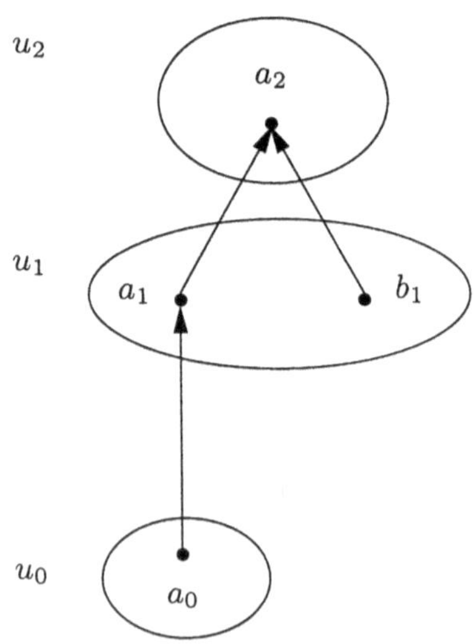

Figure 3.5.

Corollary 3.11.5

(1) $LB_1 :=$ **QS4**$^=$ $+ Ba^1$ *is* $\mathcal{K}$*-incomplete.*

(2) $\mathcal{K} \prec \mathcal{KE}$.

Remark 3.11.6 One can repeat the argument for logics without equality: Ba^1 should be replaced with

$$Ba'^1 := \Box\forall x Q(x,x) \wedge \Diamond\exists x P(x) \supset \exists x \Diamond \exists y (P(y) \wedge \Diamond Q(x,y)).$$

3.11.2 Intuitionistic case

We shall use the following intuitionistic formulas:

$$\begin{array}{l} E := \neg\neg\exists x P(x) \supset \exists x \neg\neg P(x); \\ E_1 := \exists x \forall y (\neg P(x) \supset \neg P(y)); \\ E_2 := \exists x \forall y (\neg P(y) \supset \neg P(x)); \\ F := \exists x \forall y (P(x) \supset P(y)); \\ F_0 := \forall x \neg\neg P(x) \supset \neg\neg \forall x P(x); \\ G := \exists x \forall y (P(y) \supset P(x)); \\ U := \forall x \forall y (P(x) \supset P(y)); \\ U_0 := \forall x \forall y (\neg P(x) \supset \neg P(y)); \\ C := \neg\neg p \supset p; \\ J := \neg\neg q \vee \neg q; \\ Z := (p \supset q) \vee (q \supset p). \end{array}$$

The corresponding logics were considered in [Umezawa, 1956], where many inclusions and equalities between them were proved, in particular:

(1) $\mathbf{QH} + E = \mathbf{QH} + E_1 = \mathbf{QH} + E_2$,

(2) $\mathbf{QH} + F_0 = \mathbf{QH} + \neg\neg F$.

One can easily see that

$$\mathbf{QH}^= + U_0 \subset \mathbf{QH}^= + \neg\neg U \subset \mathbf{QH}^= + U = \mathbf{QH}^= + \forall x \forall y (x = y).$$

We also have:

Lemma 3.11.7

(1) $\mathbf{QH}^{=s} + U = \mathbf{QH}^{=s} + U_0$;

(2) $\mathbf{QH}^{=s} + U_0 \vee A = \mathbf{QH}^{=s} + U \vee A$ *for every sentence A without occurrences of P.*

Proof

(1) From $R(x) \supset \forall y \neg\neg R(y)$ by substitution we obtain

$$z = x \supset \forall y \neg\neg\ z = y,$$

and

$$x = x \supset \forall y \neg\neg\ x = y.$$

Now, $x = x$ and $\neg\neg\ x = y \supset x = y$ (in $\mathbf{QH}^{=s}$) yield $\forall x \forall y (x = y)$.

(2) Similar. ■

Lemma 3.11.8 $(U \vee J) \in (\mathbf{QH}^{=s} + E)$.

Proof ('Naive argument'). Due to 3.11.7 it is sufficient to show that

$$(U_0 \vee J) \in (\mathbf{QH}^{=s} + E).$$

E_1 and E_2 provide x_1, x_2 such that

$$\forall y(\neg P(y) \supset \neg P(x_1))$$

and

$$\forall y(\neg P(x_2) \supset \neg P(y)).$$

If $x_1 = x_2$ we have U_0, so suppose $x_1 \neq x_2$. Let

$$R(x) := (\neg\neg q \wedge (x = x_1)) \vee (\neg q \wedge (x = x_2)).$$

It is easily seen that $\neg\neg\exists x R(x)$, so due to E, there exists x_0 such that $\neg\neg R(x_0)$. Since $(x_0 = x_1) \vee (x_0 \neq x_1)$, we obtain J. ■

Corollary 3.11.9 $\mathbf{QH} + E \models_{\mathcal{K}} U \vee J$.

Lemma 3.11.10 $\mathbf{QH} + E \not\models_{\mathcal{KE}} U \vee J$.

Proof The Kripke sheaf Φ depicted at Figure 3.6, does not validate $U \vee J$, as it is easily seen. If $u \models \neg\neg\exists x P(x)$ in some model on Φ then $v_1 \models P(b), V_2 \models P(c_i)$ for some i and so, $u \models \neg\neg P(a_i)$.

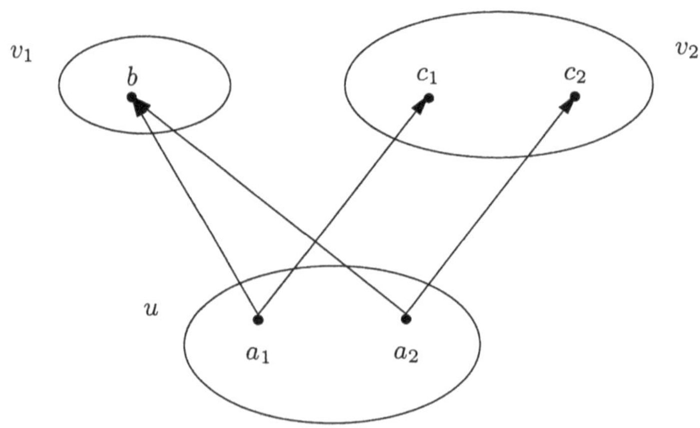

Figure 3.6.

■

Corollary 3.11.11

(1) $\mathbf{QH} + E$ *is* $\mathcal{K}$*-incomplete.*

(2) $\mathcal{K} \prec \mathcal{KE}$ *(intuitionistic case).*

(3) $\mathbf{QH}^{=c} + E$ *is not conservative w.r.t.* $\mathbf{QH} + E$.

Remark 3.11.12 Kripke-incompleteness (i.e. $\mathcal{K}$-incompleteness) of $\mathbf{QH} + E$ was first observed by H.Ono [1973]. Namely, he proved that $\mathbf{QH} + E \vDash_{\mathcal{K}} CD$ but $\mathbf{QH}+E \nvDash_{\mathcal{CT}} CD$ (the semantics $\mathcal{CT}$ will be defined in Chapter 4). Actually $\mathbf{QH} + E \nvDash_{\mathcal{KE}} CD$ also holds, and Figure 3.7 shows a Kripke sheaf needed for the proof.

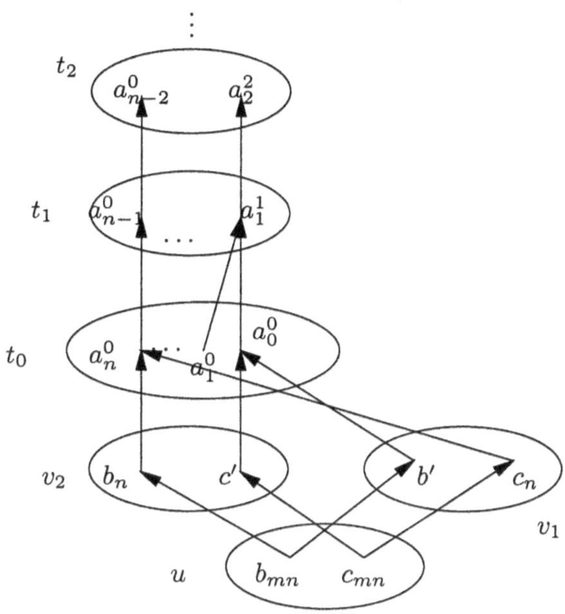

Figure 3.7.

3.12 Translations into classical logic

In Section 1.8 we showed that in many cases Kripke-complete modal or superintuitionistic propositional logics are recursively axiomatisable and complete w.r.t. countable frames. To prove these results, we used the standard translation of propositional modal formulas into classical first-order formulas. In this section we extend this translation to modal first-order formulas, cf. [van Benthem, 1983].

Let us consider the two-sorted classical predicate language $\mathcal{L}2_N$. The variables of the first sort (ranging over individuals) are taken from the same set Var as in our basic language $\mathcal{L}_N$. The second sort has a countable set of variables (ranging over worlds) $Wvar := \{w_n \mid n \in \omega\}$. $\mathcal{L}2_N$ also contains equality[21]

[21]Strictly speaking, there are two kinds of equality - for worlds and for individuals.

and binary predicate letters $R_1, \ldots, R_N, U$; R_i are of type (worlds $\times$ worlds) and U is of type (worlds $\times$ individuals); thus atomic formulas are of the form $R_i(w_m, w_n)$ or $U(w_m, v_n)$ or $w_m = w_n$ or $v_m = v_n$. $\mathcal{L}2_N^\star$ is the expansion of $\mathcal{L}2_N$ with $(n+1)$-ary predicate symbols $P_j^{n\star}$ of type (worlds $\times$ individuals^n) corresponding to n-ary predicate symbols $P_j^n(\mathbf{x})$ of our basic language $\mathcal{L}_N$. Consider the following classical $\mathcal{L}2_N^\star$-theory:

$$\Gamma_0 := \{Exd\} \cup \{A_j^n \mid n, j \geq 0\},$$

where

$$Exd := \forall w_0 \exists v_0 U(w_0, v_0) \wedge \forall v_0 \exists w_0 U(w_0, v_0) \wedge$$
$$\bigwedge_{i=1}^{N} \forall w_1 \forall w_2 \forall v_0 (R_i(w_1, w_2) \wedge U(w_1, v_0) \supset U(w_2, v_0)),$$
$$A_j^n := \forall w_0 \forall v_1 \ldots \forall v_n (P_j^{n\star}(w_0, v_1, \ldots, v_n) \supset \bigwedge_{i=1}^{n} U(w_0, v_i)).$$

The intended interpretation of the formula $U(w_0, x)$ is $x \in D_{w_0}$. So the conjuncts of Exd respectively mean that every individual domain is non-empty, every individual belongs to some domain, and the domains are expanding with respect to R_i. The formula $P_j^{n\star}(w_0, \mathbf{x})$ asserts that $w_0 \models P_j^n(\mathbf{x})$; thus A_j^n means that an atomic formula can be true at world w_0 only for individuals from D_{w_0}.

Given a Kripke frame $\mathbf{F} = (F, D)$ based on a propositional frame $F = (W, \rho_1, \ldots, \rho_N)$, we can construct the following (*associated*) $\mathcal{L}2_N$-structure $\mathbf{F}^\star$ expanding $F^\star$ and satisfying Exd, in which the universes of sorts 1 and 2 are the sets W and D^+ respectively, and for any u_1, u_2, u, a

$$\mathbf{F}^\star \models R_i(u_1, u_2) \text{ iff } u_1 \rho_i u_2,$$
$$\mathbf{F}^\star \models U(u, a) \text{ iff } a \in D_u.$$

Next, a Kripke model $M = (F, D, \xi)$ gives rise to an *associated* $\mathcal{L}2_N^\star$-structure $M^\star$ satisfying Γ_0 and expanding $\mathbf{F}^\star$ such that for any u, $\mathbf{a}$

$$M^\star \models P_j^{n\star}(u, \mathbf{a}) \text{ iff } \mathbf{a} \in D_u^n \ \&\ M, u \models P_j^n(\mathbf{a}).$$

Then we have

Lemma 3.12.1 *Every classical model of Exd is associated with some predicate Kripke frame. Every classical model of Γ_0 is associated with some predicate Kripke model.*

Proof In fact, let μ be a model of Γ_0. To define M, let D^+ be the domain of the first sort, W the domain of the second sort. The accessibility relations ρ_i in M are taken from μ. Then put for every $u \in W$,

$$D_u := \{a \in D^+ \mid \mu \models U(u, a)\},$$

and

$$\xi^+(P_k^n(\mathbf{a})) := \{u \in W \mid \mu \models P_k^{n\star}(u, \mathbf{a})\},$$

for $\mathbf{a} \in D_u^n$. From the above remarks it follows that $\mu = M^\star$. In the same way a model of Exd can be presented as $\mathbf{F}^\star$ for some $\mathbf{F}$. ∎

Now by induction let us construct an $\mathcal{L}2_N^{\star}$-formula $A^{\star}(w_0, \mathbf{x})$, the standard translation of a first-order modal formula $A(\mathbf{x})$:

$$\begin{array}{l} P_j^n(\mathbf{x})^{\star} := P_j^{n\star}(w_0, \mathbf{x}), \\ (x_1 = x_2)^{\star} := (x_1 = x_2), \\ \bot^{\star} := \bot, \\ (B \supset C)^{\star} := B^{\star} \supset C^{\star}, \\ (\exists y B)^{\star} := \exists y (U(w_0, y) \wedge B^{\star}), \\ (\Box_i B)^{\star}(w_0, \mathbf{x}) := \forall w_1 (R(w_0, w_1) \supset B^{\star}(w_1, \mathbf{x})). \end{array}$$

Lemma 3.12.2 *(1) $M, u \vDash A(\mathbf{a})$ (modally) iff $M^{\star} \vDash A^{\star}(u, \mathbf{a})$ (classically) for any N-modal Kripke model M, N-modal formula A, $u \in M$, $\mathbf{a} \in D_u^n$.*

(2) $\mathbf{F} \vDash A$ iff $\forall \xi\ (\mathbf{F}, \xi)^{\star} \vDash \forall w_0 A^{\star}(w_0)$ for an $\mathcal{L}_N$-sentence A and an N-modal frame $\mathbf{F}$.

Proof
(1) By induction. Let us consider the case $A(\mathbf{x}) = \Box_k B(\mathbf{x})$.

$$\begin{array}{l} M, u \vDash A(\mathbf{a}) \Leftrightarrow \forall u_1 \in \varrho_k(u) \quad M, u_1 \vDash B(\mathbf{a}) \\ \Leftrightarrow \forall u_1 \in \varrho_k(u) \quad M^{\star} \vDash B^{\star}(u_1, \mathbf{a}) \text{ (by the induction hypothesis)} \\ \Leftrightarrow M^{\star} \vDash \forall w (R_k(u, w) \supset B^{\star}(w, \mathbf{a})) \Leftrightarrow M^{\star} \vDash (\Box_k B)^{\star}(\mathbf{a}). \end{array}$$

(2) $\mathbf{F} \vDash A \Leftrightarrow \forall \xi \forall u \in \mathbf{F}\ (\mathbf{F}, \xi), u \vDash A$
$\Leftrightarrow \forall \xi \forall u \in \mathbf{F}\ (F, \xi)^{\star} \vDash A^{\star}(u)$ (by (1)) $\Leftrightarrow \forall \xi\ (F, \xi)^{\star} \vDash \forall w_0 A^{\star}(w_0)$. ■

Lemma 3.12.3 *(1) $M, u \Vdash A(\mathbf{a})$ iff $M^{\star} \vDash (A^T)^{\star}(u, \mathbf{a})$ for any intuitionistic Kripke model M, an intuitionistic formula A, $u \in M$, $\mathbf{a} \in D_u^n$.*

*(2) $\mathbf{F} \Vdash A$ iff for any intuitionistic ξ, $(\mathbf{F}, \xi)^{\star} \vDash \forall w_0 (A^T)^{\star}(w_0)$ for an intuitionistic sentence A and an **S4**-based frame $\mathbf{F}$.*

Proof
(1) $M, u \Vdash A(\mathbf{a})$ iff $M, u \vDash A^T(\mathbf{a})$ (by Lemma 3.2.15)
iff $M^{\star} \vDash (A^T)^{\star}(u, \mathbf{a})$ (by Lemma 3.12.2).
(2) Easily follows from (1). ■

The next two definitions are predicate analogues of Definition 1.8.2.

Definition 3.12.4 *Let C be a class of N-modal predicate Kripke frames. We say that C is* Δ-elementary *(respectively,* R-elementary*) if the class of associated $\mathcal{L}2_N$-structures $C^{\star} := \{\mathbf{F}^{\star} \mid \mathbf{F} \in C\}$ is Δ-elementary (respectively, R-*elementary*).*

Definition 3.12.5 *A modal or superintuitionistic predicate logic L is called* Δ-elementary *(respectively,* R-elementary*) if the class $\mathbf{V}(L)$ of all L-frames is Δ-elementary (respectively, R-elementary).*

Definition 3.12.6 *A predicate Kripke frame* $\mathbf{F} = (F, D)$ *is called* countable *if the set of its worlds and the set of its individuals* (D^+) *are both countable. We say that* $\mathbf{F}$ *has a* countable base *if the set of worlds is countable and that* $\mathbf{F}$ *has a* countable domain *if* D^+ *is countable.*

Definition 3.12.7 *Let* $\mathcal{C}$ *be a class of Kripke frames,* $L = \mathbf{ML}^{(=)}(\mathcal{C})$ *its modal logic. We say that* L *has*

- *the* countable frame property (c.f.p.) in $\mathcal{C}$ *if*

$$L = \mathbf{ML}^{(=)}(\{\mathbf{F} \mid \mathbf{F} \in \mathcal{C},\ \mathbf{F} \text{ is countable } \}),$$

- *the* countable domain property (c.d.p.) in $\mathcal{C}$ *if*

$$L = \mathbf{ML}^{(=)}(\{\mathbf{F} \mid \mathbf{F} \in \mathcal{C},\ \mathbf{F} \text{ has a countable domain}\}),$$

- *the* countable base property (c.b.p.) in $\mathcal{C}$*:*

$$L = \mathbf{ML}^{(=)}(\{\mathbf{F} \mid \mathbf{F} \in \mathcal{C},\ \mathbf{F} \text{ has a countable base}\}).$$

and similarly for the intuitionistic case.

Obviously, the c.f.p. implies the c.d.p. and the c.b.p.

Proposition 3.12.8

(1) If a class of Kripke frames $\mathcal{C}$ *is R-elementary, then its modal predicate logic* $\mathbf{ML}^{(=)}(\mathcal{C})$ *is recursively axiomatisable (i.e. RE). Similarly, the intermediate logic* $\mathbf{IL}^{(=)}(\mathcal{C})$ *is recursively axiomatisable for any R-elementary class* $\mathcal{C}$ *of intuitionistic Kripke frames.*

(2) If $\mathcal{C}$ *is a* Δ*-elementary class of Kripke frames, then* $\mathbf{ML}^{(=)}(\mathcal{C})$ *has the c.f.p. in* $\mathcal{C}$*, and similarly for the intuitionistic case.*

Proof Similar to 1.8.5. Let $\mathcal{C}$ be the class of models of an $\mathcal{L}2_N$-theory Σ, $\mathcal{C}_1$ the class of all countable models of Σ. Then by Lemma 3.12.1, the $\mathcal{L}2_N^\star$-models of $\Sigma \cup \Gamma_0$ are exactly the structures of the form $(\mathbf{F}, \xi)^\star$, where $\mathbf{F} \vDash \Sigma$. Now by Lemma 1.8.4 and Gödel's completeness theorem we obtain for any sentence A:

$$A \in \mathbf{ML}(\mathcal{C}) \Leftrightarrow \Sigma \cup \Gamma_0 \vDash \forall w_0 A^\star(w_0) \text{ (in the classical sense)}$$
$$\Leftrightarrow \Gamma_0 \cup \Sigma \vdash \forall w_0 A^\star(w_0) \text{ (in classical predicate logic).}$$

Hence the set of all sentences in $\mathbf{ML}(\mathcal{C})$ is RE. Since for any formula A, $A \in \mathbf{ML}(\mathcal{C})$ iff $\overline{\forall} A \in \mathbf{ML}(\mathcal{C})$, the whole logic is reducible to this set, and thus (1) follows.

For the intuitionistic case, it suffices to note that $\mathbf{IL}^{(=)}(\mathcal{C}) = {}^T\mathbf{ML}^{(=)}(\mathcal{C})$, by 3.2.29, thus $\mathbf{IL}^{(=)}(\mathcal{C})$ is reducible to $\mathbf{ML}^{(=)}(\mathcal{C})$.

Similarly to the above, we obtain that for any sentence A:
$A \in \mathbf{ML}(\mathcal{C}_1)$ iff for any countable $\mathcal{L}2_N^\star$-structure μ, $(\mu \models \Sigma \cup \Gamma_0 \Rightarrow \mu \models \forall w_0 A^\star(w_0))$.

By Lemma 3.12.1 and the Löwenheim-Skolem theorem, the latter is equivalent to $\Sigma \cup \Gamma_0 \models \forall w_0 A^\star(w_0)$, and thus (as we have shown) to $A \in \mathbf{ML}(\mathcal{C})$. Therefore $\mathbf{ML}(\mathcal{C}) = \mathbf{ML}(\mathcal{C}_1)$, which proves (2).

In the intuitionistic case note that

$$\mathbf{IL}(\mathcal{C}) = {}^T(\mathbf{ML}(\mathcal{C})) = {}^T(\mathbf{ML}(\mathcal{C}_1)) = \mathbf{IL}(\mathcal{C}_1).$$

■

The following refinement of (2) is also useful.

Proposition 3.12.9 *Let L be a Δ-elementary modal or superintuitionistic predicate logic, $\mathbf{F}$ an L-frame, M a Kripke model over $\mathbf{F}$, S a countable set of worlds in M. Then there exists a countable reliable submodel of M containing S, whose frame also validates L.*

Proof Let $\mathbf{V}(L)^\star$ be the class of models of an $\mathcal{L}2_N$-theory Σ. Then $M^\star \models \Sigma$. By the Löwenheim–Skolem–Tarski theorem, $M^\star$ has a countable elementary substructure μ containing S. Then $\mu \models \Gamma_0 \cup \Sigma$, so by Lemma 3.12.1 it follows that $\mu = M_1^\star$ for some Kripke model M_1. If $\mathbf{F}_1$ is the frame of M_1, we obtain that $\mathbf{F}_1^\star \models \Sigma$, i.e. $\mathbf{F}_1 \in \mathbf{V}(L)$.

Since $M_1^\star \prec M^\star$, we have for any $u \in M$, $A(\mathbf{a}) \in \mathcal{L}(u)$:

$$M^\star \models A(u, \mathbf{a}) \text{ iff } M_1^\star \models A(u, \mathbf{a}).$$

Hence by Lemma 3.12.3

$$M, u \models A(\mathbf{a}) \text{ iff } M_1, u \models A(\mathbf{a}),$$

which means that M_1 is reliable. ■

Remark 3.12.10 In the intuitionistic case we can define a slightly different translation. Namely, we can include the axioms of quasi-ordering (or partial ordering) for R in *Exd* and add the following truth-preservation clause to A_j^n:

$$\forall z_1 \forall z_2 \forall \mathbf{x}\, (R(z_1, z_2) \wedge P_j^{n\star}(z_1, \mathbf{x}) \supset P_j^{n\star}(z_2, \mathbf{x})).$$

For an intuitionistic predicate formula $A(\mathbf{x})$, the $\mathcal{L}2_0^\star$-formula $\bar{A}(z_0, \mathbf{x})$ is constructed as follows:

$$\begin{aligned}
&\overline{P_j(\mathbf{x})} = P_j^\star(z_0, \mathbf{x});\\
&\overline{(x_1 = x_2)} = (x_1 = x_2);\\
&\overline{(A_1 \sigma A_2)} = \bar{A}_1 \sigma \bar{A}_2 \text{ for } \sigma = \wedge, \vee;\\
&\overline{(A_1 \supset A_2)}(z_0, \mathbf{x}) = \forall z_1 (R(z_0, z_1) \wedge \bar{A}_1(z_1, \mathbf{x}) \supset \bar{A}_2(z_1, \mathbf{x}));\\
&\overline{\exists x_0 A_0(x_0, \mathbf{x})} = \exists x_0 (U(z_0, x_0) \wedge \overline{A_0(x_0, \mathbf{x})});\\
&\overline{\forall x_0 A_0(x_0, \mathbf{x})} = \forall z_1 \forall x_0 (R(z_0, z_1) \wedge U(z_1, x_0) \supset \bar{A}_0(z_1, x_0, \mathbf{x})).
\end{aligned}$$

Recall (Section 3.2) that for a class of propositional Kripke frames $\mathcal{Z}$, $\mathcal{K}(Z)$ denotes the class of all Kripke frames based on frames from $\mathcal{Z}$. By $\mathcal{CK}(Z)$ we denote the class of all frames with constant domains from $\mathcal{K}(Z)$.

Corollary 3.12.11 *Let L be a Δ-elementary predicate logic, M a Kripke model over an L-frame $\mathbf{F}$, $u_0 \in M$. Then there exists a countable reliable submodel $M_0 \subseteq M$ over an L-frame containing u_0 such that $\mathbf{MT}(M) = \mathbf{MT}(M_0)$.*

Proof For any sentence $A \notin \mathbf{MT}(M)$ (in the language of L) there exists a world $u_A \in M$ such that $M, u_A \not\models A$. Put

$$S := \{u_0\} \cup \{u_A \mid A \notin \mathbf{MT}(M)\}$$

and apply the previous proposition. The resulting submodel M_0 is reliable, so $\mathbf{MT}(M) \subseteq \mathbf{MT}(M_0)$.

On the other hand, if $A \notin \mathbf{MT}(M)$, then $M, u_A \not\models A$, so $M_0, u_A \not\models A$ by reliability of M_0. Therefore $\mathbf{MT}(M_0) \subseteq \mathbf{MT}(M)$. ∎

Corollary 3.12.12 *Let $\mathcal{Z}$ be a class of propositional Kripke frames.*

(1) If $\mathcal{Z}$ is R-elementary, then the logics determined over Z $\mathbf{ML}^{(=)}(\mathcal{K}(Z))$ and $\mathbf{ML}^{(=)}(\mathcal{CK}(Z))$, are recursively axiomatisable.

(2) If $\mathcal{Z}$ is Δ-elementary, then $\mathbf{ML}^{(=)}(\mathcal{K}(Z))$ has the c.f.p. in $\mathcal{K}(Z)$, and $\mathbf{ML}^{(=)}(\mathcal{CK}(Z))$ has the c.f.p. in $\mathcal{CK}(Z)$.

Analogous properties hold for the intuitionistic case.

Proof For the case of constant domains we have to add the axiom

$$\forall z_1 \forall z_2 \forall x\ (U(z_1, x) \equiv U(z_2, x)),$$

or, if one prefers,

$$\forall z \forall x\, U(z, x).$$

Equivalently, we can drop the conjuncts containing U in the definition of $\bar{A}$. ∎

Examples

Consider the following R-elementary classes:

- $\mathcal{W}_n$, the class of all posets of width $\leq n$;
- $\mathcal{H}_n$, the class of all posets of height $\leq n$;
- $\mathcal{B}_n$, the class of all posets of branching $\leq n$.

So the superintuitionistic predicate logics of the following classes of posets are recursively axiomatisable (and satisfy the countable frame property):

(a) $\mathbf{IL}^{(=)}(\mathcal{K}(\mathcal{W}_m))$,

(b) $\mathbf{IL}^{(=)}(\mathcal{K}(\mathcal{P}_n \cap \mathcal{W}_m))$, etc.

If $m \geq 2$, explicit axiom systems for these intermediate logics are unknown. On the other hand, explicit recursive axiomatisations are known for the logics $\mathbf{IL}^{(=)}(\mathcal{K}(\mathcal{P}_n))$, $\mathbf{IL}^{(=)}(\mathcal{K}(\mathcal{P}_n \cap \mathcal{B}_m))$, $\mathbf{IL}^{(=)}(\mathcal{K}(\mathcal{B}_m))$, see Chapter 7.

In particular, we have:

$$\mathbf{IL}^{(=)}(\mathcal{K}(\mathcal{B}_1)) = \mathbf{QH}^{(=)} + Z;$$

$$\mathbf{IL}^{(=)}(\mathcal{K}(\mathcal{B}_m)) = \mathbf{QH}^{(=)} \text{ for } m \geq 2.$$

In this connection note that the logic

$$\mathbf{IL}\left(\mathcal{K}\left(\mathcal{B}_m \cap \bigcup_{n<\omega} \mathcal{P}_n\right)\right)$$

of frames of finite heights and of finite branching bounded by m in all finite heights is not RE if $m \geq 2$, see Chapter 11; recall that the corresponding propositional logic is $\mathbf{H} + B_m$ (Chapter 1). The same holds for the logics $\mathbf{IL}(\mathcal{K}(\mathcal{W}_m \cap \bigcup\limits_{n<\omega} \mathcal{P}_n))$, $\mathbf{IL}(\mathcal{K}(\mathcal{P}_m \cap \bigcup\limits_{n<\omega} \mathcal{W}_n))$, etc.

Also note that explicit finite axiomatisations of the classes of frames with constant domains in the cases (a), (b) are known, see Chapter 8.

Corollary 3.12.13 *Every predicate logic determined over a finite propositional Kripke frame F is recursively axiomatisable, and satisfies the c.d.p. (in $\mathcal{K}(F)$).*

Proof (cf. [Skvortsov, 1995], [Skvortsov, 1991]). It is well known that every finite classical structure is finitely axiomatisable. So the class $\{F\}$ is R-elementary for a finite propositional Kripke frame F; if $F = (\{a_1, \ldots, a_m\}, R)$, then up to isomorphism, F is defined by the following $\mathbf{C}_1$-sentence:

$$\exists z_1, \ldots, z_m \left(\bigwedge_{a_iRa_j} R(z_i, z_j) \wedge \bigwedge_{\neg a_iRa_j} \neg R(z_i, z_j) \wedge \forall z_0 \left(\bigvee_{i=1}^{m} (z_0 = z_i) \right) \wedge \bigwedge_{i \neq j} (z_i \neq z_j) \right)$$

∎

However in many cases the predicate logics of R-elementary classes of frames are not recursively axiomatisable. For example, for a class $\mathcal{Z}$ of rooted posets, $\mathbf{IL}(\mathcal{Z})$ is RE iff $\mathcal{Z}$ is finite, see Chapter 11.

Remark 3.12.14 We have defined a translation from a modal logic language into a classical language based on Kripke frame semantics. Similar translations can be defined for Kripke sheaves and for other Kripke-style semantics considered in further chapters. Recursive axiomatisability and countable frame (or countable domain) property also can be established for logics complete in these semantics. The constructions are rather straightforward, and we do not discuss them in detail.

We have established the countable domain property for any finite propositional Kripke frame. It turns out that the c.d.p. can be extended to any countable propositional Kripke frame as well. H. Ono proved this result for the intuitionistic case by a straightforward argument (see Theorem 1.1 in [Ono, 1972/73]). Here we give a model-theoretic proof for the modal case which is based again on the translation into a classical language.

Proposition 3.12.15 *Let F be a propositional Kripke frame of infinite cardinality κ. Then*

(1) $\mathbf{ML}^{(=)}(\mathcal{K}(F)) = \mathbf{ML}^{(=)}\{(F,D) \mid \forall u \in F\ |D_u| \leq \kappa\}$, *and similarly for the logic* $\mathbf{IL}^{(=)}(F)$ *of any intuitionistic propositional Kripke frame F of cardinality κ.*

(2) $\mathbf{ML}^{(=)}(\mathcal{CK}(F)) = \mathbf{ML}^{(=)}\{(F \odot V) \mid |V| \leq \kappa\}$.

Corollary 3.12.16 *Let $\mathcal{Z}$ be a class of countable propositional frames. Then*

$$\mathbf{ML}^{(=)}(\mathcal{K}(\mathcal{Z})) = \mathbf{ML}^{(=)}\{(F,D) \mid F \in \mathcal{Z}\ \&\ \forall u \in F\ |D_u| \leq \aleph_0\},$$

and similarly for $\mathcal{CK}(\mathcal{Z})$.

Proof We prove Proposition 3.12.15 for the modal case. For the intuitionistic case one can repeat the proof or just apply Gödel–Tarski translation.

(i) We add a set of individual constants of sort 1 of cardinality κ

$$T = \{c_\alpha \mid \alpha < \kappa\}$$

to $\mathcal{L}2_N^\bigstar$ and fix a bijection

$$\nu : T \to F.$$

We denote the resulting language by $\mathcal{L}2_{N\kappa}^\bigstar$.

Given a Kripke model $((F,D),\vDash)$, we can construct a model $\mathcal{M}_\kappa^+$ of $\mathcal{L}2_{N\kappa}^\bigstar$ by taking an $\mathcal{L}2_N^\bigstar$-model $\mathcal{M}^+$ and putting

$$\mathcal{M}_\kappa^+ \vDash u = c_\alpha \text{ iff } u = \nu(c_\alpha).$$

Due to the downward Löwenheim–Skolem–Tarski theorem, it has an elementary substructure of cardinality κ, which we denote by $\mathcal{M}_\kappa^{+'}$. Obviously, its universe of sort 1 must coincide with F, and so its $\mathcal{L}2_N^\bigstar$-reduct corresponds to some Kripke model $((F,D),\vDash')$ based on F. Since D'^+ is the universe of sort 2 in $\mathcal{M}_\kappa^{+'}$, we have $|D'^+| \leq \kappa$, and thus $\forall u\ |D'_u| \leq \kappa$. For any modal predicate formula $A(\mathbf{x})$ we have

$$(F,D), u \vDash A(\mathbf{x}) \text{ iff } \mathcal{M}_\kappa^+ \vDash \bar{A}(u,\mathbf{x}) \text{ iff } \mathcal{M}_\kappa^{+'} \vDash \bar{A}(u,\mathbf{x}) \text{ iff } (F,D'), u \vDash' A(\mathbf{x}).$$

Therefore, if $A(\mathbf{x}) \notin \mathbf{ML}^{(=)}(\mathcal{K}(F))$, it is refuted in a frame (F,D') such that $|D'_u| \leq \kappa$ for any $u \in F$.

(ii) Note that if (F,D) has a constant domain, then $\mathcal{M}_\kappa^+ \vDash \forall z \forall x U(z,x)$ and thus $\mathcal{M}_\kappa^{+'} \vDash \forall z \forall x U(z,x)$ and (F,D') has a constant domain. ∎

On the other hand, let F be a propositional frame which is well-ordered of the type κ, κ being an infinite cardinal, cf. [Skvortsov, 1989]. Consider the formula

$$KF = \neg\neg\forall x(P(x) \vee \neg P(x)),$$

see Section 2.3. Then $KF \notin \mathbf{IL}^{(=)}(\mathcal{K}(F))$, since $KF \notin \mathbf{IL}(F \odot V)$ if $|V| \geq \kappa$. In fact, let $(a_\beta \beta < \kappa)$ be a sequence of distinct elements in V, and consider the valuation ξ such that for every $\alpha < \kappa$

$$\xi_\alpha(P) = \{a_\beta \mid \beta \leq \alpha\}.$$

Then

$$\xi, 0 \Vdash \neg KF.$$

Also we have

$$KF \in \mathbf{IL}(F, D) \text{ if } |D^+| < \kappa.$$

In fact, take an arbitrary valuation ξ in (F, D); for every $a \in D^+$ take an ordinal $\beta_a \leq \kappa$ such that

$$\xi, \alpha \Vdash P(a) \text{ iff } \beta_a \leq \alpha.$$

Also take $\alpha_0 < \kappa$ such that

$$\forall a\ (\beta_a < \kappa \Rightarrow \beta_a \leq \alpha_0).$$

Then

$$\xi, \alpha_0 \Vdash \forall x(P(x) \vee \neg P(x)).$$

Also note that

$$|D^+| \leq \aleph_1 \text{ if } \forall \alpha < \kappa\, |D_\alpha| \leq \aleph_0.$$

In fact,

$$|\{\alpha < \kappa \mid D_\alpha \not\subseteq \bigcup_{\beta<\alpha} D_\beta\}| \leq \aleph_1.$$

Therefore, we cannot replace $|D_u| \leq \kappa$ with $|D_u| \leq \aleph_0$ in Proposition 3.12.15 say for $F = \aleph_2$. Let us also note that for $F = \kappa$

$$KF \notin \mathbf{IL}(F, D) \text{ if } D_\alpha = \{a_\beta \mid \beta \leq \alpha\},$$

and

$$\forall \alpha < \kappa\, |D_\alpha| < \kappa;$$

thus

$$KF \notin \mathbf{IL}\{(F, D) \mid \forall \alpha < \kappa\, |D_\alpha| \leq \aleph_0\}$$

for $F = \aleph_1$.

Let us consider another similar example. Let F be a denumerable tree (with the root 0_F), in which F^u is nonlinear (e.g. the tree ω^* of all finite sequences of natural numbers, or the binary tree $\{0, 1\}^*$ of all finite (0,1)-sequences, etc.). Let $\overline{F}$ be the coatomic tree obtained by adding maximal points w_τ above *every* branch (maximal chain) τ of F. Let us consider the following formula:

$$C^* := \forall x((\neg P(x) \supset q) \supset q) \wedge \neg\neg\exists x P(x) \supset q.$$

Then

Lemma 3.12.17 *(i)* $C^* \in \mathbf{IL}(\overline{F} \odot V)$ *for a denumerable* V*;*

(ii) $C^* \notin \mathbf{IL}(\overline{F} \odot V)$ *if* $|V| \geq |\overline{F} - F|$ *(e.g. if* $|V| \geq 2^{\aleph_0}$*).*

Proof

(i) Suppose that

$$u_0 \not\models q, u_0 \models \forall x[(\neg P(x) \supset q) \supset q], u_0 \models \neg\neg\exists x P(x), V = \{a_n : n > 0\}.$$

Then

$$\forall u \not\models q \; \forall n > 0 \; \exists v \; (uRv \,\&\, v \models \neg P(a_n) \,\&\, v \not\models q),$$

and there exists a chain $u_0 R u_1 R u_2 R \dots$ such that $\forall n > 0 \; [u_n \models \neg P(a_n) \,\&$ $, u_n \not\models q]$. Then $w_\tau \models \neg\exists x P(x)$ for a branch τ containing all u_n. This is a contradiction.

(ii) Take different elements a_τ from V for $w_\tau \in (\overline{F} - F)$, and the following valuation in $\overline{F} \odot V$:

$$u \models P(a) \text{ iff } \exists\tau \; (u = w_\tau \;\&\; a = a_\tau);$$

$$u \models q \text{ iff } \exists\tau(u = w_\tau).$$

Then $O_F \not\models q$, $O_F \models \neg\neg\exists x P(x)$. We also have

$$\forall u \in F \; \forall a \in D_0 \; u \not\models \neg P(a) \supset q$$

since for $a = a_\tau$, there exists $v \in F^u$ such that $v \notin \tau$, i.e. $v \models \neg P(a)$. Thus $O_F \models \forall x((\neg P(x) \supset q) \supset q)$.

■

Remark 3.12.18 This example is also related to the formula KF. It is known [Gabbay, 1981] that the predicate logic $\mathbf{QH} + KF$ is Kripke-complete (w.r.t. denumerable atomic trees). Moreover, $\mathbf{QH} + KF = \mathbf{IL}(\omega^*) = \mathbf{IL}(\{(\omega^*, D) : \forall u \; |D_u| \leq \aleph_0]\})$ (and similarly for the binary tree $\{0,1\}^*$). The corresponding predicate logic with constant domains is also Kripke-complete:

$$\mathbf{QH} + KF + CD = \mathbf{IL}(\{F' \odot V \mid \{0,1\}^* \subset F' \subset \omega^*, \; F' \text{ is coatomic}\})$$

for a denumerable V. On the other hand, the above mentioned example shows that

$$\mathbf{QH} + K + CD \subset \mathbf{IL}(\omega^* \odot V)$$

for a denumerable V.

Remark 3.12.19 For the case of constant domains (without equality), Proposition 3.12.15 also shows that

$$\mathbf{ML}_C(\mathcal{Q}F) = \mathbf{ML}(F \odot V)$$

for any countable F and any constant infinite domain V. This follows from 3.4.11:

$$\mathbf{ML}(F \odot V) \subseteq \mathbf{ML}(F \odot V'),$$

provided $|V| \geq |V'|$.

The translation into classical logic and the well-known generalised form of the Löwenheim–Skolem theorem shows that

$$\mathbf{ML}^{=}(F \odot V) = \mathbf{ML}^{=}(F \odot V')$$

for any finite F and infinite V, V', cf. Corollary 3.12.13.

The intuitionistic case is quite similar. On the other hand, if a poset F contains an infinite cone then

$$C^* \in (\mathbf{IL}^{=}(F \odot V) - \mathbf{IL}^{=}(F \odot V')),$$

for $|V'| = \aleph_0$, $|V| \geq 2^{|F|}$, cf. Corollary 3.12.13.

Thus we obtain the following claim (cf. Theorem 1 in [Skvortsov, 1995]):

Theorem 3.12.20 $\mathbf{IL}^{=}(F \odot V) = \mathbf{IL}^{=}(F \odot V')$ *for any infinite constant domains* V, V' *iff all the cones in* F *are finite.*

Chapter 4

Algebraic semantics

4.1 Modal and Heyting valued structures

As we know from Chapter 1, every propositional Kripke frame F corresponds to a modal algebra $MA(F)$; so Kripke semantics can be treated as a particular case of algebraic semantics. Likewise, in the predicate case Kripke frames with equality admit a straightforward algebraic generalisation.

Recall that by Lemma 3.5.2, every KFE can be represented as a propositional Kripke frame F with individuals such that for every two individuals a, b their measure of equality $E(a,b)$ is defined. $E(a,b)$ is a set of possible worlds, and thus an element of the corresponding modal algebra $MA(F)$. This suggests for the following generalization of KFEs: replace $MA(F)$ by an arbitrary modal algebra (or a Heyting algebra for the intuitionistic case) and E with a function taking values in this algebra. To make the corresponding semantics sound, E should satisfy the properties cited in Lemma 3.5.2.

Thus we come to the following two definitions.

Definition 4.1.1[1] *Let $\mathbf{\Omega}$ be a complete Heyting algebra (a 'locale'). An $\mathbf{\Omega}$-*valued set *is a triple $(\mathbf{\Omega}, D, E)$, where D is a set, $E : D \times D \longrightarrow \Omega$ is a map such that for any $a, b, c \in D$*

(E1) $E(a,b) = E(b,a)$;

(E2) $E(a,b) \wedge E(b,c) \leq E(a,c)$.

If also

(E3) $\bigvee\limits_{a \in D} E(a,a) = \mathbf{1}$,

*the triple $(\mathbf{\Omega}, D, E)$ is called an $\mathbf{\Omega}$-*valued structure *or a* Heyting valued structure (H.v.s) *over $\mathbf{\Omega}$. D is called its* individual domain, *the elements of D are called* individuals. *The elements of $\mathbf{\Omega}$ are called* truth values.

[1] Cf. [Borceaux, 1994; Fourman and Scott, 1979; Dragalin, 1988].

Obviously, (E3) implies that D is non-empty.

Definition 4.1.2 *Let* $\mathbf{\Omega} = (\Omega, \cup, \cap, -, \mathbf{0}, \mathbf{1}, \Box_1, \dots, \Box_N)$ *be a complete modal algebra. An* $\mathbf{\Omega}$-valued structure *or a* modal valued structure (m.v.s) over $\mathbf{\Omega}$ *is a triple* $(\mathbf{\Omega}, D, E)$, *where* D *is a set,* $E : D \times D \longrightarrow \Omega$ *is a map such that for any* $a, b, c \in D$, $i \in \{1, \dots, N\}$

(E1) $E(a,b) = E(b,a)$,

(E2) $E(a,b) \cap E(b,c) \leq E(a,c)$,

(E3) $\bigcup_{a\in D} E(a,a) = \mathbf{1}$,

(E4) $E(a,b) \leq \Box_i E(a,b)$.

The condition (E3) again implies the non-emptiness of D.

Since every Boolean algebra is a Heyting algebra (where $\vee$ is $\cup$, $\wedge$ is $\cap$ and $a \to b = a \supset b = -a \cup b$), an m.v.s $(\mathbf{\Omega}, D, E)$ over a modal algebra $\mathbf{\Omega}$ corresponds to the H.v.s $(\mathbf{\Omega}^b, D, E)$, where $\mathbf{\Omega}^b$ is the Boolean part of $\mathbf{\Omega}$.

The other way round, if $\mathbf{\Omega}$ is an **S4**-algebra, then an m.v.s. $(\mathbf{\Omega}, D, E)$, corresponds to the H.v.s. $(\mathbf{\Omega}^\circ, D, E)$, where $\mathbf{\Omega}^\circ$ is the pattern of $\mathbf{\Omega}$.

Definitions and results for H.v.s. and m.v.s. are often quite similar. In these cases we talk about 'structures' and denote all operations in a Heyting-algebraic style.

As in Chapter 3, we call $E(a,b)$ the *measure of equality* of individuals a, b; they are 'fully equal' if $E(a,b) = \mathbf{1}$ and 'fully different' if $E(a,b) = \mathbf{0}$. The truth value $E(a,a)$ (which in general may be not equal to $\mathbf{1}$) is called the *measure of existence* of a.

We also introduce the measure of equality for tuples $\mathbf{a} = (a_1, \dots, a_n)$, $\mathbf{b} = (b_1, \dots, b_n) \in D^n$:

$$E(\mathbf{a}, \mathbf{b}) := E(a_1, b_1) \wedge \dots \wedge E(a_n, b_n)$$

and the following abbreviations:[2]

$$E\mathbf{ab} := E(\mathbf{a}, \mathbf{b}), \ E\mathbf{a} := E(\mathbf{a}) := E\mathbf{aa}.$$

We can also include the degenerate case $n = 0$. A 0-tuple is just $\curlywedge$ (the void sequence), and we put

$$E \curlywedge \curlywedge := \mathbf{1}.$$

As we said above, Definitions 4.1.1, 4.1.2 generalise the situation in Kripke frames with equality. In that case D corresponds to the set of all individuals D^+, the conditions (E1)–(E3) in Definition 4.1.2 are exactly the same as in Lemma 3.5.2 (1), and the condition (E4) in 4.1.2 is obviously equivalent to (E4) in 3.5.2(1). Therefore we have

[2]The reader will notice that our notation is ambiguous. For example, for $a, b \in D$, Eab abbreviates both $E(a,b)$ and $E(ab) = Ea \wedge Eb$. Nevertheless we use it, when there is no confusion.

Lemma 4.1.3

(1) *Let* $\mathbf{F} = (F, D, \asymp)$ *be a KFE,* $F = (W, R_1, \ldots, R_\mu)$*, and for every two individuals* $a, b \in D^+$ *put* $E(a,b) = \{u \mid a \asymp_u b\}$*. Then* $MV(\mathbf{F}) := (MA(F), D^+, E)$ *is an m.v.s.*

(2) *Similarly, if* F *is an* **S4***-frame, the triple* $HV(\mathbf{F}) := (HA(F), D^+, E)$ *is an H.v.s.*

Proof In fact, (E1)–(E4) in 4.1.2 follow readily from 3.5.2(1). In particular, (E3) means that every individual domain D_u is non-empty. ■

The conditions (E1), (E2) in the above Definitions 4.1.1, 4.1.2 correspond to the symmetry and the transitivity of equality; more exactly, they are intended to make the corresponding first-order formulas true (see below). To prove the reflexivity, the condition (E3) is necessary. The condition (E4) corresponds to a theorem in 2.6.18(3):

$$x = y \supset \Box_i(x = y);$$

it is necessary for verifying the substitution instances of the equality axiom (Ax17).

Lemma 4.1.4 *For any structure* $(\mathbf{\Omega}, D, E)$*, individuals* $a, b \in D$ *and tuples* $\mathbf{a}, \mathbf{b}, \mathbf{c} \in D^n$*:*

(1) $E(a,b) \leq E(a,a)$*;*

(2) $E\mathbf{ab} \leq E\mathbf{aa}$*;*

(3) $E\mathbf{ab} \wedge E\mathbf{bc} \leq E\mathbf{ac}$*;*

(4) $E(\mathbf{a}) \leq \Box_i E(\mathbf{a})$ *(in the modal case);*

(5) $\bigvee\limits_{\mathbf{a} \in D^n} E(\mathbf{a}) = \mathbf{1}$.

Proof This easily follows from Definition 4.1.2. In fact,

$$E(a,b) = E(a,b) \wedge E(b,a) \leq E(a,a);$$

hence

$$E\mathbf{ab} = \bigwedge_{i=1}^{n} Ea_ib_i \leq \bigwedge_{i=1}^{n} Ea_ia_i = E\mathbf{aa};$$

$$E\mathbf{ab} \wedge E\mathbf{bc} = \bigwedge_{i=1}^{n} (Ea_ib_i \wedge Eb_ic_i) \leq \bigwedge_{i=1}^{n} Ea_ic_i = E\mathbf{ac};$$

$$\bigvee_{\mathbf{a} \in D^n} E(\mathbf{a}) = \bigvee\{Ea_1a_1 \wedge \ldots \wedge Ea_na_n \mid a_1, \ldots, a_n \in D\} \geq \bigvee_{a \in D} \bigwedge_{i=1}^{n} Eaa = \mathbf{1}.$$

In the modal case we also have

$$E(\mathbf{a}) = Ea_1a_1 \cap \ldots \cap Ea_na_n \leq \Box_i Ea_1a_1 \cap \ldots \cap \Box_i Ea_na_n = \Box_i E(\mathbf{a}).$$

■

Definition 4.1.5 *A structure* $F = (\mathbf{\Omega}, D, E)$ has a constant domain *(or briefly, F is a* CD-m.v.s. *or a* CD-H.v.s. *or a* CD-structure*) if the equality is trivial, i.e. for any* $a, b \in D$

$$E(a,b) = \begin{cases} \mathbf{1} & \textit{iff} \quad a = b, \\ \mathbf{0} & \textit{iff} \quad a \neq b. \end{cases}$$

Such a structure is denoted simply by $(\mathbf{\Omega}, D)$.

Definition 4.1.6 *A structure* $(\mathbf{\Omega}, D, E)$ *is called* flabby *if* $E(a,a) = \mathbf{1}$ *for every* $a \in D$ *and* keen *if* $E(a,b) = \mathbf{0}$ *whenever* $a \neq b$.

So a CD-structure is both flabby and keen.

Every structure $F = (\mathbf{\Omega}, D, E)$ gives rise to a flabby structure (of the same type) $F_{fl} := (\mathbf{\Omega}, D, E_{fl})$, where

$$E_{fl}(a,b) := \begin{cases} \mathbf{1} & \text{if } a = b, \\ E(a,b) & \text{otherwise,} \end{cases}$$

to a keen structure $F_{ke} := (\mathbf{\Omega}, D, E_{ke})$, where

$$E_{ke}(a,b) := \begin{cases} E(a,a) & \text{if } a = b, \\ \mathbf{0} & \text{otherwise,} \end{cases}$$

and to a CD-structure

$$F_{cd} := (\mathbf{\Omega}, D).$$

Definition 4.1.7 *For a structure* $F = (\mathbf{\Omega}, D, E)$*, an n-ary* F-predicate *(or an* $\mathbf{\Omega}$-valued predicate on D*) is a map* $\mathcal{A} : D^n \longrightarrow \Omega$*. Such a predicate is called* strict *(or* Scott*) if* $\mathcal{A}(\mathbf{a}) \leq E(\mathbf{a}, \mathbf{a})$ *for any* $\mathbf{a} \in D^n$*, and* congruential *if for any* $\mathbf{a}, \mathbf{b} \in D^n$*, for any* $i \in I_n$

$$\forall j \neq i \; a_j = b_j \Rightarrow Ea_ib_i \wedge \mathcal{A}(\mathbf{a}) \leq \mathcal{A}(\mathbf{b}).$$

For $\mathbf{a}, \mathbf{b} \in D^n$ we will use the notation

$$\mathbf{a} =_i \mathbf{b} := \forall j \neq i \; a_j = b_j.$$

Lemma 4.1.8 *For any congruential predicate* $\mathcal{A} : D^n \longrightarrow \Omega$*, for any* $\mathbf{a}, \mathbf{b} \in D^n$

(1) $E\mathbf{ab} \wedge \mathcal{A}(\mathbf{a}) \leq \mathcal{A}(\mathbf{b})$,

(2) $E\mathbf{ab} \leq \mathcal{A}(\mathbf{a}) \leftrightarrow \mathcal{A}(\mathbf{b})$.

Proof We prove (1) by induction on n.

If $n = 1$, the condition $\mathbf{a} =_1 \mathbf{b}$ holds trivially, so the claim is obvious.

Suppose (1) holds for n for any congruential $\mathcal{A}$ and consider $\mathbf{a}, \mathbf{b} \in D^{n+1}$. Let $\mathbf{a} = a_1\mathbf{a}'$, $\mathbf{b} = b_1\mathbf{b}'$. Then

$(*)$ $\mathcal{A}(\mathbf{a}) \wedge E\mathbf{ab} = \mathcal{A}(\mathbf{a}) \wedge Ea_1b_1 \wedge E\mathbf{a}'\mathbf{b}' \leq \mathcal{A}(b_1\mathbf{a}') \wedge E\mathbf{a}'\mathbf{b}'$

since $\mathcal{A}$ is congruential. The n-ary predicate $\mathcal{B} : \mathbf{c} \mapsto \mathcal{A}(b_1\mathbf{c})$ is also congruential. In fact, for $\mathbf{c}, \mathbf{d} \in D^n$ the condition $\mathbf{c} =_i \mathbf{d}$ implies $b_1\mathbf{c} =_{i+1} b_1\mathbf{d}$, thus

$$\mathcal{B}(\mathbf{c}) \wedge Ec_id_i = \mathcal{A}(b_1\mathbf{c}) \wedge Ec_id_i \leq \mathcal{A}(b_1\mathbf{d}) = \mathcal{B}(\mathbf{d}),$$

since $\mathcal{A}$ is congruential.

By the induction hypothesis,

$$\mathcal{A}(b_1\mathbf{a}') \wedge E\mathbf{a}'\mathbf{b}' = \mathcal{B}(\mathbf{a}') \wedge E\mathbf{a}'\mathbf{b}' \leq \mathcal{B}(\mathbf{b}') = \mathcal{A}(\mathbf{b}),$$

therefore by (*), we obtain (1):

$$\mathcal{A}(\mathbf{a}) \wedge E\mathbf{ab} \leq \mathcal{A}(\mathbf{b}).$$

Now (1) implies

$$E\mathbf{ab} \leq \mathcal{A}(\mathbf{a}) \to \mathcal{A}(\mathbf{b})$$

hence by symmetry

$$E\mathbf{ba} \leq \mathcal{A}(\mathbf{b}) \to \mathcal{A}(\mathbf{a}),$$

which eventually implies (2). ■

A 0-ary predicate $\mathcal{A} : \{\curlywedge\} \longrightarrow \mathbf{\Omega}$ can be treated just as an element $\mathcal{A}(\curlywedge)$ of $\mathbf{\Omega}$. This predicate is always congruential and strict, since $E(\curlywedge, \curlywedge) = \mathbf{1}$.

Lemma 4.1.9

(1) A structure F is flabby iff all F-predicates are strict.

(2) A structure F is keen iff all F-predicates are congruential.

Proof

(1) (Only if.) $E(a_i, a_i) = \mathbf{1}$ implies $\mathcal{A}(\mathbf{a}) \leq E(\mathbf{a}, \mathbf{a})$.
(If). Suppose $Ea_0 \neq \mathbf{1}$ for some $a_0 \in D$. Then the predicate $\mathcal{A}$ sending every a to $\mathbf{1}$ is not strict, since $\mathcal{A}(a_0) \not\leq Ea_0$.

(2) (Only if.) If $\mathbf{a} \neq \mathbf{b}$, then $E\mathbf{ab} = \mathbf{0}$, and obviously $Ea_ib_i \wedge \mathcal{A}(\mathbf{a}) \leq \mathcal{A}(\mathbf{b})$. If $a = b$, then $E\mathbf{ab} \wedge \mathcal{A}(\mathbf{a}) \leq \mathcal{A}(\mathbf{a}) = \mathcal{A}(\mathbf{b})$.
(If). Suppose $E(a_0, b_0) = \alpha \neq \mathbf{0}$ for some $a_0 \neq b_0$ and consider the unary F-predicate $\mathcal{A}$ sending a_0 to α and every $a \neq a_0$ to $\mathbf{0}$. Then

$$Ea_0b_0 \wedge \mathcal{A}(a_0) = \alpha \not\leq \mathbf{0} = \mathcal{A}(b_0),$$

thus $\mathcal{A}$ is not congruential.

■

Definition 4.1.10 *For n-ary F-predicates we introduce the ordering*

$$\mathcal{A} \preceq \mathcal{B} := \forall \mathbf{a} \in D^n \; \mathcal{B}(\mathbf{a}) \leq \mathcal{A}(\mathbf{a}).$$

Lemma 4.1.11 *For an n-ary F-predicate $\mathcal{A}$ put*

$$\mathcal{A}^s(\mathbf{a}) := \mathcal{A}(\mathbf{a}) \wedge E\mathbf{a}.$$

Then

(1) $\mathcal{A}^s \preceq \mathcal{A}$;

(2) $\mathcal{A}^s$ *is strict;*

(3) for any n-ary strict F-predicate $\mathcal{B}$

$$\mathcal{B} \preceq \mathcal{A} \Rightarrow \mathcal{B} \preceq \mathcal{A}^s;$$

(4) if $\mathcal{A}$ *is congruential, then* $\mathcal{A}^s$ *is congruential.*

Thus $\mathcal{A}^s$ is the greatest strict predicate "below" $\mathcal{A}$; we call it the *strict version* of $\mathcal{A}$.

Proof (1), (2), (3) are obvious.
(4) Suppose $\mathbf{a}, \mathbf{b} \in D^n$ and $\mathbf{a} =_i \mathbf{b}$. Then

(♯) $\mathcal{A}^s(\mathbf{a}) = \mathcal{A}(\mathbf{a}) \wedge E\mathbf{a} \wedge Ea_ib_i \leq \mathcal{A}(\mathbf{b}),$

since $\mathcal{A}$ is congruential. Also

$$Ea_ib_i \leq Eb_i$$

by 4.1.4(4), hence

(♯♯) $E\mathbf{a} \wedge Ea_ib_i \leq E\mathbf{b},$

and thus

$$\mathcal{A}^s(\mathbf{a}) \wedge Ea_ib_i \leq \mathcal{A}(\mathbf{b}) \wedge E\mathbf{b} = \mathcal{A}^s(\mathbf{b}).$$

by (♯), (♯♯). Therefore $\mathcal{A}^s$ is congruential. ∎

Note that trivially $\mathcal{A}^s = \mathcal{A}$ for 0-ary $\mathcal{A}$.
Let us now prove a dual to Lemma 4.1.11.

Lemma 4.1.12 *For an n-ary F-predicate $\mathcal{A}$ put*

$$\mathcal{A}^c(\mathbf{a}) := \mathcal{A}(\mathbf{a}) \vee \bigvee_{i=1}^{n} \bigvee_{\mathbf{d}=_i\mathbf{a}} (\mathcal{A}(\mathbf{d}) \wedge Ea_id_i).$$

Then

(1) $\mathcal{A} \preceq \mathcal{A}^c$;

(2) $\mathcal{A}^c$ *is congruential;*

(3) for any n-ary congruential F-predicate $\mathcal{B}$

$$\mathcal{A} \preceq \mathcal{B} \Rightarrow \mathcal{A}^c \preceq \mathcal{B};$$

(4) $\mathcal{A}$ *is strict iff* $\mathcal{A}^c$ *is strict.*

So $\mathcal{A}^c$ is the least congruential predicate 'above' $\mathcal{A}$; we call it the *congruential version* of $\mathcal{A}$.

Proof (1), (3) are obvious.

(2) Suppose $\mathbf{a} =_k \mathbf{b}$. Then

$$\mathcal{A}^c(\mathbf{a}) \wedge Ea_kb_k = \mathcal{A}(\mathbf{a}) \wedge Ea_kb_k \vee \bigvee_{i=1}^{n} \bigvee_{\mathbf{d}=_i\mathbf{a}} (\mathcal{A}(\mathbf{d}) \wedge Ea_id_i \wedge Ea_kb_k).$$

by well-distributivity (Section 1.2).

Now if $i = k$, then

$$Ea_id_i \wedge Ea_kb_k = Ea_kd_k \wedge Ea_kb_k \leq Eb_kd_k$$

by 4.1.1(1), (2). If $i \neq k$, then $a_i = b_i$, so

$$Ea_id_i \wedge Ea_kb_k = Eb_id_i \wedge Ea_kb_k \leq Eb_id_i.$$

Thus

$$\bigvee_{i=1}^{n} \bigvee_{\mathbf{d}=_i\mathbf{a}} (\mathcal{A}(\mathbf{d}) \wedge Ea_id_i \wedge Ea_kb_k) \leq \bigvee_{i=1}^{n} \bigvee_{\mathbf{d}=_i\mathbf{b}} (\mathcal{A}(\mathbf{d}) \wedge Eb_id_i).$$

Also note that

$$\mathcal{A}(\mathbf{a}) \wedge Ea_kb_k = \mathcal{A}(\mathbf{a}) \wedge Eb_id_i$$

for $\mathbf{d} = \mathbf{a}$, $i = k$. Hence

$$\mathcal{A}^c(\mathbf{a}) \wedge Ea_kb_k \leq \bigvee_{i=1}^{n} \bigvee_{\mathbf{d}=_i\mathbf{b}} (\mathcal{A}(\mathbf{d}) \wedge Eb_id_i) \leq \mathcal{A}^c(\mathbf{b}).$$

(4) Since $\mathcal{A} \preceq \mathcal{A}^c$, it follows that $\mathcal{A}^c(\mathbf{a}) \leq E\mathbf{a}$ implies $\mathcal{A}(\mathbf{a}) \leq E\mathbf{a}$.

The other way round, assume that $\mathcal{A}$ is strict. Then for any i and $\mathbf{d} =_i \mathbf{a}$,

$$\mathcal{A}(\mathbf{d}) \wedge Ea_id_i \leq E\mathbf{d} \wedge Ea_id_i = \bigwedge_{j\neq i} Ea_i \wedge Ea_id_i \leq E\mathbf{a},$$

since $Ea_id_i \leq Ea_i$ by 4.1.4(1). By assumption $\mathcal{A}(\mathbf{a}) \leq E\mathbf{a}$, so all the disjuncts in $\mathcal{A}^c(\mathbf{a})$ are bounded by $E\mathbf{a}$, and thus $\mathcal{A}^c(\mathbf{a}) \leq E\mathbf{a}$.

∎

Remark 4.1.13 Note that the disjuncts $\mathcal{A}(\mathbf{a}) \wedge Ea_i a_i$ corresponding to $\mathbf{d} = \mathbf{a}$ are obviously redundant in the definition of $\mathcal{A}^c$. On the other hand, for a strict $\mathcal{A}$ we have

$$\mathcal{A}^c(\mathbf{a}) = \bigvee_i \bigvee_{\mathbf{d}=_i\mathbf{a}} (\mathcal{A}(\mathbf{d}) \wedge Ea_i d_i),$$

since $\mathcal{A}(\mathbf{a}) = \mathcal{A}(\mathbf{a}) \wedge Ea_i a_i$.

Remark 4.1.14 For an arbitrary predicate $\mathcal{A}$, we can construct a strict congruential version

$$(\star) \quad (\mathcal{A}^s)^c(\mathbf{a}) = (\mathcal{A}^c)^s(\mathbf{a}) = \bigvee_i \bigvee_{\mathbf{d}=_i\mathbf{a}} (\mathcal{A}(\mathbf{d}) \wedge Ea_i d_i).$$

In fact, $\mathcal{A}^s$ is strict, so by Remark 4.1.13

$$\begin{aligned}(\mathcal{A}^s)^c(\mathbf{a}) &= \bigvee_i \bigvee_{\mathbf{d}=_i\mathbf{a}} (\mathcal{A}^s(\mathbf{d}) \wedge Ea_i d_i) = \bigvee_i \bigvee_{\mathbf{d}=_i\mathbf{a}} (\mathcal{A}(\mathbf{d}) \wedge E\mathbf{d} \wedge Ea_i d_i) \\ &= \bigvee_i \bigvee_{\mathbf{d}=_i\mathbf{a}} (\mathcal{A}(\mathbf{d}) \wedge E\mathbf{a} \wedge Ea_i d_i).\end{aligned}$$

The latter equality holds since $Ea_i \wedge Ea_i d_i = Ed_i \wedge Ea_i d_i$ (which easily follows from 4.1.4 (1)), and thus for $\mathbf{d} =_i \mathbf{a}$

$$E\mathbf{d} \wedge Ea_i d_i = \bigwedge_{j \neq i} Ea_j \wedge Ed_i \wedge Ea_i d_i = \bigwedge_{j \neq i} Ea_j \wedge Ea_i \wedge Ea_i d_i = E\mathbf{a} \wedge Ea_i d_i.$$

On the other hand,

$$(\mathcal{A}^c)^s(\mathbf{a}) = \mathcal{A}^c(\mathbf{a}) \wedge E\mathbf{a} = \mathcal{A}(\mathbf{a}) \wedge E\mathbf{a} \vee \bigvee_i \bigvee_{\mathbf{d}=_i\mathbf{a}} (\mathcal{A}(\mathbf{d}) \wedge E\mathbf{a} \wedge Ea_i d_i)$$

by well-distributivity. The first disjunct is redundant as it equals $\mathcal{A}(\mathbf{a}) \wedge E\mathbf{a} \wedge Ea_i a_i$, hence $(\star)$ follows.

Note that

$$\mathcal{A}^s \preceq (\mathcal{A}^s)^c \preceq \mathcal{A}^c,$$

but in general $\mathcal{A}$ and $(\mathcal{A}^s)^c$ are $\preceq$-incomparable. E.g. consider a unary $\mathcal{A}$ such that $\mathcal{A}(a) \not\leq Eaa$ for some a. Then $\mathcal{A}(a) \not\leq (\mathcal{A}^s)^c(a)$, since $(\mathcal{A}^s)^c(a) \leq Eaa$. And if $\mathcal{A}(d) \wedge Ebd \not\leq \mathcal{A}(b)$ for some b, d, then $(\mathcal{A}^s)^c(b) \not\leq \mathcal{A}(b)$, since $\mathcal{A}(d) \wedge Ebd \leq (\mathcal{A}^s)^c(b)$.

Exercise 4.1.15 Show that $(\mathcal{A}^s)^c = \mathcal{A}$ iff $\mathcal{A}$ is strict and congruential.

Remark 4.1.16 As we are interested in algebraic models satisfying the axioms of equality, atomic D^+-sentences should be evaluated by congruential predicates. This requirement also makes sense for logics without equality. A similar situation is in Kripke semantics — as we know from Chapter 3, including equality in semantics is crucial for completeness.

Remark 4.1.17 The strictness property is quite natural for evaluation of formulas — the underlying idea is that a D^+-sentence may be true only within the "life-zone" of all occurring individuals. But as we shall see, this does not matter for semantics, because replacing every $\mathcal{A}$ with $\mathcal{A}^s$ does not change the notion of validity.

However evaluating formulas with arbitrary predicates simplifies the inductive truth definition, see Definition 4.2.4. This happens because strict predicates are not closed under all Boolean operations (for example, under negation, since $\mathcal{A}(\mathbf{a}) \leq E\mathbf{a}$ does not imply $-\mathcal{A}(\mathbf{a}) \leq E\mathbf{a}$ if $E\mathbf{a} \neq \mathbf{1}$).

4.2 Algebraic models

Definition 4.2.1 *A* valuation *is an m.v.s. (H.v.s.)* $F = (\mathbf{\Omega}, D, E)$ *is a map* $\varphi : AF_D \longrightarrow \mathbf{\Omega}$ *sending every D-sentence to* $\mathbf{\Omega}$ *such that*

$$\varphi(P(\mathbf{a})) \wedge Ea_ib_i \leq \varphi(P(\mathbf{b}))$$

whenever $P \in PL^n$, $\mathbf{a}, \mathbf{b} \in D^n$, $\mathbf{a} =_i \mathbf{b}$. *Then the n-ary F-predicate* $\varphi_P : \mathbf{a} \mapsto \varphi(P(\mathbf{a}))$ *is called* associated *to* φ *and P. The pair* (F, φ) *is called an* algebraic model over F.

From the definitions it follows that all the predicates φ_P are congruential.

Definition 4.2.2 *A valuation* φ *is called* strict *if every predicate associated to* φ *is strict, i.e. if*

$$\varphi(P(\mathbf{a})) \leq E\mathbf{a}$$

for any $P \in PL^n$, $\mathbf{a} \in D^n$.

Definition 4.2.3 *For an arbitrary valuation* φ *we define its* strict version φ^s *such that*

$$\varphi^s(P(\mathbf{a})) := \varphi(P(\mathbf{a})) \wedge E\mathbf{a}.$$

for any $\mathbf{a} \in D^n$, $P \in PL^n$.

Thus $(\varphi^s)_P = (\varphi_P)^s$, and so the definition is sound, since $(\varphi_P)^s$ is congruential by Lemma 4.1.11.

Certainly $\varphi^s = \varphi$ if φ is strict.

Definition 4.2.4 *For a valuation* φ *in an m.v.s. (respectively, H.v.s.)* $F = (\mathbf{\Omega}, D, E)$ *we define its (unique) 'large' extension to all modal (respectively, intuitionistic) D-sentences in the natural way:*

(1) $\varphi(\bot) := \mathbf{0}$;

(2) $\varphi(a = b) := E(a, b)$;

(3) $\varphi(A \vee B) := \varphi(A) \vee \varphi(B)$;

(4) $\varphi(A \wedge B) := \varphi(A) \wedge \varphi(B)$;

(5) $\varphi(A \supset B) := \varphi(A) \rightarrow \varphi(B)$;

(6) $\varphi(\Box_i A) := \Box_i \varphi(A)$;

(7) $\varphi(\exists x A) := \bigvee_{d \in D} (Ed \wedge \varphi([d/x]A)))$.

(8) $\varphi(\forall x A) := \bigwedge_{d \in D} (Ed \rightarrow \varphi([d/x]A))$.

Then for any D-formula A and a distinct list of variables $\mathbf{x} \supseteq FV(A)$ of length n, we may define the *associated F-predicate*

$$\varphi_{A,\mathbf{x}} : D^n \longrightarrow \mathbf{\Omega}$$

such that

$$\varphi_{A,\mathbf{x}}(\mathbf{a}) := \varphi([\mathbf{a}/\mathbf{x}]A)$$

for any $\mathbf{a} \in D^n$. In particular, $\varphi_{P(\mathbf{x}),\mathbf{x}} = \varphi_P$.

To simplify notation, we sometimes write φ_A rather than $\varphi_{A,\mathbf{x}}$.

Let us check that every predicate $\varphi_{A,\mathbf{x}}$ is congruential. We begin with a simple observation.

Lemma 4.2.5

(1) *φ_P remains congruential after fixing some parameters. In precise terms: let $P \in PL^n$, $\mathbf{c} \in D^n$, let $\mathbf{x}$, $\mathbf{y}$ be disjoint distinct list of variables of length n and m respectively, let $\sigma : I_m \longrightarrow I_n$ be an injection, and let $B = [\mathbf{c}/\mathbf{x}][\mathbf{y}/\mathbf{x} \cdot \sigma]P(\mathbf{x})$. Then $\varphi_{B,\mathbf{y}}$ is congruential.*

(2) *Let A be an atomic D-formula without equality with $FV(A) \subseteq \{x\}$. Then for any algebraic model (F, φ) with the individual domain D, for any $a, b \in D$*

$$\varphi([a/x]A) \wedge Eab \leq \varphi([b/x]A).$$

Proof (1) First note that

$$\varphi([\mathbf{a}/\mathbf{y}]B) = \varphi(P(\mathbf{a}')),$$

where

$$a'_i := \begin{cases} c_i & \text{if } i \notin r(\sigma), \\ a_{\sigma^{-1}(i)} & \text{otherwise.} \end{cases}$$

This is clear, since

$$B[\mathbf{y} \mapsto \mathbf{a}] = P(\mathbf{x})[\mathbf{x} \cdot \sigma \mapsto \mathbf{a}][\mathbf{x} \mapsto \mathbf{c}].$$

Similarly we have

$$\varphi([\mathbf{b}/\mathbf{y}]B) = \varphi(P(\mathbf{b}')),$$

where $\mathbf{b}'$ is constructed from $\mathbf{b}$ in the same way as $\mathbf{a}'$. Now $\mathbf{a} =_i \mathbf{b}$ implies $\mathbf{a}' =_{\sigma(i)} \mathbf{b}'$, and thus, since φ_P is congruential,

$$\varphi([\mathbf{a}/\mathbf{y}]B) \wedge E\mathbf{ab} = \varphi(P(\mathbf{a}')) \wedge E\mathbf{ab} \leq \varphi(P(\mathbf{b}')) = \varphi([\mathbf{b}/\mathbf{y}]B),$$

as required.

(2) If A is a D-sentence, the claim is trivial. Otherwise we have

$$A = [x^m/\mathbf{y}]B,$$

for some B as in (1), if A contains P and x occurs m times in A.

By (1), $\varphi_{B,\mathbf{y}}$ is congruential, hence by Lemma 4.1.8,

$$E(a^m, b^m) \wedge \varphi(B(a^m)) \leq \varphi(B(b^m)).$$

Now since $E(a^m, b^m) = Eab$, $B(a^m) = A(a)$ and $B(b^m) = A(b)$, we obtain (2). ∎

Lemma 4.2.6 *The predicate $\varphi_{A,x}$ is congruential for any D-formula A with $FV(A) \subseteq \{x\}$.*

Proof By induction on the complexity of A we prove

$$\varphi([a/x]A) \wedge Eab \leq \varphi([b/x]A).$$

(1) If A is $\perp$, there is nothing to prove.

(2) If A is $x = c$ or $c = x$, then by 4.1.1 we have:

$$\varphi(A(a)) \wedge Eab = Eac \wedge Eab \leq Ebc = \varphi(A(b)).$$

(3) If A is $x = x$, then

$$\varphi(A(a)) \wedge Eab = Eaa \wedge Eab \leq Ebb = \varphi(A(b)).$$

(4) If A is atomic without equality, the claim follows from Lemma 4.2.5.

(5) In the modal case, if A is $\Box_i B$, then by Definition 4.1.2(4) and the induction hypothesis
$\varphi(A(a)) \cap Eab \leq \Box_i \varphi(B(a)) \cap \Box_i Eab = \Box_i(\varphi(B(a)) \cap Eab) \leq \Box_i \varphi(B(b)) = \varphi(A(b))$.

(6) If A is $B \supset C$ and φ_B, φ_C are congruential, then

$\varphi(A(a)) \wedge Eab = (\varphi(B(a)) \to \varphi(C(a))) \wedge Eab$;
thus
$\varphi(A(a)) \wedge Eab \wedge \varphi(B(b)) =$
$(\varphi(B(a)) \to \varphi(C(a))) \wedge \varphi(B(b)) \wedge Eab \leq$
$(\varphi(B(a)) \to \varphi(C(a))) \wedge \varphi(B(a)) \wedge Eab$ (since φ_B is congruential)
$\leq \varphi(C(a)) \wedge Eab$ (by properties of Heyting algebras)
$\leq \varphi(C(b))$ (since φ_C is congruential).

Therefore

$$\varphi(A(a)) \wedge Eab \leq \varphi(B(b)) \to \varphi(C(b)) = \varphi(A(b)).$$

(7) If $A(x)$ is $\exists y\, B(y,x)$, then we may assume that $y \neq x$ (otherwise A is a sentence), and thus

$$\varphi(A(a)) = \bigvee_{d \in D} \varphi(B(d,a))).$$

So by well-distributivity, and the induction hypothesis applied to $B(d,x)$, we obtain

$$\varphi(A(a)) \wedge Eab = \bigvee_{d \in D} (\varphi(B(d,a)) \wedge Eab) \leq$$
$$\bigvee_{d \in D} \varphi(B(d,b)) = \varphi(A(b)).$$

(8) If $A = B \vee C$, and the claim is proved for B, C, we have

$$\varphi(A(a)) \wedge Eab \leq (\varphi(B(a)) \vee \varphi(C(a))) \wedge Eab =$$
$$= (\varphi(B(a)) \wedge Eab) \vee (\varphi(C(a)) \wedge Eab) \leq \varphi(B(b)) \vee \varphi(C(b)) = \varphi(A(b)).$$

(9) The simple case $A = B \wedge C$ is left to the reader.

(10) Let $A = \forall y B(y,x)$, and assume that $y \in FV(B)$, and that the claim holds for B. Then

$$\varphi(A(a)) \wedge Eab \leq \varphi(A(b))$$

is equivalent to

$$Eab \wedge \bigwedge_{c \in D} (Ec \to \varphi(B(c,a))) \leq \bigwedge_{c \in D} (Ec \to \varphi(B(c,b))).$$

It suffices to show that for any $c \in D$,

$$Eab \wedge (Ec \to \varphi(B(c,a))) \leq Ec \to \varphi(B(c,b)),$$

i.e.

$$Eab \wedge (Ec \to \varphi(B(c,a))) \wedge Ec \leq \varphi(B(c,b)).$$

The latter follows by properties of Heyting algebras and the induction hypothesis:

$$Eab \wedge (Ec \to \varphi(B(c,a))) \wedge Ec \leq Eab \wedge \varphi(B(c,a)) \leq \varphi(B(c,b)).$$

■

Lemma 4.2.7 *The predicate $\varphi_{A,\mathbf{x}}$ is congruential for any D-formula A and $r(\mathbf{x}) \supseteq FV(A)$.*

Proof Assume $\mathbf{a} =_i \mathbf{b}$ for $\mathbf{a}, \mathbf{b}$ of the same length as $\mathbf{x}$. Fix all the parameters of A but x_i, so put

$$B(x_i) := A(a_1, \ldots, a_{i-1}, x_i, a_{i+1}, \ldots, a_n).$$

Then

$$\varphi(A(\mathbf{a})) \wedge Ea_ib_i = \varphi(B(a_i)) \wedge Ea_ib_i \leq \varphi(B(b_i)) = \varphi(A(\mathbf{b}))$$

by Lemma 4.2.6. ■

However, as we mentioned in Remark 4.1.17, the predicate φ_A is not necessarily strict, even for a strict φ. So we introduce another extension of φ to D-sentences.

Let $F = (\mathbf{\Omega}, D, E)$ be a structure, A a D-sentence of the corresponding type. Then we put

$$E(A) := \bigwedge\{Ea \mid a \in D,\ a \text{ occurs in } A\}.$$

In particular, $E(A) = \mathbf{1}$ if A is a usual sentence (without constants from D).

Definition 4.2.8 *For a valuation φ in a structure $F = (\mathbf{\Omega}, D, E)$ we define an $\mathbf{\Omega}$-valued function $\varphi^\natural$ on corresponding D-sentences as follows.*

(1) $\varphi^\natural(\bot) := \mathbf{0}$;

(2) $\varphi^\natural(a = b) := E(a, b)$;

(3) $\varphi^\natural(A) := \varphi(A) \wedge E(A)$ *for all other atomic* A;

(4) $\varphi^\natural(A \vee B) := E(A \vee B) \wedge (\varphi^\natural(A) \vee \varphi^\natural(B))$;

(5) $\varphi^\natural(A \wedge B) := \varphi^\natural(A) \wedge \varphi^\natural(B)$;

(6) $\varphi^\natural(A \supset B) := E(A \supset B) \wedge (\varphi^\natural(A) \to \varphi^\natural(B))$;

(7) $\varphi^\natural(\Box_i A) := E(\Box_i A) \cap \Box_i \varphi^\natural(A)$ *(in the modal case)*;

(8) $\varphi^\natural(\exists x A) := \bigvee_{d \in D} \varphi^\natural([d/x]A)$.

(9) $\varphi^\natural(\forall x A) := E(\forall x A) \wedge \bigwedge_{d \in D} (Ed \to \varphi^\natural([d/x]A))$.

We also define another associated predicate

$$\varphi^\natural_{A,\mathbf{x}} : D^n \longrightarrow \mathbf{\Omega}$$

such that for any $\mathbf{a} \in D^n$

$$\varphi^\natural_{A,\mathbf{x}}(\mathbf{a}) := \varphi^\natural([\mathbf{a}/\mathbf{x}]A).$$

Again we often abbreviate $\varphi^\natural_{A,\mathbf{x}}$ to $\varphi^\natural_A$.

Lemma 4.2.9 *Let (F, φ) be an algebraic model, with the domain D, A a corresponding D-sentence. Then*

$$\varphi^\sharp(A) = \varphi(A) \wedge E(A).$$

Proof By induction.

(1) If A is $\bot$, the claim is trivial.

(2) If A is $a = b$, then by 4.1.4 (1) and symmetry,

$$\varphi(A) = Eab \leq Ea \wedge Eb = E(A).$$

(3) If A is $a = a$, then $\varphi(A) = Eaa = E(A)$.

(4) If A is atomic without equality, the claim holds by definition.

(5) If A is a modal formula $\Box_i B$, then by the induction hypothesis,

$$\varphi^\sharp(A) = \Box_i \varphi^\sharp(B) \cap E(A) = \Box_i \varphi(B) \cap \Box_i E(B) \cap E(A).$$

Next,

$$\Box_i E(B) \cap E(A) = \Box_i E(A) \cap E(A) = E(A)$$

by 4.1.4 (4), and thus

$$\varphi^\sharp(A) = \Box_i \varphi(B) \cap E(A) = \varphi(A) \cap E(A).$$

(6) If A is $B \vee C$, then by the induction hypothesis,

$$\varphi^\sharp(A) = (\varphi^\sharp(B) \vee \varphi^\sharp(C)) \wedge E(A) =$$
$$(\varphi(B) \wedge E(B) \wedge E(A)) \vee (\varphi(C) \wedge E(C) \wedge E(A)).$$

Since $E(A) \leq E(B), E(C)$, we obtain

$$\varphi^\sharp(A) = (\varphi(B) \wedge E(A)) \vee (\varphi(C) \wedge E(A)) = \varphi(A) \wedge E(A).$$

(7) If A is $B \wedge C$ and the claim holds for B, C, we have

$$\varphi^\sharp(A) = \varphi(B) \wedge \varphi(C) = \varphi(B) \wedge E(B) \wedge \varphi(C) \wedge E(C) =$$
$$(\varphi(B) \wedge \varphi(C)) \wedge (E(B) \wedge E(C)) = \varphi(A) \wedge E(A).$$

(8) If A is $B \supset C$, then by the induction hypothesis

$$\varphi^\sharp(A) = E(A) \wedge (\varphi^\sharp(B) \to \varphi^\sharp(C)) =$$
$$E(A) \wedge (\varphi(B) \wedge E(B) \to \varphi(C) \wedge E(C)).$$

Let us show that this equals

$$E(A) \wedge \varphi(A) = E(A) \wedge (\varphi(B) \to \varphi(C)).$$

In fact,

$$\varphi(B) \wedge E(A) \leq \varphi(B) \wedge E(B),$$

hence

$$\varphi(B) \wedge E(A) \wedge (\varphi(B) \wedge E(B) \to \varphi(C) \wedge E(C)) \leq$$
$$\varphi(B) \wedge E(B) \wedge (\varphi(B) \wedge E(B) \to \varphi(C) \wedge E(C)) \leq$$
$$\varphi(C) \wedge E(C) \leq \varphi(C).$$

So

$$E(A) \wedge (\varphi(B) \wedge E(B) \to \varphi(C) \wedge E(C)) \leq \varphi(B) \to \varphi(C).$$

The other way round

$$(\varphi(B) \to \varphi(C)) \wedge \varphi(B) \leq \varphi(C),$$

hence,

$$E(A) \wedge (\varphi(B) \to \varphi(C)) \wedge \varphi(B) \wedge E(B) \leq E(A) \wedge \varphi(C) \leq \varphi(C) \wedge E(C)$$

thus

$$E(A) \wedge (\varphi(B) \to \varphi(C)) \leq \varphi(B) \wedge E(B) \to \varphi(C) \wedge E(C).$$

(9) If $A = \exists x B(x)$ and x is a parameter of $B(x)$, then by the induction hypothesis and Definition 4.2.4 we have

$$\varphi^\sharp(A) = \bigvee_{d\in D} \varphi^\sharp(B(d)) = \bigvee_{d\in D} (\varphi(B(d)) \wedge E(B(d))) =$$
$$\bigvee_{d\in D} (\varphi(B(d)) \wedge Ed \wedge E(A)) = E(A) \wedge \bigvee_{d\in D} (\varphi(B(d)) \wedge Ed) = E(A) \wedge \varphi(A).$$

If $A = \exists x B(x)$ and $B(x)$ is a D-sentence, then $B(x) = B(d)$ for any $d \in D$, $E(B(d)) = E(A)$. So

$$\varphi^\sharp(A) = \bigvee_{d\in D} \varphi^\sharp(B(d)) = \bigvee_{d\in D} (\varphi(B(d)) \wedge E(B(d))) = E(A) \wedge \bigvee_{d\in D} \varphi(B(d))$$
$$= E(A) \wedge \varphi(A).$$

(10) Finally let $A = \forall x B(x)$ and first assume that $x \in FV(B(x))$. Then

$$\varphi^\sharp(A) = E(A) \wedge \bigwedge_{d\in D} (Ed \to \varphi^\sharp(B(d)) =$$
$$E(A) \wedge \bigwedge_{d\in D} (Ed \to E(A) \wedge Ed \wedge \varphi(B(d)))$$

by the induction hypothesis; thus

$$\varphi^\sharp(A) \leq E(A) \wedge \bigwedge_{d\in D} (Ed \to \varphi(B(d))) = E(A) \wedge \varphi(A).$$

The other way round,

$$E(A) \wedge (Ed \to \varphi(B(d))) \wedge Ed \leq E(A) \wedge Ed \wedge \varphi(B(d)),$$

$$E(A) \wedge (Ed \to \varphi(B(d))) \wedge Ed \leq E(A) \wedge Ed \wedge \varphi(B(d)),$$

hence

$$(*) \quad E(A) \wedge (Ed \to \varphi(B(d))) \leq Ed \to E(A) \wedge Ed \wedge \varphi(B(d)),$$

which implies

$$E(A) \wedge \varphi(A) \leq \varphi^\sharp(A).$$

If $B(x)$ is closed, the argument is slightly different, because now $B(d) = B(x)$ and $E(B(d)) = E(A)$. By the induction hypothesis we have

$$\varphi^\sharp(A) = E(A) \wedge \bigwedge_{d\in D} (Ed \to E(A) \wedge \varphi(B(d)))$$
$$\leq E(A) \wedge \bigwedge_{d\in D} (Ed \to \varphi(B(d))) = E(A) \wedge \varphi(A).$$

The other way round, $(*)$ implies

$$E(A) \wedge (Ed \to \varphi(B(d))) \leq Ed \to E(A) \wedge \varphi(B(d)),$$

so

$$E(A) \wedge \varphi(A) = E(A) \wedge \bigwedge_{d \in D} (Ed \to \varphi(B(d))) \leq$$
$$E(A) \wedge \bigwedge_{d \in D} (Ed \to E(A) \wedge \varphi(B(d))) = \varphi^{\sharp}(A).$$

■

Lemma 4.2.10 *Let (F, φ) be an algebraic model with a domain D. Then for any D-formula A with $FV(A) \subseteq r(\mathbf{x})$ the predicate $\varphi^{\natural}_{A,\mathbf{x}}$ is congruential; it is also strict if $FV(A) = r(\mathbf{x})$.*

Proof Strictness readily follows from the previous lemma, since

$$\varphi^{\natural}_{A,\mathbf{x}}(\mathbf{a}) = \varphi^{\natural}(A(\mathbf{a})) \leq E(A(\mathbf{a})) = E\mathbf{a}.$$

To prove the congruentiality, assume that $|\mathbf{x}| = n$ and consider $\mathbf{a} =_i \mathbf{b}$ in D^n.
If $x_i \notin FV(A)$, then $[\mathbf{a}/\mathbf{x}]A = [\mathbf{b}/\mathbf{x}]A$, so the inequality

$$\varphi^{\sharp}([\mathbf{a}/\mathbf{x}]A) \wedge Ea_ib_i \leq \varphi^{\sharp}([\mathbf{b}/\mathbf{x}]A)$$

is trivial.
If $x_i \in FV(A)$, then by the previous lemma and Lemma 4.2.7,

$$\begin{aligned}\varphi^{\sharp}([\mathbf{a}/\mathbf{x}]A) \wedge Ea_ib_i &= \varphi([\mathbf{a}/\mathbf{x}]A) \wedge E([\mathbf{a}/\mathbf{x}]A) \wedge Ea_ib_i \\ &\leq \varphi([\mathbf{b}/\mathbf{x}]A) \wedge E([\mathbf{a}/\mathbf{x}]A) \wedge Ea_ib_i.\end{aligned}$$

Since $\mathbf{a} =_i \mathbf{b}$ and $Ea_ib_i \leq Eb_i$, we have

$$E([\mathbf{a}/\mathbf{x}]A) \wedge Ea_ib_i \leq E([\mathbf{b}/\mathbf{x}]A);$$

thus

$$\varphi^{\sharp}([\mathbf{a}/\mathbf{x}]A) \wedge Ea_ib_i \leq \varphi^{\sharp}([\mathbf{b}/\mathbf{x}]A)$$

by 4.2.7. ■

Lemma 4.2.11 *Let (F, φ) be an algebraic model with a domain D, φ^s the corresponding strict valuation such that for any $P \in PL^n$, $\mathbf{a} \in D^n$,*

$$\varphi^s(P(\mathbf{a})) = \varphi(P(\mathbf{a})) \wedge E\mathbf{a},$$

i.e. (Lemma 4.1.11)

$$\varphi^s_P = (\varphi_P)^s.$$

Then

$$(\varphi^s)^{\natural}(A) = \varphi^{\natural}(A)$$

for any D-sentence A.

Proof By the choice of φ^s, for any $A \in AF_D$

$$\varphi^s(A) = \varphi(A) \wedge E(A) = \varphi^\natural(A),$$

so

$$(\varphi^s)^\natural(A) = \varphi^s(A) \wedge E(A) = \varphi^\natural(A).$$

Since the maps $(\varphi^s)^\natural$, $\varphi^\natural$ coincide on AF_D and they are uniquely prolonged according to 4.2.8, they also coincide on all D-sentences. ∎

Lemma 4.2.12 *Let (F, φ) be an algebraic model, A a sentence of the corresponding type. Then*

$$\varphi(A) = \varphi^s(A) = \varphi^\natural(A) = (\varphi^s)^\natural(A).$$

Proof By 4.2.11,

$$\varphi^\natural(A) = (\varphi^s)^\natural(A).$$

Since $E(A) = \mathbf{1}$, by Lemma 4.2.9 we also have

$$\varphi^\natural(A) = \varphi(A), \ (\varphi^s)^\natural(A) = \varphi^s(A).$$

∎

Lemma 4.2.13 *Let $F = (\mathbf{\Omega}, D, E)$ be a structure, (F, φ) an algebraic model, $A(\mathbf{x})$ a D-formula of the corresponding type with $r(\mathbf{x}) = FV(A(\mathbf{x}))$, $|\mathbf{x}| = n$, $\mathbf{x}$ distinct. Then*

(1) $\varphi(\forall \mathbf{x} A(\mathbf{x})) = \bigwedge\limits_{\mathbf{a} \in D^n} (E\mathbf{a} \to \varphi(A(\mathbf{a})))$.

(2) $\varphi(\forall \mathbf{x} A(\mathbf{x})) = \mathbf{1}$ *iff* $\forall \mathbf{a} \in D^n \ E\mathbf{a} \leq \varphi(A(\mathbf{a}))$.

Thus $\varphi(\forall \mathbf{x} A(\mathbf{x}))$ does not depend on the ordering of $\mathbf{x}$, so we may use the notation $\varphi(\bar{\forall} A)$.

Proof (1) By induction on n. The base is trivial. Consider the step from n to $n+1$. If $\mathbf{x} = \mathbf{y}z$, $|\mathbf{y}| = n$, then $\forall \mathbf{x} A(\mathbf{x}) = \forall \mathbf{y} \forall z A(\mathbf{y}, z)$, so by the induction hypothesis and Lemma 1.2.3,

$$\begin{aligned}
\varphi(\forall \mathbf{x} A(\mathbf{x}) &= \bigwedge_{\mathbf{a} \in D^n} (E\mathbf{a} \to \varphi(\forall z A(\mathbf{a}, z))) \\
&= \bigwedge_{\mathbf{a} \in D^n} (E\mathbf{a} \to \bigwedge_{c \in D} (Ec \to \varphi(A(\mathbf{a}, c)))) = \bigwedge_{\mathbf{a} \in D^n} \bigwedge_{c \in D} (E\mathbf{a} \to (Ec \to \varphi(A(\mathbf{a}, c)))) \\
&= \bigwedge_{(\mathbf{a}, c) \in D^{n+1}} (E\mathbf{a} \wedge Ec \to \varphi(A(\mathbf{a}, c))) = \bigwedge_{\mathbf{b} \in D^{n+1}} (E\mathbf{b} \to \varphi(A(\mathbf{b})))
\end{aligned}$$

as required.

(2) Readily follows from (1). ∎

Definition 4.2.14 *A closed formula A is called* true *in an algebraic model (F, φ) (notation: $(F, \varphi) \models A$ or $(F, \varphi) \Vdash A$ in the intuitionistic case) if $\varphi(A) = \mathbf{1}$; an arbitrary formula A is called* true *in (F, φ) (with the same notation) if $\bar{\forall} A$ is true.*

Definition 4.2.15 *A predicate $\mathcal{A} : D^n \longrightarrow \mathbf{\Omega}$ is called* true *in a structure $F = (\mathbf{\Omega}, D, E)$ (notation: $F \models \mathcal{A}$ or $F \Vdash \mathcal{A}$) if $E\mathbf{a} \leq \mathcal{A}(\mathbf{a})$ for any $\mathbf{a} \in D^n$.*

Thus

$$F \models \mathcal{A} \text{ iff } \forall \mathbf{a} \in D^n \; E\mathbf{a} = \mathcal{A}(\mathbf{a}).$$

for a strict $\mathcal{A}$,

$$F \models \mathcal{A} \text{ iff } \mathcal{A}(\curlywedge) = \mathbf{1}.$$

for a 0-ary $\mathcal{A}$, and

$$F \models \mathcal{A} \text{ iff } F \models \mathcal{A}^s.$$

Lemma 4.2.16 *The following conditions are equivalent (where $\models$ stands for $\Vdash$ in the intuitionistic case)*

(1) $(F, \varphi) \models A$,

(2) $(F, \varphi^s) \models A$,

(3) $F \models \varphi_A$, *i.e.* $\forall \mathbf{a} \in D^n \; E\mathbf{a} \leq \varphi(A(\mathbf{a}))$,

(4) $F \models \varphi^s_A$, *i.e.* $\forall \mathbf{a} \in D^n \; \varphi^s(A(\mathbf{a})) = E\mathbf{a}$,

(5) $F \models \varphi^\natural_A$, *i.e.* $\forall \mathbf{a} \in D^n \; E\mathbf{a} \leq \varphi^\natural(A(\mathbf{a}))$,

(6) $F \models (\varphi^s)^\natural_A$, *i.e.* $\forall \mathbf{a} \in D^n \; (\varphi^s)^\natural(A(\mathbf{a})) = E\mathbf{a}$.

Proof By 4.2.12 and 4.2.13. ∎

Definition 4.2.17 *A formula A is* valid *in a structure F (of the corresponding kind) if A is true in every model over F (or equivalently, in every strict model over F).*

Validity is denoted again by $\models$ in the modal case, $\Vdash$ in the intuitionistic case.

Remark 4.2.18 If in the definition of validity we also allow for 'valuations' φ in $F = (\mathbf{\Omega}, D, E)$, for which φ_P is not congruential (they are just valuations in the corresponding D-structure $(\mathbf{\Omega}, D)$), then some $\mathbf{QK}^{=}_N$- (or $\mathbf{QH}^{=}$-) theorems become nonvalid. For instance, if $P \in PL^1$ and φ_P is not congruential, then the formula

$$A := \forall x \forall y (x = y \wedge P(x) \supset P(y))$$

is not true in $(\mathbf{\Omega}, \varphi)$ (in the sense of Definition 4.2.14).

In fact, suppose

$$Eab \wedge \varphi(P(a)) \not\leq \varphi(P(b))$$

for some $a, b \in D$. Then by Lemma 4.2.13 (the proof of which does not use congruentiality),

$$\begin{aligned}\varphi(A) &\leq Ea \wedge Eb \to (Eab \wedge \varphi(P(a)) \to \varphi(P(b)))\\ &= Ea \wedge Eb \wedge Eab \wedge \varphi(P(a)) \to \varphi(P(b))\\ &= Eab \wedge \varphi(P(a)) \to \varphi(P(b)) \neq \mathbf{1},\end{aligned}$$

by our assumption.

Lemma 4.2.19 *For any valuation φ in an m.v.s.,*

$$\begin{aligned}&(1) \qquad \varphi^\natural(\neg A) = E(A) - \varphi^\natural(A);\\ &(2) \qquad \varphi^\natural(\Diamond_i A) = E(A) \cap \Diamond_i \varphi^\natural(A).\end{aligned}$$

Proof The equality (1) is checked easily, so let us check (2). In fact, by (1), we have:

$$\begin{aligned}&\varphi^\natural(\Diamond_i A) = \varphi^\natural(-\Box_i - A) = E(A) \cap -\Box_i(E(A) \cap -\varphi^\natural(A)) =\\ &E(A) \cap (-\Box_i E(A) \cup -\Box_i - \varphi^\natural(A)) = E(A) \cap \Diamond_i \varphi^\natural(A)\end{aligned}$$

since $E(A) \leq \Box_i E(A)$. ■

4.3 Soundness

We begin with an analogue of Lemma 3.2.23.

Lemma 4.3.1 *Let F be a modal or Heyting-valued structure with a set of individuals D, and let $A(\mathbf{x})$, $B(\mathbf{x})$ be congruent formulas of the corresponding type, $|\mathbf{x}| = n$. Then for any $\mathbf{a} \in D^n$, for any valuation ϕ in F;*

(I) $\varphi(A(\mathbf{a})) = \varphi(B(\mathbf{a}))$, $\varphi^\sharp(A(\mathbf{a})) = \varphi^\sharp(B(\mathbf{a}))$

(II) for A, B without constants, $F \vDash (\Vdash)A \Longleftrightarrow F \vDash (\Vdash)B$.

Proof

(I) Along the same lines as 3.2.22. Consider the equivalence relation on modal (or intuitionistic) formulas:

$A \sim B$ iff $FV(A) = FV(B)$ and for any distinct list $\mathbf{x}$ such that $r(\mathbf{x}) = FV(A)$, for any $a \in D^{|\mathbf{x}|}$

$$\varphi([\mathbf{a}/\mathbf{x}]A) = \varphi([\mathbf{a}/\mathbf{x}]B).$$

It is sufficient to show that $\sim$ has the properties 2.3.17(1)–(4).

$$(1) \qquad QyA \sim Qz(A[y \mapsto z]) \text{ for } y \notin BV(A),\ z \notin V(A).$$

We only consider the case $Q = \forall$. If $FV(\forall yA) = r(\mathbf{x})$ for a distinct $\mathbf{x}$, we have two options:

(i) If $y \notin FV(A)$, then $y \notin V(A)$, $A[y \mapsto z] = A$,

$$\forall z(A[y \mapsto z]) = \forall zA, \ [\mathbf{a}/\mathbf{x}]\forall yA = \forall y[\mathbf{a}/\mathbf{x}]A,$$
$$[\mathbf{a}/\mathbf{x}]\forall zA = \forall z[\mathbf{a}/\mathbf{x}]A.$$

By Definition 4.2.4,

$$\varphi(\forall y[\mathbf{a}/\mathbf{x}]A) = \bigwedge_{d\in D}(Ed \to \varphi([\mathbf{a}/\mathbf{x}]A)),$$

since y does not occur in $[\mathbf{a}/\mathbf{x}]A$. Hence by Lemma 1.2.4 and 4.1.1(E3),

$$\varphi(\forall y[\mathbf{a}/\mathbf{x}]A) = \bigvee_{d\in D} Ed \to \varphi([\mathbf{a}/\mathbf{x}]A) = \varphi([\mathbf{a}/\mathbf{x}]A).$$

By the same reason

$$\varphi(\forall z[\mathbf{a}/\mathbf{x}]A) = \varphi([\mathbf{a}/\mathbf{x}]A),$$

so (1) holds in this case.

(ii) If $y \in FV(A)$, then

$$[\mathbf{a}/\mathbf{x}]\forall yA = \forall y[\mathbf{a}/\mathbf{x}]A,$$
$$[\mathbf{a}/\mathbf{x}]\forall z(A[y \mapsto z]) = \forall z[\mathbf{a}/\mathbf{x}](A[y \mapsto z]),$$

and so

$$\varphi([\mathbf{a}/\mathbf{x}]\forall yA) = \bigwedge_{d\in D}(Ed \to \varphi([d/y][\mathbf{a}/\mathbf{x}]A)),$$
$$\varphi([\mathbf{a}/\mathbf{x}]\forall z(A[y \mapsto z)) = \bigwedge_{d\in D}(Ed \to \varphi([d/z][\mathbf{a}/\mathbf{x}](A[y \mapsto z]))).$$

It remains to note that

$$[d/y][\mathbf{a}/\mathbf{x}]A = [d/z][\mathbf{a}/\mathbf{x}](A[y \mapsto z]),$$

since $y \notin BV(A)$, $z \notin V(A)$.

(2) Supposing $A \sim B$, let us show that

$$\forall yA \sim \forall yB.$$

We have $FV(A) = FV(B)$, $FV(\forall yA) = FV(\forall yB)$; let $\mathbf{r}(x) = FV(\forall yA)$. Then

$$\varphi([\mathbf{a}/\mathbf{x}]\forall yA) = \bigwedge_{d\in D}(Ed \to \varphi([\mathbf{a}d/\mathbf{x}y]A)),$$
$$\varphi([\mathbf{a}/\mathbf{x}]\forall yB) = \bigwedge_{d\in D}(Ed \to \varphi([\mathbf{a}d/\mathbf{x}y]B)).$$

Now $A \sim B$ implies

$$\varphi([\mathbf{a}d/\mathbf{x}y]A) = \varphi([\mathbf{a}d/\mathbf{x}y]B),$$

and the claim follows.

The argument for $\exists$ is quite similar.

(3) $A \sim A' \mathbin{\&} B \sim B' \Rightarrow (A * B) \sim (A' * B')$.

Consider the case $* = \supset$. We argue similarly to 3.2.22. If $r(\mathbf{x}) = FV(A \supset B)$, then $FV(A) = \mathbf{x} \cdot \sigma$, $FV(B) = \mathbf{x} \cdot \tau$ for injections σ, τ. Now $A \sim A' \mathbin{\&} B \sim B'$ implies $FV(A \supset B) = FV(A' \supset B')$ and

$$\begin{aligned} & \varphi([\mathbf{a}/\mathbf{x}](A \supset B)) = \varphi([\mathbf{a} \cdot \sigma/\mathbf{x}/ \cdot \sigma]A \supset [\mathbf{a} \cdot \tau/\mathbf{x} \cdot \tau]B) = \\ = \ & \varphi([\mathbf{a} \cdot \sigma/\mathbf{x} \cdot \sigma]A) \to \varphi([\mathbf{a} \cdot \tau/\mathbf{x} \cdot \tau]B) \\ = \ & \varphi([\mathbf{a} \cdot \sigma/\mathbf{x} \cdot \sigma]A') \to \varphi([\mathbf{a} \cdot \tau/\mathbf{x} \cdot \tau]B') \\ = \ & \varphi([\mathbf{a}/\mathbf{x}](A' \supset B')). \end{aligned}$$

The proof of (4) is trivial.

Since $E(A(\mathbf{a})) = E(B(\mathbf{a}))$, by Lemma 4.2.9 it follows that $\varphi^\sharp(A(\mathbf{a})) = \varphi^\sharp(B(\mathbf{a}))$.

(II) is an obvious consequence of (I).

■

Theorem 4.3.2 (Soundness theorem)

(1) For an N-m.v.s. F, the set

$$\mathbf{ML}^{(=)}(F) := \{A \in MF_N^{(=)} \mid F \vDash A\}$$

is an m.p.l. (=).

(2) For an H.v.s. F, the set

$$\mathbf{IL}^{(=)}(F) := \{A \in IF^{(=)} \mid F \Vdash A\}$$

is an s.p.l. (=).

Proof First let us show that the set of valid formulas is substitution closed. The argument resembles the proof of Lemma 3.2.16, but the detail is slightly different. Assume that $F = (\mathbf{\Omega}, D, E) \vDash A$. Let $S = [C(\mathbf{x}, \mathbf{y})/P(\mathbf{x})]$ be a simple formula substitution, where $\mathbf{x}$, $\mathbf{y}$ are distinct lists such that $r(\mathbf{y}) \subseteq FV(C(\mathbf{x}, \mathbf{y})) \subseteq r(\mathbf{xy})$. We may also assume that P occurs in A (otherwise $SA = A$) and $\mathbf{y} \cap FV(A) = \varnothing$, due to Lemma 4.3.2 (cf. the proof of Lemma 3.2.16). Since SA is defined up to congruence and validity respects congruence (Lemma 4.3.1), we may further assume that A is clean and $BV(A) \cap \mathbf{y} = \varnothing$; then SA is obtained by replacing every subformula of the form $P(\mathbf{x}')$ with $C(\mathbf{x}', \mathbf{y})$. By Lemma 2.5.23 we also have $FV(SA) = \mathbf{y} \cup FVe(S, A)$, where $FVe(S, A) \subseteq FV(A)$ is the set of essential parameters, see Definition 2.5.22. Let

$$FV(A) = r(\mathbf{z}),\ FVe(S, A) = r(\mathbf{z}'),\ \mathbf{z} = \mathbf{z}'\mathbf{z}''.$$

For an algebraic model (F, φ) let us show that $(F, \varphi) \vDash (\Vdash)\ SA$, i.e. (Lemma 4.2.16) for any $\mathbf{b} \in D^m$, $\mathbf{c} \in D^j$ (where $m = |\mathbf{y}|$, $j = |\mathbf{z}'|$)

$$(\sharp) \qquad \varphi([\mathbf{c}, \mathbf{b}/\mathbf{z}', \mathbf{y}]SA) \geq E\mathbf{b} \wedge E\mathbf{c}.$$

Given tuples $\mathbf{b}$, $\mathbf{c}$, we construct a valuation η in F such that for any $\mathbf{a} \in D^n$ (where $n = |\mathbf{x}|$)

- $\eta(P(\mathbf{a})) := \varphi(C(\mathbf{a}, \mathbf{b}))$;
- $\eta(B) := \varphi(B)$ for any other $B \in AF_D$.

Let us check that η is a valuation. In fact, let $\mathbf{a} =_i \mathbf{d}$; then by Lemma 4.2.7

$$Ea_i d_i \wedge \eta(P(\mathbf{a})) = Ea_i d_i \wedge \varphi(C(\mathbf{d}, \mathbf{b})) \leq \varphi(C(\mathbf{a}, \mathbf{b})) = \eta(P(\mathbf{d})).$$

Then we claim that for any formula $B(\mathbf{u})$, with a distinct $\mathbf{u}$ such that

$$FV(B(\mathbf{u})) \subseteq r(\mathbf{u}),\ r(\mathbf{u}) \cap r(\mathbf{y}) = r(\mathbf{u}) \cap BV(B(\mathbf{u})) = r(\mathbf{y}) \cap V(B(\mathbf{u})) = \varnothing,$$

and for any $\mathbf{d} \in D^l$ (where $l = |\mathbf{u}|$)

$$(1) \quad \eta(B(\mathbf{d})) = \varphi([\mathbf{d}, \mathbf{b}/\mathbf{u}, \mathbf{y}]SB),$$

or in a simpler notation,

$$\eta(B(\mathbf{d})) = \varphi(SB(\mathbf{d}, \mathbf{b})).$$

Note that $SB(\mathbf{d}, \mathbf{b})$ is a D-sentence, since $FV(SB) \subseteq \mathbf{uy}$, by Lemma 2.5.23.

The claim (1) is proved by induction.

The case when B is atomic and does not contain P, is trivial.

Assume that B is atomic, $B = P(\mathbf{x}')$, $\mathbf{x}' = \mathbf{u} \cdot \sigma$, where $\sigma : I_n \longrightarrow I_l$. Then we have:

$$B(\mathbf{d}) = [\mathbf{d}/\mathbf{u}][\mathbf{u} \cdot \sigma/\mathbf{x}]P(\mathbf{x}) = [\mathbf{d} \cdot \sigma/\mathbf{x}]P(\mathbf{x}) = P(\mathbf{a}),$$

where $\mathbf{a} = \mathbf{d} \cdot \sigma$. On the other hand, $SB = C(\mathbf{u} \cdot \sigma, \mathbf{y})$, and so

$$[\mathbf{d}, \mathbf{b}/\mathbf{u}, \mathbf{y}]SB = [\mathbf{d}, \mathbf{b}/\mathbf{u}, \mathbf{y}]C(\mathbf{u} \cdot \sigma, \mathbf{y}) = C(\mathbf{d} \cdot \sigma, \mathbf{b}) = C(\mathbf{a}, \mathbf{b}).$$

Thus by definition of η,

$$\eta(B(\mathbf{d})) = \varphi(SB(\mathbf{d}, \mathbf{b})).$$

Let $B = B_1 * B_2$, $SB = SB_1 * SB_2$, where $*$ is $\vee$, $\wedge$ or $\supset$. Then by Definition 4.2.4 and the induction hypothesis

$$\begin{aligned} \eta(B(\mathbf{d})) &= (\eta(B_1(\mathbf{d})) \star \eta(B_2(\mathbf{d}))) = \varphi(SB_1(\mathbf{d}, \mathbf{b})) \star \varphi(SB_2(\mathbf{d}, \mathbf{b})) \\ &= \varphi(SB_1(\mathbf{d}, \mathbf{b})) * SB_2(\mathbf{d}, \mathbf{b})) = \varphi(SB(\mathbf{d}, \mathbf{b})), \end{aligned}$$

where $\star$ is the corresponding operation in $\mathbf{\Omega}$.

If $B = \forall x B_1$, then by our assumption $x \notin \mathbf{uy}$, so $SB = \forall x\, SB_1$, and thus by Definition 4.2.4 and the induction hypothesis

$$\begin{aligned} \eta(B(\mathbf{d})) &= \bigwedge_{a \in D} (Ea \to \eta(B_1(a, \mathbf{d}))) = \bigwedge_{a \in D} (Ea \to \varphi(SB_1(\mathbf{b}, a, \mathbf{d}))) \\ &= \varphi(\forall x SB_1(\mathbf{b}, \mathbf{d})) = \varphi(SB(\mathbf{b}, \mathbf{d})). \end{aligned}$$

The case $B = \exists x B_1$ is similar and is left to the reader.

Now let us verify ($\sharp$). Take an arbitrary $\mathbf{e} \in D^{k-j}$ and the corresponding $\mathbf{d} = \mathbf{ce} \in D^k$. Then by (1) we have:

$$(2) \quad \varphi([\mathbf{d}, \mathbf{b}/\mathbf{z}, \mathbf{y}]SA) = \eta([\mathbf{d}/\mathbf{z}]A).$$

But

$$(3) \quad [\mathbf{d}, \mathbf{b}/\mathbf{z}, \mathbf{y}]SA = [\mathbf{c}, \mathbf{b}/\mathbf{z}', \mathbf{y}]SA, \ [\mathbf{d}/\mathbf{z}]A = [\mathbf{c}/\mathbf{z}']A,$$

since $FV(A) = r(\mathbf{z}),\ FV(SA) = r(\mathbf{yz}')$.

By our assumption, $F \vDash A$, and so by (2), (3), and Lemma 4.2.16, we obtain

$$\varphi([\mathbf{c}, \mathbf{b}/\mathbf{z}', \mathbf{y}]SA) = \eta([\mathbf{d}/\mathbf{z}]A) = \eta([\mathbf{c}/\mathbf{z}']A) \geq E\mathbf{c} \geq E\mathbf{b} \wedge E\mathbf{c},$$

i.e. ($\sharp$) holds.

The remaining properties $(m0)$–$(m3)$, $(m5)^{=}$, (s1), (s3) from Definitions 2.6.1, 2.6.2, 2.6.3 are verified in a standard way. We check only some of them.

Note that the value $\varphi(A)$ of a propositional formula A in a structure $F = (\mathbf{\Omega}, D, E)$ is exactly the same as in the algebra $\mathbf{\Omega}$. So, since all propositional axioms are valid in $\mathbf{\Omega}$ (Lemma 1.2.6), they are also valid in F.

Let us check the validity of (Ax12): $\forall x P(x) \supset P(y)$. In fact, consider an algebraic model (F, φ). We have

$$\begin{aligned} &Ea \wedge \varphi(\forall x P(x)) = Ea \wedge \bigwedge_{d \in D} (Ed \to \varphi(P(d))) \\ &\leq Ea \wedge (Ea \to \varphi(P(a))) \leq \varphi(P(a)), \end{aligned}$$

hence

$$Ea \leq \varphi(\forall x P(x)) \to \varphi(P(a)) = \varphi(\forall x P(x)) \to \varphi(P(a)) = \varphi(\forall x P(x) \supset P(a)),$$

i.e. $(F, \varphi) \vDash \forall x P(x) \supset P(y)$, by Lemma 4.2.16.

Let us also show the validity of (Ax14): $\forall x(P(x) \supset q) \supset (\exists x P(x) \supset q)$. In fact, $\varphi(\forall x(P(x) \supset q) \supset (\exists x P(x) \supset q)) = \mathbf{1}$
iff $\varphi(\forall x(P(x) \supset q)) \leq \varphi(\exists x P(x) \supset q) = \varphi(\exists x P(x)) \to \varphi(q)$
iff $\varphi((\forall x(P(x) \supset q)) \wedge \varphi(\exists x P(x)) \leq \varphi(q)$.

Transforming the left part of the latter inequality by well-distributivity, we obtain

$$\begin{aligned} &\bigwedge_{d \in D} (\varphi(P(d)) \to \varphi(q)) \wedge \bigvee_{a \in D} \varphi(P(a)) \\ &= \bigvee_{a \in D} [\varphi(P(a)) \wedge \bigwedge_{d \in D} (\varphi(P(d)) \to \varphi(q))] \\ &\leq \bigvee_{a \in D} [\varphi(P(a)) \wedge (\varphi(P(a)) \to \varphi(q))] \leq \varphi(q) \end{aligned}$$

as required.

And finally let us consider modus ponens. Suppose $F \vDash A(\mathbf{x}, \mathbf{y}),\ A(\mathbf{x}, \mathbf{y}) \supset B(\mathbf{y}, \mathbf{z})$, where $FV(A(\mathbf{x}, \mathbf{y})) = r(\mathbf{x}, \mathbf{y}),\ FV(B(\mathbf{y}, \mathbf{z})) = r(\mathbf{yz}),\ r(\mathbf{x}) \cap r(\mathbf{z}) = \varnothing$ and the lists $\mathbf{xy}, \mathbf{yz}$ are distinct (and some of them may be empty). Let l, m, n be the lengths of $\mathbf{x}, \mathbf{y}, \mathbf{z}$ respectively. By 4.2.16, for any $\mathbf{a} \in D^l$, $\mathbf{b} \in D^m$, $\mathbf{c} \in D^n$ and valuation φ

$$E(\mathbf{ab}) \leq \varphi(A(\mathbf{a},\mathbf{b})),\ E(\mathbf{abc}) \leq \varphi(A(\mathbf{a},\mathbf{b}) \supset B(\mathbf{b},\mathbf{c})).$$

Hence

$$E(\mathbf{abc}) = E(\mathbf{ab}) \wedge E(\mathbf{abc}) \leq \varphi(A(\mathbf{a},\mathbf{b})) \wedge \varphi(A(\mathbf{a},\mathbf{b}) \supset B(\mathbf{b},\mathbf{c})) \leq \varphi(B(\mathbf{b},\mathbf{c})).$$

As this happens for any $\mathbf{a} \in D^l$, we obtain

$$\bigvee_{\mathbf{a}\in D^l} E(\mathbf{abc}) \leq \varphi(B(\mathbf{b},\mathbf{c})).$$

But

$$\bigvee_{\mathbf{a}\in D^l} E(\mathbf{abc}) = \bigvee_{\mathbf{a}\in D^l} (E(\mathbf{a}) \wedge E(\mathbf{bc})) \leq (\bigvee_{\mathbf{a}\in D^l} E(\mathbf{a})) \wedge E(\mathbf{bc}) = E(\mathbf{bc}),$$

by 4.1.4(5). Thus

$$E(\mathbf{bc}) \leq \varphi(B(\mathbf{b},\mathbf{c})),$$

which implies $F \vDash B(\mathbf{y},\mathbf{z})$, by 4.2.16. ■

Let us now consider a particular kind of H.v.s. — those arising from **S4**-m.v.s.

Definition 4.3.3 *The* pattern *of an* **S4**-*m.v.s.* $F = (\mathbf{\Omega}, D, E)$ *is the H.v.s.* $F^\circ = (\mathbf{\Omega}^\circ, D, E)$.

An H.v.s. of this form is called 'basic'. More generally:

Definition 4.3.4 *A locale is said to be* basic *if it is isomorphic to the pattern of a complete* **S4**-*algebra. An H.v.s. over a basic locale is also called* basic.

For basic H.v.s. the intuitionistic truth definition matches with the modal definition. In precise terms, this means the following.

Lemma 4.3.5 *Let* φ *be a valuation in an* **S4**-*m.v.s* $F = (\mathbf{\Omega}, D, E)$, *and put*

$$\psi(A) := \Box\varphi(A) \text{ for every } A \in AF_D.$$

Then ψ *is a valuation in* F° *and* $\psi(A) = \varphi(A^T)$ *for every* $A \in IF_D^{=}$.

Proof By Lemma 4.2.7, ψ is congruential. The required equality is easily proved by induction. Let us consider the induction step for $A = \forall x B(x)$, where $B(x)$ is a D-formula, $FV(B(x)) \subseteq \{x\}$. Then according to Definitions 4.2.4, 2.11.1 and Proposition 1.2.7, we have:

$$\begin{aligned}\psi(A) &= \bigwedge_{d\in D} (Ed \to \psi(B(d))) = \Box \bigcap_{d\in D} \Box(Ed \supset \varphi(B^T(d))) \\ &\leq \Box \bigcap_{d\in D} (Ed \supset \varphi(B^T(d))) = \varphi(A^T).\end{aligned}$$

On the other hand,

$$\bigcap_{d\in D}(Ed \supset \varphi(B^T(d))) \leq Ed \supset \varphi(B^T(d)),$$

hence

$$\varphi(A^T) = \Box \bigcap_{d\in D}(Ed \supset \varphi(B^T(d))) \leq \Box(Ed \supset \varphi(B^T(d))),$$

and thus

$$\varphi(A^T) \leq \bigcap_{d\in D} \Box(Ed \supset \varphi(B^T(d))),$$

which implies

$$\varphi(A^T) = \Box\varphi(A^T) \leq \Box \bigcap_{d\in D} \Box(Ed \supset \varphi(B^T(d))) = \psi(A),$$

and eventually

$$\varphi(A^T) = \psi(A).$$

The case $A = \exists x B(x)$ is left to the reader. ∎

Lemma 4.3.6 *Let $F = (\mathbf{\Omega}, D, E)$ be an* **S4**-*m.v.s., ψ a valuation in F°, $\psi^\sim$ the same valuation in F. Then $\psi(A) = \psi^\sim(A^T)$ for every $A \in IF_D^=$.*

Proof Apply Lemma 4.3.5 to the case when $\varphi = \psi^\sim$. ∎

Proposition 4.3.7 *For any* **S4**-*m.v.s. F and intuitionistic formula $A(\mathbf{x})$, $F^\circ \vDash A(\mathbf{x})$ iff $F \vDash A^T(\mathbf{x})$, i.e. $\mathbf{IL}^{(=)}(F^\circ) = s(\mathbf{ML}^{(=)}(F))$.*

Proof We may assume that $\mathbf{x} = FV(A)$.

($\Leftarrow$) Suppose $F \vDash A^T$ and consider a valuation ψ in F°. Let $\psi^\sim$ be the same valuation in F; then $\psi^\sim(\forall\mathbf{x}A^T(\mathbf{x})) = \mathbf{1}$, and thus $\psi^\sim(\Box\forall\mathbf{x}A^T(\mathbf{x})) = \mathbf{1}$. By Lemma 2.11.7,

$$\mathbf{QS4} \vdash \Box\forall\mathbf{x}A^T(\mathbf{x}) \equiv (\forall\mathbf{x}A(\mathbf{x}))^T;$$

hence by Lemmas 4.3.2, 4.3.6

$$\psi^\sim(\Box\forall\mathbf{x}A^T(\mathbf{x})) = \psi^\sim((\forall\mathbf{x}A(\mathbf{x}))^T) = \psi(\forall\mathbf{x}A(\mathbf{x})) = \mathbf{1}.$$

Since ψ is arbitrary, it follows that $F^\circ \vDash A$.

($\Rightarrow$) Suppose $F^\circ \vDash A$. For an arbitrary valuation φ in F, let us show that $\varphi(\forall\mathbf{x}A^T(\mathbf{x})) = \mathbf{1}$. Let ψ be a valuation in F° described in Lemma 4.3.5. Then by Lemmas 2.11.7, 4.3.2 and since $F^\circ \vDash A$, we obtain

$$\varphi(\Box\forall\mathbf{x}A^T(\mathbf{x})) = \varphi((\forall\mathbf{x}A(\mathbf{x}))^T) = \psi(\forall\mathbf{x}A(\mathbf{x})) = \mathbf{1},$$

and therefore $\varphi(\forall\mathbf{x}A^T(\mathbf{x})) = \mathbf{1}$. ∎

Definition 4.3.8 *We introduce* general algebraic semantics *for our four types of logics* $(\mathcal{M}, \mathcal{M}^=, \mathcal{S}, \mathcal{S}^=)$*:*

$$\begin{aligned}\mathcal{AE}_{m=} &:= \{\mathbf{ML}^=(F) \mid F \text{ is an m.v.s}\},\\ \mathcal{AE}_m &:= \{\mathbf{ML}(F) \mid F \text{ is an m.v.s}\},\\ \mathcal{AE}_{s=} &:= \{\mathbf{IL}^=(F) \mid F \text{ is an H.v.s}\},\\ \mathcal{AE}_s &:= \{\mathbf{IL}(F) \mid F \text{ is an H.v.s}\}.\end{aligned}$$

An algebraic semantics *is a semantics generated by a class of m.v.s (or H.v.s.).*

Let us also introduce some particular cases of algebraic semantics.

Definition 4.3.9 *For superintuitionistic logics the semantics*

$$\mathcal{AE}^-_{s(=)} := \{\mathbf{IL}^{(=)}(F) \mid F \text{ is a basic H.v.s.}\}.$$

is called basic general algebraic.

The question, whether every locale is basic, seems open, and so we do not know if $\mathcal{AE}^-_{s(=)}$ and $\mathcal{AE}_{s(=)}$ are equivalent.

Neighbourhood frames generate a special kind of modal algebras, so we can define the associated algebraic semantics.

Definition 4.3.10 *A* neighbourhood frame with equality *is a triple* $\Phi = (F, D, E)$ *such that* F *is a neighbourhood frame and* $(MA(F), D, E)$ *is an m.v.s. The latter m.v.s. is denoted by* $MV(\Phi)$*. If* F *is a topological space, we call* Φ *a* topological frame with equality *and define*

$$HV(\Phi) := (HA(F), D, E) \ (= MV(\Phi)^\circ).$$

The corresponding logics are defined in an obvious way:

$$\mathbf{ML}^{(=)}(\Phi) := \mathbf{ML}^{(=)}(MV(\Phi));$$

$$\mathbf{IL}^{(=)}(\Phi) := \mathbf{IL}^{(=)}(HV(\Phi)).$$

Definition 4.3.11 *We define* general neighbourhood/topological semantics *as follows:*

$$\begin{aligned}\mathcal{TE}_{m(=)} &:= \{\mathbf{ML}^{(=)}(\Phi) \mid \Phi \text{ is a neighbourhood frame with equality}\};\\ \mathcal{TE}_{s(=)} &:= \{\mathbf{IL}^{(=)}(\Phi) \mid \Phi \text{ is a topological frame with equality }\}.\end{aligned}$$

From the definitions and since Kripke frames correspond to a special kind of neighbourhood frames, we have:

Lemma 4.3.12

$$\mathcal{KE}_{m(=)} \preceq \mathcal{TE}_{m(=)} \preceq \mathcal{AE}_{m(=)};$$

$$\mathcal{KE}_{s(=)} \preceq \mathcal{TE}_{s(=)} \preceq \mathcal{AE}^-_{s(=)} \preceq \mathcal{AE}_{s(=)}.$$

4.4 Morphisms of algebraic structures

In this section we consider maps between algebraic structures preserving validity; they are analogues of p-morphisms used in propositional logic. Let us begin with the intuitionistic case.

Definition 4.4.1 *Let* $F_1 = (\mathbf{\Omega}, D_1, E_1)$ *and* $F_2 = (\mathbf{\Omega}, D_2, E_2)$ *be H.v.s. A* p-morphism $\gamma : F_1 \longrightarrow F_2$ *is a map* $\alpha : D_1 \times D_2 \longrightarrow \Omega$ *(an '*$\mathbf{\Omega}$*-valued graph') satisfying the following conditions for any* $a \in D_1,\ b \in D_2$*:*

(E1) $\alpha(a, b) \wedge E_1(a) \leq E_2(b)$;

(E2) $\alpha(a, b) \wedge E_2(b) \leq E_1(a)$;

(Q1) $E_1(a) \leq \bigvee\limits_{b \in D_2} \alpha(a, b)$ *('totality');*

(Q2) $E_2(b) \leq \bigvee\limits_{a \in D_1} \alpha(a, b)$ *('surjectivity').*

If instead of (E1), (E2), α *satisfies the conditions*

(I1) $\alpha(a_1, b_1) \wedge \alpha(a_2, b_2) \wedge E_1(a_1, a_2) \leq E_2(b_1, b_2)$ *('functionality');*

(I2) $\alpha(a_1, b_1) \wedge \alpha(a_2, b_2) \wedge E_2(b_1, b_2) \leq E_1(a_1, a_2)$ *('injectivity');*

it is called a p$^=$-morphism.

A p$^{(=)}$*-morphism* α *is called a* p$^{(=)}$-embedding *if for any* $a, a' \in D_1,\ b, b' \in D_2$*:*

(ε*)* $E_2(b', b) \wedge \alpha(a, b) \leq \alpha(a, b')$.

It is obvious that (I1) implies (E1), and (I2) implies (E2), so every p$^=$-morphism is a p-morphism. It is also clear that every p$^{(=)}$-morphism $\alpha : F_1 \longrightarrow F_2$ gives rise to the converse p$^{(=)}$-morphism $\alpha^{-1} : F_2 \longrightarrow F_1$, where $\alpha^{-1}(a, b) = \alpha(b, a)$. This is because the conditions (E1), (E2), (Q1), (Q2), (I1), (I2) are 'symmetrical'.

Definition 4.4.2 *A* p$^{(=)}$*-morphism* α *is called a* p$^{(=)}$-equivalence *if both* α, α^{-1} *are* p$^{(=)}$*-embeddings, i.e.* α *satisfies (*ε*) and*

$$(\varepsilon') \quad E_1(a', a) \wedge \alpha(a, b) \leq \alpha(a', b).$$

The diagram below shows the correlation between different kinds of morphisms:

$$\begin{array}{ccc}
p^= - morphism & \Rightarrow & p - morphism \\
\Uparrow & & \Uparrow \\
p^= - embedding & \Rightarrow & p - embedding \\
\Uparrow & & \Uparrow \\
p^= - equivalence & \Rightarrow & p - equivalence
\end{array}$$

For $\mathbf{a} = (a_1, \ldots, a_n) \in D_1^n,\ \mathbf{b} = (b_1, \ldots, b_n) \in D_2^n$, let $\alpha(\mathbf{a}, \mathbf{b}) := \bigwedge\limits_{i=1}^{n} \alpha(a_i, b_i)$.

Lemma 4.4.3 *If $\alpha : F_1 \longrightarrow F_2$ is a p-morphism of H.v.s., then for any* $\mathbf{a},\ \mathbf{c} \in D_1^n;\ \mathbf{b} \in D_2^n$,

$$\alpha(\mathbf{a}, \mathbf{b}) \wedge \alpha(\mathbf{c}, \mathbf{b}) \wedge E_1(\mathbf{c}) \leq E_1(\mathbf{a}, \mathbf{c}).$$

Proof In fact, the condition (E1) yields:

$$\alpha(c_i, b_i) \wedge E_1(c_i) \leq E_2(b_i).$$

On the other hand, from (I2) we have:

$$\alpha(a_i, b_i) \wedge \alpha(c_i, b_i) \wedge E_2(b_i) \leq E_1(a_i, c_i).$$

These two inequalities imply

$$\alpha(a_i, b_i) \wedge \alpha(c_i, b_i) \wedge E_1(c_i) \leq E_1(a_i, c_i),$$

whence the statement follows easily. ■

Lemma 4.4.4 *Let $\alpha : F_1 \longrightarrow F_2$ be a $p^{(=)}$-morphism, $\mathbf{x} = (x_1, \dots, x_n)$, $A(\mathbf{x}) \in IF^{(=)})$, $FV(A) \subseteq \mathbf{x}$, $\mathbf{a} \in D_1^n$, $\mathbf{b} \in D_2^n$. Then $\alpha(\mathbf{a}, \mathbf{b}) \wedge E_1(A(\mathbf{a})) \leq E_2(A(\mathbf{b}))$.*

Proof Easy by (E1). ■

We will use the following abbreviations:

$$\varphi A := \varphi(A), \quad \psi^\sim A := \psi^\sim(A).$$

Definition 4.4.5 *Let $\alpha : F_1 \longrightarrow F_2$ be a p-morphism of H.v.s. Valuations φ_1 in F_1 and φ_2 in F_2 are said to be* matching *if*

$$\alpha(\mathbf{a}, \mathbf{b}) \leq \varphi_1 P(\mathbf{a}) \leftrightarrow \varphi_2 P(\mathbf{b}),$$

for any $P \in PL^n$, $\mathbf{a} \in D_1^n$, $\mathbf{b} \in D_2^n$.

The above condition is equivalent to

$$\alpha(\mathbf{a}, \mathbf{b}) \wedge \varphi_1 P(\mathbf{a}) = \alpha(\mathbf{a}, \mathbf{b}) \wedge \varphi_2 P(\mathbf{b})$$

and can be replaced with the following two:

$$\alpha(\mathbf{a}, \mathbf{b}) \wedge \varphi_1 P(\mathbf{a}) \leq \varphi_2 P(\mathbf{b});$$
$$\alpha(\mathbf{a}, \mathbf{b}) \wedge \varphi_2 P(\mathbf{b}) \leq \varphi_1 P(\mathbf{a}).$$

Lemma 4.4.6 *Let $\alpha : F_1 \longrightarrow F_2$ be a $p^{(=)}$-morphism, $\mathbf{x} = (x_1, \dots, x_n)$, $A(\mathbf{x}) \in IF^{(=)}$, $FV(A) \subseteq \mathbf{x}$, $\mathbf{a} \in D_1^n$, $\mathbf{b} \in D_2^n$.*
Then

$$\alpha(\mathbf{a}, \mathbf{b}) \leq \varphi_1^\sim A(\mathbf{a}) \leftrightarrow \varphi_2^\sim A(\mathbf{b}),$$

whenever valuations φ_1 in F_1 and φ_2 in F_2 are matching.

Proof By induction on A we check the following two properties:

(*) $\alpha(\mathbf{a},\mathbf{b}) \wedge \varphi_1^{\sim} A(\mathbf{a}) \leq \varphi_2^{\sim} A(\mathbf{b})$;

(**) $\alpha(\mathbf{a},\mathbf{b}) \wedge \varphi_2^{\sim} A(\mathbf{b}) \leq \varphi_1^{\sim} A(\mathbf{a})$.

- The atomic case is obvious; if A is $(x_1 = x_2)$, apply the conditions (I1), (I2).
- The case $A = B \wedge C$ is trivial.
- Let $A = B \vee C$. Then

  $$\alpha(\mathbf{a},\mathbf{b}) \wedge \varphi_1^{\sim} A(\mathbf{a}) = \alpha(\mathbf{a},\mathbf{b}) \wedge E_1(A(\mathbf{a})) \wedge (\varphi_1^{\sim} B(\mathbf{a}) \vee \varphi_1^{\sim} C(\mathbf{a})) \leq E_2(A(\mathbf{b})) \wedge (\varphi_2^{\sim} B(\mathbf{b}) \vee \varphi_2^{\sim} C(\mathbf{b})) = \varphi_2^{\sim} A(\mathbf{b})$$

  by the inductive hypothesis and Lemma 4.4.4; similarly we obtain (**).
- Let $A = B \supset C$. Then we have:

  $$\alpha(\mathbf{a},\mathbf{b}) \wedge \varphi_1^{\sim} A(\mathbf{a}) = \alpha(\mathbf{a},\mathbf{b}) \wedge E_1(A(\mathbf{a})) \wedge (\varphi_1^{\sim} B(\mathbf{a}) \rightarrow \varphi_1^{\sim}(C(\mathbf{a})) \leq \alpha(\mathbf{a},\mathbf{b}) \wedge E_2(A(\mathbf{b})) \wedge (\varphi_1^{\sim} B(\mathbf{a}) \rightarrow \varphi_1^{\sim}(C(\mathbf{a}))$$

  by Lemma 4.4.4.

  Next,

  $$\begin{aligned} &\alpha(\mathbf{a},\mathbf{b}) \wedge (\varphi_1^{\sim} B(\mathbf{a}) \rightarrow \varphi_1^{\sim} C(\mathbf{a})) \wedge \varphi_2^{\sim} B(\mathbf{b}) \\ &\leq \alpha(\mathbf{a},\mathbf{b}) \wedge (\varphi_1^{\sim} B(\mathbf{a}) \rightarrow \varphi_1^{\sim} C(\mathbf{a})) \wedge \varphi_1^{\sim} B(\mathbf{a}) \\ &\qquad\qquad \text{by the induction hypothesis for } B \\ &\leq \alpha(\mathbf{a},\mathbf{b}) \wedge \varphi_1^{\sim} C(\mathbf{a}) \text{ by the property of Heyting algebras} \\ &\leq \varphi_2^{\sim} C(\mathbf{b}) \end{aligned}$$

  by the inductive hypothesis for C, so we obtain:

  $$\alpha(\mathbf{a},\mathbf{b}) \wedge (\varphi_1^{\sim} B(\mathbf{a}) \rightarrow \varphi_1^{\sim}(C(\mathbf{a})) \leq \varphi_2^{\sim} B(\mathbf{b}) \rightarrow \varphi_2^{\sim} C(\mathbf{b}).$$

  Thus

  $$\alpha(\mathbf{a},\mathbf{b}) \wedge \varphi_1^{\sim} A(\mathbf{a}) \leq E_2(A(\mathbf{b})) \wedge (\varphi_2^{\sim} B(\mathbf{b}) \rightarrow \varphi_2^{\sim} C(\mathbf{b})) = \varphi_2^{\sim} A(\mathbf{b}).$$

  Similarly one can check (**).
- Let $A(\mathbf{x}) = \exists y B(y,\mathbf{x})$, $y \in FV(B)$. Then by Definition 4.2.1, we have:

  $$\begin{aligned} \alpha(\mathbf{a},\mathbf{b}) \wedge \varphi_1^{\sim} A(\mathbf{a}) &= \alpha(\mathbf{a},\mathbf{b}) \wedge \bigvee_{c \in D_1} \varphi_1^{\sim} B(c,\mathbf{a}) \\ &= \bigvee_{c \in D_1} (\alpha(\mathbf{a},\mathbf{b}) \wedge E_1(c) \wedge \varphi_1^{\sim} B(c,\mathbf{a})). \end{aligned}$$

  Now, since by $(Q1)$,

  $$E_1(c) \leq \bigvee_{d \in D_2} \alpha(c,d)$$

and by the inductive hypothesis,

$$\alpha(\mathbf{a},\mathbf{b}) \wedge \alpha(c,d) \wedge \varphi_1^{\sim} B(c,\mathbf{a}) \leq \varphi_2^{\sim} B(d,\mathbf{b}),$$

we obtain:

$$\bigvee_{c\in D_1} (\alpha(\mathbf{a},\mathbf{b}) \wedge E_1(c) \wedge \varphi_1^{\sim} B(c,\mathbf{a})) \leq \bigvee_{d\in D_2} \varphi_2^{\sim} B(d,\mathbf{b}) = \varphi_2^{\sim} A(\mathbf{b}).$$

So A satisfies (*). To check (**), note:

$$\alpha(\mathbf{a},\mathbf{b}) \wedge \varphi_2^{\sim} A(\mathbf{b}) = \alpha(\mathbf{a},\mathbf{b}) \wedge \bigvee_{d\in D_2} \varphi_2^{\sim} B(d,\mathbf{b})$$
$$= \bigvee_{d\in D_2} (\alpha(\mathbf{a},\mathbf{b}) \wedge E_2(d) \wedge \varphi_2^{\sim} B(d,\mathbf{b})) \leq \bigvee_{c\in D_1} \varphi_1^{\sim} B(c,\mathbf{a}) = \varphi_1^{\sim} A(\mathbf{a}),$$

by the inductive hypothesis and since $E_2(d) \leq \bigvee_{c\in D_1} \alpha(c,d)$ by (Q2).

- Let $A(\mathbf{x}) = \exists y B(\mathbf{x}),\ y \notin FV(B)$; then

$$\alpha(\mathbf{a},\mathbf{b}) \wedge \varphi_1^{\sim} A(\mathbf{a}) = \alpha(\mathbf{a},\mathbf{b}) \wedge \bigvee_{c\in D_1} \varphi_1^{\sim} B(\mathbf{a})$$
$$= \alpha(\mathbf{a},\mathbf{b}) \wedge \varphi_1^{\sim} B(\mathbf{a}) \leq \varphi_2^{\sim} B(\mathbf{b}) = \bigvee_{d\in D_2} \varphi_2^{\sim} B(\mathbf{b}) = \varphi_2^{\sim} B(\mathbf{b}).$$

- Let $A(\mathbf{x}) = \forall \mathbf{y} B(y,\mathbf{x})$. For any $d \in D_2$ we have:

$$\alpha(\mathbf{a},\mathbf{b}) \wedge \bigwedge_{c\in D_1} (E_1(c) \to \varphi_1^{\sim} B(c,\mathbf{a})) \wedge E_2(d)$$
$$= \alpha(\mathbf{a},\mathbf{b}) \wedge \bigwedge_{c\in D_1} (E_1(c) \to \varphi_1^{\sim} B(c,\mathbf{a})) \wedge E_2(d) \wedge \bigvee_{c\in D_1} \alpha(c,d) \text{ (by (Q2))}$$
$$\leq \bigvee_{c\in D_1} (\alpha(\mathbf{a},\mathbf{b}) \wedge E_2(d) \wedge \alpha(c,d) \wedge (E_1(c) \to \varphi_1^{\sim} B(c,\mathbf{a})))$$
(by distributivity)
$$\leq \bigvee_{c\in D_1} (\alpha(\mathbf{a},\mathbf{b}) \wedge \alpha(c,d) \wedge E_1(c) \wedge (E_1(c) \to \varphi_1^{\sim} B(c,\mathbf{a})))$$
by (E2)
$$\leq \bigvee_{c\in D_1} (\alpha(\mathbf{a},\mathbf{b}) \wedge \alpha(c,d) \wedge \varphi_1^{\sim} B(c,\mathbf{a}))$$
(since in Heyting algebras $x \wedge (x \to y) \leq y$)
$$= \bigvee_{c\in D_1} (\alpha(c\mathbf{a}, d\mathbf{b}) \wedge \varphi_1^{\sim} B(c,\mathbf{a})) \leq \varphi_2^{\sim} B(d,\mathbf{b})$$
(by IH).

Hence we obtain $(*)$:

$$\alpha(\mathbf{a},\mathbf{b}) \wedge \varphi_1^{\sim} A(\mathbf{a}) = \alpha(\mathbf{a},\mathbf{b}) \wedge E_1(A(\mathbf{a})) \wedge \bigwedge_{c\in D_1} (E_1(c) \to \varphi_1^{\sim} B(c,\mathbf{a})) \leq$$
$$\leq E_2(A(\mathbf{b})) \wedge \bigwedge_{d\in D_2} (E_2(d) \to \varphi_2^{\sim} B(d,\mathbf{b})) = \varphi_2^{\sim} A(\mathbf{b}).$$

Similarly, for any $c \in D_1$ we have:

$$\alpha(\mathbf{a},\mathbf{b}) \wedge \bigwedge_{d\in D_2} (E_2(d) \to \varphi_2^{\sim} B(d,\mathbf{b})) \wedge E_1(c) \leq \varphi_1^{\sim} B(c,\mathbf{a}),$$

and thus we obtain (**):

$$\alpha(\mathbf{a},\mathbf{b}) \wedge \varphi_2^{\sim} A(\mathbf{b}) = \alpha(\mathbf{a},\mathbf{b}) \wedge E_2(A(\mathbf{b})) \wedge \bigwedge_{d\in D_2} (E_2(d) \to \varphi_2^{\sim} B(d,\mathbf{b})) \leq$$
$$\leq E_1(A(\mathbf{a})) \wedge \bigwedge_{c\in D_1} (E_1(c) \to \varphi_1^{\sim} B(c,\mathbf{a})) = \varphi_1^{\sim} A(\mathbf{a}).$$

■

Lemma 4.4.7 *If $\alpha : F_1 \longrightarrow F_2$ is a $p^{(=)}$-embedding, then*

(1) every valuation φ_1 in F_1 matches with some valuation φ_2 in F_1;

(2) $\mathbf{IL}^{(=)}(F_2) \subseteq \mathbf{IL}^{(=)}(F_1)$.

Proof

(1) Given φ_1, we define
$\varphi_2 P(\mathbf{b}) := \bigvee_{\mathbf{c}\in D_1^n} (\alpha(\mathbf{c},\mathbf{b}) \wedge \varphi_1 P(\mathbf{c}))$
for every $\mathbf{b} \in D_2^n$. Then

$$\begin{aligned}\alpha(\mathbf{a},\mathbf{b}) \wedge \varphi_2 P(\mathbf{b}) &= \bigvee_{\mathbf{c}\in D_1^n} (\alpha(\mathbf{a},\mathbf{b}) \wedge \alpha(\mathbf{c},\mathbf{b}) \wedge \varphi_1 P(\mathbf{c}))\\ &= \bigvee_{\mathbf{c}\in D_1^n} (\alpha(\mathbf{a},\mathbf{b}) \wedge \alpha(\mathbf{c},\mathbf{b}) \wedge E_1(\mathbf{c}) \wedge \varphi_1 P(\mathbf{c}))\\ &\text{(since } \varphi_1 P(\mathbf{c}) \leq E_1(\mathbf{c}) \text{ for the valuation } \varphi_1)\\ &\leq \bigvee_{\mathbf{c}\in D_1^n} (E_1(\mathbf{a},\mathbf{c}) \wedge \varphi_1 P(\mathbf{c})) \leq \varphi_1 P(\mathbf{a}),\end{aligned}$$

by Lemma 4.4.3 and Definition 4.2.1.

This φ_2 is a valuation in F_2. Indeed,

$$E_2(\mathbf{d},\mathbf{b}) \wedge \varphi_2 P(\mathbf{b}) = \bigvee_{\mathbf{c}\in D_1^n} (\alpha(\mathbf{c},\mathbf{b}) \wedge E_2(\mathbf{d},\mathbf{b}) \wedge E_1(\mathbf{c}) \wedge \varphi_1 P(\mathbf{c})) \leq$$
$$\bigvee_{\mathbf{c}\in D_1^n} (\alpha(\mathbf{c},\mathbf{d}) \wedge \varphi_1 P(\mathbf{c})) = \varphi_2 P(\mathbf{d}),$$

by (ε). Also $\varphi_2 P(\mathbf{b}) \leq E_2(\mathbf{b})$, since by (E1), $\alpha(\mathbf{c},\mathbf{b}) \wedge E_1(\mathbf{c}) \leq E_2(\mathbf{b})$ for any $\mathbf{c} \in D_1^n$.

(2) Let us show that $A \notin \mathbf{IL}^{(=)}(F_1)$ implies $A \notin \mathbf{IL}^{(=)}(F_2)$. We may assume that A is a sentence. Let φ_1 be a valuation in F_1 such that $\varphi_1^{\sim} A \neq \mathbf{1}$, and let φ_2 be a valuation in F_2 matching with φ_1. By Lemma 4.4.6 (for $n = 0$), $\varphi_1^{\sim} A \leftrightarrow \varphi_2^{\sim} A = \mathbf{1}$, and thus $\varphi_2^{\sim} A \neq \mathbf{1}$.

■

Corollary 4.4.8 *If $\alpha : F_1 \longrightarrow F_2$ is a $p^{(=)}$-equivalence, then $\mathbf{IL}^{(=)}(F_1) = \mathbf{IL}^{(=)}(F_2)$.*

Definition 4.4.9 *Let $F_1 = (\mathbf{\Omega}, D_1, E_1)$ and $F_2 = (\mathbf{\Omega}, D_2, E_2)$ be two H.v.s. A map $g : D_1 \longrightarrow D_2$ is called a* strong p-morphism *if for any $a \in D_1$, $b \in D_2$*

(1) $E_1(a) = E_2(g(a))$;

(2) $E_2(b) \leq \bigvee_{a \in D_1} E_2(b, g(a))$.

A strong p$^=$-morphism *is a map g satisfying (ii) and*

(3) $E_1(a_1, a_2) = E_2(g(a_1), g(a_2))$.

for any $a_1, a_2 \in D_1$. An isomorphism *is a bijection $g : D_1 \longrightarrow D_2$ satisfying (iii).*

Note that (ii) obviously holds if g is surjective and that (iii) implies (i). So a surjective g is a strong p$^=$-morphism iff (iii) holds. It is also obvious that every isomorphism is a strong p$^=$-morphism and that every strong p$^=$-morphism is a strong p-morphism.

Lemma 4.4.10 *For H.v.s. $F_1 = (\mathbf{\Omega}, D_1, E_1)$, $F_2 = (\mathbf{\Omega}, D_2, E_2)$ and a map $g : D_1 \longrightarrow D_2$, consider $\alpha : D_1 \times D_2 \longrightarrow \mathbf{\Omega}_2$ such that*

$$\alpha(a, b) := E_2(b, g(a))$$

for $a \in D_1, b \in D_2$. Then

- *g is a strong p-morphism iff α is a p-morphism;*
- *g is a strong p$^=$-morphism iff α is a p$^=$-morphism (or a p-embedding, p-equivalence, p$^=$-embedding, p$^=$-equivalence).*

Proof First note that α satisfies

$$(E1) \quad \alpha(a, b) \wedge E_1(a) \leq E_2(b).$$

In fact, this means

$$E_2(b, g(a)) \wedge E_1(a) \leq E_2(b),$$

which obviously holds, by Lemma 4.1.4(1).

Let us show that $(E2)$ for α and (i) for g are equivalent. The condition

$$(E2) \quad \alpha(a, b) \wedge E_2(b) \leq E_1(a)$$

is equivalent to

$$E_2(b, g(a)) \leq E_1(a);$$

this follows from (i) and implies $E_2(g(a)) \leq E_1(a)$ if we take $b = g(a)$.

The condition

$$(Q1) \quad E_1(a) \leq \bigvee_{b \in D_2} \alpha(a, b)$$

is equivalent to

$$E_1(a) \leq \bigvee_{b \in D_2} E_2(b, g(a));$$

this follows from (i), since $E_2(g(a)) = E_2(g(a), g(a)) \leq \bigvee_{b \in D_2} E_2(b, g(a))$. On the other hand, this condition implies $E_1(a) \leq E_2(g(a))$, since $E_2(b, g(a)) \leq E_2(g(a))$ for any $b \in D_2$.

Next, the condition

$$(Q2) \quad E_2(b) \leq \bigvee_{a \in D_1} \alpha(a, b)$$

is equivalent to

$$E_2(b) \leq \bigvee_{a \in D_1} E_2(b, g(a)),$$

which is the same as the condition (ii).

Therefore g is a strong p-morphism iff α is a p-morphism.

Next, let us show that g satisfies (iii) if α satisfies (I1) and (I2). In fact, the condition

$$\text{(I1)} \quad \alpha(a_1, b_1) \wedge \alpha(a_2, b_2) \wedge E_1(a_1, a_2) \leq E_2(b_1, b_2)$$

means

$$E_2(b_1, g(a_1)) \wedge E_2(b_2, g(a_2)) \wedge E_1(a_1, a_2) \leq E_2(b_1, b_2),$$

and

$$\text{(I2)} \quad \alpha(a_1, b_1) \wedge \alpha(a_2, b_2) \wedge E_2(b_1, b_2) \leq E_1(a_1, a_2)$$

means

$$E_2(b_1, g(a_1)) \wedge E_2(b_2, g(a_2)) \wedge E_2(b_1, b_2) \leq E_1(a_1, a_2).$$

These two conditions follow from (iii) and taken together, imply (iii) if we put $b_1 = g(a_1)$, $b_2 = g(a_2)$.

The condition

$$(\varepsilon) \quad E_2(b', b) \wedge \alpha(a, b) \wedge E_1(a) \leq \alpha(a, b')$$

means

$$E_2(b', b) \wedge E_2(b, g(a)) \wedge E_1(a) \leq E_2(b', g(a))$$

and holds by Definition 4.1.2.

The condition

$$(\varepsilon') \quad E_1(a', a) \wedge \alpha(a, b) \wedge E_2(b) \leq \alpha(a', b)$$

means

$$E_1(a', a) \wedge E_2(b, g(a)) \wedge E_2(b) \leq E_2(b, g(a'))$$

and thus follows from (iii)

This proves the second part of the Lemma. ∎

Remark 4.4.11 One can consider a more natural, but slightly more restrictive definition of $p^{(=)}$-morphism. Viz., the conditions (E1), (E2) can be replaced with a stronger one:

(E′) $\alpha(a,b) \leq E_2(b) \wedge E_1(a)$.

In this case the conjunct $E_1(a)$ in (ε), etc. should be omitted. Note that a strong $p^{(=)}$-morphism g gives rise to the $p^{(=)}$-morphism α in this stronger sense.

Problem 4.4.12 *Does the existence of a $p^{(=)}$-morphism ($p^{(=)}$-embedding in our form) imply the existence of a $p^{(=)}$-embedding satisfying (E′)?*

Now let us consider the modal case. It is clear that if $F = (\mathbf{\Omega}, D, E)$ is an m.v.s., then its Boolean part $F^\flat = (\mathbf{\Omega}^\flat, D, E)$ is an H.v.s. So we can give

Definition 4.4.13 *A* p$^{(=)}$-morphism *(respectively,* p$^{(=)}$-embedding, $p^{(=)}$- equivalence*)* $\alpha : F_1 \longrightarrow F_2$ *from a μ-m.v.s. $F_1 = (\mathbf{\Omega}, D_1, E_1)$ to an m.v.s. $F_2 = (\mathbf{\Omega}, D_2, E_2)$ is a $p^{(=)}$-morphism (respectively, $p^{(=)}$-embedding, $p^{(=)}$-equivalence) of the Boolean parts $F_1^\flat \longrightarrow F_2^\flat$ satisfying the condition*

(E0) $\alpha(a,b) \leq \Box_i \alpha(a,b)$

for all $i \in I_\mu$.

Thus a p-morphism is a map $\alpha : D_1 \times D_2 \to \Omega$ satisfying (E0) and the conditions (for any $a \in D_1,\ b \in D_2$):

(E1) $\alpha(a,b) \cap E_1(a) \leq E_2(b)$,

and so on.

The following is rather trivial.

Lemma 4.4.14 *If F_1, F_2 are m.v.s. over the same* **S4**-*algebra, F_1°, F_2° are the corresponding H.v.s., $\alpha : F_1 \longrightarrow F_2$ is a $p^{(=)}$-morphism (respectively, $p^{(=)}$-embedding, $p^{(=)}$-equivalence), then the same map α is a $p^{(=)}$-morphism (respectively, $p^{(=)}$-embedding, $p^{(=)}$-equivalence) $F_1^\circ \longrightarrow F_2^\circ$ of the corresponding H.v.s.*

The definition of matching valuations in the modal case is the same as Definition 4.4.5:

Definition 4.4.15 *Let $\alpha : F_1 \longrightarrow F_2$ be a p-morphism of m.v.s. Valuations φ_1 in F_1 and φ_2 in F_2 are* matching *if*

$$\alpha(\mathbf{a},\mathbf{b}) \leq (\varphi_1 P(\mathbf{a}) \triangleq \varphi_2 P(\mathbf{b})),$$

for any $P \in PL^n$, $\mathbf{a} \in D_1^n$, $\mathbf{b} \in D_2^n$.

Lemma 4.4.16 *Let $\alpha : F_1 \longrightarrow F_2$ be a $p^{(=)}$-morphism, $A(\mathbf{x}) \in MF^{(=)}$, $FV(A) \subseteq \mathbf{x} = (x_1, \ldots, x_n), \mathbf{a} \in D_1^n$, $\mathbf{b} \in D_2^n$. Then*

$$\alpha(\mathbf{a},\mathbf{b}) \leq (\varphi_1 A(\mathbf{a}) \triangleq \varphi_2 A(\mathbf{b}))$$

whenever valuations φ_1 in F_1 and φ_2 in F_2 are matching.

Proof By induction on A.

The atomic case and the cases of $\wedge$, $\vee$, $\supset$, $\exists$, $\forall$ were considered in the proof of Lemma 4.4.6.

Consider the case $A = \Box_i B$. By induction hypothesis, we have:

$$\alpha(\mathbf{a}, \mathbf{b}) \cap \varphi_1 B(\mathbf{a})) \leq \varphi_2 B(\mathbf{b}),$$

and thus

$$\Box_i \alpha(\mathbf{a}, \mathbf{b}) \cap \Box_i \varphi_1 B(\mathbf{a}) \leq \Box_i (\alpha(\mathbf{a}, \mathbf{b}) \cap \varphi_1 B(\mathbf{a})) \leq \Box_i \varphi_2 B(\mathbf{b}).$$

Hence, by (E0),

(1) $\alpha(\mathbf{a}, \mathbf{b}) \cap \Box_i \varphi_1 B(\mathbf{a}) \leq \Box_i \varphi_2 B(\mathbf{b})$.

By Lemma 4.4.4 (which obviously holds in the modal case),

(2) $\alpha(\mathbf{a}, \mathbf{b}) \cap E_1(A(\mathbf{a})) \leq E_2(A(\mathbf{b}))$.

So

$$\begin{aligned} \alpha(\mathbf{a}, \mathbf{b}) \cap \varphi_1 A(\mathbf{a}) &= \alpha(\mathbf{a}, \mathbf{b}) \cap E_1(A(\mathbf{a})) \cap \Box_i \varphi_1 B(\mathbf{a}) \\ &\leq E_2(A(\mathbf{b})) \cap \Box_i \varphi_2 B(\mathbf{b}) \end{aligned}$$

by (1) and (2)

$$= \varphi_2 A(\mathbf{b}).$$

∎

Let us also define some auxiliary kinds of morphism.

Definition 4.4.17 *Let* $F_i = (\mathbf{\Omega}, D_i, E_i)$, $i = 1, 2$, *be two H.v.s (respectively, m.v.s) over the same algebra* $\mathbf{\Omega}$. *A* regular morphism $\gamma : F_1 \longrightarrow F_2$ *is a mapping* $\gamma : D_1 \longrightarrow D_2$ *such that for any* $a, b \in D_1$

(1) $E_1(a, b) \leq E_2(\gamma(a), \gamma(b))$,

(2) $E_1(a, a) = E_2(\gamma(a), \gamma(a))$.

$\mathbf{\Omega}$-Hvs *(respectively,* $\mathbf{\Omega}$-mvs*) denotes the category of H.v.s. (respectively, m.v.s.) over* $\mathbf{\Omega}$ *and regular morphisms.*

In particular, every strong p$^=$-morphism in the sense of Definition 4.4.9 is a regular morphism.

Remark 4.4.18 For H.v.s. there also exists another kind of morphisms, which we call *weak morphisms* [Goldblatt, 1984; Borceaux, 1994]. A weak morphism is defined as a mapping $\alpha : D_1 \times D_2 \rightarrow \Omega$ with the following properties:

- $E_2(b, b') \wedge \alpha(a, b') \leq \alpha(a, b)$;
- $\alpha(a, b) \wedge E_1(a, a') \leq \alpha(a', b)$;

- $\alpha(a,b) \wedge \alpha(a,b') \leq E_2(b,b')$;
- $E_1(a,a) = \bigvee\{\alpha(a,b) \mid b \in D_2\}$.

It is clear that every regular morphism γ corresponds to a weak morphism α, such that

$$\alpha(a,b) = E(b,\gamma(a)).$$

Moreover, the category of $\mathbf{\Omega}$-sets and weak morphisms is equivalent to its full subcategory of complete $\mathbf{\Omega}$-sets, and in the latter subcategory all weak morphisms correspond to regular morphisms [Borceaux, 1994]. However, weak morphisms are not logically faithful (see the discussion below). Thus the above equivalence is not sufficient for showing the equivalence between the semantics of H.v.s. and the semantics of complete H.v.s.

4.5 Presheaves and $\mathbf{\Omega}$-sets

In this section we consider an equivalent representation of algebraic semantics using presheaves. The connection between presheaves and $\mathbf{\Omega}$-sets is well known in topos theory. Every presheaf corresponds to an $\mathbf{\Omega}$-set in a standard way [Fourman and Scott, 1979]; this construction can be used to define logics of presheaves. On the other hand, for the case of sheaves there exist a connection in the opposite direction; Higgs's theorem [Higgs, 1984], also cf. [Borceaux, 1994; Goldblatt, 1984; Makkai and Reyes, 1977; Fourman and Scott, 1979], states the equivalence between the category of sheaves and sheaf morphisms and some full subcategory of $\mathbf{\Omega}$-sets (so called 'complete' $\mathbf{\Omega}$-sets). It also turns out that the latter subcategory is 'representative', i.e. equivalent to the whole category of $\mathbf{\Omega}$-sets [Goldblatt, 1984]. Unfortunately, this result does not help for our purpose, because the $\mathbf{\Omega}$-set isomorphisms used in this theorem are not logically faithful. So in this section we describe another construction showing that the logic of any m.v.s. can be presented as the logic of some presheaf.

First, let us recall (Chapter 3) that every poset (F,R) corresponds to a category $\mathrm{Cat}(F,R)$, in which (u,v) is the unique morphism from u to v whenever uRv. For every locale $\mathbf{\Omega}$ let $\mathrm{Cat}(\mathbf{\Omega}) := \mathrm{Cat}(\Omega, \geq)$.

Definition 4.5.1 *A* presheaf (of sets) over a locale $\mathbf{\Omega}$ *is defined as a functor*

$$\mathcal{F} : \mathrm{Cat}(\mathbf{\Omega}) \rightsquigarrow \mathrm{SET}.$$

This means that we have a family of sets $(\mathcal{F}(u) : u \in \Omega)$ and a family of functions (restriction maps*)*

$$\mathcal{F}(u,v) : \mathcal{F}(u) \longrightarrow \mathcal{F}(v),$$

where $u, v \in \Omega$, $u \geq v$, and the following conditions hold:

(1) $\mathcal{F}(u,u) = \mathrm{id}_{\mathcal{F}(u)}$ (the identity function);

(2) $\mathcal{F}(v,w) \circ \mathcal{F}(u,v) = \mathcal{F}(u,w)$, *provided* $u \geq v \geq w$.

A presheaf $\mathcal{F}$ *is called* inhabited *if* $\bigvee\{u \mid \mathcal{F}(u) \neq \varnothing\} = \mathbf{1}$.

The set $\mathcal{F}(u)$ *is called the* (domain) *of* $\mathcal{F}$ *at* u*; its elements are called* individuals *(or* sections*) over* u*. The set of all the individuals*

$$\mathcal{F}^* = \bigcup\{\mathcal{F}(u) \mid u \in \Omega\}$$

is called the total domain *of* $\mathcal{F}$.

Definition 4.5.2 *A* morphism $f : \mathcal{F}_1 \longrightarrow \mathcal{F}_2$ *of presheaves over* $\mathbf{\Omega}$ *is a functor morphism (natural transformation) i.e. a family of mappings*

$$f_u : \mathcal{F}_1(u) \to \mathcal{F}_2(u),$$

such that the following diagram commutes (provided $u \geq v$*).*

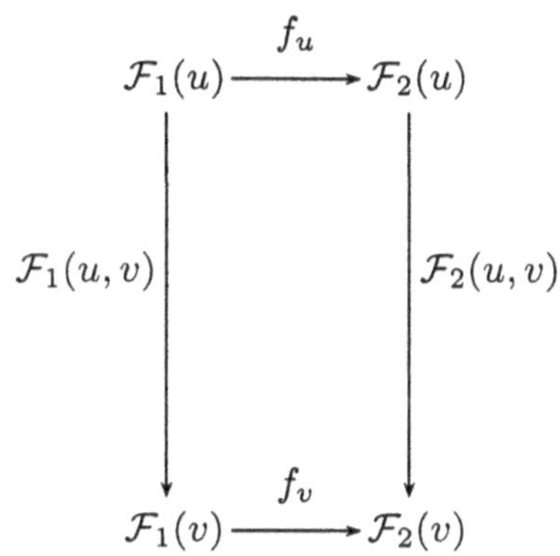

An isomorphism *is an invertible morphism, as usual; this is equivalent to bijectivity of every* f_u.

Presheaves over $\mathbf{\Omega}$ and morphisms constitute a category $\mathsf{Psh}(\mathbf{\Omega})$.

Definition 4.5.3 *A presheaf* $\mathcal{F}$ *over* $\mathbf{\Omega}$ *is* disjoint *if*

$$\forall u, v \in \Omega \; (u \neq v \Rightarrow \mathcal{F}(u) \cap \mathcal{F}(v) = \varnothing).$$

Lemma 4.5.4 *Every presheaf is isomorphic to some disjoint presheaf.*

Proof Almost obvious. Given a presheaf $\mathcal{F}$, let

$$\mathcal{F}'(u) := \mathcal{F}(u) \times \{u\},$$
$$\mathcal{F}'(u,v)(a,u) := (\mathcal{F}(u,v)(a), v).$$

Then $\mathcal{F}'$ is a disjoint presheaf, and there exists an isomorphism $f : \mathcal{F} \to \mathcal{F}'$ such that

$$f_u(a) = (a,u).$$

In fact, f is obviously bijective, and the corresponding diagram commutes, since

$$\mathcal{F}'(u,v) \circ f_u : a \mapsto (a,u) \mapsto (\mathcal{F}(u,v)(a), v)$$

and also

$$f_v \circ \mathcal{F}(u,v) : a \mapsto (\mathcal{F}(u,v)(a), v).$$

■

Thanks to the above lemma, we can assume that all presheaves are disjoint.

In a disjoint presheaf $\mathcal{F}$ every individual belongs to a unique $\mathcal{F}(u)$; the corresponding u is called the *extent* of a and denoted by $|a|$.

Definition 4.5.5 *If* $v \leq |a|$, *the* restriction of a to v *is*

$$a|v := \mathcal{F}(|a|, v)(a).$$

Definition 4.5.1 for disjoint presheaves can be reformulated as follows.

Lemma 4.5.6 *Let* $\mathcal{F}$ *be a disjoint presheaf,* $a \in \mathcal{F}^*$, $w \leq v \leq |a|$. *Then*

(1) $a \mid (|a|) = a$;

(2) $(a|v)|w = a|w$.

The following simple lemma yields an equivalent definition of a morphism of disjoint presheaves as a map of domains preserving extents and commuting with restrictions:

Lemma 4.5.7 *Let* $f : \mathcal{F}_1 \longrightarrow \mathcal{F}_2$ *be a morphism of disjoint presheaves. Consider a map* $g : \mathcal{F}_1^* \longrightarrow \mathcal{F}_2^*$ *such that*

$$g(a) := f_{|a|}(a).$$

Then g *has the following properties.*

(1) $|g(a)| = |a|$;

(2) $g(a|v) = g(a) \mid v$.

The other way round, given a map g *satisfying (1), (2), we can define a morphism* $f : \mathcal{F}_1 \longrightarrow \mathcal{F}_2$ *by putting*

$$f_{|a|}(a) := g(a).$$

Proof Straightforward. ■

Proposition 4.5.8 *Let $\mathcal{F}$ be an inhabited disjoint presheaf over a locale $\mathbf{\Omega}$, and for $a, b \in \mathcal{F}^*$ let*

$$X(a,b) = \{u \in \Omega \mid u \leq |a| \wedge |b|,\ a|u = b|u\},$$

$$E(a,b) = \bigvee X(a,b).$$

Then the triple

$$HV(\mathcal{F}) = (\mathbf{\Omega}, \mathcal{F}^*, E)$$

is an H.v.s.

Proof The equality $E(a,b) = E(b,a)$ is trivial, since $X(a,b) = X(b,a)$.
$\bigvee\{E(a,a) \mid a \in \mathcal{F}^*\} = \mathbf{1}$ holds, because $E(a,a) = |a|$ and $\mathcal{F}$ is inhabited.
Since by Lemma 4.5.7,

$$w \leq v \leq \mid a \mid \Rightarrow (a \mid v) \mid w = a \mid w,$$

$X(a,b)$ is $\geq$-stable, that is

$$v \in X(a,b) \ \&\ v \ \geq w \ \Rightarrow \ w \in X(a,b).$$

So for any $u \in X(a,b),\ v \in X(b,c)$ we have

$$(u \wedge v) \in X(a,b) \cap X(b,c) \subseteq X(a,c).$$

Hence it follows that

$$E(a,b) \wedge E(b,c) = \bigvee\{u \wedge v \mid u \in X(a,b), v \in X(b,c)\} \leq \bigvee X(a,c) = E(a,c).$$

■

Obviously, if the locale $\mathbf{\Omega}$ is basic, isomorphic to Ξ^0, for some **S4**-algebra Ξ, and E is the same as in Proposition 4.5.8, then

$$MV(\mathcal{F}) = (\Xi, \mathcal{F}^*, E)$$

is an **S4**-m.v.s.

Lemma 4.5.9 *If $\mathcal{F}_1$, $\mathcal{F}_2$ are isomorphic disjoint presheaves over $\mathbf{\Omega}$, then the corresponding H.v.s. $HV(\mathcal{F}_1)$, $HV(\mathcal{F}_2)$ are isomorphic; similarly for m.v.s. (and basic $\mathbf{\Omega}$).*

Proof Let X_1, X_2 be the functions corresponding to $\mathcal{F}_1$, $\mathcal{F}_2$ described in Proposition 4.5.8. If $g : \mathcal{F}_1^* \longrightarrow \mathcal{F}_2^*$ is a bijective map satisfying Lemma 4.5.7 (i), (ii), then one can easily check that $X_1(a,b) = X_2(g(a), g(b))$. Thus $E_1(a,b) = E_2(g(a), g(b))$, which means that g is an isomorphism of H.v.s. ■

Definition 4.5.10 *If $\mathcal{F}$ is an inhabited presheaf over a locale $\mathbf{\Omega}$, we define its* superintuitionistic logic *(with or without equality) via the corresponding H.v.s.:*

$$\mathbf{IL}^{(=)}(\mathcal{F}) = \mathbf{IL}^{(=)}(HV(\mathcal{F}_1)),$$

where $\mathcal{F}_1$ is a disjoint presheaf isomorphic to $\mathcal{F}$.

If $\mathbf{\Omega}$ is basic, we also define the modal logic

$$\mathbf{ML}^{(=)}(\mathcal{F}) = \mathbf{ML}^{(=)}(MV(\mathcal{F}_1)).$$

These logics are well-defined. In fact, by Lemma 4.5.9, an isomorphism of disjoint presheaves $\mathcal{F}_1$, $\mathcal{F}_2$ gives rise to an isomorphism of the associated H.v.s. (m.v.s.), and thus the corresponding logics are equal, by Corollary 4.4.8.

Proposition 4.5.11

(1) For any H.v.s. G over a Heyting algebra $\mathbf{\Omega}$ there exists a disjoint presheaf $\mathcal{G}$ over $\mathbf{\Omega}$ such that $\mathbf{IL}^{=}(\mathcal{G}) = \mathbf{IL}^{=}(G)$.

(2) For any m.v.s. G over an $\mathbf{S4}$-algebra $\mathbf{\Omega}$ there exists a disjoint presheaf $\mathcal{G}$ over $\mathbf{\Omega}^0$ such that $\mathbf{ML}^{=}(\mathcal{G}) = \mathbf{ML}^{=}(G)$.

Proof We prove only (1), leaving (2) for the reader.

For every $u \in \mathbf{\Omega}$ let

$$\mathcal{G}_0(u) = \{(u, a) \mid a \in D,\ u \leq Eaa\},$$

and let $\approx_u$ be the following equivalence relation in $\mathcal{G}_0(u)$:

$$(u, a) \approx_u (u, b) \Leftrightarrow u \leq Eab$$

We denote the equivalence class $(u, a)/\approx_u$ by $[u, a]$.

Now we define the presheaf $\mathcal{G}$ as follows:

(*) $\mathcal{G}(u) := \mathcal{G}_0(u)/\approx_u$;

(**) $\mathcal{G}(u, v)([u, a]) := [v, a]$ (provided $v \leq u$).

The map $\mathcal{G}(u, v)$ is well-defined, since $[u, a] = [u, b]$ means that $u \leq Eab$, which implies $v \leq Eab$, i.e. $[v, a] = [v, b]$.

Now let us show that $\mathbf{IL}^{=}(\mathcal{G}) = \mathbf{IL}^{=}(G)$.

It is clear that (*), (**), really define a presheaf, and we may assume that $\mathcal{G}$ is disjoint. $\mathcal{G}$ is inhabited, since

$$\begin{aligned}&\bigvee\{u \mid \mathcal{G}(u) \neq \varnothing\} = \bigvee\{u \mid \mathcal{G}_0(u) \neq \varnothing\}\\ &= \bigvee\{u \mid \exists a \in D\ u \leq Eaa\} = \bigvee\{Eaa \mid a \in D\} = \mathbf{1}.\end{aligned}$$

Recall that $\mathbf{IL}^{=}(\mathcal{G}) = \mathbf{IL}^{=}(HV(\mathcal{G}))$, where $HV(\mathcal{G}) = (\mathbf{\Omega}, \mathcal{G}^*, E^*)$ is the H.v.s., in which

$$\begin{aligned}&\mathcal{G}^* = \bigcup_{u \in \Omega} \mathcal{G}(u) = \{[u, a] \mid a \in D,\ u \leq Eaa\},\\ &E^*([u, a], [v, b]) = \bigvee\{w \mid w \leq u \wedge v,\ [w, a] = [w, b]\}\\ &= \bigvee\{w \mid w \leq u \wedge v,\ w \leq Eab\} = u \wedge v \wedge Eab.\end{aligned}$$

In particular, we have:

$$E^*([u,a]) = u \wedge E(a) = u.$$

Consider the map $\alpha : \mathcal{G}^* \times D \longrightarrow \Omega$ such that

$$\alpha([u,a],b) := u \wedge Eab.$$

This map is well-defined, since if $[u,a] = [u,a']$, i.e. if $u \leq Eaa'$, then $u \wedge Eab = u \wedge Ea'b$. Let us show that

$$\alpha : G \rightarrow HV(\mathcal{G})$$

is a p$^=$-equivalence (Definitions 4.4.1, 4.4.2).

(Q1) $E(b) \leq \bigvee_{c \in \mathcal{G}^*} \alpha(c,b)$, since $E(b) = \alpha([E(b),b],b)$.

(Q2) $EE^*([u,a]) \leq \bigvee_{b \in D} \alpha([u,a],b)$, since $E^*([u,a]) = u = u \wedge Eaa = \alpha([u,a],a)$.

(I1) $\alpha([u_1,a_1],b_1) \wedge \alpha([u_2,a_2],b_2) \wedge Eb_1b_2 =$
$= u_1 \wedge Ea_1b_2 \wedge u_2 \wedge Ea_2b_2 \wedge Eb_1b_2 \leq u_1 \wedge u_2 \wedge Ea_1a_2 = E^*([u_1a_1],[u_2,a_2])$.

(I2) $\alpha([u_1,a_1],b_1) \wedge \alpha([u_2,a_2],b_2) \wedge E^*([u_1,a_1],[u_2,a_2]) =$
$= u_1 \wedge Ea_1b_1 \wedge u_2 \wedge Ea_2b_2 \wedge u_1 \wedge u_2 \wedge Ea_1a_2 \leq Eb_1b_2$.

(ε) $E^*([u',a'],[u,a]) \wedge \alpha([u,a],b) =$
$u' \wedge u \wedge Ea'a \wedge u \wedge Eab \leq u' \wedge Ea'b = \alpha([u',a'],b)$.

(ε') $Eb'b \wedge \alpha([u,a],b) =$
$Eb'b \wedge u \wedge Eab \leq u \wedge Eab' = \alpha([u,a],b')$.

Therefore $\mathbf{IL}^=(G) = \mathbf{IL}^=(HV(\mathcal{G}))$, by Corollary 4.4.8. ∎

Corollary 4.5.12 *Every algebraic semantics can be obtained from a class of presheaves. In particular, the general algebraic semantics $\mathcal{AE}$ is generated by the class of all presheaves.*

4.6 Morphisms of presheaves

Lemma 4.6.1 *Let $\mathcal{F}_1, \mathcal{F}_2$ be two inhabited disjoint presheaves over a locale $\mathbf{\Omega}$, $g : \mathcal{F}_1 \rightarrow \mathcal{F}_2$ a morphism (in the sense of Lemma 4.5.7). Then g is a strong morphism from $HV(\mathcal{F}_1)$ to $HV(\mathcal{F}_2)$ (in the sense of Definition 4.4.17).*

Proof

Let $HV(\mathcal{F}_i) = (\mathbf{\Omega}, \mathcal{F}_i^*, E_i)$. Then

$$E_1(a,b) = \bigvee\{u \mid u \leq |a| \wedge |b|,\ a|u = b|u\},$$
$$E_2(g(a),g(b)) = \bigvee\{u \mid u \leq |g(a)| \wedge |g(b)|,\ g(a)|u = g(b)|u\}$$

(see Proposition 4.5.8). Since g is a morphism, we have:

$$|g(a)| = |a|, \quad g(a)|u = g(a|u), \quad g(b)|u = g(b|u).$$

So $u \in X_1(a,b)$ implies

$$g(a)|u = g(a|u) = g(b|u) = g(b)|u,$$

and thus $u \in X_2(g(a), g(b))$. Hence $X_1(a,b) \subseteq X_2(g(a), g(b))$, and so

$$E_1(a,b) \leq E_2(g(a), g(b)).$$

On the other hand,

$$X_1(a,a) = \{u \mid u \leq |a|\},$$

and thus

$$E_1(a,a) = |a|;$$

similarly,

$$E_2(g(a), g(a)) = |g(a)|;$$

hence

$$E_1(a,a) = E_2(g(a), g(a))$$

holds, and the proof is completed. ■

So we obtain

Proposition 4.6.2 *There exist functors*
HV: $\mathsf{Psh}(\mathbf{\Omega}) \rightsquigarrow \mathbf{\Omega}\text{–}\mathsf{Hvs}$
for any locale $\mathbf{\Omega}$*;*
MV: $\mathsf{Psh}(\mathbf{\Omega}) \rightsquigarrow \mathbf{\Omega}\text{–}\mathsf{mvs}$
for any basic locale $\mathbf{\Omega}$*,*
such that $HV(\mathcal{F})$*,* $MV(\mathcal{F})$ *are defined as above, and for a strong morphism* f*,* $HV(f) = MV(f) = f$*.*

The converse to Proposition 4.6.2 is in general false.

Example 4.6.3 Let $\mathcal{F}_1$ and $\mathcal{F}_2$ be inhabited disjoint presheaves shown at Fig. 4.1. Let $g : \mathcal{F}_1^* \longrightarrow \mathcal{F}_2^*$ be the map such that

$$g(a) = a, \ g(a|u) = b, \ g(a|x) = a|x$$

for $x = v$ or w. Then g is not a morphism of presheaves. On the other hand, g is a strong morphism of the corresponding H.v.s. (Definition 4.4.17), since $E_2(a,b) = u = E_1(a, a|u)$.

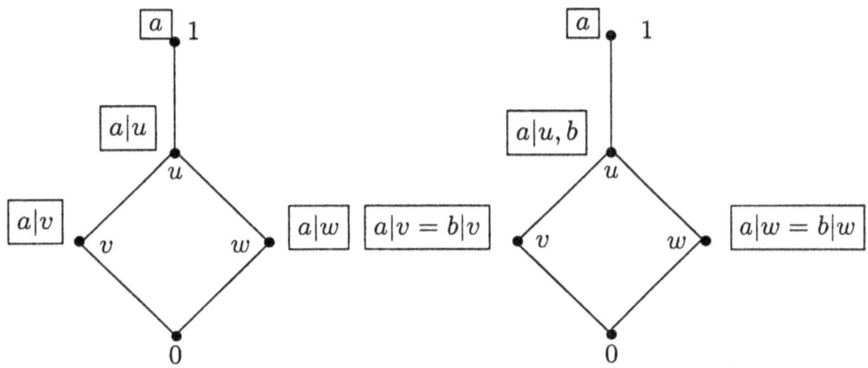

Figure 4.1.

Proposition 4.6.4 *Let $\mathcal{F}_1$ and $\mathcal{F}_2$ be two inhabited disjoint presheaves over a locale $\mathbf{\Omega}$. Every strong morphism $f : HV(\mathcal{F}_1) \longrightarrow HV(\mathcal{F}_2)$ (Definition 4.4.17) is a morphism of presheaves if $\mathcal{F}_2$ satisfies the following condition:*

$$\forall v \in \Omega\ \forall b, c \in \mathcal{F}_2(v)\ (E_2(b,c) = v \Rightarrow b = c).$$

Moreover, every strong H.v.s.-morphism $f : HV(\mathcal{F}_1) \longrightarrow HV(\mathcal{F}_2)$ is a morphism of presheaves $\mathcal{F}_1 \longrightarrow \mathcal{F}_2$ iff for any strong H.v.s.-morphism $g : HV(\mathcal{F}_1) \longrightarrow HV(\mathcal{F}_2)$ the following holds:

$$(*)\quad \forall u > v\ \forall a \in g(\mathcal{F}_1(u)) \forall b \in \mathcal{F}_2(v)\ (E_2(a,b) = v \Rightarrow a|v = b).$$

Proof (If.) Let f be a morphism of H.v.s. Let us show that $f(x|v) = f(x)|v$ for $x \in F_1(u), v < u$. Take $a = f(x)$, $b = f(x|v)$. Then $E_2(a,b) \geq E_1(x, x|v) = v$, and thus $E_2(a,b) = v$ and $a|v = b$, by the condition (*) applied to f.

(Only if.) Assume that every strong H.v.s.-morphism $HV(\mathcal{F}_1) \longrightarrow HV(\mathcal{F}_2)$ is a morphism of presheaves $\mathcal{F}_1 \longrightarrow \mathcal{F}_2$. Suppose there exists an H.v.s.-morphism $g : HV(\mathcal{F}_1) \longrightarrow HV(\mathcal{F}_2)$ that does not satisfy (*), i.e. for some $u > v$, $x \in \mathcal{F}_1(u)$, $b \in \mathcal{F}_2(v)$ we have $E_2(a,b) = v$, $a|v \neq b$ for $a = f(x) \in \mathcal{F}_2(u)$. By our assumption, g is a morphism of presheaves, and thus $g(x|v) = a|v$. Consider the map $f : \mathcal{F}_1^* \to \mathcal{F}_2^*$ defined as follows

$$f(y) := \begin{cases} b & \text{if } y = x|v, \\ g(y) & \text{otherwise.} \end{cases}$$

Then f is not a morphism of presheaves, since $f(x|v) = b$, but $f(x)|v = a|v \neq b$. On the other hand, f is a strong morphism of H.v.s.

In fact, let us show that $E_1(y,z) \leq E_2(f(y), f(z))$ for $y, z \in \mathcal{F}_1^*$. It is sufficient to consider the case $y \neq z = x|v$. Then

$$E_1(y,z) \leq E_2(g(y), a|v) =$$
$$E_2(g(y), a|v) \wedge E_2(a|v, b) \leq E_2(g(y), b) = E_2(f(y), f(z)),$$

since

$$E_2(g(u), a|v) \leq E_2(a|v, a|v) = v = E_2(a, b) \wedge E_2(a|v, a) \leq E_2(a|v, b).$$

■

Definition 4.6.5 *Let $\mathcal{F}_1$, $\mathcal{F}_2$ be (inhabited) disjoint presheaves over a locale $\mathbf{\Omega}$. A* pre-morphism *from $\mathcal{F}_1$ to $\mathcal{F}_2$ is a map $g : \mathcal{F}_1^* \longrightarrow \mathcal{F}_2^*$ preserving extents, i.e. such that*
(1) $|g(a)| = |a|$ *for any a.*
A pre-morphism g is called a p-morphism *if it satisfies*

(3) *for any $u \in \Omega$, $b \in \mathcal{F}_2(u)$*

$$u = \bigvee\{v \leq u \mid \exists a \in \mathcal{F}_1^* \ (v \leq |g(a)| \ \& \ b|v = g(a)|v)\}$$

A p$^=$-morphism *is a p-morphism satisfying*

(4) *if $a_1 \in \mathcal{F}_1(u_1)$, $a_2 \in \mathcal{F}_1(u_2)$, then for any w*
$w \leq u_1 \wedge u_2 \ \& \ g(a_1)|w = g(a_2)|w$
$\Rightarrow w \leq \bigvee\{u \mid u \leq u_1 \wedge u_2 \ \& \ a_1|u = a_2|u\}$;

(5) *if $a \in \mathcal{F}_1(u)$, $v \leq u$, then*
$v = \bigvee\{w \leq v \mid g(a)|w = g(a|v)|w\}$.

A p-embedding *is a p-morphism satisfying (4).*

Note that

$$(2) \quad g(a|u) = g(a)|u$$

implies (5); (3) holds for every surjective pre-morphism; (4) holds for every injective morphism.

The definition of a p$^{(=)}$-morphism of presheaves is rather complicated, but it exactly corresponds to the definition of a p-morphism of H.v.s. (Definition 4.4.9). More precisely, the following holds.

Lemma 4.6.6

(1) A map $g : \mathcal{F}_1^* \longrightarrow \mathcal{F}_2^*$ is a p$^{(=)}$-morphism of presheaves iff g is a strong p$^{(=)}$-morphism of corresponding H.v.s. $g : HV(\mathcal{F}_1) \longrightarrow HV(\mathcal{F}_2)$.

(2) If there exists a p-embedding $g : \mathcal{F}_1 \longrightarrow \mathcal{F}_2$ (respectively, p$^=$-morphism) of disjoint presheaves, then $\mathbf{IL}(\mathcal{F}_1) \subseteq \mathbf{IL}(\mathcal{F}_2)$ (respectively, $\mathbf{IL}^=(\mathcal{F}_1) \subseteq \mathbf{IL}^=(\mathcal{F}_2)$).

(Cf. Lemma 4.4.7 stating the converse inclusion: $\mathbf{IL}^{(=)}(F_2) \subseteq \mathbf{IL}^{(=)}(F_1)$.)

Proof Let us check items (i)–(iii) from Definition 4.4.9.

(i) $E_1(a) = E_2(g(a)) \Leftrightarrow |a| = |g(a)| \Leftrightarrow (1)$.

(ii) $E_2(b) \leq \bigvee_{a \in D_1} E_2(b, g(a))$
$\Leftrightarrow |b| \leq \bigvee_{a \in D_1} \bigvee\{v \mid v \leq |b| \wedge |g(a)|,\ b|v = g(a)|v\} \Leftrightarrow (3).$

(iii) $E_2(g(a_1), g(a_2)) \leq E_1(a_1, a_2) \Leftrightarrow$
$\Leftrightarrow \bigvee\{v \mid v \leq |g(a_1)| \wedge |g(a_2)|,\ g(a_1)|v = g(a_2)|v\} \leq$
$\leq \bigvee\{u \mid u \leq |a_1| \wedge |a_2|, a_1|u = a_2|u\}$
$\Leftrightarrow \forall v\ (v \leq |g(a_1)| \wedge |g(a_2)|\ \&\ g(a_1)|v = g(a_2)|v) \Rightarrow$
$\Rightarrow v \leq \bigvee\{u \mid u \leq |a_1| \wedge |a_2|,\ a_1|u = a_2|u\}) \Leftrightarrow (4).$
$E_1(a_1, a_2) \leq E_2(g(a_1), g(a_2)) \Leftrightarrow$
$\Leftrightarrow \bigvee\{u \mid u \leq |a_1| \wedge |a_2|,\ a_1|u = a_2|u\} \leq$
$\leq \bigvee\{v \mid v \leq |g(a_1)| \wedge |g(a_2)|,\ g(a_1)|v = g(a_2)|v\} \Leftrightarrow$

($\sharp$) $\forall u\ (u \leq |a_1| \wedge |a_2|\ \&\ a_1|u = a_2|u \Rightarrow$
$\Rightarrow u \leq \bigvee\{v \mid v \leq |g(a_1)| \wedge |g(a_2)|,\ g(a_1)|v = g(a_2)|v\}).$

This condition ($\sharp$) implies (5). In fact, let $a \in \mathcal{F}_1(u),\ v \leq u$. Take $a_1 = a,\ a_2 = a|v)$; then $a_1|v = a_2|v$. Thus

$$v \leq \bigvee\{w \mid w \leq v,\ g(a)|w = g(a|w)|v\}$$

(recall that $|g(a)| = u \geq v = |g(a|v)|$, by (1)).

On the other hand, (5) implies ($\sharp$). Indeed, assume that $a_1 \in \mathcal{F}_1(u_1), a_2 \in \mathcal{F}_2(u_2),\ u \leq u_1 \wedge u_2,\ a_1|u = a_2|u$. Take the sets

$$X_i := \{w \mid w \leq u,\ g(a_i)|w = g(a_i|u)|w\}$$

for $i = 1, 2$. Then by (5),

$$u = \left(\bigvee X_2\right) \wedge \left(\bigvee X_2\right) = \bigvee\{w_1 \wedge w_2 \mid w_1 \in X_1,\ w_2 \in X_2\}$$

(recall that $\mathbf{\Omega}$ is a complete Heyting algebra). And if $v = w_1 \wedge w_2,\ w_i \in X_i$, then

$$\begin{aligned}
&v \leq u \leq u_1 \wedge u_2 = |g(a_1)| \wedge |g(a_2)|,\\
&(a_1)|v = (g(a_1)|w_1)|v = (g(a_1|u)|w_1)|v = g(a_1|u)|v = g(a_2|u)|v\\
&= (g(a_2|u)|w_2)|v = (g(a_2)|w_2)|v = g(a_2)|v.
\end{aligned}$$

Note that a p$^{(=)}$-morphism is not necessarily a morphism (because the condition (2) does not always hold). On the other hand, (1) and (2) implies (5); in fact, $g(a)|v = g(a|v) = g(a|v)|v$.

Also note that every surjective pre-morphism is a p-morphism: the condition (3) holds if $b = g(a)$ for some $a \in \mathcal{F}_1^*$. Every isomorphism (i.e. a bijective morphism) is a $p^{=}$-morphism; in fact, (4) follows from (2) and the bijectivity. ∎

4.7 Sheaves

In this and the next section we look more closely at topological semantics. First let us show that in this case presheaves can be replaced with sheaves.

Recall (cf. [Godement, 1958]) that a *sheaf* (over a locale $\mathbf{\Omega}$) is a presheaf $\mathcal{F}$ satisfying two conditions:

(F1) if $u = \bigvee_{i\in I} u_i$, $a, b \in \mathcal{F}(u)$ and $(a|u_i) = (b|u_i)$ for all $i \in I$, then $a = b$;

(F2): if $u = \bigvee_{i\in I} u_i$, $a_i \in \mathcal{F}(u_i)$ and $(a_i|u_i \wedge u_j) = (a_j|u_i \wedge u_j)$ for all $i, j \in I$, then

$$\exists a \in \mathcal{F}(u)\ \forall i \in I\ (a|u_i) = a_i.$$

Every disjoint presheaf $\mathcal{F}$ over a topological space $(W, \Box)$ gives rise to the *canonical sheaf* $\tilde{\mathcal{F}}$ defined as follows.

For $x \in W$ consider the set

$$\mathcal{F}_0(x) := \{a \in \mathcal{F}^* \mid x \in |a|\}$$

with the equivalence relation

$$(a \equiv_x b) := x \in E(a, b).$$

The equivalence class $a_x := (a/\equiv_x)$ is called the *germ of a at x*. The set $\mathcal{F}(x) := (\mathcal{F}_0(x)/\equiv_x)$ of all the germs at x is called the *stalk* (or the *fibre*) of $\mathcal{F}$ at x. We define $\tilde{\mathcal{F}}(u)$ as the set of maps from u to $\mathcal{F}^*$ sending every $x \in u$ to some germ at x. More precisely,

$$\tilde{\mathcal{F}}(u) := \{f \in \prod_{x\in u} \mathcal{F}(x) \mid \forall x \in u\ \exists v \subseteq u\ \exists a \in \mathcal{F}(v)\ (x \in v\ \&\ \forall y \in v\ f(y) = a_y)\},$$

and for $v \subseteq u$, let

$$\tilde{\mathcal{F}}(u, v)(f) := (f \restriction v),$$

the restriction of the map f from u to v.

The following fact is well-known [Godement, 1958]:

Proposition 4.7.1

- *$\tilde{\mathcal{F}}$ is a sheaf;*
- *the family of mappings $\alpha_u : \mathcal{F}(u) \longrightarrow \tilde{\mathcal{F}}(u)$ such that*

$$\alpha_u(a) = f \Leftrightarrow \forall x \in u\ f(x) = a_x$$

  is a morphism of presheaves;
- *α is isomorphic iff $\mathcal{F}$ is a sheaf.*

For $a \in \mathcal{F}^*$ let us denote
$$\tilde{a} := \alpha_{|a|}(a).$$

Lemma 4.7.2 *Let $\mathcal{F}$ be a presheaf over a topological space, $\tilde{\mathcal{F}}$ its canonical sheaf. Let $HV(\mathcal{F}) = (\mathbf{\Omega}, \mathcal{F}^*, E), HV(\tilde{\mathcal{F}}) = (\mathbf{\Omega}, \tilde{\mathcal{F}}^*, \tilde{E})$.*

Then for any $a, b \in \mathcal{F}^*$ $E(a,b) = \tilde{E}(\tilde{a}, \tilde{b})$.

Proof Let $\mathbf{\Omega}$ be the Heyting algebra of the given topological space. Then by Definition 4.3.4,

$$\begin{aligned} x \in E(a,b) &\Leftrightarrow \exists u \in \mathbf{\Omega}\ (x \in u \ \&\ u \leq |a| \cap |b| \ \&\ a|u = b|u) \\ &\Leftrightarrow \exists u \in \mathbf{\Omega}\ (x \in u \ \&\ u \leq |a| \cap |b| \ \&\ \forall z \in u\ a_z = b_z) \\ &\Leftrightarrow \exists u \in \mathbf{\Omega}\ (x \in u \ \&\ u \leq |a| \cap |b| \ \&\ \tilde{a}|u = \tilde{b}|u) \\ &\Leftrightarrow x \in \tilde{E}(\tilde{a}, \tilde{b}). \end{aligned}$$

■

Lemma 4.7.3 *Consider the same $\mathcal{F}$, $\tilde{\mathcal{F}}$ as in the previous lemma. Let $g : \mathcal{F}^* \longrightarrow \tilde{\mathcal{F}}^*$ be the map such that $g(a) = \tilde{a}$ for $a \in \mathcal{F}^*$. Then g is a strong $\mathrm{p}^=$-morphism $HV(\mathcal{F}) \longrightarrow HV(\tilde{\mathcal{F}})$.*

Proof By Lemma 4.7.2 and because g is surjective (see remarks after Definition 4.4.9). ■

Proposition 4.7.4 $\mathbf{IL}^=(\mathcal{F}) = \mathbf{IL}^=(\tilde{\mathcal{F}})$, $\mathbf{ML}^=(\mathcal{F}) = \mathbf{ML}^=(\tilde{\mathcal{F}})$

Proof Follows from the previous lemma, by Lemma 4.4.10 and Corollary 4.4.8; recall that in the modal **S4**-case $\mathrm{p}^{(=)}$-morphisms are just the same as in the intuitionistic case. ■

Corollary 4.7.5 *$\mathcal{TE}$ is generated by the class of all sheaves over topological spaces.*

Question 4.7.6 Is the semantics $\mathcal{AE}$ generated by the class of all sheaves over locales (or complete modal algebras)?

4.8 Fibrewise models

In this section we show that validity in a presheaf $\mathcal{F}$ over a topological space can be defined via a forcing relation between points and formulas, similarly to Kripke semantics. Instead of Kripke models, we can use 'fibrewise models', which are collections of classical models on fibres of $\mathcal{F}$. As in Definition 3.2.2, a *system of domains* over a non-empty set W (whose elements are called *possible worlds*, or *points*) is a pair (W, D), where D is a family of non-empty sets $D = (D_u)_{u \in W}$. A *classical valuation* in a non-empty set V is a map ξ from PL to relations on V such that $\xi(P) \subseteq V^n$, whenever $P \in PL^n$ (V^0 is treated as a certain singleton). A *(modal) valuation* in a system of domains (W, D) is a family $(\eta_u)_{u \in W}$, where

η_u is a classical valuation in D_u such that $\eta_u(p) \subseteq \{u\}$ for every proposition letter p. A *(modal) model* over (W, D) is a triple (W, D, η), where η is a valuation in (W, D).

Definition 4.8.1 *Let $\mathcal{F}$ be a presheaf over a topological space $(W, \Box)$. We associate with $\mathcal{F}$ the system of domains $\mathcal{F}^+ := (W, D)$ with $D_u = \mathcal{F}(u)$ (the fibre at u). A* fibrewise model *over $\mathcal{F}$ is a model over $\mathcal{F}^+$.*

Definition 4.8.2 *For a fibrewise model $M = (W, D, \eta)$ we define the* (modal) forcing relation *$M, u \vDash A$ (or briefly: $u \vDash A$) between $u \in W$ and a D_u-sentence A:*

- $M, u \vDash P_k^0$ *iff* $u \in \eta_u(P_k^0)$;
- $M, u \vDash P_k^m(\mathbf{a})$ *iff* $\mathbf{a} \in \eta_u(P_k^m)$ *(for* $m > 0$*);*
- $M, u \vDash a = b$ *iff* a *equals* b*;*
- $M, u \nvDash \bot$*;*
- $M, u \vDash B \vee C$ *iff* $M, u \vDash B$ *or* $M, u \vDash C$*;*
- $M, u \vDash B \wedge C$ *iff* $M, u \vDash B \,\&\, M, u \vDash C$*;*
- $M, u \vDash B \supset C$ *iff* $M, u \nvDash B$ *or* $M, u \vDash C$*;*
- $M, u \vDash \exists x A$ *iff* $\exists a \in D_u \; M, u \vDash [a/x]A$*;*
- $M, u \vDash \forall x A$ *iff* $\forall a \in D_u \; M, u \vDash [a/x]A$*;*
- $M, u \vDash \Box A((a_1)_u, \ldots, (a_n)_u)$ *iff there exist an open* U *such that* $u \in U$, $U \subseteq |a_1| \cap \ldots \cap |a_n|$ *and* $\forall v \in U \; M, v \vDash A((a_1)_v, \ldots, (a_n)_v)$.

Let us check that forcing is well-defined. This has to be done only for the last item. More precisely, we have to show that if

Lemma 4.8.3 *Let $M = (\mathcal{F}, D, \eta)$ be a fibrewise model. Consider the algebra $\mathbf{\Omega} = MA(W, \Box)$ and define the map $\hat{\eta} : AF_{\mathcal{F}^*} \longrightarrow \Omega$ as follows:*

(*) $\hat{\eta}(A(\mathbf{a})) = \{u \in E(\mathbf{a}) \mid u \vDash A(\mathbf{a})\}$,

where $FV(A) \subseteq \mathbf{x}$, $A(\mathbf{a}) = [\mathbf{a}/\mathbf{x}]A$. Then $\hat{\eta}$ is a valuation on $\mathcal{F}$ and (*) is true for every $A(\mathbf{a}) \in MF^{=}_{\mathcal{F}^*}$.

Proof $\hat{\eta}$ is a valuation since

$$\hat{\eta}(P(\mathbf{a})) \cap E(\mathbf{a}, \mathbf{b})) = \{w \in E(\mathbf{a}, \mathbf{b}) \mid M, w \vDash P(\mathbf{a})\} \subseteq$$
$$\{w \in E(\mathbf{b}) \mid w \models P(\mathbf{b})\} = \hat{\eta}(P(\mathbf{b}))$$

(here $a_u = b_u$ for $u \in E(a, b)$, by the definition of germs). The second claim is proved by induction on the length of A. Let us consider two cases.

(i) $w \in \hat{\eta}(a = b) \Leftrightarrow w \in E(a, b) \Leftrightarrow \exists U \ni w(a|U = b|U) \Leftrightarrow a_w = b_w \Leftrightarrow w \models a = b$;

(ii) $w \in \hat{\eta}(\Box A(\mathbf{a}))$ iff $\exists U \ni w\ (U \subseteq \hat{\eta}(A(\mathbf{a})))$
iff $\exists U \ni w\ \forall v \in U\ (v \models A(\mathbf{a}))$ $(*)$
$w \models \Box A(\mathbf{a})$ iff
$\exists U \ni w\ \exists \mathbf{b} \in \mathcal{F}(u)^n (\mathbf{a}_w = \mathbf{b}_w\ \&\ \forall v \in U(v \models A(\mathbf{b})))$ $(**)$

It is clear that (*)⇒(**). Conversely, assume (**). Since $(a_w = b_w)$, there exists an open $V \subseteq U$ containing w such that $(\forall i\ (a_i|V) = (b_i|V))$; thus $(\forall v \in V\ \mathbf{a}_v = \mathbf{b}_v)$, and so $v \models A(\mathbf{b}_v)$, by (**). ■

Proposition 4.8.4 *A modal formula is valid in $\mathcal{F}$ iff it is true in every fibrewise model over $\mathcal{F}$.*

Proof It is sufficient to show that every valuation φ in $\mathcal{F}$ equals $\hat{\eta}$ for some fibrewise valuation η. In fact, define η by

$$\eta_u(P) := \{\mathbf{a}_u \mid u \in \varphi(P(\mathbf{a}))\}.$$

η is well-defined, since $a_u = b_u$ implies $x \in E(\mathbf{a}, \mathbf{b})$, and therefore $u \in \varphi(P(\mathbf{a}))$ iff $u \in \varphi(P(\mathbf{b}))$, by Definition 4.2.1. ■

4.9 Examples of algebraic semantics

Definition 4.9.1 *A presheaf $\mathcal{F}$ over a locale $\mathbf{\Omega}$ is called* constant *if all its domains are the same (usually, non-empty) and all its restriction maps are identity functions. A constant presheaf over $\mathbf{\Omega}$ with the domain D is denoted by $\mathcal{C}(\mathbf{\Omega}, D)$. A presheaf $\mathcal{F}$* has a constant domain *(or briefly, is a* CD-presheaf*) if every $\mathcal{F}(u, v)$ is bijective (for $u \geq v$).*

It is clear that every constant presheaf is a sheaf. It is also clear that every CD-presheaf $\mathcal{F}$ over $\mathbf{\Omega}$ is isomorphic to the constant sheaf $\mathcal{C}(\mathbf{\Omega}, \mathcal{F}(\mathbf{1}))$.

One can easily see that for a CD-m.v.s. $(\mathbf{\Omega}, D)$ the corresponding presheaf constructed in Proposition 4.5.11 is isomorphic to $\mathcal{C}(\mathbf{\Omega}, D)$. Using these observations, we obtain:

Proposition 4.9.2 *The following classes correspond to equal semantics:*

(1) the class of constant presheaves;

(2) the class of CD-presheaves;

(3) the class of CD-m.v.s. (or H.v.s.).

Remark 4.9.3 Note that $HV(\mathcal{C}(\mathbf{\Omega}, D)) \neq (\mathbf{\Omega}, D)$. In fact, the domain of $HV(\mathcal{C}(\mathbf{\Omega}, D)$ contains not only the individuals $a \in D$ with $E(a) = \mathbf{1}$, but also the restrictions $a|u$ such that $E(a|u) = E(a, a|u)$.

Thus we obtain the *algebraic semantics with constant domains* denoted by $\mathcal{A}$. Historically, it was introduced earlier than the general semantics $\mathcal{AE}$ [Rasiowa and Sikorski, 1963, Ch. X, §15, Ch XI], [Takano, 1987].

Similarly to Lemma 3.10.25 we have:

Lemma 4.9.4 *The semantics $\mathcal{T}$ and $\mathcal{A}$ do not satisfy (CP).*

Proof Consider the Kripke sheaf Θ_0 from the proof of Proposition 3.10.24. From the proof of Lemma 3.10.25 we know that $\mathbf{ML}(\Theta_0)$ is an intersection of $\mathcal{CK}$-complete logics. So it suffices to show that $\mathbf{ML}(\Theta_0) \notin \mathcal{A}$; in fact, it is easily seen that $F \vDash (C' \vee K')$ implies $F \vDash C'$ or $F \vDash K'$ for every CD-m.v.s. F, similarly to Proposition 3.10.24. ∎

Definition 4.9.5 *A presheaf $\mathcal{F}$ is called* monic *if all $\mathcal{F}(u,v)$ (for $u \geq v$) are injective. An m.v.s. $F = (\mathbf{\Omega}, D, E)$ is called* monic *if*

$$E(a,b) \in \{\mathbf{0}, (E(a) \cap E(b))\} \text{ for all } a, b \in D; \tag{4.1}$$

similarly for H.v.s.

Lemma 4.9.6 *If an m.v.s. F is monic, then the associated presheaf $\mathcal{F}$ (Proposition 4.5.11) is monic.*

Proof It is sufficient to show that $[v,a] = [v,b]$ implies $[u,a] = [u,b]$, whenever $v \leq u \leq E(a) \cap E(b)$. But $[v,a] = [v,b]$ (i.e. $v \leq E(a,b)$) only if $E(a,b) \neq \mathbf{0}$, and so, $E(a,b) = E(a) \cap E(b)$, and thus $u \leq E(a,b)$ by our assumption. ∎

Lemma 4.9.7 *Let $\mathcal{F}$ be an inhabited disjoint monic presheaf. Then the m.v.s. $MV(\mathcal{F})$ is monic.*

Proof Assume that $a, b \in \mathcal{F}^*$, $E(a,b) \neq \mathbf{0}$, $u = |a| \cap |b|$. Then $a|v = b|v$ for some $v \leq u$, and $a|u = b|u$ (since $\mathcal{F}$ is monic). Thus

$$E(a) \cap E(b) = u = \bigcup \{v \leq u \mid (a|v) = (b|v)\} = E(a,b).$$

∎

From Lemmas 4.9.6 and 4.9.7 we obtain:

Proposition 4.9.8 *The classes of monic m.v.s and monic presheaves generate the same semantics.*

This semantics is denoted by $\mathcal{MA}$.

Note that the semantics $\mathcal{MK}$ introduced in Section 3.4 is generated by monic presheaves over Kripke spaces.

Let us state some simple logical properties of monic m.v.s, which will be used later on. Recall that CE denotes the formula $(x \neq y) \supset \Box(x \neq y)$, DE denotes the formula $(x = y) \vee (x \neq y)$.

Lemma 4.9.9 *CE is valid in every monic m.v.s.*

Proof For a valuation φ we have:

$$\varphi(a \neq b) = (E(a) \cap E(b)) - E(a, b),$$

so $\varphi(a \neq b)$ is open, being either $E(a) \cap E(b)$ or $\mathbf{0}$. ■

Corollary 4.9.10 *DE is valid in every monic H.v.s. (see 1.4 and Lemma 2.2.2).*

The converse is not true: an H.v.s. validating *DE* may be not monic.

Corollary 4.9.11 $\mathbf{\Lambda}^{=c} \subseteq C_{\mathcal{MA}}(\mathbf{\Lambda})$ *if* $\mathbf{\Lambda}$ *is an m.p.l.,*
$\mathbf{\Lambda}^{=d} \subseteq C_{\mathcal{MA}}(\mathbf{\Lambda})$ *if* $\mathbf{\Lambda}$ *is an s.p.l.*

Proposition 4.9.12 *Let* $\mathbf{\Lambda}$ *be a propositional logic,* $\mathbf{Q\Lambda}$ *its quantified version (see. Section 2.4). Then* $C_{\mathcal{MA}}(\mathbf{Q\Lambda}) = C_{\mathcal{A}}(\mathbf{Q\Lambda})$.

Proof Assume that an m.v.s. $F = (\mathbf{\Omega}, D, E)$ separates some sentence $B \in MF$ from $\mathbf{Q\Lambda}$; let us find a monic m.v.s. with the same property. Consider $F' = (\mathbf{\Omega}, D, E')$, where

$$E'(a,b) = \begin{cases} E(a) & \text{if } a = b, \\ \mathbf{0} & \text{if } a \neq b. \end{cases}$$

It is clear that F and F' validate the same propositional formulas, that is $\mathbf{Q\Lambda} \subseteq \mathbf{ML}(F')$. There exists a valuation φ in F such that $\varphi(B) \neq \mathbf{1}$, and φ is also a valuation in F, since $\mathbf{a} \neq \mathbf{b}$ implies $\varphi(P(\mathbf{b})) \cap E'(\mathbf{a}, \mathbf{b}) = \mathbf{0} \leq \varphi(P(\mathbf{a}))$, and $\varphi(P(\mathbf{a})) \leq E(\mathbf{a}) = E'(\mathbf{a})$. ■

Corollary 4.9.13 *If the quantified version* $\mathbf{Q\Lambda}$ *of a modal propositional logic* $\mathbf{\Lambda}$ *is* $\mathcal{A}$*-complete, then* $\mathbf{Q\Lambda}^{=c}$ *is a conservative extension of* $\mathbf{Q\Lambda}$*, i.e.* $\mathbf{Q\Lambda} = (\mathbf{Q\Lambda}^{=c})^{\circ}$ *(cf. Section 2.9).*

We can also define corresponding restrictions of neighbourhood and Kripke semantics.

Definition 4.9.14 $\mathcal{T} := \mathcal{A} \cap \mathcal{TE}$ *(neighbourhood semantics with constant domains),*
$\mathcal{MT} := \mathcal{MA} \cap \mathcal{TE}$ *(mono-neighbourhood semantics),*
$\mathcal{MK} := \mathcal{MA} \cap \mathcal{KE}$ *(mono-Kripke semantics).*

Note that $\mathcal{MT}$ corresponds to monic presheaves over neighbourhood frames.

Chapter 5

Metaframe semantics

5.1 Preliminary discussion

In this chapter we study some further variations of Kripke semantics. In all these semantics we keep to the main idea that the truth value of a formula depends on a possible world and a formula $\Box A$ is true at some world iff A is true at every accessible world.

Kripke bundles

Let us first consider **S4**-frames. Our starting point is Kripke sheaf semantics. Recall (Section 3.5) that a Kripke sheaf over a propositional **S4**-frame F is a system of domains $D = (D_u \mid u \in F)$, together with transition maps $\rho_{uv} : D_u \longrightarrow D_v$ parametrised by pairs $(u, v) \in R$; $\rho_{uv}(a)$ is an 'inheritor' of an individual $a \in D_u$ in the world v. In Kripke sheaves inheritors of different individuals can collapse in some accessible world. But can we give more freedom to individuals allowing them to have several inheritors in the same accessible world (perhaps, in their own)? Following this idea we come to *Kripke bundles* [Shehtman and Skvortsov, 1990], where transition *maps* ρ_{uv} are replaced with 'inheritance' (or 'counterpart') *relations* $\rho_{uv} \subseteq D_u \times D_v$; $a\rho_{uv}b$ is read as 'b is an *inheritor* (or a *counterpart*) of a in v' (or a v-*inheritor* of a). If the domains D_u are disjoint, we can equivalently use the global inheritance relation $\rho \subseteq D^+ \times D^+$ on the set of all individuals D^+ and define ρ_{uv} as $\rho \cap (D_u \times D_v)$.

Here we assume that the relation ρ is reflexive and transitive — which is analogous to the properties of transition maps from Definition 3.6.1.

Models in Kripke bundles are defined in the same way as in Kripke sheaves. Forcing is defined by induction, but with a special care about the $\Box$-case.

The first idea is to extend the definition of forcing from Kripke sheaves to Kripke bundles as follows. For a D_u-sentence A put

(∇) $\quad M, u \vDash \Box A$ iff for any $v \in R(u)$ every v-version of A is true at M, v.

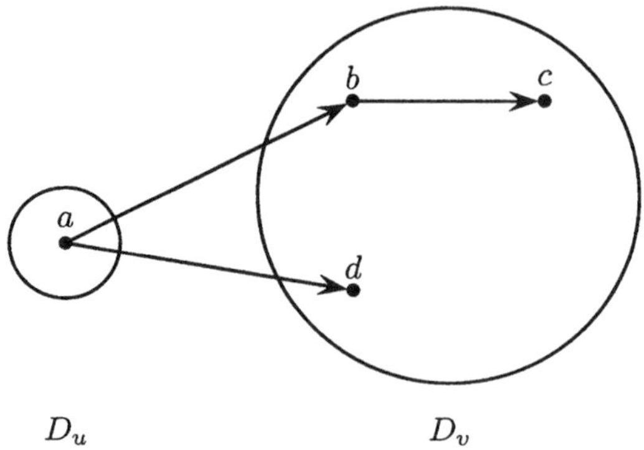

Figure 5.1. An example of a Kripke bundle: ρ is the smallest partial ordering containing the indicated arrows; $R = \{(u, v)\}$.

In Kripke sheaves v-versions of D_u-sentences are obtained by replacing every occurrence of every $a \in D_u$ with $a|v$, the unique inheritor of a in v. In Kripke bundles inheritors may be not unique, so we can define a v-version of A as a result of replacing every occurrence of every individual with some of its v-inheritor. But this definition gives too much freedom, because then we can refute some theorems of $\mathbf{QK}^=$ (and $\mathbf{QK}$).

To see this, consider the formula $B := \forall x \Box (x = x)$ and the following Kripke bundle.

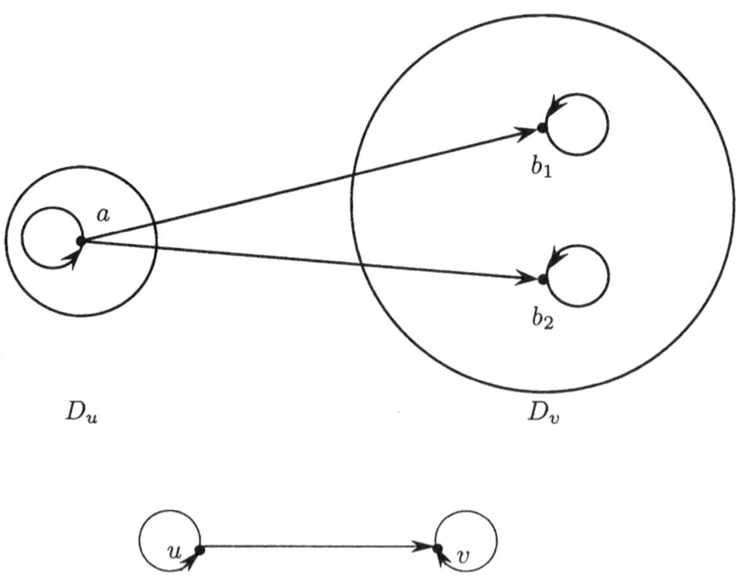

Figure 5.2.

Then (for any model M)

$$M, u \not\models \Box(a = a),$$

because $a = a$ has a false v-version $b_1 = b_2$. Hence $M, u \not\models B$.

The same argument applies to the formula without equality

$$C := \forall x \Box (P(x) \supset P(x)).$$

It is refuted at M, u for any model M such that $M, v \models P(b_1)$ and $M, v \not\models P(b_2)$. In fact, then $P(a) \supset P(a)$ has a false v-version $P(b_1) \supset P(b_2)$.

In these examples undesirable v-versions of D_u-sentences are obtained by replacing *different occurrences* of the same individual a with its *different inheritors* b_1, b_2. There is the simplest 'way out' — to forbid such a splitting of individuals within the same D_u-sentence. So we adopt the following definition:

($\nabla\nabla$) a v-version of a D_u-sentence A s a D_v-sentence obtained by replacing *all* occurrences of every $a \in D_u$ in A with some its v-inheritor a'.

In other words, v-versions of A are of the form $f \cdot A$ for maps $f : \mathrm{Const}(A) \longrightarrow D_v$, where $\mathrm{Const}(A)$ is the set of all constants from A, such that $a\rho f(a)$ for any $a \in \mathrm{Const}(A)$.

These maps are 'local analogues' of the maps ρ_{uv} in Kripke sheaves; the latter are defined on the whole D_u.

It seems natural to write the condition (∇) (which is analogue of (...)) in a form similar to Definition 3.6.4.

(∇')
$$M, u \models \Box A(a_1, \ldots, a_n) \text{ iff}$$
$$\forall v \in R(u)\, \forall b_1, \ldots, b_n \left(\mathop{\&}_{i=1}^{n} a_i \rho_{uv} b_i \Rightarrow M, v \models A(b_1, \ldots, b_n) \right).$$

But such a 'definition' is ambiguous. In fact, as we know (Section 2.4) the same D_u-sentence can be presented as $A(a_1, \ldots, a_n)$ $(= [a_1, \ldots, a_n/x_1, \ldots, x_n]\, A)$ for different formulas $A(x_1, \ldots, x_n)$. For example,

$$(a = a) = [a/x]\ (x = x) = [a, a/x, y]\ (x = y).$$

Now consider the same Kripke bundle model M as above. By applying (∇') we obtain

$$M, u \not\models [a, a/x, y]\, \Box(x = y),$$

but

$$M, u \models [a/x]\, \Box(x = x).$$

However the ambiguity of (∇') disappears after fixing one of the presentations of a D_u-sentence. For example, if we use only maximal generators,[1] $\Box(a = a)$ is regarded as $[a/x]\Box(x = x)$, so $M, u \models (a = a)$ from this viewpoint. This

[1] Recall that a maximal generator is obtained by replacing all occurrences of a constant with the same variable, cf. Section 2.4.

restricted version of (∇') exactly corresponds to the combination of (∇) and ($\nabla\nabla$).

Although the use of maximal generators and the condition ($\nabla\nabla$) seem ad hoc, later on we shall see that this is sufficient for (and closely related to) soundness.

There is another natural way to avoid ambiguity in definition (∇') — instead of evaluating D_u-sentences, we can evaluate formulas under variable assignments. This approach is well-known in classical logic. In full detail and in a more general context it will be considered in Section 5.9. Here we only sketch the main idea.

The truth value of a formula $A(\mathbf{x})$ with a list of parameters $\mathbf{x}$ at a world u is defined under a 'finite assignment' (or a D_u-transformation) $[\mathbf{x} \mapsto \mathbf{a}]$, where $\mathbf{a}$ is a tuple in D_u (cf. 2.4). Then (∇') changes to

$$(\nabla'') \quad \begin{array}{l} M, u \vDash \Box A(\mathbf{x})\ [\mathbf{x} \mapsto \mathbf{a}] \text{ iff} \\ \forall v \in R(u)\ \forall b_1 \dots b_n\ (\forall i\ a_i \rho_{uv} b_i \Rightarrow M, v \vDash A(\mathbf{x})\ [\mathbf{x} \mapsto \mathbf{b}]). \end{array}$$

Now the formulas $\Box(x = x)$, $\Box(P(x) \supset P(x))$ become true — in fact, e.g.

$$M, u \vDash \Box(x = x)\ [x \mapsto a]$$

is equivalent to

$$\forall v \in R(u)\ \forall b\ (a \rho_{uv} b \Rightarrow M, v \vDash x = x\ [x \mapsto b]).$$

But soundness still remains a problem. Viz., consider the same Kripke bundle on Fig. 5.2. Then according to (∇''), the $\mathbf{QK}^{=}$-theorem

$$A := (x = y \supset \Box(x = y))$$

is refuted. In fact (for any model M)

$$M, u \vDash x = y\ [x, y \mapsto a, a],$$

but

$$M, u \nvDash \Box(x = y)\ [x, y \mapsto a, a],$$

since

$$M, v \nvDash \Box(x = y)\ [x, y \mapsto b_1, b_2].$$

A similar example exists in modal logics without equality. Consider the following $\mathbf{QK}$-theorem:

$$\Box P(x, x) \supset \exists y\, \Box P(x, y),$$

which is a substitution instance of axiom (Ax13) from 2.6.10. This formula is refuted in the same Kripke bundle in the model M such that

$$\begin{array}{l} M, u \vDash P(x, x)\ [x \mapsto a], \\ M, v \vDash P(x, y)\ [x, y \mapsto b_i, b_j] \text{ iff } i = j. \end{array}$$

In fact,

$$M, u \models \Box P(x, x)\ [x \mapsto a],$$

since

$$M, u \models P(x, x)\ [x \mapsto a],$$
$$M, v \models P(x, x)\ [x \mapsto b_i].$$

But

$$M, u \not\models \Box P(x, y)\ [x, y \mapsto a, a],$$

since

$$M, v \not\models P(x, y)\ [x, y \mapsto b_1, b_2].$$

These counterexamples, together with the soundness theorem (Proposition 5.2.12), justify the truth definition stated in $(\nabla), (\nabla\nabla)$ or its equivalent (∇') using maximal generators.

Functor semantics

$\mathcal{C}$-sets are another generalisation of Kripke sheaves, where we allow for many transition maps between the same possible worlds. This happens, because a quasi-ordered set is replaced with an arbitrary category $\mathcal{C}$. More precisely, let $\mathcal{C}$ be a category with the set of objects W. Consider the frame $F = (W, R)$, where uRv iff there exists a morphism from u to v, i.e. iff the set $\mathcal{C}(u, v)$ of morphisms from u to v is non-empty. F is called the *frame representation* of $\mathcal{C}$. A $\mathcal{C}$-set is defined as a functor from $\mathcal{C}$ to SET; it can be regarded as a system of domains (F, D) together with a family of functions

$$\rho = (\rho_\alpha : D_u \longrightarrow D_v)_{\alpha \in \mathcal{C}(u,v)}$$

respecting composition and identity, cf. Definition 3.6.1. Obviously, Kripke sheaves over F are nothing but Cat$F$-sets.

In a $\mathcal{C}$-set, the v-inheritors of an individual $d \in D_u$ are $\rho_\alpha d$ for all $\alpha \in \mathcal{C}(u, v)$.

Given a model M in a $\mathcal{C}$-set (F, D, ρ), we define forcing, so that

$$M, u \models \Box B(d_1, \ldots, d_n) \text{ iff}$$
$$\forall v \in R(u)\ \forall \alpha \in \mathcal{C}(u, v)\ M, v \models B(\rho_\alpha d_1, \ldots, \rho_\alpha d_n).$$

Similarly to Kripke sheaves, in $\mathcal{C}$-sets we may not worry about different presentation of a D_u-sentence as $B(d_1, \ldots, d_n)$, because ρ_α are functions. A v-inheritor of a D_u-sentence is obtained by replacing all individuals $a \in D_u$ with their images $\rho_\alpha a$ under the same transition map ρ_α, $\alpha \in \mathcal{C}(u, v)$.

Validity

Validity in a Kripke bundle or in a $\mathcal{C}$-set is defined again as the truth in all models. But there is another problem: the set of valid formulas may be not substitution closed! To show this, consider the category $\mathcal{C}$ with a single object[2]

[2] I.e. its frame representation F is a reflexive singleton.

u and two arrows: id and γ such that $\gamma \circ \gamma = \gamma$ (and certainly, $id \circ \gamma = \gamma \circ id = \gamma$; C is just a two-element monoid). Let $\mathbb{F}$ be a C-set in Fig. 5.3 with $D_u = \{a, b\}$, $\rho_\gamma(a) = \rho_\gamma(b) = b$.

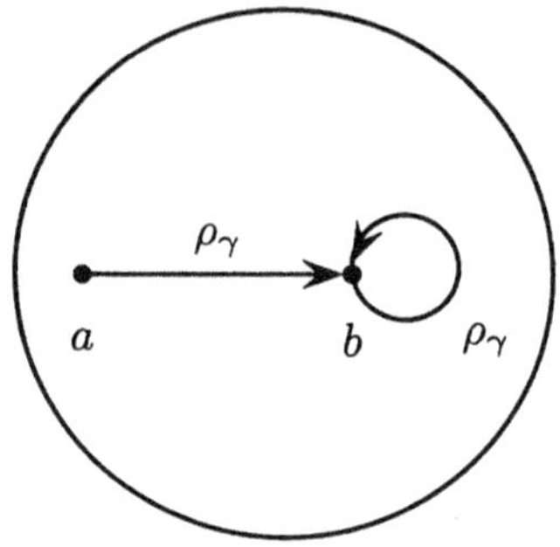

Figure 5.3.

Then the formula $p \supset \Box p$ is valid in $\mathbb{F}$ (since $F \vDash p \supset \Box p$) , but $P(x) \supset \Box P(x)$ is not. In fact, consider a model in which $u \vDash P(a), u \nvDash P(b)$; then $u \nvDash \Box P(a)$, because $u \nvDash P(\rho_\gamma a)$.

Exercise 5.1.1 Let us call a formula A *intuitionistically valid* in a C-set (or in a Kripke bundle) if A^T is valid. Show that the formula $p \vee \neg p$ is intuitionistically valid in the same C-set as above, whereas $P(x) \vee \neg P(x)$ is not intuitionistically valid.

Exercise 5.1.2 Consider the Kripke bundle $\mathbb{G}$ with the single domain $D_u = \{a, b\}$ such that $\rho_{uu} = \{(a, a), (b, b), (a, b)\}$, see Fig. 5.4. Show that the formula $P(x) \supset \Box P(x)$ is not valid and $P(x) \vee \neg P(x)$ is not intuitionistically valid in $\mathbb{G}$.

So it turns out that the set of all formulas (modal or intuitionistic) that are valid in a given Kripke bundle or a C-set, is not necessarily a predicate logic. To obtain a sound semantics, we replace validity with the notion of *strong validity* — a formula is said to be strongly valid if all its substitution instances are valid.

At first glance, two new semantics seem independent. But actually the semantics of C-sets is stronger. Furthermore, their natural combination (the 'C-bundle semantics') is strongly equivalent to C-sets. C-bundles are defined analogously to C-sets, with the following difference — we replace functions $\rho_\alpha : D_u \longrightarrow D_v$ with relations $\rho_\alpha \subseteq D_u \times D_v$ parametrised by morphisms $\alpha \in C(u, v)$; the clause (∇') is modified as follows:

$$(\nabla^+) \quad \begin{array}{l} M, u \vDash \Box A(a_1, \ldots, a_n) \text{ iff} \\ \forall v \in R(u)\ \forall \gamma \in C(u, v)\ \forall b_1, \ldots, b_n\ \ (\forall_i\ a_i \rho_\gamma b_i \Rightarrow M, v \vDash A(b_1, \ldots, b_n)), \end{array}$$

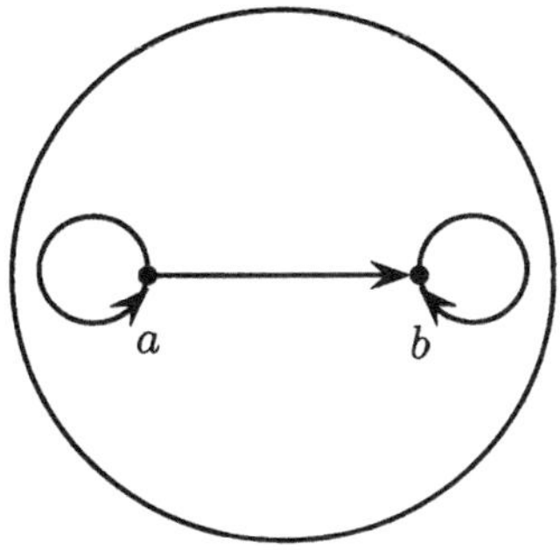

Figure 5.4.

with the same requirement about A. It turns out that for any $\mathcal{C}$-bundle there exists a $\mathcal{C}$-set with the same valid formulas; therefore every Kripke bundle also corresponds to some $\mathcal{C}$-set. This is not very surprising, because 'functionality' of $\mathcal{C}$-bundles is hidden in the fact that v-versions of A are of the form $f \cdot A$ for 'local functions' f, cf. ($\nabla\nabla$).

5.2 Kripke bundles

Now let us turn to precise definitions for the polymodal case.

Definition 5.2.1 *An* (N-modal) Kripke bundle *over an N-modal propositional frame $F = (W, R_1, \ldots, R_N)$ is a triple $\mathbb{F} = (F, D, \rho)$, in which $D = (D_u \mid u \in F)$ is a system of domains over F and $\rho = (\rho_{iuv} \mid uR_iv,\ 1 \leq i \leq N)$ is a family of relations $\rho_{iuv} \subseteq D_u \times D_v$ such that for any i, u, v*

($\sharp_1$) $uR_iv \Rightarrow \forall a \in D_u\ \exists b \in D_v\ a\rho_{iuv}b$,

i.e. $\mathrm{dom}(\rho_{iuv}) = D_u$*; this means that every individual has inheritors in all accessible worlds. The frame F is called the* base *of $\mathbb{F}$, and the domain D_u the* fibre at u.

For the 1-modal case we use the notation ρ_{uv} rather than ρ_{1uv}.

Definition 5.2.2 *An* intuitionistic Kripke bundle *is a 1-modal Kripke bundle (F, D, ρ) over an* **S4**-*frame $F = (W, R)$ satisfying the following two conditions:*

($\sharp_2$) every ρ_{uu} is reflexive;

($\sharp_3$) if $uRvRw$, then $\rho_{uv} \circ \rho_{vw} \subseteq \rho_{uw}$.

Definition 5.2.3 *An intuitionistic Kripke bundle* (F, D, ρ) *is called a* Kripke quasi-sheaf *if every* ρ_{uu} *is the identity function on* D_u, *i.e. if*

$$\forall a, b \in D_u \ (a\rho_{uu}b \Leftrightarrow a = b).$$

Obviously, every Kripke sheaf is a 1-modal Kripke bundle, in which all ρ_{uv} are functions. It is also clear that in the intuitionistic (and in the **S4**-) case every Kripke sheaf is a Kripke quasi-sheaf.

The other way round, Lemma 3.6.3 shows that every Kripke bundle, in which ρ_{iuv} are functions satisfying 'coherence conditions', gives rise to a Kripke sheaf. In a sense, Kripke sheaves are exactly Kripke bundles of this kind.

As mentioned in the previous section, a Kripke bundle can be presented in another equivalent form.

Lemma 5.2.4

(1) Let $\mathbb{F} = (F, D, \rho)$ *be a Kripke bundle, in which the domains* D_u *are disjoint.*[3] *Consider the total domain*

$$D^1 := \bigcup\{D_u \mid u \in W\}$$

with relations

$$\rho_i := \bigcup\{\rho_{iuv} \mid uR_iv\}$$

for $1 \leq i \leq N$.

Then we obtain the propositional frame of individuals (also called the total frame *of* $\mathbb{F}$*):*

$$F_1 = (D^1, \rho_1, \ldots, \rho_N).$$

Let $\pi : D^1 \longrightarrow W$ *be the map such that* $\pi(a) = u$ *for* $a \in D_u$. *Then* π *is a p-morphism* $F_1 \twoheadrightarrow F$. *Moreover, for* $N = 1$, $\mathbb{F}$ *is an intuitionistic Kripke bundle, iff* F_1 *is an* **S4**-*frame.*

(2) Conversely, a p-morphism of modal frames

$$\pi : F_1 = (W', \rho_1, \ldots, \rho_N) \twoheadrightarrow F = (W, R_1, \ldots, R_N)$$

gives rise to a system of disjoint domains $D_u = \pi^{-1}(u)$ *for* $u \in W$ *and a family of relations* $\rho_{iuv} = \rho_i \cap (D_u \times D_v)$ *for* uR_iv, $1 \leq i \leq N$, *which forms a modal Kripke bundle.*

For the case $N = 1$, *the corresponding Kripke bundle is intuitionistic iff both* F_1, F *are* **S4**-*frames.*

Sometimes it is convenient to denote the base $F = (W, R_1, \ldots, R_N)$ of a Kripke bundle $\mathbb{F}$ by $F_0 = (D^0, R_1^0, \ldots, R_N^0)$ and the frame of individuals F_1 by $(D^1, R_1^1, \ldots, R_N^1)$.

Thus we come to the following equivalent definition of Kripke bundles:

[3] As in the case of Kripke sheaves, cf. Section 3.5.

Definition 5.2.5 *A* Kripke bundle *(modal or intuitionistic) is a p-morphism between two propositional frames (resp., modal or intuitionistic).*

The choice between two equivalent definitions of Kripke bundles is a matter of convenience.

It is clear that quasi-sheaves correspond to p-morphisms of **S4**-frames $\pi : F_1 \longrightarrow F_0$ such that $F_1 \upharpoonright \pi^{-1}(u)$ is a discrete frame, i.e. a quasi-sheaf is a Kripke bundle with discrete fibres.

Exercise 5.2.6 Show that Kripke sheaves correspond to coverings, i.e. to Kripke bundles with the unique lift property:

$$\pi(a)Rv \Rightarrow \exists! b\,(a\rho b \ \&\ \pi(b) = v).$$

Definition 5.2.7 *A* valuation *ξ in a (modal) Kripke bundle $\mathbb{F} = (F, D, \rho)$ is just a (modal) valuation in its system of domains $D = (D_u \mid u \in W)$, cf. Definition 3.2.4. The pair $M = (\mathbb{F}, \xi)$ is called a* Kripke bundle model.

The inductive definition of forcing is more delicate than for Kripke frames or Kripke sheaves, because (as explained in Section 5.1) in the clause for $\Box$ we should explicitly indicate all the individuals occurring in A and use the presentation of D_u-sentences described in Lemma 2.4.4.

Forcing in Kripke bundle models is quire similar to Kripke frames or Kripke sheaves; the only difference is in the $\Box$-clause:

Definition 5.2.8 *Let $\mathbb{F} = (F, D, \rho)$ be a Kripke bundle, $F = (W, R_1, \dots, R_N)$, $M = (\mathbb{F}, \xi)$ a Kripke bundle model. We define the forcing relation $M, u \vDash A$ between worlds $u \in F$ and D_u-sentences (with equality) by induction:*

- $M, u \vDash P_k^0$ *iff* $u \in \xi_u(P_k^0)$;
- $M, u \vDash P_k^m(\mathbf{a})$ *iff* $\mathbf{a} \in \xi_u(P_k^m)$ *(for* $m > 0$*)*;
- $M, u \vDash a = b$ *iff* a *equals* b;
- $M, u \nvDash \bot$;
- $M, u \vDash B \vee C$ *iff* $(M, u \vDash B$ *or* $M, u \vDash C)$;
- $M, u \vDash B \wedge C$ *iff* $M, u \vDash B \ \&\ M, u \vDash C$;
- $M, u \vDash B \supset C$ *iff* $(M, u \nvDash B$ *or* $M, u \vDash C)$;
- $M, u \vDash \exists x A$ *iff* $\exists a \in D_u\ M, u \vDash [a/x]\,A$;
- $M, u \vDash \forall x A$ *iff* $\forall a \in D_u\ M, u \vDash [a/x]\,A$.
- $M, u \vDash \Box_i\,[a_1, \dots, a_n/x_1, \dots, x_n]\,B_0$ *iff*
  $\forall v \in R_i(u)\ \forall b_1, \dots, b_n \in D_v\ (\forall j\ a_j \rho_{iuv} b_j \Rightarrow M, v \vDash [b_1,\dots,b_n/x_1,\dots,x_n]B_0)$
  where $FV(B_0) = \{x_1, \dots, x_n\}$ *and* $a_1, \dots, a_n \in D_u$ *are* distinct.

So in the latter case we present a D_u-sentence $\Box_i B$ as $\Box_i[a/x]B_0$ for a formula B_0 and an injective $[x \mapsto a]$, recall that B_0 is called a maximal generator of B and it is unique up to renaming of x. This clause can be equivalently presented in the form discussed in 5.1:

(∇) $\quad M, u \vDash \Box A$ iff for any $v \in R(u)$ every v-version of A is true at M, v.

In particular, for a sentence B we have:

$$M, u \vDash \Box_i B \text{ iff } \forall v \in R_i(u)\ M, v \vDash B,$$

exactly as in standard Kripke semantics.

Definition 5.2.9 *A (modal) predicate formula is called* true *in a Kripke bundle model if its universal closure $\bar{\forall} A$ is true at every world of this model. A formula is called* valid *in a Kripke bundle $\mathbb{F}$ if it is true in every model over $\mathbb{F}$.*

As above, the sign $\vDash$ denotes the truth in a model and the validity in a Kripke bundle.

The following is trivial (cf. Lemma 3.2.20(1)):

Lemma 5.2.10 *For a Kripke bundle model M and a modal formula $A(x_1, \dots, x_n)$ $M \vDash A(x_1, \dots, x_n)$ iff $\forall u \in M\, \forall a_1, \dots, a_n \in D_u \quad M, u \vDash A(a_1, \dots, a_n)$.*

Recall that $A(a_1, \dots, a_n)$ denotes $[a_1, \dots, a_n/x_1, \dots, x_n]A(x_1, \dots, x_n)$, cf. 2.4.

Consider the set of valid formulas

$$\mathbf{ML}_{-}^{(=)}(\mathbb{F}) := \{A \in MF_N^{(=)} \mid \mathbb{F} \vDash A\}.$$

This set is closed under MP, Gen and $\Box$-introduction, cf. 3.2.27. But it may be not substitution closed (cf. Exercise 5.1.2), so we introduce the following notion.

Definition 5.2.11 *A formula $A \in MF_N^{(=)}$ is called* strongly valid *in a Kripke bundle $\mathbb{F}$ (respectively,* strongly valid with equality*) in $\mathbb{F}$ if all its MF_N (respectively, $MF_N^{=}$) -substitution instances are valid in $\mathbb{F}$; notation: $\mathbb{F} \vDash^{+} A$ (resp., $\mathbb{F} \vDash^{+=} A$).*

Let

$$\mathbf{ML}^{(=)}(\mathbb{F}) := \{A \in MF_N^{(=)} \mid \mathbb{F} \vDash^{+(=)} A\}.$$

Then we have

Proposition 5.2.12

(1) $\mathbf{ML}^{(=)}(\mathbb{F})$ is an m.p.l.(=);

(2) The strong validity of a formula A follows from the validity of all its m-shifts A^m, i.e.

$$\mathbf{ML}^{(=)}(\mathbb{F}) = \{A \in MF_N^{(=)} \mid \forall m\ \mathbb{F} \vDash A^m\}.$$

Speaking informally, Proposition 5.2.12(2) means that A is strongly valid iff it is valid with arbitrarily many extra parameters.

The proof is postponed until Section 5.13.

More exactly, this is a particular case of a general soundness theorem for metaframe semantics, cf. Section 5.12. However, let us sketch a general plan of the proof.

At the first stage we prove (2). Recall that substitution instances of a formula A are the strict substitution instances of its m-shifts A^m. So it is sufficient to check the following

Claim. $\mathbf{ML}_-^{(=)}(\mathbb{F})$ is closed under strict substitutions.

The proof is very similar to the substitution closedness of Kripke sheaves (Proposition 3.6.17) in the particular case of strict substitutions.

In the proof of (1), the strong validity of axioms, due to (2), follows from the validity of their shifts. The latter is proved in a straightforward way. For propositional axioms it is easier to use proposition 5.3.7 (see below), which in its turn, follows from (2).

The substitution closedness of $\mathbf{ML}^{(=)}(\mathbb{F})$ is obvious. The closedness under M, Gen, $\Box$-introduction follows from the same property of $\mathbf{ML}_-^{(=)}(\mathbb{F})$ (again,due to (2)). For example, for MP note that $(A \supset B)^n = A^n \supset B^n$, so $B^n \in \mathbf{ML}_-^{(=)}(\mathbb{F})$ whenever $A^n, (A \supset B)^n \in \mathbf{ML}_-^{(=)}(\mathbb{F})$.

Thanks to 5.2.12(1), we may call $\mathbf{ML}^{(=)}(\mathbb{F})$ the *modal logic of* $\mathbb{F}$. Also note that $\mathbf{ML}^{(=)}(\mathbb{F})$ is the largest modal logic contained in $\mathbf{ML}_-^{(=)}(\mathbb{F})$ (so to say, the *substitution interior* of $\mathbf{ML}_-^{(=)}(\mathbb{F})$).

Let us point out again that for Kripke sheaves the notions of validity and strong validity are equivalent, because the set of valid formulas is already substitution closed. Exercise 5.1.2 shows that this is not always the case in Kripke bundles.

Exercise 5.2.13 Give an example of a 1-modal Kripke bundle, in which ρ_{uv} are functions, but the set of valid formulas is not substitution-closed.

We also remark that the notion of strong validity a priori depends on the language. In fact, it might happen that $\mathbb{F} \models SA$ for any MF_N-substitution S, but not for any $MF_N^=$-substitution. But 5.2.12(2) shows that this is impossible, so $\mathbf{ML}^=(\mathbb{F})$ is conservative over $\mathbf{ML}(\mathbb{F})$.

Definition 5.2.14 *For a class $\mathcal{C}$ of Kripke bundles we define*

$$\mathbf{ML}^{(=)}(\mathcal{C}) := \bigcap\{\mathbf{ML}^{(=)}(\mathbb{F}) \mid \mathbb{F} \in \mathcal{C}\}.$$

Definition 5.2.15 Kripke bundle semantics $\mathcal{KB}_N$ *is generated by the class of all N-modal Kripke bundles.* $\mathcal{KB}_N$*-complete logics are called* Kripke bundle complete.

Definition 5.2.16 *(Cf. Definition 3.2.32) If F is a propositional Kripke frame, $\mathcal{KB}(F)$ denotes the class of all Kripke bundles based on F. The logic* $\mathbf{ML}^{(=)}$ *$\mathcal{KB}(F)$) is called the* Kripke bundle modal logic (with equality) determined over *F. Similarly we define $\mathcal{KB}(\mathcal{C})$ for a class of propositional frames $\mathcal{C}$ and the* Kripke bundle modal logic determined over *$\mathcal{C}$.*

5.3 More on forcing in Kripke bundles

Now let us present an arbitrary Kripke bundle as a collection of propositional Kripke frames. This presentation is more convenient from the technical viewpoint, and it will lead us to a more general notion of a metaframe.

Definition 5.3.1 *Let $D = (D_u \mid u \in W)$ be a system of disjoint domains over a propositional modal frame $F = (W, R_1, \ldots, R_N)$. For $n \geq 0$ we define the nth* level *of D as the set $D^n := \bigcup\{D_u^n \mid u \in W\}$.*

We also identify D_u^1 and D_u (strictly speaking, they are not equal), so that our new definition of D^1 corresponds to that in Lemma 5.2.4. Also note that $D^0 = W$, since $D_u^0 = \{u\}$, see Definition 3.2.4.

Note that every valuation $\xi = (\xi_u)_{u \in W}$ in a system of disjoint domains is associated with the following function ξ^+ sending n-ary predicate letters to subsets of D^n:

$$\xi^+(P_k^n) := \bigcup_{u \in W} \xi_u(P_k^n).$$

In particular, we have

$$\xi^+(P_k^0) \subseteq W.$$

Recall that for any function θ such that $\theta(P_k^n) \subseteq D^n$, there exists a valuation ξ such that $\theta = \xi^+$:

$$\xi_u(P_k^n) := \theta(P_k^n) \cap D_u^n.$$

To define accessibility relations on the sets D^n, we first recall the notation from Lemma 5.2.4:

$$\rho_i := \bigcup\{\rho_{iuv} \mid uR_iv\}.$$

Recall the subordination relation between n-tuples (see Introduction):

$$\mathbf{a}\ sub\ \mathbf{b} := \forall j, k\, (a_j = a_k \Rightarrow b_j = b_k).$$

Definition 5.3.2 *Let $\mathbb{F} = (F, D, \rho)$ be a modal Kripke bundle over $F = (W, R_1, \ldots, R_N)$. For $n > 0$, $1 \leq i \leq N$, we define the relation on D^n:*

$$\mathbf{a} R_i^n \mathbf{b} \text{ iff } \forall j\ a_j \rho_i b_j\ \&\ \mathbf{a}\ sub\ \mathbf{b}.$$

So in particular, $R_i^1 = \rho_i$. For $n = 0$ we put $R_i^0 := R_i$.

The frame $F_n := (D^n, R_1^n, \ldots, R_N^n)$ is called the nth level *of $\mathbb{F}$.*

Exercise 5.3.3 Show that if ρ_i is transitive, then R_i^n is also transitive.

Again in the 1-modal case we write R^n rather than R_1^n.

Now the inductive clause for $\Box_i B$ from the definition of forcing can be rewritten in the following form.

Lemma 5.3.4 *Under the conditions of Definition 5.2.8, let B be an N-modal formula with $FV(B) \subseteq r(\mathbf{x})$, $|\mathbf{x}| = n$. Then for any $u \in F$ and $\mathbf{a} \in D_u^n$*

$$M, u \vDash \Box_i B(\mathbf{a}) \qquad \textit{iff}$$
$$(*) \quad \forall v \in R_i(u)\ \forall \mathbf{b} \in D_v^n\ (\mathbf{a} R_i^n \mathbf{b} \Rightarrow M, v \vDash B(\mathbf{b}))$$

where as usual, $B(\mathbf{a})$ denotes $[\mathbf{a}/\mathbf{x}]\,B$.

variables

Proof ('Only if'.) Suppose $M, u \vDash \Box_i B(\mathbf{a})$. As explained after Definition 5.2.8, this means that every v-version of $B(\mathbf{a})$ is true at M, v (for any $v \in R_i(u)$), and v-versions are obtained according to $(\nabla\nabla)$. To check (*), let us show that for $\mathbf{a} R_i^n \mathbf{b}$, $B(\mathbf{b})$ is a v-version of $B(\mathbf{a})$.

In fact, every occurrence of a_k in $B(\mathbf{a})$ either replaces an occurrence of x_k in B or replaces an occurrence of x_j in B for $j \neq k$ — if $a_j = a_k$. In the first case an occurrence of a_k in $B(\mathbf{a})$ corresponds to an occurrence of b_k in $B(\mathbf{b})$. Similarly, in the second case an occurrence of $a_j = a_k$ in $B(\mathbf{a})$ corresponds to an occurrence of b_j in $B(\mathbf{b})$, and $b_j = b_k$, since $\mathbf{a} R_i^n \mathbf{b}$.

('If'.) Suppose (*), and consider a presentation of $B(\mathbf{a})$ as $[\mathbf{c}/\mathbf{x}]B_0$ with a distinct $\mathbf{c}$. Then B_0 is a maximal generator of $B(\mathbf{a}) = [\mathbf{a}/\mathbf{y}]B(\mathbf{y})$, so by 2.4.5(2), B_0 is obtained by a variable substitution from B, i.e. $B_0 = [\mathbf{x} \cdot \sigma/\mathbf{y}]B$ for some $\sigma \in \Sigma_{mn}$, where $|\mathbf{x}| = n$, $|\mathbf{y}| = m$.

To show that $M, u \vDash \Box_i [\mathbf{c}/\mathbf{x}]B_0$, consider $v \in R_i(u)$ and $d \in D_v^n$ such that $\forall k\ c_k \rho_i d_k$. Then

$$[\mathbf{d}/\mathbf{x}]B_0 = [\mathbf{d}/\mathbf{x}][\mathbf{x} \cdot \sigma/\mathbf{y}]B = [\mathbf{d} \cdot \sigma/\mathbf{y}]B. \tag{**}$$

Also $(\mathbf{c} \cdot \sigma)$ sub $(\mathbf{d} \cdot \sigma)$, since $c_{\sigma(i)} = c_{\sigma(j)}$ implies $\sigma(i) = \sigma(j)$ and thus $d_{\sigma(i)} = d_{\sigma(j)}$ (remember that $\mathbf{c}$ is distinct). Therefore $(\mathbf{c} \cdot \sigma) R_i^m (\mathbf{d} \cdot \sigma)$; hence $M, v \vDash [\mathbf{d} \cdot \sigma/\mathbf{y}]B$ by (*), i.e. $M, v \vDash [\mathbf{d}/\mathbf{x}]B_0$ by (**). Eventually $M, u \vDash \Box_i [\mathbf{c}/\mathbf{x}]B_0$ by Definition 5.2.8. ∎

Since the world u is fully determined by a tuple $\mathbf{a} \in D^n$, we can drop u from the notation of forcing and just write $M \vDash A(\mathbf{a})$ (or $M \vDash [\mathbf{a}/\mathbf{x}]A$, to be more precise). Now $(*)$ becomes almost the same as in the propositional case:

$$M \vDash \Box_i B(\mathbf{a}) \text{ iff } \forall \mathbf{b}\ (\mathbf{a} R_i^n \mathbf{b} \Rightarrow M \vDash B(\mathbf{b})).$$

Since by definition of R_i^n, $\mathbf{a}$ is subordinate to $\mathbf{b}$, the truth value of $\Box_i B(\mathbf{a})$ is well-defined for any tuple of individuals $\mathbf{a} = (a_1, \ldots, a_n)$ in the same world; some of a_i's may coincide.

So $\mathbb{F}$ corresponds to the sequence of propositional frames F_n, in which F_0 is the basic propositional frame. A Kripke bundle model $M = (\mathbb{F}, \xi)$ is then associated with the family of propositional Kripke models $(M_n)_{n\in\omega}$, where $M_n = (F_n, \xi^n)$ and for any k, $\xi^n(P_k^0) := \xi^+(P_k^n)$. F_n (respectively, M_n) is called the *n-level* of $\mathbb{F}$ (respectively, M).

This presentation of Kripke bundles allows us to describe the propositional fragment $\mathbf{ML}_\pi^{(=)}(\mathbb{F})$ of the predicate logic $\mathbf{ML}^{(=)}(\mathbb{F})$.

Lemma 5.3.5 *Let $M = (\mathbb{F}, \xi)$ be an N-modal Kripke bundle model, and let $M_n = (F_n, \xi^n)$ be the n-level of M. Then for any N-modal propositional formula A and for any $\mathbf{a} \in D^n$*

$$M_n, \mathbf{a} \vDash A \textit{ iff } M \vDash A^n(\mathbf{a}).$$

Proof Easy, by induction on the length of A. The case $A = \Box_i B$ readily follows from Lemma 5.3.4. ■

Lemma 5.3.6 *Let $\mathbb{F}$ be an N-modal Kripke bundle, A a propositional N-modal formula. Then for any $n \geq 0$*

$$\mathbb{F} \vDash A^n \textit{ iff } F_n \vDash A.$$

Proof Lemma 5.3.5 implies that for any model M over $\mathbb{F}$,

$$M_n \vDash A \text{ iff } M \vDash A^n.$$

In fact, note that by 5.2.10,

$$M \vDash A^n \text{ iff } \forall u \in M\ \forall \mathbf{a} \in D_u^n\ M, u \vDash A^n(\mathbf{a})$$

iff $M_n \vDash A$ by 5.3.5.

Hence

$$\mathbb{F} \vDash A^n \text{ iff } \forall\xi\ (\mathbb{F}, \xi) \vDash A^n \text{ iff } \forall\xi\ (F_n, \xi^n) \vDash A.$$

Finally note that every valuation θ in F_n is ξ^n for some valuation ξ in $\mathbb{F}$. In fact, one can put $\xi^+(P_k^n) := \theta(P_0^k)$ for any k and define arbitrary values of ξ^+ on other predicate letters. Hence $\forall\xi\ (F_n, \xi^n) \vDash A \Leftrightarrow F_n \vDash A$. ■

Proposition 5.3.7 *For an N-modal Kripke bundle $\mathbb{F}$ and an N-modal propositional formula A, $\mathbb{F} \vDash^+ A$ iff $\forall n\, F_n \vDash A$, i.e.*

$$\mathbf{ML}_\pi^{(=)}(\mathbb{F}) = \bigcap_{n\in\omega} \mathbf{ML}(F_n).$$

Proof Follows from Lemma 5.3.6 and Proposition 5.2.12. ■

5.4 Morphisms of Kripke bundles

Definition 5.4.1 *Let* $\mathbb{F} = (F, D, \rho)$, $\mathbb{G} = (G, D', \rho')$ *be Kripke bundles. A pair* $f = (f_0, f_1)$ *is called an* equality-morphism *(briefly,* =-morphism*) from* $\mathbb{F}$ *to* $\mathbb{G}$ *(notation:* $f : \mathbb{F} \longrightarrow^{=} \mathbb{G}$*) if*

(1) $f_0 : F \longrightarrow G$ *is a morphism of propositional frames,*

(2) $f_1 : F_1 \longrightarrow G_1$ *is a morphism of propositional frames,*

(3) f_1 *is a* fibrewise bijection, *i.e. every* $f_{1u} := f_1 \restriction D_u$ *is a bijection between* D_u *and* $D'_{f_0(u)}$.

f_0 *is called the* world component, f_1 *the* individual component *of* (f_0, f_1). *An =-morphism* (f_0, f_1) *is called*

- *a* p=-morphism *if* f_0 *is surjective (and thus, a p-morphism);*
- *an* isomorphism *if* f_0, f_1 *are isomorphisms.*

It is clear that Definition 3.3.1 for Kripke frames is a particular case of 5.4.1. From the definition one can easily see that the following diagram commutes (where π, π' are the p-morphisms corresponding respectively to $\mathbb{F}$, $\mathbb{G}$).

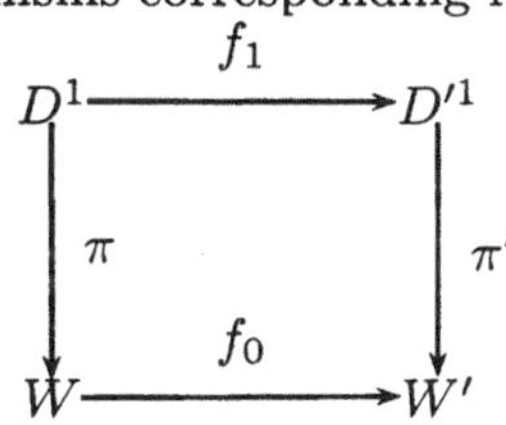

Exercise 5.4.2 Show that in Definition 5.4.1 the condition (i) follows from (ii) and (iii).

Exercise 5.4.3 Show that if (f_0, f_1) is an =-morphism and f_0 is an isomorphism, then f_1 is also an isomorphism.

Lemma 5.4.4 *If* $f : \mathbb{F} \longrightarrow^{=} \mathbb{F}'$ *is an =-morphism of Kripke bundles, then the map* f_n *sending* $\mathbf{a}$ *to* $f \cdot \mathbf{a}$ *is a morphism* $F_n \longrightarrow F'_n$.

Proof By definition, the statement holds for $n = 0, 1$.

Let us show that f_n respects subordination. In fact, suppose $\mathbf{a}\,sub\,\mathbf{b}$; then $f_1(a_i) = f_1(a_j)$ implies $a_i = a_j$ (since f is a fibrewise bijection), and hence $b_i = b_j$ (due to $\mathbf{a}\,sub\,\mathbf{b}$), $f_1(b_i) = f_1(b_j)$. Thus $(f \cdot \mathbf{a}) sub (f \cdot \mathbf{b})$.

The other way round, $(f \cdot \mathbf{a})\,sub\,(f \cdot \mathbf{b})$ implies $\mathbf{a}\,sub\,\mathbf{b}$; the reader can show this in a similar way.

Now the monotonicity of f_n $(n > 1)$ follows easily. In fact, if $\mathbf{a}R_k^n\mathbf{b}$, then for every i, $a_i R_k b_i$, and thus $f_1(a_i) R'_k f_1(b_i)$, which implies $(f \cdot \mathbf{a}) R'^n_k (f \cdot \mathbf{b})$, since f respects subordination.

To show the lift property for f_n ($n > 1$), assume $(f \cdot \mathbf{a})R'^n_k\mathbf{c}$. Then for every i, $f_1(a_i)R'_k c_i$, and thus by the lift property of f_1, there exists b_i such that $a_i R_k b_i$ and $c_i = f_1(b_i)$. Put $\mathbf{b} := (b_1, \ldots, b_n)$; then $\mathbf{a}\, sub\, \mathbf{b}$. In fact, since f is a fibrewise bijection, we have

$$a_i = a_j \text{ iff } f_1(a_i) = f_1(a_j)$$

and

$$b_i = b_j \text{ iff } c_i = c_j.$$

Since $(f \cdot \mathbf{a})\, sub\, \mathbf{c}$, we also have

$$f_1(a_i) = f_1(a_j) \Rightarrow c_i = c_j.$$

Thus $a_i = a_j$ implies $b_i = b_j$. So we obtain $\mathbf{a}R^n_k\mathbf{b}$ and $f_n(\mathbf{b}) = \mathbf{c}$. ∎

Definition 5.4.5 *Let $f = (f_0, f_1) : \mathbb{F} \longrightarrow^{=} \mathbb{G}$. For Kripke bundle models $M = (\mathbb{F}, \xi)$, $M' = (\mathbb{F}', \xi')$ we say that f is an* =-morphism from M to M' *(notation: $f : M \longrightarrow^{=} M'$) if*

$$M, u \vDash P(\mathbf{a}) \iff M', f_0(u) \vDash P(f_1 \cdot \mathbf{a}).$$

for any $P \in PL^n$, $\mathbf{a} \in D^n_u$, $n > 0$; or in another notation:[4]

$$\mathbf{a} \in \xi^+(P) \textit{ iff } (f_1 \cdot \mathbf{a}) \in \xi'^+(P),$$

and also

$$M, u \vDash P \iff M', f_0(u) \vDash P$$

for any $P \in PL^0$.

Then we obtain an analogue of Lemma 3.3.11

Lemma 5.4.6 *If $(f_0, f_1) : M \longrightarrow^{=} M'$ then for any $u \in M$ and for any D_u-sentence A*

$$(*) \qquad M, u \vDash A \textit{ iff } M', f_0(u) \vDash f_1 \cdot A,$$

where $f_1 \cdot A$ is obtained from A by replacing every $c \in D_u$ with $f_1(c)$. D_u-

Proof By induction. Let us consider only the $\Box$-case, i.e. $A = \Box_i B(\mathbf{a})$, where $\mathbf{a} \in D^n_u$. We also assume that $n > 0$; the easy case $n = 0$ is left for the reader.

So suppose $M, u \nvDash \Box_i B(\mathbf{a})$, $\mathbf{a} \in D^n_u$, and let us show that

$$M', f_0(u) \nvDash \Box_i B(f_1 \cdot \mathbf{a}).$$

We have $v \in R_i(u)$, $\mathbf{b} \in D^n_v$ such that $M, v \nvDash B(\mathbf{b})$ and $\mathbf{a}R^n_i\mathbf{b}$. By the induction hypothesis, we obtain $M', f_0(v) \nvDash B(f_1 \cdot \mathbf{b})$. We also have $(f_1 \cdot \mathbf{a})R^n_i(f_1 \cdot \mathbf{b})$ by Lemma 5.4.4, and $f_0(u)R_i f_0(v)$, since f_0 is a morphism. Hence $M', f_0(u) \nvDash \Box_i B(f_1 \cdot \mathbf{a})$.

[4] Cf. Section 5.3.

For the converse, suppose $M', f_0(u) \not\models \Box_i B(f_1 \cdot \mathbf{a})$. Then there exist w and $\mathbf{c} \in D^n_w$ such that

$$(f_1 \cdot \mathbf{a}) R^n_i \mathbf{c}, \ f_0(u) R'_i w, \ M', w \not\models B(\mathbf{c}).$$

By Lemma 5.4.4, $\mathbf{c} = f_1 \cdot \mathbf{b}$ for some $\mathbf{b} \in R^n_i(\mathbf{a})$, and (say, from $a_1 R_i b_1$) it easily follows that $u R_i v$, where v is the world of $\mathbf{b}$. Hence $M, v \not\models B(\mathbf{b})$ by the induction hypothesis, and thus $M, u \not\models \Box_i B(\mathbf{a})$. ■

Lemma 5.4.7 *Let $(f_0, f_1) : \mathbb{F} \longrightarrow^= \mathbb{G}$, and let M' be a Kripke bundle model over $\mathbb{G}$. Then there exists a model M over $\mathbb{F}$ such that $(f_0, f_1) : M \longrightarrow^= M'$.*

Proof If $M' = (\mathbb{G}, \psi')$, one can take

$$\psi^+(P) := \{\mathbf{a} \mid f_1 \cdot \mathbf{a} \in \psi'^+(P)\}$$

for $P \in PL^n$, $n > 0$, and

$$\psi^+(P) := \{u \mid f_0(u) \in \psi'^+(P)\}$$

for $P \in PL^0$. ■

Hence we obtain an analogue of Proposition 3.3.13.

Proposition 5.4.8 *If there exists a p=-morphism $f : \mathbb{F} \twoheadrightarrow^= \mathbb{F}'$, then $\mathbf{ML}^=_-(\mathbb{F}) \subseteq \mathbf{ML}^=_-(\mathbb{F}')$ and thus, $\mathbf{ML}^=(\mathbb{F}) \subseteq \mathbf{ML}^=(\mathbb{F}')$.*

Proof Let $(f_0, f_1) : \mathbb{F} = (F, D, \rho) \twoheadrightarrow^= \mathbb{F}' = (F', D', \rho')$, and assume that $\mathbb{F}' \not\models A(x_1, \dots, x_n)$. Then for some $u' \in F'$, $a_1, \dots, a_n \in D'_{u'}$, for some model M' over $\mathbb{F}'$ we have

$$M', u' \not\models A(a'_1, \dots, a'_n).$$

By the previous lemma, there exists a model M over $\mathbb{F}$ such that

$$(f_0, f_1) : \ M \twoheadrightarrow M'.$$

Since f_0 and all f_{1u} are surjective, there exist $u, a_1, \dots, a_n$, such that

$$u' = f_0(u), \ a'_1 = f_{1u}(a_1), \dots, \ a'_n = f_{1u}(a_n),$$

and thus by Lemma 5.4.6, we obtain $M, u \not\models A(a_1, \dots, a_n)$.

Therefore, $\mathbb{F}' \not\models A$ implies $\mathbb{F} \not\models A$. ■

A particular case of an =-morphism is an *inverse image morphism* obtained by the pullback construction as follows.

Lemma 5.4.9 *Let $\mathbb{F} = (F, D, \rho)$ be an N-modal Kripke bundle, $f : \ G \longrightarrow F$ a morphism of propositional frames, and put*

$$X := \{(u', a) \mid u' \in G, \ a \in D^1, \ f(u') = \pi(a)\},$$

where $\pi : F_1 \twoheadrightarrow F_0$ *corresponds to* $\mathbb{F}$. *Consider the relations on* X *defined as follows.*

$$(u', a)\rho'_i(v', b) \quad \text{iff} \quad u'R'_iv' \ \& \ a\rho_i b,$$

where ρ_i *and* R'_i *are the relations respectively in* F^1 *and* G. *Let also*

$$\pi'(u', a) := u', \ \varphi(u', a) := a.$$

Then

(1) $\pi' : (X, \rho'_1, \ldots, \rho'_N) \twoheadrightarrow F'$;

(2) $(f, \varphi) : \mathbb{G} \longrightarrow^{=} \mathbb{F}$, *where* $\mathbb{G}$ *corresponds to* π'.

Proof In all the cases the monotonicity is obvious.

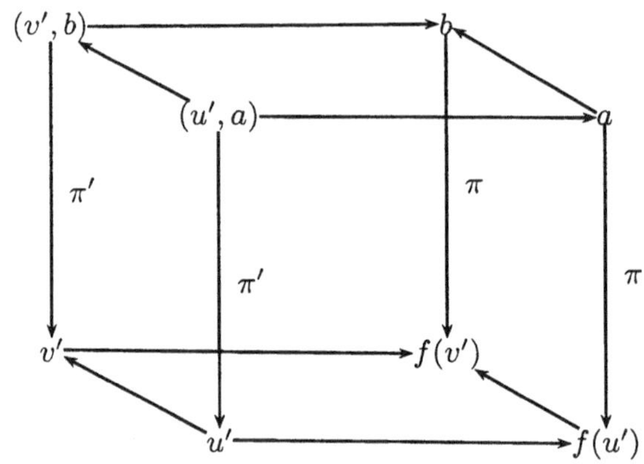

To check the lift property for π', suppose $u'R'_iv'$. Then $f(u')R_if(v')$ by monotonicity, and thus there exists b such that $a\rho_i b$ and $\pi(b) = f(v')$. Hence $(v', b) \in X$, $\pi'(v', b) = b$, $(u', a)\rho'_i(v', b)$ as required.

It is clear that φ is a fibrewise bijection; in fact,

$$a \in D_{f(u')} \text{ iff } \pi(a) = f(u') \text{ iff } (u', a) \in X \text{ iff } (u', a) \in D'_{u'}.$$

Checking the lift property for φ is left to the reader. ∎

For the Kripke bundle $\mathbb{G}$ constructed in Lemma 5.4.9 we shall use the notation $f_*\mathbb{F}$ and say that it is obtained by *change of the base along* f.

A particular case of this construction is a generated subbundle. Explicitly it is defined as follows.

Definition 5.4.10 *Let* $F \restriction V$ *be a generated subframe of a propositional frame* F, $\mathbb{F} = (F, D, \rho)$ *a Kripke bundle over* F. *Then we define*

$$\mathbb{F} \restriction V := (F', D', \rho'),$$

where $F' = F \restriction V$, $D' = D \restriction V$, ρ' *is an appropriate restriction of* ρ, *i.e.* $\rho'_{iuv} = \rho_{iuv}$ *for* $(u, v) \in R_i \cap (V \times V)$. $\mathbb{F} \restriction V$ *is called a* generated subbundle *of* $\mathbb{F}$ *(more precisely, the* restriction of $\mathbb{F}$ to V*).*

Lemma 5.4.11 *Let $j : F \restriction V \longrightarrow F$ be the inclusion morphism, $\varphi : D'^1 \longrightarrow D^1$ the inclusion map. Then $(j, \varphi) : \mathbb{F} \restriction V \longrightarrow^{=} \mathbb{F}$.*

Proof An easy exercise. ■

Exercise 5.4.12 Show that $\mathbb{F} \restriction V$ is isomorphic to $j_*\mathbb{F}$.

In particular, if $F \restriction V$ is a cone $F \uparrow u$, then $\mathbb{F} \restriction V$ is also called the *cone of* $\mathbb{F}$ and denoted by $\mathbb{F} \uparrow u$.

Lemma 5.4.13

$$\mathbf{ML}^{(=)}(\mathbb{F}) = \bigcap\{\mathbf{ML}^{(=)}(\mathbb{F} \uparrow u) \mid u \in F\}.$$

Proof Follows readily from Lemmas 5.4.6 and 5.4.7; an exercise for the reader (cf. the proof of Lemma 1.3.26). ■

Lemma 5.4.14 *Let $\mathbb{F}$ be a Kripke bundle, $\mathbb{F} \restriction V$ its generated subbundle. Then*

$$\mathbf{ML}^{(=)}(\mathbb{F}) \subseteq \mathbf{ML}^{(=)}(\mathbb{F} \restriction V).$$

Proof Follows for the previous lemma and the observation that every cone in $\mathbb{F} \restriction V$ is a cone in $\mathbb{F}$. ■

Finally let us show that $\mathbb{G} = f_*\mathbb{F}$ gives rise to the pullback (or 'coamalgam') diagram:

$$\begin{array}{ccc} G_1 & \xrightarrow{\quad \varphi \quad} & F_1 \\ {\scriptstyle \pi'}\downarrow & & \downarrow{\scriptstyle \pi} \\ G & \xrightarrow[\quad f \quad]{} & F \end{array}$$

In precise terms, this means the following

Lemma 5.4.15 *Let $f : G \to F$ be a morphism of propositional frames, $\mathbb{F}$ a Kripke bundle over F, $\pi : F_1 \twoheadrightarrow F$ the associated p-morphism, $\mathbb{F} = f_*\mathbb{F}$ and $(f, \varphi) : \mathbb{G} \to^{=} \mathbb{F}$ the $=$-morphism described in 5.4.9. Also let $(f, \psi) : \mathbb{H} \to^{=} \mathbb{F}$ be an arbitrary $=$-morphism from a Kripke bundle $\mathbb{H}$ over G to $\mathbb{F}$.*

Then

(1) there exists a unique map $g : G_1 \to F_1$ such that the following diagram (where τ is associated with $\mathbb{H}$) commutes.

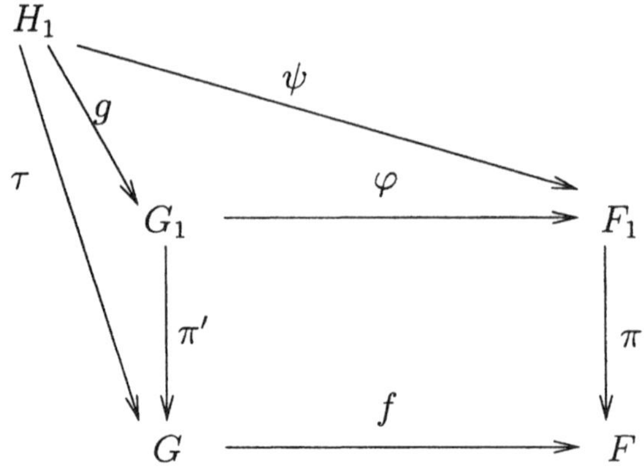

(2) g is an isomorphism between H_1 and G_1.

Proof (i) The diagram commutes iff for any $x \in H$, $\pi'(g(x)) = \tau(x)$ and $\varphi(g(x)) = \psi(x)$, i.e. iff $g(x) = (\tau(x), \psi(x))$; note that $(\tau(x), \psi(x)) \in G_1$ since $\pi \cdot \psi = f \cdot \tau$. So g is well-defined.

(ii) We can consider the category $\mathcal{P}$ of Kripke bundle morphisms of the form $(f, \psi) : \mathbb{H} \to \mathbb{F}$, where $\mathbb{H}$ is an arbitrary Kripke bundle over G. A morphism (in $\mathcal{P}$) from $(f, \psi) : \mathbb{H} \to \mathbb{F}$ to $(f, \psi') : \mathbb{H}' \to \mathbb{F}$ is defined as a map h, for which the following diagram (where $\tau' : H_1' \twoheadrightarrow G$ is associated with $\mathbb{H}'$) commutes.

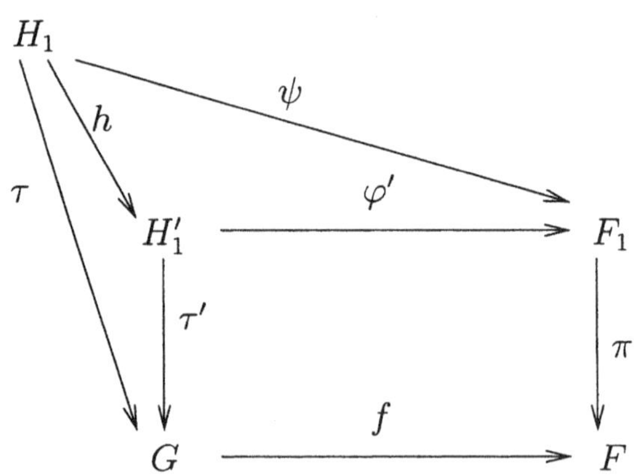

The assertion (i) shows that our specific (f, φ) is a terminal object of $\mathcal{P}$. But actually the above argument applies to any $(f, \varphi) : \mathbb{G} \to \mathbb{F}$ from $\mathcal{P}$.

In fact, we only need to find a unique $y \in G_1$ such that $\pi'(y) = \tau(x)$ & $\varphi(y) = \psi(x)$. So it suffices to show that

$$\exists! y \in G_1(\pi'(y) = u \ \& \ \varphi(y) = a)$$

whenever $f(u) = \pi(a)$. But this holds since φ is a fibrewise bijection between $\pi'^{-1}[u]$ and $\pi^{-1}[f(u)]$.

Thus all objects in $\mathcal{P}$ are terminal, and all morphisms are invertible. So the map g from (i) is bijective. To complete the proof of (ii), it remains to show the monotonicity of g; the monotonicity g^{-1} follows by the same argument. So let S_i be the relations in H_1. Then for any $y, z \in H_1$

$$yS_iz \text{ implies } \psi(y)\,\rho_i\,\psi(z), \text{ i.e. } \varphi(g(y))\,\rho_i\,\varphi(g(z)),$$

and similarly

$$yS_iz \Rightarrow \pi'(g(y))R_i'\pi'(g(z)).$$

Thus

$$yS_iz \Rightarrow g(y)\rho_i'g(z)$$

by definition (see Lemma 5.4.9). ∎

5.5 Intuitionistic Kripke bundles

Now let us consider the intuitionistic case. First note the following:

Lemma 5.5.1 *If $\mathbb{F}$ is an intuitionistic Kripke bundle, then all the F_n are* **S4**-*frames.*

Proof An exercise. ∎

Definition 5.5.2 *A valuation ξ in an intuitionistic Kripke bundle $\mathbb{F}$ is called* intuitionistic *if every set $\xi^+(P_k^n) = \bigcup\{\xi_u(P_k^n) \mid u \in W\}$ is stable in F_n, i.e. if $R^n(\xi^+(P_k^n)) \subseteq \xi^+(P_k^n)$. The model $(\mathbb{F}, \xi)$ is also called* intuitionistic *in this case.*

Intuitionistic forcing in intuitionistic Kripke bundle models is defined via modal forcing similarly to the case of Kripke frames (Definition 3.2.12):

Definition 5.5.3 *Let M be an intuitionistic model over an intuitionistic Kripke bundle $\mathbb{F} = (F, D, \rho)$, $F = (W, R)$, $u \in M$, A an intuitionistic D_u-sentence. Then we put*

$$M, u \Vdash A := M, u \vDash A^T.$$

Now we obtain an analogue of Lemma 3.2.13:

Lemma 5.5.4 *Under the conditions of Definition 5.5.3*

(1) $M, u \Vdash B$ iff $M, u \vDash B$ (for B atomic);

(2) $M, u \Vdash B \wedge C$ iff $(M, u \Vdash B \;\&\; M, u \Vdash C)$;

(3) $M, u \Vdash B \vee C$ iff $(M, u \Vdash B$ or $M, u \Vdash C)$;

(4) $M, u \Vdash B(\mathbf{a}) \supset C(\mathbf{a})$ iff
$\forall v \in R(u)\, \forall \mathbf{b} \in D_v^n\, (\mathbf{a}R^n\mathbf{b} \;\&\; M, v \Vdash B(\mathbf{b}) \Rightarrow M, v \Vdash C(\mathbf{b}))$;

(5) $M, u \Vdash \exists x B$ *iff* $\exists a \in D_u\ M, u \Vdash [a/x]\, B$*;*

(6) $M, u \Vdash \forall x B(x, \mathbf{a})$ *iff*
$\forall v \in R(u)\, \forall d \in D_v\, \forall \mathbf{b} \in D_v^n\, (\mathbf{a}R^n\mathbf{b} \Rightarrow M, v \Vdash B(d, \mathbf{b}))$*;*

(7) $M, u \Vdash \neg B(\mathbf{a})$ *iff* $\forall v \in R(u)\, \forall \mathbf{b} \in D_v^n\ (\mathbf{a}R^n\mathbf{b} \Rightarrow M, v \not\Vdash B(\mathbf{b}))$*;*

(8) $M, u \Vdash a \neq b$ *iff* a *does not equal* b*.*

Here we assume that in all cases we evaluate D_u*-sentences at* u*. In (4), (6) and (7) we also assume that* $\mathbf{a} \in D_u^n$ *is a list of individuals, which replaces a list of variables containing the parameters respectively of* $B \supset C$ *or* $\forall x\, B$*,* B*.*

Proof Let us consider only the case (4); the others are an exercise for the reader. In the fixed model M we have

$u \Vdash B(\mathbf{a}) \supset C(\mathbf{a})$ iff
$u \vDash (B(\mathbf{a}) \supset C(\mathbf{a}))^T = \Box(B(\mathbf{a})^T \supset C(\mathbf{a})^T)$ by Definitions 5.5.3, 2.11.1
iff $\forall v \in R(u)\, \forall \mathbf{b} \in D_v^n\, (\mathbf{a}R^n\mathbf{b} \Rightarrow v \vDash B(\mathbf{b})^T \supset C(\mathbf{b})^T)$ by Lemma 5.3.4
iff $\forall v \in R(u)\, \forall \mathbf{b} \in D_v^n\, (\mathbf{a}R^n\mathbf{b}\ \&\ v \Vdash B(\mathbf{b}) \Rightarrow$
$v \Vdash C(\mathbf{b}))$ by Definitions 5.5.3, 5.2.8.

■

As in the modal case, we can drop u from the notation; so e.g. (4), (6) are written as follows:

(4) $M \Vdash (B \supset C)(\mathbf{a})$ iff $\forall \mathbf{b}(\mathbf{a}R^n\mathbf{b}\ \&\ M \Vdash B(\mathbf{b}) \Rightarrow M \Vdash C(\mathbf{b}))$;

(6) $M \Vdash \forall x B(x, \mathbf{a})$ iff $\forall d \forall \mathbf{b}\ ((d\mathbf{b}) \in D^{n+1}\ \&\ \mathbf{a}R^n\mathbf{b} \Rightarrow M \Vdash B(d, \mathbf{b}))$.

Now we obtain some facts similar to those proved in Section 3.2.

Lemma 5.5.5 *Let* M *be the same as in Lemma 5.5.4. Then for any* D^1*-sentence* $A(\mathbf{a})$

$$M \Vdash A(\mathbf{a})\ \&\ \mathbf{a}R^n\mathbf{b} \Rightarrow M \Vdash A(\mathbf{b}).$$

Proof By Definition 5.5.3 and Lemma 5.3.4. ■

Definition 5.5.6 *Let* $\mathbb{F}$ *be an intuitionistic Kripke bundle;* M *a Kripke bundle model over* $\mathbb{F}$*. The* pattern *of* M *is the Kripke bundle model* M_0 *over* $\mathbb{F}$ *such that for any* $u \in \mathbb{F}$ *and any atomic* D_u*-sentence* A

$$M_0, u \vDash A \text{ iff } M, u \vDash \Box A.$$

It is clear that the model M_0 is intuitionistic.

Lemma 5.5.7 *Under the conditions of Definition 5.5.6, we have for any* $u \in \mathbb{F}$*, for any* D_u*-sentence* A

$$M_0, u \Vdash A \text{ iff } M, u \vDash A^T.$$

Proof By induction, similar to Lemma 3.2.15. Let us consider only the case $A = B(\mathbf{a}) \supset C(\mathbf{a})$. We have

$M_0, u \Vdash A$ iff
$\forall v \in R(u)\, \forall \mathbf{b} \in D_v^n\, (\mathbf{a}R^n\mathbf{b} \;\&\; M_0, v \Vdash B(\mathbf{b}) \Rightarrow M_0, v \Vdash C(\mathbf{b}))$ (by Lemma 5.5.4)
iff $\forall v \in R(u)\, \forall \mathbf{b} \in D_v^n\, (\mathbf{a}R^n\mathbf{b} \;\&\; M, v \vDash B(\mathbf{b})^T \Rightarrow M, v \vDash C(\mathbf{b})^T)$
(by the induction hypothesis)
iff $M, u \vDash \Box(B(\mathbf{a})^T \supset C(\mathbf{a})^T)$ (by Lemma 5.3.4 and Definition 5.2.8).

The latter formula is A^T by Definition 2.11.1. ■

Lemma 5.5.8 *Let M be an intuitionistic Kripke bundle model with the accessibility relation R, $A(\mathbf{x})$ an intuitionistic formula with all its parameters in the list $\mathbf{x}$, $|\mathbf{x}| = n$. Then for any $u \in M$*

$$M, u \Vdash \forall \mathbf{x} A(\mathbf{x}) \text{ iff } \forall v \in R(u)\, \forall \mathbf{a} \in D_v^n\; M, u \Vdash A(\mathbf{a}).$$

Proof The claim is similar to Lemma 3.2.18, but the proof is different; now it is based on soundness (Proposition 5.2.12).

By Definition 3.2.12, in the given model M we have:

$$u \Vdash \forall \mathbf{x}\, A(\mathbf{x}) \text{ iff } u \vDash (\forall \mathbf{x}\, A(\mathbf{x}))^T.$$

By Lemma 2.11.7, in **QS4** the latter formula is equivalent to $\Box\forall \mathbf{x} A(\mathbf{x})^T$, and thus

$u \vDash (\forall \mathbf{x} A(\mathbf{x}))^T$ iff $u \vDash \Box\forall \mathbf{x} A(\mathbf{x})^T$ (by Proposition 5.2.12)
iff $\forall v \in R(u)\, \forall \mathbf{a} \in D_v^n\; v \vDash A(\mathbf{a})^T$ (by Definition 5.2.8 and Lemma 5.2.10)
iff $\forall v \in R(u)\, \forall \mathbf{a} \in D_v^n\; v \Vdash A(\mathbf{a})$ (by Definition 5.5.3).

■

Exercise 5.5.9 Let M be an intuitionistic Kripke bundle model with the accessibility relation R, $u \in M$, $A(\mathbf{x}, \mathbf{y})$ an intuitionistic formula with $FV(A) \subseteq \mathbf{xy}$, $|\mathbf{x}| = n$, $|\mathbf{y}| = m$. Then for any $u \in M$, $\mathbf{c} \in D_u^m$

$$M, u \Vdash \forall \mathbf{x} A(\mathbf{x}, \mathbf{c}) \text{ iff } \forall v \in R(u)\, \forall \mathbf{a} \in D_v^n\, \forall \mathbf{d} \in D_v^m\, (\mathbf{c}R^m\mathbf{d} \Rightarrow M, u \Vdash A(\mathbf{a}, \mathbf{d})).$$

Definition 5.5.10 *An intuitionistic predicate formula A is called*

- true *in an intuitionistic Kripke bundle model M (notation: $M \Vdash A$) if $\bar{\forall} A$ is true at every world of M;*
- valid *in an intuitionistic Kripke bundle $\mathbb{F}$ (notation: $\mathbb{F} \Vdash A$) if it is true in all intuitionistic Kripke models over $\mathbb{F}$;*
- strongly valid *in an intuitionistic Kripke bundle $\mathbb{F}$ (notation: $\mathbb{F} \Vdash^+ A$) if all its substitution instances are valid in $\mathbb{F}$.*

Lemma 5.5.11 *$M \Vdash A(x_1, \ldots, x_n)$ iff $\forall u \in M \forall a_1, \ldots, a_n \in D_u \quad M, u \Vdash A(a_1, \ldots, a_n)$.*

Proof Similar to Lemma 3.2.20. ∎

Proposition 5.5.12 *Let $\mathbb{F}$ be an* **S4***-based Kripke bundle, $A \in IF^=$. Then*

(1) $\mathbb{F} \Vdash A$ iff $\mathbb{F} \vDash A^T$;

(2) the following three assertions are equivalent:

(a) $\mathbb{F} \Vdash^+ A$,

(b) $\forall m\ \mathbb{F} \Vdash A^m$,

(c) $\mathbb{F} \vDash^+ A^T$.

Proof
(1) Let $A = A(\mathbf{x})$, $|\mathbf{x}| = n$.

(Only if.) Assume $\mathbb{F} \Vdash A$. Let M be a Kripke bundle model over $\mathbb{F}$. By Lemma 5.5.11, we have to show $M, u \vDash A^T(\mathbf{a})(= A(\mathbf{a})^T)$ for any $u \in \mathbf{F}$, $\mathbf{a} \in D_u^n$. By Lemma 5.5.7 we have:

$$M_0, u \Vdash A(\mathbf{a}) \text{ iff } M, u \vDash A(\mathbf{a})^T.$$

By Lemma 5.5.11, $\mathbb{F} \Vdash A$ implies $M_0, u \Vdash A(\mathbf{a})$, and thus we obtain $\mathbb{F} \vDash A^T$.

(If.) Assume $\mathbb{F} \vDash A^T$. Then for any intuitionistic model M, any $u \in M, \mathbf{a} \in D_u^n$ we have $M, u \vDash A(\mathbf{a})^T (= A^T(\mathbf{a}))$ by Lemma 5.5.11, i.e. $M, u \Vdash A(\mathbf{a})$. Hence $\mathbb{F} \Vdash A$, by Lemma 5.5.11.

(2) (a)⇒(b) is obvious.

(b)⇒(c). Assume $\mathbb{F} \Vdash A^m$, i.e. $\mathbb{F} \vDash (A^m)^T$. Since $(A^m)^T = (A^T)^m$, we obtain $\mathbb{F} \vDash^+ A^T$ by Proposition 5.2.12.

(c)⇒(a). Assume $\mathbb{F} \vDash^+ A^T$, and for an arbitrary $IF^=$-substitution S let us show $\mathbb{F} \Vdash SA$, i.e. $\mathbb{F} \vDash (SA)^T$. By Lemma 2.11.5, $(SA)^T \equiv S^T(A^T)$ is a $\mathbf{QS4}^=$-theorem, and thus it is valid in $\mathbb{F}$, by Proposition 5.2.12. By assumption, $\mathbb{F} \vDash S^T(A^T)$; thus we obtain $\mathbb{F} \vDash (SA)^T$. Therefore $\mathbb{F} \Vdash A^+$. ∎

Proposition 5.5.13 *For an intuitionistic Kripke bundle $\mathbb{F}$ the set of strongly valid formulas*

$$\mathbf{IL}^{(=)}(\mathbb{F}) := \{A \in IF^{(=)} \mid \mathbb{F} \Vdash^+ A\}$$

is a superintuitionistic predicate logic (with or without equality), moreover,

$$\mathbf{IL}^{(=)}(\mathbb{F}) = {}^T\mathbf{ML}^{(=)}(\mathbb{F}).$$

Proof Similar to Proposition 3.2.29. By Proposition 5.5.12, $\mathbb{F} \Vdash^+ A$ iff $\mathbb{F} \vDash^+ A^T$, which means $A \in \mathbf{IL}^{(=)}(\mathbb{F})$ iff $A^T \in \mathbf{ML}^{(=)}(\mathbb{F})$. Thus $\mathbf{IL}^{(=)}(\mathbb{F}) = s(\mathbf{ML}^{(=)}(\mathbb{F}))$. ∎

Definition 5.5.14 *The set* $\mathbf{IL}^{(=)}(\mathbb{F})$ *is called the* superintuitionistic logic of $\mathbb{F}$ *(respectively, with or without equality).*

Now we easily obtain intuitionistic analogues of some assertions from the previous section.

Lemma 5.5.15 *Let* $(f_0, f_1) : M \longrightarrow^{=} M'$ *for intuitionistic Kripke bundle models* M, M'. *Then for any* $u \in M$ *and for any intuitionistic* D_u*-sentence* B

$$M, u \Vdash B \text{ iff } M', f_0(u) \Vdash f_1 \cdot B,$$

Proof From Lemma 5.4.6, Definition 5.5.3 and the observation that $f_1 \cdot B^T = (f_1 \cdot B)^T$. ■

Proposition 5.5.16 *If* $\mathbb{F}$ *and* $\mathbb{G}$ *are intuitionistic Kripke bundles and* $\mathbb{F} \twoheadrightarrow^{=} \mathbb{G}$, *then* $\mathbf{IL}_{-}^{=}(\mathbb{F}) \subseteq \mathbf{IL}_{-}^{=}(\mathbb{G})$ *and* $\mathbf{IL}^{=}(\mathbb{F}) \subseteq \mathbf{IL}^{=}(\mathbb{G})$.

Proof The first statement is an exercise for the reader. For the second, note that $\mathbb{F} \twoheadrightarrow^{=} \mathbb{G}$ implies $\mathbf{ML}^{=}(\mathbb{F}) \subseteq \mathbf{ML}^{=}(\mathbb{G})$ by Proposition 5.4.8; hence $s(\mathbf{ML}^{=}(\mathbb{F})) \subseteq s(\mathbf{ML}^{=}(\mathbb{G}))$, i.e. $\mathbf{IL}^{=}(\mathbb{F}) \subseteq \mathbf{IL}^{=}(\mathbb{G})$, by Proposition 5.5.13. ■

Lemma 5.5.17 *For an intuitionistic Kripke bundle* $\mathbb{F}$

$$\mathbf{IL}^{(=)}(\mathbb{F}) = \bigcap\{\mathbf{IL}^{(=)}(\mathbb{F} \uparrow u) \mid u \in F\}.$$

Proof Follows readily from Lemma 5.4.13, Proposition 5.5.13, and the equality $s\left(\bigcap_i L_i\right) = \bigcap_i s(L_i)$. ■

Lemma 5.5.18 *Let* $\mathbb{F}$ *be an intuitionistic Kripke bundle,* $\mathbb{F} \restriction V$ *its generated subbundle. Then*

$$\mathbf{IL}^{(=)}(\mathbb{F}) \subseteq \mathbf{IL}^{(=)}(\mathbb{F} \restriction V).$$

Proof Similar to Lemma 5.4.14. ■

For the intuitionistic case the notion of a Kripke bundle can be slightly extended.

Definition 5.5.19 *An* intuitionistic Kripke quasi-bundle *is a quasi-p-morphism* $\pi : F_1 \longrightarrow F_0$ *between* **S4***-frames, cf. Definition 1.4.18.*[5]

A Kripke quasi-bundle can be presented equivalently as a triple $\mathbb{F} = (F_0, D, \rho)$ with a system of domains $D = (D_u \mid u \in W)$ and a family of relations $\rho = (\rho_{uv} \mid uRv)$ satisfying ($\sharp_2$), ($\sharp_3$) from Definition 5.2.2 and

($\sharp_1'$) $uRv \Rightarrow \forall a \in D_u\ \exists v' \approx_R v\ \exists b \in D_v\ \ a\rho_{uv'}b.$

[5] Recall that quasi-p-morphisms are intuitionistic analogues of p-morphisms, cf. Chapter 1.

However one should be careful about using quasi-bundles. Since an intuitionistic quasi-bundle is not necessarily a modal bundle, it is not a surprise that the semantics of quasi-bundles *is not sound* for modal logics. But it is not sound for intuitionistic logic either! To obtain soundness, we have to reduce the class of quasi-bundles, see Section 5.17 for details.

Exercise 5.5.20 Try to give an inductive definition of intuitionistic forcing in a Kripke quasi-bundle (intuitionistic valuations are defined similarly to Kripke bundles). Note that the definition of forcing is not so straightforward.

A familiar truth definition for a formula $\exists x A$ is irrelevant in this case, as the following counterexample shows.

Example 5.5.21 Consider a quasi-bundle $\mathbb{F}$ pictured at Fig. 5.5. $b_1 \in D_{u_1}$ does not have inheritors in u_2, but b_1 is its own inheritor in $u_1 \approx_R u_2$. So $\mathbb{F}$ is still a quasi-bundle.

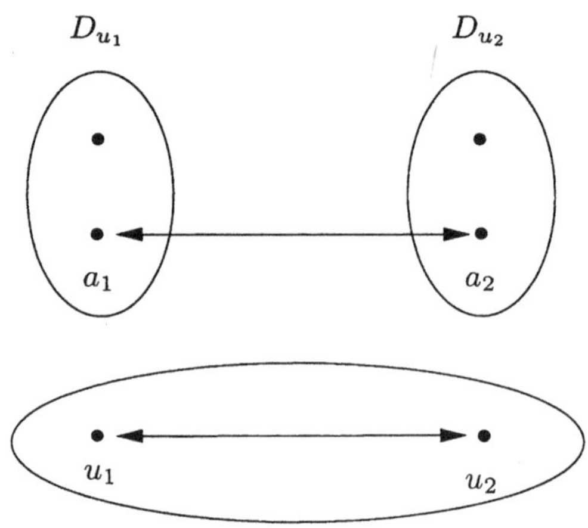

Figure 5.5. R is universal; ρ is universal on $\{a_1, a_2\}$.

Consider an intuitionistic model $M = (\mathbb{F}, \xi)$ such that $\xi^+(P) = \{b\}$. Putting

$$M, u \Vdash \exists x P(x) \Leftrightarrow \exists c \in D_u \; M, u \Vdash P(c)$$

we obtain

$$M, u_1 \Vdash \exists x P(x),$$

but

$$M, u_2 \nVdash \exists x P(x).$$

So since $u_1 R u_2$, there is no truth-preservation for $\exists x P(x)$.

But as we shall soon see (cf. Lemma 5.14.15), truth-preservation (or 'monotonicity') is necessary for soundness.

So here is an alternative definition:

$$M, u \Vdash \exists x B(x, \mathbf{a}) \text{ iff } \exists v \approx u\ \exists d \in D_v\ \exists \mathbf{b} \in D_v^n(\mathbf{b} \approx^n \mathbf{a}\ \&\ M, v \Vdash B(d, \mathbf{b})). \quad (*)$$

Due to the quasi-lift property (1.4.18) of quasi-bundles, (*) can be rewritten as

$$M, u \Vdash \exists x B(x, \mathbf{a}) \text{ iff } \exists v R u\ \exists d \in D_v\ \exists \mathbf{b} \in D_v^n(\mathbf{b} R^n \mathbf{a}\ \&\ M, v \Vdash B(d, \mathbf{b})). \quad (**)$$

Note that this definition for the $\exists$-case is dual to the clause for the $\forall$-case in Lemma 5.5.4(6).

The modified definition provides the desired truth-preservation property, as we shall see later on.

Lemma 5.5.22 *Let $M = (\mathbb{F}, \xi)$ be an intuitionistic Kripke bundle model, and let ξ^n be a propositional valuation in its nth level F_n, $n \geq 0$ such that $\xi^n(p_k) = \xi^+(P_k^n)$ for any $k \geq 0$. Let $M_n = (F_n, \xi^n)$ be the corresponding propositional Kripke model. Then every M_n is intuitionistic, and for any intuitionistic propositional formula A and for any $\mathbf{a} \in M_n$*

$$M_n, \mathbf{a} \Vdash A \text{ iff } M \Vdash A^n(\mathbf{a}).$$

Proof M_n is intuitionistic, according to Definition 5.5.2.

$M_n, \mathbf{a} \Vdash A$ iff $M_n, \mathbf{a} \vDash A^T$ (by Definition 1.4.1)
iff $M \vDash (A^T)^n(\mathbf{a})$ $(= (A^n(\mathbf{a}))^T)$ (by Lemma 5.3.5)
iff $M \Vdash A^n(\mathbf{a})$ (by Definition 5.5.3).

■

Lemma 5.5.23 *Let $\mathbb{F}$ be an intuitionistic Kripke bundle, A a propositional intuitionistic formula. Then for any $n \geq 0$*

$$\mathbb{F} \Vdash A^n \text{ iff } F_n \Vdash A.$$

Proof $\mathbb{F} \Vdash A^n$ iff $\mathbb{F} \vDash (A^n)^T (= (A^T)^n)$ (by Proposition 5.5.12) iff $F_n \vDash A^T$ (by Lemma 5.3.6) iff $F_n \Vdash A$ (by Lemma 1.4.7). ■

Proposition 5.5.24 *For an intuitionistic Kripke bundle $\mathbb{F}$,*

$$\mathbf{IL}_\pi^{(=)}(\mathbb{F}) = \bigcap_{n \in \omega} \mathbf{IL}(F_n).$$

Proof $\forall n\, F_n \Vdash A$ iff $\forall n\, F_n \Vdash A^T$ (by Lemma 1.4.7) iff $\mathbb{F} \vDash^+ A^T$ (by Proposition 5.3.7) iff $\mathbb{F} \Vdash^+ A$ (by Proposition 5.5.12). ■

Let us also make some remarks on Kripke quasi-sheaves.

Lemma 5.5.25 *Let $\mathbb{F}$ be a quasi-sheaf over an (intuitionistic) frame F. Assume that $u \approx_R v$ in F, i.e. u, v are in the same cluster. Then ρ_{uv} is a bijection between D_u and D_v with the converse ρ_{vu}.*

Proof First, note that $a\rho_{uv}b$ implies $b\rho_{vu}a$. In fact, by Definition 5.2.1, there exists a' such that $b\rho_{vu}a'$. Then by transitivity, $a\rho_{uu}a'$, and thus $a = a'$, since $\mathbb{F}$ is a quasi-sheaf. Obviously, the same argument shows that $b\rho_{vu}a$ implies $a\rho_{uv}b$, and thus $\rho_{vu} = \rho_{uv}^{-1}$. It remains to show that both ρ_{uv}, ρ_{vu} are functions, and again we can check this only for ρ_{uv}. So assume $a_0\rho_{uv}a_1, a_0\rho_{uv}a_2$; then $a_1\rho_{vu}a_0$, and thus by transitivity, $a_1\rho_{vv}a_2$, which implies $a_1 = a_2$, by Definition 5.2.3. ∎

So, similarly to the case of Kripke sheaves (Section 3.5), we can factorise a quasi-sheaf $\mathbb{F}$ and obtain a quasi-sheaf over the partially ordered frame $F^\sim$:

Lemma 5.5.26 *Let $\mathbb{F} = (F, D, \rho)$ be the same as in Lemma 5.5.25, with $F = (W, R)$ and consider the corresponding frame of individuals $F^1 = (D^1, R^1)$; let $\pi : F^1 \twoheadrightarrow F$ be the associated p-morphism. Let $F^\sim = (D^\sim, R^\sim)$, $(F^1)^\sim = ((D^1)^\sim, (R^1)^\sim)$ be their skeletons (Definition 1.3.39). Then*

(1) The map $\pi^\sim : a^\sim \mapsto \pi(a)^\sim$ is a p-morphism (and thus defines a Kripke bundle $\mathbb{F}^\sim$).

(2) $\mathbb{F}^\sim$ is a Kripke quasi-sheaf.

(3) There exists a p=-morphism $\mathbb{F} \twoheadrightarrow \mathbb{F}^\sim$.

Proof (i) π is well-defined. In fact, $a \approx_{R^1} b$ only if $\pi(a) \approx_R \pi(b)$, due to the monotonicity of π.

Next, $a^\sim(R^1)^\sim b^\sim$ iff aR^1b (by Definition 1.3.39), only if $\pi(a)R\pi(b)$ (since π is p-morphism), iff $\pi(a)^\sim R^\sim \pi(b)^\sim$ (by Definition 1.3.39).

To check the lift property for $\pi^\sim$, suppose $\pi^\sim R^\sim v^\sim$, i.e. $\pi(a)Rv$, by Definition 1.3.39. Since π is a p-morphism, there exists b such that aR^1b and $\pi(a) = v$, and thus $a^\sim(R^1)^\sim b^\sim$, $v^\sim = \pi(b)^\sim = \pi^\sim(b^\sim)$.

(ii) Suppose $\pi^\sim(a^\sim) = \pi^\sim(b^\sim)$, i.e. $\pi(a) \approx_{R^1} \pi(b)$ and also $(a^\sim)(R^1)^\sim(b^\sim)$, i.e. aR^1b. To obtain $a^\sim = b^\sim$, we have to show that bR^1a.

By the lift property, there exists a' such that bR^1a' and $\pi(a') = \pi(a)$. But then aR^1a' by transitivity, and since $\mathbb{F}$ is a Kripke quasi-sheaf, it follows that $a' = a$. Hence bR^1a.

(iii) Let us show that $(f_0, f_1) : \mathbb{F} \longrightarrow^= \mathbb{F}^\sim$, where $f_0(u) := u^\sim$, $f^1(a) := a^\sim$.

In fact, these maps are propositional p-morphisms, by Lemma 1.3.40, so it remains to show that (f_0, f_1) is a fibrewise bijection, i.e. that f_1 restricted to D_u is a bijection.

To show the injectivity, suppose $\pi(a) = \pi(b) = u$, but $a \neq b$. Then $a^\sim \neq b^\sim$, since in Kripke quasi-sheaves different individuals in the same fibre are not related.

For the surjectivity, note that an arbitrary element of $(\pi^\sim)^{-1}(u^\sim)$ is of the form $b^\sim$, with $u^\sim = \pi^\sim(b^\sim) = \pi(b)^\sim$. Let us find $a \in D_u$ such that $a^\sim = b^\sim$.

Let $v := \pi(b)$; then $u \approx_R v$, and by Lemma ρ_{vu} is a bijection between D_u and D_v, whose converse is ρ_{uv}. Now take $a := \rho_{vu}$; it follows that $a \approx_{R^1} b$, and $a \in D_u$. ∎

Definition 5.5.27 $\mathbb{F}^\sim$ *is called the* skeleton *of* $\mathbb{F}$.

So we obtain an analogue of Lemma 3.7.18:

Lemma 5.5.28 $\mathbf{IL}^{(=)}(\mathbb{F}) = \mathbf{IL}^{(=)}(\mathbb{F}^{\sim})$ *for a Kripke quasi-sheaf* $\mathbb{F}$.

Proof The inclusion $\mathbf{IL}^{(=)}(\mathbb{F}) \subseteq \mathbf{IL}^{(=)}(\mathbb{F}^{\sim})$ follows from Lemma 5.5.26 and Proposition 5.5.6. To prove the converse, suppose $M = (\mathbb{F}, \varphi)$ is an intuitionistic model, A is an intuitionistic sentence and $M \not\Vdash A$, and let us show $\mathbb{F}^{\sim} \not\Vdash A$.

Let $f : \mathbb{F} \twoheadrightarrow \mathbb{F}^{\sim}$ (Lemma 5.5.26), and let us define a valuation ψ in $\mathbb{F}^{\sim}$ such that

$$\psi^+(P) = \{f \cdot \mathbf{a} \mid \mathbf{a} \in \varphi^+(P)\}.$$

for any predicate letter P. Then ψ is an intuitionistic valuation. In fact, suppose $\mathbf{c} = f \cdot \mathbf{a} \in \psi^+(P)$, $\mathbf{c}(R^{\sim})^n\mathbf{d}$, and let us show that $\mathbf{d} \in \psi^+(P)$.

By Lemma 5.4.4, there exists $\mathbf{b} \in R^n(\mathbf{a})$ such that $\mathbf{d} = f \cdot \mathbf{b}$ (Fig. 5.6). Then from $\mathbf{a}R^n\mathbf{b}$ we obtain $\mathbf{b} \in \varphi^+(P)$. Thus by definition, $\mathbf{d} \in \psi^+(P)$.

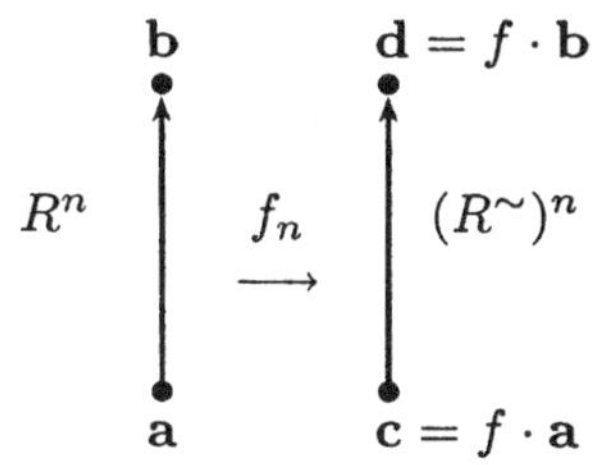

Figure 5.6.

So ψ is intuitionistic, and we also have $f : (\mathbb{F}, \varphi) \longrightarrow^{=} (\mathbb{F}^{\sim}, \psi)$.

In fact, by definition, $\mathbf{a} \in \varphi^+(P)$ only if $f \cdot \mathbf{a} \in \psi^+(P)$, and it remains to show the converse.

Suppose $f \cdot \mathbf{a} \in \psi^+(P)$, i.e. $f \cdot \mathbf{a} = f \cdot \mathbf{b}$ for some $\mathbf{b} \in \varphi^+(P)$. Then $\mathbf{a} \approx \mathbf{b}$ (by the definition of f), and thus $\mathbf{a} \in \varphi^+(P)$, since φ is intuitionistic.

By Lemma 5.4.6 we obtain (for a given sentence A and for any world u):

$$M, u \Vdash A \text{ iff } (\mathbb{F}^{\sim}, \psi), u^{\sim} \Vdash A.$$

Since $M \not\Vdash A$ by our assumption, we conclude that $\mathbb{F}^{\sim} \not\Vdash A$. ∎

Thus in the intuitionistic case it is sufficient to consider Kripke quasi-sheaves only over posets.

On the other hand, for Kripke bundles analogous reductions are not always possible, because the domains of equivalent worlds may be of different cardinality. So we have to consider Kripke bundles over arbitrary quasi-ordered sets. Lemma 5.5.16 shows that we can use only cones, but this does not simplify anything, because a quasi-ordered cone may still have several equivalent roots with non-equivalent domains.

5.6 Functor semantics

In Chapter 4 we defined validity in a presheaf over a locale and showed that the presheaf semantics is equivalent to algebraic semantics. Functor semantics is based on an alternative definition of validity in a presheaf (over an arbitrary category), which is mainly due to Ghilardi, Makkai and Reyes. In this section we define this semantics in a more general polymodal setting.

Let us begin with a notion of a precategory. This is actually a coloured multigraph, or a labelled transition system. Similarly to a category (cf. 3.4.3), it consists of objects (or points) and morphisms (arrows), but a priori without composition, and without special identity morphisms. Every morphism is coloured.

Definition 5.6.1 *An N-*precategory *is a tuple $C = (\mathbf{X}, \mathbf{Y}, \mathbf{o}, \mathbf{t}, \mathbf{cr})$, where $\mathbf{X}$, $\mathbf{Y}$ are non-empty classes; $\mathbf{o}, \mathbf{t} : \mathbf{Y} \longrightarrow \mathbf{Y}$ and $\mathbf{cr} : \mathbf{Y} \longrightarrow \{1, \ldots, N\}$ are functions. $\mathbf{X}$ is called the class of* objects *(notation: $Ob\,C$), and $\mathbf{Y}$ is the class of* morphisms *(notation: $Mor\,C$). $\mathbf{o}(f), \mathbf{t}(f), \mathbf{cr}(f)$ are called respectively the* origin, *the* target, *and the* colour *of f. An i-*morphism *is a morphism of colour i. We also use the following notation:*

$Mor_i(C) := \mathbf{cr}^{-1}(i)$ *(the class of all i-morphisms),*
$C(u, v) := \{f \mid \mathbf{o}(f) = u,\ \mathbf{t}(f) = v\}$ *(the class of all morphisms from u to v),*
$C_i(u, v) := Mor_i(C) \cap C(u, v)$ *(the class of all i-morphisms from u to v).*

We shall usually consider small *precategories, where $\mathbf{X}$, $\mathbf{Y}$ are sets. In this case the Kripke frame $(W, R_1, \ldots, R_N)$ is called the* frame representation *of C (notation: $FR(C)$) if $W = Ob\,C$ and $R_i = \{(u, v) \mid C_i(u, v) \neq \varnothing\}$.*

Note that a 1-precategory is nothing but a multigraph. It is also clear that a category is a 1-precategory expanded by identity morphisms and composition of consecutive morphisms (cf. Section 3.6); its frame representation is an **S4**-frame.

Every 1-precategory C can be extended to a category C^*; this construction is similar to reflexive transitive closure of a binary relation. Namely, morphisms in C^* are paths (sequences of consecutive morphisms) in C:

$$u_0 \xrightarrow{\alpha_1} u_1 \xrightarrow{\alpha_2} u_2 \longrightarrow \ldots \longrightarrow u_{n-1} \xrightarrow{\alpha_n} u_n$$

(where $u_0, \ldots, u_n$ may be not all different), and we also add all identity morphisms 1_u (as new).

The composition of morphisms in C^* is defined as the join of sequences:

$$(u \xrightarrow{\alpha_1} \ldots \xrightarrow{\alpha_n} v) \circ (v \xrightarrow{\beta_1} \ldots \xrightarrow{\beta_n} w) := u \xrightarrow{\alpha_1} \ldots \xrightarrow{\beta_n} w$$

and of course we define

$$1_u \circ \varphi := \varphi \circ 1_v := \varphi$$

for any $\varphi \in C^*(u, v)$.

If a (small) category C has a single object, then $Mor\,C^*$ is the free monoid generated by the set $Mor\,C$.

Definition 5.6.2 *Let $\mathcal{C}$ be a (small) N-precategory with the set of objects W. A $\mathcal{C}$-*preset *is a triple $\mathbb{F} = (\mathcal{C}, D, \rho)$, in which $D = (D_u \mid u \in W)$ is a system of domains on W and $\rho = (\rho_\alpha \mid \alpha \in Mor\,\mathcal{C})$ is a family of functions indexed by morphisms of $\mathcal{C}$, such that $\rho_\alpha : D_u \longrightarrow D_v$ for $\alpha \in \mathcal{C}(u, v)$.*

The objects of $\mathcal{C}$ are called possible worlds *of $\mathbb{F}$, and the elements of D_u are called* individuals *at u. If $d \in D_u$, $\alpha \in \mathcal{C}(u, v)$, the image $\rho_\alpha(d)$ is called the* α-inheritor *of the individual d in the world v.*

Note that an individual $d \in D_u$ can have different α-inheritors in the same world v for different morphisms α.

Definition 5.6.3 *Let $\mathcal{C}$ be a category based on a precategory $\mathcal{C}_0$, i.e. $\mathcal{C} = (\mathcal{C}_0, \circ, 1)$. A $\mathcal{C}_0$-preset $\mathbb{F}$ is called a* $\mathcal{C}$-set *(more precisely, an* intuitionistic $\mathcal{C}$-set*) if it preserves composition and identity morphisms:*

(iv) $\rho_{\alpha\circ\beta} = \rho_\alpha \circ \rho_\beta$;

(v) $\rho_{1_u} = id_{D_u}$;

for any $\alpha, \beta \in Mor\,\mathcal{C}$, $u \in Ob\,\mathcal{C}$.

Let also give an alternative definition of $\mathcal{C}$-presets.

Definition 5.6.4 *A* prefunctor $\mathbf{F} : \mathcal{C} \rightsquigarrow \mathcal{C}'$ *from an N-precategory $\mathcal{C}$ to an N-precategory $\mathcal{C}'$ is a map sending objects of $\mathcal{C}$ to objects of $\mathcal{C}'$ and morphisms of $\mathcal{C}$ to morphisms of $\mathcal{C}'$*[6] *and preserving colours, origins and targets: if $f \in \mathcal{C}_i(u, v)$, then $\mathbf{F}(f) \in \mathcal{C}'_i(\mathbf{F}(u), \mathbf{F}(v))$.*

We also extend this definition to the case when $\mathcal{C}$ is an N-precategory and $\mathcal{C}'$ is a 1-precategory: if $f \in \mathcal{C}_i(u, v)$, then $\mathbf{F}(f) \in \mathcal{C}'(\mathbf{F}(u), \mathbf{F}(v))$.

So we see that a $\mathcal{C}$-preset $(\mathcal{C}, D, \rho)$ is nothing but a prefunctor $\mathbf{F} : \mathcal{C} \rightsquigarrow SET$ such that $\mathbf{F}(u) = D_u$, $\mathbf{F}(\alpha) = \rho_\alpha$. Conversely, every prefunctor $\mathbf{F} : \mathcal{C} \rightsquigarrow SET$ such that every $\mathbf{F}(u)$ is non-empty, can be considered as a preset.

On the other hand, a preset $\mathbf{F} = (\mathcal{C}, D, \rho)$ is also associated with the *precategory of individuals* $\mathcal{C}^1(\mathbf{F})$ such that

$$Ob\,\mathcal{C}^1(\mathbf{F}) := D^1 := \bigcup_{u \in Ob\mathcal{C}} D_u,$$
$$\mathcal{C}^1(\mathbf{F})_i(a, b) := \{(a, b, \alpha) \mid b = \rho_\alpha(a),\ \alpha \in Mor_i\mathcal{C}\}.$$

We also obtain an *etale prefunctor* $\mathbf{E} : \mathcal{C}^1(\mathbf{F}) \rightsquigarrow \mathcal{C}$ such that $\mathbf{E}(a) = u$ whenever $a \in D_u$, and $\mathbf{E}(a, b, \alpha) = \alpha$.

Proposition 5.6.5

(1) Let $\mathbf{F} : \mathcal{C} \rightsquigarrow SET$ be a $\mathcal{C}$-preset, and let $\mathbf{E} : \mathcal{C}^1(\mathbf{F}) \rightsquigarrow \mathcal{C}$ be the corresponding etale prefunctor. Then $\mathbf{E}$ has the following properties

[6] We may assume that $Ob\,\mathcal{C} \cap Mor\,\mathcal{C} = \varnothing$; otherwise, $\mathcal{C}$ can be replaced with an isomorphic category having this property.

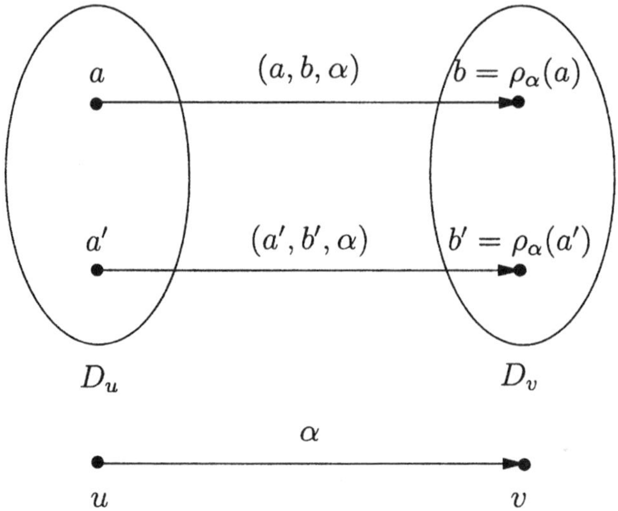

Figure 5.7. Etale prefunctor.

(i) $\forall u \in Ob\,\mathcal{C}\ \exists a \in Ob\,\mathcal{C}^1(\mathbf{F})\ \mathbf{E}(a) = u$ *(surjectivity on objects);*

(ii) $\forall a \in Ob\,\mathcal{C}^1(\mathbf{F})\ \forall \alpha \in Mor\,\mathcal{C}\ (\mathbf{E}(a) = \mathbf{o}(\alpha) \Rightarrow \exists! f\ (\mathbf{F}(f) = \alpha\ \&\ \mathbf{o}(f) = a))$ *(the unique lift property).*

(2) Let $\mathbf{G} : \mathcal{C}' \rightsquigarrow \mathcal{C}$ *be a prefunctor which is surjective on objects and has the unique lift property. Then* $\mathcal{C}'$ *is isomorphic to* $\mathcal{C}^1(\mathbf{F})$ *for some* $\mathcal{C}$*-preset* $\mathbf{F}$.

Proof (1) (i) $\mathbf{F}$ is surjective on objects since every set $\mathbf{F}(u)$ is non-empty.

(ii) Suppose $\alpha \in \mathcal{C}(u, v)$ and $a \in D_u$. Then $\mathbf{E}(f) = \alpha\ \&\ \mathbf{o}(f) = a$ holds iff $f = (a, b, \alpha)$, where $b = \rho_\alpha(a)$.

(2) For an object u put $D_u := \mathbf{F}(u) := \mathbf{G}^{-1}[u]$.

For a morphism $\alpha \in \mathcal{C}(u, v)$, put

$$\mathbf{F}(\alpha) := \rho_\alpha := \{(a, b) \mid \exists f\,(\mathbf{o}(f) = a\ \&\ \mathbf{t}(f) = b\ \&\ \mathbf{G}(f) = \alpha)\}.$$

The precategories $\mathcal{C}'$ and $\mathcal{C}^1(\mathbf{F})$ are isomorphic. In fact, they have the same objects, and there is a bijection between $\mathcal{C}^1(\mathbf{F})(a, b)$ and $\mathcal{C}'(a, b)$ sending (a, b, α) to the the unique lift of α beginning at a, i.e. to f such that $\mathbf{G}(f) = \alpha$, $\mathbf{o}(f) = a$. ∎

If $\mathcal{C}$ is category, $\mathbf{F} : \mathcal{C} \rightsquigarrow SET$ is a $\mathcal{C}$-set, then we can also make $\mathcal{C}^1(\mathbf{F})$ a category by putting

$$(a, b, \alpha) \circ (b, c, \beta) := (a, c, \alpha \circ \beta);\ 1_a := (a, a, 1_u) \text{ for } a \in D_u.$$

Lemma 5.6.6 *If* $\mathbf{F}$ *is a* $\mathcal{C}$*-set, then* $\mathbf{E} : \mathcal{C}^1(\mathbf{F}) \rightsquigarrow \mathbf{F}$ *is a functor.*

Proof In fact, $\mathbf{E}((a, b, \alpha) \circ (b, c, \beta)) = \alpha \circ \beta = \mathbf{E}(a, b, \alpha) \circ \mathbf{E}(b, c, \beta)$, and $\mathbf{E}(1_a) = 1_{\mathbf{E}(a)}$. ∎

Definition 5.6.7 *For a propositional N-frame $F = (W, R_1, \dots, R_N)$ we define the* associated N-precategory $\mathcal{C} = \mathrm{Cat}_0\mathrm{F}$ *as follows:*

$$Ob\,\mathcal{C} := W,$$

$$\mathcal{C}_i(u, v) := \begin{cases} \{(i, u, v)\} & \text{if } uR_iv, \\ \varnothing & \text{otherwise.} \end{cases}$$

From definitions it is clear that every Kripke sheaf based on a frame F is a Cat_0F-preset; if F is an **S4**-frame, we obtain a $\mathrm{Cat}F$-preset (where $\mathrm{Cat}F$ is the category defined in Section 3.5.3).

Remark 5.6.8 Note that if F is not an **S4**-frame, it may happen that a Cat_0F-preset is not a Kripke sheaf, because it may not satisfy the coherence conditions $(1)^*$, $(2)^*$ from Lemma 3.6.3.

Definition 5.6.9 *(cf. Definitions 5.2.7, 5.2.8). A (modal)* valuation *in a $\mathcal{C}$-preset* $\mathbf{F}$ *is a valuation in the system of domains $D = (\mathbf{F}(u) \mid u \in Ob\,\mathcal{C})$. A* model *over* $\mathbf{F}$ *is a pair $(\mathbf{F}, \xi)$, where ξ is a valuation in* $\mathbf{F}$.

The forcing relation $M, u \vDash A$ between a world $u \in W$ and a D_u-sentence (with equality) A is defined by induction:

- $M, u \vDash P_k^0$ *iff* $u \in \xi_u(P_k^0)$;
- $M, u \vDash P_k^m(\mathbf{a})$ *iff* $\mathbf{a} \in \xi_u(P_k^m)$ *(for $m > 0$)*;
- $M, u \vDash a = b$ *iff a equals b*;
- $M, u \not\vDash \bot$;
- $M, u \vDash B \vee C$ *iff* $(M, u \vDash B$ *or* $M, u \vDash C)$;
- $M, u \vDash B \wedge C$ *iff* $M, u \vDash B \,\&\, M, u \vDash C$;
- $M, u \vDash B \supset C$ *iff* $(M, u \not\vDash B$ *or* $M, u \vDash C)$;
- $M, u \vDash \Box_i B(\mathbf{a})$ *iff* $\forall v \in R_i(u)\ \forall \alpha \in \mathcal{C}_i(u, v)\ M, v \vDash B(\rho_\alpha \cdot \mathbf{a})$;
- $M, u \vDash \exists x A$ *iff* $\exists a \in D_u\ M, u \vDash [a/x]\, A$;
- $M, u \vDash \forall x A$ *iff* $\forall a \in D_u\ M, u \vDash [a/x]\, A$.

Here as usual in the $\Box$-case $B(\mathbf{a})$ means $[\mathbf{a}/\mathbf{x}]\, B$, and we assume that $FV(B) = \mathbf{x}$.

Note that the truth condition for $\Box_i[\mathbf{a}/\mathbf{x}]\, B$ obviously extends to the case when $FV(B) \subseteq \mathbf{x}$.

The following definition is similar to the case of Kripke bundles.

Definition 5.6.10 *A (modal) predicate formula A is called*

- true *in a* $\mathcal{C}$*-preset model* M *if its universal closure* $\bar{\forall}A$ *is true at every world of* M;
- valid *in a* $\mathcal{C}$*-preset* $\mathbf{F}$ *if* A *is true in every model over* $\mathbf{F}$;
- strongly valid *in a* $\mathcal{C}$*-preset* $\mathbf{F}$ *if all substitution instances of* A *are valid in* $\mathbf{F}$.

We use the same signs as above: $\vDash$ for truth and validity, $\vDash^+$ for strong validity.

The following is a trivial consequence of the definitions:

Lemma 5.6.11 *For a* $\mathcal{C}$*-preset model* M *and a modal formula* $A(\mathbf{x})$

$$M \vDash A(\mathbf{x}) \text{ iff } \forall u \in M \, \forall \mathbf{a} \in D_u^n \;\; M, u \vDash A(\mathbf{a}).$$

For a $\mathcal{C}$-preset $\mathbf{F}$ let

$$\mathbf{ML}_-^{(=)}(\mathbf{F}) := \{A \in MF_N^{(=)} \mid \mathbf{F} \vDash A\},$$

$$\mathbf{ML}^{(=)}(\mathbf{F}) := \{A \in MF_N^{(=)} \mid \mathbf{F} \vDash^+ A\}.$$

The proof of the following soundness result is postponed until Section 5.13 (Proposition 5.13.2).

Proposition 5.6.12 *For a* $\mathcal{C}$*-preset* $\mathbf{F}$:

(1) $\mathbf{ML}^{(=)}(\mathbf{F})$ *is a modal predicate logic;*

(2) $\mathbf{ML}^{(=)}(\mathbf{F}) = \{A \in MF_N^{(=)} \mid \forall m \; \mathbf{F} \vDash A^m\}$.

Definition 5.6.13 *For* N*-modal predicate logics we introduce the* functor semantics $\mathcal{FS}_N$ *generated by presets over* N*-precategories.*

Similarly to Kripke bundles, we introduce levels for $\mathcal{C}$-presets.

Definition 5.6.14 *Let* $\mathbf{F} = (\mathcal{C}, D, \rho)$ *be a* $\mathcal{C}$*-preset over an* N*-precategory* $\mathcal{C}$*. Let us define relations* R_i^n *on* D^n *for* $n > 0, 1 \leq i \leq N$:

$$\mathbf{a} R_i^n \mathbf{b} \text{ iff } \exists \gamma \in Mor_i \mathcal{C} \;\; \rho_\gamma \cdot \mathbf{a} = \mathbf{b}.$$

Also let $R_i^0 := R_i$, $F_n := (D^n, R_1^n, \ldots, R_N^n)$.

The nth level *of* $\mathbf{F}$ *is the frame* $F_n := (D^n, R_1^n, \ldots, R_N^n)$*. We again abbreviate* R_1^n *to* R^n *in the 1-modal case.*

From the definition of forcing 5.6.9 we readily obtain

Lemma 5.6.15 *Let* $\mathcal{C}$ *be an* N*-precategory,* $M = (\mathbf{F}, \xi)$ *a model over a* $\mathcal{C}$*-preset* $\mathbf{F} = (\mathcal{C}, D, \rho)$*. Let* B *be an* N*-modal formula with* $FV(B) \subseteq \mathbf{x}$, $|\mathbf{x}| = n$*. Then for any* $u \in F$ *and* $\mathbf{a} \in D_u^n$

$$M, u \vDash \Box_i B(\mathbf{a}) \text{ iff } \forall v \in R_i(u) \; \forall \mathbf{b} \in D_v^n \; (\mathbf{a} R_i^n \mathbf{b} \Rightarrow M, v \vDash B(\mathbf{b})).$$

Lemma 5.6.16 *If $\mathcal{C}$ is a category and* $\mathbf{F}$ *is a* $\mathcal{C}$*-set, then every* F_n *is an intuitionistic propositional frame.*

Proof An exercise. ■

Definition 5.6.17 *A valuation* ξ *in a* $\mathcal{C}$*-set* $\mathbf{F}$ *is called* intuitionistic *if every set* $\xi^+(P_k^n) = \bigcup\{\xi_u(P_k^n) \mid u \in W\}$ *is stable in* F_n*, i.e. if* $R^n(\xi^+(P_k^n)) \subseteq \xi^+(P_k^n)$*. The model* $(\mathbb{F}, \xi)$ *is also called* intuitionistic.

Definition 5.6.18 *Let* M *be an intuitionistic model over a* $\mathcal{C}$*-set* $\mathbf{F} = (\mathcal{C}, D, \rho)$*,* $u \in M$*,* A *an intuitionistic* D_u*-sentence. Then we put*

$$M, u \Vdash A := M, u \models A^T.$$

Lemma 5.6.19 *Under the conditions of Definition 5.6.18*

(1) $M, u \Vdash B$ *iff* $M, u \models B$ *(for* B *atomic);*

(2) $M, u \Vdash B \wedge C$ *iff* $(M, u \Vdash B \,\&\, M, u \Vdash C)$*;*

(3) $M, u \Vdash B \vee C$ *iff* $(M, u \Vdash B$ *or* $M, u \Vdash C)$*;*

(4) $M, u \Vdash B(\mathbf{a}) \supset C(\mathbf{a})$ *iff*
$\forall v \in R(u) \forall \mu \in \mathcal{C}(u, v) \, (M, v \Vdash B(\rho_\mu \cdot \mathbf{a}) \Rightarrow M, v \Vdash C(\rho_\mu \cdot \mathbf{a})$*;*

(5) $M, u \Vdash \forall x B(x, \mathbf{a})$ *iff*
$\forall v \in R(u) \forall \mu \in \mathcal{C}(u, v) \forall c \in D_v \; M, v \models B(c, \rho_\mu \cdot \mathbf{a})$*;*

(6) $M, u \Vdash \exists x B$ *iff* $\exists a \in D_u \; M, u \Vdash [a/x] B$*;*

(7) $M, u \Vdash \neg B(\mathbf{a})$ *iff* $\forall v \in R(u) \; \forall \mu \in \mathcal{C}(u, v) \, M, v \nVdash B(\rho_\mu \cdot \mathbf{a})$*;*

(8) $M, u \Vdash a \neq b$ *iff* a *does not equal* b*.*

Proof Similar to 5.5.4, with obvious changes. ■

Lemma 5.6.20 *Let* M *be the same as in Lemma 5.6.19. Then for any* D^1*-sentence* $A(\mathbf{a})$

$$M \Vdash A(\mathbf{a}) \;\&\; \mathbf{a} R^n \mathbf{b} \Rightarrow M \Vdash A(\mathbf{b}).$$

Definition 5.6.21 *Let* $\mathbf{F}$ *be a* $\mathcal{C}$*-set,* M *a model over* $\mathbf{F}$*. The* pattern *of* M *is the model* M_0 *over* $\mathbf{F}$ *such that for any* $u \in \mathbf{F}$ *and any atomic* D_u*-sentence* A

$$M_0, u \models A \text{ iff } M, u \models \Box A.$$

The model M_0 is intuitionistic, and similarly to Lemma 5.5.7, we obtain

Lemma 5.6.22 *For any* $u \in \mathbf{F}$*, for any* D_u*-sentence* A

$$M_0, u \Vdash A \text{ iff } M, u \models A^T.$$

Lemma 5.6.23 *Let M be an intuitionistic model over a C-set with the accessibility relation R, $A(\mathbf{x})$ an intuitionistic formula, $|\mathbf{x}| = n$. Then for any $u \in M$*

$$M, u \Vdash \forall \mathbf{x} A(\mathbf{x}) \text{ iff } \forall v \in R(u) \forall \mathbf{a} \in D_v^n \; M, u \Vdash A(\mathbf{a}).$$

Proof Similar to Lemma 5.5.8 using soundness (5.6.12). ■

Definition 5.6.24 *A formula $A \in IF^{(=)}$ is called*

- true *in an intuitionistic C-set model M (notation: $M \Vdash A$) if $\bar{\forall} A$ is true at every world of M;*
- valid *in a C-set $\mathbf{F}$ (notation: $\mathbf{F} \Vdash A$) if it is true in all intuitionistic models over $\mathbf{F}$;*
- strongly valid *in a C-set $\mathbf{F}$ (notation: $\mathbf{F} \Vdash^+ A$) if all its $IF^{(=)}$-substitution instances are valid in $\mathbf{F}$.*

Lemma 5.6.25 *For an intuitionistic C-set model M and an intuitionistic formula A*

$$M \Vdash A(\mathbf{x}) \quad \text{iff} \quad \forall u \in M \forall \mathbf{a} \in D_u^n \; M, u \Vdash A(\mathbf{a}).$$

Proposition 5.6.26 *Let $\mathbf{F}$ be a C-set, $A \in IF^=$. Then*

(1) $\mathbf{F} \Vdash A$ *iff* $\mathbf{F} \vDash A^T$.

(2) The following three assertions are equivalent:

(a) $\mathbf{F} \Vdash^+ A$;

(b) $\forall m \; \mathbf{F} \Vdash A^m$;

(c) $\mathbf{F} \vDash^+ A^T$.

Proof Cf. Proposition 5.5.12. ■

Proposition 5.6.27 *For a C-set $\mathbf{F}$ the set*

$$\mathbf{IL}^{(=)}(\mathbf{F}) := \{A \in IF^{(=)} \mid \mathbb{F} \Vdash^+ A\}$$

is a superintuitionistic predicate logic (with or without equality), and

$$\mathbf{IL}^{(=)}(\mathbf{F}) =^T \mathbf{ML}^{(=)}(\mathbf{F}).$$

Definition 5.6.28 *The set $\mathbf{IL}^{(=)}(\mathbf{F})$ is called the* superintuitionistic logic of $\mathbf{F}$.

5.7 Morphisms of presets

Now let us define truth-preserving morphisms of presets.

Definition 5.7.1 *(Cf. Proposition 5.6.5.) A prefunctor $\Phi : \mathcal{D} \rightsquigarrow \mathcal{E}$ has the* lift property *if*

$$\forall a \in Ob\mathcal{D}\, \forall \mu \in Mor\mathcal{E}\, (\Phi(a) = \mathbf{o}(\mu) \Rightarrow \exists f \in Mor\mathcal{D}\, \Phi(f) = \mu).$$

The reader can see that this is a generalisation of the lift property for propositional frame morphisms.

Definition 5.7.2 *Let $\mathbf{F}$ be a $\mathcal{C}$-preset, $\mathbf{F}'$ a $\mathcal{C}'$-preset. A pair $\gamma = (\Phi, \Psi)$ is called an* =-morphism *from $\mathbf{F}$ to $\mathbf{F}'$ (notation: $\gamma : \mathbf{F} \longrightarrow^{=} \mathbf{F}'$) if*

(1) $\Phi : \mathcal{C} \rightsquigarrow \mathcal{C}'$ is a prefunctor with the lift property;

(2) $\Psi = (\Psi_u \mid u \in Ob\mathcal{C})$ is a family of bijections

$$\Psi_u : \mathbf{F}(u) \longrightarrow \mathbf{F}'(\Phi(u));$$

(3) Ψ respects morphisms, i.e. for any $\mu \in \mathcal{C}(u, v)$ the following diagram commutes:

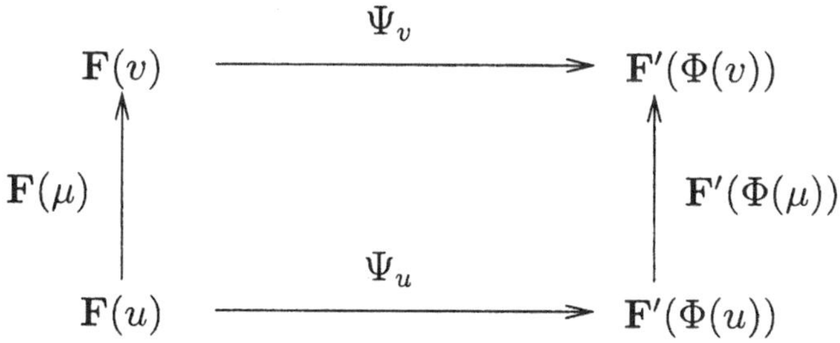

γ is called a p=-morphism *if it also has the property*

(4) Φ is surjective on objects.

The above conditions (2), (3) mean that Ψ is a functor morphism ('natural transformation') from $\mathbf{F}$ to $\mathbf{F}' \cdot \Phi$.

Definition 5.7.3 *For $\gamma = (\Phi, \Psi) : \mathbf{F} \longrightarrow^{=} \mathbf{F}'$, $n \geq 0$, we define the map $\gamma_n : F_n \longrightarrow F'_n$ as follows:*

$$\gamma_0(u) := \Phi(u), \;\; \gamma_n(\mathbf{a}) := \Psi_u \cdot \mathbf{a}$$

whenever $\mathbf{a} \in \mathbf{F}(u)^n$, $n > 0$.

Lemma 5.7.4 *$\gamma_n : F_n \longrightarrow^{=} F'_n$.*

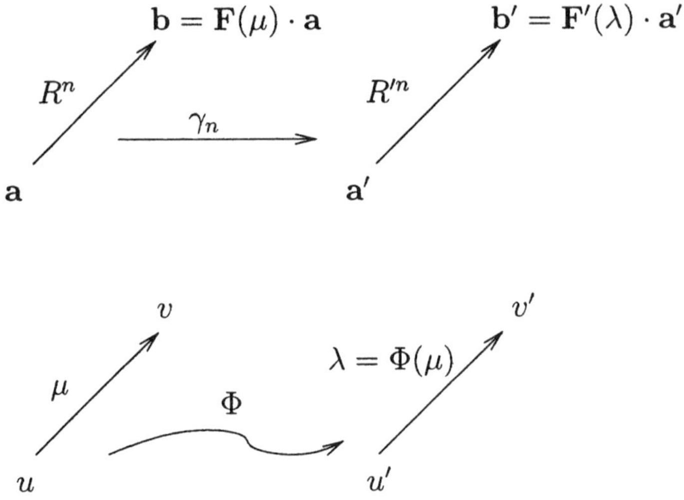

Figure 5.8.

Proof We consider only the case $n > 0$, and assume $N = 1$ to simplify notation. To prove the monotonicity, suppose $\mathbf{a}R^n\mathbf{b}$, i.e.

$$\exists \mu \in \mathcal{C}(u,v)\, \forall i\ b_i = \mathbf{F}(\mu)(a_i).$$

Then by 5.7.2(3),

$$\Psi_v(b_i) = \Psi_v(\mathbf{F}(\mu)(a_i)) = \mathbf{F}'(\Phi(\mu))(\Psi_u(a_i)).$$

Thus

$$\Psi_v \cdot \mathbf{b} = \mathbf{F}'(\Phi(\mu)) \cdot (\Psi_u \cdot \mathbf{a}),$$

and so by definition, $\gamma_n(\mathbf{a})(R')^n\gamma_n(\mathbf{b})$.

To check the lift property, suppose $\mathbf{a} \in F_n$, $\mathbf{a}' = \Psi_u \cdot \mathbf{a}$, $\mathbf{a}'R'^n\mathbf{b}'$, $\mathbf{a}' \in \mathbf{F}'(u')$, $\mathbf{b}' \in \mathbf{F}'(v')$. Then

$$\exists \lambda \in \mathcal{C}'(u',v')\ \mathbf{b}' = \mathbf{F}'(\lambda) \cdot \mathbf{a}'.$$

Since Φ has the lift property, there exist $v \in Ob\mathcal{C}$ and $\mu \in \mathcal{C}(u,v)$ such that $\lambda = \Phi(\mu)$. Then put

$$\mathbf{b} := \mathbf{F}(\mu) \cdot \mathbf{a}.$$

By definition, $\mathbf{a}R^n\mathbf{b}$, and we also claim that $\mathbf{b}' = \gamma_n(\mathbf{b}) = \Psi_v \cdot \mathbf{b}$ (Fig. 5.8).

In fact, let $\mathbf{b}'' = \Psi_v \cdot \mathbf{b}$. By 5.7.1, the following diagram commutes:

$$\begin{array}{ccc} b_i & \xrightarrow{\Psi_v} & b'_i \\ {\scriptstyle F(\mu)}\uparrow & & \uparrow{\scriptstyle F'(\lambda)} \\ a_i & \xrightarrow{\Psi_u} & a'_i \end{array}$$

Thus $\mathbf{b}'' = \mathbf{F}'(\lambda) \cdot \mathbf{a}' = \mathbf{b}'$. ■

Definition 5.7.5[7] *Let $\gamma : \mathbf{F} \longrightarrow^{=} \mathbf{G}$ be an $=$-morphism of presets. γ is called an* $=$-morphism from a model $M = (\mathbf{F}, \xi)$ to $M' = (\mathbf{F}', \xi')$ *(notation: $f : M \longrightarrow^{=} M'$) if*

$$M, u \vDash P(\mathbf{a}) \iff M', \gamma_0(u) \vDash P(\gamma_n(\mathbf{a}))$$

for any $P \in PL^n$, $\mathbf{a} \in \mathbf{F}(u)^n$, $n > 0$; and also

$$M, u \vDash P \iff M', \gamma_0(u) \vDash P$$

for any $P \in PL^0$.

Lemma 5.7.6 *If $\gamma : M \longrightarrow^{=} M'$ for preset models $M = (\mathbf{F}, \xi)$, $M' = (\mathbf{F}', \xi')$, then for any $u \in M$ and for any $\mathbf{F}(u)$-sentence A*

$$M, u \vDash A \text{ iff } M', \gamma_0(u) \vDash \gamma_1 \cdot A.$$

If the models are intuitionistic, then for any intuitionistic $\mathbf{F}(u)$-sentence A

$$M, u \Vdash A \text{ iff } M', \gamma_0(u) \Vdash \gamma_1 \cdot A.$$

Proof By an obvious modification of the proofs of 5.4.6, 5.5.15 based on Lemma 5.7.4; an exercise for the reader. ■

Lemma 5.7.7[8] *Let $\gamma : \mathbf{F} \longrightarrow^{=} \mathbf{G}$, and let M' be a model over $\mathbf{G}$. Then there exists a model M over $\mathbf{F}$ such that $\gamma : M \longrightarrow^{=} M'$.*

Proof The same as for Lemma 5.4.7. ■

Proposition 5.7.8 *If there exists a p=-morphism $\mathbf{F} \twoheadrightarrow^{=} \mathbf{F}'$, then $\mathbf{ML}^{=}_{-}(\mathbf{F}) \subseteq \mathbf{ML}^{=}_{-}(\mathbf{F}')$ and thus, $\mathbf{ML}^{=}(\mathbf{F}) \subseteq \mathbf{ML}^{=}(\mathbf{F}')$. Similarly, for the intuitionistic case: $\mathbf{IL}^{=}_{-}(\mathbf{F}) \subseteq \mathbf{IL}^{=}_{-}(\mathbf{F}')$ and $\mathbf{IL}^{=}(\mathbf{F}) \subseteq \mathbf{IL}^{=}(\mathbf{F}')$.*

Proof Cf. Propositions 5.4.8, 5.5.16. ■

[7] Cf. Definition 5.4.5.
[8] Cf. 5.4.7.

Let us now construct inverse image morphisms.

Lemma 5.7.9[9] *Let* $\mathbf{F}$ *be a* $\mathcal{C}$*-preset,* $\mathbf{E} : \mathcal{C}' \rightsquigarrow \mathcal{C}$ *the corresponding etale prefunctor,* $\mathbf{G} : \mathcal{D} \rightsquigarrow \mathcal{C}$ *an arbitrary prefunctor with the lift property. Consider a precategory* $\mathcal{D}'$ *with the class of objects*

$$\{(u, a) \mid u \in Ob\mathcal{D},\ a \in Ob\mathcal{C}',\ \mathbf{G}(u) = \mathbf{E}(a)\}$$

and with

$$\mathcal{D}'((u, a), (v, b)) := \{(\mu, f) \mid \mu \in \mathcal{D}(u, v),\ f \in \mathcal{C}'(a, b),\ \mathbf{G}(\mu) = \mathbf{E}(f)\}.$$

Then

(1) There exists an etale prefunctor $\mathbf{E}' : \mathcal{D}' \rightsquigarrow \mathcal{D}$ *such that* $\mathbf{E}'(u, a) = u$, $\mathbf{E}'(\mu, f) = \mu$.

(2) There exists a prefunctor with the lift property. $\mathbf{G}' : \mathcal{D}' \rightsquigarrow \mathcal{C}'$ *such that* $\mathbf{G}'(u, a) = a$, $\mathbf{G}'(\mu, f) = f$.

(3) The following diagram commutes:

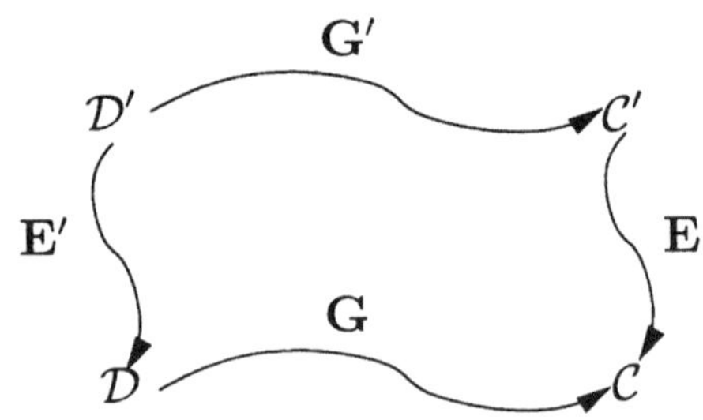

(4) If $\mathbf{F}'$ *is a* $\mathcal{D}$*-preset corresponding to* $\mathbf{E}'$*, then there exists* $(\mathbf{G}, \Psi) : \mathbf{F}' \longrightarrow^{=} \mathbf{F}$ *such that* $\Psi_u(u, a) = a$.

(5) If $\mathcal{C}$, $\mathcal{D}$ *are categories,* $\mathbf{F}$ *is a* $\mathcal{C}$*-set and* $\mathbf{G}$ *is a functor, then* $\mathcal{D}'$ *becomes a category with the composition*

$$(\mu, f) \circ (\nu, g) := (\mu \circ \nu, f \circ g)$$

and $\mathbf{F}'$ *becomes a* $\mathcal{D}$*-set.*

Proposition 5.7.10 *Every* $=$*-morphism of presets* $(\Phi, \Psi) : \mathbf{F}' \longrightarrow^{=} \mathbf{F}$ *is the composition of the inverse image morphism* $\Theta : \Phi_*\mathbf{F} \longrightarrow^{=} \mathbf{F}$ *and a prefunctor isomorphism (over the same precategory)* $\Xi : \mathbf{F}' \longrightarrow \Phi_*\mathbf{F}$.

Definition 5.7.11 *Let* $\mathcal{C}$, $\mathcal{D}$ *be* N*-precategories.* $\mathcal{D}$ *is called a* full subprecategory *of* $\mathcal{C}$ *(notation:* $\mathcal{D} \subseteq_0 \mathcal{C}$*) if*

[9] Cf. Lemma 5.4.9.

(1) $Ob\mathcal{D} \subseteq Ob\mathcal{C}$,

(2) $\forall u, v \in Ob\mathcal{D}\, \forall i \leq N\ \mathcal{C}_i(u,v) = \mathcal{D}_i(u,v)$.

If $\mathcal{C}$, $\mathcal{D}$ *are categories and* $\mathcal{D} \subseteq_0 \mathcal{C}$, *then* $\mathcal{D}$ *is called a* full subcategory *of* $\mathcal{C}$ *(notation:* $\mathcal{D} \subseteq \mathcal{C}$*) if*

(3) for any consecutive morphisms f, g *in* $\mathcal{D}$

$$f \circ g \ (in\ \mathcal{D}) = f \circ g \ (in\ \mathcal{C}).$$

Definition 5.7.12 *A full sub(pre)category* $\mathcal{D}$ *of a (pre)category* $\mathcal{C}$ *is called a* conic sub(pre)category *(notation:* $\mathcal{D} \sqsubseteq_{(0)} \mathcal{C}$*) if*

$$\forall u \in Ob\mathcal{D}\, \forall v\, (\mathcal{C}^*(u,v) \neq \varnothing \Rightarrow v \in Ob\mathcal{D}).$$

Here is a typical example.

Definition 5.7.13 *For a (pre)category* $\mathcal{C}$ *and* $u \in Ob\mathcal{C}$ *the* cone generated by u *(notation:* $\mathcal{C}{\uparrow}u$*) is defined as the full sub(pre)category with the class of objects* $\{v \mid \mathcal{C}^*(u,v) \neq \varnothing\}$.

Thus every object of $\mathcal{C}{\uparrow}u$ is accessible from u via a path of morphisms.

Definition 5.7.14 *Let* $\mathbf{F}$ *be a* $\mathcal{C}$*-preset,* $\mathcal{D} \subseteq_0 \mathcal{C}$*. The* restriction *of* $\mathbf{F}$ *to* $\mathcal{D}$ *(notation:* $\mathbf{F} \restriction \mathcal{D}$*) is the* $\mathcal{D}$*-preset taking the same values as* $\mathbf{F}$ *on objects and morphisms of* $\mathcal{D}$.

For $u \in Ob\mathcal{C}$ *we define the* cone of $\mathbf{F}$ generated by u*:* $\mathbf{F}{\uparrow}u := \mathbf{F} \restriction (\mathcal{C}{\uparrow}u)$.

Note that if $\mathbf{F}$ is a $\mathcal{C}$-set and $\mathcal{D} \subseteq \mathcal{C}$, then obviously, $\mathbf{F} \restriction \mathcal{D}$ is a $\mathcal{D}$-set.

Lemma 5.7.15 *Let* $\mathbf{F}$ *be a* $\mathcal{C}$*-preset,* $\mathcal{D} \subseteq_0 \mathcal{C}$*. Let* $\mathbf{J} : \mathcal{D} \rightsquigarrow \mathcal{C}$ *be the inclusion prefunctor (sending every object and every morphism of* $\mathcal{D}$ *to itself). Also let* $\Psi = (id_{\mathbf{F}(u)} \mid u \in Ob\mathcal{D})$*. Then*

$$(\mathbf{J}, \Psi) : \mathbf{F} \restriction \mathcal{D} \longrightarrow^{=} \mathbf{F}.$$

Proof Obviously, $\mathbf{J}$ has the lift property and Ψ respects morphisms. ∎

Lemma 5.7.16 *(1) For a* $\mathcal{C}$*-preset* $\mathbf{F}$

$$\mathbf{ML}^{(=)}(\mathbf{F}) = \bigcap \{\mathbf{ML}^{(=)}(\mathbf{F} \uparrow u) \mid u \in F\}.$$

(2) For a $\mathcal{C}$*-set* $\mathbf{F}$

$$\mathbf{IL}^{(=)}(\mathbf{F}) = \bigcap \{\mathbf{IL}^{(=)}(\mathbf{F} \uparrow u) \mid u \in F\}.$$

Proof Along the same lines as Lemmas 5.4.13, 5.5.17. ∎

Lemma 5.7.17 *(1) If* $\mathbf{F}$ *is a* $\mathcal{C}$*-preset,* $\mathcal{D} \sqsubseteq_0 \mathcal{C}$*, then*

$$\mathbf{ML}^{(=)}(\mathbf{F}) \subseteq \mathbf{ML}^{(=)}(\mathbf{F} \restriction \mathcal{D}).$$

(2) If $\mathbf{F}$ *is a* $\mathcal{C}$*-set,* $\mathcal{D} \sqsubseteq \mathcal{C}$*, then*

$$\mathbf{IL}^{(=)}(\mathbf{F}) \subseteq \mathbf{IL}^{(=)}(\mathbf{F} \restriction \mathcal{D}).$$

Proof Easy, from 5.7.16. ∎

5.8 Bundles over precategories

Let us now combine two generalisations of Kripke sheaves: Kripke bundles and presets.

Definition 5.8.1 *A* modal $\mathcal{C}$-bundle *over an N-precategory $\mathcal{C}$ is a triple $\mathbb{F} = (\mathcal{C}, D, \rho)$, in which*

- *$D = (D_u \mid u \in W)$ is a system of domains;*
- *$\rho = (\rho_\alpha \mid \alpha \in Mor\mathcal{C})$ is a family of relations parametrised by morphisms of $\mathcal{C}$;*
- *$\rho_\alpha \subseteq D_u \times D_v$, dom $\rho_\alpha = D_u$ for $\alpha \in \mathcal{C}(u, v)$.*

An intuitionistic $\mathcal{C}$-bundle *over a category $\mathcal{C}$ is a modal $\mathcal{C}$-bundle satisfying two extra conditions:*

(1) ρ_{1_u} is reflexive (on D_u) for $u \in W$,

(2) $\rho_\alpha \circ \rho_\beta \subseteq \rho_{\alpha\circ\beta}$,

cf. the corresponding conditions for intuitionistic Kripke bundles in Definition 5.2.2.

The notion of a $\mathcal{C}$-bundle generalises both Kripke bundles and $\mathcal{C}$-sets. In fact, $\mathcal{C}$-sets are just $\mathcal{C}$-bundles with *functions* ρ_α, and Kripke bundles based on a frame F are exactly Cat_0F-bundles (where Cat_0F is the precategory from Definition 5.6.7).

But it turns out that the semantics of $\mathcal{C}$-bundles is strongly equivalent (in the terminology of Section 2.16) to the semantics of $\mathcal{C}$-sets, as we show below. Hence it follows that the semantics of Kripke bundles $\mathcal{KB}$ is included in the functor semantics $\mathcal{FS}$. But actually, $\mathcal{FS}$ is stronger than $\mathcal{KB}$, as we shall see later on (Section 5.10).

To define forcing in $\mathcal{C}$-bundles, we first extend Definition 5.3.2 to this case:

Definition 5.8.2 *For a $\mathcal{C}$-bundle $\mathbb{F} = (\mathcal{C}, D, \rho)$ put*

$$\mathbf{a}R_i^n\mathbf{b} \textit{ iff } \quad \mathbf{a} \textit{ sub } \mathbf{b} \;\&\; \exists f \in Mor_i\mathcal{C}\ \forall j\, a_j \rho_f b_j$$

for $\mathbf{a},\ \mathbf{b} \in D^n,\ n > 0$.

Also put $R_i^0 := R_i$, the corresponding relation in $FR(\mathcal{C})$.

Definition 5.8.3 *A (modal)* valuation *in a $\mathcal{C}$-bundle $\mathbb{F} = (\mathcal{C}, D, \rho)$ is a valuation in the system of domains D. A* model *over $\mathbb{F}$ is a pair $(\mathbb{F}, \xi)$, where ξ is a valuation in $\mathbb{F}$.*

Now let us give a preliminary definition of forcing.

Definition 5.8.4 *(cf. Definitions 5.2.7, 5.2.8, 5.6.9). The forcing relation $M, u \vDash A$ between a world $u \in W$ and a D_u-sentence (with equality) A is defined by induction:*

- $M, u \vDash P_k^0$ *iff* $u \in \xi_u(P_k^0)$;
- $M, u \vDash P_k^m(\mathbf{a})$ *iff* $\mathbf{a} \in \xi_u(P_k^m)$ *(for* $m > 0$*)*;
- $M, u \vDash a = b$ *iff* a *equals* b;
- $M, u \nvDash \bot$;
- $M, u \vDash B \vee C$ *iff* $(M, u \vDash B$ *or* $M, u \vDash C)$;
- $M, u \vDash B \wedge C$ *iff* $M, u \vDash B \,\&\, M, u \vDash C$;
- $M, u \vDash B \supset C$ *iff* $(M, u \nvDash B$ *or* $M, u \vDash C)$;
- $M, u \vDash \Box_i B(\mathbf{a})$ *iff* $\forall v \in R_i(u)\ \forall \mathbf{b} \in D_v^n\ (\mathbf{a} R_i^n \mathbf{b} \Rightarrow M, v \vDash B(\mathbf{b}))$ *(for* $|\mathbf{a}| = n$*)*;
- $M, u \vDash \exists x A$ *iff* $\exists a \in D_u\ M, u \vDash [a/x]\, A$;
- $M, u \vDash \forall x A$ *iff* $\forall a \in D_u\ M, u \vDash [a/x]\, A$.

Again in the $\Box$-case $B(\mathbf{a})$ means $[\mathbf{a}/\mathbf{x}]\, B$, and we assume that $FV(B) = \mathbf{x}$.

But this definition should be justified. In fact, as we know, the presentation of the same formula in a form $B(\mathbf{a})$ is not unique. So we should prove that forcing does not really depend on this presentation. To show this, we find a precategory corresponding to the same 'metaframe' relations R_i^n.

Proposition 5.8.5 *Let $\mathbb{F} = (\mathcal{C}, D, \rho)$ be a $\mathcal{C}$-bundle over an N-precategory $\mathcal{C}$. Then there exists an N-precategory $\mathcal{C}'$ and a $\mathcal{C}'$-preset $\mathbf{F}'$ such that*

- $FR(\mathcal{C}) = FR(\mathcal{C}')$;
- $D_u = \mathbf{F}'(u)$ *for any* $u \in Ob\,\mathcal{C}$;
- $R_i^n = R_i'^n$ *for all* $n \in \omega$, $1 \leq i \leq N$ *(where R_i^n and $R_i'^n$ are the nth level accessibility relations in $\mathbb{F}$ and $\mathbf{F}'$ respectively).*

Moreover, if $\mathbb{F}$ is an intuitionistic $\mathcal{C}$-bundle over a category $\mathcal{C}$ (in particular a Kripke bundle over an **S4***-frame, then $\mathcal{C}'$ is also a category and $\mathbf{F}'$ is a $\mathcal{C}'$-set.*

Proof Put

$$\mathcal{C}'_i(u, v) := \{f \text{ is a function } D_u \longrightarrow D_v \mid \exists \gamma \in \mathcal{C}_i(u, v)\ f \subseteq \rho_\gamma\},$$

and let $\rho'_f := f$ in $\mathbb{F}'$. So we replace every *relation* $\rho_\gamma \subseteq D_u \times D_v$ with the set of all *functions* from D_u to D_v included in this relation.

Let us show that $R_i'^n = R_i^n$. In fact, suppose $\mathbf{a}R_i'^n\mathbf{b}$; then $\mathbf{a}\,sub\,\mathbf{b}$, and there exists $f \in C_i'(u,v)$ such that $\mathbf{b} = \rho_f' \cdot \mathbf{a} = f \cdot \mathbf{a}$, i.e. for all j, $b_j = f(a_j)$. Since $f \in C_i'(u,v)$, there exists $\gamma \in C_i(u,v)$ such that $f \subseteq \rho_\gamma$; thus $a_j\rho_\gamma b_j$, which implies $\mathbf{a}R_i^n\mathbf{b}$.

The other way round, suppose $\mathbf{a}R_i^n\mathbf{b}$. Then the set $g = \{(a_1,b_1),\ldots,(a_n,b_n)\}$ is a function (since $\mathbf{a}\ sub\ \mathbf{b}$), and $g \subseteq \rho_\gamma$ for some $\gamma \in C_i(u,v)$. This g can be prolonged to a total function $f : D_u \longrightarrow D_v$ such that $f \subseteq \rho_\gamma$, since $\mathrm{dom}\rho_\gamma = D_u$ by Definition 5.8.1. Hence $\mathbf{b} = f \cdot \mathbf{a} = \rho_f' \cdot \mathbf{a}$, which implies $\mathbf{a}R_i'^n\mathbf{b}$.

If C is category, then C' also becomes a category after we define $\circ$ as the composition of functions and 1_{D_u} as the identity function id_{D_u}. In fact, if $f \subseteq \rho_\alpha$ and $g \subseteq \rho_\beta$, then $f \circ g \subseteq \rho_\alpha \circ \rho_\beta \subseteq \rho_{\alpha\circ\beta}$ (by Definition 5.8.1); $id_{D_u} \subseteq \rho_{1_u}$, since ρ_{1_u} is reflexive by 5.8.1. ∎

Remark 5.8.6 The crucial point of this construction is 'local functionality' in the definition of forcing for a Kripke bundle. This corresponds to the choice of distinct individuals $a_1,\ldots,a_n$ in the inductive clause for $\Box$, or to the conjunct $\mathbf{a}\,sub\,\mathbf{b}$ in the definition of R_i^n.

Remark 5.8.7 Note that the argument fails for intuitionistic Kripke quasi-bundles, because the totality of ρ_{uv} is not guaranteed in this case.

Therefore a model $M = (\mathbb{F},\xi)$ corresponds to the model $M' = (\mathbf{F}',\xi)$, and we can *define* $M,u \vDash A$ as $M',u \vDash A$. Then we obtain the properties described in Definition 5.8.4, and we can further define validity in $\mathbb{F}$ and the logic

$$\mathbf{ML}^{(=)}(\mathbb{F}) := \mathbf{ML}^{(=)}(\mathbf{F}')$$

in the modal case and

$$\mathbf{IL}^{(=)}(\mathbb{F}) := \mathbf{IL}^{(=)}(\mathbf{F}')$$

in the intuitionistic case.

Hence we obtain the semantics of C-bundles, which is obviously strongly equivalent to the functor semantics, and the following

Corollary 5.8.8 $\mathcal{KB}_N \preceq \mathcal{FS}_N$, $\mathcal{KB}_{int} \preceq \mathcal{FS}_{int}$.

Proof In fact, for a Kripke bundle $\mathbb{F}$ with the base F there exists a Cat_0F-preset $\mathbf{F}'$ such that $\mathbf{ML}^{(=)}(\mathbb{F}) = \mathbf{ML}^{(=)}(\mathbf{F}')$. ∎

5.9 Metaframes

The definition of forcing in Kripke bundles and C-sets via forcing in propositional frames F_n motivates a further generalisation of Kripke semantics. So far the relations R_i^n were derived from other relations or functions. Now let us consider arbitrary metaframes; they are defined as sequences of propositional frames of tuples, without special requirements for accessibility relations.

Definition 5.9.1 *Let $F_0 = (W, R_1, \dots, R_N)$ be a propositional Kripke frame, D a system of disjoint (nonempty) domains over W.*

*Let $F_n = (D^n, R_1^n, \dots, R_N^n)$, $n > 0$, be propositional Kripke frames such that $D^n = \bigcup_{u \in W} D_u^n$. Then the pair $\mathbb{F} = ((F_n)_{n\in\omega}, D)$ is called an N-*metaframe *based on F_0. As usual, W is called the set of* possible worlds *and D^1 the set of individuals of $\mathbb{F}$. F_n is called the n-*level *of $\mathbb{F}$.*

For $\mathbf{a} \in D_u^n$ we denote the domain D_u by $D(\mathbf{a})$.

We shall often identify a metaframe $\mathbb{F}$ with the sequence $(F_n)_{n\in\omega}$, i.e. consider it as a graded propositional frame. Note that the worlds of F_n are n-tuples of individuals, and the original worlds from W are '0-tuples'. Thus we may define

$$D_u^0 := \{u\},\ D^0 := W,\ R_i^0 := R_i.$$

We say that an n-tuple $\mathbf{b}$ is an *i-inheritor* (*counterpart*) of an n-tuple $\mathbf{a}$ if $\mathbf{a}R_i^n\mathbf{b}$.

As usual, $i(= 1)$ is not indicated in the notation if $N = 1$.

By default we assume that an arbitrary metaframe and all its components are denoted as in 5.9.1.

Definition 5.9.2 *A (modal)* valuation *in an N-metaframe $\mathbb{F}$ is a valuation in its system of domains. A (modal) metaframe model over $\mathbb{F}$ is a pair $M = (\mathbb{F}, \xi)$, where ξ is a (modal) valuation in $\mathbb{F}$.*

So a valuation in $\mathbb{F}$ is a graded propositional valuation; n-ary predicate letters are evaluated in $\mathbb{F}$ as propositional letters in F_n.

Our goal is now to define forcing in metaframe models. This definition is not a straightforward generalisation of the corresponding definitions in functor and Kripke bundle semantics. The reason is that in arbitrary metaframes it is impossible to define forcing for D_u-sentences in the natural way — this will be shown later on.

Remember that every D_u-sentence is obtained by replacing parameters with D_u-individuals in a formula $A(\mathbf{x})$, or by applying a D_u-transformation $[\mathbf{x} \mapsto \mathbf{a}]$ to $A(\mathbf{x})$, cf. Definition 2.4.1.

This transformation itself (not only its result $A(\mathbf{a})$ 'forgetting' the original $A(\mathbf{x})$ and $[\mathbf{x} \mapsto \mathbf{a}]$) is essential in our definition. Moreover, we define forcing for a formula $A(\mathbf{x})$ with respect to an 'ordered assignment' $(\mathbf{x}, \mathbf{a})$ (arranging a transformation in a certain order).

In the classical case (and in all our earlier Kripke-type semantics) the two versions of the truth definition (with D-sentences or variable assignments) are equivalent. The original Tarski's definition is given in terms of assignments.

So we begin with

Definition 5.9.3 *An* ordered assignment *(at a world u) in a metaframe $\mathbb{F} = ((F_n)_{n\in\omega}, D)$ is a pair $(\mathbf{x}, \mathbf{a})$, where $\mathbf{x}$ is a distinct list of variables, $\mathbf{a} \in D_u^n$. If the list $\mathbf{x}$ is empty, then by definition, $\mathbf{a}$ is just the world u.*

We define forcing in a model M as a relation $M, (\mathbf{x}, \mathbf{a}) \vDash A$ between ordered assignments and formulas. It can also be regarded as a relation between tuples $\mathbf{a} \in D^n$ and pairs $(A, \mathbf{x})$, so we use the alternative notation $M, \mathbf{a} \vDash A\,[\mathbf{x}]$. This is read as '$\mathbf{a}$ forces a pair $(A, \mathbf{x})$', or '$\mathbf{a}$ formal expression $A[\mathbf{x}]$'.[10] It is convenient to give such a definition for the case when $FV(A) \subseteq r(\mathbf{x})$.

Let us make some remarks about notation. Remember that according to the Introduction, for a tuple $\mathbf{a} = (a_1, \dots, a_n)$ and $\sigma : I_m \longrightarrow I_n$, we denote

$$\mathbf{a} \cdot \sigma = (a_{\sigma(1)}, \dots, a_{\sigma(m)})$$

and

$$\mathbf{a} - a_i = \widehat{\mathbf{a}}_i = \mathbf{a} \cdot \delta_i^n = (a_1, \dots, a_{i-1}, a_{i+1}, \dots, a_n).$$

The same notation applies to lists of variables; thus in Definition 5.9.4, $\mathbf{x} \cdot \sigma$ means $(x_{\sigma(1)}, \dots, x_{\sigma(m)})$ and $\mathbf{x} - x_i$ means $\mathbf{x} \cdot \delta_i^n$ for a list of variables $\mathbf{x} = (x_1, \dots, x_n)$, $\sigma : I_m \longrightarrow I_n$.

However the case $m = 0$ is special; then $\sigma = \varnothing_n$ is the empty map, and we define

$\mathbf{x} \cdot \varnothing_n$ as the empty list,
$\mathbf{a} \cdot \varnothing_n$ as the world of $\mathbf{a}$ (i.e. $\mathbf{a} \cdot \varnothing_n = u \Leftrightarrow \mathbf{a} \in D_u^n$).

Note that every atomic formula $P_k^m(\mathbf{y})$ with $r(\mathbf{y}) \subseteq r(\mathbf{x})$, $|\mathbf{x}| = n$, has the form $P_k^m(x_{\sigma(1)}, \dots, x_{\sigma(m)})$, or $P_k^m(\mathbf{x} \cdot \sigma)$, for some $\sigma \in \Sigma_{mn}$.

Definition 5.9.4 *For a metaframe model M and a modal formula (with equality) A we define forcing $M, \mathbf{a} \vDash A\,[\mathbf{x}]$ under an ordered assignment $(\mathbf{x}, \mathbf{a})$ (such that $FV(A) \subseteq r(\mathbf{x})$)*[11] *by induction:*

(1) $M, \mathbf{a} \nvDash \bot\,[\mathbf{x}]$;

(2) $M, \mathbf{a} \vDash P_j^m(\mathbf{x} \cdot \sigma)\,[\mathbf{x}]$ *iff* $(\mathbf{a} \cdot \sigma) \in \xi^+(P_j^m)$ *(for $m > 0$);*

(3) $M, \mathbf{a} \vDash P_j^0\,[\mathbf{x}]$ *iff* $u \in \xi^+(P_j^0)$ *(for $a \in D_u^n$);*

(4) $M, \mathbf{a} \vDash (x_i = x_j)\,[\mathbf{x}]$ *iff* $a_i = a_j$*;*

(5) $M, \mathbf{a} \vDash B \supset C\,[\mathbf{x}]$ *iff* $(M, \mathbf{a} \nvDash B[\mathbf{x}]$ *or* $M, \mathbf{a} \vDash C\,[\mathbf{x}])$*;*

(6) $M, \mathbf{a} \vDash B \vee C\,[\mathbf{x}]$ *iff* $(M, \mathbf{a} \vDash B\,[\mathbf{x}]$ *or* $M, \mathbf{a} \vDash C\,[\mathbf{x}])$*;*

(7) $M, \mathbf{a} \vDash B \wedge C\,[\mathbf{x}]$ *iff* $(M, \mathbf{a} \vDash B\,[\mathbf{x}]\ \&\ M, \mathbf{a} \vDash C\,[\mathbf{x}])$*;*

(8) $M, \mathbf{a} \vDash \Box_i B\,[\mathbf{x}]$ *iff* $\forall \mathbf{b}\,(\mathbf{a} R_i^n \mathbf{b} \Rightarrow M, \mathbf{b} \vDash B\,[\mathbf{x}])$*;*

(9) if $y \notin \mathbf{x}$, $\mathbf{a} \in D_u^n$, *then*
$M, \mathbf{a} \vDash \exists y B\,[\mathbf{x}]$ *iff* $\exists c \in D_u\ M, (\mathbf{a}c) \vDash B\,[\mathbf{x}y]$,
$M, \mathbf{a} \vDash \forall y B\,[\mathbf{x}]$ *iff* $\forall c \in D_u\ M, (\mathbf{a}c) \vDash B\,[\mathbf{x}y]$;

[10] The notation $A[\mathbf{x}]$ should not be mixed up with $A(\mathbf{x})$ denoting a formula with parameters in $\mathbf{x}$.

[11] If $\mathbf{x}$ is empty, $\mathbf{a}$ is a possible world u, this is written as $M, u \vDash A\,[\,]$, or just $M, u \vDash A$.

(10) $M, \mathbf{a} \models \exists x_i B[\mathbf{x}]$ *iff* $M, \mathbf{a} - a_i \models \exists x_i B[\mathbf{x} - x_i]$;
$M, \mathbf{a} \models \forall x_i B[\mathbf{x}]$ *iff* $M, \mathbf{a} - a_i \models \forall x_i B[\mathbf{x} - x_i]$.

So by Definition 5.9.4, we obtain:

- $M, \mathbf{a} \models \neg B[\mathbf{x}]$ iff $M, \mathbf{a} \not\models B[\mathbf{x}]$,
- $M, \mathbf{a} \models \Diamond_i B[\mathbf{x}]$ iff $\exists v \in W\, \exists \mathbf{b} \in D_v^n\, (\mathbf{a} R_i^n \mathbf{b}\ \&\ M, \mathbf{b} \models B[\mathbf{x}])$.

Also note that if D is a system of domains over W, $m, n > 0$, then every map $\sigma \in \Sigma_{mn}$ gives rise to the map $\pi_\sigma : D^n \longrightarrow D^m$ sending $\mathbf{a}$ to $\mathbf{a} \cdot \sigma$, and it is clear that π_σ maps every D_u^n to D_u^m exactly as in the Introduction.

For the empty map $\varnothing_n \in \Sigma_{0n}$ we can define the map $\pi_{\varnothing_n} : D^n \longrightarrow D^0 = W$ sending every tuple to its world; sometimes we use a simpler notation and write $\pi_\varnothing$ rather than $\pi_{\varnothing_n}$.

In a metaframe the maps π_σ are called *jections*. They have the properties mentioned in Lemma 0.0.1:

$$\pi_\sigma \cdot \pi_\tau = \pi_{\tau \cdot \sigma} \text{ for } \sigma \in \Sigma_{km},\ \tau \in \Sigma_{mn};$$

if $\sigma \in \Upsilon_n$, then π_σ is a permutation of D^n and $\pi_{\sigma^{-1}} = (\pi_\sigma)^{-1}$.

Lemma 5.9.5 *Let* $\mathbb{F} = ((F_n)_{n \in \omega}, D)$ *be an* N*-metaframe,* $\sigma \in \Sigma_{mn}$. *Then* $\pi_\sigma[D^n] = \{\mathbf{a} \in D^m \mid \sigma\, sub\, \mathbf{a}\}$.[12]

Proof Follows from Lemma 0.0.2 applied to every D_u. ∎

Lemma 5.9.6 *Let* $\mathbb{F}$ *be a metaframe, in which not all domains are one-element. Then for* $\sigma \in \Sigma_{mn}$, σ *is injective iff* $\pi_\sigma : D^n \longrightarrow D^m$ *is surjective.*

Proof Follows from Lemma 0.0.3. ∎

We also use notation from the Introduction:

$$\pi_i^n := \pi_{\delta_i^n},\ \pi_-^n := \pi_{\sigma_-^n},\ \pi_+^n := \pi_{\sigma_+^n},$$

recall that

$\pi_-^n(a_1, \ldots, a_n) = (a_1, \ldots, a_n, a_n)$ for $n > 0$,
$\pi_i^n(\mathbf{a}) = \mathbf{a} - a_i$ for $\mathbf{a} \in D^n$, $n > 0$, $\pi_+^n(\mathbf{a}) = \mathbf{a} - a_{n+1}$ for $\mathbf{a} \in D^{n+1}$, $n \geq 0$.

Remembering that $\sigma_+^0 = \varnothing_1 \in \Sigma_{01}$, we also have $\pi_+^0(a_1) = u$ for $a_1 \in D_u$.

We shall use the same notation π_σ for lists of variables; in particular, $\pi_\varnothing(x_1 \ldots x_n)$ is the empty list.

Note that in all the clauses except (8), (10), Definition 5.9.4 is essentially the same as in earlier Kripke-type semantics. For example, consider Kripke bundles. Our new definition (5.9.4) corresponds to the old one (5.2.8) as follows.

[12] Here similarly to 5.3.2, $\sigma\, sub\, \mathbf{a}$ denotes the property $\forall i, j\ (\sigma(i) = \sigma(j) \Rightarrow a_i = a_j)$.

$M, \mathbf{a} \models A[\mathbf{x}]$ (in the new sense) is equivalent to $M, u \models [\mathbf{a}/\mathbf{x}]A$ (in the old sense), where u is the world of $\mathbf{a}$. So 5.9.4(2)

$$M, \mathbf{a} \models P_k^m(\mathbf{x} \cdot \sigma)\ [\mathbf{x}] \Leftrightarrow \mathbf{a} \cdot \sigma \in \xi^+(P_k^m)$$

becomes

$$M, u \models [\mathbf{a}/\mathbf{x}]P_k^m(\mathbf{x} \cdot \sigma)(= P_k^m(\mathbf{a} \cdot \sigma)) \Leftrightarrow \mathbf{a} \cdot \sigma \in \xi^+(P_k^m)$$

exactly as in 5.2.8.

For evaluating a formula A we use assignments $(\mathbf{x}, \mathbf{a})$ with $r(\mathbf{x}) \supseteq FV(A)$. The requirement $r(\mathbf{x}) = FV(A)$ is insufficient, e.g. in clause (5), because it may be that $FV(B \supset C) \neq FV(B), FV(C)$.

We have to be careful about the quantifier clauses. In fact, if $r(\mathbf{x}) \supset FV(\exists yB)$, then either $y \notin r(\mathbf{x})$[13] or $y \in r(\mathbf{x})$. In the first case we obtain the clause (9), which is clear. But if $y \in r(\mathbf{x})$, the clause (9) is inapplicable, because $(\mathbf{x}y, \mathbf{a}c)$ is not a correct assignment — y is repeated twice. So before adding y to $\mathbf{x}$, we should eliminate it from $\mathbf{x}$ and its value from $\mathbf{a}$. This leads us to clause (10).

Sometimes it is convenient to join (9) and (10) in a single clause. To do this, let us introduce some more notation.

For a (distinct) list of variables $\mathbf{x}$ and $y \in Var$ we put

$$\mathbf{x} - y := \begin{cases} \mathbf{x} - x_i & \text{if } y = x_i, \\ \mathbf{x} & \text{if } y \notin \mathbf{x}, . \end{cases}$$

and

$$\mathbf{x}||y := (\mathbf{x} - y)y.$$

These lists can be obtained by corresponding transformations, viz. put

$$\varepsilon_{\mathbf{x}-y} := \begin{cases} \delta_i^n & \text{if } y = x_i, \\ id_n & \text{if } y \notin \mathbf{x}, \end{cases}$$

$$\varepsilon_{\mathbf{x}||y} := (\varepsilon_{\mathbf{x}-y})^+ := \begin{cases} (\delta_i^n)^+ & \text{if } y = x_i, \\ id_{n+1} & \text{if } y \notin \mathbf{x}, \end{cases}$$

provided $|\mathbf{x}| = n$;[14] then

$$\mathbf{x} \cdot \varepsilon_{\mathbf{x}-y} = \mathbf{x} - y,\ (\mathbf{x}y) \cdot \varepsilon_{\mathbf{x}||y} = \mathbf{x}||y.$$

The associated jections are denoted as follows:

$$\pi_{\mathbf{x}-y} := \pi_{\varepsilon_{\mathbf{x}-y}},\ \pi_{\mathbf{x}||y} := \pi_{\varepsilon_{\mathbf{x}||y}}.$$

[13] This is always the case if $r(\mathbf{x}) = FV(\exists yB)$.

[14] $+$ means the simple extension, see the Introduction.

So we can combine (9), (10) from Definition 5.9.4 in a single clause:

$$M, \mathbf{a} \models \exists y B\,[\mathbf{x}] \text{ iff } \exists c \in D_u\ M, \pi_{\mathbf{x}||y}(\mathbf{a}c) \models B\,[\mathbf{x}||y], \qquad (9+10)$$

and similarly for $\forall$. In fact, if $y \notin \mathbf{x}$, then $\pi_{\mathbf{x}||y}(\mathbf{a}c) = \mathbf{a}c$ and $\mathbf{x}||y = \boldsymbol{x}y$. And if $y = x_i$, then

$$\pi_{\mathbf{x}||y}(\mathbf{a}c) = (\mathbf{a}c) \cdot (\delta_i^n)^+ = (\mathbf{a} - a_i)c,\ \mathbf{x}||y = (\mathbf{x} - x_i)x_i.$$

The clause (8) generalises a version of the truth definition for $\Box B(\mathbf{a})$ in Kripke bundles given in 5.3.4. A similar version

for functor semantics was considered in 5.6.15.

In fact, in the case of Kripke bundles

(8) $\quad M, \mathbf{a} \models \Box B\ [\mathbf{x}] \Leftrightarrow \forall \mathbf{b}(\mathbf{a}R^n\mathbf{b} \Rightarrow M, \mathbf{b} \models B\ [\mathbf{x}])$

corresponds to

(8.1) $\quad M, u \models [\mathbf{a}/\mathbf{x}]\,\Box B \Leftrightarrow \forall v \in R(u)\,\forall \mathbf{b} \in D_v^n\ (\mathbf{a}R^n\mathbf{b} \Rightarrow M, v \models [\mathbf{b}/\mathbf{x}]B),$

i.e. to (*) in 5.3.4.

Let us also recall another reading of the $\Box$-clause in Kripke-type semantics considered so far:

(8.2) $\quad M, u \models \Box C$ iff for any $v \in R(u)$, for any v-version C' of C in v, $M, v \models C'$,

where C is a D_u-sentence.

(8.1) transforms to (8.2) if we define v-versions of a D_u-sentence $[\mathbf{a}/\mathbf{x}]B$ as D_v-sentences $[\mathbf{b}/\mathbf{x}]B$ for $\mathbf{b} \in R^n(\mathbf{a}) \cap D_v^n$ (i.e. for $\mathbf{b}$ that are v-inheritors of $\mathbf{a}$).

In a Kripke sheaf $\mathbf{a} \in D_u^n$ has a unique v-inheritor $\rho_{uv}{\cdot}\mathbf{a}$. In functor semantics inheritors also result from map actions: $\mathbf{b} = \rho_\mu \cdot \mathbf{a}$, for $\mu \in \mathcal{C}(u, v)$. In Kripke bundles we define the relation R^n in a special way.

In metaframes we can also call R^n the 'inheritance relation' between n-tuples and read (8) as

(8′) $\quad M, \mathbf{a} \models \Box B\ [\mathbf{x}]$ iff for any inheritor $\mathbf{b}$ of $\mathbf{a}$, $\ M, \mathbf{b} \models B\ [\mathbf{x}]$.

But now we cannot always define *inheritance relations* between *D_u-sentences* and transform (8) to (8.2) or (8.1). This happens,

because a D_u-sentence can be presented as $[\mathbf{a}/\mathbf{x}]A$ in different ways, as mentioned in 2.4. Different presentations may lead to different sets of inheritors for the same D_u-sentence. The following example shows that this is crucial.

Example 5.9.7 Consider a D_u-sentence $C := P(a, a)$. Then

$$C = [aa/xy]B_1 = [a/x]B_2,$$

where $B_1 := P(x, y)$, $B_2 := P(x, x)$.

Put

$$\begin{aligned} X_1 &:= R^2(aa) = \{\mathbf{b} \in D^2 \mid (aa)R^2\mathbf{b}\}, \\ X_2 &:= \{(dd) \in D^2 \mid aR^1 d\}, \\ \mathcal{B}_1 &:= \{[\mathbf{b}/xy]B_1 \mid \mathbf{b} \in X_1\} = \{P(\mathbf{b}) \mid \mathbf{b} \in X_1\}, \\ \mathcal{B}_2 &:= \{[d/x]B_1 \mid aR^1 d\} = \{P(\mathbf{b}) \mid \mathbf{b} \in X_2\}. \end{aligned}$$

Since X_1 consists of all inheritors of aa, we may call the members of $\mathcal{B}_1$ 'inheritors of $[aa/xy]B_1$'. Similarly, the members of $\mathcal{B}_2$ are 'inheritors of $[a/x]B_2$'. But it may happen that $X_1 \neq X_2$,[15] so $\mathcal{B}_1 \neq \mathcal{B}_2$, and thus we cannot properly define inheritors of C. In this case (8) cannot be rewritten as (8.1). In fact, in a metaframe model $M = (\mathbb{F}, \xi)$

$$\begin{aligned} M, aa \vDash \Box P(x,y)\ [xy] &\Leftrightarrow X_1 \subseteq \xi^+(P), \\ M, a \vDash \Box P(x,x)\ [x] &\Leftrightarrow X_2 \subseteq \xi^+(P). \end{aligned}$$

So there exists a model, in which

$$M, aa \vDash \Box P(x,y)\ [xy] \not\Leftrightarrow M, a \vDash \Box P(x,x)\ [x]$$

— just put

$$\xi^+(P) := \begin{cases} X_2 \text{ if } X_1 \not\subseteq X_2, \\ X_1 \text{ if } X_2 \not\subseteq X_1. \end{cases}$$

So we see that in arbitrary metaframes there is no reasonable way to define the relation $M, u \vDash A$ for D_u-sentences A. That is why we define metaframe forcing in the form $M, \mathbf{a} \vDash A[\mathbf{x}]$.

Now let us define the truth in a metaframe model.

Definition 5.9.8 *A modal formula A is called* true *in a metaframe model $M = (\mathbb{F}, \xi)$ (notation: $M \vDash A$) if $M, \mathbf{a} \vDash A[\mathbf{x}]$ for any ordered assignment $(\mathbf{x}, \mathbf{a})$ in $\mathbb{F}$ such that $FV(A) \subseteq r(\mathbf{x})$.*

Definition 5.9.9 *A modal formula A is called* valid *in a metaframe $\mathbb{F}$ (notation: $\mathbb{F} \vDash A$) if A is true in all models over $\mathbb{F}$.*

Let

$$\mathbf{ML}_{-}^{(=)}(\mathbb{F}) := \{A \in MF_N^{(=)} \mid \mathbb{F} \vDash A\}.$$

Definition 5.9.10 *A formula $A \in MF_N^{(=)}$ is called* strongly valid *(respectively,* strongly valid with equality*) in a metaframe $\mathbb{F}$ if all its MF_N- (respectively, $MF_N^=$-) substitution instances are valid in $\mathbb{F}$ (notation: $\mathbb{F} \vDash^+ A$ or respectively, $\mathbb{F} \vDash^{+=} A$). Let*

$$\mathbf{ML}(\mathbb{F}) := \{A \in MF_N \mid \mathbb{F} \vDash^+ A\}.$$

$$\mathbf{ML}^=(\mathbb{F}) := \{A \in MF_N^= \mid \mathbb{F} \vDash^{+=} A\}.$$

Definition 5.9.11 *Let $\mathbf{F} = (\mathcal{C}, D, \rho)$ be a preset over an N-precategory $\mathcal{C}$. Then its* associated metaframe *is $\mathbb{M}\mathbf{f}(\mathbf{F}) := ((F_n), D)$, where F_n is the n-level of $\mathbf{F}$ (Definition 5.6.14). In the same way we define $\mathbb{M}\mathbf{f}(\mathbf{F})$ for a Kripke bundle $\mathbf{F}$.*

[15] The reader can easily construct such an example.

So a Kripke bundle or a $\mathcal{C}$-preset model $M = (\mathbf{F}, \xi)$ corresponds to a metaframe model $\mathbb{M}\mathtt{f}(M) := (\mathbb{M}\mathtt{f}(\mathbf{F}), \xi)$, and the two definitions of forcing 5.9.4, 5.6.9 are obviously the same:

Lemma 5.9.12 *Let* $\mathbf{F}$ *be a preset over an* N*-precategory (or a Kripke bundle),* M *a model over* $\mathbf{F}$. *Then for any formula* $A \in MF_N^{(=)}$ *for any ordered assignment* $(\mathbf{x}, \mathbf{a})$ *in* $\mathbb{M}\mathtt{f}(\mathbf{F})$ *with* $FV(A) \subseteq r(\mathbf{x})$

(1) $M \vDash [\mathbf{a}/\mathbf{x}]\, A$ *iff* $\mathbb{M}\mathtt{f}(M), \mathbf{a} \vDash A\ [\mathbf{x}]$.

(2) $M \vDash A$ *iff* $\mathbb{M}\mathtt{f}(M) \vDash A$.

Proof (1) By straightforward induction on the length of A, according to 5.9.4, 5.6.9.

(2) follows from (1), 5.6.9, 5.2.10. ∎

Proposition 5.9.13 *For a* $\mathcal{C}$*-preset or a Kripke bundle* $\mathbf{F}$*:*

$$\mathbf{ML}_{-}^{(=)}(\mathbf{F}) = \mathbf{ML}_{-}^{(=)}(\mathbb{M}\mathtt{f}(\mathbf{F})),$$
$$\mathbf{ML}^{(=)}(\mathbf{F}) = \mathbf{ML}^{(=)}(\mathbb{M}\mathtt{f}(\mathbf{F})).$$

Proof Follows readily from 5.9.12. ∎

Remark 5.9.14 One can also define a metaframe $\mathbb{M}\mathtt{f}(\Phi)$ for any Kripke quasi-bundle Φ (cf. 5.5.19) such that

$$\Phi \vDash A \Leftrightarrow \mathbb{M}\mathtt{f}(\Phi) \vDash A$$

for any modal formula A. The counterexample 5.5.19(a) shows that the set $\mathbf{ML}(\mathbb{M}\mathtt{f}(\Phi))$ is not always a modal predicate logic.

Definition 5.9.15 *A metaframe* $\mathbb{F}$ *is called* modally sound *(respectively* modally sound with equality*), in brief,* $m^{(=)}$-sound, *if* $\mathbf{ML}^{(=)}(\mathbb{F})$ *is an m.p.l.(=).*

Our goal in the next sections (5.10–5.12) will be to describe the class of m-sound metaframes; as we shall see, $m^=$-soundness is equivalent to m-soundness (Theorem 5.12.13).

But first, let us point out some peculiarities of forcing in arbitrary metaframes. This notion is still too broad, because $M, \mathbf{a} \vDash A\ [\mathbf{x}]$ may depend on the ordering of $\mathbf{x}$ and $\mathbf{a}$ and also on the variables x_i that do not occur in A.

Example 5.9.16 Let $F = (W, R)$ be a reflexive singleton $\{u\}$. Consider a metaframe $\mathbb{F}$ over F such that $D_u = \{c_1, c_2\}$,

$$R^1 := \{(c_i, c_j) \mid i \leq j\},\ R^2 := \{(\mathbf{a}, \mathbf{a}) \mid \mathbf{a} \in D_u^2\} \cup \{(c_1c_2, c_2c_2), (c_2c_1, c_2c_2)\}.$$

So R^2 is the same as the corresponding relation in the Kripke bundle in Fig. 5.9, but without the pair (c_1c_2, c_2c_2).

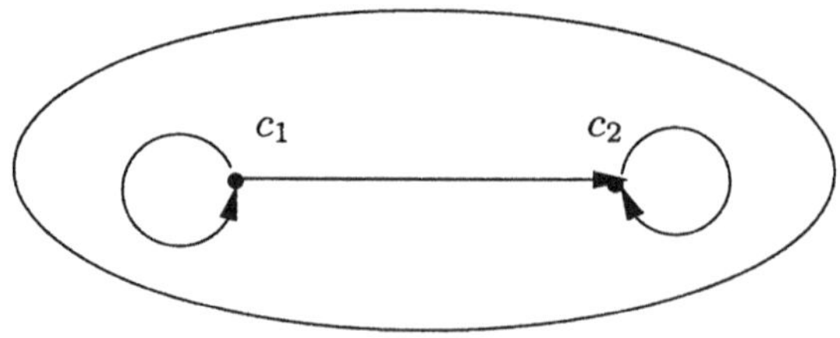

Figure 5.9.

We do not specify the frames F_n for $n > 2$; for example, R^n can be universal on D_u^n. Consider a metaframe model $M = (\mathbb{F}, \xi)$ such that

$$\xi^+(P) := \{c_2c_2\},\ \xi^+(Q) := \{c_2\} \text{ for certain } P \in PL^2,\ Q \in PL^1.$$

Let $A := \Diamond P(x_1, x_2)$, then in M we have

$$c_2c_1 \vDash A\ [x_2x_1]\ (\text{since } (c_2c_1)R^2(c_2c_2)),$$

but

$$c_1c_2 \vDash A\ [x_1x_1]\ (\text{since } (c_1c_2)\ \not R^2(c_2c_2)).$$

So $M, \mathbf{a} \vDash A\ [\mathbf{x}]$ depends on the ordering of $\mathbf{x}$ and $\mathbf{a}$.

Now put $B := \Diamond Q(x)$, then (in M)

$$c_1 \vDash B\ [x]\ (\text{since } a_1R^1c_2),$$

$$c_1c_1 \vDash\ B[xy]\ (\text{since } (c_1c_1)R^2(c_2c_1)),$$

but

$$c_1c_2 \nvDash\ B[xy]\ (\text{since } R^2(c_1c_2) = \{c_1c_2\} \text{ and } c_1c_2 \nvDash Q(x)\ [xy]).$$

So $M, \mathbf{a} \vDash B\ [\mathbf{x}]$ depends on the variables in $\mathbf{x}$ that do not occur in B.

Remark 5.9.17 The above counterexample may cause doubts about the clause (10) in Definition 5.9.4. In fact, as $M, \mathbf{a} \vDash A\ [\mathbf{x}]$ may depend on the values of x_i that do not occur in A, it seems improper to make $M, \mathbf{a} \vDash \exists x_i B\ [\mathbf{x}]$ equivalent to $M, \mathbf{a} - a_i \vDash \exists x_i B\ [\mathbf{x} - x_i]$.

Instead we can define an alternative forcing relation $\vDash^*$ by the same clauses (1)–(9) and the following modification of (10).

(10*) $M, \mathbf{a} \vDash^* \exists x_i B\ [\mathbf{x}]$ iff $\exists \mathbf{b} \in D^n\ (\mathbf{b} \cdot \delta_i^n = \mathbf{a} \cdot \delta_i^n\ \&\ M, \mathbf{b} \vDash^* B\ [\mathbf{x}])$

iff $\exists c \in D(\mathbf{a})\ M, [c/a_i]\mathbf{a} \vDash^* B\ [\mathbf{x}]$,

where $[c/a_i]\mathbf{a}$ is obtained by replacing a_i with c at position i;[16] and similarly

$$M, \mathbf{a} \vDash^* \forall x_i B[x] \text{ iff } \forall \mathbf{b} \in D^n\ (\mathbf{b} \cdot \delta_n^i = \mathbf{a} \cdot \delta_n^i \Rightarrow M, \mathbf{b} \vDash^* B\ [\mathbf{x}])$$

$$\text{iff } \forall c \in D(\mathbf{a})\ M, [c/a_i]\mathbf{a} \vDash^* B\ [\mathbf{x}].$$

In arbitrary metaframes the relations $\vDash$ and $\vDash^*$ may not be equivalent:

[16]So $\mathbf{b} \cdot \delta_i^n = \mathbf{a} \cdot \delta_i^n$ iff $\mathbf{b} = [c/a_i]\mathbf{a}$ for some $c \in D(\mathbf{a})$.

Exercise 5.9.18 Consider a metaframe model M from Example 5.9.16 and formula $C = \neg Q(x_1) \wedge A$, where $A = \Diamond P(x_1, x_2)$. Show that

$$M, c_1c_2 \vDash \exists x_1 C\ [x_1x_2] \text{ and } M, c_1c_2 \not\vDash^* \exists x_1 C\ [x_1x_2].$$

Hint: recall that $M, c_1c_1 \vDash A\ [x_2x_1]$ and $M, c_1c_2 \not\vDash A\ [x_1x_2]$.

On the other hand, later on we will show (see the end of Section 5.12) that $\vDash$ and $\vDash^*$ are equivalent in logically sound metaframes.

5.10 Permutability and weak functoriality

Now let us describe a class of metaframes without pecularities mentioned at the end of Section 5.9. This class will be used in the further description of m-sound metaframes.

First we give the following

Definition 5.10.1 *An N-metaframe $\mathbb{F}$ is called* permutable *if π_σ is monotonic for every permutation σ:*

$$\forall \sigma \in \Upsilon_n\, \forall i \in I_N\, \forall \mathbf{a}, \mathbf{b}\, (\mathbf{a}R_i^n\mathbf{b} \Rightarrow \pi_\sigma(\mathbf{a})R_i^n\pi_\sigma(\mathbf{b})),$$

Note that for $n = 0, 1$ this condition holds trivially, since in these cases $\Upsilon_n = \{id_n\}$.

Also note that to show permutability, it is sufficient to check monotonicity of jections corresponding to (simple) transpositions σ_k^n, where $2 \leq k \leq n$, which are generators of the symmetric group.

Since for any permutation σ, $\pi_{\sigma^{-1}} = (\pi_\sigma)^{-1}$, we obtain the following equivalent condition:

Lemma 5.10.2 *A metaframe $\mathbb{F}$ is permutable iff for any permutation $\sigma \in \Upsilon_n$, π_σ is an automorphism of the n-level F_n.*

Proof (Only if.) π_σ is a bijection for any $\sigma \in \Upsilon_n$ by Lemma 0.0.1. $(\pi_\sigma\mathbf{a})R^n(\pi_\sigma\mathbf{b}) \Rightarrow \mathbf{a}R^n\mathbf{b}$ follows from 5.10.1 by applying $\pi_{\sigma^{-1}} = (\pi_\sigma)^{-1}$. ∎

Exercise 5.10.3 Show that in a 1-modal permutable metaframe, where not all domains are one-element, the relation R^2 cannot be a linear ordering.

Definition 5.10.4 *A metaframe $\mathbb{F}$ is called* weakly functorial *(*w-functorial*) if for any injection $\sigma \in \Upsilon_{mn}$ (where $m \leq n$), π_σ is a p-morphism $F_n \twoheadrightarrow F_m$.*

From 5.9.6 we already know that π_σ is surjective for $\sigma \in \Upsilon_{mn}$.

To show w-functoriality, it suffices to check that π_σ is a (p-)morphism, whenever σ is a simple transposition $\sigma_i^n \in \Upsilon_n$ or a simple embedding $\sigma_+^n \in \Sigma_{n,n+1}$, because every injection is a composition of functions of this kind.[17]

Remembering that $\pi_+^n\mathbf{a} = \mathbf{a} - a_{n+1}$ for $n \neq 0$, and $\pi_+^0 a$ is the world of a (for $a \in D^1$), we obtain:

[17] See Introduction.

Lemma 5.10.5 *A metaframe* $\mathbb{F} = ((F_n)_{n\in\omega}, D)$ *is w-functorial iff* $\mathbb{F}$ *is permutable and for any* $n > 0$, $i \in I_N$, *the following conditions hold:*

(1) $\forall n > 1\, \forall \mathbf{a}, \mathbf{b} \in D^n\ (\mathbf{a}R_i^n\mathbf{b} \Rightarrow (\mathbf{a} - a_n)R_i^{n-1}(\mathbf{b} - b_n))$ *(the monotonicity for* π_+^{n-1}*);*

(2) $\forall n > 0\, \forall \mathbf{a}, \mathbf{b} \in D^n\, \forall c \in D^1\ (\mathbf{a}R_i^n\mathbf{b}\ \&\ (\mathbf{a}c) \in D^n \Rightarrow \exists d\ (\mathbf{a}c)R_i^{n+1}(\mathbf{b}d))$ *(the lift property for* π_+^n*);*

(3) $\forall a \in D_u\, \forall b \in D_v\, (aR_i^1 b \Rightarrow uR_i v)$ *(the monotonicity for* π_+^0*);*

(4) $\forall u, v\ (uR_i v \Rightarrow \forall a \in D_u \exists b \in D_v\ aR_i^1 b)$ *(the lift property for* π_+^0*).*

Lemma 5.10.6 *Let* $\mathbb{F}$ *be a metaframe.*

(1) $\mathbb{F}$ *is permutable iff for any model* M *over* $\mathbb{F}$, *for any ordered assignment* $(\mathbf{x}, \mathbf{a})$, *for any formula* A *with* $FV(A) \subseteq \mathbf{x}$ *and for any permutation* $\sigma \in \Upsilon_n$, *where* $n = |\mathbf{x}|$,

$$M, \mathbf{a} \vDash A\,[\mathbf{x}] \Leftrightarrow M, \mathbf{a}\cdot\sigma \vDash A\,[\mathbf{x}\cdot\sigma]. \qquad (*)$$

(2) $\mathbb{F}$ *is w-functorial iff (*) holds for any model* M *over* $\mathbb{F}$, *for any ordered assignment* $(\mathbf{x}, \mathbf{a})$, *for any injection* $\sigma \in \Upsilon_{mn}$, *where* $|\mathbf{x}| = n \geq m \geq 0$, *and for any formula* A *with* $FV(A) \subseteq r(\mathbf{x}\cdot\sigma)$.

Note that if σ is injective and $(\mathbf{x}, \mathbf{a})$ is an ordered assignment, then the list $\mathbf{x}\cdot\sigma$ is distinct, so $(\mathbf{x}\cdot\sigma, \mathbf{a}\cdot\sigma)$ is also an ordered assignment.

Proof The case $n = 0$ (when $\mathbf{a}, \mathbf{x}$ are empty and A is a sentence) is trivial.

(I) 'Only if' is proved by induction on A, and the proofs of (1) and (2) can be made in parallel. So we assume that $\mathbb{F}$ is respectively permutable or transformable and check the equivalence (*). The model M is fixed, so we drop it from the notation. Let $\mathbf{y} := \mathbf{x}\cdot\sigma$; then $r(\mathbf{y}) \subseteq r(\mathbf{x})$.

- Let A be atomic; then since $FV(A) \subseteq r(\mathbf{y})$, it follows that

$$A = P_j^k(\mathbf{y}\cdot\tau) = P_j^k(\mathbf{x}\cdot(\tau\cdot\sigma))$$

for some $j, k \geq 0$, $\tau \in \Sigma_{km}$ (not necessarily injective). Hence by definition,

$$\mathbf{a} \vDash A\,[\mathbf{x}] \text{ iff iff } \mathbf{a}\cdot(\tau\cdot\sigma) \in \xi^+(P_j^k).$$

On the other hand,

$$(\mathbf{a}\cdot\sigma) \vDash A\,[\mathbf{y}] \text{ iff } (\mathbf{a}\cdot\sigma)\cdot\tau \in \xi^+(P_j^k),$$

and since $(\mathbf{a}\cdot\sigma)\cdot\tau = \mathbf{a}\cdot(\tau\cdot\sigma)$, (*) follows.

- The cases when $A = \bot$, $B \wedge C$, $B \vee C$, $B \supset C$, are trivial.

- Let $A = \Box_i B$. Then $\mathbf{a} \vDash A\ [\mathbf{x}]$ iff

(3) $\forall \mathbf{b} \in R_i^n(\mathbf{a})\ \mathbf{b} \vDash B\ [\mathbf{x}]$

which by induction hypothesis, is equivalent to

(4) $\forall \mathbf{b} \in R_i^n(\mathbf{a})\ \pi_\sigma(\mathbf{b}) \vDash B\ [\pi_\sigma(\mathbf{x})]$.

On the other hand, $\pi_\sigma(\mathbf{a}) \vDash A\,[\pi_\sigma(\mathbf{x})]$ iff

(5) $\forall \mathbf{b}' \in R_i^n(\pi_\sigma(\mathbf{a}))\ \mathbf{b}' \vDash B\ [\pi_\sigma(\mathbf{x})]$.

But $R_i^n(\pi_\sigma(\mathbf{a})) = \pi_\sigma[R_i^n(\mathbf{a})]$, since π_σ is a p-morphism. So (4) is equivalent to (5).

- Let $A = \exists y B$.

Put $\tau := \varepsilon_{\mathbf{x}-y}$, $\rho := \varepsilon_{\mathbf{x}\cdot\sigma-y}$. Then $\varepsilon_{\mathbf{x}\|y} = \tau^+$, $\varepsilon_{(\mathbf{x}\cdot\sigma)\|y} = \rho^+$.

So according to our definitions and notation (see condition (9+10) in Section 5.9), for any appropriate assignment $(\mathbf{x}, \mathbf{a})$ we obtain[18]

(6) $\mathbf{a} \vDash \exists y B\,[\mathbf{x}]$ iff $\exists c \in D_u\ (\mathbf{a}c)\cdot\tau^+ \vDash B\ [(\mathbf{x}y)\cdot\tau^+]$,

and

(7) $\mathbf{a}\cdot\sigma \vDash \exists y B\,[\mathbf{x}\cdot\sigma]$ iff $\exists c \in D_u\ ((\mathbf{a}\cdot\sigma)c)\cdot\rho^+ \vDash B\ [((\mathbf{x}\cdot\sigma)y)\cdot\rho^+]$.

In view of (6), (7), to apply the induction hypothesis, it suffices to find an injective λ' such that

(8) $((\mathbf{x}y)\cdot\tau^+)\cdot\lambda' = ((\mathbf{x}\cdot\sigma)y)\cdot\rho^+$;

(9) $((\mathbf{a}c)\cdot\tau^+)\cdot\lambda' = ((\mathbf{a}\cdot\sigma)c)\cdot\rho^+$;

i.e.

(10) $(\mathbf{x}y)\cdot(\tau^+\cdot\lambda') = (\mathbf{x}y)\cdot(\sigma^+\cdot\rho^+)$;

(11) $(\mathbf{a}c)\cdot(\tau^+\cdot\lambda') = (\mathbf{a}c)\cdot(\sigma^+\cdot\rho^+)$.

Note that $\mathbf{x}\cdot\sigma - y$ is a distinct list of variables from $\mathbf{x} - y$. So by Lemma 0.0.8, $\mathbf{x}\cdot\sigma - y = (\mathbf{x} - y)\cdot\lambda$ for some injection λ.

Claim $\tau\cdot\lambda = \sigma\cdot\rho$.

For the proof note that

$$\mathbf{x}\cdot\tau = \mathbf{x} - y,\ (\mathbf{x}\cdot\sigma)\cdot\rho = \mathbf{x}\cdot\sigma - y,$$

and thus

$$\mathbf{x}\cdot(\tau\cdot\lambda) = (\mathbf{x}\cdot\tau)\cdot\lambda = (\mathbf{x} - y)\cdot\lambda = \mathbf{x}\cdot\sigma - y = (\mathbf{x}\cdot\sigma)\cdot\rho = \mathbf{x}\cdot(\sigma\cdot\rho).$$

[18] Remember that $(\mathbf{x}y)\cdot\tau^+ = \mathbf{x}\|y$, $((\mathbf{x}\cdot\sigma)y)\cdot\rho^+ = (\mathbf{x}\cdot\sigma)\|y$.

Since $\mathbf{x}$ is distinct, the claim now follows by Lemma 0.0.5.

Hence by Lemma 0.0.7, $\tau^+ \cdot \lambda^+ = \sigma^+ \cdot \rho^+$, so (10), (11) hold for $\lambda' := \lambda^+$.

(II) Now let us prove the 'if' part in both assertions, (1) and (2). We assume (*) and check the p-morphism properties of π_σ (respectively, for $\sigma \in \Upsilon_{mn}$ or $\sigma \in \Upsilon_n$).

To check the monotonicity, we also assume $\mathbf{a}R_i^n\mathbf{b}$, and show that $\pi_\sigma(\mathbf{a})R_i^m \pi_\sigma(\mathbf{b})$. Let $A = \Diamond_i P(\pi_\sigma(\mathbf{x}))$, where $\mathbf{x} = (x_1, \dots, x_n)$ is distinct. Consider the model $M = (\mathbb{F}, \xi)$ such that $\xi^+(P) = \{\pi_\sigma(\mathbf{b})\}$. Then by Definition 5.9.4

$$M, \mathbf{b} \models P(\pi_\sigma(\mathbf{x}))\,[\mathbf{x}],$$

and so

$$M, \mathbf{a} \models A\,[\mathbf{x}].$$

Hence by (*),

$$M, \pi_\sigma(\mathbf{a}) \models A\,[\pi_\sigma(\mathbf{x})],$$

which means by Definition 5.9.4, $\pi_\sigma(\mathbf{a})R_i^m\mathbf{b}'$ for some $\mathbf{b}' \in \xi^+(P)$. By the choice of ξ, $\mathbf{b}' = \pi_\sigma(\mathbf{b})$. Therefore $\pi_\sigma(\mathbf{a})R_i^m\pi_\sigma(\mathbf{b})$.

Note that the same argument is valid in the particular case $m = 0$; then we have $\sigma = \varnothing_n$, $A = \Diamond_i P$ for $P \in PL^0$, $\xi^+(P) = \{\pi_\varnothing(\mathbf{b})\}$.

To check the lift property, we assume $\pi_\sigma(\mathbf{a})R_i^m\mathbf{b}'$ and find $\mathbf{b} \in R_i^n(\mathbf{a})$ such that $\pi_\sigma(\mathbf{b}) = \mathbf{b}'$.

Consider the same formula $A = \Diamond_i P(\pi_\sigma(\mathbf{x}))$, and the model $M = (\mathbb{F}, \xi)$ such that $\xi^+(P) = \{\mathbf{b}'\}$. Then

$$M, \mathbf{b}' \models P(\pi_\sigma \mathbf{x})\,[\pi_\sigma \mathbf{x}],$$

and thus

$$M, \pi_\sigma(\mathbf{a}) \models A\,[\pi_\sigma(\mathbf{x})].$$

Hence by (*),

$$M, \mathbf{a} \models A\,[\mathbf{x}],$$

i.e. by 5.9.4,

$$M, \mathbf{b} \models P(\pi_\sigma(\mathbf{x}))\,[\mathbf{x}]$$

for some $\mathbf{b} \in R_i^n(\mathbf{a})$. But then $\pi_\sigma(\mathbf{b}) \in \xi^+(P)$, according to 5.9.4, thus $\pi_\sigma(\mathbf{b}) = \mathbf{b}'$, by the choice of ξ. ■

Lemma 5.10.6(1) shows that in permutable metaframes forcing $M, \mathbf{a} \models A[\mathbf{x}]$ is invariant under simultaneous permutations of $\mathbf{x}$ and $\mathbf{a}$. So we can also define forcing under *unordered assignments*. Such an assignment (at a world u) is nothing but a D_u-substitution (Section 2.2), i.e. a function $[\mathbf{a}/\mathbf{x}]$ sending every x_i to a_i. An ordered assignment $(\mathbf{x}, \mathbf{a})$ corresponds to the unordered assignment $[\mathbf{a}/\mathbf{x}] = \{(x_1, a_1), \dots, (x_n, a_n)\}$, where $n = |\mathbf{x}|$. So in a model M over a permutable metaframe we can define

$$M, [\mathbf{a}/\mathbf{x}] \models A := M, \mathbf{a} \models A\,[\mathbf{x}].$$

This definition is sound; in fact, $[\mathbf{a}/\mathbf{x}] = [\mathbf{b}/\mathbf{y}]$ iff $(\mathbf{y}, \mathbf{b})$ is obtained from $(\mathbf{x}, \mathbf{a})$ by some permutation, i.e. iff $\mathbf{y} = \mathbf{x}\cdot\sigma,\ \mathbf{b} = \mathbf{a}\cdot\sigma$ for some $\sigma \in \Upsilon_n$. So by Lemma 5.10.6(1),

$$M, \mathbf{a} \vDash A\ [\mathbf{x}] \text{ iff } M, \mathbf{b} \vDash A\ [\mathbf{y}].$$

Furthermore, in w-functorial metaframes the truth value of a formula A depends only on individuals assigned to parameters of A (or only on a possible world if A is a sentence). Viz., suppose $r(\mathbf{y}) \supseteq FV(A) = r(\mathbf{x})$. Then $\mathbf{x} = \mathbf{y}\cdot\sigma$ for some injection σ. So by Lemma 5.10.6(2),

$$M, \mathbf{a} \vDash A\ [\mathbf{y}] \text{ iff } M, \mathbf{a}\cdot\sigma \vDash\ A[\mathbf{x}].$$

Thus forcing $M, \mathbf{a} \vDash A\ [\mathbf{y}]$ reduces to $M, \mathbf{b} \vDash A\ [\mathbf{x}]$ for an appropriate $\mathbf{b}$. In particular:

Lemma 5.10.7 *If $r(\mathbf{x}) = FV(A)$, $\mathbf{z}$ is a distinct list of other variables and $\mathbf{c}$ is a tuple of individuals (from the world of $\mathbf{a}$) of the same length, then*

$$M, \mathbf{a} \vDash A\ [\mathbf{x}] \textit{ iff } M, \mathbf{ac} \vDash A\ [\mathbf{xz}].$$

Since in w-functorial metaframes we may use only the forcing relation $M, \mathbf{a} \vDash A\ [\mathbf{x}]$ for $r(\mathbf{x}) = FV(A)$, in Definition 5.9.4 we need only the clause (9) for $A = \forall yB$ or $\exists yB$, because clearly $y \notin FV(A) = r(\mathbf{x})$.

Exercise 5.10.8 Show that in a permutable metaframe for any model M

$$M, \mathbf{a} \vDash \exists x_i B\,[\mathbf{x}] \text{ iff } \exists c \in D(\mathbf{a})\ M, [c/a_i]\mathbf{a} \vDash B\,[\mathbf{x}],$$

where the tuple $[c/a_i]\mathbf{a}$ is obtained from $\mathbf{a}$ by replacing a_i with c; cf. Remark 5.9.17.

Hint: $(\mathbf{x} - x_i)x_i = \mathbf{x}\cdot\sigma$ and $(\mathbf{a} - a_i)c = ([c/a_i]\mathbf{a})\cdot\sigma$ for $\sigma \in \Upsilon_n$.

So by induction on $|A|$, one can easily show that

$$M, \mathbf{a} \vDash A[\mathbf{x}] \Leftrightarrow M, \mathbf{a} \vDash^* A[\mathbf{x}]$$

for a model M over a permutable metaframe for any $A \in MF^=$, where $\vDash^*$ denotes the modified forcing from Remark 5.9.17.

To show that forcing in w-functorial metaframes is congruence independent, we begin with an auxiliary lemma.

Lemma 5.10.9 *Let M be model over an arbitrary metaframe, A a modal formula, $(\mathbf{x}, \mathbf{a})$ an assignment in M such that $FV(A) \subseteq r(\mathbf{x})$. Also let $y' \notin V(A)$, $y \notin BV(A)$, $\mathbf{x}' = [y'/y]\,\mathbf{x}$, and $A' = A[y \mapsto y']$. Then*

$$M, \mathbf{a} \vDash A\ [\mathbf{x}] \textit{ iff } M, \mathbf{a} \vDash A'\ [\mathbf{x}'].$$

Proof This statement is quite obvious — it means that the truth value of a formula does not depend on the name of a certain free variable. Formally we argue by induction on the complexity of A. Let us consider three cases.

(1) If $A = P(\mathbf{x} \cdot \sigma)$, then $A' = P(\mathbf{x}' \cdot \sigma)$. So

$$M, \mathbf{a} \vDash A\ [\mathbf{x}] \text{ iff } \pi_\sigma \mathbf{a} \in \xi^+(P) \text{ iff } M, \mathbf{a} \vDash A'\ [\mathbf{x}'].$$

(2) Let $A = \exists z B$. Then $z \neq y, y'$ by the assumption of our lemma, so $A' = \exists z B'$, $(\mathbf{x}\|z)' = \mathbf{x}'\|z$, $\pi_{\mathbf{x}'\|z} = \pi_{\mathbf{x}\|z}$. Then by the modified clause (9+10) of Definition 5.9.4 and the induction hypothesis,

$$M, \mathbf{a} \vDash A\ [\mathbf{x}] \Leftrightarrow \exists c \in D(\mathbf{a})\ M, \pi_{\mathbf{x}'\|z}(\mathbf{a}c) \vDash B\ [\mathbf{x}\|z]$$
$$\Leftrightarrow \exists c \in D(\mathbf{a})\ M, \pi_{\mathbf{x}'\|z}(\mathbf{a}c) \vDash B'\ [\mathbf{x}'\|z] \Leftrightarrow M, \mathbf{a} \vDash \exists z B' (= A')\ [\mathbf{x}'].$$

(3) Let $A = \Box_i B$, $\mathbf{a} \in D^n$. By Definition 5.9.4 and the induction hypothesis,

$$M, \mathbf{a} \vDash A[\mathbf{x}] \Leftrightarrow \forall \mathbf{b} \in R_i^n(\mathbf{a})\ M, \mathbf{b} \vDash B\ [\mathbf{x}]$$
$$\Leftrightarrow \forall \mathbf{b} \in R_i^n(\mathbf{a})\ M, \mathbf{b} \vDash B'\ [\mathbf{x}'] \Leftrightarrow M, \mathbf{a} \vDash \Box_i B'\ (= A')\ [\mathbf{x}'].$$

All other cases are rather trivial. ∎

Lemma 5.10.10 *Let M be model over a w-functorial metaframe, and let A, A^1 be congruent modal formulas. Then for any assignment $[\mathbf{a}/\mathbf{x}]$ with $FV(A) \subseteq r(\mathbf{x})$*

$$(**) \qquad M, [\mathbf{a}/\mathbf{x}] \vDash A \Leftrightarrow M, [\mathbf{a}/\mathbf{x}] \vDash A^1,$$

and thus

$$M \vDash A \Leftrightarrow M \vDash A^1.$$

Proof We apply Proposition 2.3.17. Consider the equivalence relation between formulas:

$$A \sim B := \text{ for any assignment } [\mathbf{a}/\mathbf{x}] \text{ such that } FV(A), FV(B) \subseteq r(\mathbf{x}),$$
$$M, [\mathbf{a}/\mathbf{x}] \vDash A \Leftrightarrow M,]\mathbf{a}/\mathbf{x}] \vDash B.$$

It suffices to check that $\sim$ satisfies the conditions (1)–(4) from 2.3.17.

(1) $M, [\mathbf{a}/\mathbf{x}] \vDash \mathcal{Q}yA \Leftrightarrow M, [\mathbf{a}/\mathbf{x}] \vDash \mathcal{Q}z(A[y \mapsto z])$ provided $y \notin BV(A)$, $z \notin V(A)$, $\mathcal{Q} \in \{\forall, \exists\}$ and $FV(\mathcal{Q}yA) \subseteq r(\mathbf{x})$.

(Obviously, $FV(\mathcal{Q}z(A[y \mapsto z])) = FV(\mathcal{Q}yA)$.)

Consider e.g. the case $\mathcal{Q} = \exists$. We may assume $r(\mathbf{x}) = FV(\mathcal{Q}yA)$, since the metaframe is w-functorial. Then $y, z \notin r(\mathbf{x})$, so we have

$$M, [\mathbf{a}/\mathbf{x}] \vDash \exists y A \Leftrightarrow \exists c \in D(\mathbf{a})\ M, [\mathbf{a}c/\mathbf{x}y] \vDash A$$
$$\Leftrightarrow \exists c \in D(\mathbf{a})\ M, [\mathbf{a}c/\mathbf{x}z] \vDash A[y \mapsto z] \text{ (by Lemma 5.10.9)}$$
$$\Leftrightarrow M, [\mathbf{a}/\mathbf{x}] \vDash \exists z(A[y \mapsto z]).$$

(2) Assuming $A \sim B$ let us prove $QyA \sim QyB$ (for $Q = \exists$). We again assume $r(\mathbf{x}) = FV(A)$, so $y \notin r(\mathbf{x})$, and thus we obtain

$$M, [\mathbf{a}/\mathbf{x}] \models \exists y A \Leftrightarrow \exists c \in D(\mathbf{a})\ M, [\mathbf{a}c/\mathbf{x}y] \models A \Leftrightarrow \text{ (since } A \sim B)$$
$$\exists c \in D(\mathbf{a})\ M, [\mathbf{a}c/\mathbf{x}y] \models B \Leftrightarrow M, [\mathbf{a}/\mathbf{x}] \models \exists y B.$$

(3) This property holds trivially, by the definition of forcing.

(4) Assuming $A \sim B$ let us show $\Box_i A \sim \Box_i B$:

$$M, [\mathbf{a}/\mathbf{x}] \models \Box_i A \Leftrightarrow \forall \mathbf{b} \in R_i^n(\mathbf{a})\ M, [\mathbf{b}/\mathbf{x}] \models A$$
$$\Leftrightarrow \forall \mathbf{b} \in R_i^n(\mathbf{a})\ M, [\mathbf{b}/\mathbf{x}] \models B \Leftrightarrow M, [\mathbf{a}/\mathbf{x}] \models \Box_i B.$$

■

Hence we readily obtain

Lemma 5.10.11 *Let $\mathbb{F}$ be a w-functorial metaframe. Then for any congruent modal formulas A, A^1*

$$\mathbb{F} \models A \Leftrightarrow \mathbb{F} \models A^1.$$

Remark 5.10.12 We cannot extend Lemma 5.10.10 to arbitrary (not w-functorial) metaframes. For example, suppose $r(\mathbf{x}z) = FV(B)$, $z \notin BV(B)$ and y, y' are distinct variables that do not occur in B. Then the formulas $\exists y(B[z \mapsto y])$ and $\exists y'(B[z \mapsto y'])$ are congruent. However by 5.9.4 and 5.10.10

$$M, \mathbf{a}b \models \exists y(B[z \mapsto y])\ [\mathbf{x}y] \Leftrightarrow \exists c \in D(\mathbf{a})\ M, \mathbf{a}c \models B[z \mapsto y]\ [\mathbf{x}y] \Leftrightarrow$$

$$(*1) \qquad \exists c \in D(\mathbf{a})\ M, \mathbf{a}c \models B\,[\mathbf{x}z],$$

while

$$M, \mathbf{a}b \models \exists y'(B[z \mapsto y'])\ [\mathbf{x}y] \Leftrightarrow \exists c \in D(\mathbf{a})\ M, \mathbf{a}bc \models B[z \mapsto y']\ [\mathbf{x}yy'] \Leftrightarrow$$

$$(*2) \qquad \exists c \in D(\mathbf{a})\ M, \mathbf{a}bc \models B\ [\mathbf{x}yz].$$

Now, e.g. if $B = \Diamond P(\mathbf{x}, z)$, then $(*1)$ means

$$\exists c\, \exists \mathbf{d}\, \exists e\, ((\mathbf{a}c)R^{n+1}(\mathbf{d}e)\ \&\ \mathbf{d}e \in \xi^+(P)),$$

and $(*2)$ means

$$\exists c\, \exists \mathbf{d}\, \exists e\, \exists e'\, ((\mathbf{a}bc)R^{n+1}(\mathbf{d}e'e)\ \&\ \mathbf{d}e \in \xi^+(P)).$$

If a metaframe is permutable, but not w-functorial, these conditions may be not equivalent, cf. Example 5.9.16.

On the other hand, we do not know if w-functoriality is *necessary* for Lemma 5.10.10. One can try to construct counterexamples explicitly, as in the proof of Lemma 5.10.6.

5.11 Modal metaframes

Definition 5.11.1 *An (N)-*modal metaframe *is a w-functorial metaframe satisfying*

(mm$_2$) $\forall i \in I_N\ \forall a, b_1, b_2\ ((a,a)R_i^2(b_1,b_2) \Rightarrow b_1 = b_2)$.

An equivalent form of (mm$_2$) is $R_i^2(\Delta) \subseteq \Delta$, where

$$\Delta := \{(a,a) \mid a \in D^1\}.$$

For the next lemma recall that

$$\mathbf{a}\, sub\, \mathbf{b} \Leftrightarrow \forall j,k\, (a_j = a_k \Rightarrow b_j = b_k),$$
$$\pi_-^n \mathbf{a} = \mathbf{a}a_n \text{ for } \mathbf{a} \in D^n.$$

Lemma 5.11.2 *For an N-modal metaframe for any $n > 0$, $i \in I_N$ we have*

(mm$_n$) $\forall \mathbf{a} \forall \mathbf{b}\ (\mathbf{a}R_i^n\mathbf{b} \Rightarrow \mathbf{a}\, sub\, \mathbf{b})$;

(mm$_n^+$) $\forall \mathbf{a} \forall \mathbf{b}\ (\mathbf{a}R_i^n\mathbf{b} \Rightarrow (\pi_-^n\mathbf{a})R_i^{n+1}(\pi_-^n\mathbf{b}))$.

Proof In fact, (mm$_n$) follows from (mm$_2$), since $(a_j, a_k) = \mathbf{a} \cdot \lambda_{jk}^n$ (see the Introduction). So for a w-functorial metaframe $\mathbf{a}R_i^n\mathbf{b}$ implies $(a_j,a_k)R_i^2(b_j,b_k)$, and we can apply (mm$_2$) to (a_j,a_k) and (b_j,b_k).

To check (mm$_n^+$), note that $\pi_+^n : F_{n+1} \twoheadrightarrow F_n$ in a w-functorial metaframe. Thus by the lift property,

$$\mathbf{a}R_i^n\mathbf{b} \text{ implies } \exists c\ (\mathbf{a}a_n)R_i^{n+1}(\mathbf{b}c),$$

and by (mm$_{n+1}$), it follows that $c = b_n$. Hence

$$\pi_-^n\mathbf{a} = (\mathbf{a}a_n)R_i^{n+1}(\mathbf{b}b_n) = \pi_-^n\mathbf{b}.$$

∎

The next lemma will be mainly used later on, in the intuitionistic case (section 5.14).

Lemma 5.11.3 *If an N-metaframe satisfies (mm$_m$), $i \in I_N$ and $\sigma \in \Sigma_{mn}$, then $\pi_\sigma[D^n]$ is R_i^m-stable.*

Proof If $\pi_\sigma(\mathbf{a})R_i^m\mathbf{b}$, then $\pi_\sigma(\mathbf{a})\, sub\, \mathbf{b}$, and the latter yields $\sigma\, sub\, \mathbf{b}$, i.e. $\mathbf{b} \in \pi_\sigma[D^n]$, by 5.9.5. ∎

Exercise 5.11.4 Show that for a certain i, (mm$_m$) is equivalent to R_i^m-stability of all sets $\pi_\sigma[D^n]$ for $\sigma \in \Sigma_{mn}$, $n > 0$.

Hint: $\mathbf{a} \in D^m$ can be presented as $\pi_\sigma\mathbf{c}$ for some tuple $\mathbf{c}$ with different components, where $\sigma \in \Sigma_{mn}$, $n = |\mathbf{c}|$.

Definition 5.11.5 *A metaframe* $\mathbb{F}$ *is called* functorial *if every its jection* $\pi_\sigma : D^n \to D^m$ *is a morphism* $F_n \longrightarrow F_m$.

Lemma 5.11.6 *A metaframe is modal iff it is functorial.*

Proof (Only if.) Assume that a metaframe $\mathbb{F}$ is modal. Recall that every function $\sigma \in \Sigma_{mn}$ is a composition of simple injections and simple projections.[19] Since $\mathbb{F}$ is w-functorial, it is sufficient to show that every jection $\pi^n_-(= \pi_{\sigma^n_-})$ is a p-morphism. Its monotonicity is already stated by (mm_n). To check the lift property, assume $\pi^n_-(\mathbf{a}) = (\mathbf{a}a_n)R^{n+1}_i(\mathbf{b}c)$. Then $(\mathbf{a}a_n)\,sub\,(\mathbf{b}c)$ by (mm_{n+1}), and hence $b_n = c$. Thus $\mathbf{b}c = \mathbf{b}b_n = \pi^n_-(\mathbf{b})$. We also have $\mathbf{a}R^n_i\mathbf{b}$ by Lemma 5.10.5(1), since $\mathbb{F}$ is w-functorial.

(If.) We assume that $\mathbb{F}$ is functorial and check (mm_2). In fact, suppose $(a,a) = \pi^1_-(a)R^2_i(b_1,b_2)$. Since π^1_- is a p-morphism, $(b_1,b_2) = \pi^1_-(b)$ for some b, i.e. $b_1 = b_2 = b$. ∎

The next lemma shows that forcing in functorial metaframes 'respects variable substitutions'.

Lemma 5.11.7 *Let* $M = (\mathbb{F}, \xi)$ *be a model over a functorial metaframe. Then, for any modal formula* A

(♮) *for any* $\sigma \in \Sigma_{mn}$, *for any distinct lists of variables* $\mathbf{x} = (x_1,\ldots,x_n)$, $\mathbf{y} = (y_1,\ldots,y_m)$ *such that* $FV(A) \subseteq r(\mathbf{y})$, *and for any* $\mathbf{a} \in D^n$:

$$M, [\mathbf{a}/\mathbf{x}] \vDash [\mathbf{x}\cdot\sigma/\mathbf{y}]\,A \Leftrightarrow M, [\mathbf{a}\cdot\sigma/\mathbf{y}] \vDash A.$$

Recall that $[\mathbf{x}\cdot\sigma/\mathbf{y}]\,A$ is defined up to congruence. But this does not matter in functorial metaframes, by 5.10.10.

Proof By Lemma 5.10.10, the claim (♮) does not change if we replace A (in both sides) with any $B \doteq A$. So we may replace A with its clean version A° such that $BV(A^\circ) \cap r(\mathbf{xy}) = \varnothing$ (and $FV(A^\circ) = FV(A) \subseteq r(\mathbf{y})$); then we have $[\mathbf{x}\cdot\sigma/\mathbf{y}]\,A \doteq A^\circ[\mathbf{y} \mapsto \mathbf{x}\cdot\sigma]$.

To simplify notation, we now assume $A^\circ = A$ and proceed by induction on A.

- Let $A = P^k_j(\mathbf{y}\cdot\tau)$, $\tau \in \Sigma_{km}$. Then $[\mathbf{x}\cdot\sigma/\mathbf{y}]A = P^k_j((\mathbf{x}\cdot\sigma)\cdot\tau) = P^k_j(\mathbf{x}\cdot(\sigma\cdot\tau))$. So by Definition 5.9.4, we have:

  $M, [\mathbf{a}/\mathbf{x}] \vDash [\mathbf{x}\cdot\sigma/\mathbf{y}]\,A \Leftrightarrow \mathbf{a}\cdot(\sigma\cdot\tau) = ((\mathbf{a}\cdot\sigma)\cdot\tau) \in \xi(P^k_j) \Leftrightarrow M, [\mathbf{a}\cdot\sigma/\mathbf{y}] \vDash A.$

- Let $A = (y_j = y_k)$, then $[\mathbf{x}\cdot\sigma/\mathbf{y}]\,A = (x_{\sigma(j)} = x_{\sigma(k)})$. So we have

  $$M, [\mathbf{a}/\mathbf{x}] \vDash [\mathbf{x}\cdot\sigma/\mathbf{y}]\,A \Leftrightarrow a_{\sigma(j)} = a_{\sigma(k)} \Leftrightarrow M, [\mathbf{a}\cdot\sigma/\mathbf{y}] \vDash (y_j = y_k).$$

[19] See Introduction.

- Let $A = \exists z B$, then $z \notin r(\mathbf{xy})$ by the choice of A, and so

$$[\mathbf{x} \cdot \sigma/\mathbf{y}]\, A = \exists z\, [\mathbf{x} \cdot \sigma/\mathbf{y}]\, B = \exists z\, [(\mathbf{x}z) \cdot \sigma^+/\mathbf{y}z]\, B.$$

Hence

$$M, [\mathbf{a}/\mathbf{x}] \vDash [\mathbf{x} \cdot \sigma/\mathbf{y}]A \Leftrightarrow \exists c\ M, [\mathbf{a}c/\mathbf{x}z] \vDash [(\mathbf{x}z) \cdot \sigma^+/\mathbf{y}z]\, B \Leftrightarrow$$
$$\exists c\ M, [(\mathbf{a}c) \cdot \sigma^+/\mathbf{y}z] \vDash B \text{ (by the induction hypothesis) } \Leftrightarrow$$
$$\exists c\ M, [(\mathbf{a} \cdot \sigma)c/\mathbf{y}z] \vDash B \Leftrightarrow M, [\mathbf{a} \cdot \sigma/\mathbf{y}] \vDash A.$$

- Let $A = \Box_i B$. By Definition 5.9.4

$$M, [\mathbf{a}/\mathbf{x}] \vDash [\pi_\sigma(\mathbf{x})/\mathbf{y}]\, A \Leftrightarrow \forall \mathbf{b} \in R_i^n(\mathbf{a})\ M, [\mathbf{b}/\mathbf{x}] \vDash [\pi_\sigma(\mathbf{x})/\mathbf{y}]\, B$$
$$\Leftrightarrow \forall \mathbf{b} \in R_i^n(\mathbf{a})\ M, [\pi_\sigma(\mathbf{b})/\mathbf{y}] \vDash B \text{ (by the induction hypothesis)}$$

$$\Leftrightarrow \forall \mathbf{b}' \in \pi_\sigma[R_i^n(\mathbf{a})]\ M, [\mathbf{b}'/\mathbf{y}] \vDash B. \tag{1}$$

On the other hand, $M, [\pi_\sigma(\mathbf{a})/\mathbf{y}] \vDash A$ is equivalent to

$$\forall \mathbf{b}' \in R_i^m(\pi_\sigma(\mathbf{a}))\ M, [\mathbf{b}'/\mathbf{y}] \vDash B. \tag{2}$$

Now, since π_σ is a p-morphism, we have $R_i^m(\pi_\sigma(\mathbf{a})) = \pi_\sigma[R_i^n(\mathbf{a})]$. Thus (1) $\Leftrightarrow$ (2).

- All other cases are obvious.

■

Hence it follows that variable substitutions preserve validity.

Lemma 5.11.8 *Let $\mathbb{F}$ be a functorial metaframe, A a modal formula such that $\mathbb{F} \vDash A$. Then for any variable substitution $[\mathbf{y}/\mathbf{x}]$, $\mathbb{F} \vDash [\mathbf{y}/\mathbf{x}]A$.*

Proof Since $[\mathbf{y}/\mathbf{x}]A$ does not depend on the variables beyond $FV(A)$, we may assume that $r(\mathbf{x}) = FV(A)$; then $r(\mathbf{y}) = FV([\mathbf{y}/\mathbf{x}]A)$. Let $\mathbf{z}$ be a distinct list such that $r(\mathbf{z}) = r(\mathbf{y})$; then $\mathbf{y} = \mathbf{z} \cdot \sigma$ for some transformation σ. By Definitions 5.9.8, 5.9.9 and Lemma 5.10.7, $\mathbb{F} \vDash [\mathbf{y}/\mathbf{x}]A$ iff for any model M over $\mathbb{F}$, for any assignment $[\mathbf{a}/\mathbf{z}]$ in $\mathbb{F}$

$$M, [\mathbf{a}/\mathbf{z}] \vDash [\mathbf{y}/\mathbf{x}]A.$$

By 5.11.7, the latter is equivalent to

$$M, [\mathbf{a} \cdot \sigma/\mathbf{x}] \vDash A,$$

which follows from $\mathbb{F} \vDash A$. Therefore, $[\mathbf{y}/\mathbf{x}]A$ is valid in $\mathbb{F}$. ■

The next lemma can be considered as a converse to 5.11.7 in a stronger form.

Lemma 5.11.9 *Let $\mathbb{F}$ be a metaframe such that for any model M over $\mathbb{F}$, for any quantifier-free formula A*

($\sharp$) *for any* $\sigma \in \Sigma_{mn}$, *for any distinct lists* $\mathbf{x}$ *of length* n *and* $\mathbf{y}$ *of length* m *such that* $FV(A) \subseteq r(\mathbf{y})$, *for any* $\mathbf{a} \in F_n$

$$M, \mathbf{a} \models ([\mathbf{x} \cdot \sigma/\mathbf{y}] A) [\mathbf{x}] \Leftrightarrow M, \mathbf{a} \cdot \sigma \models A [\mathbf{y}].$$

Then $\mathbb{F}$ *is functorial.*

Proof Given an arbitrary $\sigma \in \Sigma_{mn}$, let us check the p-morphism properties for π_σ.

(1) $\mathbf{a}R_i^n\mathbf{b} \Rightarrow (\pi_\sigma\mathbf{a})R_i^m(\pi_\sigma\mathbf{b})$.
In fact, assume $\mathbf{a}R_i^n\mathbf{b}$. Consider the formula $A := \Diamond_i P(\mathbf{y})$, $P \in PL^m$ and the model $M = (\mathbb{F}, \xi)$ such that $\xi^+(P) = \{\pi_\sigma\mathbf{b}\}$. Then $[\pi_\sigma\mathbf{x}/\mathbf{y}] A = \Diamond_i P(\pi_\sigma\mathbf{x})$,

$$M, \mathbf{b} \models P(\pi_\sigma\mathbf{x}) [\mathbf{x}],$$

and so

$$M, \mathbf{a} \models [\pi_\sigma\mathbf{x}/\mathbf{y}] A [\mathbf{x}].$$

Hence by assumption ($\sharp$) we have $M, \pi_\sigma\mathbf{a} \models A [\mathbf{y}]$, and thus

$$M, \mathbf{b}' \models P(\mathbf{y}) [\mathbf{y}]$$

for some $\mathbf{b}' \in R_i^m(\pi_\sigma\mathbf{a})$. Hence $\mathbf{b}' \in \xi^+(P)$, i.e. $\mathbf{b}' = \pi_\sigma\mathbf{b}$, by the choice of ξ, so we obtain $(\pi_\sigma\mathbf{a})R_i^m(\pi_\sigma\mathbf{b})$.

(2) $(\pi_\sigma\mathbf{a})R_i^m\mathbf{b}' \Rightarrow \exists\mathbf{b} \in R_i^n(\mathbf{a})\ \pi_\sigma\mathbf{b} = \mathbf{b}'$.
Assume $(\pi_\sigma\mathbf{a})R_i^m\mathbf{b}'$. Consider the same formula $A = \Diamond_i P(\mathbf{y})$ and the model $M = (\mathbb{F}, \theta)$ such that $\theta^+(P) = \{\mathbf{b}'\}$. Then

$$M, \mathbf{b}' \models P(\mathbf{y}) [\mathbf{y}],$$

and so

$$M, \pi_\sigma\mathbf{a} \models A [\mathbf{y}].$$

Thus by assumption ($\sharp$),

$$M, \mathbf{a} \models [\pi_\sigma\mathbf{x}/\mathbf{y}] A (= \Diamond_i P(\pi_\sigma\mathbf{x})) [\mathbf{x}],$$

i.e. $\mathbf{a}R_i^n\mathbf{b}$ for some $\mathbf{b}$ such that

$$M, \mathbf{b} \models P(\pi_\sigma\mathbf{x}) [\mathbf{x}].$$

Hence $\pi_\sigma\mathbf{b} \in \theta^+(P)$, i.e. $\pi_\sigma\mathbf{b} = \mathbf{b}'$.

Obviously, if $m = 0$, $\sigma = \varnothing_n$, then we can use the formula $A = \Diamond_i p$ (with $p \in PL^0$) both in (1) and (2). ∎

Remark 5.11.10 Note that in the above proof we use the condition ($\sharp$) only for formulas of the form $\Diamond_i P(\mathbf{y})$ or $\Diamond_i p$.

Hence we obtain

Proposition 5.11.11 *For a metaframe $\mathbb{F}$ the following properties are equivalent:*

(1) $\mathbb{F}$ is functorial;

(2) 5.11.9 ($\sharp$) holds for any formula A;

(3) 5.11.9 ($\sharp$) holds for any quantifier-free A.

Actually ($\sharp$) in (2) should be formulated as the equivalence

$$M, \mathbf{a} \Vdash ([\mathbf{x} \cdot \sigma/\mathbf{y}]A)\ [\mathbf{x}] \Leftrightarrow M, \mathbf{a} \cdot \sigma \Vdash A\ [\mathbf{y}]$$

for any congruent version of $[\mathbf{x} \cdot \sigma/\mathbf{y}]A$.

Note that variable substitutions are defined up to congruence, while in non-functorial metaframes forcing may be sensitive to congruence. But for a quantifier-free A, the formula $[\mathbf{x} \cdot \sigma/\mathbf{y}]\, A$ is unique, so there is no ambiguity in (3).

Proposition 5.11.12 *Let M be a model over a functorial metaframe, $u \in M$, and let A, A^* be modal formulas, $[\mathbf{a}/\mathbf{x}]$, $[\mathbf{a}^*/\mathbf{x}^*]$ assignments giving rise to equal D_u-sentences: $[\mathbf{a}/\mathbf{x}]\, A = [\mathbf{a}^*/\mathbf{x}^*]\, A^*$. Then*

($\sharp\sharp$) $\quad M, \mathbf{a} \vDash A\,[\mathbf{x}] \Leftrightarrow M, \mathbf{a}^* \vDash A\,[\mathbf{x}^*]$.

Proof Recall that (Lemma 2.4.2) a D_u-sentence $[\mathbf{a}/\mathbf{x}]A$ can be presented in the form $[\mathbf{b}/\mathbf{y}]B$, where $\mathbf{b}$ is a list of distinct individuals (from D_u) and $r(\mathbf{y}) = FV(B)$; this presentation is unique up to the choice of a distinct list $\mathbf{y}$. So let us show that

$$(1) \quad M, [\mathbf{a}/\mathbf{x}] \vDash A \text{ iff } M, [\mathbf{b}/\mathbf{y}] \vDash B$$

and

$$(1^*) \quad M, [\mathbf{a}^*/\mathbf{x}^*] \vDash A \text{ iff } M, [\mathbf{b}/\mathbf{y}] \vDash B.$$

Since both these assertions are similar, it is sufficient to check (1). Let $\mathbf{x}'$ be a sublist of $\mathbf{x}$ such that $FV(A) = r(\mathbf{x}')$, $\mathbf{a}'$ the corresponding sublist of $\mathbf{a}$ (i.e. if $\mathbf{x}' = \mathbf{x} \cdot \tau$ for an injection τ, then $\mathbf{a}' = \mathbf{a} \cdot \tau$). Thus

$$(2) \quad M, [\mathbf{a}/\mathbf{x}] \vDash A \text{ iff } M, [\mathbf{a}'/\mathbf{x}'] \vDash A$$

by Lemma 5.10.6, since $\mathbb{F}$ is w-functorial. Here $\mathbf{b}$ is a list of distinct individuals from $\mathbf{a}'$, and B is obtained from A' by identifying those variables x_j in $\mathbf{x}'$, which correspond to equal individuals a_j in $\mathbf{a}'$ (cf. the proof of Lemma 2.4.2). Thus if $m = |\mathbf{x}'| = |\mathbf{a}'|$ and $n = |\mathbf{y}| = |\mathbf{b}|$ is the number of distinct individuals in $\mathbf{a}'$, then for some function $\sigma \in \Sigma_{mn}$, we have $\mathbf{a}' = \mathbf{b} \cdot \sigma$, $B = [\mathbf{y} \cdot \sigma/\mathbf{x}']A$, Now by 5.11.7($\sharp$),

$$(3) \quad M, [\mathbf{b}/\mathbf{y}] \vDash B\,(= [\mathbf{y} \cdot \sigma/\mathbf{x}']\, A) \Leftrightarrow M, [\mathbf{b} \cdot \sigma/\mathbf{x}'] \vDash A.$$

Since $\mathbf{b} \cdot \sigma = \mathbf{a}'$, (1) follows from (2) and (3). ■

Again we have the converse to the above proposition in the following stronger form.

Corollary 5.11.13 *Let $\mathbb{F}$ be a metaframe such that for any model M over $\mathbb{F}$, for any $u \in M$, quantifier-free formulas A, A^* and assignments $[\mathbf{a}/\mathbf{x}]$, $[\mathbf{a}^*/\mathbf{x}^*]$ in $\mathbb{F}$ such that $[\mathbf{a}/\mathbf{x}]A = [\mathbf{a}^*/\mathbf{x}^*]A^*$.*

(♯♯) $M, \mathbf{a} \vDash A\,[\mathbf{x}] \Leftrightarrow M, \mathbf{a}^* \vDash A^*\,[\mathbf{x}^*]$.

Then $\mathbb{F}$ is functorial.

Proof Note that (♯♯) implies 5.11.9 (♯) for a quantifier-free A, since $[\mathbf{a}/\mathbf{x}]([\mathbf{x} \cdot \sigma/\mathbf{y}]A) = [\mathbf{a} \cdot \sigma/\mathbf{y}]A$ by 2.4.2(4). So we can apply Lemma 5.11.9.[20] ∎

Therefore modal (i.e. functorial) metaframes are exactly those in which forcing can be defined for D_u-sentences, cf. Section 5.9.

5.12 Modal soundness

In this section we show that modal metaframes are exactly $\mathrm{m}^{(=)}$-sound metaframes

Recall that by Definition 5.9.8, for a metaframe model $M = (\mathbb{F}, \xi)$, $M \vDash A$ if $M, \mathbf{a} \vDash A\,[\mathbf{x}]$ for any ordered assignment $(\mathbf{x}, \mathbf{a})$ in $\mathbb{F}$ with $FV(A) \subseteq r(\mathbf{x})$. For a w-functorial metaframe we can fix the list $\mathbf{x}$ as the following simple lemma shows.

Lemma 5.12.1 *Let $M = (\mathbb{F}, \xi)$ be a model over a w-functorial N-metaframe $\mathbb{F}$, A an N-modal formula. Then the following conditions are equivalent.*

(1) $M \vDash A$;

(2) $M, \mathbf{a} \vDash A\,[\mathbf{x}]$ *for any ordered assignment* $(\mathbf{x}, \mathbf{a})$ *in* $\mathbb{F}$ *such that* $FV(A) = r(\mathbf{x})$;

(3) there exists a list of distinct variables $\mathbf{y}$ *containing* $FV(A)$ *such that* $M, \mathbf{b} \vDash A\,[\mathbf{y}]$ *for any ordered assignment* $(\mathbf{y}, \mathbf{b})$ *in* $\mathbb{F}$.

Proof Obviously, (1) implies (2), and (2) implies (3).

The other way round, assuming (2), let us show (1), i.e. $M, [\mathbf{b}/\mathbf{y}] \vDash A$ for any assignment $[\mathbf{b}/\mathbf{y}]$ with $r(\mathbf{y}) \supseteq FV(A)$. By 0.0.7, we have $\mathbf{x} = \mathbf{y} \cdot \sigma$ for some injection σ, hence by Lemma 5.10.6, $M, [\mathbf{b}/\mathbf{y}] \vDash A$ is equivalent to $M, [\mathbf{b} \cdot \sigma/\mathbf{x}] \vDash A$, which holds by our assumption.

Finally, assuming (3), let us check (2). In fact, again we have $\mathbf{x} = \pi_\sigma(\mathbf{y})$ for an injection σ, and $\mathbf{a} = \pi_\sigma(\mathbf{b})$ for some $\mathbf{b}$, due to the surjectivity of π_σ (see Lemma 5.9.6); thus 5.10.6 can be applied. ∎

Recall that $\mathbf{ML}^{(=)}_{-}(\mathbb{F})$ denotes the set of all formulas valid in $\mathbb{F}$ (Definition 5.9.9).

[20] This equality holds for a quantifier-free A; for an arbitrary A it should be replaced by congruence.

Lemma 5.12.2 *For any metaframe $\mathbb{F}$:*

(1) $\mathbf{ML}_{-}^{(=)}(\mathbb{F})$ *is closed under necessitation and generalisation.*

(2) If $\mathbb{F}$ is w-functorial, then $\mathbf{ML}_{-}^{(=)}(\mathbb{F})$ *is closed under modus ponens.*

(3) If $\mathbb{F}$ is w-functorial, then for any $B \in MF_N^=$

$$\mathbb{F} \vDash B \Leftrightarrow \mathbb{F} \vDash \forall y B \Leftrightarrow \mathbb{F} \vDash \overline{\forall} B.$$

(4) If $\mathbb{F}$ is functorial, then $\mathbf{ML}_{-}^{(=)}(\mathbb{F})$ *is closed under strict substitutions.*

Proof
(1) Rather trivial. Note that by definition (cf. (8) and (9+10) in Section 5.9)

$$M, \mathbf{a} \vDash \Box_i B[\mathbf{x}] \Leftrightarrow \forall \mathbf{b} \in R_i^n(\mathbf{a})\ M, \mathbf{b} \vDash B\,[\mathbf{x}];$$

and

$$M, \mathbf{a} \vDash \forall y B\,[\mathbf{x}] \Leftrightarrow \forall c \in D_u\ M, \pi_{\mathbf{x}||\mathbf{y}}(\mathbf{a}c) \vDash B\,[\mathbf{x}||y].$$

(if $\mathbf{a} \in D_u^n$).

(2) If $M, [\mathbf{a}/\mathbf{x}] \vDash B \supset C$ and $M, [\mathbf{a}/\mathbf{x}] \vDash B$, for any assignment $(\mathbf{x}, \mathbf{a})$ with a fixed $\mathbf{x}$ containing $FV(B \supset C) \supseteq FV(C)$, then we have $M, [\mathbf{a}/\mathbf{x}] \vDash C$. Hence $M \vDash C$ by Lemma 5.12.1.

(3) Obviously, it suffices to check the first equivalence. ($\Rightarrow$) follows readily from (1).
($\Leftarrow$) Suppose $\mathbb{F} \vDash \forall y B$ and consider a model M over $\mathbb{F}$. Let $r(\mathbf{x}) = FV(\forall y B)$, then for any assignment $(\mathbf{x}, \mathbf{a})$ we have $M, \mathbf{a} \vDash \forall y B\,[\mathbf{x}]$. Since $y \notin \mathbf{x}$, this means that $M, \mathbf{a}c \vDash B\,[\mathbf{x}y]$ for any assignment $(\mathbf{x}y, \mathbf{a}c)$. Hence $M \vDash B$ by Lemma 5.12.1.

(4) Suppose $\mathbb{F} \vDash A$. Consider a strict substitution $S = [C/P(\mathbf{y})]$, where P is an m-ary predicate letter, $\mathbf{y} = (y_1, \ldots, y_m)$ is a distinct list containing $FV(C)$, and let us show that $\mathbb{F} \vDash SA$.

Recall that a substitution instance $[C/P(\mathbf{y})]A$ is defined up to congruence and can be obtained from a clean version A° of A by replacing every subformula of the form $P(\mathbf{y}')$ with $[\mathbf{y}'/\mathbf{y}]C$ ($\mathbf{y}'$ may be not distinct). By Lemma 5.10.10, for any model M over $\mathbb{F}$ we have

$$M \vDash A \text{ iff } M \vDash A^\circ,$$

and thus

$$\mathbb{F} \vDash A \text{ iff } \mathbb{F} \vDash A^\circ.$$

So we may assume that A is clean (i.e. $A = A^\circ$).

For a model $M = (\mathbb{F}, \xi)$ and a list of distinct variables $\mathbf{x}$ such that $r(\mathbf{x}) = FV(A) \supseteq FV([C/P(\mathbf{y})]A)$, we have to show that $M \vDash SA$, i.e. that

(4.1) $M, [\mathbf{a}/\mathbf{x}] \vDash [C/P(\mathbf{y})]A$ for any assignment $[\mathbf{a}/\mathbf{x}]$.

Since A is clean, $BV(A) \cap r(\mathbf{x}) = \varnothing$.

Consider the model $N = (\mathbb{F}, \eta)$ such that

$$\eta^+(P) = \{\mathbf{b} \mid M, [\mathbf{b}/\mathbf{y}] \vDash C\},$$
$$\eta^+(Q) = \xi^+(Q) \text{ for any other predicate letter } Q.$$

We claim that

(4.2) $N, [\mathbf{a}/\mathbf{x}] \vDash A$ iff $M, [\mathbf{a}/\mathbf{x}] \vDash SA$.

By our assumption, $\mathbb{F} \vDash A$, thus $N, [\mathbf{a}/\mathbf{x}] \vDash A$, and so (4.2) implies (4.1). To show (4.2), let us prove

(4.3) $N, [\mathbf{a}/\mathbf{z}] \vDash B$ iff $M, [\mathbf{a}/\mathbf{z}] \vDash SB$

for any subformula B of A and for any assignment $[\mathbf{a}/\mathbf{z}]$ with $r(\mathbf{z}) \supseteq FV(B)$, $r(\mathbf{z}) \cap BV(B) = \varnothing$.

The proof is by induction on the complexity of B.

If $B = P(\mathbf{z} \cdot \sigma)$ for some map σ, then by Definition 5.9.4 and the choice of N,

$$N, [\mathbf{a}/\mathbf{z}] \vDash B \text{ iff } (\mathbf{a} \cdot \sigma) \in \eta^+(P) \text{ iff } M, [\mathbf{a} \cdot \sigma/\mathbf{y}] \vDash C.$$

By Lemma 5.11.7 in a functorial metaframe we have

$$M, [\mathbf{a} \cdot \sigma/\mathbf{y}] \vDash C \text{ iff } M, [\mathbf{a}/\mathbf{z}] \vDash [\mathbf{z} \cdot \sigma/\mathbf{y}]C \; (= SB),$$

and thus B satisfies (4.3).

If $B = B_1 \wedge B_2$ and (4.3) holds for B_1, B_2, we obtain it for B:

$$N, [\mathbf{a}/\mathbf{z}] \vDash B \text{ iff } N, [\mathbf{a}/\mathbf{z}] \vDash B_1 \;\&\; N, [\mathbf{a}/\mathbf{z}] \vDash B_2 \text{ iff}$$
$$M, [\mathbf{a}/\mathbf{z}] \vDash SB_1 \;\&\; M, [\mathbf{a}/\mathbf{z}] \vDash SB_2 \text{ iff } M, [\mathbf{a}/\mathbf{z}] \vDash SB_1 \wedge SB_2 \; (= SB).$$

The cases $B = B_1 \supset B_2$, $B = B_1 \vee B_2$ are similar to the above, and the cases $B = \bot$, $B = Q(z \cdot \sigma)$ $(Q \neq P)$ are trivial.

If $B = \Box_i B_1$, and (4.3) holds for B_1, then for an n-tuple $\mathbf{a}$ we have:

$$N, [\mathbf{a}/\mathbf{z}] \vDash B \text{ iff } \forall \mathbf{b} \in R_i^n(\mathbf{a}) \; N, [\mathbf{b}/\mathbf{z}] \vDash B_1$$
$$\text{iff } \forall \mathbf{b} \in R_i^n(\mathbf{a}) \; M, [\mathbf{b}/\mathbf{z}] \vDash SB_1 \text{ iff } M, [\mathbf{a}/\mathbf{z}] \vDash \Box_i(SB_1) \; (= SB),$$

i.e. (4.3) holds for B.

If $B = \exists v B_1$, then by our assumption, $v \notin r(\mathbf{z})$. Also $r(\mathbf{z}v) \cap BV(B_1) = \varnothing$, since B is clean. So by Definition 5.9.4 and the induction hypothesis we have

$$N, [\mathbf{a}/\mathbf{z}] \vDash B \text{ iff } \exists c\, (\mathbf{a}c \in D^{n+1} \;\&\; N, [\mathbf{a}c/\mathbf{z}v] \vDash B_1)$$
$$\text{iff } \exists c\, (\mathbf{a}c \in D^{n+1} \;\&\; M, [\mathbf{a}c/\mathbf{z}v] \vDash SB_1) \text{ iff } M, [\mathbf{a}/\mathbf{z}] \vDash \exists v(SB_1) \; (= SB).^{21}$$

So (4.3) holds for B.

The case $B = \forall v B_1$ is left to the reader. ■

[21]Note that in our case $S\exists v B_1 = \exists v S B_1$.

Lemma 5.12.2 (3), (4) yields the following convenient description of the set of strongly valued formulas $\mathbf{ML}^{(=)}(\mathbb{F})$ for a functorial metaframe $\mathbb{F}$.

Proposition 5.12.3 *Let $\mathbb{F}$ be a functorial N-metaframe. Then*

$$\begin{array}{rcl}\mathbf{ML}^{(=)}(\mathbb{F}) & = & \{A \in MF_N^{(=)} \mid \mathbb{F} \vDash A^n \text{ for any } n \in \omega\} \\ & = & \{A \in MF_N^{(=)} \mid \mathbb{F} \vDash \overline{A^n} \text{ for any } n \in \omega\}.\end{array}$$

Proof Recall that $\overline{A^n} = \overline{\forall} A^n$, so the second equality follows from 5.12.2(3). Since A^n is a substitution instance of A, we have

$$\{A \in MF_N^{(=)} \mid \forall n\, \mathbb{F} \vDash A^n\} \supseteq \mathbf{ML}^{(=)}(\mathbb{F}).$$

The other way round, suppose $\forall n\, \mathbb{F} \vDash A^n$. By Lemma 2.5.30, every substitution instance SA of A is congruent to a formula of the form $[\mathbf{y}/\mathbf{x}]S_1 A^n$, where S_1 is a strict substitution. By 5.12.2(4) and 5.11.8, the latter formula is valid in $\mathbb{F}$. Thus SA is valid, by 5.10.11. ∎

Corollary 5.12.4 *If $\mathbb{F}$ is a functorial metaframe, then $\mathbf{ML}^{=}(\mathbb{F})$ is conservative over $\mathbf{ML}(\mathbb{F})$.*

Remark 5.12.5 For an arbitrary N-metaframe $\mathbb{F}$, we can only state that $\mathbf{ML}^{=}(\mathbb{F})^{\circ} \subseteq \mathbf{ML}(\mathbb{F})$. In fact, if all $MF_N^{=}$-instances of a formula $A \in MF_N$ are valid in $\mathbb{F}$, then all its MF_N-instances are also valid, but not the other way round. So we cannot claim that $\mathbf{ML}(\mathbb{F}) \subseteq \mathbf{ML}^{=}(\mathbb{F})$.

The next two lemmas are full analogues of 5.3.5 and 5.3.6, so we leave their proofs to the reader.

Lemma 5.12.6 *Let $\mathbb{F}$ be an arbitrary N-metaframe, $M = (\mathbb{F}, \xi)$ a model over $\mathbb{F}$, and let ξ^n be a propositional valuation in its n-th level F_n, $n \geq 0$ such that $\xi^n(p_k) = \xi^+(P_k^n)$ for any $k \geq 0$. Let $M_n = (F_n, \xi^n)$ be the corresponding propositional Kripke model. Then for any N-modal propositional formula A, for any $\mathbf{a} \in F_n$ and for any assignment $(\mathbf{x}, \mathbf{a})$*

$$M_n, \mathbf{a} \vDash A \text{ iff } M, \mathbf{a} \vDash A^n\, [\mathbf{x}].$$

Lemma 5.12.7 *Let $\mathbb{F}$ be a w-functorial N-metaframe, A an N-modal propositional formula. Then*

$$\mathbb{F} \vDash A^n \text{ iff } F_n \vDash A.$$

Remark 5.12.8 The w-functoriality of $\mathbb{F}$ is essential in Lemma 5.12.2, which is used in the proof of 5.12.7.

By 5.12.3 and 5.12.7 we obtain:

Proposition 5.12.9 *For a modal metaframe $\mathbb{F}$*

$$\mathbf{ML}_{\pi}^{(=)}(\mathbb{F}) = \bigcap_{n \in \omega} \mathbf{ML}(F_n).$$

Corollary 5.12.10 *All theorems of $\mathbf{K}_N$ are strongly valid in every N-modal metaframe $\mathbb{F}$.*

Proof Obvious, since they are valid in all propositional frames F_n. ■

Exercise 5.12.11 Show that Proposition 5.12.9 (and Corollary 5.12.10) holds for any N-metaframe $\mathbb{F}$ (not necessarily functorial). Hint: for a substitution instance A' of a propositional formula A and for an assignment $(\mathbf{s}, \mathbf{a})$ (where $FV(A') \subseteq r(\mathbf{x})$ and $\mathbf{a} \in D^n$) the truth of $M, \mathbf{a} \vDash A'[\mathbf{x}]$ can be checked in the propositional frame F_n.

However the claim of the above exercise is not so important, because as we will show in Theorem 5.12.13, only modal (i.e. functorial) N-metaframes validate $\mathbf{QK}_N^{(=)}$. Therefore only modal metaframes are interesting from the logical point of view.

Lemma 5.12.12 *If a metaframe $\mathbb{F}$ is functorial, then the set $\mathbf{ML}^{(=)}(\mathbb{F})$ is closed under necessitation, generalisation, modus ponens, and arbitrary substitutions.*

Proof For a functorial $\mathbb{F}$, $\mathbf{ML}^{(=)}(\mathbb{F})$ is substitution closed, by Lemma 2.5.27. (Note that $\mathbf{ML}^{(=)}(\mathbb{F})$ is the largest substitution closed subset, the 'substitution interior' of $\mathbf{ML}_{-}^{(=)}(\mathbb{F})$.)

By Lemma 5.12.2 and Proposition 5.12.3, it follows that $\mathbf{ML}^{(=)}(\mathbb{F})$ is closed under necessitation and generalisation (for a functorial $\mathbb{F}$) as well as under modus ponens (for an *arbitrary* metaframe $\mathbb{F}$).

For example, consider generalisation. Suppose $A \in \mathbf{ML}^{(=)}(\mathbb{F})$. Then $A^n \in \mathbf{ML}_{-}^{(=)}(\mathbb{F})$ for any n, and hence $(\forall x A)^n = \forall x A^n \in \mathbf{ML}_{-}^{(=)}(\mathbb{F})$ by 5.12.2. Hence $\forall x A \in \mathbf{ML}^{(=)}(\mathbb{F})$ by 5.12.3. ■

Theorem 5.12.13 (Soundness theorem) *Let $\mathbb{F}$ be an N-metaframe. Then the following properties are equivalent:*

(1) *$\mathbb{F}$ is modal;*

(2) *$\mathbf{ML}(\mathbb{F})$ is a predicate N-modal logic without equality;*

(3) *$\mathbf{ML}^{=}(\mathbb{F})$ is a predicate N-modal logic with equality;*

(4) $\mathbf{QK}_N \subseteq \mathbf{ML}(\mathbb{F})$;

(5) $\mathbf{QK}_N^{=} \subseteq \mathbf{ML}^{=}(\mathbb{F})$.

Proof (1) $\Rightarrow$ (3), (1) $\Rightarrow$ (2). For a modal metaframe $\mathbb{F}$, let us check the strong validity of the predicate axioms and the axioms of equality, i.e. the validity of their n-shifts (for $n \geq 0$). We fix a model $M = (\mathbb{F}, \xi)$ and do not indicate it in the notation of forcing.

- $A_1 := \forall y P(\mathbf{x}, y) \supset P(\mathbf{x}, z)$, for $y, z \notin r(\mathbf{x})$, $y \neq z$.

  Suppose $[\mathbf{a}b/\mathbf{x}z] \models \forall y P(\mathbf{x}, y)$, $\mathbf{a}b \in D_u^{n+1}$; then by 5.9.4, for any $c \in D_u$, $[\mathbf{a}bc/\mathbf{x}zy] \models P(\mathbf{x}, y)$, which is equivalent to $\forall c \in D_u\ \mathbf{a}c \in \xi^+(P)$. Hence $\mathbf{a}b \in \xi^+(P)$, and thus $[\mathbf{a}b/\mathbf{x}z] \models P(\mathbf{x}, z)$. Therefore $[\mathbf{a}b/\mathbf{x}z] \models A_1$.

- $A_2 := P(\mathbf{x}, z) \supset \exists y P(\mathbf{x}, y)$, where $y, z \notin r(\mathbf{x})$, $y \neq z$.

  Similarly to the previous case, assume $[\mathbf{a}b/\mathbf{x}z] \models P(\mathbf{x}, z)$, i.e. $\mathbf{a}b \in \xi^+(P)$. Then for some $c \in D_u$ (viz. for $c = b$), $[\mathbf{a}bc/\mathbf{x}zy] \models P(\mathbf{x}, y)$, and thus $[\mathbf{a}b/\mathbf{x}z] \models \exists y P(\mathbf{x}, y)$. So $[\mathbf{a}b/\mathbf{x}z] \models A_2$.

- $A_3 := \forall y(Q(\mathbf{x}) \supset P(\mathbf{x}, y)) \supset (Q(\mathbf{x}) \supset \forall y P(\mathbf{x}, y))$, where $y \notin r(\mathbf{x})$.

  Assume
  $$[\mathbf{a}/\mathbf{x}] \models \forall y(Q(\mathbf{x}) \supset P(\mathbf{x}, y)),$$
  $\mathbf{a} \in D_u^n$. Then for any $b \in D_u$, we have
  $$[\mathbf{a}b/\mathbf{x}y] \models Q(\mathbf{x}) \supset P(\mathbf{x}, y).$$
  If also $[\mathbf{a}/\mathbf{x}] \models Q(\mathbf{x})$, i.e. $\mathbf{a} \in \xi^+(Q)$, then for any $b \in D_u$, $[\mathbf{a}b/\mathbf{x}y] \models Q(\mathbf{x})$. Hence $[\mathbf{a}b/\mathbf{x}y] \models P(\mathbf{x}, y)$ for any b, and thus we obtain
  $$[\mathbf{a}/\mathbf{x}] \models \forall y\, P(\mathbf{x}, y).$$
  Therefore $M \models A_3$.

- $A_4 := \forall y(P(\mathbf{x}, y) \supset Q(\mathbf{x})) \supset (\exists y P(\mathbf{x}, y) \supset Q(\mathbf{x}))$, $y \notin \mathbf{x}$.

  Assume
  $$[\mathbf{a}/\mathbf{x}] \models \forall y(P(\mathbf{x}, y) \supset Q(\mathbf{x}))$$
  and
  $$[\mathbf{a}/\mathbf{x}] \models \exists y P(\mathbf{x}, y).$$
  Then
  $$[\mathbf{a}b/\mathbf{x}y] \models P(\mathbf{x}, y)$$
  for some $b \in D_u$, and also
  $$[\mathbf{a}b/\mathbf{x}y] \models P(\mathbf{x}, y) \supset Q(\mathbf{x}).$$
  Thus $[\mathbf{a}b/\mathbf{x}y] \models Q(\mathbf{x})$, i.e. $\mathbf{a} \in \xi^+(Q)$, which implies $[\mathbf{a}/\mathbf{x}] \models Q(\mathbf{x})$. Therefore $M \models A_4$.

- $A_5 := (x = x)$.

  $[a/x] \models x = x$ holds trivially, since $a = a$.

- $A_6 := (y = z \supset (P(\mathbf{x}, y) \supset P(\mathbf{x}, z))$, $y, z \notin r(\mathbf{x})$.

  We may assume that y, z are distinct, since otherwise $M \models A_6$ is trivial.

  If $[\mathbf{a}bc/\mathbf{x}yz] \models y = z$, and $[\mathbf{a}bc/\mathbf{x}yz] \models P(\mathbf{x}, y)$, then by Definition 5.9.4, $b = c$ and $\mathbf{a}b \in \xi^+(P)$, which implies $[\mathbf{a}bc/\mathbf{x}yz] \models P(\mathbf{x}, z)$ by the same definition. Therefore $M \models A_6$.

The implications (2) $\Rightarrow$ (4), (3) $\Rightarrow$ (5) are obvious. (5) $\Rightarrow$ (4) follows by Remark 5.12.5.

(4) $\Rightarrow$ (1). We assume $\mathbf{QK}_N \subseteq \mathbf{ML}(\mathbb{F})$ and show that $\mathbb{F}$ is functorial.

(I) $\mathbb{F}$ is permutable, i.e.

$$\mathbf{a}R_i^n\mathbf{b} \Rightarrow \pi_\sigma(\mathbf{a})R_i^n\pi_\sigma(\mathbf{b}) \text{ for any } \sigma \in \Upsilon_n,\ n > 0.$$

Suppose $\mathbf{a}R_i^n\mathbf{b}$, $\mathbf{a} \in D_u^n$. Consider the $\mathbf{QK}_N$-theorem

$$B_1 := \exists\mathbf{x}(P(\pi_\sigma(\mathbf{x})) \wedge \Diamond_i Q(\pi_\sigma(\mathbf{x}))) \supset \exists\mathbf{x}(P(\mathbf{x}) \wedge \Diamond_i Q(\mathbf{x})),\ |\mathbf{x}| = n,$$

and a model $M = (\mathbb{F}, \xi)$ such that $\xi^+(P) = \{\pi_\sigma(\mathbf{a})\}$, $\xi^+(Q) = \{\pi_\sigma(\mathbf{b})\}$. Then $M, \mathbf{b} \models Q(\pi_\sigma(\mathbf{x}))\,[\mathbf{x}]$ and $M, \mathbf{a} \models (P(\pi_\sigma(\mathbf{x}))) \wedge \Diamond_i Q(\pi_\sigma(\mathbf{x}))\,[\mathbf{x}]$.

Hence

$$M, u \models \exists\mathbf{x}(P(\pi_\sigma(\mathbf{x})) \wedge \Diamond_i Q(\pi_\sigma(\mathbf{x}))).$$

Since $M, u \models B_1$, we have

$$M, u \models \exists\mathbf{x}(P(\mathbf{x}) \wedge \Diamond_i Q(\mathbf{x})),$$

i.e.

$$M, \mathbf{c} \models (P(\mathbf{x}) \wedge \Diamond_i Q(\mathbf{x}))\,[\mathbf{x}]$$

for some $\mathbf{c} \in D_u^n$. Thus $\mathbf{c} \in \xi^+(P)$, i.e. $\mathbf{c} = \pi_\sigma(\mathbf{a})$ and $\mathbf{c}R_i^n\mathbf{d}$ for some $\mathbf{d} \in \xi^+(Q)$, i.e. for $\mathbf{d} = \pi_\sigma(\mathbf{b})$. Therefore $\pi_\sigma(\mathbf{a})R_i^n\pi_\sigma(\mathbf{b})$.

(II) $\mathbb{F}$ is w-functorial.

We apply Lemma 5.10.6.

(IIa) $aR_i^1b \Rightarrow uR_iv$, whenever $a \in D_u$, $b \in D_v$.

In fact, assume aR_i^1b. Consider the $\mathbf{QK}_N$-theorem

$$B_2 := \exists y \Diamond_i p \supset \Diamond_i p,$$

where $p \in PL^0$, and a model $M = (\mathbb{F}, \xi)$ such that $\xi^+(p) = \{v\}$. Then (by Definition 5.9.4) $M, b \models p\,[y]$, (since $\pi_\varnothing(b) = v$), and so

$$M, a \models \Diamond_i p\ [y],$$

since aR_i^1b. Thus

$$M, u \models \exists y \Diamond_i p.$$

Since $M, u \models B_2$, we have $M, u \models \Diamond_i p$, which implies uR_iv.

(IIb) $(\mathbf{a}c)R_i^{n+1}(\mathbf{b}d) \Rightarrow \mathbf{a}R_i^n\mathbf{b}$ for $n > 0$.

Suppose $(\mathbf{a}c)R_i^{n+1}(\mathbf{b}d)$. Consider the $\mathbf{QK}_N$-theorem

$$B_3 := \exists y \Diamond_i P(\mathbf{x}) \supset \Diamond_i P(\mathbf{x}),$$

where $|\mathbf{x}| = n$, $y \notin r(\mathbf{x})$, and a model $M = (\mathbb{F}, \xi)$ such that $\xi^+(P) = \{\mathbf{b}\}$. Then

$$M, \mathbf{b}d \models P(\mathbf{x})\,[\mathbf{x}y].$$

Since $(\mathbf{a}c)R_i^{n+1}(\mathbf{b}d)$, we have

$$M, \mathbf{a}c \vDash \Diamond_i P(\mathbf{x})\,[\mathbf{x}y],$$

and so

$$M, \mathbf{a} \vDash \exists y \Diamond_i P^n(\mathbf{x})\,[\mathbf{x}].$$

Now from $M, \mathbf{a} \vDash B_3\ [\mathbf{x}]$, we obtain $M, \mathbf{a} \vDash \Diamond_i P(\mathbf{x})\,[\mathbf{x}]$, and thus $\mathbf{a}R_i^n\mathbf{b}$ by the choice of M.

(IIc) $\mathbf{a}c \in D^{n+1}\ \&\ \mathbf{a}R_i^n\mathbf{b} \Rightarrow \exists d\,(\mathbf{a}c)R_i^{n+1}(\mathbf{b}d)$ for $n > 0$.

In fact, assume $\mathbf{a}c \in D_u^{n+1}$, $\mathbf{a}R_i^n\mathbf{b}$, and consider the $\mathbf{QK}_N$-theorem

$$B_4 := \Diamond_i P(\mathbf{x}) \supset \forall y \Diamond_i P(\mathbf{x}),$$

where $|\mathbf{x}| = n$, $y \notin \mathbf{x}$, and a model $M = (\mathbb{F}, \xi)$ such that $\xi^+(P) = \{\mathbf{b}\}$. Then $M, \mathbf{b} \vDash P(\mathbf{x})\,[\mathbf{x}]$, and so

$$M, \mathbf{a} \vDash \Diamond_i P(\mathbf{x})\,[\mathbf{x}].$$

Since $M, \mathbf{a} \vDash B_4\ [\mathbf{x}]$, we also have

$$M, \mathbf{a} \vDash \forall y \Diamond_i P(\mathbf{x})\,[\mathbf{x}],$$

i.e. $M, \mathbf{a}e \vDash \Diamond_i P(\mathbf{x})\,[\mathbf{x}y]$ for any $e \in D_u$. In particular, $M, \mathbf{a}c \vDash \Diamond_i P(\mathbf{x})\,[\mathbf{x}y]$. So there exists $\mathbf{g} = (g_1, \ldots, g_{n+1}) \in D^{n+1}$ such that

$$(\mathbf{a}c)R_i^{n+1}\mathbf{g}\ \&\ M, \mathbf{g} \vDash P(\mathbf{x})\,[\mathbf{x}y].$$

Hence $(g_1, \ldots, g_n) = \mathbf{b}$, i.e. $(\mathbf{a}c)R_i^{n+1}(\mathbf{b}g_{n+1})$, and so we can take $d = g_{n+1}$.

(IId) $uR_iv \Rightarrow \forall a \in D_u\ \exists b \in D_v\ aR_i^1b$.

Consider the $\mathbf{QK}_N$-theorem

$$B_5 := \Diamond_i p \supset \forall y \Diamond_i p,$$

for $p \in PL^0$ and a model $M = (\mathbb{F}, \xi)$ such that $\xi^+(p) = \{v\}$. Then we obviously have $M, v \vDash p$, and $M, u \vDash \Diamond_i p$. Since $M, u \vDash B_5$, it follows that $M, u \vDash \forall y \Diamond_i p$. Thus

$$\forall a \in D_u\ M, a \vDash \Diamond_i p\ [y],$$

which yields $M, b \vDash p\ [y]$ for some $b \in R_i^1(a)$. By definition, the latter means that $\pi_\varnothing(b) \in \xi^+(p)$, i.e. $\pi_\varnothing(b) = v$, or $b \in D_v$.

(III) $(a, a)R_i^2(b_1, b_2) \Rightarrow b_1 = b_2$, for $a \in D_u$, $b_1, b_2 \in D_v$.

This property (mm_2) easily follows from the validity of the $\mathbf{QK}_N^=$-theorem $x = y \supset \Box_i(x = y)$. In fact, obviously, $aa \vDash x = y\ [xy]$. Thus $aa \vDash x =$

$y \supset \Box_i(x = y)\,[xy]$ implies $aa \vDash \Box_i(x = y)\,[xy]$, which is equivalent to (III).

So we have proved (5) $\Rightarrow$ (1).

To check (III) for the case without equality (i.e. for the proof of (4)$\Rightarrow$(1)) we need a longer argument.

Let $(a,a)R_i^2(b_1,b_2)$, $a \in D_u$. Consider the $\mathbf{QK}_N$-theorem

$$B_6 := \exists x(P(x) \wedge \Box_i Q(x,x)) \supset \exists x_1 \exists x_2(P(x_1) \wedge P(x_2) \wedge \Box_i Q(x_1,x_2)).$$

Take a model $M = (\mathbb{F}, \xi)$ such that

$$\xi^+(P) = \{a\},\ \xi^+(Q) = \{(b,b) \mid b \in D^1\}.$$

Then $M, a \vDash \Box_i Q(x,x)\,[x]$, since $M, b \vDash Q(x,x)\,[x]$ for any b. We also have $M, a \vDash P(x)\,[x]$; hence $M, u \vDash \exists x(P(x) \wedge \Box_i Q(x,x))$. Thus from $M, u \vDash B_6$ we obtain:

$$M, u \vDash \exists x_1 \exists x_2(P(x_2) \wedge P(x_1) \wedge \Box_i Q(x_1,x_2)),$$

i.e.

$$M, a_1a_2 \vDash P(x_1) \wedge P(x_2) \wedge \Box_i Q(x_1,x_2)\,[x_1x_2]$$

for some $a_1, a_2 \in D_u$. Then $M, a_1a_2 \vDash P(x_1) \wedge P(x_2)\,[x_1x_2]$, which implies $a_1 = a_2 = a$, since $\xi^+(P) = \{a\}$. Now we have

$$M, aa \vDash \Box_i Q(x_1,x_2)\,[x_1x_2]$$

and so

$$M, b_1b_2 \vDash Q(x_1,x_2)\,[x_1x_2],$$

since $(a,a)R_i^2(b_1,b_2)$. Thus $(b_1,b_2) \in \xi^+(Q)$, i.e. $b_1 = b_2$.

Therefore $\mathbb{F}$ is modal. ■

Theorem 5.12.13 means that modal metaframes are exactly m-sound (and also $m^=$-sound) metaframes. This allows us to introduce the metaframe semantics $\mathcal{M}_N^=$ and $\mathcal{M}_N$ for modal predicate logics (with or without equality) generated by the class of N-modal (or functorial) metaframes. This is the *largest sound* modal Kripke-type semantics generated by metaframes. Therefore we obtain a precise criterion of logical soundness in metaframes. In the next section this criterion will be applied to Kripke bundles and $\mathcal{C}$-sets.

Now let us make some remarks on terminology. According to general definitions, a metaframe validating a modal logic L (predicate or propositional), should be called an 'L-metaframe'. If L is a propositional logic, 'L-metaframes' are those, for which every F_n is a propositional L-frame (Proposition 5.12.9); such metaframes are not necessarily modal. But if L is a predicate logic, then $\mathbb{F}$ is an L-metaframe iff $L \subseteq \mathbf{ML}^{(=)}(\mathbb{F})$; so every L-metaframe is modal (Theorem 5.12.13).

In particular, for a propositional logic $\mathbf{\Lambda}$, $\mathbf{Q\Lambda}$-metaframes are just modal $\mathbf{\Lambda}$-metaframes. So e.g. $\mathbb{F}$ is an **S4**-metaframe iff every F_n is reflexive and transitive. An **S4**-metaframe is also called *propositionally intuitionistic.* On the other hand, a **QS4**-*metaframe*, is a 1-modal **S4**-metaframe.

Finally let us explain why the modified forcing $\models^*$ described in 5.9.17 does not really change the notion of logical soundness. To see this we consider an arbitrary modification $\models'$ of forcing satisfying the clauses (1)–(9) from Definition 5.9.4. Then we define the notions of truth, validity and strong validity for $\models'$ in the natural way. Let $\mathbf{ML}'^{(=)}(\mathbb{F})$ be the corresponding set of strongly valid formulas in a metaframe $\mathbb{F}$. If this set is an m.p.l. (=), we say that $\mathbb{F}$ is *modally sound for* $\models'$ (respectively, without or with equality). This notion is described as follows.

Proposition 5.12.14 *Consider a modified forcing relation $\models'$ satisfying 5.9.4(1)-(9).*

(1) If $\mathbf{QK}_N \subseteq \mathbf{ML}'(\mathbb{F})$, then $\mathbb{F}$ is a modal metaframe;

(2) If $\models'$ is equivalent to $\models$ in modal metaframes, i.e.

$$M, \mathbf{a} \models' A\ [\mathbf{x}] \Leftrightarrow M, \mathbf{a} \models A\ [\mathbf{x}]$$

for any model M over a modal metaframe, for any assignment $(\mathbf{x}, \mathbf{a})$ in M and formula A with $FV(A) \subseteq r(\mathbf{x})$, then for any metaframe $\mathbb{F}$ the following conditions are equivalent:

(i) $\mathbb{F}$ is modally sound for $\models'$ (with or without equality);

(ii) $\mathbb{F}$ is modally sound;

(iii) $\mathbb{F}$ is modal.

Moreover, $\mathbf{ML}'^{(=)}(\mathbb{F}) = \mathbf{ML}^{(=)}(\mathbb{F})$ for modal (i.e. modally sound) $\mathbb{F}$.

Proof

(1) We can repeat the part (4) $\Rightarrow$ (1) from the proof of 5.12.13. The argument does not use the clause (1°), because it does not involve forcing $M, \mathbf{a} \models \exists y B[\mathbf{x}]$ with $y \in r(\mathbf{x})$.[22]

(2) Obvious.

∎

Thus a 'reasonable' modification of 5.9.4(10) does not affect logical soundness. This applies to the forcing $\models^*$ from 5.9.17, because it is equivalent to $\models$ in modal metaframes. In fact, by 5.9.4 we have

$$\begin{aligned} M, \mathbf{a} \models \exists x_i B[\mathbf{x}] &\Leftrightarrow M, \mathbf{a} - a_i \models \exists x_i B\ [\mathbf{x} - x_i] \\ &\Leftrightarrow \exists c \in D(\mathbf{a})\ M, (\mathbf{a} - a_i)c \models B\ [(\mathbf{x} - x_i)x_i)]. \end{aligned}$$

[22] One can easily rewrite the proof using only clean formulas and forcing $M, \mathbf{a} \models A\ [\mathbf{x}]$ with $r(\mathbf{x}) = FV(A)$.

Now let $|\mathbf{x}| = n$, and let $\sigma \in \Upsilon_n$ be a permutation

$$\begin{pmatrix} 1\dots i-1 & i & \dots & n-1 & n \\ 1\dots i-1 & i+1 & & n & i \end{pmatrix};$$

then $((\mathbf{x} - x_i)x_i) \cdot \sigma = \mathbf{x}$. By 5.10.6 we can further write

$$M, \mathbf{a} \vDash \exists x_i B\ [\mathbf{x}] \Leftrightarrow \exists c \in D(\mathbf{a})\ M, ((\mathbf{a} - a_i)c) \cdot \sigma \vDash B\ [\mathbf{x}].$$

But $((\mathbf{a} - a_i)c) \cdot \sigma = (a_1, \dots, a_{i-1}, c, a_{i+1}, \dots, a_n)$, so the existence of c is equivalent to the existence of $\mathbf{b} \in D^n$ such that $\mathbf{b} - b_i = \mathbf{a} - a_i$ and

$$M, \mathbf{b} \vDash B\ [\mathbf{x}].$$

Thus $\vDash$ satisfies the inductive clause (10*) from 5.9.17 for $\exists$ (and similarly for $\forall$). Hence the equivalence of $M, \mathbf{a} \vDash A\ [\mathbf{x}]$ and $M, \mathbf{a} \vDash^* A\ [\mathbf{x}]$ easily follows by induction on $|A|$.

5.13 Representation theorem for modal metaframes

Recall that a preset $\mathbf{F} = (\mathcal{C}, D, \rho)$ over an N-precategory $\mathcal{C}$ gives rise to the metaframe $\mathbb{M}\mathrm{f}(\mathbf{F}) = (F_n)_{n\in\omega}$ with the relations

$$\mathbf{a} R_i^n \mathbf{b} \Leftrightarrow \exists \gamma \in Mor_i\mathcal{C}\ \rho_\gamma \cdot \mathbf{a} = \mathbf{b},$$

cf. Definition 5.6.14. By 5.9.13, $\mathbf{ML}^{(=)}(\mathbb{M}\mathrm{f}(\mathbf{F})) = \mathbf{ML}^{(=)}(\mathbf{F})$.

We also know that every Kripke bundle $\mathbf{F} = (F, D, \rho)$ is associated with a metaframe $\mathbb{M}\mathrm{f}(\mathbf{F}) = (F_n)_{n\in\omega}$ (cf. Definition 5.3.2) as well as with a preset $\mathbf{F}'$ (over some precategory), cf. 5.8.5. By Proposition 5.8.5 these constructions are coherent, i.e. $\mathbb{M}\mathrm{f}(\mathbf{F}) = \mathbb{M}\mathrm{f}(\mathbf{F}')$. Recall (Definition 5.3.2) that the relations in $\mathbb{M}\mathrm{f}(\mathbf{F})$ are

$$\mathbf{a} R_i^n \mathbf{b} \text{ iff } \forall j\ a_j \rho_i b_j\ \&\ \mathbf{a}\ sub\ \mathbf{b}.$$

By 5.9.13, $\mathbf{ML}^{(=)}(\mathbf{F}) = \mathbf{ML}^{(=)}(\mathbb{M}\mathrm{f}(\mathbf{F}))$; hence $\mathbf{ML}^{(=)}(\mathbf{F}) = \mathbf{ML}^{(=)}(\mathbf{F}')$.

Lemma 5.13.1

(1) If $\mathbf{F}$ is a preset over an N-precategory $\mathcal{C}$ (or an N-modal Kripke bundle), then $\mathbb{M}\mathrm{f}(\mathbf{F})$ is an N-modal metaframe.

(2) If $\mathbf{F}$ is a $\mathcal{C}$-set over a category $\mathcal{C}$ (or an intuitionistic Kripke bundle), then $\mathbb{M}\mathrm{f}(\mathbf{F})$ is a $\mathbf{QS4}$-metaframe.

Proof

(1) By the above remark, it is sufficient to consider only presets. To show the w-functoriality, consider $\sigma \in \Upsilon_{mn}$, and show that $\pi_\sigma : F_n \twoheadrightarrow F_m$. To check the

monotonicity, suppose $\mathbf{a}R_i^n\mathbf{b}$, $u = \pi_\varnothing(\mathbf{a})$, $v = \pi_\varnothing(\mathbf{b})$. Then $\rho_\gamma \cdot \mathbf{a} = \mathbf{b}$ for some $\gamma \in \mathcal{C}_i(u, v)$.

Now note that

(3) $\rho_\gamma \cdot (\mathbf{a} \cdot \sigma) = (\rho_\gamma \cdot \mathbf{a}) \cdot \sigma$.

In fact,

$$(\rho_\gamma \cdot (\mathbf{a} \cdot \sigma))_j = \rho_\gamma((\mathbf{a} \cdot \sigma)_j) = \rho_\gamma(a_{\sigma(j)}) = (\rho_\gamma \cdot \mathbf{a})_{\sigma(j)} = ((\rho_\gamma \cdot \mathbf{a}) \cdot \sigma)_j.$$

So by (3) we have:

$$\rho_\gamma \cdot (\mathbf{a} \cdot \sigma) = (\rho_\gamma \cdot \mathbf{a}) \cdot \sigma = \mathbf{b} \cdot \sigma.$$

Thus $(\mathbf{a} \cdot \sigma)R_i^n(\mathbf{b} \cdot \sigma)$.

Next, let us check the lift property for π_σ. If $(\mathbf{a} \cdot \sigma)R_i^n\mathbf{b}'$, $u = \pi_\varnothing(\mathbf{a})$, $v = \pi_\varnothing(\mathbf{b}')$, then $\rho_\gamma \cdot (\mathbf{a} \cdot \sigma) = \mathbf{b}'$ for some $\gamma \in \mathcal{C}_i(u, v)$. By (1) it follows that $(\rho_\gamma \cdot \mathbf{a}) \cdot \sigma = \mathbf{b}'$, i.e. $\mathbf{b}' = \pi_\sigma\mathbf{b}$, where $\mathbf{b} = \rho_\gamma \cdot \mathbf{a}$, and thus $\mathbf{a}R_i^n\mathbf{b}$.

Therefore $\pi_\sigma : F_n \twoheadrightarrow F_m$.

To check the property 5.11.1 (mm_2), suppose $(a, a)R_i^2(b_1, b_2)$, $a \in D_u$, $b_1, b_2 \in D_v$. Then for some $\gamma \in \mathcal{C}_i(u, v)$

$$\rho_\gamma \cdot (a, a) = (b_1, b_2),$$

i.e. $b_1 = \rho_\gamma(a) = b_2$. Thus ($\mathrm{mm}_2$) holds.

(2) If $\mathbf{F}$ is a $\mathcal{C}$-set, then every F_n is an **S4**-frame by 5.6.16. The same holds for intuitionistic Kripke bundles by 5.5.1 or by the observation that if $\mathbb{F}$ is an intuitionistic Kripke bundle, then the corresponding $\mathcal{C}$-preset $\mathbf{F}'$ is a $\mathcal{C}$-set (Proposition 5.8.5), and $\mathbb{M}\mathbf{f}(\mathbb{F}) = \mathbb{M}\mathbf{f}(\mathbf{F}')$. ∎

Therefore by Proposition 5.9.13, Theorem 5.12.13, Proposition 5.12.3, and Corollary 5.12.4 we obtain

Proposition 5.13.2 *Let* $\mathbf{F}$ *be a preset over an* N*-precategory* $\mathcal{C}$ *(or an* N*-modal Kripke bundle). Then*

(1) $\mathbf{ML}^{(=)}(\mathbf{F}) = \mathbf{ML}^{(=)}(\mathbb{M}\mathbf{f}(\mathbf{F}))$ *is an N-m.p.l.(=);*

(2) $\mathbf{ML}^{(=)}(\mathbf{F}) = \{A \in MF_N^{(=)} \mid \forall m\ \mathbf{F} \vDash A^m\} = \{A \in MF_N^{(=)} \mid \forall m\ \mathbf{F} \vDash \overline{A^m}\}$*;*

(3) $\mathbf{ML}^{=}(\mathbf{F})$ *is conservative over* $\mathbf{ML}(\mathbf{F})$.

(4) if $\mathbf{F}$ *is* C*-set or an intuitionistic Kripke bundle, then* $\mathbf{ML}^{(=)}(\mathbf{F}) \supseteq \mathbf{QS4}^{(=)}$.

Let us now show that all countable modal metaframes can be represented by C-sets.

Definition 5.13.3 *We say that* $D = (D_u : u \in W)$ *is a* system of countable domains *if every* D_u *is countable.*[23] *In this case we also say that a metaframe with a system of domains* D has countable domains.

[23] Recall that according to our terminology, a countable set may be finite.

Theorem 5.13.4 (Representation theorem)

(1) Let $\mathbb{F}$ be an N-modal metaframe with countable domains. Then $\mathbb{F} = \mathbb{M}\mathbf{f}(\mathbf{F}')$ for some preset $\mathbf{F}'$ over an N-precategory $\mathcal{C}$. Moreover, if $\mathbb{F}$ is a **QS4***-metaframe, then $\mathcal{C}$ is a category and $\mathbf{F}'$ is a $\mathcal{C}$-set.*

*(2) There exists a modal (***QS4***-) metaframe $\mathbb{F}$, for which (1) does not hold, and all but one domains are countable.*

Proof

(1) Let $\mathbb{F} = ((F_n)_{n\in\omega}, D)$. Consider the N-precategory $\mathcal{C}$ with $Ob\,\mathcal{C} = W$ and

(1.1) $\mathcal{C}_i(u, v) = \{f : D_u \to D_v \mid \forall n > 0\ \forall \mathbf{a} \in D_u^n\ \mathbf{a}\, R_i^n\, (f \cdot \mathbf{a})\}$.

Note that if $\mathbb{F}$ is a **QS4**-metaframe ($N = 1$), then $\mathcal{C}$ is a concrete category, i.e. the composition of morphisms is the composition of functions and identity morphisms are identity functions. Recall that R^n are reflexive and transitive in this case; thus $f \circ g \in \mathcal{C}(u, w)$ whenever $f \in \mathcal{C}(u, v)$, $g \in \mathcal{C}(v, w)$, $uRvRw$, and also $id_{D_u} \in \mathcal{C}(u, u)$ for $u \in W$.

Consider the $\mathcal{C}$-set $\mathbf{F}' = (F, D, \rho)$, in which $\rho_f = f$ for $f \in \mathcal{C}_i(u, v)$ and $F = FR(\mathcal{C})$.

Let us ensure that $\mathbb{F} = \mathbb{M}\mathbf{f}(\mathbf{F}')$ according to Definition 5.6.14, i.e.

(1.2) $\mathbf{a}R_i^n\mathbf{b} \Leftrightarrow \exists f \in \mathcal{C}_i(u, v)\ f \cdot \mathbf{a} = \mathbf{b}$

for $\mathbf{a} \in D_u^n$, $\mathbf{b} \in D_v^n$, (recall that $\rho_f = f$).

In fact, if $f \cdot \mathbf{a} = \mathbf{b}$, $f \in \mathcal{C}_i(u, v)$, then $\mathbf{a}R_i^n\mathbf{b}$ by (1.1).

The other way round, suppose $\mathbf{a}R_i^n\mathbf{b}$. Let $a_1, \dots, a_n, a_{n+1} \dots$ be an enumeration of D_u starting at our $\mathbf{a} = (a_1, \dots, a_n)$. Using Lemma 5.10.5 (2), by induction we construct a sequence $b_1, \dots, b_n, b_{n+1} \dots$ in D_v such that $\mathbf{b} = (b_1, \dots, b_n)$ and $(a_1, \dots, a_k)R_i^k(b_1, \dots, b_k)$ for all $k \geq n$. Then the function f sending each a_k to b_k belongs to $\mathcal{C}_i(u, v)$. In fact, $(a_1, \dots, a_k)R_i^k(b_1, \dots, b_k)$ for $k \geq n$, by our construction, and this is also true for $k < n$, by monotonicity of π_σ, where $\sigma : I_k \longrightarrow I_n$ is the inclusion map. Thus (1.2) holds.

Finally note that (1.2) implies $F(= FR(\mathcal{C})) = F_0$:

$$uR_iv \Leftrightarrow \mathcal{C}_i(u, v) \neq \varnothing.$$

In fact, suppose $uR_iv, a \in D_u$. By 5.10.5(4), aR_i^1b for some $b \in D_v$, thus by (1.2), $f(a) = b$ for some $f \in \mathcal{C}_i(u, v)$ and so $\mathcal{C}_i(u, v) \neq \varnothing$.

The other way round, if $f \in \mathcal{C}_i(u, v)$, then $aR_i^1f(a)$ and $f(a) \in D_v$. Hence uR_iv by 5.10.5(3).

(2) Consider a **QS4**-metaframe $\mathbb{F} = ((F_n)_{n\in\omega}, D)$, such that

- $F_0 = (W, R)$ is a 2-element chain:
  $W = \{u_0, u_1\}$, $R = W \times W - \{(u_1, u_0)\}$;
- $|D_{u_0}| > |D_{u_1}| = \aleph_0$;

- $\mathbf{a}R^n\mathbf{b} \Leftrightarrow \mathbf{a}\, sub\, \mathbf{b}$ & $\mathbf{b}\, sub\, \mathbf{a} \Leftrightarrow \forall_{i,j}(a_i = a_j \Leftrightarrow b_i = b_j)$ for $\mathbf{a} \in D^n_{u_j}$, $\mathbf{b} \in D^n_{u_k}$, $j \leq k$.

To check that $\mathbb{F}$ is a modal metaframe, we apply 5.10.2, 5.10.5.

First, for a permutation $\sigma \in \Upsilon_n$, π_σ is an automorphism of F_n. In fact, $\mathbf{a}R^n\mathbf{b}$ iff $\forall i,j\,(a_i = a_j \Leftrightarrow b_i = b_j)$ iff $\forall i,j\,(a_{\sigma(i)} = a_{\sigma(j)} \Leftrightarrow b_{\sigma(i)} = b_{\sigma(j)})$ iff $(\mathbf{a}\cdot\sigma)R^n(\mathbf{b}\cdot\sigma)$.

A similar argument shows that π_σ is monotonic for any $\sigma \in \Upsilon_{mn}$.

Second, π^n_+ has the lift property. In fact, suppose $\mathbf{a}R^{n-1}\mathbf{b}$, i.e. $\mathbf{a}\, sub\, \mathbf{b}$ and $\mathbf{b}\, sub\, \mathbf{a}$. Let u, v be the worlds of $\mathbf{a}$, $\mathbf{b}$, respectively. Then for any $c \in D_u$ there is $d \in D_v$ such that $(\mathbf{a}c)R^n(\mathbf{b}d)$: if $c = a_i$ for some i, take $d = b_i$; otherwise take $d \neq b_i$ for any i (since D_v is infinite, such an individual d always exists).

The case $n = 1$ described in 5.10.5(4) is trivial. By definition, the property (mm_2) also holds: $(a,a)R^2(b_1,b_2) \Rightarrow b_1 = b_2$.

All the F_n are **S4**-frames, since the relation *sub* is reflexive and transitive.

Now suppose $\mathbb{F}$ corresponds to a $\mathcal{C}$-set $\mathbf{F}' = (F, D, \rho)$. Let $\mu \in \mathcal{C}(u_0, u_1)$; then ρ_μ is a function from D_{u_0} to D_{u_1}. But $|D_{u_0}| > |D_{u_1}|$, so there exist $a_1, a_2 \in D_{u_0}$, $b \in D_{u_1}$ such that $a_1 \neq a_2$ and $\rho_\mu(a_1) = \rho_\mu(a_2) = b$. Then $(a_1,a_2)R^2(b,b)$ in $\mathbb{M}\mathtt{f}(\mathbf{F}')$, but not in $\mathbb{F}$.

One can easily construct similar examples for any $N > 1$. ∎

Therefore modal metaframes *with countable domains* are nothing but $\mathcal{C}$-sets. However we do not know if the semantics of modal metaframes is stronger than functor semantics (either in the case of $\mathcal{C}$-presets or $\mathcal{C}$-sets). For the intuitionistic case (discussed below) this question is also open.

5.14 Intuitionistic forcing and monotonicity

5.14.1 Intuiutionistic forcing

Now let us consider the intuitionistic case. We shall begin with a simple observation that **QS4**-metaframes generate a sound semantics for superintuitionistic logics via Gödel–Tarski translation (Proposition 5.14.7). But this semantics is probably not maximal in the intuitionistic case, so our goal will be to identify a larger class of 'intuitionistic sound metaframes'.

Let us first give a definition of intuitionistic forcing in **S4**-metaframes.

Definition 5.14.1 *A valuation ξ in an* **S4**-*metaframe $\mathbb{F}$ is called* intuitionistic *if it is intuitionistic in every F_n, i.e. for any predicate letter P^n_j, $n \geq 0$*

$$R^n(\xi^+(P^n_j)) \subseteq \xi^+(P^n_j).$$

The pair $(\mathbb{F}, \xi)$ is called an intuitionistic metaframe model.

Henceforth until Section 5.19, we shall consider mainly **S4**-metaframes.

Definition 5.14.2 *An intuitionistic model* $M = (\mathbb{F}, \xi)$ *gives rise to the forcing relation* $M, \mathbf{a} \Vdash A[\mathbf{x}]$ *(where* $\mathbf{a} \in D^n$, $|\mathbf{x}| = n$, $A \in IF^=$, $FV(A) \subseteq r(\mathbf{x})$*) defined by induction:*

(I) $M, \mathbf{a} \nVdash \bot\, [\mathbf{x}]$;

(II) $M, \mathbf{a} \Vdash P_k^n(\mathbf{x} \cdot \sigma)\, [\mathbf{x}]$ *iff* $(\mathbf{a} \cdot \sigma) \in \xi^+(P_k^n)$;

(III) $M, \mathbf{a} \Vdash (x_j = x_k)\, [\mathbf{x}]$ *iff* $a_j = a_k$;

(IV) $M, \mathbf{a} \Vdash (B \wedge C)\, [\mathbf{x}]$ *iff* $M, \mathbf{a} \Vdash B\, [\mathbf{x}]$ *and* $M, \mathbf{a} \Vdash C\, [\mathbf{x}]$;

(V) $M, \mathbf{a} \Vdash (B \vee C)\, [\mathbf{x}]$ *iff* $M, \mathbf{a} \Vdash B\, [\mathbf{x}]$ *or* $M, \mathbf{a} \Vdash C\, [\mathbf{x}]$;

(VI) $M, \mathbf{a} \Vdash (B \supset C)\, [\mathbf{x}]$ *iff* $\forall \mathbf{b} \in R^n(\mathbf{a})\ (M, \mathbf{b} \Vdash B\, [\mathbf{x}] \Rightarrow M, \mathbf{b} \Vdash C\, [\mathbf{x}])$;

(VII) $M, \mathbf{a} \Vdash \forall y B\, [\mathbf{x}]$ *iff* $\forall \mathbf{b} \in R^n(\mathbf{a})\ \forall c \in D(\mathbf{b})\ M, \pi_{\mathbf{x}||y}(\mathbf{b}c) \Vdash B\, [\mathbf{x}||y]$;
$M, \mathbf{a} \Vdash \exists y B\, [\mathbf{x}]$ *iff* $\exists \mathbf{b} \in (R^n)^{-1}(\mathbf{a})\ \exists c \in D(\mathbf{b})\ M, \pi_{\mathbf{x}||y}(\mathbf{b}c) \Vdash B\, [\mathbf{x}||y]$.

This definition of forcing is motivated by Gödel–Tarski translation. E.g. the intuitionistic forcing for $\forall y B$ resembles the modal forcing for $\Box \forall y B$. In more detail the connection between modal and intuitionistic forcing will be discussed later on. However note that for the $\exists$-case the intuitionistic and the modal definition differ, unlike Kripke frame or Kripke sheaf semantics. This situation was already discussed for quasi-bundles in section 5.5.

Definition 5.14.3 *Let* $\mathbb{F}$ *be an* **S4**-*metaframe. An intuitionistic formula* A *is called*

- true *in an intuitionistic model* M *over* $\mathbb{F}$ *(notation:* $M \Vdash A$*) if* $M, \mathbf{a} \Vdash A[\mathbf{x}]$ *for any assignment* $(\mathbf{x}, \mathbf{a})$ *with* $r(\mathbf{x}) \supseteq FV(A)$;
- valid *in* $\mathbb{F}$ *(notation:* $\mathbb{F} \Vdash A$*) if it is true in all intuitionistic models over* $\mathbb{F}$;
- strongly valid *(respectively,* strongly valid with equality*) in* $\mathbb{F}$ *if all its* IF*-(respectively,* $IF^=$*)-substitution instances are valid in* $\mathbb{F}$*. Strong validity is denoted by* $\Vdash^+$ *(respectively,* $\Vdash^{+=}$*).*

Similarly to the modal case, we use the notation

$$\mathbf{IL}_-^{(=)}(\mathbb{F}) := \{A \in IF^{(=)} \mid \mathbb{F} \Vdash A\},$$
$$\mathbf{IL}^{(=)}(\mathbb{F}) := \{A \in IF^{(=)} \mid \mathbb{F} \Vdash^{+(=)} A\}.$$

Note that $\mathbf{IL}^=(\mathbb{F})$ is the largest substitution closed subset of $\mathbf{IL}_-^{(=)}(\mathbb{F})$.

As we have pointed out for Kripke bundles, it may be the case that $\mathbb{F} \Vdash SA$ for any IF-substitution S, but not for any $IF^=$-substitution; so $\mathbf{IL}^=(\mathbb{F})$ may be not conservative over $\mathbf{IL}(\mathbb{F})$, cf. Remark 5.12.5 for the modal case. Similarly to the modal case we have

$$\mathbf{IL}^=(\mathbb{F}) \cap IF \subseteq \mathbf{IL}(\mathbb{F}).$$

Definition 5.14.4 *A metaframe* $\mathbb{F}$ *is called* intuitionistic$^{(=)}$ sound *($i^{(=)}$-sound, for short) if* $\mathbf{IL}^{(=)}(\mathbb{F})$ *is an s.p.l.(=).*

As we shall see in Section 5.16, i$^=$-soundness implies i-soundness.

Now let us show that all **QS4**-metaframes are i$^=$-sound. This happens because intuitionistic forcing for **QS4**-metaframes corresponds to modal forcing via Gödel–Tarski translation.

Definition 5.14.5 *Let* $\mathbb{F}$ *be a* **QS4**-*metaframe;* $M = (\mathbb{F}, \xi)$ *a metaframe model. The* pattern *of* M *is the model* $M_0 = (\mathbb{F}, \xi_0)$ *such that*

$$\xi_0^+(P_j^n) = \{\mathbf{a} \in D^n \mid R^n(\mathbf{a}) \subseteq \xi^+(P_j^n)\}$$

for any $j, n \geq 0$.

It is clear that M_0 is an intuitionistic model and $M_0 = M$ if M itself is intuitionistic.

Lemma 5.14.6 *Let* M *be the same as in Definition 5.14.5. Then for any* $A \in IF^=$ *and any assignment* $(\mathbf{x}, \mathbf{a})$ *with* $FV(A) \subseteq \mathbf{x}$

$$M_0, \mathbf{a} \Vdash A\,[\mathbf{x}] \Leftrightarrow M, \mathbf{a} \Vdash A^T\,[\mathbf{x}].$$

Proof By induction, cf. Lemmas 3.2.15, 5.5.7.

If $A = P(\pi_\sigma \mathbf{x})$, $\sigma \in \Sigma_{mn}$, $|\mathbf{x}| = n$, then

$$M_0, \mathbf{a} \Vdash A\,[\mathbf{x}] \Leftrightarrow \pi_\sigma \mathbf{a} \in \xi_0^+(P) \Leftrightarrow R^m(\pi_\sigma \mathbf{a}) \subseteq \xi^+(P),$$
$$M, \mathbf{a} \Vdash \Box A\,[\mathbf{x}] \Leftrightarrow \forall \mathbf{b} \in R^n(\mathbf{a})\ \pi_\sigma(\mathbf{b}) \in \xi^+(P) \Leftrightarrow \pi_\sigma(R_n(\mathbf{a})) \subseteq \xi^+(P).$$

But $R_m(\pi_\sigma \mathbf{a}) = \pi_\sigma(R_n(\mathbf{a}))$ for a morphism $\pi_\sigma : F_n \longrightarrow F_m$ (see the remark after Definition 1.3.30), so the statement is true in this case.

For the induction step let us only check the quantifier cases.

Suppose $A = \forall y B$ and the statement holds for B. Then

$$M_0, \mathbf{a} \Vdash A \Leftrightarrow \forall \mathbf{b} \in R^n(\mathbf{a})\ \forall c \in D(\mathbf{b})\ M_0, \pi_{\mathbf{x}||y}(\mathbf{b}c) \Vdash B\,[\mathbf{x}||y] \Leftrightarrow$$
$$\forall \mathbf{b} \in R^n(\mathbf{a})\ \forall c \in D(\mathbf{b})\ M, \pi_{\mathbf{x}||y}(\mathbf{b}c) \vDash B^T\,[\mathbf{x}||y] \Leftrightarrow M, \mathbf{a} \vDash \Box \forall y B^T (= A^T)\,[\mathbf{x}].$$

The argument for the $\exists$-case is slightly longer as the intuitionistic and the modal definitions differ in this case.

Suppose $A = \exists y B$ and the statement holds for B. We have

$$M_0, \mathbf{a} \Vdash A \Leftrightarrow \exists \mathbf{b} \in (R^n)^{-1}(\mathbf{a})\ \exists c \in D(\mathbf{b})\ M_0, \pi_{\mathbf{x}||y}(\mathbf{b}c) \Vdash B\,[\mathbf{x}||y] \Leftrightarrow$$
$$\exists \mathbf{b} \in (R^n)^{-1}(\mathbf{a})\ \exists c \in D(\mathbf{b})\ M, \pi_{\mathbf{x}||y}(\mathbf{b}c) \vDash B^T\,[\mathbf{x}||y] \Leftrightarrow$$
$$\exists \mathbf{b} \in (R^n)^{-1}(\mathbf{a})\ M, \mathbf{b} \vDash \exists y B^T (= A^T)\,[\mathbf{x}].$$

The latter holds if $M, \mathbf{a} \vDash A^T$ — just take $\mathbf{b} = \mathbf{a}$. The other way round, suppose $M, \mathbf{b} \vDash A^T\,[\mathbf{x}]$. As we know, $\mathbf{QS4}^{(=)} \vdash A^T \equiv \Box A^T$ (Lemma 2.11.2), so $M \vDash A^T \equiv \Box A^T$ by soundness (5.12.13), and thus $M, \mathbf{b} \vDash \Box A^T\,[\mathbf{x}]$, which implies $M, \mathbf{a} \vDash A^T$.

The case $A = \exists x_i B$ is reduced to the above:

$M_0, \mathbf{a} \Vdash A \Leftrightarrow M_0, \mathbf{a} - a_i \Vdash \exists x_i B\,[\mathbf{x} - x_i] \Leftrightarrow M, \mathbf{a} - a_i \models \exists x_i B^T\,[\mathbf{x} - x_i]$
$\Leftrightarrow M, \mathbf{a} \models \exists x_i B^T (= A^T)\,[\mathbf{x}]$.

All other cases are left to the reader. ∎

Proposition 5.14.7 *Let $\mathbb{F}$ be an* **QS4***-metaframe, $A \in IF^=$. Then*

(1) $\mathbb{F} \Vdash A$ iff $\mathbb{F} \models A^T$.

(2) The following three assertions are equivalent:

(a) $\mathbb{F} \Vdash^+ A$;

(b) $\forall m\ \mathbb{F} \Vdash A^m$;

(c) $\mathbb{F} \models^+ A^T$.

(3) $\mathbf{IL}^{(=)}(\mathbb{F}) = {}^T\mathbf{ML}^{(=)}(\mathbb{F})$ and therefore $\mathbf{IL}^{(=)}(\mathbb{F})$ is an s.p.l.(=).

Proof Similar to 5.5.12, 5.5.13.

(1) (Only if.) Assume $\mathbb{F} \Vdash A$. For a metaframe model M over $\mathbb{F}$ let us show $M \models A^T$, i.e. $M, \mathbf{a} \models A^T\,[\mathbf{x}]$ for any appropriate assignment $(\mathbf{x}, \mathbf{a})$. By Lemma 5.14.6, the latter is equivalent to $M_0, \mathbf{a} \Vdash A\,[\mathbf{x}]$, which follows from our assumption.

(If.) Assume $\mathbb{F} \models A^T$. Let M be an intuitionistic metaframe model over $\mathbb{F}$, and let us show $M \Vdash A$, i.e. $M, \mathbf{a} \Vdash A\,[\mathbf{x}]$ for any appropriate assignment $(\mathbf{x}, \mathbf{a})$. Since $M_0 = M$, by Lemma 5.14.6, $M, \mathbf{a} \Vdash A\,[\mathbf{x}]$ iff $M, \mathbf{a} \models A^T[\mathbf{x}]$, and the latter follows from $\mathbb{F} \models A^T$.

(2) The proof is completely analogous to 5.5.12. Use 5.12.3 for (b)⇒(c) and soundness (5.12.13) for (c)⇒(a).

(3) follows readily from the equivalence (a)⇔(c) in (2). ∎

Proposition 5.14.7 shows that **QS4**-metaframes are $\mathrm{i}^{(=)}$-sound. As we shall see later on, the class of i-sound metaframes is larger, but the question, whether i-sound metaframes generate a stronger semantics, is open. The same happens to $\mathrm{i}^=$-soundness.

Corollary 5.14.8 *If* **F** *is a C-set (or an intuitionistic Kripke bundle), then* $\mathbf{IL}^{(=)}(\mathbf{F})$ *is an s.p.l.(=); moreover,* $\mathbf{IL}^{(=)}(\mathbf{F}) = {}^T\mathbf{ML}^{(=)}(\mathbf{F})$.

Exercise 5.14.9 Using 5.14.6, check the following properties of intuitionistic forcing in **QS4**-metaframes analogous to the properties of modal forcing (cf. Lemma 5.10.6(2), Lemma 5.10.10, Lemma 5.11.7, Proposition 5.11.12).

Let $\mathbb{F}$ be a **QS4**-metaframe, M an intuitionistic model over $\mathbb{F}$. Then

(1) for any ordered assignment $(\mathbf{x}, \mathbf{a})$, for any $\sigma \in \Upsilon_{mn}$, where $|\mathbf{x}| = n \geq m$ and for any formula $A \in IF^=$ with $FV(A) \subseteq r(\mathbf{x} \cdot \sigma)$

$$M, \mathbf{a} \Vdash A\,[\mathbf{x}] \Leftrightarrow M, \mathbf{a} \cdot \sigma \Vdash A\,[\mathbf{x} \cdot \sigma];$$

(2) for any congruent intuitionistic formulas A, B and any ordered assignment $(\mathbf{x}, \mathbf{a})$ with $FV(A) \subseteq r(\mathbf{x})$

$$M, \mathbf{a} \Vdash A\,[\mathbf{x}] \Leftrightarrow M, \mathbf{a} \Vdash B\,[\mathbf{x}];$$

(3) for any $\sigma \in \Sigma_{mn}$, for any distinct lists of variables $\mathbf{x} = (x_1, \ldots, x_n)$, $\mathbf{y} = (y_1, \ldots, y_m)$, for any intuitionistic formula A with $FV(A) \subseteq r(\mathbf{y})$, for any $\mathbf{a} \in D^n$

$$M, \mathbf{a} \Vdash ([\mathbf{x} \cdot \sigma/\mathbf{y}]\, A)\,[\mathbf{x}] \Leftrightarrow M, \mathbf{a} \cdot \sigma \Vdash A\,[\mathbf{y}];$$

(4) if A, A^* are intuitionistic formulas, $[\mathbf{a}/\mathbf{x}]$, $[\mathbf{a}^*/\mathbf{x}^*]$ assignments giving rise to equal D_u-sentences: $[\mathbf{a}/\mathbf{x}]\, A = [\mathbf{a}^*/\mathbf{x}^*]\, A^*$, then

$$M, \mathbf{a} \Vdash A\,[\mathbf{x}] \Leftrightarrow M, \mathbf{a}^* \Vdash A\,[\mathbf{x}^*].$$

As we shall see later on, all the claims in 5.14.9 extend to arbitrary $\mathrm{i}^=$-sound metaframes.

Now let us consider arbitrary **S4**-metaframes. Let us first note that the quantifier clauses in Definition 5.14.2 are analogues of the combined clause (9+10) for modal forcing from Section 5.9. Now if we consider two options in the intuitionistic case, we obtain

(VII.1) if $y \notin \mathbf{x}$, then
$M, \mathbf{a} \Vdash \forall y B\,[\mathbf{x}]$ iff $\forall \mathbf{b} \in R^n(\mathbf{a})\, \forall c \in D(\mathbf{b})\; M, \mathbf{b}c \Vdash B\,[\mathbf{x}y]$,
$M, \mathbf{a} \Vdash \exists y B[\mathbf{x}]$ iff $\exists \mathbf{b} \in (R^n)^{-1}(\mathbf{a})\, \exists c \in D(\mathbf{b})\; M, \mathbf{b}c \Vdash B\,[\mathbf{x}y]$;

(VII.2) for $1 \leq i \leq n$
$M, \mathbf{a} \Vdash \forall x_i B\,[\mathbf{x}]$ iff $\forall \mathbf{b} \in R^n(\mathbf{a})\; \forall c \in D(b)\; M, (\mathbf{b} - b_i)c \Vdash B\,[(\mathbf{x} - x_i)x_i]$
iff $\forall \mathbf{d} \in \pi_i^n(R^n(\mathbf{a}))\; \forall c \in D(\mathbf{d})\; M, \mathbf{d}c \Vdash B\,[(\mathbf{x} - x_i)x_i]$.
$M, \mathbf{a} \Vdash \exists x_i B\,[\mathbf{x}]$ iff $\exists \mathbf{b} \in (R^n)^{-1}(\mathbf{a})\; \exists c \in D(\mathbf{b}) M, \mathbf{d}c \Vdash B\,[(\mathbf{x} - x_i)x_i]$.

Note that (VII.1) corresponds to the clause (9) from Definition 5.9.4. Let us show that in **QS4**-frames (VII.2) corresponds to the clause 5.9.4(10), i.e.
$M, \mathbf{a} \Vdash \forall x_i B\,[\mathbf{x}]$ iff $M, \widehat{\mathbf{a}}_i \Vdash \forall x_i B[\widehat{\mathbf{x}}_i]$,
$M, \mathbf{a} \Vdash \exists x_i B\,[\mathbf{x}]$ iff $M, \widehat{\mathbf{a}}_i \Vdash \exists x_i B\,[\widehat{\mathbf{x}}_i]$.

In fact, by (VII.1)

$M, \mathbf{a} - a_i \Vdash \forall x_i B\,[\mathbf{x} - x_i$ iff $\forall \mathbf{d} \in R^{n-1}(\mathbf{a} - a_i)\, \forall c \in D(\mathbf{d})\, M, \mathbf{d}c \Vdash B\,[(\mathbf{x} - x_i)x_i]$
iff $\forall \mathbf{d} \in R^n(\pi_i^n \mathbf{a})\; \forall c \in D(\mathbf{d})\; M, \mathbf{d}c \Vdash B\,[(\mathbf{x} - x_i)x_i]$.

Now in **QS4**-metaframes π_i^n is a morphism, so the latter is equivalent to (VII.2). But in arbitrary **S4**-metaframes (VII.2) may not be true, because it may happen that $\pi_i^n(R^n(\mathbf{a})) \neq R^n(\pi_i^n(\mathbf{a}))$, cf. Section 5.9.[24] The reader can construct a counterexample as an exercise, cf. 5.9.16, 5.9.18.

[24] Or note that

$$M, \mathbf{a} \Vdash \forall x_i B\,[\mathbf{x}] \Leftrightarrow M, \mathbf{a} - a_i \Vdash \forall x_i B\,[\mathbf{x} - x_i]$$

may not imply

$$M, \mathbf{a} \Vdash \Box \forall x_i B\,[\mathbf{x}] \Leftrightarrow M, \mathbf{a} - a_i \Vdash \Box \forall x_i B\,[\mathbf{x} - x_i].$$

Later on we shall see that (VII.2) and its analogue for the $\exists$-case hold in all $\mathrm{i}^{(=)}$-sound metaframes.

Remark 5.14.10 These comments motivate an alternative definition of forcing ($\Vdash^{\star}$) for the quantifier case instead of (VII):

This can be rewritten as follows:
There also exists an altrnative version ($\Vdash^{*}$) resembling 5.9.17:
All these definitions are equivalent in $\mathrm{i}^{(=)}$-sound metaframes.

Now we obtain an analogue to Lemma 5.12.2 (1), (2) and Corollary 5.12.12.

Proposition 5.14.11 *Let $\mathbb{F}$ be an* **S4**-*metaframe. Then*

(1) $\mathbf{IL}_{-}^{(=)}(\mathbb{F})$ *is closed under generalisation and modus ponens;*

(2) $\mathbf{IL}^{(=)}(\mathbb{F})$ *is closed under formula substitutions;*

(3) the following conditions are equivalent:

(a) $\mathbf{IL}^{(=)}(\mathbb{F})$ *is an s.p.l. (=);*

(b) $\mathbf{QH}^{(=)} \subseteq \mathbf{IL}^{(=)}(\mathbb{F})$;

(c) $\mathbf{QH}^{(=)} \subseteq \mathbf{IL}_{-}^{(=)}(\mathbb{F})$.

Proof (1) (I) Let us first consider generalisation. Let M be an intuitionistic model over $\mathbb{F}$, $B \in IF^{(=)}$. Assuming $M \Vdash B$, let us prove $M \Vdash \forall y B$, i.e. $M, \mathbf{a} \Vdash \forall y B\,[\mathbf{x}]$ for any assignment $(\mathbf{x}, \mathbf{a})$ with $r(\mathbf{x}) \supseteq FV(\forall y B)$. Suppose $|\mathbf{x}| = n$.

If $y \notin \mathbf{x}$, then

$$M, \mathbf{a} \Vdash \forall y B\,[\mathbf{x}] \text{ iff } \forall \mathbf{b} \in R^n(\mathbf{a})\, \forall c \in D(\mathbf{b})\ M, \pi_{\mathbf{x}||y}(\mathbf{b}c) \Vdash B\,[\mathbf{x}||y].$$

But $M, \pi_{\mathbf{x}||y}(\mathbf{b}c) \Vdash B\,[\mathbf{x}||y]$ holds, since $M \Vdash B$.

(II) Now let us consider modus ponens.

Let M be an intuitionistic model M over $\mathbb{F}$. Assuming $M \Vdash B,\ B \supset C$, let us check that $M \Vdash C$.

Let $\mathbf{x} \subseteq FV(C)$, $\mathbf{y} = FV(B) - \mathbf{x}$. By our assumption, for any assignment $(\mathbf{xy}, \mathbf{ab})$,

$$M, \mathbf{ab} \Vdash B\,[\mathbf{xy}];\ M, \mathbf{ab} \Vdash (B \supset C)[\mathbf{xy}].$$

Since every relation R^n is reflexive, we readily obtain (for any assignment $(\mathbf{xy}, \mathbf{ab})$)

(3.1) $M, \mathbf{ab} \Vdash C\,[\mathbf{xy}]$.

Now let us show that

(3.2) $M, \mathbf{a} \Vdash C\,[\mathbf{x}]$.

In fact, (3.1) implies

(3.3) $M, \mathbf{a} \Vdash \forall \mathbf{y} C\, [\mathbf{x}]$,

as one can easily check. By (c), we also have

(3.4) $M, \mathbf{a} \Vdash (\forall \mathbf{y} C \supset C)[\mathbf{x}]$,

since $\forall \mathbf{y} C \supset C$ is an intuitionistic theorem. Now (3.2) follows from (3.3), (3.4), and the reflexivity of R^n.

(2) Note that the composition of substitutions is a substitution, cf. the argument in the modal case in 5.12.12.

(3) The implications (a) $\Rightarrow$ (b), (b) $\Rightarrow$ (c) are obvious.

(c) $\Rightarrow$ (b) is also obvious, $\mathbf{IL}^{(=)}(\mathbb{F})$ is substitution closed.

So assuming (c), let us check (a). Due to (2), it remains to consider modus ponens and generalisation.

First note that $\mathbf{IL}^{(=)}_{-}(\mathbb{F})$ is closed under congruence. In fact, suppose $A \in \mathbf{IL}^{=}_{-}(\mathbb{F})$ and $A \doteq B$. Then $(A \equiv B) \in \mathbf{QH}^{(=)}$ by 2.6.9, and hence $(A \supset B) \in \mathbf{QH}^{(=)}$ (by (Ax3), MP and substitution). So by the assumption (c), $(A \supset B) \in \mathbf{IL}^{(=)}_{-}(\mathbb{F})$.

Since by (1) $\mathbf{IL}^{(=)}_{-}(\mathbb{F})$ is closed under MP, it follows that $B \in \mathbf{IL}^{(=)}_{-}(\mathbb{F})$.

Now the closedness of $\mathbf{IL}^{(=)}(\mathbb{F})$ under modus ponens follows easily.

In fact, suppose

$$A,\ A \supset B \in \mathbf{IL}^{(=)}(\mathbb{F}).$$

Then $S, SA, S(A \supset B) \in \mathbf{IL}^{(=)}_{-}(\mathbb{F})$ for any formula substitution. By Lemma 2.5.13, $S(A \supset B) \doteq (SA \supset SB)$. Since $\mathbf{IL}^{(=)}_{-}(\mathbb{F})$ is congruence closed and MP-closed (by (1)), this implies $SB \in \mathbf{IL}^{(=)}_{-}(\mathbb{F})$. Eventually $B \in \mathbf{IL}^{=}(\mathbb{F})$.

Finally, let us consider generalisation. We suppose $A \in \mathbf{IL}^{(=)}(\mathbb{F})$ and show that $\forall x A \in \mathbf{IL}^{(=)}(\mathbb{F})$. This does not follow directly from (1), because a substitution instance of $\forall x A$ is not always of the form $\forall x A'$ for a substutition instance A' of A (cf. Lemma 2.5.13), and so we need a little detour.

So for a substitution S, let us prove that $\mathbb{F} \Vdash S \forall x A$.

Let y be a new variable, $y \notin V(\forall x A) \cup FV(S)$; then by 2.3.25(13)

$$\forall x A \doteq \forall y [y/x] A.$$

Hence by 2.5.12 and 2.5.13(3),

$$S \forall x A \doteq S \forall y [y/x] A \doteq \forall y S [y/x] A.$$

As we already know, the validity in $\mathbb{F}$ is closed under congruence and generalisation. So it is sufficient to show that $\mathbb{F} \Vdash S[y/x]A$.

By Lemma 2.5.14 (for complex substitutions) we can rename the bound variables of S; so

(♯1) $$S[y/x]A \doteq S_1[y/x]A,$$

with $x \notin BV(S_1)$. Now we introduce yet another new variable x'. Then we obviously have

$$(\sharp 2) \qquad S_1[y/x]A \stackrel{\circ}{=} [x/x'][x'/x]S_1[y/x]A.$$

By 2.5.17, there exists a formula substitution S_0 such that $x \notin FV(S_0)$ and

$$(\sharp 3) \qquad [x'/x]S_1[y/x]A \stackrel{\circ}{=} S_0[y/x]A.$$

But then by 2.5.15 we obtain

$$(\sharp 4) \qquad S_0[y/x]A \stackrel{\circ}{=} [y/x]S_0A$$

Eventually from ($\sharp$1)–($\sharp$4) and 2.3.24 it follows that

$$(\sharp 5) \qquad S[y/x]A \stackrel{\circ}{=} [x/x'][y/x]S_0A.$$

It remains to notice that validity in $\mathbb{F}$ respects variable substitutions. In fact, if $\mathbb{F} \Vdash B$, then $\mathbb{F} \Vdash \forall \mathbf{x} B$ by (1).

By 2.6.15(xxv), $\forall \mathbf{x} B \supset [\mathbf{y}/\mathbf{x}]B$ is an intutionistic theorem, so it is valid in $\mathbb{F}$ by our assumption (c). By (1), we can apply modus ponens, thus $\mathbb{F} \Vdash [\mathbf{y}/\mathbf{x}]B$.

Therefore, since $A \in \mathbf{IL}^{(=)}(\mathbb{F})$, we obtain $\mathbb{F} \Vdash [x/x'][y/x]S_0A$; thus $\mathbb{F} \Vdash S[y/x]A$ by ($\sharp$5), since $\mathbf{IL}_{-}^{(=)}(\mathbb{F})$ is closed under congruence. ∎

Remark 5.14.12 The above proof cannot be simplified by concluding $\mathbb{F} \Vdash S_0[y/x]A$ directly from $A \in \mathbf{IL}^{(=)}(\mathbb{F})$, before we know that $[y/x]A \in \mathbf{IL}^{(=)}(\mathbb{F})$. Since S_0 does not always commute with $[y/x]$ for $x \in FV(S_0)$, we had to replace x with x' and S_0 with S_1.

Corollary 5.14.13 *$i^{=}$–-soundness implies i-soundness.*

Proof $\mathbf{QH}^{=} \subseteq \mathbf{IL}^{=}(\mathbb{F})$ implies $\mathbf{QH} = (\mathbf{QH}^{=})^{\circ} \subseteq (\mathbf{IL}^{=}(\mathbb{F}))^{\circ} \subseteq \mathbf{IL}(\mathbb{F})$, which implies i-soundness by 5.14.11(3). ∎

5.14.2 Monotonic metaframes

Now let us consider an important property of intuitionistic sound metaframes.

Definition 5.14.14 *An* **S4**-*metaframe* $\mathbb{F}$ *is called* monotonic$^{(=)}$ *if it satisfies the condition*

(im) $M, \mathbf{a} \Vdash A[\mathbf{x}]$ & $\mathbf{a}R^n\mathbf{b} \Rightarrow M, \mathbf{b} \Vdash A[\mathbf{x}]$

for any $A \in IF^{(=)}$, *for any intuitionistic model* $M = (\mathbb{F}, \xi)$, *a distinct list* $\mathbf{x}$ *of length* n *containing* $FV(A)$ *and any* $\mathbf{a}, \mathbf{b} \in D^n$.

We already mentioned this property in Section 5.5 when quasi-bundles were discussed. The next lemma proves that monotonicity is really necessary.

Lemma 5.14.15 *If the formula $B := p \supset (\top \supset p)$ is strongly valid in an* **S4**-*metaframe* $\mathbb{F}$ *(in the language with or without equality, respectively), then* $\mathbb{F}$ *is monotonic*$^{(=)}$.

Proof Let us check (im) for an arbitrary formula A. Consider the following substitution instance $B' := A \supset (\top \supset A)$ of B.

Assume $\mathbf{a}R^n\mathbf{b}$ and $M, \mathbf{a} \Vdash A[\mathbf{x}]$. Then $M, \mathbf{a} \Vdash \top \supset A[\mathbf{x}]$ since $M, \mathbf{a} \Vdash B'[\mathbf{x}]$ and R^n is reflexive. Obviously, $M, \mathbf{b} \Vdash \top(= \bot \supset \bot)[\mathbf{x}]$, and thus $M, \mathbf{b} \Vdash A[\mathbf{x}]$. ∎

The idea of the previous proof is quite clear; B' expresses the property (im) for A — it states that if A is true now, then $\top \supset A$ is also true, and thus A will always be true in the future, by the definition of forcing for implication.

Hence we obtain

Proposition 5.14.16 *Let* $\mathbb{F}$ *be an* **S4**-*metaframe such that* $\mathbf{H} \subseteq \mathbf{IL}^{(=)}(\mathbb{F})$. *Then* $\mathbb{F}$ *is monotonic*$^{(=)}$.

Let us now describe monotonic metaframes in terms of accessibility relations.

Definition 5.14.17 *An* **S4**-*metaframe* $\mathbb{F}$ *is called* semi-functorial *(or* s-functorial, *for short) if for any* $\sigma \in \Sigma_{mn}$; $m, n \geq 0$

(0σ) $\forall \mathbf{a}, \mathbf{b} \in D^n(\mathbf{a}R^n\mathbf{b} \Rightarrow (\pi_\sigma\mathbf{a})R^m(\pi_\sigma\mathbf{b}))$,

i.e. if every π_σ *is monotonic.*

Lemma 5.14.18 *An* **S4**-*metaframe* $\mathbb{F}$ *is semi-functorial iff the following holds.*

(I1) $\mathbb{F}$ *is permutable (see Definition 5.10.1), i.e.* (0σ) *holds for all (simple) permutations* σ.[25]

(I2.1) $\forall n > 0\ \forall u, v \in W\ \forall \mathbf{a} \in D_u^n\ \forall \mathbf{b} \in D_v^n\ (\mathbf{a}R^n\mathbf{b} \Rightarrow uRv)$.

(I2.2) $\forall n > 0\ \forall u, v \in W\ \forall \mathbf{a} \in D_u^n\ \forall \mathbf{b} \in D_v^n\ \forall c \in D_u\ \forall d \in D_v\ ((\mathbf{a}c)R^{n+1}(\mathbf{b}d) \Rightarrow \mathbf{a}R^n\mathbf{b})$.

(I3) $\forall n > 0\ \forall \mathbf{a}, \mathbf{b} \in D^n\ (\mathbf{a}R^n\mathbf{b} \Rightarrow (\mathbf{a}a_n)R^{n+1}(\mathbf{b}b_n))$.

(Also note that (I2.1) can be replaced by its particular case with $n = 1$.)

Proof Every $\sigma \in \Sigma_{mn}$ is a composition of permutations, simple embeddings and simple projections. So it suffices to check the monotonicity of π_σ for $\sigma = \sigma_+^n$ $(n \geq 0)$, σ_-^n $(n > 0)$ and for all permutations. Now recall that $\sigma_+^0 = \phi_1, \pi_{\varnothing_n}(\mathbf{a}) = u$ for $\mathbf{a} \in D_u^n$, $\pi_{\sigma_-^n}(\mathbf{a}) = \mathbf{a}a_n$, $\pi_{\sigma_+^n}(\mathbf{a}c) = \mathbf{a}$. ∎

Definition 5.14.19 *A semi-functorial metaframe is called* semi-functorial with equality *(briefly,* s$^=$-functorial*) if it satisfies the condition (cf. 5.11.1)*

[25] Cf. Definition 5.10.1, Lemma 5.10.5.

(mm_2) $\forall a, b_1, b_2((a,a)R^2(b_1,b_2) \Rightarrow b_1 = b_2)$.

Lemma 5.14.20 *In an s-functorial metaframe all properties*

(mm_n) $\forall \mathbf{a}, \mathbf{b}\ (\mathbf{a}R^n\mathbf{b} \Rightarrow \mathbf{a}\ sub\ \mathbf{b})$

are equivalent for $n \geq 2$. So (mm_n) holds in every $s^=$-functorial metaframe.

Proof (mm_2) implies (mm_n), since $\mathbf{a}R^n\mathbf{b}$ implies $(a_i, a_j)R^2(b_i, b_j)$ by (I1), (I2.2); cf. Lemma 5.11.2.

The other way round, suppose (mm_n). Let $\mathbf{a} = (a,a)R^2(b_1,b_2) = \mathbf{b}$, and consider the projection $\sigma \in \Sigma_{n2}$ such that

$$\sigma(i) = \begin{cases} 1 & \text{for } i = 1, \\ 2 & \text{for } i > 1. \end{cases}$$

Then $\pi_\sigma\mathbf{a} = a^n$, $\pi_\sigma\mathbf{b} = b_1 b_2^{n-1}$.

By s-functoriality $(\pi_\sigma\mathbf{a})R^n(\pi_\sigma\mathbf{b})$, hence by ($mm_n$), $(\pi_\sigma\mathbf{a}) sub (\pi_\sigma\mathbf{b})$, and thus $b_1 = b_2$. ■

An inductive argument shows that $s^=$-functoriality implies monotonicity. The steps of the proof are almost trivial, due to the definition of forcing 5.14.2, so the only problem is the atomic case.

Recalling that

$$(\mathbb{F}, \xi), \mathbf{a} \Vdash P(\mathbf{x} \cdot \sigma)[\mathbf{x}] \Leftrightarrow \pi_\sigma\mathbf{a} \in \xi^+(P)$$

and $\xi^+(P)$ is R^n-stable, we see that monotonicity exactly corresponds to (0σ).

Now let us give more details.

Lemma 5.14.21 *A metaframe is $s^{(=)}$-functorial iff it is monotonic$^{(=)}$.*

Proof (Only if.) By induction on the complexity of A.

If $A = P_j^m(\pi_\sigma\mathbf{x})$, then (im) holds, since $\mathbf{a} \Vdash P_j^m(\pi_\sigma\mathbf{x})\,[\mathbf{x}]$ iff $\pi_\sigma\mathbf{a} \in \xi^+(P_j^m)$. By monotonicity of $\mathbb{F}$, $\mathbf{a}R^n\mathbf{b}$ implies $(\pi_\sigma\mathbf{a})R^m(\pi_\sigma\mathbf{b})$, and since M is intuitionistic, we obtain $\pi_\sigma\mathbf{b} \in \xi^+(P_j^m)$, i.e. $\mathbf{b} \Vdash A\,[\mathbf{x}]$.

If $A = (x_i = x_j)$, $\mathbf{a}R^n\mathbf{b}$, we have

$$\mathbf{a} \Vdash x_i = x_j\,[\mathbf{x}] \text{ iff } a_i = a_j,$$
$$\mathbf{b} \Vdash x_i = x_j\,[\mathbf{x}] \text{ iff } b_i = b_j.$$

By (I4), $\mathbf{a}$ sub $\mathbf{b}$, so $a_i = a_j$ implies $b_i = b_j$.

The induction step for $\wedge, \vee$ is obvious.

Let $A = B \supset C$. Suppose $\mathbf{a}R^n\mathbf{b}$ and $M, \mathbf{b} \not\Vdash A[\mathbf{x}]$, i.e. $M, \mathbf{c} \Vdash B\,[\mathbf{x}]$, $M, \mathbf{c} \not\Vdash C\,[\mathbf{x}]$ for some $\mathbf{c} \in R^n(\mathbf{b})$. By transitivity, $\mathbf{a}R^n\mathbf{c}$ and thus $M, \mathbf{a} \not\Vdash A\,[\mathbf{x}]$.

Let $A = \forall y B$. Suppose $y \notin \mathbf{x}$, $\mathbf{a}R^n\mathbf{b}$ and $M, \mathbf{b} \not\Vdash A\,[\mathbf{x}]$. Then by Definition 5.14.2, $M, \mathbf{c}d \not\Vdash B\,[\mathbf{x}y]$ for some $\mathbf{c} \in R^n(\mathbf{b})$ and some $d \in D(\mathbf{c})$. By transitivity, $\mathbf{a}R^n\mathbf{c}$, and thus $M, \mathbf{a} \not\Vdash A\,[\mathbf{x}]$.

Now suppose $A = \forall x_i B$, $\mathbf{a}R^n\mathbf{b}$, $M, \mathbf{b} \not\Vdash A[\mathbf{x}]$. Then by Definition 5.14.2, $M, \widehat{\mathbf{b}}_i, \not\Vdash A[\widehat{\mathbf{x}}_i]$. Recall that $\widehat{\mathbf{a}}_i = \pi_\sigma \mathbf{a}$ for $\sigma = \delta^i_n$. So since π_σ is monotonic, $\mathbf{a}R^n\mathbf{b}$ implies $\widehat{\mathbf{a}}_i R^n \widehat{\mathbf{b}}_i$, and thus $M, \widehat{\mathbf{a}}_i \not\Vdash A[\widehat{\mathbf{x}}_i]$, as we have already proved. Therefore $M, \mathbf{a} \not\Vdash A[\mathbf{x}]$, by Definition 5.14.2.

Let $A = \exists y B$, $y \notin \mathbf{x}$, and suppose $\mathbf{a}R^n\mathbf{b}$ and $M, \mathbf{a} \Vdash A[\mathbf{x}]$. Then $M, \mathbf{c}d \Vdash B[\mathbf{x}y]$ for some $\mathbf{c} \in (R^n)^{-1}(\mathbf{a})$ and some $d \in D(\mathbf{c})$. By transitivity, we have $\mathbf{c}R^n\mathbf{b}$, and hence we obtain $M, \mathbf{b} \Vdash A[\mathbf{x}]$.

The case $A = \exists x_i B$ is similar to $A = \forall x_i B$ and is left to the reader.

(If.) Assuming (im), let us show that $\mathbb{F}$ is monotonic$^{(=)}$. Note that $(\mathrm{mm})_2$ readily follows from (im) for $A = (x_1 = x_2)$.

If $\mathbf{a}R^n\mathbf{b}$, then $M, \mathbf{a} \Vdash P^m(\pi_\sigma \mathbf{x})[\mathbf{x}]$ in the intuitionistic model $M = (\mathbb{F}, \xi)$ such that $\xi^+(P^m) = R^m(\pi_\sigma \mathbf{a})$. Thus by (im), $M, \mathbf{a} \Vdash P^m(\pi_\sigma \mathbf{x})[\mathbf{x}]$, i.e. $\pi_\sigma \mathbf{b} \in \xi^+(P^m)$, and so $(\pi_\sigma \mathbf{a})R^m(\pi_\sigma \mathbf{b})$. ∎

Now we obtain

Proposition 5.14.22 *Every $i^{(=)}$-sound* **S4***-metaframe is $s^{(=)}$-functional.*

Remark 5.14.23 Monotonicity is an intrinsic property of intuitionistic forcing, thus Lemma 5.14.8 explains why we confine ourselves to s-functorial metaframes in our further considerations.

5.15 Intuitionistic soundness

In this section we prove intuitionistic soundness for a certain class of **S4**-metaframes. Its description is more complicated than in the modal case. As explained in the previous section, we certainly need s-functoriality corresponding to monotonicity. We also need the 'quasi-lift property', an intuitionistic analogue of the lift property for jections π_σ. But unlike the modal case, this is yet insufficient, and we need two extra conditions related to interpretation of quantifiers.

In **S4**-metaframes we use the notation $\approx^n$ for the corresponding equivalence relation $\approx_{R^n}$ on D^n; it is called *n-equivalence.*

Definition 5.15.1 *An* **S4***-metaframe $\mathbb{F}$ is called* quasi-functorial (i-functorial, *for short) if all its jections π_σ are quasi-morphisms.*

So we have

Lemma 5.15.2 *A metaframe $\mathbb{F} = ((F_n)_{n\in\omega}, D)$ is i-functorial iff it is s-functorial (i.e. all π_σ are monotonic) and all π_σ have the following* quasi-lift property*:*

(1σ) $\forall \mathbf{a} \in D^n\ \forall \mathbf{b}' \in D^m((\pi_\sigma \mathbf{a})R^m\mathbf{b}' \Rightarrow \exists \mathbf{b} \in D^n(\mathbf{a}R^n\mathbf{b}\ \&\ \pi_\sigma \mathbf{b} \approx^m \mathbf{b}'))$.

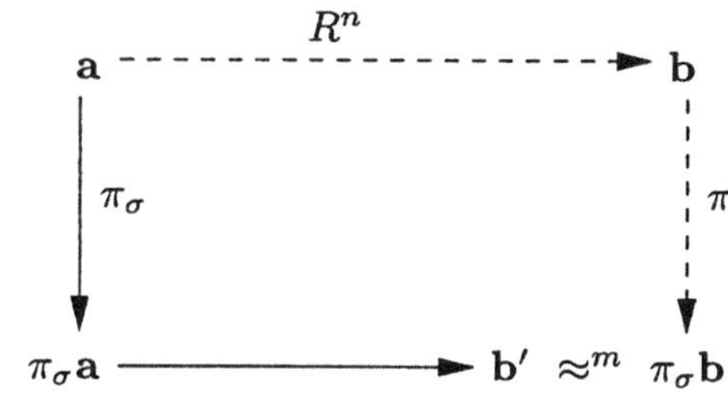

Recall that in the modal case we defined functorial metaframes, where all π_σ are morphisms, and also weakly functorial metaframes, where π_σ are morphisms for injective σ. The intuitionistic analogue of functoriality is quasi-functoriality, but as we shall see now, there is no need in the intuitionistic analogue of weak functoriality.

Lemma 5.15.3 *Let M be an intuitionistic model over an i-functorial wi-metaframe $\mathbb{F}$. Then we can replace R^n with $\approx^n$ in the truth definition for $\exists$:*

$$M, \mathbf{a} \Vdash \exists y B\,[\mathbf{x}] \textit{ iff } \exists \mathbf{b} \approx^n \mathbf{a}\ \exists c \in D(\mathbf{b})\ M, \pi_{\mathbf{x}||y}(\mathbf{b}c) \Vdash B\,[\mathbf{x}||y].$$

Proof In fact, if there exists $\mathbf{b}c \in D^{n+1}$ such that $\mathbf{b}R^n\mathbf{a}$ and $M, \pi_{\mathbf{x}||y}(\mathbf{b}c) \Vdash B\,[\mathbf{x}||y]$, then, by the quasi-lift property $(1\sigma^n_+)$, there exists $\mathbf{b}'c' \in D^{n+1}$ such that $(\mathbf{b}c)R^{n+1}(\mathbf{b}'c')$, $\mathbf{b}' \approx^n \mathbf{a}$, (recall that $\pi_{\sigma^n_+}(\mathbf{b}'c') = \mathbf{b}'$). Now by $(0\delta^i_n)$,

$$\pi_{\mathbf{x}||y}(\mathbf{b}c)R^n\pi_{\mathbf{x}||y}(\mathbf{b}'c'),$$

and thus by monotonicity 5.14.14 (im),

$$M, \pi_{\mathbf{x}||y}(\mathbf{b}'c') \Vdash B\,[\mathbf{x}||y].$$

■

The modified definition of forcing in the $\exists$-case resembles the familiar definition for intuitionistic Kripke models, but with $\approx^n$ replacing '='. So in our semantics, a 'witness' for $\exists yB$ does not always exist in the present world, but should exist in an equivalent world (cf. section 5.5). This reflects the basic idea that intuitionistic models[26] 'distinguish' worlds up to $\approx^0$, individuals up to $\approx^1$, n-tuples of individuals up to $\approx^n$.

Remark 5.15.4 One can see that in the above proof monotonicity and quasi-lift property are used explicitly only for injections σ^n_+ and $\varepsilon_{\mathbf{x}||y}$. But we also use the monotonicity of forcing relying on the monotonicity of π_σ for arbitrary (perhaps, non-injective) σ, cf. Lemma 5.14.21.

Therefore in further definitions we suppose monotonicity of all π_σ.

The next lemma shows that in functorial metaframes the intuitionistic truth definition for the $\exists$-case is quite analogous to the modal one. This simplifies the proof of Lemma 5.14.6.

[26]Due to the monotonicity of forcing, cf. 5.14.14.

Lemma 5.15.5 *Let M be an intuitionistic model over a functorial metaframe. Then for any $B \in IF^=, y \notin \mathbf{x}$*

$$M, \mathbf{a} \Vdash \exists y B[\mathbf{x}] \textit{ iff } \exists c \in D(\mathbf{a})\ M, \mathbf{a}c \Vdash B[\mathbf{x}y].$$

Proof The same as in 5.15.3, but now $\pi_{\sigma_+^n}$ is a morphism, so we can take $\mathbf{b}' = \mathbf{a}$. Note that forcing is monotonic, because in a functorial metaframe every π_σ is monotonic. ■

Remark 5.15.6 In a similar way one can show that the equivalence

$$M, \mathbf{a} \Vdash \exists x_i B\ [\mathbf{x}] \Leftrightarrow \exists c \in D(\mathbf{a})\ M, \hat{\mathbf{a}}_i c \Vdash B\ [\hat{\mathbf{x}}_i x_i].$$

holds in functorial metaframes. Then we readily have

$$M, \mathbf{a} \Vdash \exists x_i B\ [\mathbf{x}] \Leftrightarrow M, \hat{\mathbf{a}}_i \Vdash \exists x_i B\ [\hat{\mathbf{x}}_i].$$

The same equivalence for $\forall$ was already proved in 5.14, cf. the condition (VII.2).

Intutionistic metaframes

As we already know, functoriality is equivalent to modal soundness (Lemma 5.11.6 and Theorem 5.12.13). But i-functoriality does not yet imply intuitionistic soundness as we shall see in section 5.16. Now let us consider two other soundness conditions specific for the intuitionistic case.

Definition 5.15.7 *An* intuitionistic metaframe *(*ı-metaframe, *for short) is an i-functorial metaframe satisfying*

(2σ) (forward 2-lift property)
$\forall \mathbf{a} \in D^n\ \forall \mathbf{b} \in D^{m+1}((\pi_\sigma \mathbf{a}) R^m (\pi_+^m \mathbf{b}) \Rightarrow \exists \mathbf{c} \in D^{n+1}(\mathbf{a} R^n (\pi_+^n \mathbf{c})\ \&\ (\pi_{\sigma^+} \mathbf{c}) R^{m+1} \mathbf{b}))$.

(3σ) (backward 2-lift property)
$\forall \mathbf{a} \in D^n\ \forall \mathbf{b} \in D^{m+1}((\pi_+^m \mathbf{b}) R^m (\pi_\sigma \mathbf{a}) \Rightarrow \exists \mathbf{c} \in D^{n+1}((\pi_+^n \mathbf{c}) R^n \mathbf{a}\ \&\ \mathbf{b} R^{m+1} (\pi_{\sigma^+} \mathbf{c})))$.

Recall that $\sigma_+^m \in \Upsilon_{m,m+1}$, $\sigma_+^n \in \Upsilon_{n,n+1}$ are simple embeddings and $\sigma^+ \in \Sigma_{m+1,n+1}$ is the simple extension of $\sigma \in \Sigma_{mn}$, see the Introduction.

So a metaframe is intuitionistic iff it satisfies $(0\sigma), (1\sigma), (2\sigma), (3\sigma)$ for any σ.

Definition 5.15.8 *A* weak intuitionistic (wi-) metaframe *is an s-functorial metaframe satisfying* $(1\sigma), (2\sigma), (3\sigma)$, *for all injections* σ.

Note that in this case (0σ) holds for arbitrary σ.

Definition 5.15.9 *An* intuitionistic metaframe with equality *(an* i$^=$-metaframe, *for short) and respectively, a* weak intuitionistic metaframe with equality *(a* wi$^=$-metaframe*) is an $s^=$-functorial i-metaframe (respectively, an $s^=$-functorial wi-metaframe).*

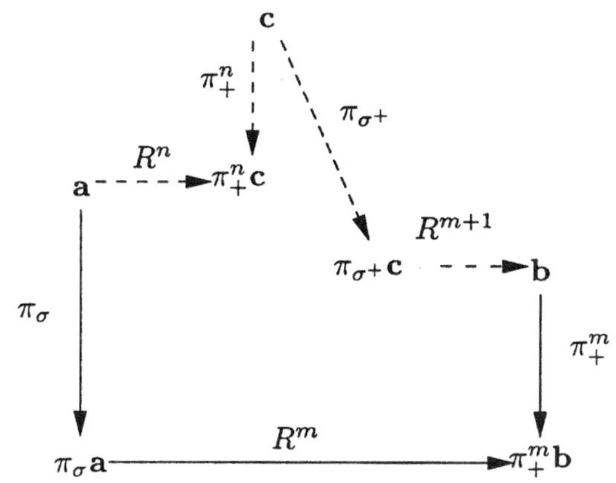

Figure 5.10. Forward 2-lift property

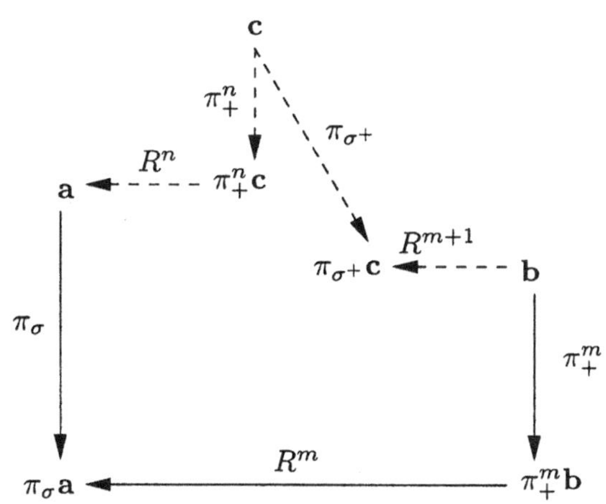

Figure 5.11. Backward 2-lift property.

Lemma 5.15.10 *$\mathrm{i}^{=}$-metaframes (respectively $\mathrm{wi}^{=}$-metaframes) are exactly i-metaframes (respectively, wi-metaframes) satisfying* (mm_2) *from 5.11.1.*

Our next goal is to show that $\mathrm{i}^{=}$-metaframes are exactly $\mathrm{i}^{=}$-sound metaframes (Theorem 5.16.13) and at the same time — $\mathrm{wi}^{=}$-metaframes (Theorem 5.16.10(3)).

Remark 5.15.11 Let us give a motivation for 2-lift properties. The starting point is the following simple observation on tuples (called the 'lift property' in [Skvortsov and Shehtman, 1993]). Let

$$\sigma \in \Sigma_{mn},\ \mathbf{a} = (a_1, \dots, a_n) \in D^n,\ b \in D^{m+1},\ \pi_\sigma \mathbf{a} = \pi_+^m \mathbf{b},$$

i.e. $\mathbf{b} = a_{\sigma(1)} \dots a_{\sigma(m)} b_{m+1}$; then there exists $\mathbf{c} \in D^{n+1}$ such that $\mathbf{a} = \pi_+^n \mathbf{c}$ and $\pi_{\sigma^+} \mathbf{c} = \mathbf{b}$; in fact, put $\mathbf{c} = \mathbf{a} b_{m+1}$. This shows ($2\sigma$), ($3\sigma$) for the case when R^k are the equality relations. See Fig. 5.12 below.

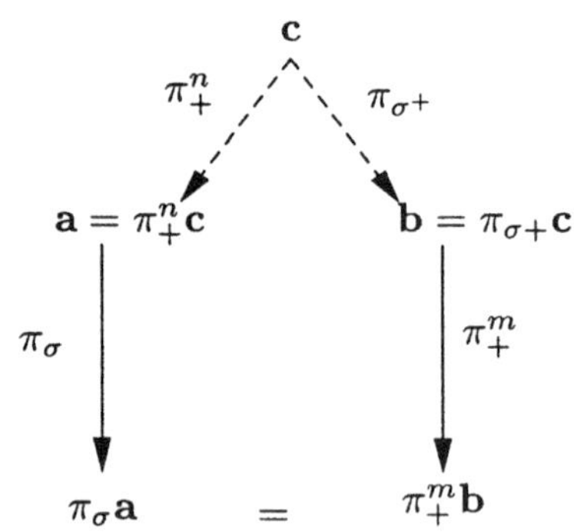

Figure 5.12. $\pi_\sigma \mathbf{a}$ is a 'common part' of $\mathbf{b}$ and $\mathbf{a}$; $\mathbf{c}$ is their 'join'.

We can also check these properties for a metaframe $\mathbf{Mf}(\mathbb{F})$ associated with a $\mathcal{C}$-set $\mathbf{F} = (C, D, \rho)$ over a category $\mathcal{C}$. In fact, if for $\mathbf{a} \in D_u^n$ and $\mathbf{b} \in D_v^{m+1}$ we have

$$\pi_\sigma \mathbf{a} = (a_{\sigma(1)}, \dots, a_{\sigma(m)}) R^m (b_1, \dots, b_m) = \pi_+^m \mathbf{b},$$

then there exists a morphism $\mu \in \mathcal{C}(u, v)$ such that $\pi_+^m \mathbf{b} = \rho_\mu \cdot (\pi_\sigma \mathbf{a})$, i.e. $\forall i\ b_i = \rho_\mu(a_{\sigma(i)})$. Then consider

$$\mathbf{c}' := \rho_\mu \cdot \mathbf{a} = (\rho_\mu(a_1), \dots, \rho_\mu(a_n)) \in D_v^n$$

and extend it to

$$\mathbf{c} := \mathbf{c}' b_{m+1} \in D_v^{n+1}.$$

It follows that

$$\mathbf{a} R^n \mathbf{c}' = \pi_+^n \mathbf{c},\ \pi_{\sigma^+} \mathbf{c} = (\pi_\sigma \mathbf{c}') b_{m+1} = (\rho_\mu \cdot (\pi_\sigma \mathbf{a})) b_{m+1} = \mathbf{b}$$

and thus obviously

$$(\pi_{\sigma^+} \mathbf{c}) R^{m+1} \mathbf{b}.$$

So (2σ) holds; (3σ) is checked in a dual way.

We see that in the first example the condition (2σ) expresses the 'co-amalgamation property' of tuples (Fig. 5.12).

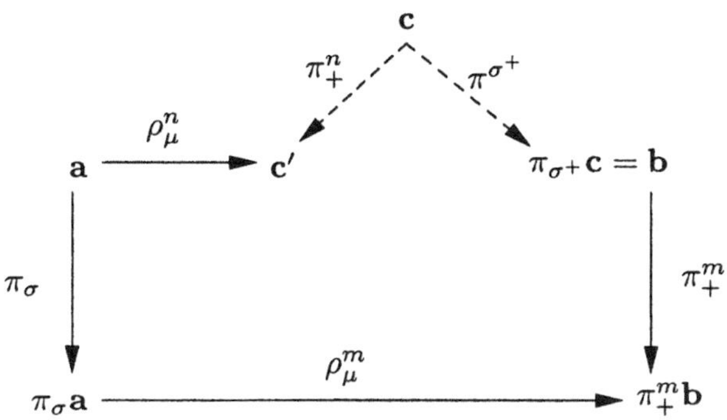

Figure 5.13.

The second example corresponds to Fig. 5.14. Here ρ_μ^n sends $\mathbf{a}$ to $\rho_\mu \cdot \mathbf{a}$; similarly for ρ_μ^m. Again $\mathbf{c}$ is the 'join' of $\mathbf{c}'$ and $\mathbf{b}$; it exists, because they have a 'common part' $\pi_+^m\mathbf{b}$.

Lemma 5.15.12 *In i-functorial metaframes* (2σ) *is equivalent to*

$$\forall \mathbf{a} \in D^n\, \forall \mathbf{b} \in D^{m+1}(\pi_\sigma\mathbf{a} \approx^m \pi_+^m\mathbf{b} \Rightarrow \exists \mathbf{c} \in D^{n+1}(\mathbf{a}R^n(\pi_+^n\mathbf{c})\ \&\ \pi_{\sigma+}\mathbf{c} \approx^{m+1} \mathbf{b})). \tag{$2^\sim\sigma$}$$

Similarly (3σ) *is equivalent to*

$$\forall \mathbf{a} \in D^n\, \forall \mathbf{b} \in D^{m+1}(\pi_+^m\mathbf{b} \approx^m \pi_\sigma\mathbf{a} \Rightarrow \exists \mathbf{c} \in D^{n+1}((\pi_+^n\mathbf{c})R^n\mathbf{a}\ \&\ \mathbf{b} \approx^{m+1} \pi_{\sigma+}\mathbf{c}). \tag{$3^\sim\sigma$}$$

As we can see, these conditions are obtained from (2σ) and (3σ) by changing R^m, R^{m+1} to $\approx^m$, $\approx^{m+1}$.

Proof (2σ) $\Rightarrow$ ($2^\sim\sigma$). We assume (2σ) and prove a stronger claim than ($2^\sim\sigma$). Namely, for $\mathbf{a} \in D^{n+1}$, $\mathbf{b} \in D^{m+1}$ such that $(\pi_\sigma\mathbf{a})R^m(\pi_+^m\mathbf{b})$ we find $\mathbf{c}' \in D^{n+1}$ such that

$$\mathbf{a}R^n\pi_+^n\mathbf{c}'\ \&\ \pi_{\sigma+}\mathbf{c}' \approx^m \mathbf{b}.$$

First we apply (2σ) to $\mathbf{a}$, $\mathbf{b}$ and find $\mathbf{c}$ shown in Fig. 5.13.

Then using the quasi-lift property ($1\sigma^+$) we find $\mathbf{c}'$. Finally we note that $\mathbf{a}R^n(\pi_+^n\mathbf{c})R^n(\pi_+^n\mathbf{c}')$ by monotonicity ($0\pi_+^n$).

($2^\sim\sigma$) $\Rightarrow$ (2σ). Assume ($2^\sim\sigma$). Given $\mathbf{a}, \mathbf{b}$ such that $(\pi_\sigma\mathbf{a})R^m(\pi_+^n\mathbf{b})$, by the quasi-lift property (1σ), we can find $\mathbf{a}' \in R^n(\mathbf{a})$ with $\pi_\sigma\mathbf{a}' \approx^m \pi_+^m\mathbf{b}$, see Fig. 5.15. Then by applying ($2^\sim\sigma$) to $\mathbf{a}'$ and $\mathbf{b}$, we obtain $\mathbf{c}$ such that $\mathbf{a}' \approx^n \pi_+^n\mathbf{c}$, $\mathbf{b} \approx^{m+1} \pi_{\sigma+}\mathbf{c}$. Hence by transitivity $\mathbf{a}R^n\pi_+^n\mathbf{c}$. $\pi_{\sigma+}\mathbf{c}R^{m+1}\mathbf{b}$ is obvious.

The proof of (3σ) $\Leftrightarrow$ ($3^\sim\sigma$) is dual, so we skip it. ■

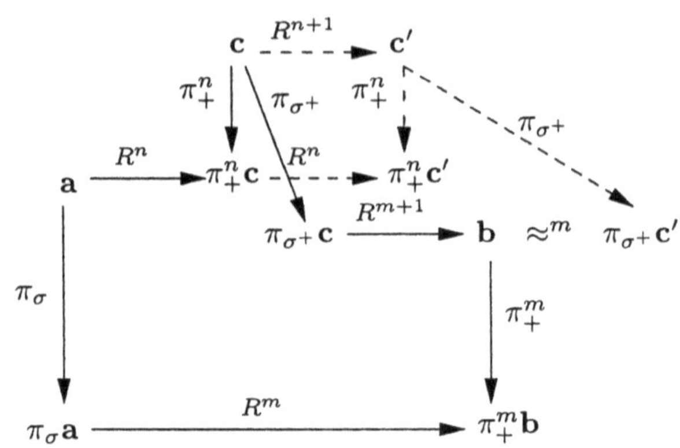

Figure 5.14.

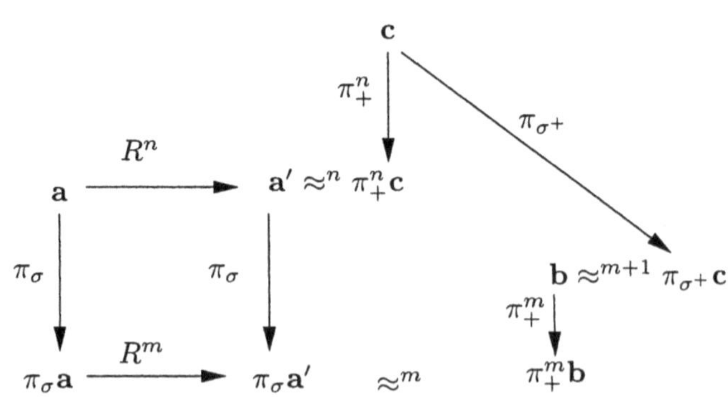

Figure 5.15.

Lemma 5.15.13 *In i-functorial metaframes* (2σ) *is equivalent to*

$$\forall \mathbf{a} \in D^n \, \forall \mathbf{b} \in D^{m+1}(\pi_\sigma \mathbf{a} \approx^m \pi_+^m \mathbf{b} \Rightarrow \exists \mathbf{c} \in D^{n+1}(\mathbf{a}R^n(\pi_+^n \mathbf{c}) \;\&\; (\pi_{\sigma^+}\mathbf{c})R^{m+1}\mathbf{b}))$$

and dually for (3σ).

Proof Similar to Lemma 5.15.12 (an exercise). ■

Now we are going to prove intuitionistic analogues of the results from Sections 5.11 and 5.12.

The property (1σ) together with s-functoriality means that π_σ is a quasi-p-morphism. In modal logic we have a stronger requirement that π_σ is a p-morphism. In fact, in intuitionistic models, due to monotonicity (im), $\approx^n$-equivalent worlds are indistinguishable. Thus equality transforms into $\approx^n$ in the intuitionistic case.

The next assertion is an intuitionistic analogue of Lemma 5.10.6.

Lemma 5.15.14 *Let* $\mathbb{F}$ *be an* $s^{(=)}$*-functorial metaframe. Then* $\mathbb{F}$ *is weakly functorial iff for any* $A \in IF^{(=)}$*, for any assignment* $(\mathbf{x}, \mathbf{a})$ *of length* n*, for any injection* $\sigma \in \Upsilon_{mn}$ *such that* $FV(A) \subseteq \mathbf{x} \cdot \sigma$*, for any intuitionistic model* $M = (\mathbb{F}, \xi)$

$$M, \mathbf{a} \Vdash A\,[\mathbf{x}] \Leftrightarrow M, \mathbf{a} \cdot \sigma \Vdash A\,[\mathbf{x} \cdot \sigma]. \tag{*i}$$

So (*i) means that $M, \mathbf{a} \Vdash A[\mathbf{x}]$ does not depend on the choice of a list $\mathbf{x} \supseteq FV(A)$ and possible renumbering of variables.

Proof $(\Rightarrow)$ is proved by induction on the complexity of A (cf. Lemma 5.10.6). The atomic cases are obvious. The cases when the main connective of A is $\vee$ or $\wedge$, are also easy, and we leave them to the reader.

Let $A = B \supset C$. Suppose $M, \mathbf{a} \nVdash A\,[\mathbf{x}]$, i.e.

$$M, \mathbf{b} \Vdash B\,[\mathbf{x}]; \;\; M, \mathbf{b} \nVdash C\,[\mathbf{x}]$$

for some $\mathbf{b} \in R^n(\mathbf{a})$. Then $(\mathbf{a} \cdot \sigma)R^m(\mathbf{b} \cdot \sigma)$ by s-functoriality, and so

$$M, \mathbf{b} \cdot \sigma \Vdash B\,[\mathbf{x} \cdot \sigma], \;\; M, \mathbf{b} \cdot \sigma \nVdash C\,[\mathbf{x} \cdot \sigma]$$

by the induction hypothesis, thus $M, \mathbf{a} \cdot \sigma \nVdash A\,[\mathbf{x} \cdot \sigma]$.

The other way round, suppose $M, \mathbf{a} \cdot \sigma \nVdash A[\mathbf{x} \cdot \sigma]$, i.e. $M, \mathbf{b}' \Vdash B[\mathbf{x} \cdot \sigma]$, $M, \mathbf{b}' \nVdash C[\mathbf{x} \cdot \sigma]$. for some $\mathbf{b}' \in R^m(\mathbf{a} \cdot \sigma)$. Then by (1σ), $\mathbf{b} \cdot \sigma \approx^m \mathbf{b}'$ for some $\mathbf{b} \in R^n(\mathbf{a})$. Now by monotonicity,

$$M, \mathbf{b} \cdot \sigma \Vdash B[\mathbf{x} \cdot \sigma]; \;\; M, \mathbf{b} \cdot \sigma \nVdash C[\mathbf{x} \cdot \sigma],$$

and thus by the induction hypothesis,

$$M, \mathbf{b} \Vdash B[\mathbf{x}]; \;\; M, \mathbf{b} \nVdash C[\mathbf{x}].$$

Hence $M, \mathbf{a} \nVdash A[\mathbf{x}]$.

Now let us consider the case $A = \forall yB$. Supposing $M, \mathbf{a}\cdot\sigma \not\Vdash A[\mathbf{x}\cdot\sigma]$, and let us prove $M, \mathbf{a} \not\Vdash A[\mathbf{x}]$. By definition, we have $\mathbf{c} \in R^m(\mathbf{a}\cdot\sigma)$, $d \in D(\mathbf{c})$ such that

$$M, \pi_{(\mathbf{x}\cdot\sigma)||y}(\mathbf{c}d) \not\Vdash B[(\mathbf{x}\cdot\sigma)||y].$$

Since $(\pi_\sigma\mathbf{a})R^m\mathbf{c} = \pi^m_+(\mathbf{c}d)$, by applying (2σ), we obtain $\mathbf{b} \in D^n$, $e \in D^1$ such that

$$\mathbf{a}R^n\pi^n_+(\mathbf{b}e) = \mathbf{b} \ \&\ (\mathbf{b}\cdot\sigma)e = \pi_{\sigma+}(\mathbf{b}e)R^{m+1}(\mathbf{c}d).$$

Hence by s-functoriality,

$$\pi_{(\mathbf{x}\cdot\sigma)||y}((\mathbf{b}\cdot\sigma)e)R^l\pi_{(\mathbf{x}\cdot\sigma)||y}(\mathbf{c}d),$$

where l is the length of $(\mathbf{x}\cdot\sigma)||y$.

Now note that

$$\pi_{(\mathbf{x}\cdot\sigma)||y}((\mathbf{x}\cdot\sigma)y) = (\mathbf{x}\cdot\sigma)||y = (\mathbf{x}||y)\cdot\tau = \pi_{\mathbf{x}||y}(\mathbf{x}y)\cdot\tau$$

for some τ, as we saw in the proof of 5.10.6. Hence by a standard argument we have the same for arbitrary tuples:

$$\pi_{(\mathbf{x}\cdot\sigma)||y}((\mathbf{b}\cdot\sigma)e) = \pi_{\mathbf{x}||y}(\mathbf{b}e)\cdot\tau.$$

Then by monotonicity we obtain

$$\pi_{\mathbf{x}||y}(\mathbf{b}e)\cdot\tau \not\Vdash B[(\mathbf{x}\cdot\sigma)||y],$$

and since $(\mathbf{x}\cdot\sigma)||y = (\mathbf{x}||y)\cdot\tau$, by the induction hypothesis, it follows that

$$\pi_{\mathbf{x}||y}(\mathbf{b}e) \not\Vdash B[\mathbf{x}||y].$$

But $\mathbf{a}R^n\mathbf{b}$, thus $\mathbf{a} \not\Vdash \forall yB[\mathbf{x}]$, as required.

The other way round, suppose $\mathbf{a} \not\Vdash A[\mathbf{x}]$. Then

$$\pi_{\mathbf{x}||y}(\mathbf{b}e) \not\Vdash B[\mathbf{x}||y]$$

for some $\mathbf{b} \in R^n(\mathbf{a})$, $e \in D(\mathbf{b})$. Hence by the induction hypothesis,

$$\pi_{\mathbf{x}||y}(\mathbf{b}e)\cdot\tau \not\Vdash B[(\mathbf{x}||y)\cdot\tau],$$

where τ is the same as above, i.e. we obtain

$$\pi_{(\mathbf{x}\cdot\sigma)||y}((\mathbf{b}\cdot\sigma)e) \not\Vdash B[(\mathbf{x}\cdot\sigma)||y].$$

But $(\mathbf{a}\cdot\sigma)R^n(\mathbf{b}\cdot\sigma)$, by s-functoriality; thus eventually,

$$\mathbf{a}\cdot\sigma \not\Vdash A[\mathbf{x}\cdot\sigma],$$

by Definition 5.14.2.

The case $A = \exists yB$ is dual to $A = \forall yB$. The proof for this case is obtained by inverting the accessibility relations R^n, changing $\not\Vdash$ to $\Vdash$ and applying (3σ) instead of (2σ).

In more detail, suppose $M, \mathbf{a} \cdot \sigma \Vdash A[\mathbf{x} \cdot \sigma]$. By definition, there exists $\mathbf{c} \in (R^m)^{-1}(\mathbf{a} \cdot \sigma)$, $d \in D(\mathbf{c})$ such that

$$\pi_{(\mathbf{x}\cdot\sigma)||y}(\mathbf{c}d) \Vdash B[(\mathbf{x} \cdot \sigma)||y].$$

From $\mathbf{c}R^m(\pi_\sigma \mathbf{a})$, by ($3\sigma$), we obtain $\mathbf{b} \in D^n$, $e \in D^1$ such that

$$\pi_+^n(\mathbf{b}e) = \mathbf{b}R^n\mathbf{a} \ \&\ (\mathbf{c}d)R^{m+1}(\mathbf{b} \cdot \sigma)e.$$

Hence by s-functoriality,

$$\pi_{(\mathbf{x}\cdot\sigma)||y}(\mathbf{c}d)R^l\pi_{(\mathbf{x}\cdot\sigma)||y}((\mathbf{b} \cdot \sigma)e) = \pi_{\mathbf{x}||y}(\mathbf{b}e) \cdot \tau,$$

where l is the length of $(\mathbf{x} \cdot \sigma)||y$. Then by monotonicity

$$\pi_{\mathbf{x}||y}(\mathbf{b}e) \cdot \tau \Vdash B[(\mathbf{x} \cdot \sigma)||y],$$

and since $(\mathbf{x} \cdot \sigma)||y = (\mathbf{x}||y) \cdot \tau$ by the induction hypothesis, it follows that

$$\pi_{\mathbf{x}||y}(\mathbf{b}e) \Vdash B[\mathbf{x}||y].$$

Since $\mathbf{a}R^n\mathbf{b}$, we obtain $\mathbf{a} \not\Vdash \forall yB[\mathbf{x}]$.

To show the converse, suppose $\mathbf{a} \Vdash A[\mathbf{x}]$. Then

$$\pi_{\mathbf{x}||y}(\mathbf{b}e) \Vdash B[\mathbf{x}||y]$$

for some $\mathbf{b} \in (R^n)^{-1}(\mathbf{a})$, $e \in D(\mathbf{b})$. Hence by the induction hypothesis,

$$\pi_{\mathbf{x}||y}(\mathbf{b}e) \cdot \tau \Vdash B[(\mathbf{x}||y) \cdot \tau].$$

This is the same as

$$\pi_{(\mathbf{x}\cdot\sigma)||y}((\mathbf{b} \cdot \sigma)e) \Vdash B[(\mathbf{x} \cdot \sigma)||y].$$

Since by s-functoriality, $(\mathbf{b} \cdot \sigma)R^n(\mathbf{a} \cdot \sigma)$, it follows that

$$\mathbf{a} \cdot \sigma \Vdash A[\mathbf{x} \cdot \sigma].$$

($\Leftarrow$) Assuming the equivalence ($*i$) for suitable formulas A, let us prove the properties (1σ), (2σ), (3σ) for an injection $\sigma \in \Upsilon_{mn}$.

(1σ) Let $\mathbf{a} \in D^n$, $\mathbf{b}' \in D^m$, and assume $(\pi_\sigma \mathbf{a})R^m\mathbf{b}'$. Consider the formula

$$A := P(\pi_\sigma \mathbf{x}) \supset Q(\pi_\sigma \mathbf{x})$$

and the intuitionistic model $M = (\mathbb{F}, \xi)$ such that

$$\xi^+(P) := R^m(\mathbf{b}'),\ \xi^+(Q) := \{\mathbf{c} \in D^m \mid \neg(\mathbf{c}R^m\mathbf{b}')\}$$

(recall that $\mathbb{F}$ is an **S4**-frame). Then $\mathbf{b}' \in \xi^+(P) - \xi^+(Q)$, thus $M, \pi_\sigma \mathbf{a} \not\Vdash A[\pi_\sigma \mathbf{x}]$. Therefore by ($*i$), $M, \mathbf{a} \not\Vdash A[\mathbf{x}]$, i.e. there exists $\mathbf{b} \in R^m(\mathbf{a})$ such that $M, \mathbf{b} \Vdash P(\pi_\sigma \mathbf{x})[\mathbf{x}]$ and $M, \mathbf{b} \not\Vdash Q(\pi_\sigma \mathbf{x})[\mathbf{x}]$. Thus $\pi_\sigma \mathbf{b} \in (\xi^+(P) - \xi^+(Q)) = \ \approx^m (\mathbf{b}')$, i.e. $\mathbf{b}' \approx^m \pi_\sigma \mathbf{b}$.

(2σ) Let $\mathbf{a} \in D^n$, $\mathbf{c} = (\pi_+^m \mathbf{c})d \in D^{m+1}$, and assume that $(\pi_\sigma \mathbf{a})R^m(\pi_+^m \mathbf{c})$. Consider the formula $A := \forall y P(\pi_\sigma \mathbf{x}, y)$ with $\mathbf{x} = (x_1, \ldots, x_n)$, $y \notin \mathbf{x}$, and the intuitionistic model $M = (\mathbb{F}, \xi)$ such that

$$\xi^+(P) = \{\mathbf{c}' \in D^{m+1} \mid \neg(\mathbf{c}' R^{m+1} \mathbf{c})\}.$$

Since $\mathbf{c} \notin \xi^+(P)$, we have

$$M, \mathbf{c} \not\Vdash P(\pi_\sigma \mathbf{x}, y)\, [(\pi_\sigma \mathbf{x})y],$$

and so $M, \pi_\sigma \mathbf{a} \not\Vdash A[\pi_\sigma \mathbf{x}]$, since $(\pi_\sigma \mathbf{a})R^m(\pi_+^m \mathbf{c})$. Thus by (∗i), $M, \mathbf{a} \not\Vdash A\,[\mathbf{x}]$, i.e. there exists $\mathbf{b} = (\pi_+^n \mathbf{b})d \in D^{n+1}$ such that

$$\mathbf{a}R^n(\pi_+^n \mathbf{b}) \ \&\ M, \mathbf{b} \not\Vdash P(\pi_\sigma \mathbf{x}, y)\, [\mathbf{x}y].$$

But $(\pi_\sigma \mathbf{x})y = \pi_{\sigma^+}(\mathbf{x}y)$, thus $\pi_{\sigma^+}(\mathbf{b}) \notin \xi^+(P)$, so by definition of M, $\pi_{\sigma^+}(\mathbf{b})R^{m+1}\mathbf{c}$, as required.

(3σ) The proof is quite similar to the previous case. Now let $\mathbf{a} \in D^n$, $\mathbf{c} = (\pi_+^m \mathbf{c})d \in D^{m+1}$, $(\pi_+^m \mathbf{c})R^m(\pi_\sigma \mathbf{a})$.
Consider the formula $A := \exists y P(\pi_\sigma \mathbf{x}, y)$, where $\mathbf{x} = (x_1, \ldots, x_n)$, $y \notin \mathbf{x}$, and the intituionistic model $M = (\mathbb{F}, \xi)$ such that $\xi^+(P) = R^{m+1}(\mathbf{c})$. Since $\mathbf{c} \in \xi^+(P)$, we have $M, \mathbf{c} \Vdash P(\pi_\sigma \mathbf{x}, y)\, [\pi_\sigma \mathbf{x}, y]$, and so $M, \pi_\sigma \mathbf{a} \Vdash A\,[\pi_\sigma \mathbf{x}]$, since $(\pi_+^m \mathbf{c})R_m(\pi_\sigma \mathbf{a})$. Thus by (∗i) $M, \mathbf{a} \Vdash A\,[\mathbf{x}]$, i.e. there exists $\mathbf{b} = (\pi_+^n \mathbf{b})d \in D^{n+1}$ such that $(\pi_+^n \mathbf{b})R^n \mathbf{a}$ and $M, \mathbf{b} \Vdash P(\pi_\sigma \mathbf{x}, y)\, [\mathbf{x}y]$, which is equivalent to

$$M, \mathbf{b} \vDash P(\pi_{\sigma^+}(\mathbf{x}y))\, [\mathbf{x}y],$$

i.e. $\pi_{\sigma^+}\mathbf{b} \in \xi^+(P)$, and hence $\mathbf{c}R^{m+1}(\pi_{\sigma^+}\mathbf{b})$.

■

As in the modal case, Lemma 5.15.14 allows us to arbitrarily change a list $\mathbf{x} \supseteq FV(A)$ in $M, \mathbf{a} \Vdash A\,[\mathbf{x}]$; e.g. we can take $\mathbf{x} = FV(A)$. So in the truth conditions for quantifiers we may always assume $y \notin \mathbf{x}$, and thus we have

$$\mathbf{a} \Vdash \forall y B\,[\mathbf{x}] \text{ iff } \forall \mathbf{b} \in R^n(\mathbf{a})\ \forall c \in D(\mathbf{b})\ \mathbf{b}c \Vdash B\,[\mathbf{x}y];$$
$$\mathbf{a} \Vdash \exists y B\,[\mathbf{x}] \text{ iff } \exists \mathbf{b} R^n \mathbf{a}\ \exists c \in D(\mathbf{b})\ \mathbf{b}c \Vdash B\,[\mathbf{x}y].$$

(or equivalently, $\mathbf{b}R^n\mathbf{a}$ may be replaced with $\mathbf{b} \approx^n \mathbf{a}$ in the clause for $\exists$, cf. Lemma 5.15.3).

Now let us prove an intuitionistic analogue of Lemma 5.10.9.

Lemma 5.15.15 *Let $M = (\mathbb{F}, \xi)$ be an intuitionistic model over an* **S4**-*metaframe, $A \in IF^{(=)}$, $FV(A) \subseteq r(\mathbf{x})$. Also let $y \notin BV(A)$, $y' \notin V(A)$, $\mathbf{x}' := [y'/y]\,\mathbf{x}$, $y' \notin r(\mathbf{x})$, and $A' := A[y \mapsto y']$. Then*

$$M, \mathbf{a} \Vdash A\,[\mathbf{x}] \textit{ iff } M, \mathbf{a} \Vdash A'[\mathbf{x}'].$$

Proof Since $\mathbb{F}$ is a wi$^{(=)}$-metaframe, due to the previous remarks, we prove the equivalence only for $\mathbf{x} = FV(A)$. We argue by induction and consider only the atomic and the quantifier cases; other cases can be easily checked by the reader.

Let $A = P(\mathbf{x} \cdot \sigma)$. Then $A' = P((\mathbf{x} \cdot \sigma)') = P(\mathbf{x}' \cdot \sigma)$, so

$$\mathbf{a} \Vdash A\,[\mathbf{x}] \text{ iff } (\mathbf{a} \cdot \sigma) \in \xi^+(P) \text{ iff } \mathbf{a} \Vdash A'\,[\mathbf{x}'].$$

Let $A = \forall z B$ and suppose the assertion holds for B. Then $z \neq y, y'$ by the assumption of the lemma, and $z \notin \mathbf{x}$, by our additional assumption; so $z \notin \mathbf{x}'$. Thus

$$\mathbf{a} \Vdash A\,[\mathbf{x}] \text{ iff } \forall \mathbf{b} \in R^n(\mathbf{a})\, \forall c \in D(\mathbf{b})\ \mathbf{b}c \Vdash B\,[\mathbf{x}z];$$

$$\mathbf{a} \Vdash A'\,[\mathbf{x}'] \text{ iff } \forall \mathbf{b} \in R^n(\mathbf{a})\, \forall c \in D(\mathbf{b})\ \mathbf{b}c \Vdash B'\,[\mathbf{x}'z].$$

Since $(\mathbf{x}z)' = \mathbf{x}'z$, the right parts of these equivalences, are also equivalent by the induction hypothesis. So the assertion holds for A.

If $A = \exists z B$, the argument is similar:

$$\mathbf{a} \Vdash A\,[\mathbf{x}] \text{ iff } \exists \mathbf{b} \approx^n \mathbf{a}\, \exists c \in D(\mathbf{b})\ \mathbf{b}c \Vdash B\,[\mathbf{x}z];$$

$$\mathbf{a} \Vdash A'\,[\mathbf{x}'] \text{ iff } \exists \mathbf{b} \approx^n \mathbf{a}\, \exists c \in D(\mathbf{b})\ \mathbf{b}c \Vdash B'\,[\mathbf{x}'z],$$

and again the right parts of these equivalences are equivalent. ∎

Let us prove an intuitionistic analogue of 5.10.10.

Lemma 5.15.16 *Let $\mathbb{F}$ be a wi$^{(=)}$-metaframe, M an intuitionistic model over $\mathbb{F}$ Then:*

(c) $$M, \mathbf{a} \Vdash A\,[\mathbf{x}] \Leftrightarrow M, \mathbf{a} \Vdash A^1\,[\mathbf{x}]$$

for any congruent formulas A, A^1 and any appropriate assignment $(\mathbf{x}, \mathbf{a})$.

Proof (Cf. Lemma 5.10.10.) We assume $\mathbf{x} = FV(A)$, $|\mathbf{x}| = n$.

It is sufficient to consider the case when A^1 is obtained from A by replacing its subformula QyB with $Qy'B'$, where Q is a quantifier, $y' \notin V(A)$, $y \notin BV(B)$, $B' = B[y \mapsto y']$. Now the proof is by induction.

(1) If $A = \forall y B$, then $A^1 = \forall y' B'$. By our assumption, $\mathbf{x} = FV(A)$, so $y \notin \mathbf{x}$, and thus

$$\begin{array}{l} \mathbf{a} \vDash \forall y B\,[\mathbf{x}] \Leftrightarrow \forall \mathbf{b} \in R^n(\mathbf{a})\, \forall d \in D(\mathbf{b})\ \mathbf{a}d \Vdash B\,[\mathbf{x}y] \Leftrightarrow \\ \forall \mathbf{b} \in R^n(\mathbf{a})\, \forall d \in D(\mathbf{b})\ \mathbf{a}d \Vdash B'\,[\mathbf{x}y'] \quad \text{(by Lemma 5.15.15)} \\ \Leftrightarrow \mathbf{a} \Vdash \forall y' B'\,[\mathbf{x}]. \end{array}$$

(2) If $A = \exists y B$, the proof is almost the same. We have $A^1 = \exists y' B'$. Suppose $y', y \notin \mathbf{x}$, and thus

$$\begin{array}{l} \mathbf{a} \vDash \exists y B\,[\mathbf{x}] \Leftrightarrow \exists \mathbf{b} \approx^n \mathbf{a}\, \exists d \in D(\mathbf{b})\ \mathbf{a}d \Vdash B\,[\mathbf{x}y] \Leftrightarrow \\ \exists \mathbf{b} \approx^n \mathbf{a}\, \exists d \in D(\mathbf{b})\ \mathbf{a}d \Vdash B'\,[\mathbf{x}y'] \quad \text{(by Lemma 5.15.15)} \\ \Leftrightarrow \mathbf{a} \Vdash \exists y' B'\,[\mathbf{x}]. \end{array}$$

(1), (2) prove the base of the induction. The step is rather routine, so we only consider the case $A = \forall z C$, $A^1 = \forall z C^1$. We have $\mathbf{x} = FV(A)$, $z \notin \mathbf{x}$, and thus

$$\mathbf{a} \Vdash A\,[\mathbf{x}] \Leftrightarrow \forall \mathbf{b} \in R^n(\mathbf{a})\, \forall d \in D(\mathbf{b})\ \mathbf{a}d \Vdash C\,[\mathbf{x}z] \Leftrightarrow$$
$$\forall \mathbf{b} \in R^n(\mathbf{a})\, \forall d \in D(\mathbf{b})\ \mathbf{a}d \Vdash B^1\,[\mathbf{x}z] \quad \text{(by induction hypothesis)}$$
$$\Leftrightarrow \mathbf{a} \Vdash \forall z B^1\,(= A^1)\,[\mathbf{x}].$$

The remaining cases are left to the reader. ■

Let us prove the intuitionistic analogue of 5.11.7.

Lemma 5.15.17 *Let $\mathbb{F}$ be an $s^{(=)}$-functorial metaframe. Then $\mathbb{F}$ is an i-metaframe iff*

(#) for any $\sigma \in \Sigma_{mn}$, distinct lists of variables $\mathbf{x} = (x_1, \ldots, x_n)$, $\mathbf{y} = (y_1, \ldots, y_m)$, and formula $A \in IF^{(=)}$ with $FV(A) \subseteq r(\mathbf{y})$, for any model M over $\mathbb{F}$, $\mathbf{a} \in D^n$:

$$M, \mathbf{a} \Vdash [(\mathbf{x} \cdot \sigma)/\mathbf{y}]\, A\ [\mathbf{x}] \Leftrightarrow M, \mathbf{a} \cdot \sigma \Vdash A\,[\mathbf{y}].$$

Similarly to 5.11.7, this equivalence is stated for all congruent versions of $[\mathbf{x} \cdot \sigma/\mathbf{y}]A$.

Proof ('Only if'.) Recall that $[(\mathbf{x} \cdot \sigma)/\mathbf{y}]A$ is defined up to congruence and can be obtained from a clean version A° of A such that $BV(A^\circ) \cap r(\mathbf{xy}) = \varnothing$ (and $FV(A^\circ) = FV(A) \subseteq \mathbf{y}$) by replacing $[\mathbf{y} \mapsto \mathbf{x} \cdot \sigma]$. By Lemma 5.15.16 we may assume that $A^\circ = A$. Now we argue similarly to 5.11.7 by induction.

(1) If A is $y_i = y_j$ or $P_j^k(\mathbf{y} \cdot \tau)$, $\tau \in \Sigma_{km}$ or A is, the proof is the same as in 5.11.7.

(2) Suppose $A = B \supset C$. If $\mathbf{a} \nVdash [(\mathbf{x} \cdot \sigma)/\mathbf{y}]A\ [\mathbf{x}]$, i.e. for some $\mathbf{b} \in R^n(\mathbf{a})$

$$\mathbf{b} \Vdash [(\mathbf{x} \cdot \sigma)/\mathbf{y}]B\ [\mathbf{x}]$$

and

$$\mathbf{b} \nVdash [(\mathbf{x} \cdot \sigma)/\mathbf{y}]C\ [\mathbf{x}],$$

then by the induction hypothesis

$$\mathbf{b} \cdot \sigma \Vdash B\,[\mathbf{y}] \text{ and } \mathbf{b} \cdot \sigma \nVdash C\,[\mathbf{y}].$$

Since $\mathbf{a}R^n\mathbf{b}$ implies $(\mathbf{a} \cdot \sigma)R^m(\mathbf{b} \cdot \sigma)$ by (0σ), we have $\mathbf{a} \cdot \sigma \nVdash (B \supset C)\,[\mathbf{y}]$.

The other way round, if

$$\mathbf{a} \cdot \sigma \nVdash (B \supset C)\,[\mathbf{y}],$$

then for some $\mathbf{c} \in R^m(\mathbf{a} \cdot \sigma)$

$$\mathbf{c} \Vdash B\,[\mathbf{y}] \text{ and } \mathbf{c} \nVdash C\,[\mathbf{y}].$$

Now by (1σ), there exists $\mathbf{b} \in R^n(\mathbf{a})$ such that $\mathbf{b}\cdot\sigma \approx^m \mathbf{c}$. Then by monotonicity

$$\mathbf{b}\cdot\sigma \Vdash B[\mathbf{y}] \text{ and } \mathbf{b}\cdot\sigma \nVdash C[\mathbf{y}].$$

So by the induction hypothesis,

$$\mathbf{b} \Vdash [(\mathbf{x}\cdot\sigma)/\mathbf{y}]B\ [\mathbf{x}] \text{ and } \mathbf{b} \nVdash [(\mathbf{x}\cdot\sigma)/\mathbf{y}]C\ [\mathbf{x}].$$

Since $\mathbf{a}R^n\mathbf{b}$, this implies

$$\mathbf{a} \nVdash [(\mathbf{x}\cdot\sigma)/\mathbf{y}]A\ [\mathbf{x}].$$

(3) Let $A = \forall z B$, then $z \notin r(\mathbf{xy})$ by the choice of A so

$$[(\mathbf{x}\cdot\sigma)/\mathbf{y}]\,A = \forall z[(\mathbf{x}z)\cdot\sigma^+/\mathbf{y}z]\,B.$$

If

$$\mathbf{a} \nVdash [(\mathbf{x}\cdot\sigma)/\mathbf{y}]A\ [\mathbf{x}],$$

then there exists $\mathbf{b} \in R^n(\mathbf{a}), d \in D(\mathbf{b})$ such that

$$\mathbf{b}d \nVdash [(\mathbf{x}z)\cdot\sigma^+/\mathbf{y}z]B\ [\mathbf{x}z].$$

So by the induction hypothesis

$$(\mathbf{b}\cdot\sigma)d = (\mathbf{b}d)\cdot\sigma^+ \nVdash B[\mathbf{y}z].$$

Since $(\mathbf{a}\cdot\sigma)R^m(\mathbf{b}\cdot\sigma)$, it follows that

$$\mathbf{a}\cdot\sigma \nVdash A[\mathbf{y}].$$

The other way round, suppose $\mathbf{a}\cdot\sigma \nVdash A[\mathbf{y}]$. Then for some $\mathbf{c}' \in R^m(\mathbf{a}\cdot\sigma)$, $d \in D(\mathbf{c}')$

$$\mathbf{c}'d \nvDash B[\mathbf{y}z].$$

So for $\mathbf{c} := \mathbf{c}'d$ we have $(\pi_\sigma\mathbf{a})R^m\mathbf{c}' = \pi^m_+\mathbf{c}$. Hence by ($2\sigma$) there exists $\mathbf{b} \in D^{n+1}$ such that

$$\mathbf{a}R^n(\pi^n_+\mathbf{b}) \text{ and } (\pi_{\sigma^+}\mathbf{b})R^{m+1}\mathbf{c}.$$

Put $\mathbf{b}' := \pi^n_+\mathbf{b}$, then

$$\mathbf{b} = \mathbf{b}'b_{n+1},\ \ \pi_{\sigma^+}\mathbf{b} = (\pi_\sigma\mathbf{b}')b_{n+1}.$$

Since

$$(\pi_{\sigma^+}\mathbf{b})R^{m+1}\mathbf{c} \nVdash B[\mathbf{y}z],$$

by monotonicity we have

$$\pi_{\sigma^+}\mathbf{b} \nVdash B[\mathbf{y}z].$$

Then by the induction hypothesis,

$$\mathbf{b} \nVdash [(\mathbf{x}z)\cdot\sigma^+/\mathbf{y}z]B\ [\mathbf{x}z].$$

But $\mathbf{b} = \mathbf{b}'b_{n+1}$, while $\mathbf{a}R^n\mathbf{b}'$. Hence by the definition of forcing

$$\mathbf{a} \not\Vdash \forall z[(\mathbf{x}z) \cdot \sigma^+/\mathbf{y}z]B\ [\mathbf{x}]$$

as required.

(4) The case $A = \exists zB$ is dual to the previous one. The proof makes use of (3σ), cf. the $\exists$-case in the proof of 5.15.14.

(5) The cases of $\wedge$, $\vee$ are almost trivial.

('If'.) (Cf. the proof of 5.15.14.) For $\sigma \in \Sigma_{mn}$ let us check the properties (1σ), (2σ), (3σ).

(1σ) Let $\mathbf{a} \in D^n$, $\mathbf{b}' \in D^m$, $(\pi_\sigma\mathbf{a})R^m\mathbf{b}'$. Consider $A := P(\mathbf{y}) \supset Q(\mathbf{y})$, $|\mathbf{y}| = m$ and the intuitionistic model $M = (\mathbb{F}, \xi)$ such that

$$\xi^+(P) = R^m(\mathbf{b}'),\ \xi^+(Q) = R^m(\mathbf{b}') - \approx^m (\mathbf{b}').$$

Then $\mathbf{b}' \in \xi^+(P) - \xi^+(Q)$, thus $M, \pi_\sigma\mathbf{a} \not\Vdash A\,[\mathbf{y}]$. Therefore by (#),

$$M, \mathbf{a} \not\Vdash [\pi_\sigma\mathbf{x}/\mathbf{y}]\, A (= P(\pi_\sigma\mathbf{x}) \supset Q(\pi_\sigma\mathbf{x}))\ [\mathbf{x}],$$

i.e. there exists $\mathbf{b} \in R^m(\mathbf{a})$ such that $M, \mathbf{b} \Vdash P(\pi_\sigma\mathbf{x})\,[\mathbf{x}]$ and $M, \mathbf{b} \not\Vdash Q(\pi_\sigma\mathbf{x})\,[\mathbf{x}]$. Thus $\pi_\sigma\mathbf{b} \in (\xi^+(P) - \xi^+(Q)) = \approx_m (\mathbf{b}')$, i.e. $\mathbf{b}' \approx^m \pi_\sigma\mathbf{b}$.

(2σ) Let $\mathbf{a} \in D^n$, $\mathbf{c} = (\pi^m_+\mathbf{c})d \in D^{m+1}$, and assume that $(\pi_\sigma\mathbf{a})R^m(\pi^m_+\mathbf{c})$. Consider the formula $A := \forall zP(\mathbf{y}, z)$ with $|\mathbf{y}| = m$, $z \notin r(\mathbf{yx})$, and the intuitionistic model $M = (\mathbb{F}, \xi)$ such that

$$\xi^+(P) = \{\mathbf{c}' \in D^{m+1} \mid \neg(\mathbf{c}'R^{m+1}\mathbf{c})\}.$$

Since $\mathbf{c} \notin \xi^+(P)$, we have

$$M, \mathbf{c} \not\Vdash P(\mathbf{y}, z)\,[\mathbf{y}z],$$

and so $M, \pi_\sigma\mathbf{a} \not\Vdash A[\mathbf{y}]$, since $(\pi_\sigma\mathbf{a})R^m(\pi^m_+\mathbf{c})$. Thus by (#), $M, \mathbf{a} \not\Vdash \forall zP(\pi_\sigma\mathbf{x}, z)\,[\mathbf{x}]$,[27] i.e. there exists $\mathbf{b} = (\pi^n_+\mathbf{b})d \in D^{n+1}$ such that

$$\mathbf{a}R^n(\pi^n_+\mathbf{b})\ \&\ M, \mathbf{b} \not\Vdash P(\pi_\sigma\mathbf{x}, z)\,[\mathbf{x}z].$$

But $(\pi_\sigma\mathbf{x})z = \pi_{\sigma^+}(\mathbf{x}z)$, thus $\pi_{\sigma^+}(\mathbf{b}) \notin \xi^+(P)$, so by definition of M, $\pi_{\sigma^+}(\mathbf{b})R^{m+1}\mathbf{c}$, as required.

(3σ) The proof is similar to (2σ). We assume $\mathbf{a} \in D^n$, $\mathbf{c} = (\pi^m_+\mathbf{c})d \in D^{m+1}$, $(\pi^m_+\mathbf{c})R^m(\pi_\sigma\mathbf{a})$ and consider the formula $A := \exists zP(\mathbf{y}, z)$, where $|\mathbf{y}| = m$, $z \notin r(\mathbf{yx})$, and the intituionistic model $M = (\mathbb{F}, \xi)$ such that $\xi^+(P) = R^{m+1}(\mathbf{c})$. Checking the details is left to the reader.

■

[27] It is essential that here we use $\forall zP(\pi_\sigma\mathbf{x}, z)$, a congruent version of $[(\mathbf{x} \cdot \sigma)/y]A$ with a new variable z (not $x_i \in r(\mathbf{x}) - r(\pi_\sigma\mathbf{x})$).

Note that unlike the modal case (Lemma 5.11.9, Proposition 5.11.11), we cannot consider only quantifier free formulas A in the proof of 'if' in the above lemma, because $(2\sigma), (3\sigma)$ are related to quantifiers.

Now we obtain an intuitionistic analogue of 5.11.12:

Proposition 5.15.18 *Let M be a model over a $wi^{(=)}$-metaframe, $u \in M$, $A, A^* \in IF^{(=)}$, and let $[\mathbf{a}/\mathbf{x}]$, $[\mathbf{a}^*/\mathbf{x}^*]$ be assignments in M giving rise to equal D_u-sentences: $[\mathbf{a}/\mathbf{x}]\,A = [\mathbf{a}^*/\mathbf{x}^*]\,A^*$. Then*

$$(\sharp\sharp) \quad M, \mathbf{a} \Vdash A\,[\mathbf{x}] \Leftrightarrow M, \mathbf{a}^* \Vdash A\,[\mathbf{x}^*].$$

Proof Similar to 5.11.12, but now we apply 5.15.14, 5.15.17 instead of 5.10.6, 5.11.7. ∎

Proposition 5.15.18 allows us to define forcing for D_u-sentences $[\mathbf{a}/\mathbf{x}]A$ in i-metaframes in the same manner as we did in Kripke sheaves, Kripke bundles and modal metaframes. In the modal case this definition is successful, due to the property

$$(mm_n) \quad \forall\mathbf{a}\forall\mathbf{b}\ (\mathbf{a}R^n\mathbf{b} \Rightarrow \mathbf{a}\,sub\,\mathbf{b}),$$

cf. Lemma 5.11.2, Proposition 5.11.12. This property also holds in $i^=$-metaframes (5.14.20), but not always in i-metaframes (cf. Example 2* in 5.16). But i-metaframes still enjoy the property (##), perhaps thanks to the following weaker version of (mm_n).

Exercise 5.15.19 Show that

$$(mm_n^{\sim}) \quad \forall\mathbf{a}\forall\mathbf{b}\,(\mathbf{a}R^n\mathbf{b} \Rightarrow \exists\mathbf{b}' \approx^n \mathbf{b}\ \mathbf{a}\,sub\,\mathbf{b}')$$

holds in i-functorial metaframes. Hint: choose σ such that $\mathbf{a}\,sub\,\sigma$ & $\sigma\,sub\,\mathbf{a}$ and apply (1σ) to $\mathbf{c}$ such that $\mathbf{a} = \pi_\sigma\mathbf{c}$.

Now we can prove intuitionistic analogues of results from Section 5.12.

The following lemma is an analogue of 5.12.1.

Lemma 5.15.20 *Let $M = (\mathbb{F}, \xi)$ be a model over an $w^{(=)}$-metaframe $\mathbb{F}$, and let $A \in IF^{(=)}$. Then the following conditions are equivalent.*

(1) $M \Vdash A$;

(2) $M, \mathbf{a} \Vdash A\,[\mathbf{x}]$ for any ordered assignment $(\mathbf{x}, \mathbf{a})$ in $\mathbb{F}$ such that $FV(A) = r(\mathbf{x})$;

(3) there exists a list of distinct variables $r(\mathbf{y}) \supseteq FV(A)$ such that
$(\natural)$ $\quad M, \mathbf{b} \Vdash A[\mathbf{y}]$ for any ordered assignment $(\mathbf{y}, \mathbf{b})$ in $\mathbb{F}$.

Thus $(\natural)$ does not depend on the choice of $\mathbf{y}$.

Proof Along the same lines as 5.12.1, but now we apply 5.15.14 instead of 5.10.6. ∎

The next lemma is an analogue of 5.12.2.

Lemma 5.15.21 *Let $\mathbb{F}$ be an $i^{(=)}$-metaframe. Then*

(1) $\mathbf{IL}_{-}^{(=)}(\mathbb{F})$ *is closed under modus ponens;*

(2) for any $B \in IF^{(=)}$

$$\mathbb{F} \Vdash B \Leftrightarrow \mathbb{F} \Vdash \forall y B \Leftrightarrow \mathbb{F} \vDash \overline{\forall} B;$$

(3) $\mathbf{IL}_{-}^{(=)}(\mathbb{F})$ *is closed under generalisation;*

(4) $\mathbf{IL}_{-}^{(=)}(\mathbb{F})$ *is closed under strict substitutions.*

Note that (1) for $B \in IF$ holds in every wi-metaframe $\mathbb{F}$.

Proof

(1) Similar to the proof of 5.12.2(2), but now we use Lemma 5.15.14 (and the reflexivity of R^n 5.15.20).

(2) Cf. the proof of Lemma 5.12.2(3); again we use 5.15.14 and the reflexivity of R^n.

(3) Follows from (2).

(4) The argument is essentially the same as in the proof of 5.12.2(4). Assuming that $S = [C/P(\mathbf{y})]$ is a strict substitution ($FV(C) \subseteq \mathbf{y}$) and $\mathbb{F} \Vdash A$, we show that $\mathbb{F} \Vdash SA$, i.e. $M, \mathbf{a} \Vdash SA\,[\mathbf{x}]$ for any intuitionistic model M and any assignment $(\mathbf{x}, \mathbf{a})$ with $r(\mathbf{x}) = FV(A) \supseteq FV(SA)$.

By Lemma 5.15.16, we may assume that A is clean, $BV(A) \cap r(\mathbf{x}) = \varnothing$.

Consider the model $N = (\mathbb{F}, \eta)$ such that

$$\eta^+(P) = \{\mathbf{b} \mid M, \mathbf{b} \Vdash C\,[\mathbf{y}]\},$$
$$\eta^+(Q) = \xi^+(Q) \text{ for any other predicate letter } Q.$$

This model is intuitionistic, by monotonicity of $\mathbb{F}$. Now it suffices to prove by induction that

($\sharp$) $N, \mathbf{a} \Vdash B\,[\mathbf{z}]$ iff $M, \mathbf{a} \Vdash SB\,[\mathbf{z}]$

for any subformula B of A and for any assignment $(\mathbf{z}, \mathbf{a})$ with $r(\mathbf{z}) \supseteq FV(B)$, $r(\mathbf{z}) \cap BV(B) = \varnothing$.

If $B = P(\mathbf{z} \cdot \sigma)$ for a map σ, then $SB = [\mathbf{z} \cdot \sigma/\mathbf{y}]\,C$. By Definition 5.14.2, the choice of N, and Lemma 5.15.17 we have

$$N, \mathbf{a} \Vdash B\,[\mathbf{z}] \text{ iff } (\mathbf{a} \cdot \sigma) \in \eta^+(P) \text{ iff } M, \mathbf{a} \cdot \sigma \Vdash C\,[\mathbf{y}] \text{ iff } M, \mathbf{a} \Vdash [\mathbf{z} \cdot \sigma/\mathbf{y}]\,C\,[\mathbf{z}],$$

and thus ($\sharp$) holds.

If $B = B_1 \supset B_2$ and (♯) holds for B_1, B_2, we obtain it for B:

$$\begin{array}{l} N, \mathbf{a} \Vdash B\,[\mathbf{z}] \text{ iff } \forall \mathbf{b} \in R^n(\mathbf{a})\ (N, \mathbf{b} \Vdash B_1\,[\mathbf{z}] \Rightarrow N, \mathbf{b} \Vdash B_2\,[\mathbf{z}]) \text{ iff} \\ \forall \mathbf{b} \in R^n(\mathbf{a})\ (M, \mathbf{b} \Vdash SB_1\,[\mathbf{z}] \Rightarrow M, \mathbf{b} \Vdash SB_2\,[\mathbf{z}]) \text{ iff} \\ M, \mathbf{a} \Vdash SB_1 \supset SB_2\,(= SB)\ [\mathbf{z}] \end{array}$$

(where $n = |\mathbf{a}|$).

Suppose $B = \forall v B_1$ and (♯) holds for B_1. By our assumption, $v \notin r(\mathbf{z})$, so we have

$$\begin{array}{l} N, \mathbf{a} \Vdash B\,[\mathbf{z}] \text{ iff } \forall \mathbf{b} \in R^n(\mathbf{a})\ \forall c \in D(\mathbf{b})\ N, \mathbf{b}c \Vdash B_1\,[\mathbf{z}v] \\ \text{iff } \forall \mathbf{b} \in R^n(\mathbf{a})\ \forall c \in D(\mathbf{b})\ M, \mathbf{b}c \Vdash SB_1\,[\mathbf{z}v] \\ \text{iff } M, \mathbf{a} \Vdash \forall v(SB_1)\,(= SB)\,[\mathbf{z}]. \end{array}$$

Thus (♯) holds for B.

The cases $B = B_1 \wedge B_2$, $B_1 \vee B_2$, $\exists v B_1$ are left to the reader, and the case $B = \bot$ is trivial.

■

Now similarly to 5.12.3 we obtain the following

Proposition 5.15.22 *Let $\mathbb{F}$ be an $i^{(=)}$-metaframe. Then*

$$\begin{array}{rcl} \mathbf{IL}^{(=)}(\mathbb{F}) & = & \{A \in IF^{(=)} \mid \mathbb{F} \Vdash A^n \text{ for any } n \in \omega\}, \\ & = & \{A \in IF^{(=)} \mid \mathbb{F} \Vdash \overline{A^n} \text{ for any } n \in \omega\}. \end{array}$$

Therefore $\mathbf{IL}^{=}(\mathbb{F})$ is conservative over $\mathbf{IL}(\mathbb{F})$.

We also obtain the following analogues of 5.12.9, 5.12.10.

Proposition 5.15.23 *For an $i^{(=)}$-metaframe $\mathbb{F}$*

$$\mathbf{IL}_{\pi}^{(=)}(\mathbb{F}) = \bigcap_{n \in \omega} \mathbf{IL}(F_n).$$

Corollary 5.15.24 *All theorems of* **H** *are strongly valid in every i-metaframe.*

Proof $\mathbf{H} \subseteq \mathbf{IL}(F_n)$ if F_n is an **S4**-frame. ■

Remark 5.15.25 Note that we cannot extend Lemma 5.15.21 and Proposition 5.15.22 to wi-metaframes, but Proposition 5.15.23 still holds (cf. Exercise 5.12.11 for the modal case). However in the next section we will show that only $\mathrm{i}^{(=)}$-metaframes can validate predicate all **QH**-axioms (and theorems).

Theorem 5.15.26 (Soundness theorem) *For any $i^{(=)}$-metaframe $\mathbb{F}$, $\mathbf{IL}^{(=)}(\mathbb{F})$ is an s.p.l.$(=)$.*

Proof Soundness of the inference rules is proved by the same argument as in 5.12.12. Now similarly to 5.12.13, we prove the strong validity of the predicate axioms (and the axioms of equality — for $\mathrm{i}^=$-metaframes); all these formulas can be supposed clean. So consider an intuitionistic model $M = (\mathbb{F}, \xi)$.

- $A_1 := \forall y P(\mathbf{x}, y) \supset P(\mathbf{x}, z)$, for $y, z \notin \mathbf{x}$, $y \neq z$, $|\mathbf{x}| = n$.

  It is sufficient to show that

  $$\mathbf{a}b \Vdash \forall y P(\mathbf{x}, y)\,[\mathbf{x}z] \Rightarrow \mathbf{a}b \Vdash P(\mathbf{x}, y)\,[\mathbf{x}z].$$

  So assume $\mathbf{a}b \Vdash \forall y P(\mathbf{x}, y)\,[\mathbf{x}z]$; then by 5.15.14 $\mathbf{a} \Vdash \forall y P(\mathbf{x}, y)\,[\mathbf{x}]$. Hence (by the reflexivity of R^n) for any $c \in D(\mathbf{a})$, $\mathbf{a}c \Vdash P(\mathbf{x}, y)\,[\mathbf{x}y]$, which is equivalent to $\mathbf{a}c \in \xi^+(P)$, and thus to $\mathbf{a}c \Vdash P(\mathbf{x}, z)\,[\mathbf{x}z]$. Therefore $\mathbf{a}b \Vdash P(\mathbf{x}, z)\,[\mathbf{x}z]$.

- $A_2 := P(\mathbf{x}, z) \supset \exists y P(\mathbf{x}, y)$, where $y, z \notin \mathbf{x}$, $y \neq z$, $|\mathbf{x}| = n$.

  Assume $\mathbf{a}b \Vdash P(\mathbf{x}, z)\,[\mathbf{x}z]$, i.e. $\mathbf{a}b \in \xi^+(P)$, which is equivalent to $\mathbf{a}b \Vdash P(\mathbf{x}, y)\,[\mathbf{x}y]$. Since $\mathbf{a} \approx_n \mathbf{a}$, by the definition of forcing it follows that $\mathbf{a} \Vdash \exists y P(\mathbf{x}, y)\,[\mathbf{x}]$; hence $\mathbf{a}b \Vdash \exists y P(\mathbf{x}, y)\,[\mathbf{x}z]$ by 5.15.14.

- $A_3 := \forall y(Q(\mathbf{x}) \supset P(\mathbf{x}, y)) \supset (Q(\mathbf{x}) \supset \forall y P(\mathbf{x}, y))$, where $y \notin \mathbf{x}$, $|\mathbf{x}| = n$.

  Assuming

  (3.1) $\mathbf{a} \Vdash \forall y(Q(\mathbf{x}) \supset P(\mathbf{x}, y))\,[\mathbf{x}]$,

  let us show

  (3.2) $\mathbf{a} \Vdash (Q(\mathbf{x}) \supset \forall y P(\mathbf{x}, y))\,[\mathbf{x}]$.

  In fact, (3.1) means

  (3.3) $\forall \mathbf{b} \in R^n(\mathbf{a})\ \forall c \in D(\mathbf{b})\ \mathbf{b}c \Vdash (Q(\mathbf{x}) \supset P(\mathbf{x}, y))\,[\mathbf{x}y]$,

  and we have to show for any $\mathbf{b} \in R^n(\mathbf{a})$

  (3.4) $\mathbf{b} \Vdash Q(\mathbf{x})\,[\mathbf{x}] \Rightarrow \mathbf{b} \Vdash \forall y P(\mathbf{x}, y)\,[\mathbf{x}]$.

  So we assume

  (3.5) $\mathbf{b} \Vdash Q(\mathbf{x})\,[\mathbf{x}]$

  and prove that

  (3.6) $\mathbf{b} \Vdash \forall y P(\mathbf{x}, y)\,[\mathbf{x}]$,

  which means

  (3.7) $\forall \mathbf{d} \in R^n(\mathbf{b})\ \ \forall c \in D(\mathbf{d})\ \mathbf{d}c \Vdash P(\mathbf{x}, y)\,[\mathbf{x}y]$.

Let $\mathbf{b}R^n\mathbf{d}$; by monotonicity, (3.5) implies

(3.8) $\mathbf{d} \Vdash Q(\mathbf{x})\,[\mathbf{x}]$;

then $\mathbf{d} \in \xi^+(Q)$, and thus

(3.9) $\forall c \in D(\mathbf{d})\ \mathbf{d}c \Vdash Q(\mathbf{x})\,[\mathbf{x}y]$.

On the other hand, R^n is transitive, so $\mathbf{d} \in R^n(\mathbf{a})$, and (3.3) implies

(3.10) $\forall c \in D(\mathbf{d})\ \mathbf{d}c \Vdash (Q(\mathbf{x}) \supset P(\mathbf{x}, y))\,[\mathbf{x}y]$.

Eventually from (3.9) and (3.10) we obtain

$$\forall c \in D(\mathbf{d})\ \mathbf{d}c \Vdash P(\mathbf{x}, y)\,[\mathbf{x}y],$$

which yields (3.6).

Therefore (3.4) holds.

- $A_4 := \forall y(P(\mathbf{x}, y) \supset Q(\mathbf{x})) \supset (\exists y P(\mathbf{x}, y) \supset Q(\mathbf{x}))$, for $y \notin \mathbf{x}$, $|\mathbf{x}| = n$.

Assuming

(4.1) $\mathbf{a} \Vdash \forall y(P(\mathbf{x}, y) \supset Q(\mathbf{x}))\,[\mathbf{x}]$

let us show

(4.2) $\mathbf{a} \Vdash (\exists y P(\mathbf{x}, y) \supset Q(\mathbf{x}))\,[\mathbf{x}]$.

Consider an arbitrary $\mathbf{b} \in R^n(\mathbf{a})$ such that

(4.3) $\mathbf{b} \Vdash \exists y P(\mathbf{x}, y)\,[\mathbf{x}]$.

We have to show that

(4.4) $\mathbf{b} \Vdash Q(\mathbf{x})\,[\mathbf{x}]$.

By (4.3) we have for some $\mathbf{b}' \approx^n \mathbf{b}$, $c \in D(\mathbf{b}')$,

(4.5) $\mathbf{b}'c \Vdash P(\mathbf{x}, y)\,[\mathbf{x}y]$.

Now $\mathbf{a}R^n\mathbf{b}'$, so by (4.1)

(4.6) $\mathbf{b}'c \Vdash (P(\mathbf{x}, y) \supset Q(\mathbf{x}))\,[\mathbf{x}y]$.

From (4.5), (4.6) we obtain

(4.7) $\mathbf{b}'c \Vdash Q(\mathbf{x})\,[\mathbf{x}y]$,

which implies

(4.8) $\mathbf{b}' \Vdash Q(\mathbf{x})\,[\mathbf{x}]$,

and the latter implies (4.4) by monotonicity, since $\mathbf{b}' \approx^n \mathbf{b}$.

- $A_5 := (x = x)$. The proof is trivial.
- $A_6 := (y = z \supset_\bullet P(\mathbf{x}, y) \supset P(\mathbf{x}, z))$, $y, z \notin \mathbf{x}$, y, z are distinct, $|\mathbf{x}| = n$.
  In fact, assume

  (6.1) $\mathbf{a}bc \Vdash y = z\,[\mathbf{x}yz]$

  and

  (6.2) $(\mathbf{a}bc)R^{n+2}\mathbf{d}$.

  From (6.1) we readily obtain $b = c$, and (6.2) implies $(\mathbf{a}bc)\,sub\,\mathbf{d}$ by 5.14.20 (I4). Hence $d_{n+1} = d_{n+2}$. Now we have to check that

  (6.3) $\mathbf{d} \Vdash P(\mathbf{x}, y)\,[\mathbf{x}yz] \Rightarrow \mathbf{d} \Vdash P(\mathbf{x}, z)\,[\mathbf{x}yz]$.

  So suppose

  (6.3) $\mathbf{d} \Vdash P(\mathbf{x}, y)\,[\mathbf{x}yz]$.

  By 5.15.18, this implies

  (6.4) $(d_1, \ldots, d_{n+1}) \Vdash P(\mathbf{x}, z)\,[\mathbf{x}z]$.

  Since $d_{n+1} = d_{n+2}$, (6.4) is the same as

  (6.5) $(d_1, \ldots, d_n, d_{n+2}) \Vdash P(\mathbf{x}, z)\,[\mathbf{x}z]$.

  which (again by 5.15.18) implies

  $$\mathbf{d} \Vdash P(\mathbf{x}, z)\,[\mathbf{x}yz].$$

  Thus (6.3) holds.

■

Recall that **QS4**-metaframe is i$^=$-functorial. So we can conclude that $\mathbf{IL}^{(=)}(\mathbb{F})$ is a superintuitionistic predicate logic for any **QS4**-metaframe $\mathbb{F}$. Thus soundness theorem 5.15.26 implies Proposition 5.14.7.

5.16 Maximality theorem

In this section we prove necessary conditions for intuitionistic soundness. To this end, we present Definitions 5.15.7, 5.15.8, 5.15.9 in a more convenient form. But first let us recall basic properties of i-metaframes and introduce some their versions.

(0σ) $\forall \mathbf{a}, \mathbf{b} \in D^n(\mathbf{a}R^n\mathbf{b} \Rightarrow (\pi_\sigma\mathbf{a})R^m(\pi_\sigma\mathbf{b}))$,

(1σ) $\forall \mathbf{a} \in D^n\ \forall \mathbf{b}' \in D^m((\pi_\sigma \mathbf{a})R^m\mathbf{b}' \Rightarrow \exists \mathbf{b} \in D^n(\mathbf{a}R^n\mathbf{b}\ \&\ \pi_\sigma\mathbf{b} \approx^m \mathbf{b}'))$.

(2σ) (forward 2-lift property)

$$\forall \mathbf{a} \in D^n\ \forall \mathbf{b} \in D^{m+1}((\pi_\sigma \mathbf{a})R^m(\pi_+^m \mathbf{b}) \Rightarrow \exists \mathbf{c} \in D^{n+1}(\mathbf{a}R^n(\pi_+^n\mathbf{c})\ \&\ (\pi_{\sigma^+}\mathbf{c})R^{m+1}\mathbf{b}));$$

(3σ) (backward 2-lift property)

$$\forall \mathbf{a} \in D^n\ \forall \mathbf{b} \in D^{m+1}((\pi_+^m \mathbf{b})R^m(\pi_\sigma \mathbf{a}) \Rightarrow \exists \mathbf{c} \in D^{n+1}((\pi_+^n\mathbf{c})R^n\mathbf{a}\ \&\ \mathbf{b}R^{m+1}(\pi_{\sigma^+}\mathbf{c})));$$

(mm_2) $\forall a, b_1, b_2\ ((a,a)R^2(b_1,b_2) \Rightarrow b_1 = b_2)$.

Recall that i-metaframes are the **S4**-metaframes satisfying (0σ), (1σ), (2σ), (3σ) for all σ; wi-metaframes satisfy the same conditions for all injective σ; (w)i$^=$-metaframes also satisfy (mm$_2$). The correlation between different kinds of metaframes is shown in the diagram below.

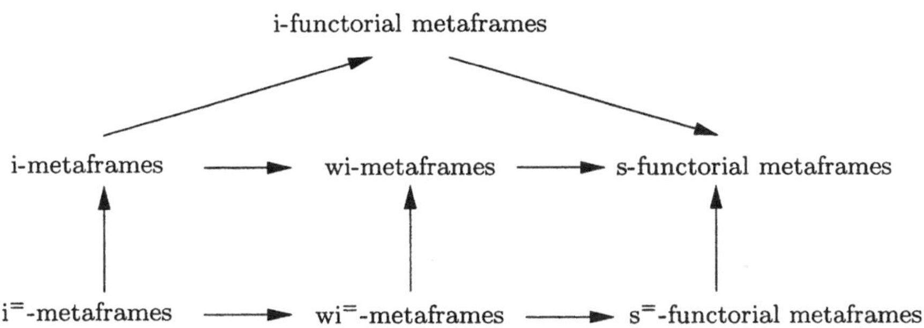

By Lemma 5.14.14 the condition $\forall\sigma$ (0σ) (i.e. s-functorality) is equivalent to the conjunction of the following properties:

(I1) (0σ) holds for all permutations σ;

(I2.1) $\forall n > 0\ \forall u, v \in W\ \forall \mathbf{a} \in D_u^n\ \forall \mathbf{b} \in D_v^n\ (\mathbf{a}R^n\mathbf{b} \Rightarrow uRv)$;

(I2.2) $\forall n > 0\ \forall u, v \in W\ \forall \mathbf{a} \in D_u^n\ \forall \mathbf{b} \in D_v^n\ \forall c \in D_u\ \forall d \in D_v\ ((\mathbf{a}c)R^{n+1}(\mathbf{b}d) \Rightarrow \mathbf{a}R^n\mathbf{b})$,

(I3) $\forall n > 0\ \forall \mathbf{a}, \mathbf{b} \in D^n\ (\mathbf{a}R^n\mathbf{b} \Rightarrow (\mathbf{a}a_n)R^{n+1}(\mathbf{b}b_n))$.

We also need the following conditions:

(I2.3) $\forall n > 0\, \forall u, v\, (uRv \Rightarrow \forall \mathbf{a} \in D_u^n\ \exists t \approx^0 v\ \exists \mathbf{b} \in D_t^n\ \mathbf{a}R^n\mathbf{b})$'

(I2.4) $\forall n, \mathbf{a}, \mathbf{b}\, (\pi_+^n(\mathbf{a})R^n\mathbf{b} \Rightarrow \exists \mathbf{c} \in D^{n+1}\ (\mathbf{a}R^{n+1}\mathbf{c}\ \&\ (\pi_+^n\mathbf{c}) \approx^n \mathbf{b}))$;

(I4.1) for any $a, b \in D^1$

$$\pi_\varnothing(a)R^0\pi_\varnothing(b) \Rightarrow \exists c_1, c_2\ (\pi_\varnothing(c_1) = \pi_\varnothing(c_2)\ \&\ aR^1c_1\ \&\ c_2R^1b);$$

(I4.2) for any $n > 0$, $\mathbf{a}, \mathbf{b} \in D^{n+1}$

$$(\pi_+^n \mathbf{a}) R^n (\pi_+^n \mathbf{b}) \Rightarrow \exists \mathbf{c} \in D^n\ \exists d, e \in D(\mathbf{c})\, (\mathbf{a} R^{n+1}(\mathbf{c}d)\ \&\ (\mathbf{c}e) R^{n+1} \mathbf{b});$$

(I5.1) $(\mathbf{a}a_n) R^{n+1} \mathbf{b}\ \&\ b_n \neq b_{n+1} \Rightarrow \exists \mathbf{c} \in R^n(\mathbf{a})\, ((\mathbf{c}c_n) \approx^{n+1} \mathbf{b})$;

(I5.2) for any $\mathbf{a} \in D^n$, $\mathbf{b} \in D^{n+2}$, $n > 0$

$$(\mathbf{a}a_n) R^{n+1} (\pi_+^{n+1} \mathbf{b})\ \&\ b_n \neq b_{n+1} \Rightarrow \exists \mathbf{c} \in D^n\ \exists d \in D(\mathbf{c})$$
$$(\mathbf{a} R^n \mathbf{c}\ \&\ (\mathbf{c}c_n d) R^{n+2} \mathbf{b});$$

(I5.3) for any $\mathbf{a} \in D^n$, $\mathbf{b} \in D^{n+2}$, $n > 0$

$$(\pi_+^{n+1} \mathbf{b}) R^{n+1} (\mathbf{a}a_n)\ \&\ b_n \neq b_{n+1} \Rightarrow$$
$$\exists \mathbf{c} \in D^n\ \exists d \in D(\mathbf{c})\, (\mathbf{c} R^n \mathbf{a}\ \&\ \mathbf{b} R^{n+2} (\mathbf{c}c_n d));$$

and also the following:[28]

(I5) for any $\mathbf{a} \in D^n$, $\mathbf{b} \in D^{n+2}$, $n > 0$
$(\mathbf{a}a_n) R^{n+1} \widehat{\mathbf{b}_1}\ \&\ b_{n+1} \neq b_{n+2} \Rightarrow \exists \mathbf{c} \in D^{n+1}\ \mathbf{b} \approx^{n+2} \mathbf{c}c_{n+1}$;

(I5′) for any $a \in D^1$, $\mathbf{b} \in D^{n+1}$, $n > 0$
$(a, a) R^2 (b_n, b_{n+1})\ \&\ b_n \neq b_{n+1} \Rightarrow \exists \mathbf{c} \in D^n\ \mathbf{b} \approx^{n+1} \mathbf{c}c_n$.

Later on we will show (Lemma 5.16.9) that each of (I5), (I5′) is equivalent to (I5.1) & (I5.2) & (I5.3).

Lemma 5.16.1 *In i-functorial metaframes (I4.1), (I4.2) are respectively equivalent to*

(I4.1~) for any $a, b \in D^1$

$$\pi_\varnothing(a) \approx^0 \pi_\varnothing(b) \Rightarrow \exists c_1, c_2\ (\pi_\varnothing(c_1) = \pi_\varnothing(c_2)\ \&\ a R^1 c_1\ \&\ c_2 R^1 b),$$

(I4.2~) for any $n > 0$, $\mathbf{a}, \mathbf{b} \in D^{n+1}$

$$\pi_+^n \mathbf{a} \approx^n \pi_+^n \mathbf{b} \Rightarrow \exists \mathbf{c} \in D^n\ \exists d, e \in D(\mathbf{c})\, (\mathbf{a} R^{n+1} (\mathbf{c}d)\ \&\ (\mathbf{c}e) R^{n+1} \mathbf{b}).$$

Proof Note that (I4.1) is $(2\varnothing_1)$. In fact $\varnothing_1 = \sigma_+^0$ (and $\pi_{\varnothing_1}(a) = \pi_+^0(a)$ is the world of $a \in D^1$), so $(2\varnothing_1)$ is

$$\forall a, b \in D^1\, (\pi_\varnothing(a) R^0 \pi_\varnothing(b) \Rightarrow \exists \mathbf{c} \in D^2 (a R^1 \pi_+^1 \mathbf{c}\ \&\ \pi_{\varnothing_1^+}(\mathbf{c}) R^1 b)).$$

Now if $\mathbf{c} = c_1 c_2$, then $\pi_+^1(\mathbf{c}) = c_1$, $\pi_{\varnothing_1^+}(\mathbf{c}) = c_2$. By the same reason, (I4.1~) is $(2^\sim\varnothing_1)$ from 5.15.12. So by 5.15.12, (I4.1) ⇔ (I4.1~).

Analogously, (I4.2) is $(2\sigma_+^n)$ and (I4.2~) is $(2^\sim\sigma_+^n)$, so (I4.1) ⇔ (I4.2~). ■

[28]Recall that $\widehat{\mathbf{c}_1}$ is obtained by eliminating c_1 from $\mathbf{c}$.

The next lemma is similar to 5.11.3 and 5.11.6.

Lemma 5.16.2 *Let* $\mathbb{F}$ *be an s-functorial metaframe, and let* $\sigma \in (\Sigma_{mn} - \Upsilon_{mn})$, $m \geq 2$. *Then* $\mathbb{F}$ *is* $s^=$*-functorial (i.e. satisfies* (mm_2)*) iff* $\pi_\sigma[D^n]$ *is* R^m*-stable.*

Recall that π_σ is surjective iff σ is injective.

Proof (Only if.) Assume that $\pi_\sigma[D^n]$ is R^m-stable, but $\mathbb{F}$ does not satisfy (mm_2). Suppose $(aa)R_2(b_1b_2)$, $b_1 \neq b_2$, and $i, j \in I_m$, $i \neq j$, $\sigma(i) = \sigma(j)$.

Now let[29] $\mathbf{a} := a^m$; then $\mathbf{a} = a^n \cdot \sigma \in \pi_\sigma[D^n]$. Next, let $\mathbf{c} := b_1^{i-1}b_2b_1^{m-i}$. Then $c_i \neq c_j$, so $\neg(\sigma \operatorname{sub} \mathbf{c})$, and thus $\mathbf{c} \notin \pi_\sigma[D^n]$, by Lemma 5.9.5.

On the other hand, consider $\tau \in \Sigma_{n2}$ such that $\tau(i) = 2$, $\tau(j) = 1$ for $j \neq i$. Then $\mathbf{a} = \pi_\tau(aa)$, $\mathbf{c} = \pi_\tau(b_1b_2)$, and thus $\mathbf{a}R^m\mathbf{c}$ by s-functoriality. Therefore $\mathbf{c} \in (R^m(\pi_\sigma[D^n]) - \pi_\sigma[D^n])$, which contradicts our assumption.

(If.) If $\mathbb{F}$ is s$^=$-functorial, it satisfies (mm_m) (Lemma 5.11.2). Then by Lemma 5.11.3, $\pi_\sigma[D^n]$ is R^m-stable. ■

Let us now reformulate the condition (1σ). Note that for an s-functorial metaframe $\mathbb{F}$, (1σ) means that π_σ is a quasi-p-morphism from (D^n, R^n) onto (D^m, R^m).

Lemma 5.16.3 *Every s-functorial metaframe satisfies* (1σ) *for any permutation* σ.

Proof s-functoriality implies permutability (cf. Definition 5.10.1), so by 5.10.2, for any $\sigma \in \Upsilon_n$, π_σ is an automorphism of F_n, and thus a (p-)morphism. ■

Lemma 5.16.4 *Let* $\mathbb{F}$ *be an s-functorial metaframe*

(1) If $\mathbb{F}$ *satisfies* $(1\sigma_+^n)$ *for any simple embedding* σ_+^n*, then it satisfies* (1σ) *for any injection* σ.

(2) If $\mathbb{F}$ *satisfies* $(1\sigma_+^n)$ *and* $(1\sigma_-^n)$ *for every* n*, then it satisfies* (1σ) *for any* σ.

Proof Follows from 5.16.3 and the observation that the composition of quasi-morphisms is a quasi-morphism. ■

Lemma 5.16.5 *Let* $\mathbb{F}$ *be an s-functorial metaframe. Then*

(1) $\mathbb{F}$ *satisfies* (1σ) *for any injection* σ *iff it satisfies (I2.3) and (I2.4);*

(2) $\mathbb{F}$ *satisfies* (1σ) *for any* σ *iff it satisfies (I2.3), (I2.4), and (I5.1);*

(3) If $\mathbb{F}$ *is* $s^=$*-functorial, then*

  (a) $\mathbb{F}$ *satisfies (I5.1);*

  (b) (1σ) *holds for any* σ *iff it holds for any injection* σ *(iff* $\mathbb{F}$ *satisfies (I2.3) and (I2.4), by (1)).*

[29] For the notation a^m, see the Introduction.

(I2.3) can also be replaced with its particular case corresponding to $n = 1$, cf. Lemma 5.14.18.

Note that in (I2.3) the worlds of $\mathbf{a}$ and $\mathbf{b}$ are $\approx^0$-equivalent, but not necessarily equal.

Proof

(1) (I2.3) is the quasi-lift property $(1\varnothing_n)$ and (I2.4) is the quasi-lift property $(1\sigma_+^n)$ for $n > 0$. Since $\varnothing_1$ is σ_+^0, we can apply Lemma 5.16.4(1).

(2) Similarly, we can apply 5.16.4(2). Now (I5.1) is $(2\sigma_-^n)$, up to the conjunct $b_n \neq b_{n+1}$ in the premise. But $b_n = b_{n+1}$ obviously implies the conclusion of (I5.1) — just put $\mathbf{c} = \mathbf{b}$; thus (I5.1) $\Leftrightarrow (1\sigma_-^n)$.

(3) (a) By (mm_{n+1}), $(\mathbf{a}a_{n+1})R^{n+1}\mathbf{b}$ implies $b_n = b_{n+1}$, so the premise of (I5.1) is false.
(b) Obviously follows from (2) and (3)(a).

■

Now let us turn to properties (2σ), (3σ). It is clear that (3σ) is dual to (2σ), so it can be analysed in a similar way.

Lemma 5.16.6 *Let $\mathbb{F}$ be an s-functorial* **S4**-*metaframe. Then*

$$(2\sigma)\ \&\ (2\tau) \Rightarrow (2(\sigma \cdot \tau))$$

and

$$(3\sigma)\ \&\ (3\tau) \Rightarrow (3(\sigma \cdot \tau)).$$

Proof We prove only the first implication; the second one is an easy exercise for the reader.

So assume (2σ) & (2τ). Let $\sigma \in \Sigma_{mn}$, $\tau \in \Sigma_{km}$; then $\sigma \cdot \tau \in \Sigma_{kn}$. Let $\mathbf{a} \in D^n$, $\mathbf{c} = \mathbf{c}'c_{k+1} \in D^{k+1}$, and assume that

$$(\pi_{\sigma\cdot\tau}\mathbf{a}) = ((\pi_\sigma \mathbf{a}) \cdot \tau)R^k(\pi_+^k \mathbf{c}) = \mathbf{c}'.$$

Then by (2τ), there exists $\mathbf{d} = \mathbf{d}'d_{m+1} \in D^{m+1}$ such that

$$(\pi_\sigma \mathbf{a})R^m(\pi_+^m \mathbf{d}) = \mathbf{d}',\ (\mathbf{d} \cdot \tau^+)R^{k+1}\mathbf{c}.$$

Now by (2σ), there exists $\mathbf{b} = \mathbf{b}'b_{n+1} \in D^{n+1}$ such that $\mathbf{a}R^n(\pi_+^n \mathbf{b}) = \mathbf{b}'$ and $(\pi_{\sigma^+}\mathbf{b}) = (\mathbf{b}' \cdot \sigma)b_{n+1}R^{m+1}\mathbf{d}$. Thus

$$\pi_{(\sigma\cdot\tau)^+}\mathbf{b} = \mathbf{b} \cdot (\sigma \cdot \tau)^+ = ((\mathbf{b}' \cdot \sigma) \cdot \tau)b_{n+1} = ((\pi_{\sigma^+}\mathbf{b}) \cdot \tau^+)R^{k+1}(\mathbf{d} \cdot \tau^+)$$

by s-functoriality. So by the transitivity of R^{k+1}, it follows that $(\pi_{(\sigma\circ\tau)^+}\mathbf{b})R^{k+1}\mathbf{c}$.
■

Lemma 5.16.7 *Let $\mathbb{F}$ be an s-functorial* **S4***-metaframe. Then* (2σ), (3σ) *hold for any permutation* $\sigma \in \Upsilon_n$.

Proof Let us check (3σ). Let

$$\mathbf{a} \in D^n,\ \mathbf{c} = \mathbf{c}'c_{n+1} \in D^{n+1},\ \pi_+^n \mathbf{c} = \mathbf{c}' R^n(\mathbf{a} \cdot \sigma).$$

Put $\mathbf{b} = (\mathbf{c}' \cdot \sigma^{-1})c_{n+1}$. Then by s-functoriality,

$$\pi_+^n \mathbf{b} = (\mathbf{c}' \cdot \sigma^{-1}) R^n((\mathbf{a} \cdot \sigma) \cdot \sigma^{-1}) = \mathbf{a}.$$

We also have:

$$\pi_{\sigma^+} \mathbf{b} = ((\mathbf{c}' \cdot \sigma^{-1}) \cdot \sigma)c_{n+1} = \mathbf{c}'c_{n+1} = \mathbf{c},$$

and thus $\mathbf{c}R^{n+1}(\pi_{\sigma^+}\mathbf{b})$ by the reflexivity of R^{n+1}. ■

So to prove $(2\sigma), (3\sigma)$ for all injective σ (respectively, for all σ), it suffices to check them only for simple embeddings σ_+^n (respectively for all σ_+^n, σ_-^n). This is the same as with (1σ) in Lemma 5.16.7.

Lemma 5.16.8 *Let $\mathbb{F}$ be an s-functorial* **S4***-metaframe. Then*

(1) (2σ) & (3σ) *holds for any* σ *iff (I4.1) & (I4.2).*

(2) (2σ) & (3σ) *holds for any* σ *iff (I4.1) & (I4.2) & (I5.2) & (I5.3).*

(3) If $\mathbb{F}$ is $s^=$-functorial, then

  (a) $\mathbb{F}$ satisfies (I5.2) & (I5.3).

  (b) (2σ) & (3σ) *holds for any* σ *iff it holds for any injection* σ *(i.e. iff $\mathbb{F}$ satisfies (I4.2) & (I4.2), by (2)).*

Note that $\mathbf{a}R^n\mathbf{b}$ iff $(\mathbf{a}a_n)R^{n+1}(\mathbf{b}b_n)$ by s-functoriality; so we can respectively change (I5.2); similarly with (I5.3).

Proof (1) As we noticed in the proof of 5.16.1, $(2\varnothing_1)$ is equivalent to (I4.1), and $(2\sigma_+^n)$ is equivalent to (I4.2).

(3σ) is obtained from (2σ) by replacing every R^k with its converse and permuting $\mathbf{a}$ with $\mathbf{c}$. But this replacement does not change (I4.1), (I4.2); so $(3\sigma_+^n)$ is equivalent to $(2\sigma_+^n)$.

(2) $(2\sigma_-^n)$ is

$$\begin{gathered}\forall \mathbf{a} \in D^n\ \forall \mathbf{b} \in D^{n+2}\ ((\mathbf{a}a_n)R^{n+1}(\pi_+^{n+1}\mathbf{b}) \Rightarrow \\ \exists \mathbf{g} \in D^{n+1}\ (\mathbf{a}R^n(\pi_+^n\mathbf{g})\ \&\ (\pi_{(\sigma_-^n)^+}\mathbf{g})R^{n+2}\mathbf{b})).\end{gathered}$$

If $\mathbf{g} = \mathbf{c}d,\ \mathbf{c} \in D^n,\ d \in D(\mathbf{c})$, then

$$\mathbf{c} = \pi_+^n\mathbf{g},\ \pi_{(\sigma_-^n)^+}\mathbf{g} = (\mathbf{c} \cdot \sigma_-^n)d = \mathbf{c}c_n d,$$

so $(2\sigma_-^n)$ is equivalent to

$$\forall \mathbf{a} \in D^n\ \forall \mathbf{b} \in D^{n+2}\,((\mathbf{a}a_n)R^{n+1}(\pi_+^{n+1}\mathbf{b}) \Rightarrow \\ \exists \mathbf{c} \in D^n\ \exists d \in D(\mathbf{c})\,(\mathbf{a}R^n\mathbf{c}\ \&\ (\mathbf{c}c_n d)R^{n+2}\mathbf{b})).$$

Note that this implication holds if $b_n = b_{n+1}$. In fact, put

$$\mathbf{c} := b_1 \dots b_n = \pi_+^n \pi_+^{n+1}\mathbf{b},\ d := b_{n+2};$$

then $(\mathbf{a}a_n)R^{n+1}(\pi_+^{n+1}\mathbf{b})$ implies $\mathbf{a}R^n\mathbf{c}$ by s-functoriality, and $\mathbf{c}c_n d = \mathbf{c}b_{n+1}d = \mathbf{b}R^{n+2}\mathbf{b}$ by reflexivity. Therefore $(2\sigma_-^n)$ is equivalent to (I5.2).

Thus an equivalent of $(3\sigma_-^n)$ is obtained from (I5.2) by replacing R^{n+1}, R^{n+2} with their converses, which is an equivalent of (I5.3).

(3)(b) By 5.14.20 s-functorial metaframes satisfy (mm_{n+1}). Thus $(\mathbf{a}a_n)R^{n+1}$ $(\pi_+^{n+1}\mathbf{b})$, implies $(\mathbf{a}a_n)\,sub\,(\pi_+^{n+1}\mathbf{b})$, and hence $b_n = b_{n+1}$. So (I5.2) holds trivially. Similarly we obtain (I5.3). ∎

Lemma 5.16.9 *Let $\mathbb{F}$ be an s-functorial metaframe satisfying. Then*

(1) (I5.1) & (I5.2) is equivalent to each of (I5), (I5′).

(2) If $\mathbb{F}$ also satisfies (I2.4), then each of these conditions implies (I5.3) and thus

(I5) ⇔ (I5′) ⇔ (I5.1) & (I5.2) & (I5.3).

Proof Let us first show that (I5) is equivalent to (I5′).

- (I5) for $n = 1 \Rightarrow$ (I5′) for $n = 1$.

Assume (I5) for $n = 1$:

$$\forall a \in D^1\ \forall \mathbf{d} \in D^3\,((a,a)R^2(d_2,d_3)\ \&\ d_2 \neq d_3 \Rightarrow \exists \mathbf{e} \in D^2\ \mathbf{d} \approx^3 \mathbf{e}e_2). \qquad (\star)$$

To check (I5′), suppose $(a,a)R^2(b_1,b_2)$, $b_1 \neq b_2$. Let us find $c \in D^1$ such that $(b_1,b_2) \approx^2 (c,c)$. Put $\mathbf{d} := b_1b_2b_1$. By $(\star)$ there exists $\mathbf{e} \in D^2$ such that $\mathbf{d} \approx^3 \mathbf{e}e_2 = e_1e_2e_2$. If $\sigma \in \Upsilon_{23}$ is such that $\sigma(1) = 2$, $\sigma(2) = 3$, then $b_1b_2b_1 \cdot \sigma = b_1b_2$, $\mathbf{e}e_2 \cdot \sigma = e_2e_2$; so by s-functoriality it follows that $b_1b_2 \approx^2 e_2e_2$, and we can put $c = e_2$.

- Suppose (I5), (I5′) hold for n, and let us prove (I5′) for $n+1$.

  So we assume $\mathbf{b} \in D^{n+2}$, $a \in D^1$, $(a,a)R^2(b_{n+1},b_{n+2})$, $b_{n+1} \neq b_{n+2}$, and find $\mathbf{c} \in D^{n+1}$ such that $\mathbf{b} \approx^{n+2} \mathbf{c}c_{n+1}$. By (I5′) for n, for $\widehat{\mathbf{b}}_1 = b_2 \dots b_{n+2}$ there exists $\mathbf{c}' \in D^{n+1}$ such that $\widehat{\mathbf{b}}_1 \approx^{n+1} \mathbf{c}'c_n'$. Then we apply (I5) to $\mathbf{a} := \mathbf{c}'$, $\mathbf{b}$ and obtain $\mathbf{c} \in D^{n+1}$ such that $\mathbf{b} \approx^{n+2} \mathbf{c}c_{n+1}$.

- (I5′) for $n+1 \Rightarrow$ (I5) for n.

  In fact, assume (I5′) for $n+1$. Suppose

  $$\mathbf{a} \in D^n,\ \mathbf{b} \in D^{n+2},\ (\mathbf{a}a_n)R^{n+1}\widehat{\mathbf{b}}_1,\ b_{n+1} \neq b_{n+2}.$$

  Then $(a_n,a_n)R^{n+1}(b_{n+1},b_{n+2})$ by s-functoriality, hence by (I5′), $\exists \mathbf{c} \in D^{n+1}\ \mathbf{b} \approx^{n+2} \mathbf{c}c_{n+1}$.

- (I5.1) & (I5.2) ⇒ (I5).

  In fact, assume (I5.1), (I5.2). Suppose

  $$\mathbf{a} \in D^n,\ \mathbf{b} \in D^{n+2},\ b_{n+1} \neq b_{n+2},$$

  $(\mathbf{a}a_n)R^{n+1}\widehat{\mathbf{b}}_1$. Now we can apply (I5.2) to $\mathbf{a}$ and $\widehat{\mathbf{b}}_1 b_1 = b_2 \dots b_{n+1} b_{n+2} b_1$ (this is possible, since $b_{n+1} \neq b_{n+2}$). So there exist $\mathbf{e} \in D^n$, $d \in D(\mathbf{e})$ such that $(\mathbf{e}e_n d)R^{n+2}(\widehat{\mathbf{b}}_1 b_1)$. Hence by s-functoriality $(d\mathbf{e}e_n)R^{n+2}\mathbf{b}$, and thus by (I5.1) there exists $\mathbf{c} \in D^{n+1}$ such that $\mathbf{b} \approx^{n+2} \mathbf{c}c_{n+1}$.

- (I5′) ⇒ (I5.1).

  In fact, assume (I5′). Suppose $\mathbf{a} \in D^n$, $(\mathbf{a}a_n)R^{n+1}\mathbf{b}$, $b_n \neq b_{n+1}$. Then by s-functoriality, $(a_n, a_n)R^2(b_n, b_{n+1})$. Hence by (I5′), there exists $\mathbf{c} \in D^n$ such that $\mathbf{b} \approx_{n+1} \mathbf{c}c_n$. Hence $(\mathbf{a}a_n)R^{n+1}(\mathbf{c}c_n)$ by transitivity, and eventually $\mathbf{a}R^n\mathbf{c}$, by s-functoriality. Therefore (I5.1) holds.

- (I5′) ⇒ (I5.2).

  In fact, assume (I5′). Suppose

  $$\mathbf{a} \in D^n,\ \mathbf{b} \in D^{n+2},\ (\mathbf{a}a_n)R^{n+1}(b_1 \dots b_{n+1}) = \pi_+^{n+1}\mathbf{b};$$

  then

  $$\mathbf{a}R^n(b_1 \dots b_n),\ (a_n, a_n)R^2(b_n, b_{n+1})$$

  by s-functoriality. Now by (I5′) applied to $\mathbf{d} := b_{n+2}b_1 \dots b_{n+1} \in D^{n+2}$, there exists $\mathbf{e} = c_0 c_1 \dots c_n \in D^{n+1}$ such that

  $$\mathbf{d} \approx^{n+2} \mathbf{e}e_{n+1} = c_0 c_1 \dots c_n c_n.$$

  Then by s-functoriality,

  $$c_1 \dots c_n c_n c_0 \approx^{n+2} \mathbf{b},\ b_1 \dots b_n \approx^n c_1 \dots c_n,$$

  and so $\mathbf{a}R^n(c_1 \dots c_n)$. Therefore for $\mathbf{c} := c_1 \dots c_n$, we have $\mathbf{a}R^n\mathbf{c}$ and $(\mathbf{c}c_n c_0)R^{n+2}\mathbf{b}$ as required.

- Finally let us show that together with (I2.4), (I5′) implies (I5.3).

  Assume (I5′). Let $\mathbf{a} \in D^n$, $\mathbf{b} \in D^{n+2}$, and $(\pi_+^{n+1}\mathbf{b})R^{n+1}(\mathbf{a}a_n)$. Then by (I2.4) there exists $\mathbf{c} \in D^{n+2}$ such that

  $$\mathbf{b}R^{n+2}\mathbf{c} \text{ and } \mathbf{a}a_n \approx^{n+1} c_1 \dots c_{n+1}.$$

  Now $(a_n, a_n)R^2(c_n, c_{n+1})$ by s-functoriality, so by (I5′) (applied to a_n, $c_1 \dots c_{n+1}$) we obtain $\mathbf{d} \in D^{n+1}$, such that $c_1 \dots c_{n+1} \approx^{n+2} \mathbf{d}d_{n+1}$. Thus

  $$\mathbf{b}R^{n+2}\mathbf{c} \approx^{n+2} d_2 \dots d_{n+1} d_n d_1$$

  again by s-functorality. By s-functoriality we also have

  $$d_2 \dots d_{n+1} \approx^n c_1 \dots c_n \approx^n \mathbf{a},$$

  so $(d_2 \dots d_{n+1})R^n\mathbf{a}$, which proves the conclusion of (I5.3) (where $d_2 \dots d_{n+1}$ stands for $\mathbf{c}$, d_1 stands for d).

■

Note that again (I5), (I5′) become obvious if $b_n \neq b_{n+1}$ is replaced with $b_n = b_{n+1}$ — take $\mathbf{c} = \mathbf{b}$.

Therefore we obtain the following equivalent form of Definitions 5.15.7, 5.15.8, 5.15.9.

Proposition 5.16.10 *Let $\mathbb{F}$ be an* **S4**-*metaframe. Then*

(1) $\mathbb{F}$ is a wi-metaframe iff $\mathbb{F}$ is an s-functorial metaframe satisfying (I2.3) & (I2.4) & (I4.1) & (I4.2);

(2) $\mathbb{F}$ is an i-metaframe iff $\mathbb{F}$ is a wi-metaframe satisfying (I5) (or equivalently, (I5′));

(3) $\mathbb{F}$ is a $wi^=$-metaframe iff $\mathbb{F}$ is an $i^=$-metaframe iff $\mathbb{F}$ is a wi-metaframe satisfying (mm_2).

Proof (1) If a metaframe is *s*-functorial, then by 5.16.5, the quasi-lift property for injections is equivalent to (I2.3) & (I2.4), and by 5.16.8, the 2-lift properties for injections are (together) equivalent to (I4.1) & (I4.2).

(2) By 5.16.5, in *s*-functorial metaframes the quasi-lift property for all maps is equivalent to (I2.3) & (I2.4) & (I5.1). By 5.16.8, the 2-lift properties for all maps are equivalent to (I4.1) & (I4.2) & (I5.2) & (I5.3).

So if $\mathbb{F}$ is quasi-functorial, it is weakly functorial and satisfies (I5.1) & (I5.2). By 5.16.9, the latter implies (I5).

The other way round, if $\mathbb{F}$ is weakly functorial and satisfies (I5), then by 5.16.9, it satisfies (I5.1), (I5.2), (I5.3). Also by (1), it satisfies (I2.3) & (I2.4) & (I4.1) & (I4.2), which yields the quasi-lift and the 2-lift properties.

(3) By 5.16.5, for monotonic$^=$ metaframes the quasi-lift property for all injections implies it for all maps. By 5.16.8, the same happens to the 2-lift properties. ■

Proposition 5.16.11 *Let $\mathbb{F}$ be an* **S4**-*metaframe such that* $\mathbf{QH} \subseteq \mathbf{IL}^{(=)}(\mathbb{F})$. *Then $\mathbb{F}$ is an $i^{(=)}$-metaframe.*

Proof By Proposition 5.14.16, $\mathbb{F}$ is monotonic$^{(=)}$. So by Proposition 5.16.10 it is sufficient to check the properties (I2.3), (I2.4), (I4.1), (I4.2), (I5):

(I2.4) $\mathbf{a}R^n\mathbf{b} \Rightarrow \forall d \in D(\mathbf{a})\ \exists \mathbf{c} \in D^{n+1}\ ((\mathbf{a}d)R^{n+1}\mathbf{c}\ \&\ (\pi^n_+\mathbf{c}) \approx^n \mathbf{b})$;

(I2.3) $uRv \Rightarrow \forall d \in D_u\ \exists t \approx^0 v\ \exists c \in D_t\ dR^1c$.

The proofs in both cases are quite similar. Suppose $\mathbf{a}R^n\mathbf{b}$, $n > 0$ $d \in D(\mathbf{a})$ (or uRv, $d \in D_u$). Consider the **QH**-theorem

$$A_1 := \exists yp \supset p$$

for $p \in PL^0$ and its substitution instance.

$$A := \exists y(P(\mathbf{x}) \supset Q(\mathbf{x})) \supset_\bullet P(\mathbf{x}) \supset Q(\mathbf{x}),$$

where $|\mathbf{x}| = n$, $y \notin \mathbf{x}$. Let $M := (\mathbb{F}, \xi)$ be an intuitionistic model such that

$$\xi^+(P) = R^n(\mathbf{b}),\ \xi^+(Q) = D^n - (R^n)^{-1}(\mathbf{b}).$$

Thus

$$\xi^+(P) - \xi^+(Q) = \approx^n(\mathbf{b}),$$

in particular $\mathbf{b} \in \xi^+(P) - \xi^+(Q)$, hence $M, \mathbf{a} \not\Vdash P(\mathbf{x}) \supset Q(\mathbf{x})\,[\mathbf{x}]$, so from the assumption $M \Vdash A$ we obtain

$$M, \mathbf{a} \not\Vdash \exists y(P(\mathbf{x}) \supset Q(\mathbf{x}))\,[\mathbf{x}],$$

and thus

$$M, \mathbf{a}d \not\Vdash\ (P(\mathbf{x}) \supset Q(\mathbf{x}))\,[\mathbf{x}y].$$

So there exists $\mathbf{c}$ such that

$$(\mathbf{a}d)R^{n+1}\mathbf{c}\ \&\ \mathbf{c} \Vdash P(\mathbf{x})\,[\mathbf{x}y]\ \&\ \mathbf{c} \not\Vdash\ Q(\mathbf{x})\,[\mathbf{x}y].$$

Since $\mathbf{x} = (\mathbf{x}y) \cdot \sigma^n_+$, by definition of forcing it follows that

$$\pi^n_+\mathbf{c} \in (\xi^+(P) - \xi^+(Q)) = \ \approx^n(\mathbf{b}),$$

which proves (I2.4). The changes for (I2.3) are now obvious: $n = 0$ and

$$\xi^+(P) = R(v),\ \xi^+(Q) = W - R^{-1}(v).$$

(I4.1) $\pi_\varnothing(a)R^0\pi_\varnothing(c) \Rightarrow \exists b_1, b_2\ (\pi_\varnothing(b_1) = \pi_\varnothing(b_2)\ \&\ aR^1b_1\ \&\ b_2R^1c)$,

(I4.2) $(\pi^n_+\mathbf{a})R^n(\pi^n_+\mathbf{c}) \Rightarrow \exists \mathbf{b} \in D^n\ \exists d, e \in D(\mathbf{b})\,(\mathbf{a}R^{n+1}(\mathbf{b}d)\ \&\ (\mathbf{b}e)R^{n+1}\mathbf{c})$.

For these properties the proofs are also similar, so we check only (I4.2). Suppose $(\pi^n_+\mathbf{a})R^n(\pi^n_+\mathbf{c})$ and consider the intuitionistic model $M := (\mathbb{F}, \xi)$ such that $\xi^+(P) = D^{n+1} - (R^{n+1})^{-1}(\mathbf{c})$. By the assumption, the **QH**-theorem (the substitution instance of the same A_1)

$$B := \exists y \forall z P(\mathbf{x}, z) \supset \forall z P(\mathbf{x}, z),$$

where $|\mathbf{x}| = n$, $y, z \notin \mathbf{x}$, $y \neq z$, is true in M.

But $M, \mathbf{c} \not\Vdash\ P(\mathbf{x}, z)\,[\mathbf{x}z]$, and so since $(\pi^n_+\mathbf{a})R^n(\pi^n_+\mathbf{c})$, we have $M, \pi^n_+\mathbf{a} \not\Vdash \forall z P(\mathbf{x}z)\,[\mathbf{x}]$. Now $M \Vdash B$ implies $M, \pi^n_+\mathbf{a} \not\Vdash \exists y \forall z P(\mathbf{x}, z)[\mathbf{x}]$; hence $M, \mathbf{a} \not\Vdash \forall z P(\mathbf{x}, z)\,[\mathbf{x}y]$. So there exists $\mathbf{b}de \in D^{n+2}$ such that $\mathbf{a}R^{n+1}(\mathbf{b}d)$ and $M, \mathbf{b}de \not\Vdash P(\mathbf{x}, z)\,[\mathbf{x}yz]$. The latter is equivalent to $\mathbf{b}e \notin \xi^+(P)$, i.e. to $(\mathbf{b}e)R^{n+1}\mathbf{c}$, by the definition of ξ. Thus (I4.2) holds.

(I5) $(\mathbf{a}a_n)R^{n+1}\widehat{\mathbf{c}}_1\ \&\ c_n \neq c_{n+1} \Rightarrow \exists \mathbf{b} \in D^{n+1}\ \mathbf{c} \approx^{n+2} \mathbf{b}b_{n+1}$.

Recall that by Proposition 5.16.10, this property is essential only for the case without equality.

Suppose $(\mathbf{a}a_n)R^{n+1}\widehat{\mathbf{c}}_1$. Take a distinct list $\mathbf{x}$ of length n, $y, z \notin \mathbf{x}$, $y \neq z$ and consider the following formulas:

$$B_1 := \forall y(P_1(y, \mathbf{x}, z) \supset P_2(y, \mathbf{x}, z)),$$
$$B_2 := Q_1(\mathbf{x}, z) \supset Q_2(\mathbf{x}),$$
$$B := B_1 \wedge B_2,\ A := [x_n/z]B \supset \exists zB.$$

Obviously, A is a **QH**-theorem (a substitution instance of the axiom $A_2 := P(x_n) \supset \exists zP(z)$). Consider the intuitionistic model $M := (\mathbb{F}, \xi)$ such that

$$\xi^+(P_1) = R^{n+2}(\mathbf{c}),$$
$$\xi^+(P_2) = D^{n+2} - (R^{n+2})^{-1}(\mathbf{c}),$$
$$\xi^+(Q_1) = D^{n+1} - (R^{n+1})^{-1}(\mathbf{a}a_n),$$
$$\xi^+(Q_2) = D^n - (R^n)^{-1}(\mathbf{a}).$$

Then obviously

(0) $\xi^+(P_1) - \xi^+(P_2) = \ \approx^{n+2}(\mathbf{c})$.

Now we claim that

(0.1) $M, \mathbf{a} \not\Vdash \exists zB\,[\mathbf{x}]$.

In fact, suppose the contrary. Then there exists $\mathbf{d}e \in D^{n+1}$ such that

(1) $\mathbf{d}R^n\mathbf{a}$ and $\mathbf{d}e \Vdash B\,[\mathbf{x}z]$.

Hence

(2) $\mathbf{d}e \Vdash (Q_1(\mathbf{x}, z) \supset Q_2(\mathbf{x}))\,[\mathbf{x}z]$.

By the choice of M, $\mathbf{d}R^n\mathbf{a}$ implies $\mathbf{d} \notin \xi^+(Q_2)$ and thus

(3) $\mathbf{d}e \not\Vdash Q_2(\mathbf{x})\,[\mathbf{x}z]$.

Now from (2), (3) we obtain

(4) $\mathbf{d}e \not\Vdash Q_1(\mathbf{x}, z)\,[\mathbf{x}z]$,

which according to the choice of M, is equivalent to

(5) $(\mathbf{d}e)R^{n+1}(\mathbf{a}a_n)$.

By (1) we also have

(6) $(\mathbf{d}e) \Vdash \forall y(P_1(y, \mathbf{x}, z) \supset P_2(y, \mathbf{x}, z))\,[\mathbf{x}z]$.

By our initial assumption, $(\mathbf{a}a_n)R^{n+1}\widehat{\mathbf{c}}_1$, hence $(\mathbf{d}e)R^{n+1}\widehat{\mathbf{c}}$, by (5). Now by (6) we obtain

(7) $\widehat{\mathbf{c}}_1 c_1 \Vdash (P_1(y, \mathbf{x}, z) \supset P_2(y, \mathbf{x}, z))\,[\mathbf{x}y]$,

On the other hand, by (0)

$$\mathbf{c} = \mathbf{c}_1\widehat{\mathbf{c}}_1 \in \xi^+(P_1) - \xi^+(P_2),$$

so

$$\widehat{\mathbf{c}}_1 c_1 \Vdash P_1(y, \mathbf{x}, z)\ [\mathbf{x}zy]$$

and

$$\widehat{\mathbf{c}}_1 c_1 \not\Vdash P_2(y, \mathbf{x}, z)\ [\mathbf{x}zy].$$

This contradicts (7) and the reflexivity of R^{n+2}.

So we have proved that $\mathbf{a} \not\Vdash \exists z B\,[\mathbf{x}]$. But $M \Vdash A$, therefore

(8) $\mathbf{a} \not\Vdash\ [x_n/z]B\,[\mathbf{x}]$.

However note that

(9) $M \Vdash [x_n/z]B_2\,[\mathbf{x}]$.

In fact, if (in M) $\mathbf{d} \not\Vdash\ Q_2(\mathbf{x})\,[\mathbf{x}]$, then by definition, $\mathbf{d}R^n\mathbf{a}$, which implies $(\mathbf{d}d_n)R^n(\mathbf{a}a_n)$ by s-functoriality. But the latter means $(\mathbf{d}d_n) \notin \xi^+(Q_1)$, i.e. $\mathbf{d} \not\Vdash\ Q_1(\mathbf{x}, x_n)\,[\mathbf{x}]$. Eventually from (8) and (9) it follows that

$$\mathbf{a} \not\Vdash\ [x_n/z]B_1\,[\mathbf{x}].$$

So there exists $\mathbf{e} \in D^{n+1}$ such that

$$\mathbf{e} \Vdash P_1(y, \mathbf{x}, x_n)\,[\mathbf{x}y] \text{ and } \mathbf{e} \not\Vdash\ P_2(y, \mathbf{x}, x_n)\,[\mathbf{x}y],$$

which is equivalent to

$$(e_{n+1}\mathbf{e}e_n) \in (\xi^+(P_1) - \xi^+(P_2)) = \ \approx^{n+2}(\mathbf{c})$$

by (0). Then $\mathbf{b} := e_{n+1}\mathbf{e}$ fits for the conclusion of (I5). ■

Remark 5.16.12 So we see that $\mathbb{F}$ is an $\mathrm{i}^{(=)}$-metaframe if $\mathbf{IL}^{(=)}(\mathbb{F})$ contains $A_0 := (p \supset .\top \supset p)$ (cf. the proof of 5.14.12) and A_1, A_2 from the proof above.

Theorem 5.16.13 (Maximality theorem) *Let $\mathbb{F}$ be an* **S4**-*metaframe. Then the following conditions are equivalent:*

(1) $\mathbf{IL}^{(=)}(\mathbb{F})$ *is an s.p.l (=);*

(2) $\mathbf{QH} \subseteq \mathbf{IL}^{(=)}(\mathbb{F})$*;*

(3) $\mathbb{F}$ *is an* $i^{(=)}$*-metaframe.*

Proof In fact, (2) ⇒ (3) by Proposition 5.16.11; (3) ⇒ (1) by Theorem 5.15.26. (1) ⇒ (2) is trivial. ■

Therefore we can introduce the intuitionistic semantics of metaframes $\mathcal{MF}_{int}^{(=)}$ generated by $\mathrm{i}^{(=)}$-metaframes.

Theorem 5.16.13 shows that (using the terminology from Section 2.12) this is the *greatest* sound semantics of **S4**-metaframes for superintuitionistic logics, because it is generated by *all* **S4**-metaframes strongly validating $\mathbf{QH}^{(=)}$-theorems. Therefore the list of properties of $\mathrm{i}^{(=)}$-metaframes, yields a precise criterion of intuitionistic soundness for metaframes. In the next section we will apply this criterion to describe intuitionistic soundness in Kripke-quasi-bundles.

Note that the classes of weak i-metaframes, i-metaframes and $\mathrm{i}^{=}$-metaframes are actually different. Let us consider two simple examples.

Example 5.16.14 Let $\mathbb{F}_1 = ((F_n)_{n\in\omega}, D)$, be an **S4**-metaframe such that

- $D^0 = \{u_0, u_1\}$, (D^0, R^0) is a two-element chain;
- $D_{u_0} = \{d_0\}$, $D_{u_1} = \{d_1, d_2\}$;
- (D^n, R^n) is a tree of height 2 with the root d_0^n, i.e. the tuple d_0^n sees all n-tuples from D_{u_1} (which are incomparable).

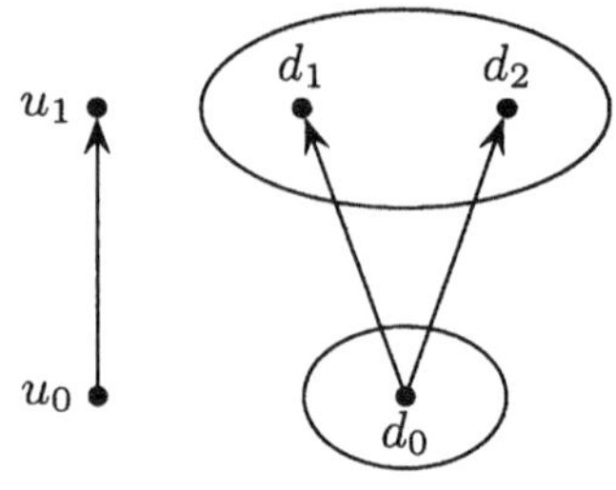

On one hand, by Proposition 5.16.10, $\mathbb{F}_1$ is a wi-metaframe: the s-functoriality, (I2.3), (I2.4), (I6.1), (I6.2) hold obviously. For example, for (I6.2): if $n > 0, (a_1, \ldots, a_n, a_{n+1}) \in (D_u)^{n+1}$, $(c_1, \ldots, c_n, c_{n+1}) \in (D_v)^{n+1}, (a_1, \ldots, a_n)$ $R_n(c_1, \ldots, c_n)$, then we can put $(b_1, \ldots, b_n, d, e) := (d_0, \ldots, d_0, d_0, c_{n+1})$ if $u = u_0$, and $(b_1, \ldots, b_n, d, e) := (a_1, \ldots, a_n, a_{n+1}, c_{n+1})$ if $(a_1, \ldots, a_n) = (c_1, \ldots, c_n)$.

On the other hand, $\mathbb{F}_1$ is not an i-metaframe, since (I5) fails for $n = 1$. In fact, $(d_0, d_0)R^2(d_1, d_2)$, but there does not exist $\mathbf{b}' = (b_0, b_1, b_1) \approx^3 (d_1, d_1, d_2)$.

Example 5.16.15 *Let $\mathbb{F}_2$ be the metaframe with the same D as $\mathbb{F}_1$ and with the universal R'^0, $R'^n = R^n \cup (R^n)^{-1}$, see the following figure:*

Again, $\mathbb{F}_2$ is a wi-metaframe, and even an i-metaframe: (I5) holds since $(d_0, d_0, d_0) \approx^3 (d_1, d_2, d_2)$, etc. On the other hand, $\mathbb{F}_2$ is not an $\mathrm{i}^{=}$-metaframe — (mm$_2$) fails, since $(d_0, d_0)R'^2(d_1, d_2)$, but $d_1 \neq d_2$.

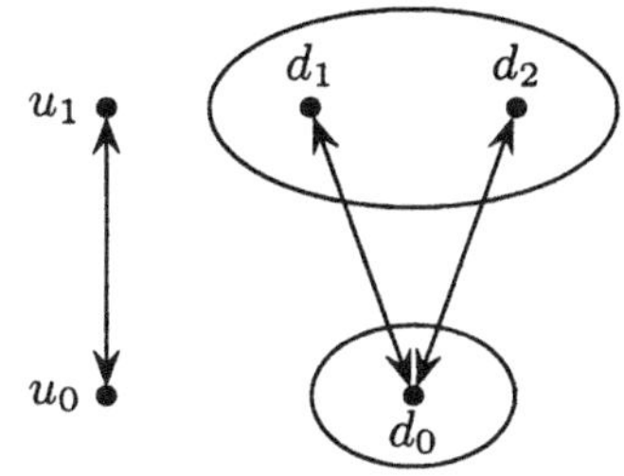

Therefore, the class of i-metaframes (which are sound for the intuitionistic logic without equality) is larger than the class of i$^=$-metaframes. Recall that in the modal case there is no difference between logics with or without equality in this respect; sound metaframes are same (see Theorem 5.12.13). Recall also that $\mathbf{IL}(\mathbb{F}) = \mathbf{IL}^=(\mathbb{F}) \cap IF$ is the fragment without equality for any i$^=$-metaframe $\mathbb{F}$ (by Proposition 5.15.22).

Let us also show that the class of i$^=$-metaframes is larger than the class of **QS4**-metaframes (which are sound for the modal case).

Example 5.16.16 Consider an **S4**-metaframe $\mathbb{F}$ such that

$$W = \{u_0, u_1\},\ u_0 \approx u_1,\ D_{u_0} = \{d_0\},\ D_{u_1} = \{d_1, d_2\},$$
$$\mathbf{a}R^n\mathbf{b} \text{ iff } \mathbf{a} = (b_1, \dots, b_n) \vee \exists i \leq 1(\mathbf{a} = (d_i, \dots, d_i)\ \&\ \mathbf{b} = (d_{1-i}, \dots, d_{1-i}))$$

Obviously, $R^n = \approx^n$. On the one hand, $\mathbb{F}$ is not a modal metaframe, because (I2.3) fails; in fact, $(a_1, a_2) \in D^2_{u_1}$, $a_1 R^1 a_0$, and $\neg\exists b \in D_{u_0}\ (a_1a_2)R^2(a_0, b)$.

On the other hand, $\mathbb{F}$ is an i$^=$-metaframe. E.g. the condition (I2.3) holds: one can take $(b'_1, \dots, b'_n, b'_0) = (a_1, \dots, a_n)$ since $R^n = \approx^n$. And (I6) also holds, because the $(n+1)$-tuple $(a_0, \dots, a_0, a_0) \in D^{n+1}_{u_0}$ has an $\approx^{n+1}$-copy $(a_1, \dots, a_1, a_1) \in D^{n+1}_{u_1}$, and we can stick together $(n+1)$-tuples in D_{u_1}.

Note that we can consider $\mathcal{MF}_m$ as a semantics for superintuitionistic logics (with or without equality) and $\mathcal{MF}^=_{int}$ also as a semantics for superintuitionistic logics without equality. We still do not know if all these semantics are equal to $\mathcal{MF}^=_m$.

5.17 Kripke quasi-bundles

Recall that according to Definition 5.5.19, a Kripke quasi-bundle is a quasi-p-morphism between **S4**-frames, i.e. a monotonic surjective map with the quasi-lift property.

Every Kripke quasi-bundle $\pi : F_1 \longrightarrow F_0$, where $F_i = (W_i, R_i)$, is associated with a system of domains $D_u := \{\pi^{-1}(u) \mid u \in F_0\}$ and a family of inheritance relations $\rho_{uv} := R_1 \cap (D_u \times D_v)$. It also corresponds to a metaframe constructed as in the case of Kripke bundles, cf. Definition 5.3.2; the n-levels are $F_n = (D^n, R^n)$, with

$$\mathbf{a}R^n\mathbf{b} \text{ iff } \forall j\ a_j R_1 b_j \ \&\ \mathbf{a}\ sub\ \mathbf{b}.$$

So $R_i = R^i$ for $i = 0, 1$.

Proposition 5.17.1 *Let $\mathbb{F} = (F, (D_n))$ be a metaframe corresponding to an intuitionistic Kripke quasi-bundle. Then $\mathbb{F}$ is an i-metaframe iff $\mathbb{F}$ is an $\mathrm{i}^=$-metaframe iff the following conditions hold:*

(1) $aR^1b \Rightarrow \exists v\, \exists c, d \in D_v\ (aR^1c\ \&\ dR^1b)$;

(2) *if $n > 0$, and $(\mathbf{a}d) \in D^{n+1}$, all a_i and d are distinct; $\mathbf{b} \in D^n$ (b_i are not necessarily distinct), and $\forall s\ a_s R^1 b_s$, then there exists $\mathbf{c} \in D^{n+1}$ such that $dR^1c_{n+1}\ \&\ \forall s \leq n\, (b_s \approx^1 c_s)\ \&\ \forall s, t \leq n\ (b_s = b_t \Leftrightarrow c_s = c_t)$;*

(3) *if $n > 0$, $\mathbf{a}, \mathbf{c} \in D^{n+1}$, all a_i are distinct, all c_i are distinct and $\forall s\ a_s R^1 c_s$, then there exists $(\mathbf{b}de) \in D^{n+2}$ such that $d, b_1, \ldots, b_n$ are distinct (but perhaps $e = d$ or $e = b_i$) such that*

$$\forall s\, (a_s R^1 b_s\ \&\ b_s R^1 c_s\ \&\ a_{n+1} R^1 d\ \&\ e R^1 c_{n+1}).$$

These conditions allow us to 'move' individuals from one world to another.

Proof First, $\mathbb{F}$ is an **S4**-metaframe: all R^n are reflexive and transitive, since the relations R^1, R^0, sub are reflexive and transitive.

Now let us check the properties of $\mathrm{i}^=$-metaframes stated in Proposition 5.16.10.

The monotonicity is almost obvious. In fact, $\forall i\ a_i R^1 b_i$ implies $\forall i\ a_{\sigma(i)} R^1 b_{\sigma(i)}$; and $\mathbf{a}\, sub\, \mathbf{b}$ implies $(\pi_\sigma \mathbf{a})\ sub\ (\pi_\sigma \mathbf{b})$.

(I2.4) is the quasi-lift property, which follows from the definition of a Kripke quasi-bundle.

(I6.1) is the same as (1).

(I2.3) follows from (2). In fact, $\mathbf{b} = (b_1, \ldots, b_n) \approx^n \mathbf{c} = (c_1, \ldots, c_n)$ iff $\forall s \leq n\, (b_s \approx^1 c_s)\ \&\ \mathbf{b}\, sub\, \mathbf{c}\ \&\ \mathbf{c}\, sub\, \mathbf{b}$; we also have $(\mathbf{a}d)R^{n+1}\mathbf{c}$, since $(a_1, \ldots, a_n)R^n(c_1, \ldots, c_n)$ and dR^1c_{n+1}. The requirement in (2) that $d, a_1, \ldots, a_n$ are distinct, is overcome due to the 'local functionality'.

It remains to check (I6.2) or its equivalent version 5.16.8 (I6.2~). So let $\mathbf{a}, \mathbf{c} \in D^{n+1}$, $\pi^n_+\mathbf{a} \approx^n \pi^n_+\mathbf{c}$. Then $\mathbf{a}\, sub\, \mathbf{c}$ and $\mathbf{c}\, sub\, \mathbf{a}$. The cases when $a_{n+1} \in \{a_1, \ldots, a_n\}$, $c_{n+1} \in \{c_1, \ldots, c_n\}$ are obvious. Thus we may assume that all a_j and all b_j are distinct. Now (3) yields us a tuple $(\mathbf{b}de) \in D^{n+2}$ such that $\mathbf{a}R^{n+1}(\mathbf{b}d)$, $(\mathbf{b}e)R^{n+1}\mathbf{c}$.

Also note that the conditions (1), (2), (3) are necessary: they definitely hold if $\mathbb{F}$ is an i-metaframe. In fact, they respectively follow from (I6.1), (I2.3), (I6.2). ∎

Therefore, the logic $\mathbf{IL}^{(=)}(\mathbb{F})$ of a Kripke quasi-bundle $\mathbb{F}$ is superintuitionistic iff the conditions (1), (2), (3) hold. We call such quasi-bundles *intuitionistic* (or, briefly, i-quasi-bundles); they generate a sound semantics for superintuitionistic logics. Every intuitionistic Kripke bundle is clearly an i-quasi-bundle.

Note that a metaframe associated with a Kripke quasi-bundle is not necessarily modal. For instance, the $i^{=}$-metaframe $\mathbb{F}$ from Example 5.16.16, which is not modal, corresponds to the Kripke quasi-bundle $(W, D, \bar{\rho}), W = \{u_0, u_1\}, u_0 \approx u_1, D_{u_0} = \{d_0\}, D_{u_1} = \{d_1, d_2\} \rho_{u_i,u_i} = id_{d_{u_i}} (i = 0, 1), \rho_{u_i,u_{1-i}} = \{(d_i, d_{1-i})\}$. This explains why the construction of the $\mathcal{C}$-set corresponding to a Kripke bundle described in the proof of Proposition 7.8.11, fails for Kripke quasi-bundles.

5.18 Some constructions on metaframes

Definition 5.18.1 *Let $\mathbb{F} = (F, (D_n))$ be an N-metaframe based on $F = (W, R_1, \ldots, R_N)$ and let $V \subseteq W$. Then the* submetaframe $\mathbb{F}|V$ *is the restriction of $\mathbb{F}$ to V, i.e. it has a system of domains is $D|V := (D_u \mid u \in V)$ with relations $R_i^n|V := R_i^n \restriction (D|V)^n$; in particular, $R^0|V = R^0 \restriction V$.*

A submetaframe $\mathbb{F}|V$ is called generated *if $V \subseteq W$ is R_i-stable for every $i = 1, \ldots, N$.*

Lemma 5.18.2

(1) If $\mathbb{F}|V$ is a generated submetaframe of $\mathbb{F}$ and all the maps $\pi_{\varnothing_n}$ in $\mathbb{F}$ are monotonic (in particular, if $\mathbb{F}$ is a modal or a wi-metaframe), then each of its n-level $(F|V)_n$ is a generated subframe of F_n.

(2) Every submetaframe of an N-modal metaframe is an N-modal metaframe

(3) Every generated submetaframe of an i- (resp., $\bar{i}^{=}$-, wi-) metaframe is also an i- (resp., $\bar{i}^{=}$-, wi-) metaframe.

Definition 5.18.3 *A* cone $\mathbb{F}{\uparrow}u$ *of a metaframe $\mathbb{F}$ (for $u \in F^0$) is its restriction to the subset (cone) $R^*(u)$ in the base F^0.*

Recall that R^* is the reflexive transitive closure of $\bigcup\limits_{i=1}^{N} R_i$.

Definition 5.18.4 *A metaframe model $M = (\mathbb{F}, \xi)$ over $\mathbb{F}$ (N-modal or intuitionistic) gives rise to the model $M|V := (\mathbb{F}|V, \xi|V)$ (the* restriction of M to V*) such that $(\xi|V)^+(P_j^n) = \xi^+(P_j^n) \cap (D|V)^n$.*

Lemma 5.18.5 *Let $\mathbb{F}$ be an N-modal metaframe satisfying (I2.1), or a wi-metaframe, let $M = (\mathbb{F}, \xi)$ be a model (N-modal or intuitionistic, respectively), $\mathbb{F}|V$ a generated submetaframe, and let $M|V$ be the corresponding submodel of M. Then*

(1) $M, \mathbf{a} \Vdash (\vDash) A\,[\mathbf{x}]$ iff $M|V, \mathbf{a} \Vdash (\vDash) A\,[\mathbf{x}]$ for any appropriate assignment $(\mathbf{a}, \mathbf{x})$ (modal or intuitionistic, respectively).

(2) $M \Vdash (\models)A \Rightarrow M|V \Vdash (\models)A$.

Proof By induction on the length of A. Note that in the intuitionistic case it is essential that in a wi-metaframe an inductive clause for $\exists$ can be rewritten with $\approx$ instead of R, see Section 5.11. Thus we cannot state (1) for an arbitrary metaframe. ■

Note that every valuation (N-modal or intuitionistic) in $\mathbb{F}|V$ is also a valuation in $\mathbb{F}$ and it coincides with its restriction to $\mathbb{F}|V$. Hence we obtain

Proposition 5.18.6

(1) $\mathbf{ML}_{-}^{(=)}(\mathbb{F}) \subseteq \mathbf{ML}_{-}^{(=)}(\mathbb{F}|V)$ *and* $\mathbf{ML}^{(=)}(\mathbb{F}) \subseteq \mathbf{ML}^{(=)}(\mathbb{F}|V)$ *for a generated submetaframe of an N-modal metaframe $\mathbb{F}$.*

(2) $\mathbf{IL}_{-}^{(=)}(\mathbb{F}) \subseteq \mathbf{IL}_{-}^{(=)}(\mathbb{F}|V)$ *and* $\mathbf{IL}^{(=)}(\mathbb{F}) \subseteq \mathbf{IL}^{(=)}(\mathbb{F}|V)$ *for a generated submetaframe of an i- (wi-, $\bar{\imath}^{=}$-) metaframe $\mathbb{F}$.*

Proposition 5.18.7

(1) $\mathbf{ML}_{-}^{(=)}(\mathbb{F}) = \bigcap\limits_{u\in W} \mathbf{ML}_{-}^{(=)}(\mathbb{F}{\uparrow}u)$ *and* $\mathbf{ML}^{(=)}(\mathbb{F}) = \bigcap\limits_{u\in W} \mathbf{ML}^{(=)}(\mathbb{F}{\uparrow}u)$ *for an N-modal metaframe $\mathbb{F}$.*

(2) $\mathbf{IL}_{-}^{(=)}(\mathbb{F}) = \bigcap\limits_{u\in W} \mathbf{IL}_{-}^{(=)}(\mathbb{F}{\uparrow}u)$ *and* $\mathbf{IL}^{(=)}(\mathbb{F}) \bigcap\limits_{u\in W} \mathbf{IL}^{(=)}(\mathbb{F}{\uparrow}u)$ *for an i- (wi-, $\bar{\imath}^{=}$-) metaframe $\mathbb{F}$.*

Proof If $M = (\mathbb{F}, \xi) \not\models A$ then $M, \mathbf{a} \not\models A\,[\mathbf{x}]$ for some $(\mathbf{a}, \mathbf{x})$, and thus $M{\uparrow}u, \mathbf{a} \not\models A\,[\mathbf{x}]$, for $u = \pi_{\varnothing}(\mathbf{a})$. ■

Definition 5.18.8 *Let $(\mathbb{F}_i)_{j\in J}$ be a family of metaframes of the same type (N-modal or intuitionistic), $\mathbb{F}_j = ((F_{jn}), D_j)$, $F_{jn} = (D_j^n, R_{j1}^n, \dots, R_{jN}^n)$. Their* disjoint sum (union) $\bigsqcup\limits_{j\in J} \mathbb{F}_j$ *is defined as the metaframe $\mathbb{F} = ((F_n), D)$ such that*

- $F_0 = \bigsqcup\limits_{j\in J} F_{j0}$;
- $D_{(u,j)} = (D_j)_u \times \{j\}$;
- $F_n = (D^n, R_1^n, \dots, R_N^n)$, *where* $D^n = \bigcup\limits_{w\in F_0} D_w^n$, *and for* $\mathbf{a}, \mathbf{b} \in D_j^n$ $((a_1, j), \dots, (a_n, j)) R_k^n ((b_1, j), \dots, (b_n, j))$ *iff* $\mathbf{a} R_{jk}^n \mathbf{b}$.

Lemma 5.18.9 *Let* $\mathbb{F} = \bigsqcup\limits_{j\in J} \mathbb{F}_j$. *Then*

(1) $F_n \cong \bigsqcup\limits_{j\in J} F_{jn}$;

(2) *$\mathbb{F}$ is an N-modal metaframe, or respectively an i- (wi-, $i^{=}$-) metaframe iff all $\mathbb{F}_j$ are.*

Proof

(1) The map $\bigsqcup_{j \in J} F_{jn} \longrightarrow F_n$ sending $(\mathbf{a}, j)$ (for $\mathbf{a} \in D_j^n$) to $((a_1, j), \ldots, (a_n, j))$, is an isomorphism.

(2) An exercise.

■

Lemma 5.18.10

(1) $\mathbf{ML}_{-}^{(=)}(\bigsqcup_j \mathbb{F}_j) = \bigcap_j \mathbf{ML}_{-}^{(=)}(\mathbb{F}_j)$ *for N-modal metaframes.*

(2) $\mathbf{IL}_{-}^{(=)}(\bigsqcup_j \mathbb{F}_j) = \bigcap_j \mathbf{IL}^{(=)}(\mathbb{F}_j)$ *for i- (wi-, i⁼)-metaframes.*

(3) Similarly for **IL**, **ML**.

Proof On one hand, each $\mathbb{F}_j$ is isomorphic to a generated submetaframe of $\bigsqcup_j \mathbb{F}_j$. On the other hand, every cone in $\bigsqcup_j \mathbb{F}_j$ is isomorphic to a cone in the corresponding $\mathbb{F}_j$. ■

Therefore the semantics $\mathcal{MF}_m^{(=)}$, $\mathcal{MF}_{int}^{(=)}$ satisfy the collection property (CP) (see Section 2.16).

5.19 On semantics of intuitionistic sound metaframes

Note that all definitions related to $\mathrm{i}^{(=)}$-soundness can be readily extended to arbitrary (not necessarily **S4**-) 1-metaframes. Namely, a valuation ξ in a 1-metaframe is called *intuitionistic* if $R^n(\xi^+(P_j^n)) \subseteq \xi^+(P_j^n)$ for any predicate letter $P_j^n, n \geq 0$, cf. 5.14.1. Now the definitions of forcing, validity, and strong validity are rewritten in a straightforward way (cf. Definitions 5.14.2, 5.14.3).

The notations $\mathbf{IL}_{-}^{(=)}$, $\mathbf{IL}^{(=)}$ are also used in this case. Finally a metaframe $\mathbb{F}$ is called *$i^{(=)}$-sound* if $\mathbf{IL}^{(=)}(\mathbb{F})$ is an s.p.l., cf. Definition 5.14.4.

To conclude our description of intuitionistic sound metaframes, we now prove the following statement.

Proposition 5.19.1 *For every $i^{(=)}$-sound 1-metaframe $\mathbb{F}$ there exists an* **S4**-*metaframe $\mathbb{F}'$ such that*

$$\mathbf{IL}^{(=)}(\mathbb{F}') = \mathbf{IL}^{(=)}(\mathbb{F}).$$

Hence $\mathbb{F}'$ is also $i^{(=)}$-sound, so it is an $i^{(=)}$-metaframe, by 5.16.13.

Therefore we can conclude that the semantics $\mathcal{MF}_{int}^{(=)}$ of arbitrary $\mathrm{i}^{(=)}$-metaframes actually equals the semantics of all $\mathrm{i}^{(=)}$-sound metaframes.

To prove Proposition 5.19.1, we need two auxiliary notions.

Definition 5.19.2 *A propositional frame $F = (W, R)$ is called* weakly reflexive *(or* w-reflexive, *for short) if the transitive closure of R is reflexive (cf. [Došen, 1993]). A 1-metaframe $\mathbb{F}$ is called* w-reflexive *if all its levels F_n are w-reflexive.*

Definition 5.19.3 *A propositional frame $F = (W, R)$ is called* co-serial *if $\forall u \exists v\ vRu$. A 1-metaframe $\mathbb{F}$ is* co-serial *if all F_n are co-serial.*

Obviously, every reflexive propositional frame (and 1-metaframe) is w-reflexive and co-serial.

Lemma 5.19.4 *Let $\mathbb{F}$ be a 1-metaframe. Consider the formulas $B_1 := \exists z \top$ and $B_2 := (\top \supset p) \supset p$.*

(1) If $B_1 \in \mathbf{IL}(\mathbb{F})$, then $\mathbb{F}$ is co-serial.

(2) If $\mathbb{F}$ is co-serial and $B_2 \in \mathbf{IL}(\mathbb{F})$, then $\mathbb{F}$ is w-reflexive.

Proof

(1) Suppose $B_1 \in \mathbf{IL}(\mathbb{F})$, $\mathbf{a} \in D^n$. Let $\mathbf{x} = (x_1, \ldots, x_n), z \notin \mathbf{x}$. Consider an arbitrary intuitionistic model $M = (\mathbb{F}, \xi)$. Then $M, \mathbf{a} \Vdash B_1[\mathbf{x}]$, so there exists $\mathbf{b} \in (R^n)^{-1}(\mathbf{a})$ satisfying the condition: $\exists c \in D(\mathbf{b})\ M, \mathbf{b}c \Vdash \top[\mathbf{x}z]$. This implies co-seriality.

(2) Suppose $\mathbb{F}$ is co-serial, $B_2 \in \mathbf{IL}(\mathbb{F})$. Given $\mathbf{a} \in D^n$, choose $\mathbf{b} \in (R^n)^{-1}(\mathbf{a})$, by co-seriality. Consider the substitution instance

$$B_2' := (\top \supset P^n(\mathbf{x})) \supset P^n(\mathbf{x})$$

of B_2, and an intuitionistic model $M = (\mathbb{F}, \xi)$ such that $\xi^+(P^n) = ((R^n)^* \circ R^n)(\mathbf{a})$. Then

$$M, \mathbf{a} \Vdash (\top \supset P^n(\mathbf{x}))[\mathbf{x}].$$

In fact, if $\mathbf{c} \in R^n(\mathbf{a})$ and $M, \mathbf{c} \Vdash \top$, then $\mathbf{c} \in \xi^+(P^n)$, and so $M, \mathbf{c} \Vdash P^n(\mathbf{x})\ [\mathbf{x}]$. Thus $M, \mathbf{a} \Vdash P^n(\mathbf{x})$, since $M, \mathbf{b} \Vdash B_2'\ [\mathbf{x}]$ and $\mathbf{b}R^n\mathbf{a}$. Thus $\mathbf{a} \in \xi(P^n)\ [\mathbf{x}]$, so $\mathbf{a}((R^n)^* \circ R^n)\mathbf{a}$. Hence the w-reflexivity of $\mathbb{F}$ follows.

∎

Definition 5.19.5 *A 1-metaframe $\mathbb{F}$ is monotonic$^{(=)}$ if it satisfies*

(im) $M, \mathbf{a} \Vdash A\,[\mathbf{x}]\ \&\ \mathbf{a}R^n\mathbf{b} \Rightarrow M, \mathbf{b} \Vdash A\,[\mathbf{x}]$

for all for $A \in IF^{(=)}$, $\mathbf{x} \supseteq FV(A)$ and $\mathbf{a}, \mathbf{b} \in D^n$, cf. 5.14.14.

Proposition 5.19.6 *Let $\mathbb{F}$ be a co-serial-metaframe, $B := p \supset (\top \supset p)$. If $B \in \mathbf{IL}^{(=)}(\mathbb{F})$, then $\mathbb{F}$ is monotonic$^{(=)}$.*

Proof [30] Let us check (im) for a formula $A \in IF^{(=)}$. Consider a substitution instance $B' := A \supset (\top \supset A)$ of B.

Assume $\mathbf{a}R^n\mathbf{b}$ and $M, \mathbf{a} \Vdash A\ [\mathbf{x}]$. Due to co-seriality, there exists $\mathbf{c}R^n\mathbf{a}$. Since $M, \mathbf{c} \Vdash B'\ [\mathbf{x}]$, it follows that $M, \mathbf{a} \Vdash (\top \supset A)\ [\mathbf{x}]$. Now $M, \mathbf{b} \Vdash \top\ [\mathbf{x}]$ implies $M, \mathbf{b} \Vdash A\ [\mathbf{x}]$. ■

Corollary 5.19.7 *If for a 1-metaframe* $\mathbb{F}$, $\mathbf{QH} \subseteq \mathbf{IL}^{(=)}(\mathbb{F})$ *and even, if* $\{p \equiv (\top \supset p), \exists z \top\} \subseteq \mathbf{IL}^{(=)}(\mathbb{F})$, *then* $\mathbb{F}$ *is w-reflexive, co-serial, and monotonic*$^{(=)}$.

Proof Note that $(p \equiv (\top \supset p)) \in \mathbf{IL}^{(=)}(\mathbb{F})$ iff $p \supset (\top \supset p), (\top \supset p) \supset p \in \mathbf{IL}^{(=)}(\mathbb{F})$. Then apply 5.19.4, 5.19.6. ■

Definition 5.19.8 *For a w-reflexive metaframe* $\mathbb{F} = ((D^n, R^n)_{n\in\omega}, D)$ *we define its* transitive closure, *the* **S4**-*metaframe* $\mathbb{F}^* = ((D^n, R^{n*})_{n\in\omega}, D)$, *where* $R^{n*} = R^{n+}$ *is the transitive closure of* R^n.

Lemma 5.19.9 *Let* $\mathbb{F}$ *be a w-reflexive monotonic*$^{(=)}$*-metaframe. Then*

$$\mathbf{IL}^{(=)}(\mathbb{F}) = \mathbf{IL}^{(=)}(\mathbb{F}^*).$$

Proof It is clear that ξ is an intuitionistic valuation in $\mathbb{F}$ iff ξ is an intuitionistic valuation in $\mathbb{F}^*$, since $\xi^+(P^n)$ is R^n-stable iff it is R^{n*}-stable.

So it is sufficient to show that for every model $M = (\mathbb{F}, \xi)$ and the corresponding model $M^* = (\mathbb{F}^*, \xi)$, for every ordered assignment $(\mathbf{x}, \mathbf{a})$ such that $\mathbf{x} \supseteq FV(A)$, the following holds:

$$M, \mathbf{a} \Vdash A\ [\mathbf{x}] \Leftrightarrow M^*, \mathbf{a} \Vdash A\ [\mathbf{x}].$$

We proceed by induction and consider only three non-trivial cases.

(1) $A = B \supset C$. If $M, \mathbf{a} \not\Vdash A\ [\mathbf{x}]$, then there exists $\mathbf{b} \in R^n(\mathbf{a})$ such that $M, \mathbf{b} \Vdash B\ [\mathbf{x}]$ and $M, \mathbf{b} \not\Vdash C\ [\mathbf{x}]$. Then $M^*, \mathbf{b} \Vdash B\ [\mathbf{x}]$ and $M^*, \mathbf{b} \not\Vdash C\ [\mathbf{x}]$ by the induction hypothesis, and obviously $\mathbf{b} \in R^{n*}(\mathbf{a})$. So $M^*, \mathbf{a} \not\Vdash A\ [\mathbf{x}]$.

The other way round, let $M^*, \mathbf{a} \not\Vdash A\ [\mathbf{x}]$. Then there exists $\mathbf{b} \in R^{n*}(\mathbf{a})$ such that $M^*, \mathbf{b} \Vdash B\ [\mathbf{x}]$ and $M^*, \mathbf{b} \not\Vdash C\ [\mathbf{x}]$. So $M, \mathbf{b} \Vdash B\ [\mathbf{x}]$ and $M, \mathbf{b} \not\Vdash C\ [\mathbf{x}]$. Take $\mathbf{d} \in D^n$ such that $\mathbf{a}R^{n*}\mathbf{d}R^n\mathbf{b}$. Then $M, \mathbf{d} \not\Vdash A\ [\mathbf{x}]$ and by monotonicity $M, \mathbf{a} \not\Vdash A\ [\mathbf{x}]$ as well.

(2) $A = \exists y B\ [\mathbf{x}]$. If $M, \mathbf{a} \Vdash A\ [\mathbf{x}]$, then there exist $\mathbf{b} \in (R^n)^{-1}(\mathbf{a})$ and $c \in D(\mathbf{b})$ such that $M, \pi_{\mathbf{x}||y}(\mathbf{b}c) \Vdash B\ [\mathbf{x}||y]$. So $M^*, \pi_{\mathbf{x}||y}(\mathbf{b}c) \Vdash B\ [\mathbf{x}||y]$ and $M^*, \mathbf{a} \Vdash A\ [\mathbf{x}]$, since $\mathbf{a}R^{n*}\mathbf{b}$.

The other way round, let $M^*, \mathbf{a} \Vdash A\ [\mathbf{x}]$. Then there exist $\mathbf{b}, \mathbf{d} \in D^n$ and $c \in D(\mathbf{b})$ such that[31] $\mathbf{b}R^n\mathbf{d}R^{n*}\mathbf{a}$ and $M^*, \pi_{\mathbf{x}||y}(\mathbf{b}c) \Vdash B\ [\mathbf{x}||y]$. Hence $M, \pi_{\mathbf{x}||y}(\mathbf{b}c) \Vdash B\ [\mathbf{x}||y]$, so $M, \mathbf{d} \Vdash A\ [\mathbf{x}]$, and thus $M, \mathbf{a} \Vdash A\ [\mathbf{x}]$ by monotonicity.

---

[30] Cf. Lemma 5.14.15.

[31] Note that $R^* \circ R = R \circ R^*$ is the transitive closure of R.

(3) The case $A = \forall y B\ [\mathbf{x}]$ is similar.

■

Lemma 5.19.9 and Corollary 5.19.7 obviously imply Proposition 5.19.1; namely we put $\mathbb{F}' = \mathbb{F}^*$.

The previous consideration in this section allows us to extend Lemma 5.14.11(3) to arbitrary 1-metaframes:

Proposition 5.19.10 *Let $\mathbb{F}$ be a 1-metaframe. Then the following conditions are equivalent:*

(1) $\mathbb{F}$ is $i^{(=)}$-sound;

(2) $\mathbf{QH} \subseteq \mathbf{IL}^{(=)}(\mathbb{F})$;

(3) $\mathbb{F}$ is w-reflexive and monotonic$^{(=)}$, and $\mathbb{F}^*$ is an $i^{(=)}$-metaframe.

Moreover, these conditions are equivalent to

(4) the following formulas are in $\mathbf{IL}^{(=)}(\mathbb{F})$:

$$p \equiv (\top \supset p),\ \exists z\top,\ A_1 := \exists yp \supset p,\ A_2 := P(x) \supset \exists zP(z).$$

Here A_1 and A_2 are the formulas used in the proof of Proposition 5.16.11. Note that all formulas in (4) are **QH**-theorems, so (2) readily implies (4).

Proof (3)⇒(1). By Lemma 5.19.9 and Theorem 5.15.26 (soundness).

(2)⇒(3). By Corollary 5.19.7, Lemma 5.19.9, and 5.16.11.

(4)⇒(3). Assume (4). Then $\mathbb{F}$ is w-reflexive and monotonic$^{(=)}$ by 5.19.7, so by 5.19.9 $\mathbf{IL}^{(=)}(\mathbb{F}) = \mathbf{IL}^{(=)}(\mathbb{F}^*)$. So formulas from (4) are in $\mathbf{IL}^{(=)}(\mathbf{F}^*)$, and therefore $\mathbb{F}^*$ is an $\mathrm{i}^{(=)}$-metaframe, cf. the proof of Proposition 5.16.11 and Lemma 5.14.15, or cf. Remark 5.16.12. ■

By the way, basing on Proposition 5.19.10 (1) ⇔ (3), we can give an explicit description of the class of all $\mathrm{i}^{(=)}$-sound metaframes. Viz., to check that $\mathbb{F}^*$ is an $\mathrm{i}^{(=)}$-metaframe, we can rewrite all conditions from the definition of an $\mathrm{i}^{(=)}$-metaframe (cf. Section 5.15 and 5.16) with R^{n*} replacing R^n. In particular, the condition (0σ) becomes

$$\mathbf{a}R^{n*}\mathbf{b} \Rightarrow (\pi_\sigma\mathbf{a})R^{m*}(\pi_\sigma\mathbf{b}),$$

or equivalently

$$\mathbf{a}R^{n}\mathbf{b} \Rightarrow (\pi_\sigma\mathbf{a})R^{m*}(\pi_\sigma\mathbf{b}).$$

Here is a description of monotonicity$^{(=)}$ for arbitrary metaframes, which we state without a proof.

Lemma 5.19.11

(1) A 1-metaframe $\mathbb{F}$ is monotonic iff it satisfies

(i) $\mathbf{a}R^n\mathbf{b} \Rightarrow (\pi_\sigma\mathbf{a})R^{m*}(\pi_\sigma\mathbf{b})$ *for any* $\sigma \in \Sigma_{mn}$*;*

(ii) $\mathbf{a}R^n\mathbf{b}R^n\mathbf{c} \Rightarrow \exists\mathbf{c}'(\mathbf{a}R^n\mathbf{c}' \ \&\ \mathbf{c}R^{n*}\mathbf{c} \ \&\ \mathbf{c}'R^{n*}\mathbf{c})$*;*

(iii) $\mathbf{a}R^n\mathbf{b}R^n\mathbf{c} \Rightarrow \forall e \in D(\mathbf{c})\ \exists\mathbf{d}\,\exists g \in D(\mathbf{c})\,(\mathbf{a}R^n\mathbf{d} \ \&\ (\mathbf{d}g)R^{*(n+1)}(\mathbf{c}e))$*;*

(iv) $\mathbf{c}R^n\mathbf{b}R^n\mathbf{a} \Rightarrow \forall e \in D(\mathbf{c})\ \exists\mathbf{d}\,\exists g \in D(\mathbf{d})\,(\mathbf{d}R^n\mathbf{a} \ \&\ (\mathbf{c}e)R^{*(n+1)}(\mathbf{d}g))$*.*

(2) $\mathbb{F}$ *is monotonic*$^=$ *iff it is monotonic and satisfies* (mm_2)*.*

Here the condition (i) corresponds to (0σ) in $\mathbb{F}^*$ and expresses the monotonicity for atomic formulas $P(\pi_\sigma\mathbf{x})$, cf. the proof of 5.14.11. The condition (ii) means that all levels F_n are weakly transitive.[32]

Actually the conditions (b), (c), (d) express the monotonicity for the implication and the quantifiers $\forall$, $\exists$ respectively, i.e. for formulas $P(\mathbf{x}) \supset Q(\mathbf{x})$, $\forall y P(\mathbf{x}, y)$, $\exists y P(\mathbf{x}, y)$.

The interested reader can try to restore the missing details. We point out that the description of $\mathrm{i}^{(=)}$-soundness in 5.19.11 is more complicated than the notion of an $i^{(=)}$-metaframe, but the semantics is the same.

5.20 Simplicial frames

Introduction

In the last section of this chapter we briefly describe the next step in generalising Kripke-type semantics.

Recall that in the usual Kripke semantics for predicate logics we have a system of nested domains $D = (D_u \mid u \in F)$, and every individual from D_u is considered as its own inheritor in the domains D_v of all worlds v accessible from u. In Kripke sheaves, Kripke bundles and functor semantics accessibility relations (or functions) between individuals are introduced; they describe inheritors of individuals in accessible worlds. Accessibility relations between n-tuples of individuals (for $n > 1$) can be derived from these accessibility relations on individuals.

At the next step, in metaframes, accessibility relations R^n (or R_i^n, in the polymodal case) between n-tuples of individuals for different $n \geq 1$ may be arbitrary, and only the requirement of soundness puts some constraints on these relations. But n-tuples $\mathbf{a} = (a_1, \ldots, a_n)$ are obtained from individuals existing in the same world. So all n-tuples are taken from the set $D^n = \bigcup\{D_u^n \mid u \in F\}$.

Now let us consider more general kind of frames, in which n-tuples are 'abstract'; so the sets D^n for $n \geq 1$ are a priori independent, and unlike the case of metaframes, D^n is not constructed from the set of 'actual' individuals $D^1 =$

[32] A propositional frame $F = (W, R)$ is called weakly transitive if

$$\forall u, v, w \in W(uRvRw \Rightarrow \exists t\,(uRt \ \&\ wR^*t \ \&\ tR^*w)).$$

$\bigcup\{D_u \mid u \in F\}$. But then for any $\sigma \in \Sigma_{mn} = (I_n)^{I_m}$, we should introduce a map $\pi_\sigma : D^n \longrightarrow D^m$ transforming 'abstract' tuples. Their metaframe analogues are 'jections' π_σ transforming 'actual' tuples: $\pi_\sigma(a_1, \dots, a_n) = (a_{\sigma(1)}, \dots, a_{\sigma(m)})$. But in the 'abstract' case π_σ are chosen arbitrarily. The definitions of valuations, forcing, validity, and strong validity can be given quite similarly to metaframe semantics.[33] And again the predicate logic $\mathbf{ML}^{(=)}(\mathbb{F})$ or $\mathbf{IL}^{(=)}(\mathbb{F})$ for such a frame $\mathbb{F}$ is the set of formulas strongly valid in $\mathbb{F}$. Next, we can find constraints corresponding to logical soundness or to other natural properties of forcing, as we did in Chapter 5 for metaframes.

An 'abstract' n-tuple $\mathbf{a} \in D^n$ corresponds to the 'real' n-tuple $(\pi_{\lambda_1^n}(\mathbf{a}), \dots, \pi_{\lambda_n^n}(\mathbf{a}))$, where $\lambda_i^n \in \Sigma_{1n}$, $\lambda_i^n(1) = i$, see the Introduction. $\pi_{\lambda_i^n}(\mathbf{a})$ can be regarded as the ith 'component' of $\mathbf{a}$. But this correspondence in general is not bijective — different tuples may have the same 'components'; moreover, for a permutation $\sigma \in \Upsilon_n$, 'n-tuples' $\mathbf{a}$ and $\pi_\sigma(\mathbf{a})$ may have different (and even disjoint!) sets of 'components'. Nevertheless, this correspondence is useful; as we shall see in Volume 2, this helps associate logically sound simplicial frames with metaframes.

As in metaframes, we can identify D^0 with the 'underlying propositional frame' $F = (W, R_1, \dots, R_N)$. And we can introduce the individual domains of a world $u \in W$ as $D_u := \{\mathbf{a} \in D^1 \mid \pi_{\varnothing_1}(\mathbf{a}) = u\}$. More generally, put $D_u^n := \pi_{\varnothing_n}^{-1}(u)$ for $n \geq 1$, where $\varnothing_n \in \Sigma_{0n}$ is the empty function. Then $D^n = \bigcup\{D_u^n \mid u \in F\}$ is a partition. But in general, an 'abstract' n-tuple $\mathbf{a}$ and its 'components' $\pi_{\sigma_{in}}(\mathbf{a})$ for $1 \leq i \leq n$ may be in domains of different worlds from F.

Forcing in simplicial frames

Now let us turn to precise definitions. We begin with the case without equality.

Definition 5.20.1 *A (formal)* simplicial N-frame *based on a propositional Kripke frame* $F = (W, R_1, \dots, R_N)$ *is a tuple* $\mathbb{F} = (F, \overrightarrow{D}, \overrightarrow{R}, \pi)$, *where* $\overrightarrow{D} = (D^n \mid n \in \omega)$ *is a sequence of (non-empty) sets,* $\overrightarrow{R} = (R_i^n \mid n \in \omega,\ 1 \leq i \leq N)$ *is a family of relations* $R_i^n \subseteq D^n \times D^n$*; so we have propositional N-modal frames:* $F_n := (D^n, R_n^1, \dots, R_n^N)$. *As usual, we assume that* $F_0 := (W, R_1, \dots, R_N)$ *is the original frame F. And finally,* $\pi = (\pi_\sigma \mid \sigma \in \bigcup_{m,n} \Sigma_{mn})$ *is a family of mappings ('abstract jections')* $\pi_\sigma : D^n \longrightarrow D^m$ *for* $\sigma \in \Sigma_{mn}$.

So D^n (for $n \geq 1$) are sets of 'abstract' n-tuples (in particular, D^1 is the set of 'individuals'), R_i^n are accessibility relations between n-tuples, and π_σ are mappings ('jections') transforming 'abstract' tuples.

Put $D_u^n := \pi_{\varnothing_n}^{-1}(u)$ for $n > 0, u \in W$, where $\varnothing_n$ is the empty function from Σ_{0n}; it may be called the *n-tuple domain* of the world u. In particular, for $n = 1$ we sometimes write D_u rather than D_u^1 and call this set the *individual domain of the world* u. Obviously $\{D_u^n \mid u \in W\}$ is a partition of D^n.

[33] Up to some details about the equality, which are briefly discussed later on.

The *ith component* of an abstract n-tuple $\mathbf{a} \in D^n$, $n > 0$ is $[\mathbf{a}]_i := \pi_{\sigma_{in}}(\mathbf{a}) \in D^1$, where $\sigma_{in} \in \Upsilon_{1n}$, $\sigma_{in}(1) = i$ (i.e. $(a_1, \dots, a_n) \cdot \sigma_{in} = a_i$). In general, we do not assume that $\mathbf{a}$ and its components 'live' in the same world, i.e. that $\pi_{\varnothing_n}(\mathbf{a}) = \pi_{\varnothing_1}([\mathbf{a}]_i)$. We do not even assume that $[\mathbf{a}]_1 = \mathbf{a}$ for $\mathbf{a} \in D^1$ — an 'individual' may be non-equal to its own 'component'; they can even 'live' in different worlds. But as we shall see later on, in natural semantics (for modal or superintuitionistic logics) we may consider only frames, in which $\pi_{id_n}(\mathbf{a}) = \mathbf{a}$ for any n; such frames are called *π-identical.* In these frames we also have $[\mathbf{a}]_1 = \mathbf{a}$ for $\mathbf{a} \in D^1$, and $\pi_\varnothing(u) = u$ for $u \in W$, i.e. $D^0_u = \pi^{-1}_\varnothing(u) = \{u\}$.

As usual, for the monomodal (and the intuitionistic) case we denote accessibility relations on n-tuples by R^n, without the subscript $i = 1$.

The notations R^{n+} (respectively, R^{n*}) for the transitive (respectively, reflexive transitive) closure of $\bigcup\limits_i R^n_i$ are usual, cf. Chapter 1. We also consider the corresponding equivalence relations $\approx_n := \approx_{R^{*n}}$ on D^n.

Obviously, every N-metaframe gives rise to a simplicial N-frame,with D^n, R^n_i, π_σ described in Section 5.9.

Definition 5.20.2 *A* valuation *in a simplicial N-frame $\mathbb{F}$ is a function ξ sending every n-ary predicate letter P^n_j to a subset $\xi^+(P^n_j) \subseteq D^n$; in particular, $\xi^+(P^0_j) \subseteq W$ (cf. Definition 5.9.2).*

A valuation ξ in a simplicial 1-frame (or more briefly, a simplicial frame) is intuitionistic *if every $\xi^+(P^n_j)$ is an R^n-stable subset of D^n:*

$$\mathbf{a}R^n\mathbf{b} \;\&\; \mathbf{a} \in \xi^+(P^n_j) \Rightarrow \mathbf{b} \in \xi^+(P^n_j),$$

cf. Definition 5.14.1. A simplicial model *is a pair $M = (\mathbb{F}, \xi)$.*

Definition 5.20.3 *An* assignment *of length n in a simplicial frame $\mathbb{F}$ is a pair $(\mathbf{x}, \mathbf{a})$, where $\mathbf{a} \in D^n$, and $\mathbf{x}$ is a distinct list of variables of length n.*

Now we can define forcing for modal formulas without equality, cf. Definition 5.9.4.

Definition 5.20.4 *A simplicial N-model $M = (\mathbb{F}, \xi)$ gives rise to the* forcing relation *$M, \mathbf{a} \vDash A$ $[\mathbf{x}]$, where $A \in MF_N$, $(\mathbf{x}, \mathbf{a})$ is an assignment in $\mathbb{F}$, $FV(A) \subseteq r(\mathbf{x})$. The definition is by induction, modifying 5.9.4 in a natural way. We assume that $(\mathbf{x}, \mathbf{a})$ is of length n.*

(At1) $M, \mathbf{a} \nvDash \bot$ $[\mathbf{x}]$;

(At2) $M, \mathbf{a} \vDash P^m_j(\mathbf{x} \cdot \sigma)$ $[\mathbf{x}]$ *iff* $\pi_\sigma \mathbf{a} \in \xi^+(P^m_j)$ *(for $\sigma \in \Sigma_{mn}$)*

($\supset$) $M, \mathbf{a} \vDash (B \supset C)$ $[\mathbf{x}]$ *iff* $M, \mathbf{a} \nvDash B$ $[\mathbf{x}]$ *or* $M, \mathbf{a} \vDash C$ $[\mathbf{x}]$;

and similarly for the other Boolean connectives;

($\Box$) $M, \mathbf{a} \vDash \Box_i B$ $[\mathbf{x}]$ *iff* $\forall \mathbf{b} \in R^n_i(\mathbf{a})$ $M, \mathbf{b} \vDash B$ $[\mathbf{x}]$;

(Q1) $M, \mathbf{a} \vDash \exists y B\ [\mathbf{x}]$ *iff* $\exists \mathbf{c} \in D^{n+1}(\pi_{\sigma_+^n}\mathbf{c} = \mathbf{a}\ \&\ M, \mathbf{c} \vDash B\ [\mathbf{x}y])$,
$M, \mathbf{a} \vDash \forall y B[\mathbf{x}]$ *iff* $\forall \mathbf{c} \in D^{n+1}(\pi_{\sigma_+^n}\mathbf{c} = \mathbf{a} \Rightarrow M, \mathbf{c} \vDash B[\mathbf{x}y])$
(for $y \notin \mathbf{x}$*);*

(Q2) $M, \mathbf{a} \vDash \exists x_i B\ [\mathbf{x}]$ *iff* $M, \pi_{\delta_i^n}\mathbf{a} \vDash \exists x_i B\ [\mathbf{x} - x_i]$,
$M, \mathbf{a} \vDash \forall x_i B[\mathbf{x}]$ *iff* $M, \pi_{\delta_i^n}\mathbf{a} \vDash \exists x_i B[\mathbf{x} - x_i]$,
where $\mathbf{x} - x_i = \mathbf{x} \cdot \delta_i^n$.

Hence we obtain the truth conditions for $\neg$ and $\Diamond_i$ similar to those in Section 5.9.

One can see that in a simplicial frame corresponding to a metaframe the clauses (At2), ($\Box$), (Q1), (Q2) become equivalent to (2), (8), (9), (10) from Definition 5.9.4. In fact, for (Q1) note that in a metaframe $\pi_{\sigma_+^n}\mathbf{c} = \mathbf{a}$ iff $\mathbf{c} = \mathbf{a}d$ for some $d \in D(\mathbf{a})$.

Let us also note that in 'π-identical' simplicial frames (where $\pi_{id_n}\mathbf{a} = \mathbf{a}$ for any $\mathbf{a} \in D^n$) both cases of the inductive clause for the quantifier can be presented in the following uniform way (cf. (9+10) in Section 5.9):

$$M, \mathbf{a} \vDash \exists y B\ [\mathbf{x}] \text{ iff } \exists \mathbf{c} \in D^{m+1}(\pi_{\sigma_+^m}\mathbf{c} = \pi_{\varepsilon_{\mathbf{x}-y}}\mathbf{a}\ \&\ M, \mathbf{c} \vDash B\ [\mathbf{x}\|y])$$

(where $m = |\mathbf{x} - y|$). In fact, if $y = x_i$, this clause is equivalent to (Q2), since $\varepsilon_{\mathbf{x}-y} = \delta_i^n$ and $\mathbf{x}\|y = (\mathbf{x} - x_i)x_i$; and if $y \notin \mathbf{x}$, then it is equivalent to (Q1), since $\pi_{\varepsilon_{\mathbf{x}-y}}\mathbf{a} = \pi_{id_n}\mathbf{a} = \mathbf{a}$ and $\mathbf{x}\|y = \mathbf{x}y$.

Remark 5.20.5 Again we can propose a reasonable alternative definition of forcing $\vDash^*$ differing in the clause (Q2):

$$M, \mathbf{a} \vDash^* \exists x_i B\ [\mathbf{x}] \text{ iff } \exists \mathbf{c} \in D^n(\pi_{\delta_i^n}\mathbf{c} = \pi_{\delta_i^n}\mathbf{a}\ \&\ M, \mathbf{c} \vDash B[\mathbf{x}]),$$

and similarly for $\forall x_i B$.

But this definition actually leads to the same *natural* semantics of modal simplicial frames as Definition 5.20.4; thus both versions are equivalent. However we do not know if these definitions give equal (or equivalent) *maximal* logically sound semantics.

Similarly we can define forcing for the intuitionistic case.

Definition 5.20.6 *An (intuitionistic) simplicial model* $M = (\mathbb{F}, \xi)$ *gives rise to* forcing *for intuitionistic formulas* A *and assignments* $(\mathbf{x}, \mathbf{a})$ *with* $r(\mathbf{x}) \supseteq FV(A)$ *defined by the following inductive clauses (cf. Definition 5.14.2):*

- $M, \mathbf{a} \not\Vdash \bot\ [\mathbf{x}]$;
- $M, \mathbf{a} \Vdash P_j^m(\pi_\sigma \mathbf{x})\ [\mathbf{x}]$ *iff* $\pi_\sigma \mathbf{a} \in \xi^+(P_j^m)$;
- $M, \mathbf{a} \Vdash (B \wedge C)\ [\mathbf{x}]$ *iff* $M, \mathbf{a} \Vdash B\ [\mathbf{x}]$ *and* $M, \mathbf{a} \Vdash C\ [\mathbf{x}]$;
- $M, \mathbf{a} \Vdash (B \vee C)\ [\mathbf{x}]$ *iff* $M, \mathbf{a} \Vdash B$ *or* $M, \mathbf{a} \Vdash C\ [\mathbf{x}]$;

- $M, \mathbf{a} \Vdash (B \supset C)$ $[\mathbf{x}]$ *iff* $\forall \mathbf{b} \in D^n(\mathbf{a}R^n\mathbf{b}$ & $M, \mathbf{b} \Vdash B$ $[\mathbf{x}] \Rightarrow M, \mathbf{b} \Vdash C$ $[\mathbf{x}])$;
- $M, \mathbf{a} \Vdash \forall y B$ $[\mathbf{x}]$ *iff* $\forall \mathbf{c} \in D^{n+1}(\mathbf{a}R^n\pi_{\sigma_+^n}(\mathbf{c}) \Rightarrow M, \mathbf{c} \Vdash B$ $[\mathbf{x}y])$ *(for $y \notin \mathbf{x}$);*
- $M, \mathbf{a} \Vdash \exists y B$ $[\mathbf{x}]$ *iff* $\exists \mathbf{c} \in D^{n+1}(\pi_{\sigma_+^n}(\mathbf{c})R^n\mathbf{a}$ & $M, \mathbf{c} \Vdash B$ $[\mathbf{x}y])$;
- $M, \mathbf{a} \Vdash Qx_iB$ $[\mathbf{x}]$ *iff* $M, \pi_{\delta_i^n}\mathbf{a} \vDash Qx_iB$ $[\mathbf{x} - x_i]$
  for a quantifier Q.

Again we have uniform presentations of quantifier clauses in π-identical simplicial frames.

Sound and natural simplicial frames

Definition 5.20.7 *An N-modal formula A is* true *in a simplicial model (notation: $M \vDash A$) if $M, \mathbf{a} \vDash A$ $[\mathbf{x}]$ for any assignment $(\mathbf{x}, \mathbf{a})$ such that $FV(A) \subseteq r(\mathbf{x})$. The definition for the intuitionistic case is similar.*

Definition 5.20.8 *An N-modal formula A is* valid *in a simplicial frame $\mathbb{F}$ (notation $\mathbb{F} \vDash A$) if it is true in all models over $\mathbb{F}$. The definition for the intuitionistic case is similar.*

$\mathbf{ML}_-(\mathbb{F})$ denotes the set of all N-modal formulas (without equality) valid in $\mathbb{F}$; the notation $\mathbf{IL}_-(\mathbb{F})$ is similar.

Definition 5.20.9 *A formula is* strongly valid *in $\mathbb{F}$ if all its substitution instances (without equality) are valid.*

The set of all formulas strongly valid in $\mathbb{F}$ is denoted by $\mathbf{ML}(\mathbb{F})$ in the modal case and by $\mathbf{IL}(\mathbb{F})$ in the intuitionistic case.

A simplicial frame $\mathbb{F}$ is *logically sound* if $\mathbf{ML}(\mathbb{F})$ or $\mathbf{IL}(\mathbb{F})$ is a modal or a superintuitionistic logic respectively; we call such a frame *m-sound* or *i-sound* respectively. Logically sound simplicial frames generate 'maximal semantics'. But unlike the case of metaframes, we do not know an explicit description of these semantics, and we may conjecture that they are rather complicated (cf. Section 5.16, 5.19 for intuitionistic sound metaframes). Therefore in Volume 2 we shall describe more convenient classes of simplicial frames, where forcing satisfies natural properties of 'logical invariance' similar to those we had for metaframes in Sections 5.10, 5.11, 5.15. These classes generate 'almost maximal' semantics (and maybe even maximal, but this is yet unknown).

Definition 5.20.10 *A simplicial frame $\mathbb{F}$ is called* modally transformable *(or* m-transformable*) if the following condition*[34]

$$M, \mathbf{a} \vDash A\ [\mathbf{x}] \Leftrightarrow M, \pi_\sigma\mathbf{a} \vDash A\ [\pi_\sigma\mathbf{x}] \qquad (trfm)$$

holds for any injection $\sigma \in \Upsilon_{mn} (n \geq m)$, for any model $M = (\mathbb{F}, \xi)$, and for any assignment $(\mathbf{x}, \mathbf{a})$ such that $FV(A) \subseteq r(\pi_\sigma\mathbf{x})$.

[34]Cf. 5.10.6(∗).

The definition of $i^{(=)}$-*transformable* frames for the intuitionistic case is similar — they satisfy the condition[35]

$$M, \mathbf{a} \Vdash A\ [\mathbf{x}] \Leftrightarrow M, \pi_\sigma \mathbf{a} \vDash A\ [\pi_\sigma \mathbf{x}]. \qquad (trfi)$$

for intuitionistic models M and formulas A.

Recall that the case $n = 0$ (when A is a sentence and $\sigma = id_0 = \Lambda_0$) is trivial.

Loosely speaking, in transformable frames forcing $M, \mathbf{a} \vDash (\Vdash)A[\mathbf{x}]$ does not depend on the choice of a list of variables $\mathbf{x}$ containing $FV(A)$. More precisely, $M, \mathbf{a} \vDash A\ [\mathbf{x}]$ with $r(\mathbf{x}) \supset FV(A)$ can be reduced to $M, \mathbf{a} \vDash A[\mathbf{x}]$ with $r(\mathbf{x}') = FV(A)$, where $\mathbf{x}' = \pi_\sigma \mathbf{x}$, σ is an injection. Moreover, the choice of $\mathbf{x}'$ is inessential, since for any enumerations $\mathbf{x}, \mathbf{x}'$ of $FV(A)$, there exists a permutation σ such that $\mathbf{x}' = \pi_\sigma \mathbf{x}$. This observation allows us to use only the first case in the inductive clause for quantifiers, cf. Section 5.6.

Lemma 5.20.11 *Let $\mathbb{F}$ be an m-transformable simplicial frame, $M = (\mathbb{F}, \xi)$ a simplicial model. Then for any congruent formulas A, A' and for any $\mathbf{x}, \mathbf{a}$ such that $FV(A) = FV(A') \subseteq r(\mathbf{x})$:*

$$M, \mathbf{a} \vDash A\ [\mathbf{x}] \Leftrightarrow M, \mathbf{a} \vDash A\ [\mathbf{x}], \qquad (*)$$

and similarly for the intuitionistic case.

Definition 5.20.12 *A simplicial frame $\mathbb{F}$ is called* modally s-transformable *(or ms-*transformable*) if for any $\sigma \in \Sigma_{mn}$, for any distinct lists $\mathbf{x}$ of length n, m respectively, for any modal formula A such that $FV(A) \subseteq r(\mathbf{y})$, for any model M over $\mathbb{F}$, and for any $\mathbf{a} \in D^n$:*

$$M, \mathbf{a} \vDash ([\pi_\sigma \mathbf{x}/\mathbf{y}]\, A)\ [\mathbf{x}] \Leftrightarrow M, \pi_\sigma \mathbf{a} \vDash A\ [\mathbf{y}]. \qquad (trfms)$$

Similarly for the intuitionistic case, we define is-transformable *simplicial frames with the following condition:*

$$M, \mathbf{a} \Vdash ([\pi_\sigma \mathbf{x}/\mathbf{y}]\, A)\ [\mathbf{x}] \Leftrightarrow M, \pi_\sigma \mathbf{a} \Vdash A\ [\mathbf{y}]. \qquad (trfis)$$

This condition expresses the invariance of forcing under variable substitutions (cf. Lemma 5.11.7 for metaframes).

Recall that variable substitutions are defined up to congruence, and the congruent versions of A are all the results of applying the identity substitution [/] to A (cf. Section 2.3).

So in ms-transformable simplicial frames forcing is congruence invariant:

$$M, \mathbf{a} \vDash A\ [\mathbf{x}] \Leftrightarrow M, \mathbf{a} \Vdash A^1\ [\mathbf{x}] \qquad (*)$$

for any congruent formulas A, A^1 (apply $(trfsm)$ with $\mathbf{y} = \mathbf{x}$, $\sigma = id_n$), and similarly in the intuitionistic case. On the other hand, we can show that forcing is congruence invariant in every transformable simplicial frame as well (cf.

[35] Cf. 5.15.14($*i$).

Lemmas 5.10.11 and 5.15.16 for metaframes in the modal and the intuitionistic cases respectively).

Anyway, if the condition (*) holds, then the choice of any congruent version of $[\pi_\sigma \mathbf{x}/\mathbf{y}]A$ in $(trfsm)$ does not matter.

Definition 5.20.13 *A simplicial frame* $\mathbb{F}$ *is called* m-natural *(respectively,* i-natural*) if it is logically sound and transformable (respectively, s-transformable).*

The classes of all m-natural and i-natural simplicial frames generate the 'natural' semantics of simplicial frames $\mathcal{SF}_m$ *and* $\mathcal{SF}_{int}$.

In Volume 2 we will also describe other classes of simplicial frames (called 'modal' and 'intuitionistic') generating the same semantics $\mathcal{SF}_m$ and $\mathcal{SF}_{int}$; this description generalises the work done in Sections 5.11 and 5.15 above.

Theorems 5.12.13, 5.16.13, along with 5.10.6, 5.11.7, 5.15.14, 5.15.17, show that for every logically sound metaframe the corresponding simplicial frame is natural; thus $\mathcal{MF}_m \preceq \mathcal{SF}_m$ and $\mathcal{MF}_{int} \preceq \mathcal{SF}_{int}$.

On the other hand, we do not know if every logically sound simplicial frame is natural and even if the semantics of natural simplicial frames equals the 'maximal' semantics of all logically sound simplicial frames. It is also unknown if the semantics of simplicial frames (natural or 'maximal;') are stronger than the semantics of metaframes.

Equality in simplicial frames

To conclude this preliminary exposition, we briefly discuss problems with interpretation of equality in simplicial frames.

Recall that in metaframes the clause for equality has the following form:[36]

$$\mathbf{a} \vDash x_j = x_k \ [\mathbf{x}] \text{ iff } a_j = a_k, \tag{=}$$

where $\mathbf{a} \in D_u^n$, $|\mathbf{x}| = n$. This definition admits several reasonable generalisations for simplicial frames; in metaframes all these versions are equivalent.

First, for the formula $x_1 = x_2$, $\mathbf{x} = (x_1, x_2)$, and $\mathbf{a} \in D^2$, we can put

$$\mathbf{a} \vDash x_1 = x_2 \ [\mathbf{x}] \text{ iff } \pi_{\sigma_{12}}(\mathbf{a}) = \pi_{\sigma_{22}}(\mathbf{a}),$$

where $\sigma_{j2}(1) = j$ (so $(a_1, a_2) \cdot \sigma_{j2} = a_j$ in metaframes).

Second, we can define

$$\mathbf{a} \vDash x_1 = x_2 \ [\mathbf{x}] \text{ iff } \mathbf{a} = \pi_{\sigma^1}(\mathbf{b}) \text{ for some } \mathbf{b} \in D^1,$$

where $\sigma^1 \in \Sigma_{21}$ is the 'diagonal' map; $\sigma^1(j) = 1$ for $j \in I_2$ (so $\pi_{\sigma^1}(b) = (b, b)$ in metaframes). These two definitions are clearly equivalent for 'real' tuples, but they differ for 'abstract' tuples. For example, in general we cannot assert that $\pi_{\sigma_{j2}}(\pi_{\sigma^1}(\mathbf{b})) = \mathbf{b}$ for $\mathbf{b} \in D^1$ or that $\pi_{\sigma^1}(\pi_{\sigma_{12}}(\mathbf{a})) = \mathbf{a}$ if $\pi_{\sigma_{12}}(\mathbf{a}) = \pi_{\sigma_{22}}(\mathbf{a})$.

[36]We do not mention a model M as equality does not depend on the valuation.

There exist other more 'exotic' versions of forcing, e.g.

$$\mathbf{a} \vDash x_1 = x_2 \ [\mathbf{x}] \text{ iff } \pi_{\sigma_{12}}(\mathbf{a}) = \pi_{\sigma_{22}}(\mathbf{a}) \ \& \ \mathbf{a} = \pi_{\sigma^1}(\pi_{\sigma_{12}}(\mathbf{a})).$$

All these interpretations are equivalent to (=) (with $n = 2,\ i = 1,\ j = 2$) in metaframes, although they are non-equivalent in simplicial frames.

To express these definitions (i.e. generalisations of (=) to simplicial frames) for arbitrary n and $i, j \in I_n$ we use projections $\pi_{\sigma_{in}}$ 'extracting' components of an n-tuple ($\pi_{\sigma_{in}}(a_1, \ldots, a_n) = a_i$) and 'pairing jections' $\pi_{\lambda^n_{jk}}$, where $\lambda^n_k \in \Sigma_{2n}$, $\lambda^n_{jk}(1) = j$, $\lambda^n_{jk}(2) = k$, i.e. $\pi_{\lambda^n_{jk}}(a_1, \ldots, a_n) = a_j a_k$.

Thus we obtain the following interpretations of equality in simplicial frames

$$\mathbf{a} \vDash^{+} x_j = x_k \ [\mathbf{x}] \text{ iff } \pi_{\sigma_{jn}}(\mathbf{a}) = \pi_{\sigma_{kn}}(\mathbf{a}), \tag{$=^+$}$$

the 'componentwise' or the *upper* interpretation,

$$\mathbf{a} \vDash^{-} x_j = x_k \ [\mathbf{x}] \text{ iff } \exists \mathbf{b} \in D^1 \ \pi_{\lambda^n_{jk}}(\mathbf{a}) = \pi_{\sigma^1}(\mathbf{b}) \tag{$=^-$}$$

the 'diagonal' or the *lower* interpretation, and also the 'combined' interpretation

$$\mathbf{a} \vDash^{\pm} x_i = x_j \ [\mathbf{x}] \text{ iff } \pi_{\sigma_{jn}}(\mathbf{a}) = \pi_{\sigma_{kn}}(\mathbf{a}) \ \& \ \pi_{\lambda^n_{jk}}(\mathbf{a}) = \pi_{\sigma^1}(\pi_{\sigma_{in}}(\mathbf{a})). \tag{$=^\pm$}$$

We suppose that the upper interpretation is the most straightforward generalisation of the condition (=) in metaframes. However, some logical properties of the lower interpretation $\vDash^-$ seem better.

In Volume 2 we will consider a general approach to interpretation of equality in simplicial frames covering all these versions — various interpretations of equality $\mathbf{a} \vDash x_i = x_j \ [\mathbf{x}]$ depend on triples of the form $(\pi_{\sigma_{in}}(\mathbf{a}), \pi_{\sigma_{jn}}(\mathbf{a}), \pi_{\lambda^n_{jk}}(\mathbf{a}))$.

But such a general approach seems too complicated, so we simplify it by using 'pairs' $\pi_{\lambda^n_{jk}}(\mathbf{a}) \in D^2$. Thus, for example, the upper interpretation has a simplified version:

$$\mathbf{a} \vDash^{+1} x_j = x_k \ [\mathbf{x}] \text{ iff } \pi_{\lambda^n_{jk}} \in I^+,$$

where $I^+ = \{\mathbf{b} \in D^2 \mid \pi_{\sigma_{12}}(\mathbf{b}) = \pi_{\sigma_{22}}(\mathbf{b})\}$.

Although this condition is not equivalent to ($=^+$) in arbitrary simplicial frames, for 'natural' simplicial frames this reduction works quite well.

Part III

Completeness

Chapter 6

Kripke completeness for varying domains

Completeness proofs in this chapter are based on various canonical model constructions. They originate from canonical models in modal propositional logic and Henkin's completeness proof in classical predicate logic. The main idea is that worlds in canonical models are consistent (or even syntactically complete) theories, and individuals are identified with individual constants of these theories.

6.1 Canonical models for modal logics

Recall (Definition 2.7.7) that for a first-order modal theory Γ, D_Γ denotes the set of individual constants occurring in Γ, $\mathcal{L}^{(=)}(\Gamma)$ is the set of all D_Γ-sentences in the language of Γ.

Definition 6.1.1 *A (first-order) modal theory Γ is called L-*consistent *if $\nvdash_L \neg \bigwedge\limits_{i=1}^{k} A_i$ for any $A_1, \ldots, A_k \in \Gamma$, or equivalently, if $\Gamma \nvdash_L \bot$. A modal theory is called L-*complete *if it is maximal (by inclusion) among L-consistent theories in the same language.*

Lemma 6.1.2 *Let Γ be a modal theory.*

(1) If Γ is L-consistent and $\Gamma \vdash_L \neg A$, then $\Gamma \nvdash_L A$.

(2) If Γ is L-complete, $A \in \mathcal{L}^{(=)}(\Gamma)$, then

$$\Gamma \vdash_L A \text{ iff } A \in \Gamma;$$

in particular, $\overline{L} \subseteq \Gamma$ (where $\overline{L}$ denotes the set of all sentences in L).

(3) If Γ is L-complete, then for any $A \in \mathcal{L}^{(=)}(\Gamma)$

$$\neg A \in \Gamma \textit{ iff } A \notin \Gamma.$$

(4) Every L-consistent Γ satisfying the equivalence (3), is L-complete.

Proof
(1) If $\Gamma \vdash_L \neg A$ and $\Gamma \vdash_L A$, then $\Gamma \vdash_L \bot$, by MP.

(2) 'If' is obvious. To show 'only if', suppose $A \notin \Gamma$. Then $\Gamma \subset \Gamma \cup \{A\}$, hence $\Gamma \cup \{A\} \vdash_L \bot$, and thus $\Gamma \vdash_L A \supset \bot (= \neg A)$ by 2.8.1. Hence $\Gamma \nvdash_L A$ by (1).

(3) 'Only if' readily follows from (1) and (2). To show 'if', suppose the contrary — that both A, $\neg A$ are not in Γ. Then as in the proof of (2), we obtain $\Gamma \vdash_L \neg A$ and similarly, $\Gamma \vdash_L \neg\neg A$, which implies the L-inconsistency of Γ.

(4) Suppose the equivalence in (3) holds for any $A \in \mathcal{L}^{(=)}(\Gamma)$, but Γ is L-incomplete. Then $\Gamma \subset \Gamma'$ for some Γ' in the same language, so there exists $A \in (\Gamma' - \Gamma)$. By (3), we have $\neg A \in \Gamma \subseteq \Gamma'$, thus Γ' is L-inconsistent. ■

Similarly to the propositional case, we have the following

Lemma 6.1.3 *Let Γ be an L-complete theory. Then for any $A, B \in \mathcal{L}^{(=)}(\Gamma)$:*

(i) $(A \wedge B) \in \Gamma$ iff ($A \in \Gamma$ and $B \in \Gamma$);

(ii) $(A \vee B) \in \Gamma$ iff ($A \in \Gamma$ or $B \in \Gamma$);

(iii) $(A \supset B) \in \Gamma$ iff ($A \notin \Gamma$ or $B \in \Gamma$).

Proof
(i) (If.) $A \wedge B \supset A$ is an instance of a tautology, so if $(A \wedge B) \in \Gamma$, we have $\Gamma \vdash_L A$ by MP. Hence $A \in \Gamma$ by 6.1.2 (2). Similarly $B \in \Gamma$.

(Only if.) If $A, B \in \Gamma$, we apply the instance of a tautology $A \supset (B \supset A \wedge B)$, MP and 6.1.2 (2).

(ii) (If.) Similarly to (i), use $A \supset A \vee B$ or $B \supset A \vee B$.

(Only if.) Suppose $A, B \notin \Gamma$. Then $\neg A, \neg B \in \Gamma$ by 6.1.2(3). Since $\neg A \supset (\neg B \supset \neg(A \vee B))$ is an instance of a tautology, we obtain $\neg A, \neg B \vdash_L \neg(A \vee B)$, hence $\neg(A \vee B) \in \Gamma$, and thus $(A \vee B) \notin \Gamma$ by 6.1.2 (3).

(iii) (Only if.) Supposing $(A \supset B), A \in \Gamma$, we obtain $B \in \Gamma$ by MP and 6.1.2 (2).

(If.) If $A \notin \Gamma$, then $\neg A \in \Gamma$ by 6.1.2 (3). Since $\neg A \supset (A \supset B)$ is an instance of a tautology, by MP and 6.1.2 (2), it follows that $(A \supset B) \in \Gamma$.

If $B \in \Gamma$, we use $B \supset (A \supset B)$ in the same way. ■

Recall that $\overline{L}$ denotes the set of all sentences in a logic L,
$\Box^*\Gamma := \{\Box_\alpha A \mid A \in \Gamma,\ \alpha \in I_N^\infty\}$ (if N is fixed).

Lemma 6.1.4 *Let L, L_1 be N-m.p.l.(=), Γ an L-complete theory. Then*

(1) Γ *is* L_1*-complete iff* $\overline{L_1} \subseteq \Gamma$.

(2) Suppose $L_1 = L + \Theta$ *for a set of sentences* Θ*. Then* Γ *an* L_1*-complete iff* $\Box^*\overline{\mathrm{Sub}}(\Theta) \subseteq \Gamma$.

Proof
(1) 'Only if' follows from Lemma 6.1.2 (2).

To show 'if', assume $\overline{L_1} \subseteq \Gamma$. By Lemma 6.1.2 (3), (4), it suffices to check that Γ is L_1-consistent. Suppose the contrary; then for a finite $X \subseteq \Gamma$, $\vdash_{L_1} A = \neg(\bigwedge X)$. Put $A := \neg(\bigwedge X)$. Then by Definition 2.7.1, A has a generator $A_1 \in L_1$. Hence $\overline{\forall} A_1 \in \overline{L_1} \subseteq \Gamma$. But then $\Gamma \vdash_L \overline{\forall} A_1$, whence $\Gamma \vdash_L A_1$, and therefore $\Gamma \vdash_L A$ implying the L-inconsistency of Γ.

(2) By (1), Γ is L_1-complete iff $\overline{L_1} \subseteq \Gamma$. Since $\Box^*\overline{\mathrm{Sub}}(\Theta) \subseteq \overline{L_1}$, the 'only if' part of (2) follows readily.

For the converse, we assume $\Box^*\overline{\mathrm{Sub}}(\Theta) \subseteq \Gamma$ and show that $\overline{L_1} \subseteq \Gamma$. In fact, by the deduction theorem 2.8.4, every $A \in \overline{L_1} \subseteq L + \Theta$ is L-provable in $\Box^*\overline{Sub}(\Theta)$, hence $\Gamma \vdash_L A$, therefore $A \in \Gamma$, by 6.1.2. ∎

Lemma 6.1.5 (Lindenbaum lemma) *Every* L*-consistent theory can be extended to an* L*-complete theory in the same language.*

Proof Similar to the propositional case. If $\mathcal{L}^{(=)}(\Gamma)$ is countable, we consider its enumeration $A_0, A_1, \ldots$ and construct a sequence of L-consistent theories $\Gamma = \Gamma_0 \subseteq \Gamma_1, \ldots$ such that $\Gamma_{n+1} = \Gamma_n \cup \{A_n\}$ or $\Gamma_{n+1} = \Gamma_n \cup \{\neg A_n\}$. This is possible, since one of $\Gamma_n \cup \{A_n\}$, $\Gamma_n \cup \{\neg A_n\}$ is L-consistent (otherwise $\Gamma_n \vdash_L \neg A_n, \neg\neg A_n$ by Lemma 2.8.1, and thus Γ_n is L-inconsistent). Eventually the union $\Gamma_\omega := \bigcup\limits_n \Gamma_n$ is L-consistent and L-complete by 6.1.2(4).

If the language is uncountable, we apply transfinite induction, but actually we do not need this case in the sequel. ∎

Definition 6.1.6 *An* L*-complete theory* Γ *is called* L-Henkin *if for any* D_Γ*-sentence* $\exists x A(x)$ *there exists a constant* $c \in D_\Gamma$ *such that*

$$(\exists x A(x) \supset A(c)) \in \Gamma.$$

We say that Γ *is* (L, S)-Henkin *if* Γ *is* L*-Henkin with* $D_\Gamma = S$.

Due to Lemma 6.1.3 (iii), every L-Henkin theory Γ has the following *existence property*:

(EP) if $\exists x A(x) \in \Gamma$, then for some $c \in D_\Gamma$, $A(c) \in \Gamma$.

Moreover, we have

Lemma 6.1.7 *Every* L*-Henkin theory* Γ *satisfies the condition*

(EP′) $\exists x A(x) \in \Gamma$ *iff for some* $c \in D_\Gamma$, $A(c) \in \Gamma$.

Proof Since $L \vdash A(y) \supset \exists x A(x)$ for a new variable y (Lemma 2.6.15(ii)), we have $\vdash_L A(c) \supset \exists x A(x)$, and so $(A(c) \supset \exists x A(x)) \in \Gamma$. Thus Lemma 6.1.3 implies the converse of (EP), therefore (EP′) holds. ■

Exercise 6.1.8 *Show that every L-complete theory with the existence property is L-Henkin.*

Lemma 6.1.9 *Let Γ be an L-consistent theory, $D_\Gamma \subset S$, $|S| = |S - D_\Gamma| = \aleph_0$. Then there exists an (L, S)-Henkin theory $\Gamma' \supseteq \Gamma$.*

Proof The set of all S-sentences is countable, so let us enumerate all S-sentences of the form $\exists x A(x)$: $\exists x_1 A_1(x_1)$, $\exists x_2 A_2(x_2), \ldots$ Choose distinct constants c_k from $(S - D_\Gamma)$ such that c_k does not occur in $\exists x_i A_i(x_i)$ for $i \leq k$, and put

$$\Gamma_\omega := \Gamma \cup \{\exists x_k A_k(x_k) \supset A_k(c_k) \mid k \in \omega\}.$$

Let us show that Γ_ω is L-consistent. Suppose the contrary. Then for some k

$$\Gamma \vdash_L \neg \bigwedge_{i \leq k} (\exists x_k A_k(x_k) \supset A_k(c_k)).$$

Take the minimal k with this property. Let

$$B := \bigwedge_{i<k} (\exists x_i A_i(x_i) \supset A_i(c_i)).$$

Then

$$\Gamma, B, \exists y A_k(y) \supset A_k(c) \vdash_L \bot$$

(where $y = x_k$, $c = c_k$). Hence by Lemma 2.8.1,

$$\Gamma, B \vdash_L \neg(\exists y A_k(y) \supset A_k(c)).$$

Then from the propositional tautology $\neg(p \supset q) \supset p \wedge \neg q$ we obtain

(1) $\Gamma, B \vdash_L \exists y A_k(y)$;

(2) $\Gamma, B \vdash_L \neg A_k(c)$.

Since c does not occur in $\Gamma \cup \{B\}$, by Lemma 2.7.11, it follows that

(3) $\Gamma, B \vdash_L \forall y \neg A_k(y)$,

and thus by 2.6.15(xiii),

(4) $\Gamma, B \vdash_L \neg \exists y A_k(y)$.

From (1) and (4) it follows that $\Gamma \vdash_L \neg B$, contrary to the choice of k. Therefore Γ_ω is L-consistent, and by Lemma 6.1.5, we can extend it to Γ'. ■

Lemma 6.1.10 *Let A be a modal sentence, $|S| = \aleph_0$.*
Then $L \vdash A$ iff ($A \in \Gamma$ for any (L, S)-Henkin theory Γ).

Proof 'Only if' follows from Lemma 6.1.3. To prove 'if', suppose $L \nvdash A$. Then $\{\neg A\}$ is an L-consistent subset of $MF_N^{(=)}(\varnothing)$, and so by Lemma 6.1.9, there exists an (L, S)-Henkin theory $\Gamma \supseteq \{\neg A\}$. ∎

For $S \subseteq S'$ and $\Gamma' \subseteq MF_N^{(=)}(S')$, let $\Gamma'|S := \Gamma \cap MF_N^{(=)}(S)$ (the *restriction* of Γ' to the domain S). Obviously, $\Gamma'|S$ is consistent (complete) if Γ' is consistent (respectively, complete), but for a Henkin (L, S')-theory Γ', $\Gamma'|S$ may be not a Henkin (L, S)-theory.

Henceforth we fix a denumerable set S^*, the 'universal set of constants'. We call a set $S \subset S^*$ *small* if $(S^* - S)$ is infinite.

Definition 6.1.11 *An L-*place *is a Henkin L-theory with a small set of constants.*

The set of all L-places is denoted by VP_L.

Lemma 6.1.12 *Let u be a world in a Kripke model M over an N-modal predicate Kripke frame $\mathbf{F} = (F, D)$ and assume that $M \vDash L$. Then the set of D_u-sentences that are true at u*

$$\Gamma_u := \{A \in MF_N^{(=)}(D_u) \mid M, u \vDash A\}$$

is a Henkin L-theory.

Proof This is an easy consequence from the definitions. Let us only give an argument for L-consistency. If $\vdash_L \neg \bigwedge_{i=1}^{k} A_i$, then $B := \bigwedge_{i=1}^{k} A_i$ can be presented as $[\mathbf{c}/\mathbf{x}]\, B_0$ for $B_0 \in L$, $\mathbf{c} \in D_u^\infty$, and then $\forall \mathbf{x} B_0 \in L$. Hence by assumption $M, u \vDash \forall \mathbf{x} B_0$, which implies $M, u \vDash [\mathbf{c}/\mathbf{x}]\, B_0 = \neg \bigwedge_{i=1}^{k} A_i$ by Lemma 3.2.18. Thus one of the A_i must be false at M, u. ∎

Definition 6.1.13 *Let M be a Kripke model for a modal logic L, in which every individual domain is a small subset of S^*. The map from (the worlds of) M to VP_L sending u to Γ_u is called* canonical *and denoted by $\nu_{M,L}$ (or by ν_M, or by ν if there is no confusion).*

Lemma 6.1.12 extends to Kripke sheaves, Kripke bundles, or other kinds of models for L. Moreover, VP_L is the set of all 'places' Γ_u of all possible Kripke models (and of metaframe models) for L with small domains. To show this, we shall now construct a certain predicate Kripke frame (the canonical frame), whose worlds are L-places and in which every 'world' Γ has the individual domain D_Γ, while Γ is the set of all D_Γ-sentences true at this world. Since all the domains are small, there are infinitely many spare constants at every world.

For an N-modal theory Γ and $i = 1, \ldots, N$, put

$$\begin{aligned} \Diamond_i\Gamma &:= \{\Diamond_i A \mid A \in \Gamma\}, \\ \Box_i^-\Gamma &:= \{B \mid \Box_i B \in \Gamma\}. \end{aligned}$$

Definition 6.1.14 *For L-places Γ, Γ' we define* canonical accessibility relations*:*
$\Gamma\, R_{Li}\, \Gamma' := \Box_i^-\Gamma \subseteq \Gamma'$.

Lemma 6.1.15 $\Gamma\, R_{Li}\, \Gamma'$ *iff* $D_\Gamma \subseteq D_{\Gamma'}$ & $\Diamond_i\Gamma' \cap \mathcal{L}(\Gamma) \subseteq \Gamma$.

Proof (Only if.) Assume $\Gamma R_{Li}\Gamma'$. For any $c \in D_\Gamma$ we have $\vdash_L \Box(P(c) \supset P(c))$, so $\Box(P(c) \supset P(c)) \in \Gamma$ by 6.1.2(2), and thus $(P(c) \supset P(c)) \in \Box_i^-\Gamma \subseteq \Gamma'$. Therefore $c \in D_{\Gamma'}$, which shows $D_\Gamma \subseteq D_{\Gamma'}$.

(If.) Assume $D_\Gamma \subseteq D_{\Gamma'}$, $\Diamond_i\Gamma' \cap \mathcal{L}(\Gamma) \subseteq \Gamma$. Next, assume $\Box_i B \in \Gamma$. Then $B \in \Gamma'$, since otherwise $\neg B \in \Gamma'$, $\Diamond_i\neg B \in \Diamond_i\Gamma' \cap \mathcal{L}(\Gamma) \subseteq \Gamma$, and thus Γ is inconsistent. Therefore $\Box_i^-\Gamma \subseteq \Gamma'$. ∎

The crucial property of canonical relations is the following

Lemma 6.1.16 *For any L-place Γ and for any D_Γ-sentence A such that $\Diamond_i A \in \Gamma$, there exists an L-place Γ' such that $\Gamma R_{Li}\Gamma'$ and $A \in \Gamma'$.*

Proof Let us show that the theory $\Gamma_0 := \Box_i^-\Gamma \cup \{A\}$ is L-consistent if $\Diamond_i A \in \Gamma$. Suppose $\vdash_L \bigwedge\limits_j B_j \supset \neg A$ for some formulas $B_j \in \Box_i^-\Gamma$. Then $\vdash_L \Box_i(\bigwedge\limits_j B_j) \supset \Box_i\neg A$, hence $L \vdash \bigwedge\limits_j \Box_i B_j \supset \neg\Diamond_i A$, $\Gamma \vdash_L \neg\Diamond_i A$, and thus by Lemma 6.1.2 (2), $\neg\Diamond_i A \in \Gamma$. This is a contradiction.

Now by Lemma 6.1.9, we can construct an L-place Γ' such that $\Gamma_0 \subseteq \Gamma'$ (remember that D_Γ is small, so there exists a small S such that $|S - D_\Gamma| = \aleph_0$). By Definition 6.1.14, $\Gamma R_{Li}\Gamma'$. ∎

Lemma 6.1.17 *For any $\Gamma \in VP_L$ and $A \in \mathcal{L}(\Gamma)$ we have*

(1) $\Diamond_i A \in \Gamma$ *iff* ($A \in \Gamma'$ *for some* $\Gamma' \in R_{Li}(\Gamma)$)*;*

(2) $\Box_i A \in \Gamma$ *iff* ($A \in \Gamma'$ *for each* $\Gamma' \in R_{Li}(\Gamma)$)*.*

Proof
(1) 'Only if' follows from 6.1.16. The other way round, if $\Diamond_i A = \neg\Box_i\neg A \notin \Gamma$, then $\Box_i\neg A \in \Gamma$, hence $\neg A \in \Gamma'$ (i.e. $A \notin \Gamma'$) for any $\Gamma' \in R_{Li}(\Gamma)$.

(2) 'Only if' follows from Definition 6.1.14. To show 'if', assume $\Box_i A \notin \Gamma$, then $\neg\Box_i A \in \Gamma$, and since $\vdash_L \neg\Box_i A \supset \Diamond_i\neg A$, this implies $\Diamond_i\neg A \in \Gamma$. Hence by (1), $\neg A \in \Gamma'$ for some $\Gamma' \in R_{Li}(\Gamma)$. ∎

Definition 6.1.18 *The* canonical Kripke frame with varying domains *of an N-modal logic L without equality is $VF_L := (VP_L, R_{L1}, \ldots, R_{LN}, D_L)$, where $(D_L)_\Gamma := D_\Gamma$.*

Definition 6.1.19 *For an N-modal logic L with equality we define the* canonical Kripke frame with equality *as*

$$VF_L^= := (VP_L, R_{L1}, \dots, R_{LN}, D_L, \asymp_L),$$

where

$$c\,(\asymp_L)_\Gamma\, d := (c = d) \in \Gamma$$

for $c, d \in D_\Gamma$.

Due to the standard properties of equality 2.6.16(i), (ii), $(\asymp_L)_\Gamma$ is an equivalence relation on D_Γ (sometimes we denote it just by $\asymp_\Gamma$). We also have

$$\Gamma R_{Li} \Gamma' \;\&\; (c = d) \in \Gamma \Rightarrow (c = d) \in \Gamma',$$

since $L \vdash (x = y) \supset \Box_i(x = y)$; thus $\asymp_L$ satisfies the R_{Li}-stability condition from Definition 3.5.1.

Definition 6.1.20 *The* canonical model *of an m.p.l.(=) L is*

$$VM_L^{(=)} := (VF_L^{(=)}, \xi_L),$$

where

$$(\xi_L)_\Gamma(P_k^m) := \{\mathbf{c} \in (D_\Gamma)^m \mid P_k^m(\mathbf{c}) \in \Gamma\}.$$

Note that $VM_L^{(=)}$ is well-defined, since $P_k^m(c_1, \dots, c_m), (c_1 = d_1), \dots, (c_m = d_m) \in \Gamma$ implies $P_k^m(d_1, \dots, d_m) \in \Gamma$, thanks to 2.6.16(v).

The main property of canonical models is the following

Theorem 6.1.21 (Canonical model theorem)

$$VM_L^{(=)}, \Gamma \vDash A \text{ iff } A \in \Gamma$$

for any L-place $\Gamma \in VP_L$ and $A \in \mathcal{L}^{(=)}(\Gamma)$.

Proof By induction on the length of A.

The base readily follows from the definitions.

The inductive step for classical connectives and quantifiers follows from 6.1.3 (i)–(iii) and 6.1.7 (EP′), properties (i)–(iv), (vi) of Henkin theories. The case $A = \Box_i B$ follows from Lemma 6.1.17 (1). ∎

Corollary 6.1.22 *For any modal formula A, $VM_L^{(=)} \vDash A$ iff $L \vdash A$.*

Proof If A is a modal sentence, then by Corollary 6.1.10, $A \in L$ implies $A \in \Gamma$ for any L-place Γ; hence $VM_L^{(=)}, \Gamma \vDash A$ by Theorem 6.1.21.

The other way round, if $A \notin L$, then by 6.1.10, $A \notin \Gamma$ for some Henkin (L, S)-theory Γ, where S is infinite and small. So Γ is an L-place, and thus $VM_L^{(=)}, \Gamma \nvDash A$ by Theorem 6.1.21.

The claim for an arbitrary formula A is reduced to the claim for $\overline{\forall} A$, since $VM_L^{(=)} \vDash A$ iff $VM_L^{(=)} \vDash \overline{\forall} A$, and $A \in L$ iff $\overline{\forall} A \in L$. ∎

Definition 6.1.23 *An m.p.l.(=) L is called V-*canonical *if* $VF_L^{(=)} \models L$.

This property is sufficient for completeness:

Corollary 6.1.24 *Every V-canonical m.p.l. is strongly Kripke complete. Every V-canonical m.p.l.= is strongly KFE (or Kripke sheaf) complete.*

Proof In fact, let L be a V-canonical m.p.l. (=), Γ_0 an L-consistent theory; by Lemma 6.1.9, there exists an L-place $\Gamma \supseteq \Gamma_0$. By 6.1.21, $VM_L^{(=)}, \Gamma \models \Gamma$, so Γ_0 is satisfied the L-frame $VF_L^{(=)}$. ∎

Now let us consider canonical models for two N-modal logics $L_1 \subseteq L_2$. Every L_2-consistent theory is clearly L_1-consistent, thus $VP_{L_1} \subseteq VP_{L_2}$. Moreover, we obtain

Lemma 6.1.25 *Let $L_1 \subseteq L_2$ be N-m.p.l.(=). Then*

(1) $VP_{L_2} = \{\Gamma \in VP_{L_1} \mid VM_{L_1}, \Gamma \models \overline{L_2}\}$.

(2) VM_{L_2} *is a generated submodel of* VM_{L_1}.

(3) If L_1 is canonical, then $VF_{L_2} \models L_1$.

Proof (1) Let Γ be an L_1-place. Then $VM_{L_1}, \Gamma \models \overline{L_2}$ iff $\overline{L_2} \subseteq \Gamma$ (by the canonical model theorem) iff Γ is L_2-complete (by 6.1.4(1)). This is equivalent to $\Gamma \in VP_{L_2}$, since $\Gamma \in VP_{L_1}$ already enjoys the Henkin property and has a small set of constants.

(2) In fact, $\Gamma \models \overline{L_2}$ implies $\Gamma \models \Box_i A$ for any $A \in \overline{L_2}$; thus $\Delta \models \overline{L_2}$ for any $\Delta \in R_{L_1 i}(\Gamma)$. So due to (1), VP_{L_2} is stable in VF_{L_1}. VM_{L_2} is a submodel of VM_{L_1}, since $VM_{L_1}, \Gamma \models P(\mathbf{a})$ iff $P(\mathbf{a}) \in \Gamma$ iff $VM_{L_2}, \Gamma \models P(\mathbf{a})$.

(3) Follows from (2) and the generation lemma 3.3.18. ∎

Proposition 6.1.26 *If L is a V-canonical N-m.p.l.(=), Γ is a set of N-modal pure equality formulas, then $L_1 = L + \Gamma$ is also V-canonical.*

Proof By Lemma 6.1.25, VF_{L_1} is a generated subframe of VF_L, so $L \subseteq \mathbf{ML}^{(=)}(VF_L) \subseteq \mathbf{ML}^{(=)}(VF_{L_1})$. On the other hand, $VM_{L_1} \models \Gamma$ by the canonical model theorem, thus since all formulas in Γ are constant, $VF_{L_1} \models \Gamma$. Therefore $VF_{L_1} \models L_1$. ∎

Corollary 6.1.22 shows that $VM_L^{(=)}$ is an exact model for L. The next proposition is a further refinement of this fact.

Proposition 6.1.27

(1) Every m.p.l. (=) L has a countable exact Kripke model (respectively, KFE-model).

(2) Moreover, if $L \supseteq L_0$ for a Δ-elementary canonical m.p.l.(=) L_0, such a model exists over an L_0-frame.

(3) Moreover, every L-consistent theory is satisfiable in some model described in (2).

Proof We can apply Corollary 3.12.11 to $L = L_0$, $M = VM_L^{(=)}$, $F = VF_L^{(=)}$ and u_0 such that $M, u_0 \models \Gamma$ (for a given L-consistent Γ).

Note that $F \models L_0$ by 6.1.33. So by 3.12.11 there exists a countable reliable $M_0 \subseteq M$ over an L_0-frame such that $\mathbf{MT}(M_0) = \mathbf{MT}(M) = L$ and $u_0 \in M_0$. Then $M_0, u_0 \models \Gamma$ by reliability. ∎

Obviously, the logics $\mathbf{QK}_N$ and $\mathbf{QK}_N^=$ are V-canonical. Let us also show V-canonicity for some other simple modal predicate logics.

For an N-modal theory Γ, $\alpha \in I_N^\infty$ let

$$\Box_\alpha^- \Gamma := \{B \mid \Box_\alpha B \in \Gamma\},$$

$$\Diamond_\alpha \Gamma := \{\Diamond_\alpha A \mid A \in \Gamma\}.$$

Also let

$$R_{Li_1\dots i_k} := R_{Li_1} \circ \dots \circ R_{Li_k},$$

and let $R_{L\lambda}$ be the equality relation Id_{VP_L}, cf. Proposition 1.11.5.

Lemma 6.1.28 *Let $\alpha \in I_N^\infty$, $\Gamma R_{L\alpha} \Delta$. Then*

(1) $\Box_\alpha^- \Gamma \subseteq \Delta$,

(2) $\Diamond_\alpha \Delta \cap \mathcal{L}(\Gamma) \subseteq \Gamma$.

Proof (1) By induction on the length of α. If $\alpha = \lambda$, everything is obvious. If (i) holds for β, $\alpha = k\beta$ and $\Gamma R_{L\alpha}\Delta$, then there exists Γ' such that $\Gamma R_{Lk} \Gamma'$, $\Gamma' R_{L\beta} \Delta$. By Definition 6.1.14 and the induction hypothesis we obtain

$$\Box_\alpha^- \Gamma = \{B \mid \Box_k \Box_\beta B \in \Gamma\} = \{B \mid \Box_\beta B \in \Box_k^- \Gamma\} \subseteq \Box_\beta^- \Gamma' \subseteq \Delta.$$

(2) Ad absurdum. Suppose $A \in \Delta$, $\Diamond_\alpha A \in (-\Gamma)$. Then $\neg\Diamond_\alpha A \in \Gamma$, and thus $\Box_\alpha \neg A \in \Gamma$. By (1), this implies $\neg A \in \Delta$, which makes Δ inconsistent. ∎

Theorem 6.1.29 *Let $\mathbf{\Lambda}$ be a propositional one-way PTC-logic. Then the logics $\mathbf{Q\Lambda}$, $\mathbf{Q\Lambda}^=$ are V-canonical.*

Proof Every constant axiom is valid in the canonical frame, since it is true in the canonical model.

If $A = \Box_k p \supset \Box_\beta p \in L = \mathbf{Q\Lambda}^{(=)}$, then $R_{L\beta} \subseteq R_{Lk}$. In fact, suppose $\Gamma R_{L\beta} \Delta$, $\Box_k B \in \Gamma$. Since $\vdash_L \Box_k B \supset \Box_\beta B$, we have $\Box_\beta B \in \Gamma$. By Lemma 6.1.28, it follows that $B \in \Delta$.

By Proposition 1.11.5 and Lemma 3.2.34, we obtain $VF_L \models A$. ∎

Example 6.1.30 The following counterexample shows that Theorem 6.1.29 does not extend to arbitrary PTC-logics, because they may be $\mathcal{KE}$-incomplete. Consider the logic **K5** $:=$ **K** $+ \Diamond\Box p \supset \Box p$; its Kripke frames are characterized by the 'Euclidian' condition

$$(\varepsilon) \qquad \forall u \forall v \forall w (uRv \wedge uRw \supset vRw).$$

Then $\mathbf{QK5}^=$ is $\mathcal{KE}$-incomplete. To see this, first note that

$$\mathbf{QK5}^= \vDash_{\mathcal{KE}} \Diamond AU_1 \supset \Box AU_1.$$

In fact, consider a Kripke sheaf model M over a Euclidian frame, and suppose $M, u \vDash \Diamond AU_1$, i.e. for some $v \in R(u)$ $M, v \vDash AU_1$, which means that $|D_v| = 1$. Then $|D_w| = 1$ for any $w \in R(u)$. In fact, by (ε) we have vRw & wRv, so there are transition functions ρ_{vw}, ρ_{wv}; and by the properties of Kripke sheaves it follows that they are bijections. So $|D_w| = 1$.

However $\mathbf{QK5}^= \nvdash \Diamond AU_1 \supset \Box AU_1$. To see this, consider a Kripke bundle $\mathbb{F}$ over the base $F = (W, R)$, where

$$W = \{u, v_1, v_2\}, \ R = W \times (W - \{u\}),$$

$$D_u = \{a\}, \ D_{v_1} = \{b\}, \ D_{v_2} = \{c_1, c_2\}, \ \rho = D^+ \times (D^+ - \{a\}).$$

It is clear that F is Euclidean. Every level $F_n = (D^n, R^n)$ is also Euclidean, because on n-tuples from D_{v_1} or D_{v_2} the relation R^n is equivalent to $\mathbf{d}\ sub\ \mathbf{c}$ & $\mathbf{c}\ sub\ \mathbf{d}$ (which is an equivalence relation) and $a^n R^n \mathbf{d}$ for any $\mathbf{d} \neq a^n$. Therefore $\mathbb{F} \vDash \mathbf{QK5}^=$ by 5.3.7 and soundness of Kripke bundle semantics. On the other hand, $\mathbb{F} \nvDash \Diamond AU_1 \supset \Box AU_1$, since $v_1 \vDash AU_1$ and $v_2 \nvDash AU_1$. Therefore $\mathbf{QK5}^= \nvdash \Diamond AU_1 \supset \Box AU_1$ by soundness.

Actually many simple and natural predicate logics are not V-canonical. So they are either Kripke-incomplete, or the completeness proof requires more work. One of the options may be to take an appropriate submodel of VM_L, for which an analogue of Lemma 6.1.21 still holds.

This leads us to the following definition.

Definition 6.1.31 *A relation $R_i \subseteq R_{Li}$ on VP_L is called* (i-)selective *if for any L-place Γ and for any $A \in \mathcal{L}(\Gamma)$ such that $\Diamond_i A \in \Gamma$, there exists an L-place Γ' such that $\Gamma R_i \Gamma'$ and $A \in \Gamma'$.*

So we obtain an analogue of Lemma 6.1.17:

Lemma 6.1.32 *For any selective R_i, for any $\Gamma \in VP_L$ and $A \in \mathcal{L}(\Gamma)$ we have*

(1) $\Diamond_i A \in \Gamma$ iff ($A \in \Gamma'$ for some $\Gamma' \in R_i(\Gamma)$);

(2) $\Box_i A \in \Gamma$ iff ($A \in \Gamma'$ for each $\Gamma' \in R_i(\Gamma)$).

Proof The same as in 6.1.17, but 'only if' in (1) now follows from 6.1.31. ■

Lemma 6.1.16 means that the relation R_{Li} is i-selective, and thus it is the greatest i-selective relation on VP_L.

Now similarly to the canonical frame, we can introduce *quasi-canonical Kripke frames* $(VP_L, R_1, \dots, R_N, D_L\ [, \asymp_L])$ with arbitrary i-selective relations R_i and the corresponding *quasi-canonical models* (with the valuation ξ_L). Then we obtain the following analogues of Lemma 6.1.21 and Corollary 6.1.22:

Lemma 6.1.33 *Let M be a quasi-canonical Kripke model for an m.p.l.(=) L. Then*

(1) for any L-place Γ and $A \in \mathcal{L}(\Gamma)$

$$M, \Gamma \vDash A \text{ iff } A \in \Gamma;$$

(2) for any formula A

$$M \vDash A \text{ iff } L \vdash A.$$

Another option is to use subsets of VP_L still satisfying Corollary 6.1.10 (and thus, Corollary 6.1.22). Various modifications will be considered in this chapter later on.

6.2 Canonical models for superintuitionistic logics

Now let us turn to the intuitionistic case. We would again like to represent possible worlds in Kripke (or other) models by sets of formulas. But now we should take false formulas into account as well, because in intuitionistic logic the falsity of A is not equivalent to the truth of $\neg A$. For this purpose we need double theories introduced in 2.7.12.

Definition 6.2.1 *If L is a superintuitionistic logic (with or without equality), a theory (Γ, Δ) is called L-*consistent *if $(\Gamma, \Delta) \nvdash \bot$, i.e.*

$$\nvdash_L \bigwedge \Gamma_1 \supset \bigvee \Delta_1$$

for any finite $\Gamma_1 \subseteq \Gamma,\ \Delta_1 \subseteq \Delta$.

It is obvious that $\Gamma \cap \Delta = \varnothing$ whenever (Γ, Δ) is L-consistent.

Definition 6.2.2 *An L-consistent theory (Γ, Δ) is called L-*complete *if $\Gamma \cup \Delta = \mathcal{L}(\Gamma, \Delta)$*

So in this case $\Delta = \mathcal{L}(\Gamma, \Delta) - \Gamma$, which is abbreviated as $(-\Gamma)$. Also note that for any $c \in \mathcal{L}(\Gamma, -\Gamma)$ we have $(P(c) \supset P(c)) \in \Gamma$, since otherwise $(P(c) \supset P(c)) \in (-\Gamma)$, while the theory $(\varnothing, \{P(c) \supset P(c)\})$ is inconsistent. So it follows that $D_{-\Gamma} \subseteq D_\Gamma$. A similar argument using $P(c) \wedge \neg P(c)$ shows that $D_\Gamma \subseteq D_{-\Gamma}$. So we can write D_Γ rather than $D_{(\Gamma, -\Gamma)}$.

Moreover, we shall often say that Γ (rather than $(\Gamma, -\Gamma)$) is an L-complete intuitionistic theory.

Definition 6.2.3 *An L-complete intuitionistic theory* Γ *is called* $L\exists$-complete *if it satisfies the existence property:*

(EP) if $\exists x A(x) \in \Gamma$, *then* $A(c) \in \Gamma$ *for some* $c \in D_\Gamma$.

More specifically, an $L\exists$-*complete* Γ *is called* $(L\exists, S)$-complete *if* $D_\Gamma = S$.

Definition 6.2.4 *An* $L\exists$-*complete (respectively, an* $(L\exists, S)$-*complete) theory* Γ *is called* $L\exists\forall$-complete *(respectively, an* $(L\exists\forall, S)$-complete*) if it satisfies the following* coexistence property*:*

(Av) if $\forall x A(x) \in (-\Gamma)$ *and* $A(d) \in \Gamma$ *for all* $d \in D_\Gamma$, *then* $\forall x(A(x) \vee C) \in \Gamma$ *for some formula* $C \in (-\Gamma)$.

The intended meaning of this property (after we prove the canonical model theorem) is the following. If at a world Γ of the canonical model $\forall x A(x)$ is false, but $A(d)$ is true for every individual d, then obviously $\forall x A(x)$ must be false at some strictly accessible world. (Av) allows us to find a formula C such that $\Gamma \Vdash \forall x(A(x) \vee C)$; so C should be true at all strongly accessible worlds, where $A(d)$ is refuted for some individual d.

For example, this property holds if C exactly defines the set of all strictly accessible worlds (and the canonical model theorem holds).

For a logic L containing the formula CD the condition (Av) may be simplified:

(Ac) if $\forall x A(x) \in (-\Gamma)$, then $A(d) \in (-\Gamma)$ for some $d \in D_\Gamma$.

In fact, let $CD \in L$. If (Av) holds and $\forall x A(x) \in (-\Gamma)$, but $A(d) \in \Gamma$ for all $d \in D_\Gamma$, then $\forall x(A(x) \vee C) \in \Gamma$ for some $C \in (-\Gamma)$. Since $CD \in \Gamma$, we obtain $(\forall x A(x) \vee C) \in \Gamma$, which leads to a contradiction. Thus (Ac) holds.

Conversely, if (Ac) holds then the premise of (Av) is false, so (Av) holds.

Note that (Ac) is an analogue to the property ($\forall$) of forcing in Kripke models with constant domains.

Since the Gödel–Tarski translation for $\exists$ is just $\exists$, the definition of intuitionistic forcing for the $\exists$-case is the same as classical. This explains, why the existence property is defined in the same way. On the other hand, the intuitionistic $\forall$ translates as modal $\Box\forall$, and thus the intuitionistic definition for the $\forall$-case consists of the 'classical' part dealing with individuals from a certain world and of the 'modal' part dealing with accessibility relations between worlds. So property (Av) is responsible for the classical part of intuitionistic universal quantification. In the case of constant domains $(\forall x B)^T$ is equivalent to $\forall x B^T$, and thus (Ac) is just the dual of (E).

$\exists\forall$-complete theories seem to be a better intuitionistic analogue of Henkin theories than $\exists$-complete theories. But the latter can be used for completeness proofs in the simplest cases (for example, for $\mathbf{QH}^{(=)}$).

Lemma 6.2.5 *Every L-complete theory* Γ *has properties 6.1.3(i), (iii), 6.1.2(2):*

- $(A \wedge B) \in \Gamma$ *iff* ($A \in \Gamma$ *and* $B \in \Gamma$);

- $(A \vee B) \in \Gamma$ *iff* $(A \in \Gamma$ *or* $B \in \Gamma)$;
- $\Gamma \vdash_L A$ *iff* $A \in \Gamma$ *(for* $A \in \mathcal{L}^{(=)}(\Gamma)$*)*.

Every $L\exists$*-complete* Γ *also satisfies 6.1.7(EP′):*

- $\exists x A(x) \in \Gamma$ *iff* $(A(c) \in \Gamma$ *for some* $c \in D_\Gamma)$.

Proof An exercise. ■

Now let us prove the crucial property of $L\exists$-complete theories analogous to Lemma 6.1.9 for Henkin theories.

For intuitionistic theories $(\Gamma, \Delta), (\Gamma', \Delta')$ put

$$(\Gamma, \Delta) \preceq (\Gamma', \Delta') := \Gamma \subseteq \Gamma' \ \&\ \Delta \subseteq \Delta'.$$

For a complete theory Θ also let

$$(\Gamma, \Delta) \preceq \Theta := \Gamma \subseteq \Theta \ \&\ \Delta \cap \Theta = \varnothing,$$

which means $(\Gamma, \Delta) \preceq (\Theta, -\Theta)$.

Lemma 6.2.6 *Let* (Γ, Δ) *be an* L*-consistent theory,* $D_{(\Gamma,\Delta)} = S \subset S'$, $|S'| = |S' - S| = \aleph_0$. *Then there exists an* $(L\exists, S')$*-complete (and even an* $(L\exists\forall, S')$*-complete) theory* Θ *such that* $(\Gamma, \Delta) \preceq \Theta$.

Proof The idea of the proof is quite standard. Namely, we can successively extend the pair (Γ, Δ) by adding S'-sentences either to Γ or to Δ. Moreover, together with adding $\exists x A(x)$ to Γ, we always add $A(c)$ for some *new* constant c from $(S' - S)$. And together with adding $\forall x A(x)$ to Δ, we add $A(c)$ for a new c — but only if this is consistent. This construction provides the existence and the coexistence properties. If we need only an $L\exists$-complete theory $\Theta \succeq (\Gamma, \Delta)$, we can use the same procedure, but without mentioning $\forall$-formulas.

Let $IF^{(=)}(S') = \{B_k \mid k \in \omega\}$. Choose distinct constants c_k from $(S' - S)$ such that c_k does not occur in $B_0, \ldots, B_k$. Next, define a sequence of L-consistent theories

$$(\Gamma, \Delta) = (\Gamma_0, \Delta_0) \preceq \ldots \preceq (\Gamma_k, \Delta_k) \preceq \ldots$$

as follows.

If $B_k = \exists x A(x)$, then

$$(\Gamma_{k+1}, \Delta_{k+1}) := \begin{cases} (\Gamma_k \cup \{B_k, A(c_k)\}, \Delta_k) \text{ if } (\Gamma_k \cup \{B_k\}, \Delta_k) \text{ is } L\text{-consistent,} \\ (\Gamma_k, \Delta_k \cup \{B_k\}) \text{ otherwise} \end{cases}$$

If B_k is not of the form $\exists x A(x)$, then

$$(\Gamma_{k+1}, \Delta_{k+1}) := \begin{cases} (\Gamma_k \cup \{B_k\}, \Delta_k) \text{ if } (\Gamma_k \cup \{B_k\}, \Delta_k) \text{ is } L\text{-consistent,} \\ (\Gamma_k, \Delta_k \cup \{B_k\}) \text{ otherwise.} \end{cases}$$

Let us show that $(\Gamma_{k+1}, \Delta_{k+1})$ is L-consistent if (Γ_k, Δ_k) is.

First, if $(\Gamma_k \cup \{B_k\}, \Delta_k)$ and $(\Gamma_k, \Delta_k \cup \{B_k\})$ are both inconsistent, then (Γ_k, Δ_k) is also inconsistent. In fact, in this case

$$\Gamma_k \vdash_L B_k \supset \bigvee \Delta', \; B_k \vee \bigvee \Delta''$$

for some finite $\Delta', \Delta'' \subseteq \Delta_k$, and thus

$$\Gamma_k \vdash_L B_k \supset \bigvee \Delta, \; B_k \vee \bigvee \Delta,$$

where $\Delta = \Delta' \cup \Delta''$. Now since $p \supset q, \; p \vee q \vdash_{\mathbf{H}} q$, we obtain $\Gamma_k \vdash_L \bigvee \Delta$, i.e. (Γ_k, Δ_k) is inconsistent.

Second, if $B_k = \exists x A(x)$ and $(\Gamma_k \cup \{\exists x A(x), A(c_k)\}, \Delta_k)$ is inconsistent, then

$$\Gamma_k \vdash_L A(c_k) \supset \bigvee \Delta$$

for some finite $\Delta \subseteq \Delta_k$; therefore by 2.7.11

$$\Gamma_k \vdash_L \forall x(A(x) \supset \bigvee \Delta),$$

since c_k does not occur in $\Gamma_k \cup \Delta_k$, which implies

$$\Gamma_k \vdash_L \exists x A(x) \supset \bigvee \Delta$$

by 2.6.10 and MP, and thus $(\Gamma_k \cup \{B_k\}, \Delta_k)$ is inconsistent.

Next, put $\Theta := \bigcup\limits_{k \in \omega} \Gamma_k$. By construction, the theory $(\Theta, -\Theta)$ is L-consistent and has the existence property.

To obtain an $L\exists\forall$-complete theory, we should slightly refine the above construction. Viz., if $B_k = \forall x A(x)$, we put

$$(\Gamma_{k+1}, \Delta_{k+1}) := \begin{cases} (\Gamma_k \cup \{B_k\}, \Delta_k) & \text{if this theory is } L\text{-consistent,} \\ (\Gamma_k, \Delta_k \cup \{B_k, A(c_k)\}) & \text{if this theory is } L\text{-consistent and} \\ & (\Gamma_k \cup \{B_k\}, \Delta_k) \text{ is } L\text{-inconsistent,} \\ (\Gamma_k, \Delta_k \cup \{B_k\}) & \text{otherwise.} \end{cases}$$

In other words, we add B_k either to Γ_k or to Δ_k keeping consistency, and also add $A(c_k)$ to Δ_k whenever this is possible.

The consistency of (Γ_k, Δ_k) again implies the consistency of $(\Gamma_{k+1}, \Delta_{k+1})$, since either $(\Gamma_k \cup \{B_k\}, \Delta_k)$ or $(\Gamma_k, \Delta_k \cup \{B_k\})$ is consistent, as we have seen above.

Finally, let us check (Av) for Θ. Suppose $B_k = \forall x A(x) \notin \Theta$ and $A(c) \in \Theta$ for all $c \in D_\Theta$. Then $B_k \in \Delta_{k+1}$, and $A(c_k) \notin \Delta_{k+1}$, so by construction, $(\Gamma_k, \Delta_k \cup \{B_k, A(c_k)\}$ is L-inconsistent, and thus,

$$\Gamma_k \vdash_L \bigvee \Delta \vee A(c_k)$$

for some finite $\Delta \subseteq \Delta_k$. Hence by 2.7.11

$$\Gamma_k \vdash_L \forall x(\bigvee \Delta \vee A(x)).$$

Since Θ is $L\exists$-complete, it satisfies 6.1.3(v), so it follows that $\forall x(C \vee A(x)) \in \Theta$, where $C = \bigvee \Delta \in (-\Theta)$. ∎

Lemma 6.2.7 *Let A be an intuitionistic sentence, $|S| = \aleph_0$. Then:*

$$L \vdash A \textit{ iff } (A \in \Gamma \textit{ for any } (L,S)\textit{-place } \Gamma)$$
$$\textit{iff } (A \in \Gamma \textit{ for any } (L\forall,S)\textit{-place } \Gamma).$$

Proof Cf. Lemma 6.1.10 for the modal case. ∎

Let $\Gamma|S := \Gamma \cap IF_S^{(=)}$ for $S \subseteq S'$ and $\Gamma \subseteq IF_{S'}^{(=)}$. We also fix a denumerable universal set of constants S^* and consider small subsets of S^* as in Section 7.1.

Definition 6.2.8 *An L-*place *(respectively, an $L\forall$-*place*) is an $L\exists$-complete (respectively, an $L\exists\forall$-complete) theory with a small set of constants.*
VP_L and UP_L denote the sets of all L-places and $L\forall$-places respectively.

Obviously, $UP_L \subseteq VP_L$. We also have an analogue of Lemma 6.1.12:

Lemma 6.2.9 *Let u be a world in an intuitionistic Kripke model M over a frame $\mathbf{F} = (F, D)$ and assume that $M \models L$. Then*

$$\Gamma_u := \{A \in IF^{(=)}(D_u) \mid M, u \Vdash A\}$$

is an $L\exists$-complete theory.

Proof Almost obvious. As for L-consistency, note that if $A_i \in \Gamma_u$ for $i = 1, \ldots, k$ and $L \vdash \bigwedge_{i=1}^{k} A_i \supset \bigvee_{j=1}^{m} B_j$, then the latter implication is true in M, and so $M, u \Vdash B_j$ for some j. Thus $(\Gamma_u, -\Gamma_u)$ is L-consistent. ∎

Note that Γ_u is not necessarily $L\exists\forall$-complete.
Similarly to the modal case we define canonical maps:

Definition 6.2.10 *Let M be a Kripke model for a superintuitionistic logic L, in which every individual domain is a small subset of S^*. The map from (the worlds of) M to VP_L sending u to Γ_u is called* canonical *and denoted by $\nu_{M,L}$ (or by ν_M, or by ν).*

Lemma 6.2.9 also extends to Kripke sheaves, bundles and to metaframe models in *modal* metaframes.[1] The intuitionistic canonical model theorem (see below) shows that VP_L is the set of all L-places Γ_u in Kripke models for L with small domains.

Now let us define some accessibility relations on VP_L and UP_L.

[1]For *intuitionistic* metaframes the situation may be different.

Definition 6.2.11 *For $\Gamma, \Gamma' \in VP_L$, we say that*

- Γ' *is* conservative *over* Γ *(notation: $\Gamma \lesssim \Gamma'$) if* $\Gamma = \Gamma'|D_\Gamma$;
- Γ' *is* proper *over* Γ *(notation: $\Gamma < \Gamma'$) if* $\Gamma \subset \Gamma'|D_\Gamma$.

Let $\leq$ be the reflexive closure of $<$, i.e.

$$\Gamma \leq \Gamma' \textit{ iff } \Gamma < \Gamma' \textit{ or } \Gamma = \Gamma'.$$

The inclusion relation $\subseteq$ on VP_L (i.e. the set $\{(\Gamma, \Gamma') \in VP_L^2 \mid \Gamma \subseteq \Gamma'\}$) is also denoted by R_L.

It is clear that $\Gamma \subseteq \Gamma'$ iff $\Gamma < \Gamma' \vee \Gamma \lesssim \Gamma'$.

Definition 6.2.12 *Let $W \subseteq VP_L$. A partial ordering $R \subseteq R_L$ is called* selective *on W if for any $\Gamma \in VP_L$ the following two conditions hold:*

($\supset$) *if $(A_1 \supset A_2) \in (-\Gamma)$, then $A_1 \in \Gamma'$ and $A_2 \notin \Gamma'$ for some $\Gamma' \in R(\Gamma) \cap W$;*

($\forall$) *if $\forall x A(x) \in (-\Gamma)$, then $A(c') \notin \Gamma'$ for some $\Gamma' \in R(\Gamma) \cap W$ and some $c' \in D_{\Gamma'}$.*

These conditions can be rewritten in an equivalent form:

($\supset'$) $(A_1 \supset A_2) \in \Gamma$ iff for any $\Gamma' \in R(\Gamma) \cap W$ $(A_1 \in \Gamma' \Rightarrow A_2 \in \Gamma')$;

($\forall'$) $\forall x A(x) \in \Gamma$ iff $A(c') \in \Gamma'$ for any $\Gamma' \in R(\Gamma) \cap W$, $c' \in D_{\Gamma'}$.

Lemma 6.2.13

(1) The relation R_L is selective on VP_L and UP_L.

(2) The relation $\leq$ is selective on UP_L.

Proof (1) ($\supset$) If $(A_1 \supset A_2) \notin \Gamma$, then $\Gamma \not\vdash_L (A_1 \supset A_2)$ (Lemma 6.2.5), i.e. the theory $(\Gamma \cup \{A_1\}, \{A_2\})$ is L-consistent. So by Lemma 6.2.6, there exists an $L\forall$-place $\Gamma' \succeq (\Gamma \cup \{A_1\}, \{A_2\})$. Thus $\Gamma \subseteq \Gamma'$, $A_1 \in \Gamma$, $A_2 \notin \Gamma'$.

($\forall$) Suppose $\forall x A(x) \notin \Gamma$, then $\Gamma \not\vdash_L \forall x A(x)$. Take $c \in (S^* - D_\Gamma)$; then the theory $(\Gamma, \{A(c)\})$ is L-consistent (otherwise $\Gamma \vdash_L A(c)$, which implies $\Gamma \vdash_L \forall x A(x)$). Hence by Lemma 6.2.6, there exists $\Gamma' \in UP_L$ such that $\Gamma' \succeq (\Gamma, \{A(c)\})$, i.e. $\Gamma' \supseteq \Gamma$ and $A(c) \notin \Gamma'$.

(2) ($\supset$) Suppose $(A_1 \supset A_2) \in (-\Gamma)$ and consider two cases.

(a) If $A_1 \notin \Gamma$, then by (1) we obtain $\Gamma' \supseteq \Gamma$ such that $A_1 \in \Gamma'$, $A_2 \notin \Gamma'$. So $A_1 \in ((\Gamma'|D_\Gamma) - \Gamma)$, and thus $\Gamma < \Gamma'$.

(b) If $A_1 \in \Gamma$, then to show (IV), we can take $\Gamma' = \Gamma$, since $A_2 \vdash_L (A_1 \supset A_2)$, and thus $(A_1 \supset A_2) \notin \Gamma$ implies $A_2 \notin \Gamma$.

($\forall$) Suppose $\forall x A(x) \in (-\Gamma)$ and consider two cases.

(a) If $A(c) \notin \Gamma$ some $c \in D_\Gamma$, there is nothing to prove: one can take $\Gamma' = \Gamma$.

(b) If $A(c) \in \Gamma$ for all $c \in D_\Gamma$, we can apply 6.2.4 (Av). So there exists a formula $C \in (-\Gamma)$ such that $\forall x(A(x) \vee C) \in \Gamma$. Now as in the proof of (1), we obtain $\Gamma' \supseteq \Gamma$ and $e \in (D_{\Gamma'} - D_\Gamma)$ such that $A(e) \notin \Gamma'$. Since $\forall x(A(x) \vee C) \in \Gamma \subseteq \Gamma'$ and $\forall x(A(x) \vee C) \vdash_L A(e) \vee C$, we also have $(A(e) \vee C) \in \Gamma'$, and thus $C \in \Gamma'$ by Lemma 6.2.5 (iii). Hence $C \in (\Gamma'|D_\Gamma) - \Gamma$, and eventually $\Gamma < \Gamma'$. ∎

Note that the previous argument fails for VP_L.

Definition 6.2.14 *The* canonical frame *of a superintuitionistic predicate logic L is defined as $VF_L := (VP_L, R_L, D_L)$, where*

$$\Gamma R_L \Gamma' := \Gamma \subseteq \Gamma', \ (D_L)_\Gamma := D_\Gamma$$

similarly to the modal case (cf. Definition 6.1.18).

The canonical frame with equality *of an s.p.l.(=) L is*

$$VF_L^= := (VP_L, R_L, D_L, \asymp_L),$$

where, as in Definition 7.1.10,

$$c\,(\asymp_L)_\Gamma\, d := (c = d) \in \Gamma.$$

The stability condition for $\asymp_L$ holds trivially, due to the definition of R_L.

Definition 6.2.15 *The* canonical model *of an s.p.l.(=) L is $VM_L^{(=)} := (VF_L^{(=)}, \xi_L)$, where*

$$(\xi_L)_\Gamma(P_k^m) := \{\mathbf{c} \in (D_\Gamma)^m \mid P_k^m(\mathbf{c}) \in \Gamma\}.$$

This valuation is obviously intuitionistic.

Definition 6.2.16 *We also introduce two kinds of* quasi-canonical frames and models*:*

$$UF_L^{(=)} := VF_L^{(=)}|UP_L, \ UM_L^{(=)} := VM_L^{(=)}|UP_L,$$

$U^{\leq}F_L^{(=)}$ (respectively, $U^{\leq}M_L^{(=)}$) is the same as $UF_L^{(=)}$ (respectively, $UM_L^{(=)}$), but with the relation $\leq$ instead of R_L.

Theorem 6.2.17 (Canonical model theorem) *Let L be an s.p.l.(=). Then for any L-place Γ and $A \in \mathcal{L}(\Gamma)$,*

$$VM_L^{(=)}, \Gamma \Vdash A \text{ iff } A \in \Gamma;$$

similarly for both quasi-canonical models $UM_L, U^{\leq}M_L$ and for all their selective submodels.

Proof By induction on the length of A (cf. Lemma 6.1.21). The inductive step for $\wedge, \vee, \exists$ uses Lemma 6.2.5. In the cases of $\supset$ and $\forall$ use 6.2.12 ($\supset$), ($\forall$). ∎

Corollary 6.2.18 *If* $M = VM_L^{(=)}$, $UM_L^{(=)}$ *or* $U^{\leq}M_L^{(=)}$, *then*

$$M \Vdash A \text{ iff } L \vdash A$$

for any $A \in IF^{(=)}$.

Now we obtain an analogue of 6.1.25:

Lemma 6.2.19 *Let* $L_1 \subseteq L_2$ *be s.p.l.(=). Then*

(1) $VP_{L_2} = \{\Gamma \in VP_{L_1} \mid VM_{L_1}, \Gamma \Vdash L_2\}$,

(2) VM_{L_2} *is a generated submodel of* VM_{L_1}.

Proof Almost the same as in the modal case. The proof of (1) uses a slightly different argument about consistency. Viz., let us show the L_2-consistency for an L_1-place $\Theta \supseteq L_2$. Suppose the contrary: $L_2 \vdash \bigwedge\Delta \supset \bigvee\nabla$ for finite $\Delta \subseteq \Theta$, $\nabla \subseteq -\Theta$; then $\Xi \vdash_{\mathbf{QH}} \bigwedge\Delta \supset \bigvee\nabla$ for a finite $\Xi \subseteq L_2 \subseteq \Theta$, and hence $\mathbf{QH} \vdash \bigwedge(\Delta \cup \Xi) \supset \bigvee\nabla$, thus $(\Theta, -\Theta)$ is L_1-inconsistent, which contradicts the assumption that Θ is an L_1-place.

For the proof of (2) note that $\Gamma \Vdash L_2$ implies $\Delta \Vdash L_2$ for any $\Delta \supseteq \Gamma$ since the model is intuitionistic. ∎

Definition 6.2.20 *An s.p.l.(=) L is called V*-canonical *if* $L \subseteq \mathbf{IL}^{(=)}(VF_L^{(=)})$. *U*-canonicity *and* $U^{\leq}$-canonicity *are defined analogously.*

Similarly to 6.1.24, we obtain

Corollary 6.2.21 *Every (quasi-) canonical superintuitionistic logic is strongly Kripke complete (Kripke sheaf complete in the case of a logic with equality).*

By definition we readily have

Proposition 6.2.22 *The logics* $\mathbf{QH}$, $\mathbf{QH}^=$ *are V-canonical.*

We also have an analogue of Proposition 6.1.26:

Proposition 6.2.23 *If L is a V-canonical s.p.l.=, Γ is a set of pure equality intuitionistic sentences, then* $L + \Gamma$ *is also V-canonical.*

Proof The same as in the modal case, but using Lemma 6.2.19. ∎

Corollary 6.2.24 *The logics with decidable and with stable equality* $\mathbf{QH}^{=d} = \mathbf{QH}^= + DE$ *and* $\mathbf{QH}^{=s} = \mathbf{QH}^= + SE$ *are V-canonical.*

Therefore, we obtain the following completeness result.

Theorem 6.2.25

(1) Intuitionistic predicate logic $\mathbf{QH}$ *is strongly Kripke frame complete.*

(2) *The logics with equality* $\mathbf{QH}^{=}, \mathbf{QH}^{=d}, \mathbf{QH}^{=s}$ *are strongly Kripke sheaf complete.*

Note that (ii) can also be deduced from (i) by applying Theorems 3.8.8, 3.8.4.

Remark 6.2.26 In Section 6.4 we shall show that $\mathbf{QH}^{=d}$ is actually Kripke-complete.

However V-canonicity is rare. In general one may expect that superintuitionistic predicate logics with simple axioms are not V-canonical. For example, classical logic **QCL** is obviously not V-canonical; in fact, every **QCL**-place (as well as every L-place for any L) is not maximal in the corresponding canonical frame, since it may be extended after adding new constants to the language. Another trivial observation is that $\mathbf{QCL} \vdash CD$, but CD is not valid in $VF_{\mathbf{QCL}}$, since as we have just noted, every **QCL**-place sees another place with a larger domain.

Moreover, the same argument shows that VF_L is of infinite depth (for any logic L). Thus every superintuitionistic logic containing a propositional finite depth formula P_k , $k \geq 1$ (see Section 1.1.2) is not V-canonical.

Nevertheless for logics with additional axioms listed in Chapter 2 we can sometimes prove $U^{\leq}$-canonicity. Let us now consider such examples.

6.3 Intermediate logics of finite depth

Let us begin with some simple, but useful remarks.

Lemma 6.3.1 *Let* L, L' *be superintuitionistic logics,* (Γ, Δ) *an intuitionistic theory.*

(1) *If* (Γ, Δ) *is* **QH**-*consistent,* $\overline{L} \subseteq \Gamma$, *then* (Γ, Δ) *is* L-*consistent.*

(2) *If* (Γ, Δ) *is* $L\exists$-*complete, then it is* $L'\exists$-*complete iff* $\overline{L'} \subseteq \Gamma$.

(3) *If* (Γ, Δ) *is* $L\exists\forall$-*complete, then it is* $L'\exists\forall$-*complete iff* $L' \subseteq \Gamma$.

Proof (i) Suppose (Γ, Δ) is L-inconsistent, $\overline{L} \subseteq \Gamma$, and let us show that (Γ, Δ) is **QH**-inconsistent. We have $\vdash_L \bigwedge \Gamma_1 \supset \bigvee \Delta_1$ for some finite $\Gamma_1 \subseteq \Gamma$, $\Delta_1 \subseteq \Delta$. So a generator A of this formula is in L; hence $\overline{\forall} A \in \overline{L} \subseteq \Gamma$. Thus $\Gamma \vdash_{\mathbf{QH}} A$, and so $\Gamma \vdash_{\mathbf{QH}} \bigwedge \Gamma_1 \supset \bigvee \Delta_1$, which implies $\Gamma \vdash_{\mathbf{QH}} \bigvee \Delta_1$, i.e. the **QH**-inconsistency of (Γ, Δ).

(ii) (If.) Assume that $\overline{L'} \subseteq \Gamma$ and (Γ, Δ) is $L\exists$-complete. Then it is **QH**-consistent, and thus L'-consistent, by (i). Now L'-completeness and the existence property readily follow from $L\exists$-completeness.

(Only if.) Suppose $\overline{L'} \not\subseteq \Gamma$, and take $A \in \overline{L'} - \Gamma$. Since $\mathbf{QH} \vdash \top \supset p$, it follows that $\vdash_L \top \supset A$, and thus (Γ, Δ) is L'-inconsistent.

(iii) The proof is similar to (ii). ∎

Remark 6.3.2 Every **QCL**-place Γ has all the properties from Lemma 6.1.3 (with $L = \mathbf{QCL}$), so it obviously satisfies the conditions 6.2.12 ($\supset$), ($\forall$). In both cases we can take $\Gamma' = \Gamma$. In fact, $(A_1 \supset A_2) \notin \Gamma$; implies $\neg(A_1 \supset A_2) \in \Gamma$, and thus $A_1 \in \Gamma$, $A_2 \notin \Gamma$. Next, if $\forall x A(x) \notin \Gamma$, then $\neg\forall x A(x) \in \Gamma$, and thus $\exists x \neg A(x) \in \Gamma$, by classical logic (cf. 2.6.17(ii)). Hence by the existence property, for some $c \in S$ we have $\neg A(c) \in \Gamma$, i.e. $A(c) \notin \Gamma$.

Now let us consider Kuroda's axiom. Recall that KF is valid in Kripke frames $F = (W, R, D)$ with the *McKinsey property*:

$$\forall u \in W \exists v \in W (uRv \ \&\ v \text{ is maximal}),$$

and similarly for Kripke sheaves. As noted at the end of the previous section, there are no maximal worlds in the canonical frame VF_L, so VF_L does not have the McKinsey property (for *any* logic L). However let us show that the McKinsey property holds in $U^{\leq}F_L^{(=)}$ whenever L contains KF.

Lemma 6.3.3 *Let L be an s.p.l.(=) containing KF, $(\Gamma, \varnothing)$ an L-consistent theory. Then $(\Gamma, \varnothing)$ is* **QCL***-consistent.*

Proof In fact, suppose $\vdash_{\mathbf{QCL}} \neg(\bigwedge \Gamma_1)$. Then $(\bigwedge \Gamma_1)$ has a generator A such that $\neg A \in \mathbf{QCL}$. By the Glivenko Theorem 2.12.1, then $\neg A \in \mathbf{QH} + KF \subseteq L$, and thus $\vdash_L \neg(\bigwedge \Gamma_1)$. So the **QCL**-inconsistency of $(\Gamma, \varnothing)$ implies its L-inconsistency. ∎

Lemma 6.3.4 *Let L be an s.p.l.(=).*

(1) Γ is a **QCL***-place iff Γ is maximal in $U^{\leq}F_L$.*

(2) If $KF \in L$, then for any L-place Γ there exists an **QCL***-place $\Gamma' \geq \Gamma$.*

Proof

(1) Every **QCL**-complete theory is maximal among consistent theories in the same language, so every extension of Γ is conservative. By 6.3.2, Γ is a **QCL**$\forall$-place, i.e. $\Gamma \in UP_L$.

(2) If already $\overline{\mathbf{QCL}} \subseteq \Gamma$, then according to 6.3.2, Γ is a **QCL**$\forall$-place, so we we can take $\Gamma' = \Gamma$,

The other way round, if an $L\forall$-place Γ is not maximal in $U^{\leq}F_L$, then $\Gamma < \Delta$ for some $\Delta \in UP_L$, so there exists $A \in ((\Delta | D_\Gamma) - \Gamma)$. Then $A = B(\mathbf{a})$ for some $B(\mathbf{x}) \in IF^{(=)}$ and a tuple $\mathbf{a}$ from D_Γ. By 6.2.17 it follows that $U^{\leq}M_L(^{(=)}, \Gamma \not\Vdash A \vee \neg A$, and thus

$$U^{\leq}M_L^{(=)}, \Gamma \not\Vdash \forall \mathbf{x}(B(\mathbf{x}) \vee \neg B(\mathbf{x})).$$

Hence $\forall \mathbf{x}(B(\mathbf{x}) \vee \neg B(\mathbf{x})) \notin \Gamma$, and thus $\overline{\mathbf{QCL}} \subseteq \Gamma$ (again by 6.2.17).

Otherwise there exists $B \in (\overline{\mathbf{QCL}} - \Gamma)$. Obviously, the theory $(\Gamma, \varnothing)$ is L-consistent, and thus **QCL**-consistent by 6.3.3. By Lemma 6.2.6, there exists a **QCL**$\forall$-place $\Gamma' \supseteq \Gamma$. We also have $\Gamma < \Gamma'$ since $B \in (\Gamma' - \Gamma) \cap IF$.

The argument for logics with equality is similar. ∎

So we readily obtain completeness:

Theorem 6.3.5 *The logic* $\mathbf{QH} + KF$ *(respectively,* $\mathbf{QH}^= + KF$*) is strongly determined by Kripke frames (respectively, Kripke sheaves) satisfying the McKinsey property.*

Now let us prove completeness for intermediate logics $\mathbf{QHP}_k^+$ introduced in Section 2.4. Recall that $\mathbf{QHP}_k^+ = \Delta^{k-1}(\mathbf{QCL})$, where the logic ΔL is axiomatised by formulas $\delta A = p \vee (p \supset A)$ for $A \in L$, p not occurring in A; $\mathbf{QHP}_k^+$-frames are the intuitionistic frames of depth $\leq k$.

Lemma 6.3.6 *If* Γ, Γ' *are* L*-places,* $\Gamma < \Gamma'$ *and* $\Delta L \subseteq \Gamma$*, then* $L \subseteq \Gamma'$.

Proof Let $B \in \mathcal{L}(\Gamma) \cap (\Gamma' - \Gamma)$ and suppose $L \not\subseteq \Gamma'$, $A \in L - \Gamma'$. Then $B \vee (B \supset A)$ is a substitution instance of δA, so $B \vee (B \supset A) \in \Delta L$.

On the other hand, $(B \supset A) \notin \Gamma$; in fact, otherwise $(B \supset A) \in \Gamma \subseteq \Gamma'$, which together with $B \in \Gamma'$, implies $A \in \Gamma'$ contradicting the choice of A. Now since also $B \notin \Gamma$, we obtain $B \vee (B \supset A) \notin \Gamma$, by Lemma 6.2.5.

Hence it follows that $\Delta L \not\subseteq \Gamma$, in a contradiction to the assumption. ∎

Lemma 6.3.7 *If* $\mathbf{QHP}_k^+ \subseteq L$*, then the frame* $U^{\leq}F_L$ *is of depth* $\leq k$.

Proof Suppose $\Gamma_1 < \ldots \Gamma_k$ is a chain in $U^{\leq}F_L$. We have $L \subseteq \mathbf{QHP}_k^+ = \Delta^{k-1}(\mathbf{QCL}) \subseteq \Gamma_1$, which yields $\mathbf{QCL} = \mathbf{QHP}_1^+ \subseteq \Gamma_k$, by applying Lemma 6.3.6 $(k-1)$ times. Thus Γ_k is maximal in $U^{\leq}F_L$, by Lemma 6.3.4(i). ∎

Theorem 6.3.8 *The logic* $\mathbf{QHP}_k^{+(=)}$ *for* $k \geq 1$ *is determined by Kripke frames (respectively, Kripke sheaves) of depth* k.

Proof We already know that $L = \mathbf{LP}_k^+$ is sound for frames of depth k, so it suffices to show that $U^{\leq}F_L$ is of depth k. By Lemma 6.3.7, $U^{\leq}F_L$ is of depth $\leq k$. On the other hand, $\mathbf{LP}_k^+ \nvdash P_{k-1}$, hence $U^{\leq}F_L \nVdash AP_{k-1}$ by Corollary 6.2.18. Therefore by Proposition 1.4.17 and Lemma 3.2.23 $U^{\leq}F_L$ is of depth $\geq k$. ∎

Remark 6.3.9 For some logics without equality canonical models allow us to prove $\mathcal{KE}$-completeness rather than $\mathcal{K}$-completeness. Indeed, let L be a non-canonical predicate logic without equality, $L^= = \mathbf{QH}^= + L$ its minimal extension with equality, and suppose the canonical model for $L^=$ validates L (unlike the canonical model for L). Then $L^=$ is $\mathcal{KE}$-complete (recall that axioms of $\mathbf{QH}^=$ are valid in all Kripke sheaves), as well as L. In fact, if a formula A without equality is not L-provable, then by conservativity (Proposition 2.9.2), A is not $L^=$-provable and thus is refuted in the canonical model of $L^=$. Although we do not know how to apply this idea for a direct construction of a Kripke incomplete logic L, later on we will show an indirect construction of this kind.

Definition 6.3.10 *Let L be a superintuitionistic logic, Γ an L-place. We say that E is a* characteristic formula *for Γ if $E \in (-\Gamma)$ and $A \vee (A \supset E) \in \Gamma$ for any $A \in \mathcal{L}^{(=)}(\Gamma)$. Or equivalently, this means that $(A \supset E) \in \Gamma$ for any $A \in (-\Gamma)$, i.e. E is the weakest D_Γ-sentence outside Γ.*

If a characteristic formula exists, it is clearly unique up to equivalence in Γ (the relation $(A \equiv B) \in \Gamma$).

The next lemma uses the operation Δ introduced in Section 2.13

Lemma 6.3.11 *If Γ is a ΔL-place, which is not an L-place, then there exists a sentence $E \in IF^{(=)}$ (without extra constants), which is characteristic for Γ.*

Proof Let $E \in (L - \Gamma)$. Then $p \vee (p \supset E) \in \Delta L$ for p not occurring in E, and thus $A \vee (A \supset E) \in \Gamma$ for any $A \in \mathcal{L}(\Gamma)$. ∎

Corollary 6.3.12 *If $\mathbf{QHP}_n^+ \subseteq L$ for some $n > 0$, then for any L-place there exists a characteristic sentence.*

Proof Recall that $\mathbf{QHP}_k^+ = \Delta(\mathbf{QHP}_{k-1}^+)$ for $k > 0$ and $\mathbf{LP}_0^+ = \mathbf{QH} + \bot$ is inconsistent. For an L-place Γ, take the least k (> 0) such that Γ is an $\mathbf{QHP}_k^+$-theory. Then apply Lemma 6.3.11. ∎

6.4 Natural models for modal and superintuitionistic logics

As we can see, canonical models work well only for some particular predicate logics. In subtler cases it is convenient to use so-called 'natural models' obtained from canonical models by a sort of 'selective filtration', together with canonical maps (cf. Definition 6.1.13).

Let us begin with some simple remarks on canonical maps.

The following is a trivial reformulation of the canonical model theorem.

Lemma 6.4.1 *Let M be a Kripke model for a modal logic L, in which every individual domain is a small subset of S^* (cf. Definition 6.1.13). Then for any world $u \in M$, for any $A \in \mathcal{L}(u)$*

$$M, u \vDash A \text{ iff } VM_L, \nu_M(u) \vDash A,$$

i.e. canonical maps are reliable.

The same holds in the intuitionistic case, with obvious changes.

Lemma 6.4.2 *Let M be the same as in the previous lemma, R_i the accessibility relations in M. Then for any $u, v \in M$*

$$uR_iv \Rightarrow \nu_M(u)R_{Li}\nu_M(v),$$

i.e. canonical maps are monotonic.

Proof Almost obvious: if $\Box_i A \in \nu_M(u)$, i.e. $M, u \models \Box_i A$, and uR_iv, then $M, v \models A$, i.e. $A \in \nu_M(v)$.

In the intuitionistic case, if uRv, then $\nu_M(u) \subseteq \nu_M(v)$, by truth-preservation. ∎

However note that in general ν_M is not a morphism of Kripke frames. The lift property does not always hold, because it may happen that $\Gamma_u R_{Li} \Gamma_v$ for incomparable $u, v \in F$, but there does not exist $w \in R_i(u)$ such that $\Gamma_w = \Gamma_v$.

Definition 6.4.3 *Let L be a predicate logic (N-modal or superintuitionistic, without or with equality), $F = (W, R_1, \ldots, R_N)$ a propositional frame of the corresponding kind. An L-*map based on F *is a monotonic map from F to (the propositional base of) $VF_L^{(=)}$, i.e. a map h such that for any $u, v \in F$*

$$uR_iv \Rightarrow h(u)R_{Li}h(v).$$

In the intuitionistic case we say that h is an $L\forall$-map *if all $h(u)$ are $L\forall$-places, i.e. h is actually a map to UF_L.*

Definition 6.4.4 *Let h be an L-map based on a propositional frame F. The* predicate Kripke frame associated with h *is $\mathbf{F}(h) := (F, D)$, where $D_u = (D_L)_{h(u)}$ for $u \in F$.*

If L is a logic with equality, the Kripke frame with equality associated with h *is $\mathbf{F}^=(h) := (F, D, \asymp)$, where*

$$c \asymp_u d \ \text{ iff } \ (c = d) \in h(u)$$

for $u \in F$ and $c, d \in D_u$.

The L-model associated with h *is $M^{(=)}(h) := (\mathbf{F}^{(=)}(h), \xi(h))$, where*

$$\xi(h)_u(P_k^m) := \{\mathbf{c} \in D_u^m \mid P_k^m(\mathbf{c}) \in h(u)\}$$

for $u \in F$.

All these frames and models are well-defined, since uR_iv implies $h(u)R_{Li}h(v)$ (which means $h(u) \subseteq h(v)$ in the intuitionistic case).

Definition 6.4.5 *An L-map $h : F \longrightarrow VF_L$ and the associated L-model $M(h)$ are called* natural *if $h = \nu_{M(h),L}$, i.e. $(h, id) : M(h) \longrightarrow VM_L$ is reliable: for any $u \in F$, $A \in \mathcal{L}(u)$*

$$M(h) \models (\Vdash)A \text{ iff } VM_L, h(u) \models (\Vdash)A \ (\Leftrightarrow A \in h(u)).$$

This definition is an analogue to the canonical model theorem for $M(h)$. In this case we readily obtain $M(h) \models L$.

Remark 6.4.6 Lemma 6.4.1 shows that every predicate Kripke model M with small domains can be presented as a natural Kripke model $M(h)$ in a unique way; namely, put

$$h(u) := \Gamma_u = \{A \in MF_N^{(=)}(D_u) \mid M, u \models A\}$$

and similarly in the intuitionistic case.

Definition 6.4.7 *A modal L-map $h : F \longrightarrow VF_L$ based on $F = (W, R_1, \ldots, R_N)$ is called* selective *if it satisfies the condition*

($\Diamond$) *if $u \in F$ and $\Diamond_i A \in h(u)$, then there exists $v \in R_i(u)$ such that $A \in h(v)$.*

Lemma 6.4.8 *An L-map $h : F \longrightarrow VF_L$ is natural iff it is selective.*

Proof The 'only if' direction is trivial, since in a natural model $M(h)$ we have

$M(h), u \vDash A$ iff $A \in h(u)$ (or $VM_L, h(u) \vDash A$, by the canonical model theorem)

for any $A \in \mathcal{L}(u)$.

To prove 'if', we check the above equivalence by induction on the length A (for any u). Let us consider the case $A = \Box_i B$; other cases easily follow from Lemma 6.1.3.

If $\Box_i B \notin h(u)$, then $h(u) \nvDash \Box_i B$, and thus $h(u) \vDash \Diamond_i \neg B$. By ($\Diamond$), there exists $v \in R_i(u)$ such that $\neg B \in h(v)$, which is equivalent to $B \notin h(v)$. Hence $M(h), v \nvDash B$ by the induction hypothesis. Since $v \in R_i(u)$, we obtain $M(h), u \nvDash \Box_i B$.

The other way round, if $h(u) \vDash \Box_i B$ and $uR_i v$, then $h(u) R_{Li} h(v)$, by monotonicity. So $h(v) \vDash B$, and thus $v \vDash B$, by the induction hypothesis. Since v is arbitrary, this implies $u \vDash \Box_i B$. ■

Note that the condition ($\Diamond$) is a generalisation of selectivity from Definition 6.1.31 (where F is a subframe of VF_L). Moreover, by ($\Diamond$) and monotonicity it follows that the image of h is a selective subframe of VF_L.

The intuitionistic analogue of 6.4.7 is the following

Definition 6.4.9 *An intuitionistic L-map $h : F \longrightarrow VF_L$ based on $F = (W, R)$ is called* selective *if it satisfies the following two conditions (cf. 6.2.12 ($\supset$), ($\forall$)).*

($\supset$) *if $(A_1 \supset A_2) \in (-h(u))$, then $A_1 \in h(v)$, $A_2 \notin h(v)$ for some $v \in R(u)$;*

($\forall$) *if $\forall x A(x) \in (-h(u))$, then $A(c) \notin h(v)$ for some $v \in R(u)$ and $c \in D_v$.*

Lemma 6.4.10 *An intuitionistic L-map $h : F \longrightarrow VF_L$ is natural iff it is selective.*

Proof Similar to Lemma 6.4.8. The 'only if' part is trivial. To prove 'if', we show by induction on the length of $A \in \mathcal{L}(u)$ (for any u) that

$$M(h), u \Vdash A \text{ iff } A \in h(u) \text{ (or } VM_L, h(u) \Vdash A).$$

Let us consider only three cases.

Let $A = B \supset C$. If $A \notin h(u)$, then by ($\supset$), there exists $v \in R(u)$ such that $B \in h(v)$, $C \notin h(v)$. By the induction hypothesis, we obtain $v \Vdash B$, $v \nVdash C$. Hence $u \nVdash B \supset C$.

The other way round, if $u \nVdash B \supset C$, then there exists $v \in R(u)$ such that $v \Vdash B$, $v \nVdash C$. Thus $h(v) \Vdash B$, $h(v) \nVdash C$ by the induction hypothesis. Since $h(u) R_L h(v)$, it follows that $h(u) \nVdash B \supset C$.

Let $A = \exists x B$. Then

$$u \Vdash A \text{ iff } \exists c \in D_u \ u \Vdash B(c) \text{ iff } \exists c \in D_u \ h(u) \Vdash B(c)$$

by the induction hypothesis. Since $h(u)R_L h(v)$, the latter is equivalent to $h(u) \Vdash A$.

Let $A = \forall x B$. If $A \notin h(u)$, then by ($\forall$), there exist $v \in R(u)$, $c \in D_v$ such that $B(c) \notin h(v)$. By the induction hypothesis, $v \nVdash B(c)$. Hence $u \nVdash A$.

The other way round, if $u \nVdash \forall x B$, then there exist $v \in R(u)$, $c \in D_v$ such that $v \nVdash B(c)$. Thus $h(v) \nVdash B(c)$ by the induction hypothesis, and therefore $h(u) \nVdash A$, since $h(u)R_L h(v)$. ■

Sometimes we specify the terminology and say, e.g. that h is an (L, R')-map if R' is a selective relation on VP_L such that uRv implies $h(u)R'h(v)$. In the intuitionistic case we also use the term *$L\forall$-map* if h is a map to UF_L; $M(h)$ is called a *natural $L\forall$-model.*

If an L-map h is injective and also

$$uR_i v \quad \text{iff} \quad h(u)R_{Li}h(v),$$

then $M(h)$ is clearly a selective submodel of the canonical model.

Also note that the naturalness condition ($\Diamond$) is a weak analogue of lift property.

In the intuitionistic case an L-map h over a p.o. set is called *proper* if

$$uR^- v \Rightarrow h(u) \neq h(v);$$

in particular, if h is $L, \leq$-natural and proper, then

$$uR^- v \Rightarrow h(u) < h(v).$$

Obviously this condition does not imply the injectivity for h. Similarly in the intuitionistic case one can consider proper natural models on a quasi-ordered F; in the modal case for extensions of **QT** proper natural models on reflexive frames can be used. Informally speaking, in these cases in natural models we do not have to 'repeat' the same L-place in strictly accessible worlds, thanks to reflexivity.

Lemma 6.4.11 *Let L be a modal or superintuitionistic logic, $M_1 \subseteq VM_L$ a selective submodel, F_1 the propositional frame of M_1, $h : F \twoheadrightarrow F_1$. Then h is a natural L-map.*

Proof By Lemmas 6.4.8, 6.4.10, it is sufficient to show that h is selective. The monotonicity of h follows from the definitions. In the modal case, to check ($\Diamond$), suppose $\Diamond_i A \in h(u)$. By selectivity of M_1, there exists $\Gamma \in M_1$ such that $h(u)R_{iL}\Gamma$ and $A \in \Gamma$. Then by the lift property for h, there exists $v \in R_i(u)$ such that $h(v) = \Gamma$, so $A \in h(v)$.

In the intuitionistic case ($\supset$), ($\forall$) are checked by a similar argument, which we leave to the reader. ■

Lemma 6.4.12

(1) Let L be an N-modal logic, $\Gamma \in VP_L$. Then there exists a Kripke model M based on a standard greedy tree F such that $\nu_M(\curlywedge) = \Gamma$ and $M \vDash L$.

(2) If L contains $\Diamond_i \top$ for $1 \leq i \leq N$, then one can take $F = F_N T_\omega$.

*(3) If L is 1-modal and $L \supseteq$ **QS4**, then the claim holds for $F = IT_\omega$.*

Proof By Proposition 3.12.9, there exists a countable reliable $M_1 \subseteq VM_L$ verifying Γ. Let F_1 be the propositional frame of M_1. We may assume that Γ is the root of M_1 (otherwise, replace M_1 with $M_1 \uparrow \Gamma$).

By Proposition 1.10.18, there exists a p-morphism h from some greedy standard tree F onto F_1 sending $\curlywedge$ to Γ. By Lemma 6.4.11, h is a natural L-map, so we can take $M = M(h)$. Thus $\nu_M(\curlywedge) = h(\curlywedge) = \Gamma$.

In the case (2) all the formulas $\Diamond_i \top$ are true in VM_L, hence they are true in its selective submodel M_1. This means that F_1 is serial, thus we can take $F = F_N T_\omega$, by Proposition 1.10.18.

In the case (3) we use Proposition 1.11.11 to construct a p-morphism $h : IT_\omega \longrightarrow F_1$ sending $\curlywedge$ to Γ. ∎

Hence we obtain

Proposition 6.4.13 *Let $\mathcal{F}_N\mathcal{T}_\omega$ be the set of all greedy standard subtrees of $F_N T_\omega$. Then*

(1) $\mathbf{QK}_N = \mathbf{ML}(\mathcal{K}\mathcal{F}_N\mathcal{T}_\omega)$, $\mathbf{QD}_N = \mathbf{ML}(\mathcal{K}F_N T_\omega)$,

(2) $\mathbf{QK}_N^= = \mathbf{ML}^=(\mathcal{K}\mathcal{E}(\mathcal{F}_N\mathcal{T}_\omega))$, $\mathbf{QD}_N^= = \mathbf{ML}^=(\mathcal{K}\mathcal{E}(F_N T_\omega))$.

Now let us modify this proposition for the logics described in Theorem 6.1.29.

Proposition 6.4.14 *Let $\mathbf{\Lambda}$ be a propositional one-way PTC-logic. Let $\mathcal{GT}(\mathbf{\Lambda})$ be the set of all greedy standard $\mathbf{\Lambda}$-trees. Then*

(1) $\mathbf{Q\Lambda} = \mathbf{ML}(\mathcal{K}\mathcal{GT}(\mathbf{\Lambda}))$;

(2) $\mathbf{Q\Lambda}^= = \mathbf{ML}^=(\mathcal{K}\mathcal{E}(\mathcal{GT}(\mathbf{\Lambda})))$.

Proof (i) By Theorem 6.1.29, the logic $L = \mathbf{Q\Lambda}$ is Kripke complete. By Proposition 1.11.5, it is also Δ-elementary. So by Proposition 3.12.8, it has the c.f.p., i.e. $L = \mathbf{ML}(\mathcal{K}\mathbf{V}_0(\mathbf{\Lambda}))$, where $\mathbf{V}_0(\mathbf{\Lambda})$ is the class of all countable $\mathbf{\Lambda}$-frames. Hence by Lemma 3.3.21, $L = \mathbf{ML}(\mathcal{K}\mathbf{V}_1(\mathbf{\Lambda}))$, where $\mathbf{V}_1(\mathbf{\Lambda})$ is the class of all countable rooted $\mathbf{\Lambda}$-frames.

Next, by Proposition 1.11.11, every frame from $\mathbf{V}_1(\mathbf{\Lambda})$ is a p-morphic image of a frame from $\mathcal{GT}(\mathbf{\Lambda})$. Therefore $\mathbf{ML}(\mathcal{K}\mathcal{GT}(\mathbf{\Lambda})) \subseteq L$, by Corollary 3.3.14. The converse inclusion is a trivial consequence of the definitions.

Of course, instead of applying 3.3.14, we could apply Lemma 6.4.12.

In the case (ii) the argument is the same, based on the corresponding facts about frames with equality. ∎

For logics with closed equality there is a similar result:

Proposition 6.4.15

(1) $\mathbf{QK}^{=}_{N} + CE = \mathbf{ML}^{=}(\mathcal{KF}_N\mathcal{T}_\omega)$, $\mathbf{QD}^{=}_{N} + CE = \mathbf{ML}^{=}(\mathcal{K}F_N T_\omega)$.

(2) Let $\mathbf{\Lambda}$ *be a propositional one-way PTC-logic.*

Then $\mathbf{Q\Lambda}^{=} + CE = \mathbf{ML}^{=}(\mathcal{KGT}(\mathbf{\Lambda}))$.

Proof (ii) We modify the argument from the previous proof. Kripke sheaf completeness follows from Theorem 6.1.29 and Proposition 6.1.26. The logics are Δ-elementary, since CE corresponds to a first-order property of KFEs.

As in the proof of Lemma 6.4.12, we obtain that every L-consistent theory Γ is satisfied at the root of a Kripke sheaf model $M = M(h)$ based on a standard greedy $\mathbf{\Lambda}$-tree. Let Φ be the Kripke sheaf of M. Since $M \vDash CE$, it follows that $\Phi \vDash CE$. Then by Lemma 3.10.8, Φ is isomorphic to a simple Kripke sheaf Φ' (in fact, to apply this lemma, one should take the corresponding Horn closure). So Γ is satisfied in a Kripke frame from $\mathcal{QGT}(\mathbf{\Lambda})$. ∎

Recall (Definition 1.10.10) that the intuitionistic universal tree is $IT_\omega := (T_\omega, \leqslant)$, where

$$\alpha \leqslant \beta \text{ iff } \exists\gamma\,(\beta = \alpha\gamma).$$

So in the intuitionistic case we have the following analogue of 6.4.12.

Lemma 6.4.16 *Let L be a superintuitionistic logic, R' a selective relation on VP_L. Then for any L-place Γ there exists a proper natural (L, R')-model on a standard (complete) subtree F of IT_ω such that $\nu_M(\lambda) = \Gamma$.*

We shall often denote $\nu_M(u)$ by Γ_u if there is no confusion.

It is clear that for $v \in \beta(u)$ one can choose an L-place Γ_v non-equal to Γ_u, due to the reflexivity.

Therefore there exists a natural (L, R')-model on IT_ω: in fact, one can copy L-places Γ_u from maximal points u of F at all points from $IT_\omega{\uparrow}u$. This natural model is in general improper. But if for example, $R' = R_L$, then one can directly construct a proper natural model on IT_ω; recall that every L-place Γ in the canonical model can be properly extended by extending its 'domain' D_Γ.

Hence we obtain completeness:

Theorem 6.4.17

(1) $\mathbf{QH} = \mathbf{IL}(\mathcal{K}(IT_\omega))$;

(2) $\mathbf{QH}^{=} = \mathbf{IL}(\mathcal{KE}(IT_\omega))$;

(3) $\mathbf{QH}^{=d} = \mathbf{IL}^{=}(\mathcal{K}(IT_\omega))$.

Now let us prove a somewhat stronger form of completeness (cf. Dragalin [1988]).

Theorem 6.4.18 *Let F be an effuse countable tree. Then*

(1) $\mathbf{IL}(\mathcal{K}F) = \mathbf{QH}$; $\mathbf{ML}(\mathcal{K}F) = \mathbf{QS4}$;

(2) $\mathbf{IL}^{=}(\mathcal{K}\mathcal{E}F) = \mathbf{QH}^{=}$; $\mathbf{ML}^{=}(\mathcal{K}F) = \mathbf{QS4}^{=}$;

(3) $\mathbf{IL}^{=}(\mathcal{K}F) = \mathbf{QH}^{=d}$, $\mathbf{ML}(\mathcal{K}F) = \mathbf{QS4}^{=c}$.

Proof (1) In the modal case the argument is similar to Proposition 1.11.21. We have $F{\uparrow}u \twoheadrightarrow IT_2 \twoheadrightarrow IT_\omega$ for some u. Hence by Proposition 3.3.14, $\mathbf{ML}(\mathcal{K}F{\uparrow}u) \subseteq \mathbf{QS4}$. Thus by the generation lemma, $\mathbf{ML}(\mathcal{K}F) \subseteq \mathbf{ML}(\mathcal{K}F{\uparrow}u) \subseteq \mathbf{QS4}$.

The remaining statements are left to the reader. ∎

Corollary 6.4.19 $\mathbf{QH}^{(=)(d)}$ *is determined by the universal n-branching tree IT_n for any $n \geq 2$.*

If a relation R' is selective on UP_L (for an intermediate logic L), then similarly to 6.4.12, for any $\Gamma \in UP_L$ one can construct a proper natural $(L\forall, R')$-model on a standard greedy subtree F of IT_ω and a natural $(L\forall, R')$-model on IT_ω (not necessarily proper), such that $\Gamma_\lambda = \Gamma$. In particular, one can apply this construction to the selective relation $\leq$ on UP_L. However unlike the case of VF_L, we cannot state the existence of a proper $(L\forall, \leq)$-model on the whole IT_ω. In particular, we obtain

Lemma 6.4.20 *If $\mathbf{QHP}_k^+ \subseteq L$ and ν is a proper $(L\forall, \leq)$-model on F, then F is a frame of depth $\leq k$.*

Lemma 6.4.21 *If $\mathbf{QHP}_k^+ \subseteq L$ then for any $\Gamma \in UP_L$ there exist a proper $(L\forall, \leq)$-model ν on a standard (greedy) subtree F of IT_ω^k and an $(L\forall, \leq)$-model on $IT^k\omega$ such that $\nu(\lambda) = \Gamma$.*

Thus we obtain the following completeness result for $\mathbf{QHP}_k^+$, $k > 0$.

Proposition 6.4.22

(1) $\mathbf{QHP}_k^+ = \mathbf{IL}(\mathcal{K}(IT_\omega^k))$,

(2) $(\mathbf{QHP}_k^{+=} = \mathbf{IL}^{=}(\mathcal{K}\mathcal{E}(IT_\omega^k))$.

In Section 6.10 we will axiomatise the superintuitonistic predicate logic of the frames IT_n^k of finite depth k and finite branching n; it is just the logic of all posets of depth $\leq k$ and branching $\leq n$ (obviously, all these posets are finite).

But first we shall give some simple remarks on natural models based on posets of finite heights. The intuitionistic conditions of naturalness for this case can be slightly modified.

Lemma 6.4.23 *Let M be an intuitionistic (L, R')-model on a poset F of finite depth. Then*

(1) *M is natural iff the following conditions hold (for any $u \in F$):*

($\supset'$) *If* $(A_1 \supset A_2) \in (-\Gamma_u)$ *and* $A_1 \notin \Gamma_u$, *then* $(A_1 \supset A_2) \notin \Gamma_u$ *for some* $v \in R^\bullet(u)$;

($\forall'$) *if* $\forall x A(x) \in (-\Gamma_u)$ *and* $A(c) \in \Gamma_u$ *for all* $c \in D_{\Gamma_u}$, *then* $\forall x A(x) \notin \Gamma_v$ *for some* $v \in R^\bullet(u)$.

(2) *If* M *is an* $(L\forall, R')$*-model, then the condition* ($\supset'$) *implies* (Av').

Proof

($\supset'$) $\Rightarrow$ ($\supset$) If $A_1 \in \Gamma_v$, $A_2 \notin \Gamma_v$ for some $v \in R^\bullet(u)$ then $(A_1 \supset A_2) \notin \Gamma_v$.

($\forall'$) $\Rightarrow$ ($\forall$) The proof is similar.

($\supset$) $\Rightarrow$ ($\supset'$) Obviously, if $(A_1 \supset A_2) \notin \Gamma_u$, then $A_1 \in \Gamma_v$ and $A_2 \notin \Gamma_v$ for a maximal $v \in R(u)$ such that $(A_1 \supset A_2) \notin \Gamma_v$.

($\supset'$) & ($\forall$) $\Rightarrow$ ($\forall$) If $\forall x A(x) \notin \Gamma_u$ and $A(c) \in \Gamma_u$ for all $c \in S_u$ then by (Av), for an $L\exists\forall$-complete Γ_u, we have $\forall x(A(x) \vee C) \in \Gamma_v$ for some $C \notin \Gamma_u$. Then $(C \supset \forall x A(x)) \notin \Gamma_u$ (otherwise $\forall x A(x) \in \Gamma_u$) and by ($\supset$)$'$, $(C \supset \forall x A(x)) \notin \Gamma_v$ for some $v \in R^\bullet(u)$; hence $\forall x A(x) \notin \Gamma_v$.

■

Let us now prove two general lemmas on natural models.

Definition 6.4.24 *A subtree* $F = (W, R_1, \ldots, R_N) \subseteq F_N T_\omega$ *is said to be* small *if*

$$\forall u \in F\ \forall i \in I_N\ (\sqsubset_i (u) - R_i(u)) \text{ is infinite.}$$

In the transitive (or intuitionistic) case we should take $\beta_F(u)$ *instead of* $R_i(u)$ *in the above condition.*

Thus every world of a small subtree F has infinitely many successors in $F_N T_\omega$ unused in F. Note that every denumerable tree, in which every point is of finite height, is isomorphic to a small subtree of $F_N T_\omega$ (or IT_ω in intuitionistic case).

Lemma 6.4.25

(1) Let L be an N-m.p.l. Then every L-map based on a small subtree F of $F_N T_\omega$ can be extended to a natural L-map based on a (small) subtree $F' \subseteq F_N T_\omega$ such that $F' \supseteq F$.

(2) The same holds for an s.p.l. L and a small subtree F of IT_ω.

Proof
(1) By induction we construct an increasing sequence of small trees

$$F = F_0 \subseteq F_1 \subseteq \ldots$$

and a sequence $h = h_0 \subseteq h_1 \subseteq \ldots$ of L-maps $h_n : F_n \longrightarrow VF_L^{(=)}$ such that $F' := \bigcup\limits_n F_n$ is small, $F_{n+1} - F_n \subset \bigcup\limits_{i=1}^{N} \sqsubset_i (F_n)$ and h_{n+1} 'cures the defects' of h_n:

$$(*) \quad \text{for any } u \in F_n, \text{ if } \Diamond_i A \in h_n(u), \text{ then } \exists v \in R_{i,n+1}(u)\ A \in h_{n+1}(v),$$

cf. Lemma 6.1.16. Actually we may assume that $F_{n+1} - F_n$ consists only of the v constructed by (*). Thus the joined map $h' := \bigcup\limits_n h_n : F' \longrightarrow VF_L^{(=)}$ is selective, i.e. natural, by Lemma 6.4.8. More explicitly this is done as follows. For any N-modal S^*-sentence $\Diamond_i A$ choose the corresponding 'Skolem function'

$$f_{\Diamond_i A} : \xi_L^+(\Diamond_i A) \longrightarrow \xi_L^+(A)$$

such that $\Gamma R_{Li} F_{\Diamond_i A}(\Gamma)$ for any $\Gamma \in \xi_L^+(\Diamond_i A)$. This function exists by the axiom of choice. Now if h_n is constructed and (*) holds for F_{n-1}, then we extend h_n to h_{n+1} by putting

$$h_{n+1}(u(2mN + i - 1)) := f_{\Diamond_i A_m}(h_n(u))$$

whenever $(A_m)_{m\geq 0}$ is the list of all N-modal S^*-sentences, $u \in F_n - F_{n-1}$ and $h_n(u) \in dom(f_{\Diamond_i A_m})$ (i.e. $\Diamond_i A_m \in h_n(u)$). So it follows that $(*)$ holds for F_n — in fact, if $\Diamond_i A_m \in h_n(u)$, the corresponding v is $u(2mN + i - 1)$. It also follows that F_{n+1} is small (if F_n is small). In fact, infinitely many $\sqsubset_i$-successors of $u \in F_n - F_{n-1}$ do not appear in F_{n+1}; they are of the form $u(kN + i - 1)$ with odd k.

(2) In the intuitionistic and in the modal transitive case the definition is slightly modified. In the transitive case one should take $\beta(u)$ instead of $\sqsubset_i (u)$. In the intuitionistic case the condition (*) is replaced with two (where $\Vdash$ denotes forcing in the canonical model):

$$(**) \quad \text{if } h_n(u) \nVdash A \supset B, \text{ then } \exists v \in \beta(u)\ (h_{n+1}(v) \Vdash A \ \&\ h_{n+1}(v) \nVdash B),$$

$$(***) \quad \text{if } h_n(u) \nVdash \forall x A(x), \text{ then } \exists v \in \beta(u)\ \exists c \in D_{h_{n+1}(v)}\ h_{n+1}(v) \nVdash A(c).$$

Similarly to the modal case, we use corresponding Skolem functions $f_{A\supset B}$, $f_{\forall x A(x)}$.

$f_{A\supset B}$ is defined on the complement of $\xi_L^+(A \supset B)$, and $\Gamma R_L f_{A\supset B}(\Gamma)$, $f_{A\supset B}(\Gamma) \in \xi_L^+(A) - \xi_L^+(B)$.

Analogously, $f_{\forall x A(x)}$ is defined on $-\xi_L^+(\forall x A(x))$. ∎

Note that in the intuitionistic case we can construct an $L\forall$-map or $L\forall^{\leq}$-map, or even $L\forall^{<}$-map h' if the original map h is of the corresponding kind. For $L\forall^{<}$ we should not repeat Γ_u in future worlds and similarly for any selective relation (in the intuitionistic or in the reflexive modal case).

Lemma 6.4.25 directly implies Lemma 6.4.12, which is its particular case for a singleton frame on $\{\curlywedge\}$, with $h(\curlywedge) = \Gamma$. A small tree F' from 6.4.25 is clearly isomorphic to a greedy one (or to the whole $F_N T_\omega$, or to IT_ω in the intuitionistic case).

Actually the proof of 6.4.25 is a more explicit version of the proof of 6.4.12. In fact, the intermediate L-maps h_n correspond to steps in calculating Skolem functions used in a well-known proof of the downward Löwenheim–Skolem theorem.

The construction in 6.4.25 extends L-maps by adding new points above the existing ones. Now we shall describe a construction allowing us to add new points (theories) *below* the existing theories. The main idea is to extend theories in a conservative way, by adding only sentences with extra constants.

Although we will further apply this construction to the linear case, we formulate it for arbitrary trees, see Lemma 6.4.30 below.

Definition 6.4.26 *The* domain *of an L-map h (and of the corresponding L-model) is* $D_h^+ := \bigcup_{u\in F} D_{h(u)}$.

Definition 6.4.27 *Let* $h : F \longrightarrow VF_L^{(=)}$ *and* $h' : F' \longrightarrow VF_L^{(=)}$ *be two L-maps. We say that h' is* conservative *over h if for every $u \in F \cap F'$:*

(1) $D_{h(u)} \subseteq D_{h'(u)}$ *and* $D_{h'(u)} \cap D_h^+ \subseteq D_{h(u)}$,

(2) $h(u) \lesssim h'(u)$ *(cf. 6.2.11).*

We say that h' is a conservative extension *of h if it is conservative and $F \subseteq F'$.*

The condition (1) means that additional constants in h'-domains are 'new' for h.

Lemma 6.4.28 *If* $h : F \longrightarrow VF_L^{(=)}$ *and* $h' : F' \longrightarrow VF_L^{(=)}$ *are two L-maps on subtrees F, F' of $F_N T_\omega$ (IT_ω in the intuitionistic case) and h' is conservative over h, then h' can be prolonged to a conservative extension* $h'' : F \cup F' \longrightarrow VF_L^{(=)}$ *of h.*

Usually this lemma is applied to the case $F' \subset F$.

Proof Consider the case when $F' \subset F$. We construct $h''(v)$ for $v \in F - F'$ by induction on its length $l(v)$. To simplify notation, we write Γ_u, Γ'_u, Γ''_u for $h(u)$, $h'(u)$, $h''(u)$ respectively; similarly D_u abbreviates $D_{h(u)}$, etc.

Assume that Γ''_u satisfying (i) and (ii) is already constructed (with $\Gamma''_u = \Gamma'_u$ for $u \in F \cap F'$) and consider $v \in R_i(u)$ (or $v \in \beta(u)$ in the intuitionistic case).

Let us begin with the intuitionistic case. Put $(\Gamma, \Delta) := (\Gamma''_u \cup \Gamma_v, -\Gamma_v)$. Then $D_{(\Gamma,\Delta)} = D''_u \cup D_v$. Let us show that (Γ, Δ) is L-consistent. In fact, suppose

$$\vdash_L A'(\mathbf{c}, \mathbf{c}'') \wedge A(\mathbf{c}, \mathbf{d}) \supset B(\mathbf{c}, \mathbf{d}),$$

where

$$\begin{array}{l} r(\mathbf{c}) \subseteq D_v \cap D''_u = D_u \text{ (by (i) for } \Gamma''_u), \ r(\mathbf{d}) \subseteq D_v - D_u, \ r(\mathbf{c}'') \subseteq D''_u - D_u, \\ A' = \bigwedge\limits_k A'_k, \ A = \bigwedge\limits_i A_i, \ B = \bigvee\limits_j B_j, \ A'_k \in \Gamma''_u, \ A_i \in \Gamma_v, \ B_j \in -\Gamma_v, \end{array}$$

so $A' \in \Gamma''_u$, $A \in \Gamma_v$, $B \in -\Gamma_v$. Put $A''(\mathbf{c}) := \exists \mathbf{x} A'(\mathbf{c}, \mathbf{x})$, where $\mathbf{x}$ is a distinct list corresponding to $\mathbf{c}''$. Then $\vdash_L A''(\mathbf{c}) \wedge A(\mathbf{c}, \mathbf{d}) \supset B(\mathbf{c}, \mathbf{d})$, and

$$A''(\mathbf{c}) \in \Gamma''_u | D_u = \Gamma_u \subseteq \Gamma_v.$$

It follows that Γ_v is L-inconsistent, which is a contradiction.

Now, by Lemma 6.2.6, we extend (Γ, Δ) to an $L\exists$-complete $\Gamma''_v \succeq (\Gamma, \Delta)$; moreover, we choose D''_v such that $D''_v \cap D^+_h \subseteq D_v \cup D''_u$; then $D''_v \cap D^+_h \subseteq D_v$, since $D''_u \cap D^+_h = D_u \subseteq D_v$. Obviously, $\Gamma''_u \subseteq \Gamma''_v$ and $\Gamma''_v | D_v = \Gamma_v$, i.e. $\Gamma_v \lesssim \Gamma''_v$. Therefore by repeating this construction for all $v \in (F - F')$, we obtain a conservative extension of F.

Now let us consider the modal case. We put

$$\Gamma := \Box_i^- \Gamma''_u \cup \Gamma_v;$$

again $D_\Gamma = D''_u \cup D_v$. Let us check the L-consistency of Γ. Suppose

$$\vdash_L \neg(A'(\mathbf{c}, \mathbf{c}'') \wedge A(\mathbf{c}, \mathbf{d})),$$

where $A' = \bigwedge\limits_k A'_k$, $A = \bigwedge\limits_j A_j$, $\Box_i A'_k \in \Gamma''_u$, $A_j \in \Gamma_v$, so $\Box_i A' \in \Gamma''_v$, $A \in \Gamma_v$ ($\mathbf{c}, \mathbf{d}, \mathbf{c}''$ are just the same as in the intuitionistic case). Similarly to that case, put $A''(\mathbf{c}) := \exists \mathbf{x} A'(\mathbf{c}, \mathbf{x})$. Then $\vdash_L \neg(A'' \wedge A)$ and $\Box_i A'' \in \Gamma''_u | D_u = \Gamma_u$. Since $\Gamma_u R_{L_i} \Gamma_v$ (in h), i.e. $\Box_i^- \Gamma_u \subseteq \Gamma_v$, we obtain $A'' \in \Gamma_v$. Hence Γ_v is L-inconsistent, which is a contradiction.

Now by Lemma 6.1.9 we extend Γ to an L-Henkin theory Γ''_v with a set of constants D''_v such that $D''_v \cap D^+_h \subseteq D_v$. Then $\Gamma''_u R_{Li} \Gamma''_v$ and $\Gamma''_v | D_v = \Gamma_v$; in fact, $A \in -\Gamma_v$ implies $\neg A \in \Gamma_v \subseteq \Gamma''_v$, so $A \notin \Gamma''_v$. ■

Lemma 6.4.29 *Under the conditions of Lemma 6.4.28, let $F \cup F'$ be a small subtree of $F_N T_\omega$ (or IT_ω). Then h' can be prolonged to a natural conservative extension h^* of h over a (small) subtree $F^* \supseteq F \cup F'$.*

Proof Apply Lemma 6.4.25 to h'' constructed in Lemma 6.4.28. ■

Lemma 6.4.30 *Let F be a subtree of $F_N T_\omega$ (or IT_ω). Let h be an L-map over $F^- := F - \{\curlywedge\}$. Let Γ be an L-place such that*

(1) $D_\Gamma \cap D_h^+ \subseteq D_u$ for all $u \in F$ (or equivalently, for all $u \in R_i(\curlywedge)$, $i \in I_N$, or for all $u \in \beta(\curlywedge)$ resp.)

(2) in the intuitionistic case: $\Gamma|(D_\Gamma \cap D_h^+) \subseteq \Gamma_u$ *for all $u \in \beta(\curlywedge) \cap F$ (hence for all $u \in F$);*
in the modal case*:* $\Box_i^-(\Gamma|(D_\Gamma \cap D_h^+)) \subseteq \Gamma_u$ *for all $u \in R_i(\curlywedge)$.*

Then there exists a conservative extension h'' of h to F such that $h''(\curlywedge) = \Gamma$.

Proof Repeat the proof of Lemma 6.4.28 word for word. For constructing $h''(v)$ for $v \in R_i(\curlywedge)$ (i.e. for $u = \curlywedge$), we use (ii) in the intuitionistic case and (i) in the modal case. ∎

Remark 6.4.31 Strictly speaking, Lemma 6.4.28 cannot be applied directly here. We cannot just prolong h to F by putting $\Gamma_\curlywedge = \Gamma|(D_\Gamma \cap D_h^+)$, since the latter does not always have the existence property. But this formal obstacle is not really essential.

6.5 Refined completeness theorem for **QH** + KF

Definition 6.5.1 *A p.o. set $F = (W, R)$ is called* coatomic *if it satisfies the* McKinsey property*:*

$$\forall u\, \exists v\, (uRv \;\&\; R(v) = \{v\}),$$

i.e. every world sees a maximal world.

Recall that formula KF is valid in all frames over coatomic p.o. sets.

Definition 6.5.2 *A p.o. set $F = (W, R)$ is called* tree-like *if it is rooted and R^{-1} is non-branching.*

Definition 6.5.3 *Let $IT_\omega^\sharp$ be the set of all ω-paths (i.e maximal chains) in the universal tree IT_ω. Let us extend IT_ω to the tree-like p.o. set $\overline{IT}_\omega = IT_\omega \cup IT_\omega^\sharp$, in which $IT_\omega^\sharp$ is the set of maximal points and*

$$u \leqslant z \Leftrightarrow u \in z$$

for $u \in IT_\omega$, $z \in IT_\omega^\sharp$. A set $G \subseteq IT_\omega^\sharp$ gives rise to the subframe

$$\overline{IT}_\omega^G := \overline{IT}_\omega \upharpoonright (IT_\omega \cup G)$$

of $\overline{IT}_\omega$ with the set of maximal elements G.

Obviously, $\overline{IT}_\omega$ is coatomic and tree-like. $\overline{IT}_\omega^G$ is coatomic iff the paths from G cover IT_ω, i.e.

$$(\alpha) \qquad \forall u \in IT_\omega \ \exists z \in G \ u \in z.$$

It is also clear that if $\overline{IT}_\omega^G$ is coatomic, then there exists a denumerable $G' \subseteq G$ such that $\overline{IT}_\omega^{G'}$ is also coatomic – in fact, for each $u \in IT_\omega$ choose a path $z_u \in G$ containing u and put $G' = \{z_u \mid u \in IT_\omega\}$.

Lemma 6.5.4 *Let L be a superintuitionistic logic containing KF. Then for any $G \subseteq IT_\omega^\sharp$ and for any L-place Γ*

(1) there exists an L-natural model M on $\overline{IT}_\omega^G$ such that $\nu_M(\curlywedge) = \Gamma$;

(2) similarly, there exists an $L^{\leq}\forall$-natural model M on $\overline{IT}_\omega^G$ with $\nu_M(\curlywedge) = \Gamma$.

Proof First, by Lemma 6.4.16 we find an L-natural model M over IT_ω. Obviously, we may assume that the set $\bigcup_u D_u$ is small, i.e. it is a subset of S^* (cf. Section 7.1). Now for every path $z = \{u_k \mid k \in \omega\} \in G$ we put $D'_z := \bigcup_k D_{u_k}$ and consider the L-consistent theory $(\Gamma, \varnothing)$, where $\Gamma := \bigcup_k \Gamma_{u_k}$. By Lemma 6.3.3, this theory is **QCL**-consistent. So by Lemma 6.2.6, there exists a **QCL**$\forall$-place $\Gamma_z \supseteq \Gamma$ with $D_{\Gamma_z} \supseteq D'_z$.

The resulting map $h : \overline{IT}_\omega^G \longrightarrow VF_L$ sending v to Γ_v, is L-natural, since the conditions $(\supset)$, $(\forall)$ for a **QCL**-place Γ_z hold trivially. It is also clear that $\Gamma_u \subseteq \Gamma_z$ whenever $u \in z$ (and $\Gamma_u \leq \Gamma_z$ in the proof of (2)). ∎

Note that this construction can be applied to an arbitrary selective relation R on VP_L (or UP_L); to obtain an (L, R)-natural model, one should only check that $\Gamma_u R \Gamma_z$ holds for $u \in z \in IT_\omega^\sharp$ in the resulting model.

Proposition 6.5.5 **QH** $+ KF = $ **IL**$(\mathcal{K}\overline{IT}_\omega) = $ **IL**$(\mathcal{K}\overline{IT}_\omega^G)$ *for any coatomic poset $\overline{IT}_\omega^G$.*

One can easily obtain the corresponding completeness results for logics with equality, in the semantics $\mathcal{KE}$ — for $\mathbf{QH}^= + KF$, and in $\mathcal{K}$ — for $\mathbf{QH}^{=d} + KF$.

6.6 Directed frames

In this section we prove completeness results from [Corsi and Ghilardi, 1989]. We consider the following formulas:

$ML := \forall x \Diamond \Box P(x) \supset \Diamond \Box \forall x P(x)$,
$ML^* := \Box \forall x \Diamond \Box P(x) \supset \Diamond \Box \forall x P(x)$,
$F := \Box \exists x P(x) \supset \Diamond \exists x \Box P(x)$.

Definition 6.6.1 *A propositional Kripke frame* $F = (W, R)$ *is called* directed *if the relation* R *is reflexive, transitive and such that*

$$\forall u, v \in W \ \exists w \in W \ (uRw \ \& \ vRw).$$

A predicate Kripke frame $\boldsymbol{F} = (W, R, D)$ *is called* directed *if its base* (W, R) *is directed.*

In this section we use specific propositional frames:

Definition 6.6.2

- IT_ω *is called the* subordination frame*;*
- $IT_\omega + 1$ *(*IT_ω *with the added top element* ∞*) is called the* subordination frame with the greatest element*;*
- $IT_\omega + C_\omega$ *(*IT_ω *with the added greatest countable cluster* C_ω*) is called the* subordination frame with the greatest cluster*; the elements of* C_ω *are denoted by* $\infty_0, \ \infty_1, \ldots$*;*
- *the ordinal product* $IT_\omega \cdot \omega$ *is called the* multiple subordination frame*; recall that this is the set* $\omega^\infty \times \omega$ *ordered by the relation*

$$(\alpha, n)R(\beta, m) \text{ iff } n < m \vee (n = m \ \& \ \alpha \leqslant \beta).$$

Definition 6.6.3 *Let* L *be a 1-m.p.l. or s.p.l. A* subordination L-map *is a natural L-map* $h : F \longrightarrow VF_L$*, where* F *is one of the subordination frames from Definition 6.6.2, with the following properties*

(1) the set $D_h^+ := \bigcup\limits_{u \in F} D_{h(u)}$ *is small;*

(2) if $\alpha, \beta, \gamma \in \omega^\infty$ *and* $\alpha \sqsubset \beta$, $\beta \not\leqslant \gamma$*, then* $(D_{h(\beta)} - D_{h(\alpha)}) \cap D_{h(\gamma)} = \varnothing$*;*

(3) if $F = IT_\omega \cdot \omega$*, then* h *is a subordination map on each copy of* IT_ω*:*

$$\forall n \in \omega \, \forall \alpha, \beta, \gamma \in \omega^\infty \, (\alpha \sqsubset \beta \ \& \ \beta \not\leqslant \gamma \Rightarrow (D_{h(\beta, n)} - D_{h(\alpha, n)}) \cap D_{h(\gamma, n)} = \varnothing).$$

In this case $M(h)$ *is called a* subordination L-model.

Lemma 6.6.4 *Let* $L \supseteq$ **QS4** *and let* Γ *be an L-place. Then there exists a subordination L-map* $g : IT_\omega \longrightarrow VF_L$ *such that* $g(\curlywedge) = \Gamma$.

Proof By 6.4.12, we can construct a natural L-map $h : IT_\omega \longrightarrow VF_L$ such that $h(\curlywedge) = \Gamma$, so now we are going to transform it into a subordination L-map g.

To satisfy the conditions 6.6.3 (1), (2), we use renaming of constants. For this purpose we should first trace all constants appearing in D_h^+. Viz., for $c \in D_h^+$ we say that h *introduces* c *at stage* α if α is minimal in the set $\{\beta \in$

$IT_\omega \mid c \in D_{h(\beta)}\}$. The same constants can be introduced at different stages, but they are all incomparable.

Now let $S^* = \{e_n \mid n \in \omega\}$ be an enumeration of our universal set of constants. We choose its small subset S and present it as a two-dimensional array:

$$S = \{c_{n,\alpha} \mid n \in \omega,\ \alpha \in \omega^\infty\}.$$

Next, we define $g : IT_\omega \longrightarrow VF_L$ as follows.

$$g(\alpha) := \{A_\alpha \mid A \in h(\alpha)\},$$

where A_α is obtained from A by replacing all occurrences of each constant e_n with $c_{n,\beta}$ such that $\beta \leqslant \alpha$ and h introduces e_n at stage β; this β is uniquely determined by n and α.

Of course we have to ensure that $g(\alpha)$ is really an L-place. First note that A can be also obtained from A_α by renaming constants. This operation respects provability, so $\vdash_L A_\alpha \Rightarrow \vdash_L A$ for any $D_{h(\alpha)}$-formula A; formally, one should argue by induction on the inference of A. Thus $\nvdash_L \neg \bigwedge_{i=1}^{k} (A_i)_\alpha$ implies $\nvdash_L \neg \bigwedge_{i=1}^{k} A_i$, i.e. the L-consistency of $h(\alpha)$ implies the L-consistency of $g(\alpha)$.

The Henkin property is almost obvious. In fact, if $\exists x A_\alpha(x) = (\exists x A(x))_\alpha \in \mathcal{L}(g(\alpha))$, then $\exists x A(x) \in \mathcal{L}(h(\alpha))$. By the Henkin property for $h(\alpha)$, there is n such that $(\exists x A(x) \supset A(e_n)) \in h(\alpha)$. Therefore $(\exists x A_\alpha(x) \supset A_\alpha(c_{n,\alpha})) \in g(\alpha)$.

Now g satisfies the condition 6.6.3 (1), since the new set of constants S is small. For (2), note that if $\alpha \sqsubset \beta$, then $D_{h(\beta)} - D_{h(\alpha)}$ consists exactly of the constants introduced at stage β, so all constants in $D_{g(\beta)} - D_{g(\alpha)}$ are of the form $c_{n,\beta}$. Thus if $\beta \not\leqslant \gamma$, they cannot appear in $g(\gamma)$. ■

Lemma 6.6.5 *Let $L \supseteq$* **QS4.2**. *If h is a subordination map such that $h(\curlywedge) \supseteq L$, then $\bigcup_{\alpha \in \omega^\infty} \Box^- h(\alpha)$ is L-consistent.*

Proof Suppose $\bigcup_{\alpha \in \omega^\infty} \Box^- h(\alpha)$ is L-inconsistent. Then for some n, $\bigcup_{|\alpha| \leq n} \Box^- h(\alpha)$ is L-inconsistent. Take the smallest such n; obviously $n > 0$.

Note that $\alpha \leqslant \beta$ implies $\Box^- h(\alpha) \subseteq \Box^- h(\beta)$.

In fact, by monotonicity of h,

$$\alpha \leqslant \beta \Rightarrow \Box^- h(\alpha) \subseteq h(\beta).$$

Now $A \in \Box^- h(\alpha)$ implies $\Box A \in h(\alpha)$, hence $\Box\Box A \in h(\alpha)$ by **S4**, so $\Box A \in \Box^- h(\alpha) \subseteq h(\beta)$, i.e. $A \in \Box^- h(\beta)$.

Hence it follows that $\bigcup_{|\alpha| = n} \Box^- h(\alpha)$ is L-inconsistent.

Now since every $\Box^- h(\alpha)$ is $\wedge$-closed, by joining conjunctions, we obtain different sequences $\alpha_1 k_1, \dots, \alpha_m k_m$ of length n and formulas $A_i \in \Box^- h(\alpha_i k_i)$ such that $\{A_i \mid 1 \leq i \leq m\}$ is L-inconsistent. (Note that α_i are not necessarily distinct.) Every A_i can be presented as $[\mathbf{c}_i/\mathbf{x}_i]B_i$, where $r(\mathbf{c}_i) \subseteq D_{h(\alpha_i)}$, $\mathbf{x}_i$ is a distinct list of variables and B_i is a $D_{h(\alpha_i)}$-sentence. Since h is a subordination map, the sets $r(\mathbf{c}_i)$ are disjoint, so we may assume that all $\mathbf{x}_i$ are disjoint. By Lemma 2.7.10

$$\vdash_L \neg \bigwedge_i A_i \text{ implies } \vdash_L \forall \mathbf{x} \neg \bigwedge_i B_i,$$

where $\mathbf{x} = \mathbf{x}_1 \dots \mathbf{x}_m$.

By 2.6.15(xvi), we obtain

$$\vdash_L \neg \exists \mathbf{x} \bigwedge_i B_i, \text{ and thus } \vdash_L \neg \bigwedge_i \exists \mathbf{x}_i B_i$$

by 2.6.15(xxvii), since $\mathbf{x}_i$ are disjoint. Hence $\vdash_L \Box\Diamond\neg \bigwedge_i \exists \mathbf{x}_i B_i$ ($\Box$-introduction is admissible, and $(\Box p \supset \Box\Diamond p) \in \mathbf{S4}$), which implies

$$\vdash_L \neg\Diamond\Box(\bigwedge_i \exists \mathbf{x}_i B_i),$$

and thus

$$(*) \qquad \vdash_L \neg \bigwedge_i \Diamond\Box\exists \mathbf{x}_i B_i,$$

by Lemmas 1.1.1, 1.1.3.

But $h(\alpha_i k_i) \vDash \Box A_i$, so $h(\alpha_i k_i) \vDash \exists \mathbf{x}_i \Box B_i$, which implies $h(\alpha_i k_i) \vDash \Box\exists \mathbf{x}_i B_i$ by 2.6.18 and soundness; thus

$$h(\alpha_i) \vDash \Diamond\Box\exists \mathbf{x}_i B_i,$$

and so

$$h(\alpha_i) \vDash \Box\Diamond\Box\exists \mathbf{x}_i B_i,$$

by 1.1.1 and soundness. This shows that $\Diamond\Box\exists \mathbf{x}_i B_i \in \Box^- h(\alpha_i)$, so eventually $(*)$ implies the L-inconsistency of $\bigcup\limits_{|\alpha|=n-1} \Box^- h(\alpha)$ contrary to the choice of n.

■

Lemma 6.6.6 *Let $L \supseteq$ **QS4.2**. Then for every L-place Γ there is an ω-subordination model M with $\nu_M(\lambda) = \Gamma$.*

Proof Let us define h as follows:

- By Lemma 6.6.4, there is a subordination model M_0 with $\nu_{M_0}(\lambda) = \Gamma$. Put $h(0, \alpha) := \nu_{M_0}(\alpha)$ for all $\alpha \in \omega^\infty$.
- By Lemmas 6.6.5 and 6.1.9, for every n there is an L-place Γ_{n+1} such that $D_{\Gamma_{n+1}} \supseteq \bigcup\limits_{\alpha\in\omega^\infty} D_{h(n,\alpha)}$ and $\Gamma_{n+1} \supseteq \bigcup\limits_{\alpha\in\omega^\infty} \Box^- h(n, \alpha)$. By Lemma 6.6.4, there is a subordination model M_{n+1} with $\nu_{M_{n+1}}(\lambda) = \Gamma_{n+1}$. Then put $h(n+1, \alpha) := \nu_{M_{n+1}}(\alpha)$ for all $\alpha \in \omega^\infty$.

■

From Lemmas 6.6.4, 6.6.6 and 6.1.9 we obtain

Theorem 6.6.7 *The logic* **QS4.2** *is determined by the ω-subordination frame.*

Definition 6.6.8 *An L-consistent theory $\Gamma \subseteq MF_S$ is called a* nearly (L,S)-Henkin *theory if $\Gamma \vdash_L \forall x\, A(x)$, whenever $\Gamma \vdash_L A(c)$ for all $c \in S$.*

Lemma 6.6.9 *(1) Let L be an 1-m.p.l., S a small set of constants. If Γ is a nearly (L,S)-Henkin theory and A is an S-sentence such that $\Gamma \cup \{A\}$ is L-consistent, then $\Gamma \cup \{A\}$ can be extended to an (L,S)-Henkin theory.*

*(2) Let $L \supseteq$ **QK** $+ Ba$. If Γ is a nearly (L,S)-Henkin theory then $\Box^-\Gamma$ is a nearly (L,S)-Henkin theory.*

Proof Standard. ■

Lemma 6.6.10 *Let $L \supseteq$ **QS4** $+ ML^*$ and let $\Gamma \subseteq MF_S$ be L-consistent and L-$\Box$-closed (i.e. $A \in \Gamma$ only if $\Gamma \vdash_L \Box A$). Let also $S \subseteq S'$, $|S'| = |S' - S| = \aleph_0$. Then there exists an L-consistent theory $\Gamma' \subseteq MF_{S'}$ such that $\Gamma \subseteq \Gamma'$ and the following conditions hold:*

(1) $\Box\Diamond\neg B \in \Gamma'$ or $\Box B \in \Gamma'$ for any $B \in MF_{S'}$;

(2) if $\forall x B(x) \in MF_{S'}$ and $\Box\Diamond\neg\forall x\, B(x) \in \Gamma'$, then $\Box\Diamond\neg B(c) \in \Gamma'$ for some $c \in S'$.

Proof Let $\{A_i \mid i \in \omega\}$ be an enumeration of $MF_{S'}$. We define the following sequence of theories:

$\Gamma_0 := \Gamma$;

$$\Gamma_{n+1} := \begin{cases} (1) & \Gamma_n \cup \{\Box\Diamond\neg A_n, \Box\Diamond\neg B(c)\}, \\ & \qquad \text{where } c \text{ does not occur in } \Gamma_n \cup \{A_n\} \\ & \quad \text{if } \Gamma_n \cup \{\Box\Diamond\neg A_n\} \text{ is } L\text{-consistent and } A_n = \forall x\, B(x); \\ (2) & \Gamma_n \cup \{\Box\Diamond\neg A_n\} \text{ if } \Gamma_n \cup \{\Box\Diamond\neg A_n\} \text{ is } L\text{-consistent and} \\ & A_n \text{ is not of the form } \forall x\, B(x); \\ (3) & \Gamma_n \cup \{\Box A_n\} \text{ otherwise.} \end{cases}$$

Each Γ_n is clearly L-$\Box$-closed. We only need to show by induction that Γ_n is L-consistent.

In fact, Γ_0 is L-consistent by assumption. For the induction step, suppose that some Γ_n is L-consistent, but Γ_{n+1} is not. Then Γ_{n+1} is not obtained by (2). Suppose Γ_{n+1} is obtained by (1). Then there are sentences $B_1, \ldots, B_m \in \Gamma_n$ such that

$$L \vdash B_1 \wedge \ldots \wedge B_m \wedge \Box\Diamond\neg\forall x\, B(x) \supset \neg\Box\Diamond\neg B(c).$$

Since c does not occur in Γ_n or in $B(x)$, it follows that

$$L \vdash B_1 \wedge \ldots \wedge B_m \wedge \Box\Diamond\neg\forall x\, B(x) \supset \forall x \neg\Box\Diamond\neg B(x).$$

Since $L \supseteq \mathbf{QS4} + ML^*$, we have

$$L \vdash \Box B_1 \wedge \ldots \wedge \Box B_m \wedge \Box\Diamond\neg\forall x\, B(x) \supset \Diamond\Box\forall x\, B(x),$$

hence

$$L \vdash \Box B_1 \wedge \ldots \wedge \Box B_m \supset \Diamond\Box\forall x\, B(x).$$

Since Γ_n is L-$\Box$-closed, we obtain

$$\Gamma_n \vdash_L \Diamond\Box\forall x\, B(x).$$

This contradicts the L-consistency of $\Gamma_n \cup \{\Box\Diamond\neg A_n\}$. If Γ_{n+1} is obtained by (3), then $\Gamma_n \cup \{\Box\Diamond\neg A_n\}$ is L-inconsistent, and so $\Gamma_n \vdash_L \neg\Box\Diamond\neg A_n$. If Γ_{n+1} is L-inconsistent, then $\Gamma_n \vdash_L \neg\Box A_n$. By Necessitation, $\Gamma_n \vdash_L \Box\Diamond\neg A_n$, and so Γ_n is L-inconsistent, contrary to hypothesis. Thus each Γ_n is L-consistent.

Therefore $\Gamma' := \bigcup\limits_{n\in\omega} \Gamma_n$ is L-consistent and L-$\Box$-closed by a standard argument. The conditions (1) and (2) hold by construction, so Γ' is an (L, S)-Henkin theory. ∎

Lemma 6.6.11 *Let $L \supseteq$ **QS4**. If $\Gamma \subseteq MF_S$ is an L-consistent theory satisfying conditions (1) and (2) of Lemma 6.6.10, then the following holds:*

(1) $\Box^-\Gamma$ is a nearly (L, S)-Henkin theory;

(2) If $\Gamma' \supseteq \Box^-\Gamma$ is L-consistent then $\Box^-\Gamma' \subseteq \Box^-\Gamma$.

(3) If $\Gamma \vdash_L A$ implies $A \in \Gamma$, then (1) and (2) above imply 6.6.10(1) and 6.6.10(2).

Proof

(1) Suppose $\Box^-\Gamma \nvdash_L \forall x\, A(x)$. Then $\forall x\, A(x) \notin \Box^-\Gamma$, and so $\Box\forall x\, A(x) \notin \Gamma$. By 6.6.10(1), then $\Box\Diamond\neg\forall x\, A(x) \in \Gamma$, and by 6.6.10(2), $\Box\Diamond\neg A(c) \in \Gamma$ for some c. Since Γ is L-consistent, $\Gamma \nvdash_L \Box A(c)$, and so $\Box^-\Gamma \nvdash_L A(c)$.

(2) Suppose $A \notin \Box^-\Gamma$. By 6.6.10(1) it follows that $\Box\Diamond\neg A \in \Gamma$, so $\Diamond\neg A \in \Box^-\Gamma$ and $\Diamond\neg A \in \Gamma'$. Since Γ' is L-consistent, $\Box A \notin \Gamma'$ and so $A \notin \Box^-\Gamma'$.

(3) We first check 6.6.10(1). If $\Box\Diamond\neg A \notin \Gamma$, then $\Diamond\neg A \notin \Box^-\Gamma$, and so $\Box^-\Gamma \nvdash_L \Diamond\neg A$. Hence $\Box^-\Gamma \cup \{\Box A\}$ is consistent. From (2) it follows that $\Box^-(\Box^-\Gamma \cup \{\Box A\}) \subseteq \Box^-\Gamma$ and so $A \in \Box^-\Gamma$ and $\Box A \in \Gamma$.

Now suppose $\Gamma \vdash_L \Box\Diamond\neg\forall x\, A(x)$, but for all $c \in S$, $\Gamma \nvdash_L \Box\Diamond\neg A(c)$. Then $\Box\Diamond\neg A(c) \notin \Gamma$, and by 6.6.10(1), $\Box A(c) \in \Gamma$ for all c. Then $A(c) \in \Box^-\Gamma$ and $\Box^-\Gamma \vdash_L \forall x\, A(x)$. It follows that $\Gamma \vdash_L \Box\forall x\, A(x)$. This contradicts the L-consistency of Γ.

∎

Lemma 6.6.12 *Let $L \supseteq$ **QS4.2** $+ ML^*$. Then for every L-place Γ there is a subordination model M with the greatest cluster such that $\nu_M(\lambda) = \Gamma$.*

Proof We construct the corresponding h as follows.

Let M_0 be a subordination model such that $\nu_{M_0}(\curlywedge) = \Gamma$. Such a model exists by Lemma 6.6.4. Let $h = \nu_{M_0}$ on ω^∞. By Lemma 6.6.5, $\Gamma' := \bigcup_{\alpha\in\omega^\infty} \Box^- h(\alpha)$ is L-consistent. One can easily to see that Γ' is L-$\Box$-closed. Let $S'' \supseteq S' = \bigcup_{\alpha\in\omega^\infty} D_{h(\alpha)}$, $|S'' - S'| = \aleph_0$. By Lemma 6.6.10 there is an L-consistent set $\Gamma'' \supseteq \Gamma'$ of S''-sentences with properties 6.6.10(1) and 6.6.10(2). Let $A_1, \dots$ be an enumeration of all S''-sentences B such that $\Box^-\Gamma'' \cup \{B\}$ is L-consistent. For any n we define Γ_{∞_n} as a Henkin (L, S'')-theory extending $\Box^-\Gamma'' \cup \{A_n\}$.

We have $\Box^-\Gamma_\alpha \subseteq \Gamma_\beta$ whenever $\alpha R\beta$ for all $\alpha, \beta \in IT_\omega + C_\omega$. In fact, for $\alpha \in \omega^\infty$ and $n \in \omega$ we have $\Box^-\Gamma_\alpha = \Box^-\Box^-\Gamma_\alpha \subseteq \Box^-\Gamma'' \subseteq \Gamma_{\infty_n}$. For $n, m \in \omega$ we have, by Lemma 6.6.11(2), $\Gamma_{\infty_n} \subseteq \Box^-\Gamma'' \subseteq \Gamma_{\infty_m}$. Therefore, $\alpha R\beta$ implies $\Box^-\Gamma_\alpha \subseteq \Gamma_\beta$.

If $\Diamond A \in \Gamma_{\infty_n}$, then $\Box^-\Gamma'' \cup \{\Diamond A\}$ is L-consistent. By Lemma 6.1.16, there is an L-place Γ^* such that $\Gamma_{\infty_n} R_L \Gamma^*$ and $A \in \Gamma^*$. Since we also have $\Box^-\Gamma_{\infty_n} \subseteq \Gamma^*$, $\Box^-\Gamma_{\infty_n} \cup \{A\}$ is L-consistent and $A = A_m$ for some m. ∎

By applying Lemmas 6.6.4, 6.6.12, 6.1.9 we obtain

Theorem 6.6.13 *The logic* **QS4.2** $+ ML^*$ *is determined by the subordination frame with the greatest cluster.*

Lemma 6.6.14 *Let* $L \supseteq$ **QS4.2.1** $+ ML^*$. *Then for every L-place Γ there is a subordination model M with the greatest element such that with $\nu_M(\curlywedge) = \Gamma$.*

Proof Let $h \restriction \omega^\infty$, Γ'' be the same as in the proof of 6.6.12.

Let us show that $\Box^-\Gamma''$ is an (L, S'')-Henkin theory. Suppose that both $A \notin \Box^-\Gamma''$ and $\neg A \notin \Box^-\Gamma''$. Then $\Box A \notin \Gamma''$ and $\Box\neg A \notin \Gamma''$. By Lemma 6.6.10(1), $\Box\Diamond\neg A \in \Gamma''$ and $\Box\Diamond\neg\neg A \in \Gamma''$, in contradiction with the L-consistency of Γ''. By Lemma 6.6.11(1), $\Box^-\Gamma''$ is a nearly (L, S'')-Henkin theory, and so it is an (L, S'')-Henkin theory as well.

For any $\alpha \in \omega^\infty$ we have $\Box^-\Gamma_\alpha = \Box^-\Box^-\Gamma_\alpha \subseteq \Box^-\Gamma''$ and $D_{\Gamma_\alpha} \subseteq S''$.

Let $\Diamond A \in \Box^-\Gamma''$. If $A \notin \Box^-\Gamma''$, then $\neg A \in \Box^-\Gamma''$ and so $\Box\neg A \in \Gamma''$ and $\Box\neg A \in \Box^-\Gamma''$ contradicting the L-consistency of $\Box^-\Gamma''$. ∎

By applying Lemmas 6.6.4, 6.6.14, 6.1.9 we obtain

Theorem 6.6.15 *The logic* **QS4.2.1**$+ML^*$ *is determined by the subordination frame with the greatest element.*

Now let us consider the intuitionistic case. Analogously to Lemma 6.6.4 we have

Lemma 6.6.16 *Let* $L \supseteq$ **QH** *and let Γ be an L-place. Then there exists a subordination map $g : F \longrightarrow VF_L$ such that $g(\curlywedge) = \Gamma$.*

Proof Same as Lemma 6.6.4, using Lemma 6.4.16 instead of Lemma 6.4.12. We need to check the existence property instead of the Henkin property, but this is trivial. ∎

We also obtain an analogue to Lemma 6.6.6:

Lemma 6.6.17 *Let $L \supseteq \mathbf{QH} + J$ and let h be a subordination map such that $L \supseteq h(\lambda)$. Then $\bigcup_{\alpha \in IT^\omega} h(\alpha)$ is L-consistent.*

Proof The argument is basically the same as in the proof of 6.6.17. By monotonicity, $\alpha \leqslant \beta \Rightarrow h(\alpha) \subseteq h(\beta)$. So the L-inconsistency of $\bigcup_{\alpha \in IT^\omega} h(\alpha)$ implies the L-inconsistency of $\bigcup_{|\alpha| \leq n} h(\alpha)$ for some n. Consider the least such n.

As in 6.6.5, we obtain $A_i \in h(\alpha_i k_i)$ with different $\alpha_i k_i$ of length n such that $\{A_{ij} \mid 1 \leq i, j \leq m\}$ is consistent. Next, we choose maximal generators B_i of A_i, so for distinct $\mathbf{c}_i$ and $\mathbf{x}_i$ $A_i = [\mathbf{c}_i/\mathbf{x}_i]B_i$.

Hence $\vdash_L \forall \mathbf{x}_i \neg \bigwedge_i B_i$, which implies $\vdash_L \neg \bigwedge_i \exists \mathbf{x}_i B_i$ by 6.6.10 (xvi, xxvii), and next

$$(*) \qquad \vdash_L \neg \bigwedge_i \neg\neg\exists \mathbf{x}_i B_i,$$

since $\mathbf{QH} \vdash p \supset \neg\neg p$. On the other hand, $h(\alpha_i k_i) \models \exists \mathbf{x}_i B_i$, thus $h(\alpha_i) \not\models \neg\exists \mathbf{x}_i B_i$, which implies

$$(**) \qquad h(\alpha_i) \models \neg\neg\exists \mathbf{x}_i B_i,$$

since $h(\alpha_i) \models \neg\exists \mathbf{x}_i B_i \vee \neg\neg\exists \mathbf{x}_i B_i$ by AJ. Now $(*), (**)$ imply the L-inconsistency of $\bigcup_{|\alpha| = n-1} h(\alpha)$. ∎

Corollary 6.6.18 *Let $L \supseteq \mathbf{QH} + J$ and let h be a subordination map such that $L \supseteq h(\lambda)$. Then $\bigcup_{\alpha \in T} h(\alpha)$ is L-consistent for any standard subtree T of IT^ω.*

Proof Obvious. ∎

Lemma 6.6.19 *Let $L \supseteq \mathbf{QH} + J$ and let Γ be an L-place. Then there exists an ω-subordination model M such that $\nu_M(\lambda) = \Gamma$.*

Proof Same as Lemma 6.6.6, replacing modal lemmas with their analogues. ∎

From Lemmas 6.6.16, 6.6.19 and 6.2.6 we obtain

Theorem 6.6.20 *The logic $\mathbf{QH}+J$ is determined by the ω-subordination frame.*

Lemma 6.6.21 *Let $L \supseteq \mathbf{QH} + KF + J$ and let Γ be an L-place. Then there exists an subordination map with the greatest element $h : IT_\omega \longrightarrow VF_L$ such that $h(\lambda) = \Gamma$.*

Proof Let g be the subordination map given by Lemma 6.6.16. By Lemmas 6.6.17, 6.2.6 and 6.3.4 there exists a **QCL**$\exists$-complete $L\forall$-place $\Gamma' \supseteq \bigcup_{\alpha \in IT^\omega} g(\alpha)$. Let $h(\alpha) = g(\alpha)$ for all $\alpha \in IT^\omega$, $h(\infty) = \Gamma'$.

Since Γ' is $\leq$-maximal and $\leq$ is selective on UP_L, we observe that $A \in \Gamma'$ and $B \notin \Gamma'$ whenever $(A \supset B) \notin \Gamma'$ and $\forall x A(x) \in \Gamma'$ whenever $A(c) \in \Gamma'$ for all $c \in D_{\Gamma'}$. Therefore h is natural. ∎

Hence we obtain

Theorem 6.6.22 *The logic* $\mathbf{QH} + KF + J$ *is determined by the subordination frame with the greatest element.*

6.7 Logics of linear frames

In this section we prove strong Kripke-completeness for the logic **QLC** w.r.t. denumerable linear Kripke frames, or moreover, w.r.t. Kripke frames over the set of nonnegative rationals $\mathbb{Q}_+$.

The first proof of this result found by Corsi [Corsi, 1992] used the 'method of finite diagrams', which is a modification of the original S. Kripke's method of semantic tableaux for the linear case, and it was rather laborious and complicated. Another proof using linear natural models was given in [Skvortsov, 2005]. Now we present a new, essentially simplified version of the latter proof basing on Lemma 6.4.28.

Let $\mathcal{C}_{\text{lin}}$ be the class of all linearly ordered sets, $\mathcal{C}^\omega_{\text{lin}}$ the class of all denumerable sets from $\mathcal{C}_{\text{lin}}$.

Recall that every set from $\mathcal{C}^\omega_{\text{lin}}$ is embeddable in $\mathbb{Q}$.

Theorem 6.7.1

(1) $\mathbf{QLC} = \mathbf{IL}(\mathcal{K}(\mathcal{C}_{\text{lin}})) = \mathbf{IL}(\mathcal{K}(\mathcal{C}^\omega_{\text{lin}}))$;

(2) $\mathbf{QLC}^= = \mathbf{IL}^=(\mathcal{KE}(\mathcal{C}_{\text{lin}})) = \mathbf{IL}^=(\mathcal{KE}(\mathcal{C}^\omega_{\text{lin}}))$;

(3) $\mathbf{QLC}^{=d} = \mathbf{IL}^=(\mathcal{K}(\mathcal{C}_{\text{lin}})) = \mathbf{IL}(\mathcal{K}(\mathcal{C}^\omega_{\text{lin}}))$.

Moreover, strong completeness holds in all these cases.

Lemma 6.7.2 [2] *If F is a countable linearly ordered set, then* $\mathbb{Q} \twoheadrightarrow F$.

Theorem 6.7.3

(1) $\mathbf{QLC} = \mathbf{IL}(\mathcal{K}(\mathbb{Q}_+)) = \mathbf{IL}(\mathcal{K}(\mathbb{Q}))$;

(2) $\mathbf{QLC}^= + \mathbf{IL}^=(\mathcal{KE}(\mathbb{Q}_+)) = \mathbf{IL}^=(\mathcal{KE}(\mathbb{Q}))$;

(3) $\mathbf{QLC}^{=d} = \mathbf{IL}^=(\mathcal{KE}(\mathbb{Q}_+)) = \mathbf{IL}^=(\mathcal{KE}(\mathbb{Q}))$.

[2] [Takano, 1987]

Proof From the previous lemma it follows that $\mathbb{Q}_+ \twoheadrightarrow F$ for any rooted $F \in C^{\omega}_{\text{lin}}$. Then apply Theorem 6.7.1 and Proposition 3.3.14. ∎

Soundness $\mathbf{QLC} \subseteq \mathbf{IL}(\mathcal{K}(\mathcal{C}_{\text{lin}}))$ follows from 3.2.34.

Let us prove strong completeness. Let $L \supseteq \mathbf{QLC}$ be a superintuitionistic predicate logic.

Definition 6.7.4 *A* finite L-chain *is an L-map over a finite linearly ordered set $W = \{0, 1, \dots, k\}$, $0 < 1 < \dots < k$, i.e. a sequence of L-places $M = (\Gamma_0, \Gamma_1, \dots, \Gamma_k)$ such that $\Gamma_i \subseteq \Gamma_{i+1}$ for $i < n$. Recall that $D^+_M := \bigcup\limits_i D_{\Gamma_i} = D_{\Gamma_m}$.*

Comment We shall construct a natural L-model step-by-step; L-chains will be finite fragments of this model at different stages of the construction. As usual, if say, $(A \supset B) \notin \Gamma_j$, then we have to insert a new point above j, i.eà new theory Γ' extending Γ_j such that $A \in \Gamma'$, $B \notin \Gamma'$. But in general we cannot put this Γ' above the whole n-chain, because it may happen that $(A \supset B) \in \Gamma_{j+1}$. Then we have to insert a new point between j and $(j+1)$, and after extending the theory $(\Gamma_j \cup \{A\}, \{B\})$ to Γ' by Lemma 6.2.6, we should extend Γ_{j+1} to a new theory Γ'_{j+2} in a conservative way, etc. and thus obtain a conservative extension $(\Gamma'_{i+1} \mid i > j)$ of $(\Gamma_i \mid i > j)$, which together with the 'old' $(\Gamma_i \mid i \leq j)$ and the 'new' Γ' makes a $(k+2)$-element L-chain extending the original $(k+1)$-element L-chain. To make this, we apply Lemma 6.4.28 (its condition is satisfied due to the linearity axiom).

Lemma 6.7.5 *Let $M = (\Gamma_0, \Gamma_1, \dots, \Gamma_k)$ be a finite L-chain and assume that for a certain $j < k$*

(1) $(A_1 \supset A_2) \in \Gamma_{j+1} - \Gamma_j$
or

(2) $\forall x A(x) \in \Gamma_{j+1} - \Gamma_j$.

Then there exists a finite L-chain $M' = (\Gamma_0, \dots, \Gamma_j, \Gamma'_{j+1}, \Gamma'_{j+2}, \dots, \Gamma'_{k+1})$ such that $D_{\Gamma'_{j+1}} \cap D^+_M = D_{\Gamma_j}$, $M' | (\{0, \dots, k+1\} - \{j+1\})$ is a conservative extension of M and respectively:

(1′) $A_1 \in \Gamma'_{j+1}$, $A_2 \notin \Gamma'_{j+1}$ *in case (1),*

(2′) $A(c) \notin \Gamma'_{j+1}$ *for some* $c \in D_{\Gamma'_{j+1}} - D_{\Gamma_j}$ *in case (2).*

Comment Between Γ_j and Γ_{j+1} we insert a new theory Γ'_{j+1} satisfying the standard falsity condition for $(A \supset B)$ or $\forall y A(y)$. This position is exact, for it is impossible to put Γ'_{j+1} after Γ_{j+1} and it is not necessary to put it before Γ_j. Along with putting Γ'_{j+1} before Γ_{j+1}, we extend the members of the former subchain $\Gamma_{j+1}, \dots, \Gamma_k$ to $\Gamma'_{j+2}, \dots, \Gamma'_{k+1}$ (by applying Lemma 6.4.28). Of course, we may not preserve the conservativity, if we use (say, in $\Gamma'_{j+1} - \Gamma_j$ or in $\Gamma'_{j+2} - \Gamma_{j+1}$, etc.) the constants that are already present in $\Gamma_{j+1}, \Gamma_{j+2}$, etc. Thus we should add only new constants.

Proof Consider the following L-consistent theory (Γ, Δ) in the case (1) or (2) respectively:

(1) $(\Gamma, \Delta) := (\Gamma_j \cup \{A_1\}, \{A_2\})$,

(2) $(\Gamma, \Delta) := (\Gamma_j, \{A(c)\})$, with $c \notin D_M^+$.

By Lemma 6.2.6, there exists an L-place Γ'_{j+1} such that $(\Gamma, \Delta) \preceq \Gamma'_{j+1}$ and $D_{\Gamma'_{j+1}} \cap D_M^+ = D_{\Gamma_j}$ (i.e. we add only new constants). Then $\Gamma_j \subseteq \Gamma'_{j+1}$ and the condition (1′) or (2′) (respectively) holds.

If $j = k$, the construction terminates. So consider the case $j < k$. Put $B_0 := (A_1 \supset A_2)$ or $B_0 := \forall x A(x)$ in case (1) or (2) respectively; then $B_0 \in (\Gamma_{j+1}|D_{\Gamma_j}) - \Gamma'_{j+1}$.

Let us show that $\Gamma'_{j+1}|D_{\Gamma_j} \subseteq \Gamma_{j+1}$. In fact, let $B \in \Gamma'_{j+1}|D_{\Gamma_j}$. By **LC**, $(B \supset B_0) \vee (B_0 \supset B) \in \Gamma_j$. Also $(B \supset B_0) \notin \Gamma_j$, since otherwise $B \in \Gamma'_{j+1}$ implies $B_0 \in \Gamma'_{j+1}$. Thus $(B_0 \supset B) \in \Gamma_j \subseteq \Gamma_{j+1}$, and so $B \in \Gamma_{j+1}$, since $B_0 \in \Gamma_{j+1}$.

Now Lemma 6.4.28 yields a conservative extension $(\Gamma'_{j+2}, \dots, \Gamma'_{k+1})$ of $(\Gamma_{j+1}, \dots, \Gamma_k)$ such that $\Gamma'_{j+1} \subseteq \Gamma'_{j+2}$. This completes the construction; obviously $M'|W$ is a conservative extension of M. ∎

Proposition 6.7.6 *Let Γ be an L-place. Then there exists a natural L-model M over a subframe (not necessarily generated) $F \subseteq \mathbb{Q}_+$ with the root 0 such that $\nu_M(0) = \Gamma$.*

Proof Rather straightforward. Consider an enumeration $B_0, B_1, B_2, \dots$ of the set $IF(S^*)$ of S^*-sentences (where S^* is the universal set of constants as usual), in which every sentence occurs infinitely many times. We construct an exhausting sequence of finite subsets of $\mathbb{Q}_+$: $\{0\} = W_0 \subseteq W_1 \subseteq \dots$ and a sequence of finite L-chains $h_n : M_n \longrightarrow VF_L$ over W_n (for $n \in \omega$) such that $h_n(0) = \Gamma$ (for all n) and every restriction $M_{n+1}|W_n$ is a conservative extensions of M_n.

The inductive step is as follows. Assume that M_n is already constructed, with $W_n = \{u_0, \dots, u_k\}$, $h_n(u_i) = \Gamma_{u_i}$, and B_n has the form $(A_1 \supset A_2)$ or $\forall x A(x)$, so that $B_n \in (-\Gamma_{u_i})$ for some $i \leq k$. Then we choose the greatest $j \leq k$ such that $B_n \notin \Gamma_{u_j}$ and construct an L-chain M_{n+1} over $W_{n+1} = \{u_0, \dots, u_j, v, u_{j+1}, \dots, u_n\}$, where $u_j < v < u_{j+1}$, according to Lemma 6.7.5; so $M_{n+1}|W_n$ is a conservative extension of M_n.

Now put $W := \bigcup\limits_n W_n$; $M := \bigcup\limits_n M_n$, with $h(u) = h_n(u)$ whenever $u \in W_n$.

Thus $h(0) = \Gamma$ and every $M|W_n$ is a conservative extension of M_n. By the construction it is clear that the L-map h is natural. In fact, if $(A_1 \supset A_2) \in -h(u)$ for some $u \in W$, then $(A_1 \supset A_2) \in -h_n(u)$ for some n such that $u \in W_n$ and $B_n = (A_1 \supset A_2)$ (recall that $(A_1 \supset A_2)$ occurs infinitely many times in our enumeration). Let $W_n = \{u_0, \dots, u_k\}$, $u = u_i$. Then $W_{n+1} = \{u_0, \dots, u_j, v, u_{j+1}, \dots, u_k\}$ for some $j \geq i$, and $A_1 \in h_{n+1}(v)$, $A_2 \notin h_{n+1}(v)$. So $A_1 \in h(v)$, $A_2 \notin h(v)$ (by conservativity) and $v > u_j \geq u_i = u$ in W.

The case $\forall x A(x) \in -h(u)$ is quite similar. ∎

Corollary 6.7.7 **QLC** *is strongly Kripke complete.*

Remark 6.7.8 The completeness proof given in [Skvortsov, 2005] follows the same scheme, but instead of Lemma 6.4.28, it uses the notion of a 'prechain' under a finite L-chain $M = (\Gamma_{u_1}, \ldots, \Gamma_{u_k})$ over $W = \{u_1, \ldots, u_k\}$. Such a prechain has the form $((\Gamma, \Delta), \Gamma_{u_1}, \ldots, \Gamma_{u_k})$ where (Γ, Δ) is an L-consistent theory. It should satisfy a special condition, when it can be extended it to a finite L-chain $M' = (\Gamma'_{u_0}, \Gamma'_{u_1}, \ldots, \Gamma'_{u_k})$ such that $(\Gamma, \Delta) \preceq \Gamma'_{u_0}$ and $M'|W$ is a conservative extension of M. But this notion makes sense only for L-chains M, where $\Gamma_{u_1} < \ldots < \Gamma_{u_k}$ (i.e. all extensions $\Gamma_{u_i} < \Gamma_{u_{i+1}}$ are proper). So for constructing a natural $L^<$-model (cf. the proof of Proposition 6.7.6) one has to take care of properness of all inclusions between L-places. But Lemma 6.4.28 allows us to avoid these complications in the above proof.

Corollary 6.7.9 *Let* $\mathbf{F}^Q := (\mathbb{Q}_+, D^Q)$ *be a Kripke frame over* $\mathbb{Q}_+$*, in which* $D^Q_u := \omega \times \{v \in \mathbb{Q}_+ \mid v \leq u\}$. *Then* $\mathbf{IL}(\mathbf{F}^Q) = \mathbf{QLC}$.

Note that here $|D_u - \bigcup\limits_{v<u} D_v| = \aleph_0$ for every $u \in \mathbb{Q}_+$.

Proof Let $\mathbf{F}' = (\mathbb{Q}_+, D')$ be a Kripke frame with denumerable domains. Then we can construct a p-morphism $f = (f_0, f_1) : \mathbf{F}^Q \twoheadrightarrow \mathbf{F}'$. Namely, we put $f_0 := id_{\mathbb{Q}_+}$, and for any $u \in \mathbb{Q}_+$ fix a surjective map $f'_u : \omega \times \{u\} \longrightarrow D'_u$. Then we define $f_u : D^Q_u \longrightarrow D'_u$ such that $f_u|\omega \times \{v\} = f'_v$ for any $v \leq u$.

Obviously, (f_0, f_1), where $f_1 = (f_u \mid u \in \mathbb{Q}_+)$, is a p-morphism. ∎

From Chapter 3 it follows that $\mathbf{QLC}^=$ is characterised by all KFEs over $\mathbf{F}^Q = (\mathbb{Q}_+, D^Q)$. On the other hand, $\mathbf{QLC}^{=d} \neq \mathbf{IL}^=(\mathbf{F}^Q)$. Moreover, there does not exist a single KFE $\mathbf{F}$ over $\mathbb{Q}_+$ such that $\mathbf{IL}^=(\mathbf{F}) = \mathbf{QLC}^=$, or a Kripke frame $\mathbf{F}$ over $\mathbb{Q}_+$ such that $\mathbf{IL}^=(\mathbf{F}) = \mathbf{QLC}^{=d}$. In fact, one cannot refute both formulas $\neg\forall xy(x = y)$ and $\neg\neg\forall xy(x = y)$, which are obviously not in $\mathbf{QLC}^{=d}$, in the same KFE over $\mathbb{Q}_+$ (e.g. since $\mathbb{Q}_+ \Vdash J$).

Now let us consider the modal case. Let us prove the result on strong Kripke completeness of **QS4.3** from [Corsi, 1989]. Our proof follows the same lines as 6.7.1. Finite L-chains are defined as in 6.7.4, i.e. $0 < 1 < \cdots < k$ corresponds to $\Gamma_0 R_L \Gamma_1 R_L \ldots R_L \Gamma_k$ (with perhaps $\Gamma_{j+1} R_L \Gamma_j$ for some j). Lemma 6.7.5 is replaced with the following

Lemma 6.7.10 *Let* $M = (\Gamma_0, \Gamma_1, \ldots, \Gamma_k)$ *be a finite* L*-chain such that for a certain* $j \leq k$

$$\Diamond A \in \Gamma_j - \Gamma_{j+1}.$$

Then there exists a finite L*-chain*

$$M' = (\Gamma_0, \ldots, \Gamma_j, \Gamma'_{j+1}, \Gamma'_{j+2}, \ldots, \Gamma'_{k+1})$$

such that $D_{\Gamma'_{j+1}} \cap D^+_M = D_{\Gamma_j}$, $M'|(\{0, \ldots, k+1\} - \{j+1\})$ *is a conservative extension of* M *and*

$$A \in \Gamma'_{j+1}. \qquad (*)$$

Note that in this case $\neg(\Gamma_{j+1} R_L \Gamma_j)$, since $\Gamma_j \models \Diamond A$, while $\Gamma_{j+1} \not\models \Diamond A$.

Proof There exists an L-place Γ'_{j+1} such that $\Box^-\Gamma_j \cup \{A\} \subseteq \Gamma'_{j+1}$ and $D_{\Gamma'_{j+1}} \cap D^+_M = D_{\Gamma_j}$ (we add only new constants); so $\Gamma_j R_L \Gamma'_{j+1}$ and $A \in \Gamma'_{j+1}$ as required. Now we can apply Lemma 6.4.28 for extending $(\Gamma_{j+1}, \ldots, \Gamma_k)$ to $(\Gamma'_{j+2}, \ldots, \Gamma'_{k+1})$ in a conservative way so that $\Gamma'_{j+1} R_L \Gamma'_{j+2}$; it is sufficient to check that

$$\Box^-(\Gamma'_{j+1}|D_{\Gamma_j}) \subseteq \Gamma_{j+1}; \qquad (**)$$

recall that $D_{\Gamma'_{j+1}} \cap D^+_M = D_{\Gamma_j}$.

In fact, let $\Box B \in \Gamma'_{j+1}$, $B \in \mathcal{L}(D_{\Gamma_j})$. Note that $\Box(\Box B \supset \neg A) \vee \Box(\Box\neg A \supset B) \in \Gamma_{u_j}$ since $\overline{\mathbf{QS4.3}} \subseteq \Gamma_j$. Also $\Box(\Box B \supset \neg A) \notin \Gamma_j$, since otherwise $(\Box B \supset \neg A) \in \Gamma'_{j+1}$ (remember that $\Gamma_j R_L \Gamma'_{j+1}$), and so $\Box B \in \Gamma'_{j+1}$ would imply $\neg A \in \Gamma'_{j+1}$. Hence $\Box(\Box\neg A \supset B) \in \Gamma_j$, so $(\Box\neg A \supset B) \in \Gamma_{j+1}$, since $\Gamma_j R_L \Gamma_{j+1}$. Now $B \in \Gamma_{j+1}$ follows from $\Box\neg A \in \Gamma_{j+1}$. ■

Proposition 6.7.11 **QS4.3** *is strongly Kripke complete.*

Proof Repeat the proof of 6.7.7 with the inductive step for $B_n = \Diamond A$. Note that $i \leq j$ & $\Diamond A \in \Gamma_j \Rightarrow \Diamond A \in \Gamma_i$. So we choose the largest j such that $\Diamond A \in \Gamma_j$ and then apply 6.7.10. ■

Now we obtain full analogues of the intuitionistic completeness results:

Theorem 6.7.12

(1) $\mathbf{QS4.3} = \mathbf{ML}(\mathcal{K}(\mathcal{C}_{\text{lin}})) = \mathbf{ML}(\mathcal{K}(\mathcal{C}^{\omega}_{\text{lin}}))$;

(2) $\mathbf{QS4.3}^{=} = \mathbf{ML}^{=}(\mathcal{KE}(\mathcal{C}_{\text{lin}})) = \mathbf{ML}^{=}(\mathcal{KE}(\mathcal{C}^{\omega}_{\text{lin}}))$;

(3) $\mathbf{QS4.3}^{=d} = \mathbf{ML}^{=}(\mathcal{K}(\mathcal{C}_{\text{lin}})) = \mathbf{ML}(\mathcal{K}(\mathcal{C}^{\omega}_{\text{lin}}))$.

Again strong completeness holds in all these cases.

Corollary 6.7.13

(1) $\mathbf{QS4.3} = \mathbf{ML}(\mathcal{K}(\mathbb{Q}_+)) = \mathbf{ML}(\mathcal{K}(\mathbb{Q}))$;

(2) $\mathbf{QS4.3}^{=}\mathbf{ML}^{=}(\mathcal{KE}(\mathbb{Q}_+)) = \mathbf{IL}^{=}(\mathcal{KE}(\mathbb{Q}))$;

(3) $\mathbf{QS4.3}^{=d} = \mathbf{ML}^{=}(\mathcal{KE}(\mathbb{Q}_+)) = \mathbf{ML}^{=}(\mathcal{KE}(\mathbb{Q}))$.

6.8 Properties of Δ-operation

In this section we use completeness to prove syntactic properties of the Δ-operation introduced in Section 2.13. First let us introduce some notation. For a poset F, let F_- be its restriction to the set of all nonminimal elements (of course, it may happen that $F_- = F$.) Then by induction we define:

$$F_{-0} := F, \; F_{(n+1)} := (F_{-n})_-.$$

If $\mathbf{F}$ is an intuitionistic Kripke frame or a KFE (or a Kripke sheaf) over F, let $\mathbf{F}_-$ be the restriction $\mathbf{F} \upharpoonright F_-$; $\mathbf{F}_{-n} := \mathbf{F} \upharpoonright F_{-n}$.

Proposition 6.8.1 *For any intuitionistic formula A and $k \geq 0$*

$$\mathbf{F} \Vdash \delta_k A \textit{ iff } \mathbf{F}_- \Vdash A.$$

Proof We write A as $A(\mathbf{x})$ for $r(\mathbf{x}) = FV(A)$.

(If.) Let us consider Kripke frames or KFEs; the case of sheaves is similar. Suppose

$$\mathbf{F} \nVdash \delta_{k,P}A(\mathbf{x}) = \forall \mathbf{y}(P(\mathbf{y}) \vee (P(\mathbf{y}) \supset A(\mathbf{x})))$$

for disjoint $\mathbf{x}$, $\mathbf{y}$. Then there exist a Kripke model M over $\mathbf{F}$, a world w, and tuples $\mathbf{a}, \mathbf{b}$ in D_w such that $M, w \nVdash P(\mathbf{a}),\ P(\mathbf{a}) \supset A(\mathbf{b})$.

So there is $v \in R(w)$ (where R is the accessibility relation in $\mathbf{F}$) such that $M, v \Vdash P(\mathbf{a})$ and $M, v \nVdash A(\mathbf{b})$. Then $w \neq v$, so v is not minimal, i.e. $v \in \mathbf{F}_-$. Let M_- be the restriction $M \upharpoonright \mathbf{F}_-$; since it is a generated submodel, we have $M_-, v \nVdash A(\mathbf{b})$ by 3.3.18. Hence $\mathbf{F}_- \nVdash A$.

(Only if.) Suppose $\mathbf{F}_- \nVdash A(\mathbf{x})$. Then by 3.3.18 there is a model M over $\mathbf{F}$ such that $M, v \nVdash A(\mathbf{a})$ for some $v \in \mathbf{F}_-$ and a tuple $\mathbf{a}$ in D_v. Then consider a model M' over $\mathbf{F}$ differing from M only on P, i.e. such that for any w, $Q \neq P$ and a tuple of individuals $\mathbf{b}$ of the corresponding length

$$M', w \Vdash Q(\mathbf{b}) \text{ iff } M, w \Vdash Q(\mathbf{b}),$$

and also such that

$$M', w \Vdash P(\mathbf{c}) \text{ iff } w \neq u_0$$

for any $\mathbf{c} \in D_w^k$, where u_0 is the root of $\mathbf{F}$. Now by induction it easily follows that

$$M', w \Vdash B \text{ iff } M, w \Vdash B$$

for any world w and D_w-sentence B that does not contain P. Thus

$$M', v \nVdash A(\mathbf{a});\ \ M', v \Vdash P(\mathbf{c}),$$

and on the other hand, $M', u_0 \nVdash P(\mathbf{c})$. Hence

$$M', u_0 \nVdash P(\mathbf{c}) \vee (P(\mathbf{c}) \supset A(\mathbf{a})),$$

and therefore $M', u_0 \nVdash \overline{\forall}\delta_k A$ implying $\mathbf{F} \nVdash \delta_k A$. ■

So the validity of δA at some world means that 'A will be true tomorrow'. All $\delta_k A$ are $\mathcal{KE}$-indistinguishable from δA; this also follows from their deductive equivalence:

$$\mathbf{QH} + \delta_k A = \mathbf{QH} + \delta A,$$

cf. Proposition 2.13.22.

Corollary 6.8.2 *Let $\mathbf{F}$ be a rooted predicate Kripke frame (or a KFE, or a Kripke sheaf).*

(1) $\mathbf{F} \Vdash \delta_k A$ *iff* $\forall v \in \mathbf{F}_- \ \mathbf{F}{\uparrow}v \Vdash A$.

(2) $\Delta L \subseteq \mathbf{IL}^{(=)}(\mathbf{F})$ *iff* $L \subseteq \mathbf{IL}^{(=)}(\mathbf{F}_-)$.

(3) $\Delta^n L \subseteq \mathbf{IL}^{(=)}(\mathbf{F})$ *iff* $L \subseteq \mathbf{IL}^{(=)}(\mathbf{F}_{-n})$, *for* $n \in \omega$.

Proof (1) By 3.3.21

$$\mathbf{F}_- \Vdash A \text{ iff } \forall v \in \mathbf{F}_- \ \mathbf{F}_-{\uparrow}v \Vdash A.$$

It remains to note that $\mathbf{F}_-{\uparrow}v = \mathbf{F}{\uparrow}v$.

(2) Readily follows from 6.8.1, the definition of ΔL and soundness.

(3) Follows from (2) by induction on n, since $\mathbf{F}_{-(n+1)} = (\mathbf{F}_{-n})_-$. ■

Lemma 6.8.3 *Let* $\mathbf{F} = (F, D)$ *be an intuitionistic Kripke frame, in which the intersection of all individual domains is non-empty, and let* $a_0 \in \bigcap_{u \in F} D_u$. *Next, let* M *be an intuitionistic Kripke model over* $\mathbf{F}$, L *an s.p.l. such that* $M \Vdash \overline{L}$. *Consider an intuitionistic Kripke model* $M' = \delta M$ *obtained by adding a new root* u_0 *with the domain* $\{a_0\}$ *below* M, *such that* $M' \restriction \mathbf{F} = M$ *and* $M', v \Vdash p$ *iff* $v = u_0$. *Then* $M' \Vdash \overline{\text{Sub}}(\delta \overline{L})$.

Proof If $A = p \vee (p \supset B) \in \delta \overline{L}$ (where $B \in \overline{L}$, and p does not occur in B), then every strict substitution instance of A has the form $C \vee (C \supset B')$ for a sentence C and $B' \in \overline{Sub}(\overline{L}) \subseteq \overline{L}$. So let us show

$$M', u_0 \Vdash C \vee (C \supset B').$$

In fact, otherwise $M', u_0 \nVdash C, C \supset B'$, and then $M', v \nVdash B'$ for some $v \in M$. Since M is a generated submodel, it follows that $M, v \nVdash B'$ contradicting $M \Vdash \overline{L}$. ■

Lemma 6.8.4 *For any intuitionistic formula* A *and an s.p.l.* L

$$\Delta L \vdash \delta A \textit{ iff } L \vdash A.$$

Proof Although this lemma is syntactic, we know only a model-theoretic proof.

(If.) Obvious, by the definition of ΔL.

(Only if.) Again we write A as $A(\mathbf{x})$. Suppose $L \nvdash A(\mathbf{x})$, or equivalently, $\overline{L} \nvdash A \times (\mathbf{x})$. Then by the canonical model theorem and the generation lemma 3.3.12, the there exists a Kripke model M over $\mathbf{F}$ with a root v_0 such that $M, v_0 \Vdash \overline{L}$ and $M, v_0 \nVdash A(\mathbf{a})$ for a tuple $\mathbf{a}$ in D_{u_0}. Consider the model $M' = \delta M$ defined in Lemma 6.8.3. By that lemma we have

$$M', u_0 \Vdash \overline{\text{Sub}}(\delta \overline{L}).$$

Since M and M' coincide on $\mathbf{F}$, an inductive argument as in the proof of 6.8.1 shows that

$$M', w \Vdash B \Longleftrightarrow M, w \Vdash B$$

for any $w \in M$ and a D_w-sentence B that does not contain p. Hence $M', u_0 \not\Vdash A(\mathbf{a})$, and thus $M', v_0 \not\Vdash \delta A(\mathbf{a})$.

Now put $\Gamma := \overline{Sub}(\delta \overline{L})$. By truth-preservation $M', v_0 \Vdash \Gamma$ implies $M' \Vdash \Gamma$; hence by Lemma 3.2.31 applied to Γ and $L = \mathbf{QH}$, we obtain

$$\overline{Sub}(\delta \overline{L}) \not\vdash_{\mathbf{QH}} \delta A(\mathbf{x}),$$

i.e.

$$\Delta L \not\vdash_{\mathbf{QH}} \delta A(\mathbf{x})$$

by the deduction theorem. ■

Proposition 6.8.5 *For any s.p.l.* L_1, L_2

(1) $L_1 \subseteq L_2$ *iff* $\Delta L_1 \subseteq \Delta L_2$*;*

(2) $L_1 = L_2$ *iff* $\Delta L_1 = \Delta L_2$.

Proof
(i) 'Only if' is trivial. For 'if' note that $A \in L_1 - L_2$ implies $\delta A \in \Delta L_1 - \Delta L_2$ by 6.8.4.

(ii) Follows from (i). ■

So as in the propositional case, Δ is monotonic.

Let us now prove an analogue to Lemma 1.16.10.

Lemma 6.8.6 *For s.p.l.s* L_1 *and* L_2 *and sentences* A_1, A_2, *if* $A_1 \in (L_1 - L_2)$ *and* $A_2 \in (L_2 - L_1)$, *then* $(\delta A_1 \lor \delta A_2) \in (\Delta L_1 \cap \Delta L_2 - \Delta(L_1 \cap L_2))$.

Proof Since $\Delta L_1 \vdash \delta A_1$ and $\Delta L_2 \vdash \delta A_2$, it follows that

$$\Delta L_1 \cap \Delta L_2 \vdash \delta A_1 \lor \delta A_2.$$

Next, $L_2 \not\vdash A_1$ implies $L_2 \not\vdash p \supset A_1$ (cf. the proof of Lemma 2.9.4). Similarly $L_1 \not\vdash p \supset A_2$. So the theories $(\overline{L}_2 \cup \{p\}, \{A_1\})$ and $(\overline{L}_1 \cup \{p\}, \{A_2\})$ are **QH**-consistent, and by strong completeness there exist Kripke models M_1, M_2 with roots u_1, u_2 such that $M_1, u_1 \Vdash (\overline{L}_2 \cup \{p\}, \{A_1\})$ and $M_2, u_2 \Vdash (\overline{L}_1 \cup \{p\}, \{A_2\})$. Obviously we may assume that there is a common individual a_0 in the domains of the worlds u_1, u_2.

Then consider the model $M' := \delta(M_1 \sqcup M_2)$ described in Lemma 6.8.3. Since $M_i \Vdash \overline{L_i}$, it follows that $M_1 \sqcup M_2 \Vdash \overline{L_1 \cap L_2}$. Hence

$$M', u_0 \Vdash \overline{\mathrm{Sub}}(\delta(\overline{L_1 \cap L_2}))$$

by Lemma 6.8.3. We also have $M_i, u_i \not\Vdash p \supset A_i$, hence $M', u_0 \not\Vdash p \supset A_i$. Since also $M', u_0 \not\Vdash p$, we obtain $M', u_0 \not\Vdash \delta A_1 \lor \delta A_2$. Therefore $\Delta(L_1 \cap L_2) \not\vdash \delta A_1 \lor \delta A_2$ by Lemma 3.2.31. ■

Let us mention the following analogue of Proposition 1.6.11.

Proposition 6.8.7 *Let* L_1 *and* L_2 *be superintuitionistic predicate logics.*

(1) $\Delta(L_1 + L_2) = \Delta L_1 + \Delta L_2$,

(2) $\Delta(L_1 \cap L_2) = \Delta L_1 \cap \Delta L_2$ *iff* L_1 *and* L_2 *are comparable by inclusion.*

6.9 Δ-operation preserves completeness

Definition 6.9.1 *Let* $F = (W, R)$ *be a non-empty poset,* $\max F$ *the set of its maximal points* $\overrightarrow{G} := (G_w \mid w \in \max F)$ *a family of rooted posets. Then* $F + \overrightarrow{G} := F + \bigsqcup_{w \in \max F} G_w$ *is a poset obtained by putting disjoint isomorphic copies of* G_w *above the corresponding points* $w \in \max F$*. Speaking precisely, if*

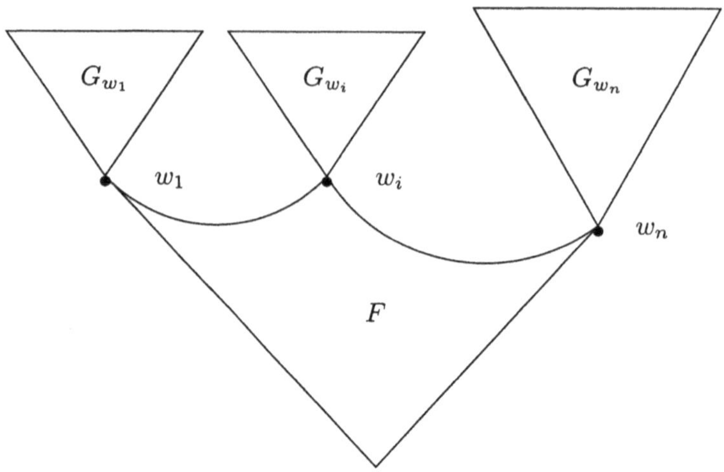

Figure 6.1. $F + \vec{G}$.

$G_w = (V_w, R_w)$, *then*

$$F + \overrightarrow{G} = (\overline{W}, \overline{R}),$$

where

$$\overline{W} := W^- \cup \bigsqcup_{w \in \max F} V_w, \ W^- := W - \max F$$

(cf. Section 1...),

$$\overline{R} := R \upharpoonright W^- \cup \{(u, (v, w)) \mid u \in W^-, \ w \in \max F \cap R(u), \ v \in V_w\}$$
$$\cup \bigcup_{w \in \max F} \{(v, w), (v', w) \mid vR_w v'\}$$

Obviously

$$V'_w := V_w \times \{w\}$$

for $w \in \max F$, are disjoint $\overline{R}$-stable subsets in $F + \overrightarrow{G}$, and $(F + \overrightarrow{G}) \upharpoonright V'_w \cong G_w$. If v_w is the root of G_w, then

$$W' := W_- \cup \{(v_w, w) \mid w \in \max F\}$$

is an $\overline{R}^{-1}$-stable subset of $F + \overrightarrow{G}$, $(F + \overrightarrow{G})\restriction W' \cong F$. It is also clear that

$$\overline{W} = W' \cup \bigcup_{w \in \max F} V'_w$$

and $W' \cap V'_w = \{(v_w, w)\}$ for $w \in \max F$.

Definition 6.9.2 *Let $\mathbf{F} = (F, D)$ be a predicate Kripke frame over a poset F and for $w \in \max F$ let $\mathbf{G}_w = (G_w, D'_w)$ be Kripke frames* [3] *over rooted posets G_w (with roots v_w) such that $D_w = D'_w(v_w)$. Then we define a Kripke frame $\mathbf{F}+\overrightarrow{\mathbf{G}} := (F+\overrightarrow{G}, \overline{D})$, in which $\overline{D}(w) = D(w)$ for $w \in W^-$ and $\overline{D}(v, w) = D'_w(v)$ for $v \in G_w$.*

Definition 6.9.3 *Similarly for Kripke sheaves $\mathbf{F} = (F, \rho, D)$ and $\mathbf{G}_w = (G_w, \rho'_w, D'_w)$, with roots v_w, where $w \in \max F$, and $D_w = D'_w(v_w)$, we define a Kripke sheaf $\mathbf{F} + \overrightarrow{\mathbf{G}} := (F + \overrightarrow{G}, \overline{\rho}, \overline{D})$, in which*

$$\overline{\rho}(u, (v, w)) = \rho(u, w) \circ \rho'_w(v_w, v)$$

for $u \in F^-$, $w \in R(u) \cap \max F$, $v \in G_w$; in all other cases $\overline{\rho}$ is inherited from $\mathbf{F}$ or $\mathbf{G}_w$ in the natural way.

Definition 6.9.4 *We can also define a KFE $\mathbf{F}' = \mathbf{F} + \overrightarrow{\mathbf{G}}$ for KFEs $\mathbf{F} = (F, D, \asymp)$ and $\mathbf{G}_w = (G_w, D'_w, \asymp_w)$, $w \in \max F$ such that $D(w) = D'_w(v_w)$ and $\asymp_w = (\asymp'_w)_{v_w}$ for $w \in \max F$. The details are left to the reader.*

Definition 6.9.5 *Let $\mathcal{C}_0$ and $\mathcal{C}$ be classes of rooted predicate Kripke frames. Then $\mathcal{C}_0 + \mathcal{C}$ denotes the class of all Kripke frames of the form $\mathbf{F} + \overrightarrow{\mathbf{G}}$, where $\mathbf{F} \in \mathcal{C}_0$ and $\overrightarrow{\mathbf{G}} = (\mathbf{G}_w \mid w \in \max \mathbf{F})$ is a family of frames from $\mathcal{C}$. Similarly we define a class $\mathcal{C}_0 + \mathcal{C}$ of KFEs or Kripke sheaves for classes $\mathcal{C}_0$ and $\mathcal{C}$ of rooted KFEs or Kripke sheaves respectively.*

Definition 6.9.6 *A tree of finite height is called* uniform *if all its maximal points are of the same height.*

Recall that IT_n^k (for $k > 0$, $n \leq \omega$) is the tree of n-sequences of length $< k$. Obviously all these trees are uniform.

Lemma 6.9.7 *Let F be a uniform tree of height k, $\overline{F} = F + \overrightarrow{G}$. Then $\overline{F}_{-k} \cong \bigsqcup \overrightarrow{G}$. Similarly, if $\mathbf{F}$ is a predicate frame over F and $\mathbf{G}_w$ are predicate frames over G_w, then*

$$(\mathbf{F} + \overrightarrow{\mathbf{G}})_{-k} \cong \bigsqcup \overrightarrow{\mathbf{G}}.$$

Proof Obvious. The levels $l < k$ in $\overline{F}$ and F are the same, and the level k in $\overline{F}$ consists of the roots of posets G_w for $w \in \max F$. ∎

[3] D'_w denotes the domain function in G_w, so we denote the domain at a world u in G_w by $D'_w(u)$; $D(u)$ (respectively, $\overline{D}(u)$) is the domain at u in F (respectively, in $\overline{F}$).

Proposition 6.9.8 *(1) Let L be an s.p.l., $\mathcal{C}$ a class of rooted predicate Kripke frames, and let F_0 be a uniform tree of height k, $\mathcal{C}_0 = \mathcal{K}(F_0)$. Then*

$$\Delta^k L \subseteq \mathbf{IL}^{(=)}(\mathcal{C}_0 + \mathcal{C}) \text{ iff } L \subseteq \mathbf{IL}^{(=)}(\mathcal{C}).$$

(2) Similarly for KFEs or Kripke sheaves, with $\mathcal{C}_0 = \mathcal{KE}(F_0)$.

Note that if $k = 0$; then F is a singleton and $\mathcal{C}_0 + \mathcal{C}$ is equivalent to $\mathcal{C}$.

Proof (If.) Let $\vec{\mathbf{G}} = (\mathbf{G}_w \mid w \in \max F_0)$ be a family of frames from $\mathcal{C}$. Then by 3.5.24

$$\mathbf{IL}(\bigsqcup \vec{\mathbf{G}}) = \bigcap_{w \in \max F} \mathbf{IL}(\mathbf{G}_w) \supseteq \mathbf{IL}(\mathcal{C}).$$

By Lemma 6.9.7, for a frame $\mathbf{F}_0$ over F_0

$$\bigsqcup \vec{\mathbf{G}} \cong (\mathbf{F}_0 + \vec{\mathbf{G}})_{-k},$$

so $L \subseteq \mathbf{IL}(\mathcal{C})$ implies $L \subseteq \mathbf{IL}((\mathbf{F}_0 + \vec{\mathbf{G}})_{-k})$, which is equivalent to

$$\Delta^k L \subseteq \mathbf{IL}(\mathbf{F}_0 + \vec{\mathbf{G}})$$

by 6.8.2. Since $\vec{\mathbf{G}}$ is arbitrary, this implies

$$\Delta^k L \subseteq \mathbf{IL}(\mathcal{C}_0 + \mathcal{C}).$$

(Only if.) Assuming $\Delta^k L \subseteq \mathbf{IL}(\mathcal{C}_0 + \mathcal{C})$, let us show that $L \subseteq \mathbf{IL}(\mathbf{F})$ for any $\mathbf{F} \in \mathcal{C}$. Given $\mathbf{F}$ with a domain V at the root, consider the constant family

$$\vec{\mathbf{G}} := (\mathbf{F} \mid v \in \max F_0).$$

Then there exists a frame $\mathbf{F}_0$ over F_0 such that copies of F can be stuck to its maximal points. For example, we can put $\mathbf{F}_0 = F_0 \odot V$, the frame with the constant domain V.

Then

$$(\mathbf{F}_0 + \vec{\mathbf{G}})_{-k} \cong \bigsqcup_{v \in \max F_0} \mathbf{F},$$

so

$$\mathbf{IL}((\mathbf{F}_0 + \vec{\mathbf{G}})_{-k}) = \bigcap_{v \in \max F_0} \mathbf{IL}(\mathbf{F}) = \mathbf{IL}(\mathbf{F}).$$

By our assumption

$$\Delta^k L \subseteq \mathbf{IL}(\mathcal{C}_0 + \mathcal{C}) \subseteq \mathbf{IL}(\mathbf{F}_0 + \vec{\mathbf{G}}),$$

hence by 6.8.2, $L \subseteq \mathbf{IL}((\mathbf{F}_0 + \vec{\mathbf{G}})_{-k}) = \mathbf{IL}(\mathbf{F})$. ∎

Proposition 6.9.9 *Let L be an intermediate predicate logic (with or without equality) strongly complete w.r.t. a class $\mathcal{C}$ of Kripke frames or KFEs. Then $\Delta^k L$ is strongly complete w.r.t. $\mathcal{C}_0 + \mathcal{C}$, where $\mathcal{C}_0 = \mathcal{K}(IT_\omega^{k+1})$ (or $\mathcal{KE}(IT_\omega^{k+1})$ respectively).*

Hence $\Delta^h L$ is Kripke-complete or Kripke sheaf complete w.r.t. $\mathcal{C}_0 + \mathcal{C}$. Note that we cannot state Kripke completeness of $\Delta^h L$ for a (strongly) Kripke complete logic L with equality, because in this case $\mathcal{C}_0$ is still $\mathcal{KE}(IT_\omega^{k+1})$.

Proof Soundness follows from 6.9.8.

Now for a $\Delta^k(L)\forall$-place Γ_0, let us construct a proper $(\Delta^h L, \forall^{\leq})$-map $(\Gamma_u \mid u \in \mathbf{IT}_\omega^{k+1})$, beginning with $\Gamma_\lambda = \Gamma_0$.

(**) if $h_j(u) \not\Vdash A \supset B$, then $\exists v \in \beta(u) \cap F_{j+1}\ (h_{j+1}(v) \Vdash A \ \&\ h_{j+1}(v) \not\Vdash B)$,

(***) if $h_j(u) \not\Vdash \forall x A(x)$, then $\exists v \in \beta(u) \cap F_{j+1}\ \exists c \in D_{h_{j+1}(v)}\ h_{j+1}(v) \not\Vdash A(c)$.

After we reach a point w of level k in IT_ω^{k+1}, we obtain a $(k+1)$-element sequence $\Gamma_\lambda < \ldots < \Gamma_w$, so $\Delta^h L \subseteq \Gamma_\lambda$ implies $L \subset \Gamma_w$ by Proposition 6.9.8 Now we can use the strong completeness of L and obtain a natural model with our Γ_w at the root of a frame from $\mathcal{C}$. So we obtain a natural model over a frame from $\mathcal{C}_0 + \mathcal{C}$. Clearly, this construction gives us a natural model over a frame from G_0^G. Naturally, for a logic with equality, we construct places Γ_w for $u \in IT_\omega^{k+1}$ in the language with equality and hence we obtain a natural KFE.

There exists one special case. Namely, if for some point $u \in IT_\omega^k$ there do not exist places $> \Gamma_u$, i.e. Γ_u is maximal in $U \leq F\Delta^k(L)$, then by Lemma 7.3.2(i) Γ_u is $\mathbf{QCL}\forall$-place. In this case the naturalness conditions hold with $\Gamma_v = \Gamma_u$ and we can put $\Gamma_v = \Gamma_u$ for all $v \in (IT_\omega^{k+1} \uparrow u)$.

And again for points w of level k we can apply the strong completeness of L, because $\Gamma_w = \Gamma_u$ is an $L\forall$-place since $L \subseteq \mathbf{QCL}$. ■

Note that the condition $L \subseteq \mathbf{QCL}^{(=)}$ is *necessary* in proposition 6.8.7. Namely:

Proposition 6.9.10 *Let L be a superintuitionistic predicate logic, $L \not\subseteq \mathbf{QCL}^{(=)}$, and $h > 0$. Then $\Delta^h L \neq \mathbf{IL}^{(=)}(C_0 + C)$ for any class $C \subseteq \mathcal{KE}$ and any $C_0 \subseteq \mathcal{KE}(IT_\omega^{h+1})$.*

Proof Consider a sentence $A \in L - \mathbf{QCL}^{(=)}$ and suppose $\Delta^h L = \mathbf{IL}^{(=)}(C_0 + C)$. Then $L \subseteq \mathbf{IL}^{(=)}(C)$, by Corollary 6.8.6, so $A \in \mathbf{IL}^{(=)}(C)$ and hence obviously $\neg\neg A \in \mathbf{IL}^{(=)}(C_0 + C)$. On the other hand, $\neg\neg A \notin \Delta^h L$ since $\Delta^h L \subseteq \Delta L \subseteq \mathbf{QCL}^{(=)}$. ■

Now let us reformulate the previous completeness result (Proposition 6.8.7) for the tree T_n^k with the uniform finite branching $n > 0$ instead of T_ω^k.

Proposition 6.9.11 *Let* **L** *be an intermediate predicate logic strongly complete w.r.t. a class* $\mathcal{C}$ *of Kripke frames. Let* $k \leq 0$, $n > 0$, *let* $L = \Delta^h L + \Theta$ *for a set* Θ *of sentences* $\forall$*-perfect in* $\Delta^h L$. *Let* $\mathcal{C}_0 = \mathcal{K}(IT_n^{k+1})$. *Let* $L' = \Delta^k L + B_n[\Theta])$, *then* L' *is strongly complete w.r.t.* $\mathcal{C} + \mathcal{C}_0$.

Recall that $\Delta^h L \subseteq L$.

Proof Soundness follows from Lemma 7.8.5.

Now, let Γ be a $L'\forall$ place. Again we construct a $L'\forall^{\leq}$-map over $IT_n^{k+1}(\Gamma_u \mid\in IT_n^{k+1})$ in such a way that $\Gamma_\lambda = \Gamma_0$ and $L \subseteq \Gamma_v$ for all $v \in max(IT_n^{k+1})$ and the naturalness condition ($\supset'$) for all $u \in IT_n^k$ with $v \in \beta(u)$ holds. Then for any $v \in \max(IT_n^{k+1})$ we apply the strong completeness for **L** and find a natural **L**-model on a frame $F_v \in C$ with Γ_v in its root; these models give us a natural model over $F + (F_v \mid v \in \max IT_n^{k+1})$, where $F \in \mathcal{C}_0$.

Let us describe the inductive step. Let Γ_u for $u \in IT_n^k$ be already constructed. If $L = \Delta^k L + \Theta \subseteq \Gamma_u$, then we put $\Gamma_v = \Gamma_u$ for all $v \in (IT_n^{k+1} \uparrow u)$. And if $\Delta^k L + \Theta \not\subseteq \Gamma_u$, then we apply Lemma 7.8.8 to obtain $\mathbf{L}'\forall$-places $\Gamma_v > \Gamma_u$ for n points $v \in \beta(u)$. Note that a characteristic formula E for Γ exists by Lemma 7.4.21. In fact, here $\Delta^k L \subseteq \Gamma$ and $L \not\subseteq \Gamma$, so there exists $l < k$ such that $\Delta^{l+1} L \subseteq \Gamma$ and $\Delta^l L \not\subseteq \Gamma$. Finally, if $n' = 0$ (in applying of Lemma 7.8.8), then Γ_u is a **QCL**-place, and again we put $\Gamma_v = \Gamma_u$ for all $(v \in (IT_n^{k+1} \uparrow u)$; here we use that $L \subseteq$ **QCL**, cf. the concluding argument in the proof of 7.10.8. ∎

A similar statement holds for KFE. Note that the completeness for $\mathbf{LP}_{k+1}^+ + Br_n = \mathbf{IL}(IT_\omega^{k+1}))$ and for $\mathbf{LP}_{k+1}^+ + Br_n = \mathbf{IL}(IT_n^{k+1})$ are the particular cases of these statements for $L =$ **QLC**.

6.10 Trees of bounded branching and depth

As we shall see in Volume 2, the s.p.l. $\mathbf{IL}(\mathcal{K}(\mathbb{F}\text{in}))$ determined by all finite posets is not recursively axiomatisable. Still we can explicitly describe 'approximations' of this logic of bounded height and width or branching.

Recall (from Chapter 1) that the propositional axiom

$$F_n := \bigvee_{i=0}^{n} \left(p_i \supset \bigvee_{j \neq i} p_j \right)$$

characterises posets of width $\leq n$. Similarly, the axiom

$$Br_n := \bigwedge_{i=0}^{n} \left(p_i \supset \bigvee_{j \neq i} p_j \;\bullet\!\supset \bigvee_{j \neq i} p_j \right) \supset \bigvee_{i=0}^{n} p_i$$

or its equivalent version:

$$Br'_n := \bigwedge_{i=0}^{n} (q \supset p_i \;\bullet\!\supset q) \wedge \bigwedge_{i<j} (q \supset p_i \vee p_j) \supset q$$

identifies posets of branching $\leq n$ among posets of finite depth. We also know that $\mathbf{H} + Br_n$ is the propositional logic of finite n-ary trees and $\mathbf{QH} + Br_1 = \mathbf{QH} + W_1$.

In this section we study the propositional axiom Br_n in predicate logics. Now it is not sufficient for axiomatising Kripke frames over finite n-ary trees, as we shall see in Volume 2. But still it works properly above $\mathbf{QHP}_k^+$, i.e. for posets of bounded height.

So consider classes $\mathcal{B}_n^k := \mathcal{B}_n \cap \mathcal{P}_k^+$ of posets of depth $\leq h$ and branching $\leq n$ (for $n, k > 0$) and also classes $\mathcal{W}_n^k := \mathcal{W}_n \cap \mathcal{P}_k^+$ of posets of depth $\leq k$ and width $\leq n$. Actually the predicate logics are determined by rooted posets, so we can deal only with the corresponding classes $\mathcal{B}_n^{k\uparrow}$ and $\mathcal{W}_n^{k\uparrow}$. Up to isomorphism, each of these classes contains finitely many finite posets. Therefore by 3.2.20, the logics $\mathbf{IL}(\mathcal{K}(\mathcal{B}_n^k))$ and $\boldsymbol{L}(\mathcal{K}(\mathcal{W}_n^k))$ are uniformly recursively axiomatisable. Note that $\mathcal{W}_n^{k\uparrow} \subseteq \mathcal{B}_n^{k\uparrow} \subseteq \mathcal{W}_{n^{k-1}}^{k\uparrow}$. Also note that every finite poset belongs to $\mathcal{W}_n^k$ (hence to $\mathcal{B}_n^k$) for some n, k; thus $\mathbb{F}\mathrm{in}^\uparrow = \bigcup\limits_{n,k>0} \mathcal{W}_n^{k\uparrow} = \bigcup\limits_{n,k>0} \mathcal{B}_n^{k\uparrow}$. It follows that the logic of finite predicate Kripke frames $\mathbf{IL}(\mathcal{K}(\mathbb{F}\mathrm{in}))$ is in Π_2^0. Now we shall find finite axiomatisations for the logics $\mathbf{IL}(\mathcal{K}(\mathcal{B}_n^k))$ and thus obtain a simpler and more natural Π_2^0-presentation for $\mathbf{IL}(\mathcal{K}(\mathbb{F}\mathrm{in}))$.

First we need some lemmas.

Lemma 6.10.1 *Let Γ be an L-place, $Br_n \in L$, and assume that there exists a characteristic sentence E for Γ such that the theories $(\Gamma \cup \{E\}, \Delta_i)$ are L-consistent for $i = 0, 1, \ldots, n$. Then $(\Gamma \cup \{E\}, \Delta_i \cup \Delta_j)$ is L-consistent for some $i \neq j$.*

Proof Suppose $(\Gamma \cup \{E\}, \Delta_i \cup \Delta_j)$ is L-inconsistent for any $i \neq j$.

Then for any pair (i, j) with $i \neq j$, we have

$$\Gamma, E \vdash_L \bigvee \Delta_{ij}^1 \vee \bigvee \Delta_{ji}^2$$

for some finite $\Delta_{ij}^1 \subseteq \Delta_i$, $\Delta_{ji}^2 \subseteq \Delta_j$.

Hence by the Deduction theorem,

$$\Gamma \vdash_L E \supset C_i \vee C_j,$$

where

$$\Delta_i^0 := \bigcup_{j \neq i} (\Delta_{ij}^1 \cup \Delta_{ij}^2),\ C_i := \bigvee \Delta_i^0.$$

On the other hand, $\vdash_L E \wedge (E \supset C_i) \supset C_i$ and $(\Gamma \cup \{E\}, \Delta_i)$ is L-consistent, so $(E \supset C_i) \notin \Gamma$.

Since $((E \supset C_i) \vee (E \supset C_i \bullet\supset E)) \in \Gamma$ (by the property of characteristic formula, see 6.3.10), it follows that $(E \supset C_i \bullet\supset E) \in \Gamma$. But

$$\vdash_L \bigwedge_{i=0}^{n} (E \supset C_i \bullet\supset E) \wedge \bigwedge_{i<j} (E \supset C_i \vee C_j) \supset E,$$

since $Br'_n \in L$. Since Γ is an L-place, we obtain $E \in \Gamma$, which contradicts Definition 6.3.10. ∎

Lemma 6.10.2 (Main lemma on n-branching) *Let Γ be an L-place, $Br_n \in L$, and let E be a characteristic formula for Γ. Assume that $D_\Gamma = S \subset S'$, $|S' - S| = |S| = \aleph_0$. Then there exist L-places $\Gamma_1, \dots, \Gamma_m$, $m \leq n$ such that*

(a) $\Gamma \cup \{E\} \subseteq \Gamma_i$ for $i = 1, \dots, m$ (so $\Gamma R_L \Gamma_i$);

(b) if $(B \supset C) \in (-\Gamma)$ and $B \notin \Gamma$, then $(B \supset C) \notin \Gamma_i$ for some i.

Proof Let $\{(B \supset C) \in (-\Gamma) \mid B \notin \Gamma\} = \{A_k \mid k \in \omega\}$. We define a family of theories $\Delta_k^i \subseteq \mathcal{L}(\Gamma)$ for $k \in \omega$, $i > 0$ such that

$$\forall i, k\ \Delta_k^i \subseteq \Delta_{k+1}^i, \tag{0}$$

and for any k there exists $n_k \leq n$ with the following properties

$$\Delta_k^i \neq \varnothing \text{ iff } 1 \leq i \leq n_k, \tag{1}$$

$$(\Gamma \cup \{E\}, \Delta_k^i) \text{ is } L\text{-consistent for } 1 \leq i \leq n_k, \tag{2}$$

$$(\Gamma \cup \{E\}, \Delta_k^i \cup \Delta_k^j) \text{ is } L\text{-inconsistent for any different } i, j \leq n_k. \tag{3}$$

Hence by Lemma 6.10.1, $n_k \leq n$.

To begin the construction, put $\Delta_0^i := \varnothing$ for all $i > 0$ (i.e. $n_0 = 0$).

Δ_{k+1}^i is constructed as follows. Consider $A_k = B \supset C$. There are two cases.

(I) If $(\Gamma \cup \{E\}, \Delta_k^i \cup \{B \supset C\})$ is L-consistent for some $i \leq n_k$, then take the minimal i with this property and put

$\Delta_{k+1}^i := \Delta_k^i \cup \{B \supset C\}$, $\Delta_{k+1}^j := \Delta_k^j$ for all $j \neq i$ (thus $n_{k+1} = n_k$).

(II) If for any $i \leq n_k$, $(\Gamma \cup \{E\}, \Delta_k^i \cup \{B \supset C\})$ is L-inconsistent, then put $\Delta_{k+1}^{n_k+1} := \{B \supset C\}$ and $\Delta_{k+1}^i := \Delta_k^i$ for all $i \leq n_k$ (thus $n_{k+1} = n_k + 1$).

The properties (1),(3) hold trivially by definition, the same with (2) in case (I). To check (2) in case (II), we have to show that $(\Gamma \cup \{E\}, \{B \supset C\})$ is L-consistent, i.e. $(E \supset_\bullet B \supset C) \notin \Gamma$.

In fact, $B \notin \Gamma$ and $B \vee (B \supset E) \in \Gamma$, since E is a characteristic formula, thus $(B \supset E) \in \Gamma$. So $(E \supset_\bullet B \supset C) \in \Gamma$ implies $(B \supset_\bullet B \supset C) \in \Gamma$, which contradicts $(B \supset C) \notin \Gamma$. Therefore $(E \supset_\bullet B \supset C) \notin \Gamma$.

Next, put $m := \max\{n_k \mid k \in \omega\}$ (thus $m \leq n$) and

$$\Delta_i := \bigcup_{k \in \omega} \Delta_k^i$$

for $1 \leq i \leq m$. Since the sequence $(\Delta_k^i)_{k \in \omega}$ is increasing, (2) implies the L-consistency of every $(\Gamma \cup \{E\}, \Delta_i)$. So there exist $L\forall$-places $\Gamma_i \succ (\Gamma \cup \{E\}, \Delta_i)$.

Thus condition (a) holds. (b) holds by our construction, since $B \supset C = A_k$ for some k, whenever $B \notin \Gamma$; so $C \in \Delta_i$. ∎

Lemma 6.10.3 *Assume that* $\mathbf{QHP}_k^+ + Br_n \subseteq L$. *Then for any* L*-place* Γ *there exists a proper natural* $(L\forall, \leq)$*-model* $N = (\Gamma_u \mid u \in F)$ *based on a greedy standard subtree of* IT_n^k *such that* $\Gamma_\lambda = \Gamma$ *and a natural* $(L\forall, \leq)$ *-model on* IT_n^k *(not necessarily proper).*

Proof We construct Γ_u using Lemma 6.10.2, which gives us the condition of naturalness from 6.4.9. If $0 < m < n$, we repeat some theories; if $m = 0$ then the construction terminates at this point. Recall that characteristic formulas used in Lemma 6.10.2 always exist by 6.3.12. The construction terminates at some height $\leq h$, due to Lemma 6.3.7. Finally we repeat the theories from the maximal points of the resulting subtree at all leaves of IT_n^k. ∎

Proposition 6.10.4 $\mathbf{IL}(\mathcal{K}(\mathcal{B}_n^k)) = \mathbf{IL}(\mathcal{K}(IT_n^k)) = \mathbf{QHP}_k^+ + Br_n$ *for any* $n, k > 0$, *where* IT_n^k *is the* n*-ary tree of height* k.

Proof Readily follows from 6.10.3. ∎

In particular, for $n = 1$ we have:

Corollary 6.10.5 $\mathbf{QHP}_k^+ + \mathbf{LC}$ *is determined by an* k*-element chain.*

Corollary 6.10.6

$$\mathbf{IL}(\mathcal{K}(\mathbb{F}\mathrm{in})) = \bigcup_{n,k}(\mathbf{QHP}_k^+ + Br_n) = \bigcap_{n,k}(\mathbf{QHP}_k^+ + W_n).$$

Proof In fact, $\mathbf{H} + W_n \vdash Br_n$ and $\mathbf{H} + AP_k \wedge Br_n \vdash W_m$ for some m (e.g. due to Kripke completeness). Thus

$$\bigcap_{n,k}(\mathbf{QHP}_k^+ + Br_n) = \bigcap_{n,k}(\mathbf{QHP}_k^+ + W_n).$$

∎

On the other hand, the logics $\mathbf{QHP}_k^+ + W_n$ for $k \geq 3$, $n \geq 2$ are Kripke incomplete [Skvortsov, 2006]; the proof will be given in Volume 2. Our conjecture (based on some results from Volume 2) is that all the corresponding logics $\mathbf{IL}(\mathcal{K}(\mathcal{W}_n^h))$ are not finitely axiomatisable.

6.11 Logics of uniform trees

As we shall see in Volume 2, generalised versions of formulas Br_n allow us to axiomatise the intermediate predicate logic of all finite trees explicitly. Recall that the logic determined by an arbitrary finite poset is recursively axiomatisable (Chapter 3), but the proof of this fact provides only an implicit axiomatisation.

In this section we make a step towards the logic of all finite trees clarifying main ideas of the whole construction; here we describe an explicit axiomatisation of an s.p.l. determined by an arbitrary 'uniform' tree of finite depth.

Definition 6.11.1 *A tree F is called* (levelwise) uniform *if all its cones $F \uparrow u$ with roots u of the same height are isomorphic.*

A levelwise uniform tree F of depth $h+1$ gives rise to an *h-sequence* $\mathbf{t} = (t_i \mid i < h)$ of non-null cardinals t_i; t_i is the branching $|\beta(u)|$ for all points $u \in F$ of height $i < h$ (obviously, the branching at a maximal point is 0).

The other way round, every h-sequence $\mathbf{t}$ of cardinals corresponds to a unique tree $T_{\mathbf{t}}$ of depth $(h+1)$ and of branching t_i at points of height i; this tree is clearly levelwise uniform. Obviously $T_{\mathbf{t}}$ is finite (respectively, countable) iff every t_i is finite (respectively, countable). The corresponding h-sequences are called h^f- (respectively, $h^\dagger$-)*sequences* .

Every denumerable uniform tree can be represented as a subtree of IT_ω:

$$T_{\mathbf{t}} = \{\alpha \in \omega^* \mid |\alpha| \leq h,\ \forall i\ \alpha_i < t_i\}.$$

Every h-sequence $\mathbf{t}$ of cardinals gives rise to the truncated $h^{\leq\omega}$-sequence $\mathbf{t}^\dagger$, in which every infinite $\mathbf{t}_i$ is replaced with ω.

We shall construct explicit axiomatisations for all denumerable levelwise uniform trees and show that

$$\mathbf{IL}(\mathcal{K}(T_{\mathbf{t}})) = \mathbf{IL}(\mathcal{K}(T_{\mathbf{t}^\dagger}))$$

for any uncountable $T_{\mathbf{t}}$. Note that for $h = 0$ there is the empty sequence corresponding to a one-element tree T_k of depth 1. A constant sequence $\mathbf{t} = (n, \dots, n)$ of length $h \geq 0$ corresponds to a uniform tree $T_{\mathbf{t}} = IT_n^{k+1}$ considered in Section 6.10 or to IT_ω^{k+1} if $n = \omega$.

Definition 6.11.2 *Let A be a predicate formula and let $q, p_0, p_1, \dots, p_n$ be different proposition letters non-occurring in A, $n > 0$. We put*

$$p_i^- := \bigvee\{p_j \mid 0 \leq j \leq n,\ j \neq i\}$$

and define the formulas

$$\begin{aligned} Br_n^A &:= \bigwedge_{i=0}^{n} ((p_i \supset p_i^-) \supset p_i^-) \wedge \left(A \supset \bigvee_{i=0}^{n} p_i\right) \supset \bigvee_{i=0}^{n} p_i, \\ Br_n'^A &:= \bigwedge_{i=0}^{n} ((q \supset p_i) \supset q) \wedge \bigwedge_{i<j} (q \supset p_i \vee p_j) \wedge (A \supset q) \supset q. \end{aligned}$$

We will also use the notation $Br_n[A]$, $Br'_n[A]$ for Br_n^A, $Br_n'^A$.

Obviously, Br_n^A and $Br_n'^A$ are equivalent to $\left(A \supset \bigvee_{i=0}^{n} p_i\right) \supset Br_n$ and $(A \supset q) \supset Br'_n$ respectively. Thus $Br_n^\perp$ and $Br_n'^\perp$ are equivalent to Br_n and Br'_n.

Now let us show that both axioms are actually deductively equivalent (cf. Lemma 2 for Br_n and Br'_n):

Lemma 6.11.3 $\mathbf{QH} + Br_n^A = \mathbf{QH} + Br_n'^A$ *(for $n > 0$).*

Proof ($\supseteq$) Let $\mathbf{p} := p_1 \dots p_n$, $\mathbf{C} := C_1 \dots C_n$, where $C_i := \bigwedge\limits_{j \neq i} p_j$ and let us show that

(1) $$\mathbf{QH} \vdash [\mathbf{C}/\mathbf{p}]\, Br_n^A \supset Br_n'^A.$$

Let B be the premise of $Br_n'^A$, $C_i^- := \bigvee\limits_{j \neq i} C_j$,

$$E := \bigwedge_{i=0}^{n} (C_i \supset C_i^- \bullet \supset C_i^-) \wedge \left(A \supset \bigvee_{i=0}^{n} C_i\right)$$

the premise of $[\mathbf{C}/\mathbf{p}]\, Br_n^A$.

It suffices to check that

(2) $$B \vdash_{\mathbf{QH}} E \wedge \left(\bigvee_{i=0}^{n} C_i \supset q\right);$$

hence we can obtain

$$B,\ [\mathbf{C}/\mathbf{p}]\, Br_n^A = \left(E \supset \bigvee_{i=0}^{n} C_i\right) \vdash_{\mathbf{QH}} \bigvee_{i=0}^{n} C_i,\ \bigvee_{i=0}^{n} C_i \supset q,$$

and thus

$$[\mathbf{C}/\mathbf{p}] Br_n^A,\ B \vdash_{\mathbf{QH}} q,$$

which implies (1).

For the proof of (2) first note that

(3) $$\bigwedge_{i<j} (q \supset p_i \vee p_j) \vdash_{\mathbf{H}} q \supset \bigvee_{i=0}^{n} C_i.$$

To show this, we argue in $\mathbf{H}$. Assume q and $\bigwedge\limits_{i<j} (q \supset p_i \vee p_j)$. Then $\bigwedge\limits_{i<j} (p_i \vee p_j)$, which is equivalent to $\bigvee\limits_{i=0}^{n} C_i$ by distributivity – in fact, this should be a disjunction of conjunctions, in which every conjunction contains either p_i or p_j from each distinct pair (i, j); so there is at most one p_i missing from that conjunction.

Thus

(4) $$B \vdash_{\mathbf{QH}} q \supset C_i \vee C_i^-.$$

We also have

(5) $$\mathbf{H} \vdash C_i^- \supset p_i,$$

since p_i is present in every C_j for $j \neq i$. Hence

(6) $$C_i \supset C_i^- \vdash_{\mathbf{H}} C_i \supset p_i$$

by (5),

(7) $$B, C_i \supset C_i^- \vdash_{\mathbf{QH}} q \supset C_i^-$$

by (4),

(8) $$B, C_i \supset C_i^- \vdash_{\mathbf{QH}} q \supset p_i$$

by (7), (5),

(9) $$B, C_i \supset C_i^- \vdash_{\mathbf{QH}} q$$

by (8), since $(q \supset p_i) \supset q$ is a conjunct in B,

(10) $$B, C_i \supset C_i^- \vdash_{\mathbf{QH}} C_i^-$$

by (9), (7). Thus

(11) $$B \vdash_{\mathbf{QH}} C_i \supset C_i^- \bullet \supset C_i^-.$$

Next,

(12) $$B \vdash_{\mathbf{QH}} A \supset \bigvee_{i=0}^{n} C_i,$$

by (3) and since $A \supset q$ is also a conjunct in B.

Next, we obviously have

$$C_i \vdash_{\mathbf{H}} p_j,$$

for $j \neq i$, hence

$$C_i \vdash_{\mathbf{H}} q \supset p_j,$$

and consequently

(13) $$B, C_i \vdash_{\mathbf{QH}} q,$$

since $B \vdash_{\mathbf{QH}} (q \supset p_j) \supset q$. Thus

(14) $$B \vdash_{\mathbf{QH}} C_i \supset q.$$

by (13). Eventually (2) follows from (11), (12), and (14).

($\subseteq$) Similarly,

(15) $$[D, p_1^-, \dots, p_n^- / q, p_1, \dots, p_n]\, Br_n'^A \vdash_{\mathbf{QH}} Br_n^A,$$

where $D := \bigvee_{i=0}^{n} p_i$.

In fact, $\vdash_{\mathbf{H}} D \equiv p_i \vee p_i^-$, hence

$$\vdash_{\mathbf{H}} (D \supset p_i^-) \equiv (p_i \supset p_i^-),$$

and thus

$$p_i \supset p_i^- \bullet \supset p_i^- \vdash_{\mathbf{H}} D \supset p_i^- \bullet \supset p_i^- .$$

Since $\vdash_{\mathbf{H}} p_i^- \supset D$, this implies

$$p_i \supset p_i^- \bullet \supset p_i^- \vdash_{\mathbf{H}} D \supset p_i^- \bullet \supset D. \tag{16}$$

It also obvious that

$$\mathbf{H} \vdash D \equiv p_i^- \vee p_j^- \tag{17}$$

for $i < j$. So by (16), (17) the premise of Br_n^A:

$$\bigwedge_{i=0}^{n} (p_i \supset p_i^- \bullet \supset p_i^-) \wedge (A \supset D)$$

implies the premise of $[D, p_1^-, \ldots, p_n^- / q, p_1, \ldots, p_n] Br_n'^A$:

$$\bigwedge_{i=0}^{n} (D \supset p_i^- \bullet \supset D) \wedge \bigwedge_{i<j} (D \supset p_i^- \vee p_j^-) \wedge (A \supset D).$$

Since the conclusions of $[D, p_1^-, \ldots, p_n^- / q, p_1, \ldots, p_n] Br_n'^A$ and Br_n^A coincide, the claim (15) follows. ■

Lemma 6.11.4 *The following formulas are* **QH**-*theorems:*

(1) $A \supset Br_n^A$, $A \supset Br_n'^A$,

(2) $(A_1 \supset A_2) \supset (Br_n^{A_1} \supset Br_n^{A_2})$, $(A_1 \supset A_2) \supset (Br_n'^{A_1} \supset Br_n'^{A_2})$,

(3) $\bigwedge_{i=1}^{r} Br_n^{A_i} \supset Br_n \left[\bigwedge_{i=1}^{r} A_i \right]$, $\bigwedge_{i=1}^{r} Br_n'^{A_i} \supset Br_n' \left[\bigwedge_{i=1}^{r} A_i \right]$.

(4) $Br_n \supset Br_n^A$, $Br_n' \supset Br_n'^A$.

Proof
(1) A trivial exercise.
(2) Note that $A_1 \supset A_2, A_2 \supset D \vdash A_1 \supset D$.
(3) By Lemma 1.1.2(5),

$$\mathbf{QH} \vdash \bigwedge_{i=1}^{r} ((A_i \supset q) \supset q) \supset \left(\left(\bigwedge_{i=1}^{r} A_i \supset q \right) \supset q \right).$$

Since $Br_n'^A$ is equivalent to $B \supset (A \supset q_\bullet \supset q)$, where B does not depend on A and q, the claim follows. For Br_n^A replace q with $\bigvee_i p_i$.

(4) Follows from (2), since $\vdash_{\mathbf{QH}} \bot \supset A$. ■

For a set of formulas Θ we define the sets

$$Br_n^\Theta := Br_n[\Theta] := \{Br_n^A \mid A \in \Theta\}, Br_n'^\Theta := Br_n'[\Theta] := \{Br_n'^A \mid A \in \Theta\}.$$

Then
$$\mathbf{QH} + Br_n^\Theta = \mathbf{QH} + Br_n'^\Theta \subseteq \mathbf{QH} + \Theta$$
by 6.11.3 and 6.11.4 (1).

We also put
$$Br_\omega := Br_\omega' := \top,\ Br_\omega^\Theta := Br_\omega'^\Theta := \varnothing.$$

Lemma 6.11.5 *Let Θ_1, Θ_2 be sets of sentences and let L be an s.p.l., $n > 0$.*

(1) If $L + \Theta_1 \subseteq L + \Theta_2$, then

(a) $L + Br_n[\Theta_1] \subseteq L + Br_n[\Theta_2]$ provided Θ_2 is $\forall$-perfect in L;

(b) $L + Br_n[\Theta_1^\forall] \subseteq L + Br_n[\Theta_2^\forall]$ for any Θ_1, Θ_2.

(2) If $L + \Theta_1 = L + \Theta_2$, then

(a) $L + Br_n[\Theta_1] = \mathbf{L} + Br_n[\Theta_2]$ provided Θ_1, Θ_2 are $\forall$-perfect in L;

(b) $L + Br_n[\Theta_1^\forall] = L + Br_n[\Theta_2^\forall]$.

Proof Cf. Lemma 2.8.7 (on δ-operation). Recall that $\Theta^\forall$ is $\forall$-perfect in **QH** for any Θ. ■

Lemma 6.11.6 *Let **F** be an intuitionistic Kripke frame of finite depth, $0 < n < \omega$. Then*

$$Br_n^A \in \mathbf{IL}(\mathbf{F}) \textit{ iff } \forall u \in \mathbf{F}\,(|\beta(u)| > n \Rightarrow \mathbf{F}\!\uparrow\! u \Vdash A).$$

So Br_n^A states that 'the branching is $\le n$ until A becomes true'.

Proof Let $\le$ be the accessibility relation in **F**.

(If.) Suppose $Br_n'^A \notin \mathbf{IL}(\mathbf{F})$, so due to the finiteness of height, there exists a Kripke model over **F** and $u \in \mathbf{F}$ of maximal height refuting $Br_n'^A$, i.e.

$$u \Vdash (q \supset p_i) \supset q,\ u \Vdash (A \supset q),\ u \Vdash q \supset p_i \vee p_j \text{ for } i \neq j,\ u \nVdash q,\ \forall v > u\ v \Vdash q.$$

Then $u \nVdash q \supset p_i$, so there exist a minimal $v_i > u$ such that $v_i \Vdash q$, $v_i \nVdash p_i$, and obviously $v_i \in \beta(u)$. Since $u \Vdash q \supset p_i \vee p_j$, it follows that $v_i \Vdash p_j$ for all $j \neq i$. So all v_i differ, and thus $|\beta(u)| > n$.

Since $u \nVdash q$, $u \Vdash A \supset q$, it follows that $u \nVdash A$, therefore $\mathbf{F}\!\uparrow\! u \nVdash A$.

(Only if.) Suppose $\mathbf{F}\!\uparrow\! u \nVdash A$ and there exist different $v_0, \ldots, v_n \in \beta(u)$. Consider an intuitionistic model over **F** such that $u \nVdash A$ and

$$v \nVdash q \Leftrightarrow v \le u,$$

$$v \nVdash p_i \Leftrightarrow v \le v_i.$$

(Such a model exists, since p_i, q do not occur in A.) Then $u \not\Vdash q$ and $u \Vdash A \supset q$, since $u \not\Vdash A$. Also $u \Vdash q \supset p_i \vee p_j$ for $i \neq j$, since

$$v > u \Rightarrow v \not\leq v_i \vee v \not\leq v_j.$$

Finally note that $v \not\Vdash q \supset p_i$; thus $u \not\Vdash q \supset p_i$ and $u \Vdash (q \supset p_i) \supset q$. Hence $\mathbf{F} \not\Vdash Br_n'^A$ ■

Remark 6.11.7 Note that $\mathbf{QH} + Br_n'^{EM} = \mathbf{QH} + Br_n'$, and so

$$\mathbf{QH} + Br_n^{EM} = \mathbf{QH} + Br_n$$

(recall that $EM = p \vee \neg p$). In fact, $Br_n'^{EM} \vdash (EM \supset q) \supset Br_n'$, whence we deduce $(q \vee \neg q \supset q) \supset Br_n'$. The latter is equivalent to $\vdash (\neg q \supset q) \supset Br_n'$, which readily implies Br_n', since $\mathbf{H} \vdash ((q \supset p_i) \supset q) \supset (\neg q \supset q)$.

From the semantical point of view, the statement we have proved means that if the branching is $\leq n$ in all nonmaximal points then it is $\leq n$ everywhere. However the equality of logics does not directly follow from this semantical argument, once we have not yet proved the completeness of $\mathbf{QH} + Br_n^{EM}$.

Now let us prove a natural analogue of the main lemma on n-branching, 6.10.2, for Br_n^{Θ}. We begin with an analogue of 6.10.1.

Lemma 6.11.8 *Let L be an s.p.l., Θ a set of sentences that are $\forall$-perfect in L. Assume that $Br_n^{\Theta} \subseteq L$, $\Gamma \in VP_L$ and $\overline{L+\Theta} \not\subseteq \Gamma$. Also let E be a characteristic formula for Γ, and assume that $(\Gamma \cup \{E\}, \Delta_i)$ are L-consistent theories in the language $\mathcal{L}(\Gamma)$ for $i = 0, 1, \ldots, n$.*

Then $(\Gamma \cup \{E\}, \Delta_i \cup \Delta_j)$ is L-consistent for some $i \neq j$.

Proof Since $\overline{L+\Theta} \not\subseteq \Gamma$, there exists $A \in Sub(\Theta)$ such that $A \in (-\Gamma)$. In fact, if $B \in \overline{L+\Theta} - \Gamma$, then by $\forall$-perfection, there exist $A_1, \ldots, A_k \in Sub(\Theta)$ such that $\vdash_L A_1 \wedge \ldots \wedge A_k \supset B$; by Lemma 2.7.11 we may assume that $A_1, \ldots, A_k \in \mathcal{L}(\Gamma)$, since D_Γ is denumerable and the parameters of A_i can be replaced with constants from D_Γ. Hence $A_i \in (-\Gamma)$ for some i.

Now we proceed as in the proof of Lemma 6.10.1 and find C_i (for $0 \leq i \leq n$) such that $((E \supset C_i) \supset E) \in \Gamma$ and $(E \supset C_i \vee C_j) \in \Gamma$ whenever $i < j$.

Finally, since $Br_n'^{\Theta} \subseteq L$,

$$\vdash_L \bigwedge_{i=0}^{n}((E \supset C_i) \supset E) \wedge \bigwedge_{i<j}(E \supset C_i \vee C_j) \wedge (A \supset E) \supset E,$$

So by L-consistency, this formula is in Γ. Hence $(A \supset E) \notin \Gamma$, which contradicts the property of characteristic formula E. ■

Lemma 6.11.9 *Let L be an s.p.l., Θ be a set of sentences that are $\forall$-perfect in L, $\Gamma \in VP_L$. Also assume that $\overline{L+\Theta} \not\subseteq \Gamma$ and $Br_n^{\Theta} \subseteq L$ for a finite n. Let E be a characteristic formula for Γ.*

Then there exist $\Gamma_1, \ldots, \Gamma_m \in VP_L$ (where $m \leq n$) such that

(1) $D_{\Gamma_1} = \cdots = D_{\Gamma_m}$;

(2) $\Gamma \cup \{E\} \subseteq \Gamma_i$ *for* $i = 1, \ldots, m$ *(and thus,* $\Gamma < \Gamma_i$*)*;

(3) if $(A_1 \supset A_2) \in (-\Gamma)$ *and* $A_1 \notin \Gamma$, *then* $(A_1 \supset A_2) \notin \Gamma_i$ *for some* i.

Proof Along the same lines as 6.10.2, using Lemma 6.11.8 instead of 6.10.1. ■

Definition 6.11.10 *Let* $\mathbf{t} = (t_i \mid i < h)$ *be an* $h^\dagger$*-sequence,* $h > 0$. *Put*

$$C^+(\mathbf{t}) := \{h+1\} \cup \{i < h \mid i \neq 0,\ t_{i-1} < t_i\},$$
$$C(\mathbf{t}) := \{0, h+1\} \cup \{i < h \mid i \neq 0,\ t_{i-1} \neq t_i\}.$$

The levels from $C^+(\mathbf{t})$ *and* $C(\mathbf{t})$ *are respectively called* +-critical *and* critical *for* $\mathbf{t}$.

For $i \in C(\mathbf{t})$, $i < h$, put

$$i^* := \min\{j > i \mid j \in C(\mathbf{t})\},$$
$$i^+ := \begin{cases} h+1 & \text{if } \forall j > i\ t_j \leq t_i, \\ \min\{j > i \mid t_j > t_i\} & \text{otherwise.} \end{cases}$$

Obviously, $i < i^* \leq i^+$ and $i^+ \in C^+(\mathbf{t})$, although i^+ may be non-equal to $\min\{j > i \mid j \in C^+(\mathbf{t})\}$ (the next element of $C^+(\mathbf{t})$).

It is also clear that for any $u \in T_{\mathbf{t}}$ of level $i < h$, the tree $T_{\mathbf{t}} \uparrow u$ is of depth $h_i := h + 1 - i$, and $T_{\mathbf{t}} \uparrow u \cong T^i_{\mathbf{t}} := T_{\mathbf{t}\uparrow i}$, where $\mathbf{t} \uparrow i := (t_i, t_{i+1}, \ldots, t_{h-1})$.

Lemma 6.11.11 *Let* $\mathbf{t} = (t_i \mid i < h)$ *be an* $h^\dagger$*-sequence, and let* $i \in C(\mathbf{t})$, $0 \leq i < h$. *Assume that the logics*

$$L_{i^*} := \mathbf{IL}(\mathcal{K}(T^{i^*}_{\mathbf{t}})),\ L_{i^+} := \mathbf{IL}(\mathcal{K}(T^{i^+}_{\mathbf{t}}))$$

are strongly complete w.r.t. the classes of frames $\mathcal{K}(T^{i^*}_{\mathbf{t}})$ *and* $\mathcal{K}(T^{i^+}_{\mathbf{t}})$ *respectively. Let* Θ_{i^+} *be a set of sentences that are* $\forall$*-perfect in* $\mathbf{QHP}^+_{k+1}$, *and assume that* $\mathbf{QHP}^+_{k+1} + \Theta_{i^+} = L_{i^+}$.

Then the logic

$$L_i := \Delta^{i^*-i}(L_{i^*}) + Br_{\mathbf{t}_i}[\Theta_{i^+}]$$

is strongly complete w.r.t. $\mathcal{K}(T^i_{\mathbf{t}})$, *and therefore* $L_i = \mathbf{IL}(\mathcal{K}(T^i_{\mathbf{t}}))$.

We may also put

$$L_{i^*} := \mathbf{QH} + \bot \text{ iff } i^* = k+1$$

and similarly,

$$\Theta_{i^+} := \{\bot\} \text{ iff } i^+ = k+1.$$

So $T^{k+1}_{\mathbf{t}}$ is an 'empty tree' and its logic is inconsistent. Also note that $\mathbf{QHP}^+_{k+1-i} \subseteq L_i$, since $\mathbf{LP}^+_{k+1-i^*} \subseteq L_{i^*}$, so $\mathbf{QHP}^+_{k+1-i} = \Delta^{i^*-1}(\mathbf{LP}^+_{k+1-i^*}) \subseteq \Delta^{i^*-i}(L_{i^*}) \subseteq L_i$.

Proof Fix a point $u_0 \in T_{\mathbf{t}}$ of height i, so $T^i_{\mathbf{t}} = T_{\mathbf{t}} \uparrow u_0$.

First let us show that $T^i_{\mathbf{t}}$ is an L_i-frame. In fact, $\Delta^{i^*-i}(L_{i^*}) \subseteq \mathbf{IL}(T^i_{\mathbf{t}})$, since L_{i^*} is valid at all points of height $\geq i^* - i$ in $T^i_{\mathbf{t}}$, i.e. at levels $\geq i^*$ in $T_{\mathbf{t}}$. Also $Br^A_{t_i} \in \mathbf{IL}(T^i_{\mathbf{t}})$ for any $A \in \Theta_{i^+}$ by Lemma 6.11.6, since $A \in \mathbf{IL}(T^j_{\mathbf{t}}){=}\mathbf{IL}(T^i_{\mathbf{t}} \uparrow v)$ for any v of level $j \geq i^+$ and the branching of $T_{\mathbf{t}}$ is $\leq t_i$ at any point of level $i \leq j < i^+$ (to minimise calculations, we consider levels in the tree $T_{\mathbf{t}}$ not in $T^i_{\mathbf{t}}$).

Now we prove the strong completeness. Let R be the accessibility relation in $T_{\mathbf{t}}$. Consider an $L_i\forall$-place Γ. We put $\Gamma_{u_0} := \Gamma$ for root u_0 of $T^i_{\mathbf{t}}$ and construct (inductively) $L_i\forall$-places Γ_u for points $u \in T^i_{\mathbf{t}}$ of levels from i till i^+ or i^*. During our construction we take care of the naturalness property $(\supset)'$ from Lemma 6.4.23 and the property of $\forall^{\leq}$-models: $\Gamma_u \leq \Gamma_v$ if uRv in T^i_t.

Let us begin with the 'trivial' case: $L_i + \Theta_{i^+} \subseteq \Gamma_{u_0}$. Then we put $\Gamma_v := \Gamma_{u_0}$ for all $v \in R(u_0)$ of levels $< i^+$. And for any w of level i^+ we obtain a natural model over $T_{\mathbf{t}} \uparrow w = T^{i^*}_{\mathbf{t}}$ with $\Gamma_{u_0} \leq \Gamma_w$ by the strong completeness of L_{i^+}; in fact, Γ_{u_0} is an L_{i^+}-place, since $L_{i^+} = \mathbf{QHP}^+_{k+1} + \Theta_{i^+} \subseteq L_i + \Theta_{i^+} \subseteq \Gamma_{u_0}$. Note that the naturalness condition $(\supset)'$ holds for all points v of levels $j < i^+$; in fact, if $(A_1 \supset A_2) \in (-\Gamma_v)$, then $(A_1 \supset A_2) \notin \Gamma_w$ for any $w \in R(v)$ of level i^+.

Now consider the 'nontrivial' case $L_i + \Theta_{i^+} \not\subseteq \Gamma_{u_0}$. Then if $t_i < \omega$, we apply Lemma 6.11.9 (for $n = t_i$, $\Theta = \Theta_{i^+}$) and find $L_i\forall$-places Γ_v for $v \in \beta(u_0)$ such that $\Gamma_{u_0} < \Gamma_v$ for all v and the naturalness condition $(\supset)'$ for Γ_{u_0} holds, cf. the proof of 6.10.3 (recall that a characteristic formula E exists by Proposition 6.3.12, since $\mathbf{LP}^+_{n+1} \subseteq L_i$).

Then we repeat the same construction for points $v \in \beta(u_0)$ and so on, till level i^*; note that $t_j = t_i$ for all levels j such that $i \leq j < i^*$. If for a certain v, $L_i + \Theta_{i^+} \subseteq \Gamma_v$ and Lemma 6.11.9 is inapplicable, then we proceed as in the 'trivial' case, and repeat this $L\forall$-place Γ_v at all points above v till level i^+. If we successfully arrive at a point w of level i^*, then we obtain a sequence of $L_i\forall$-places $\Gamma_{u_0} < \ldots < \Gamma_w$. By Lemma 6.3.1 we can conclude that $\overline{L_{i^*}} \subseteq \Gamma_w$ (i.e. Γ_w is an L_{i^*}-place), since Γ_{u_0} is a $\Delta^{i^*-i}(L_{i^*})$-place. Now we apply the strong completeness of L_{i^*} and obtain a natural model over $T_{\mathbf{t}} \uparrow w = T^{i^*}_{\mathbf{t}}$.

There also exists a degenerate case, when at point u_0 (or at some intermediate point) Lemma 6.11.9 should be applied to $m = 0$, so it does not produce any $L\forall$-place $\Gamma_v > \Gamma_{u_0}$. But in this case the naturalness condition $(\supset)$ trivially holds for Γ_{u_0}, so we can repeat the place Γ_{u_0} at every $v \in T_{\mathbf{t}} \uparrow u_0$ and thus obtain a required natural model.

Finally let us mention the case $t_i = \omega$. Now if $(A_1 \supset A_2) \in (-\Gamma_{u_0})$ and $A_1 \notin \Gamma_{u_0}$, we construct an $L\forall$-place Γ_v such that $\Gamma_{u_0} \cup \{A_1\} \subseteq \Gamma_v$, $A_2 \notin \Gamma_v$ (thus $(A_1 \supset A_2) \notin \Gamma_v$ and $\Gamma_{u_0} < \Gamma_v$), cf. the proof of Lemma 6.2.13(2). So we obtain infinitely many $L\forall$-places Γ_v for $v \in \beta(u_0)$ — if there are only finitely many, then we can repeat one of them at all other points $v \in \beta(u_0)$. We apply this construction to all points from $\beta(u_0)$ and further on, till level i^*. Eventually we obtain $L_{i^*}\forall$-places Γ_w for all points w of level i^* and use the strong completeness of L_{i^*} as in the previous case. ■

It is clear that Lemma 6.11.11 yields a $\forall$-perfect axiomatisation of $\mathbf{IL}(\mathcal{K}(T^i_{\mathbf{t}}))$

for any $h^\dagger$-sequence $\mathbf{t}$ and $i \in C(\mathbf{t})$; in particular, for $\mathbf{IL}(\mathcal{K}(T_\mathbf{t})) = \mathbf{IL}(\mathcal{K}(T_\mathbf{t}^0))$.

Proposition 6.11.12 *Let $\mathbf{t} = (t_i \mid i < h)$ be an $h^\dagger$-sequence. Consider the sequence Θ_i for $i \in C(\mathbf{t})$ constructed by induction on $h+1-i$:*

$$\Theta_{h+1} := \{\bot\},$$

$$\Theta_i := \delta^0_{i^*-i}(\Theta_{i^*}) \cup (Br_{t_i}[\Theta_{i^+}])^\forall \textit{ for } i < h.$$

Then all Θ_i are $\forall$-perfect in $\mathbf{LP}^+_{h+1}$,

$$\mathbf{IL}(\mathcal{K}(T^i_\mathbf{t})) = \mathbf{LP}^+_{h+1} + \Theta_i,$$

and these logics are strongly Kripke-complete w.r.t. $\mathcal{K}(T^i_\mathbf{t})$. In particular,

$$\mathbf{IL}(\mathcal{K}(T_\mathbf{t})) = \mathbf{LP}^+_{h+1} + \Theta_0$$

is strongly Kripke-complete w.r.t. $\mathcal{K}(T_\mathbf{t})$.

Proof Every Θ_i is $\forall$-perfect in $\mathbf{LP}^+_{h+1}$ and

$$\mathbf{LP}^+_{h+1} + \delta^0_{i^*-i}(\Theta_{i^*}) = \mathbf{LP}^+_{h+1} + \Delta^{i^*-i}(\mathbf{LP}^+_{h+1} + \Theta_{i^*}) = \Delta^{i^*-i}(\mathbf{LP}^+_{h+1} + \Theta_{i^*})$$

by Proposition 2.13.33. Recall that $\mathbf{LP}^+_{h+1} \subseteq \mathbf{IL}(\mathcal{K}(T_\mathbf{t})) \subseteq \mathbf{IL}(\mathcal{K}(T^i_\mathbf{t})$ and also that every set of the form $\Theta^\forall$ is $\forall$-perfect in $\mathbf{QH}$ and $\forall$-perfection is $\bigcup$-additive. So we can apply Lemma 6.11.11 at the induction step i. ■

Note that $AP_{h+1-i} = \delta^0_{h+1-i}\bot \in \Theta_i$ for any i, since $\bot \in \Theta_{h+1}$ and $\delta^0_{i^*-i}(\delta^0_{h+1-i^*}\bot) = \delta^0_{h+1-i}\bot$.

Recall that $\mathbf{LP}^+_{h+1} + AP_{h+1-i} = \mathbf{LP}^+_{h+1-i}$ by Proposition 2.13.31, so $\mathbf{LP}^+_{h+1-i} \subseteq \mathbf{LP}^+_{h+1} + \Theta_i$ for all $i \in C(\mathbf{t})$.

Also note that for $h = 0$ and the empty h-sequence $\mathbf{t}$ we have $C(\mathbf{t}) = \{0, 1\}$, $\Theta_1 = \{\bot\}$, $\Theta_0 = \{\delta\bot\} = \{AP_1\}$, and this axiom generates classical logic $\mathbf{QCL} = \mathbf{IL}(\mathcal{K}(T_k))$.

For a constant sequence $\mathbf{t} = (n, \dots, n)$ of length $k > 0$ (where $0 < n \leq \omega$) we have $C(\mathbf{t}) = \{0, k+1\}$, $\Theta_{k+1} = \{\bot\}$. If n is finite, then $\Theta_0 = \{\delta^0_{k+1}\bot\} \cup \{Br^\bot_n\} = \{AP_{k+1}, Br_n\}$, so we obtain the axiomatisation $\mathbf{LP}^+_{k+1} + Br_n$ for the uniform tree T^{k+1}_n described in Section 6.10.

If $n = \omega$, then $\Theta_0 = \{AP_{k+1}\}$, so we obtain a standard axiomatisation $\mathbf{LP}^+_{k+1}$ of IT^{k+1}_ω; recall that $Br^\Theta_\omega = \varnothing$.

In some cases the axiomatisation described in Proposition 6.11.12 can be simplified. Note that by Lemmas 2.9.2 and 6.11.5, δ^0_k and $Br_n[(\dots)^\forall]$ preserve deductive equivalence, so every Θ_i can be replaced with its simpler deductive equivalent. E.g. if $A \in \Theta_{i^*}$ and $A \in \Theta_i$ for some $i < h$, then we can eliminate $\delta^0_{i^*-i}A$ from $\delta^0_{i^*-i}(\Theta^*_i)$, as this formula follows from A.

Proposition 6.11.12 can be strengthened as follows.

Proposition 6.11.13 *Let* $\mathbf{t} = (t_i \mid i < h)$ *be an* $h^\dagger$*-sequence,* L_0 *an s.p.l. such that* $\mathbf{LP}^+_{h+1} \subseteq L_0 \subseteq \mathbf{IL}(\mathcal{K}(T_\mathbf{t}))$*. For each* $i \in C[\mathbf{t}]$ *let* Θ_i *be a set of sentences, which is* $\forall$*-perfect in* L_0 *whenever* $i \in C^+(\mathbf{t})$*. Assume that* $L_0 + \Theta_{h+1} = \mathbf{QH} + \bot$*, and for any* $i < h$ *in* $C(\mathbf{t})$

$$\mathbf{L}_0 + \Theta_i = \mathbf{L}_0 + \Delta^{i^*-i}(\mathbf{L}_0 + \Theta_{i^*}) + Br_{t_i}[\Theta_{i^+}].$$

Then $\mathbf{L}_0 + \Theta_i = \mathbf{IL}(\mathcal{K}(T^i_\mathbf{t}))$ *for all* $i \in C(\mathbf{t})$*; in particular,*

$$\mathbf{L}_0 + \Theta_0 = \mathbf{IL}(\mathcal{K}(T_\mathbf{t})).$$

Proof Again we proceed by induction on $h + 1 - i$, and the argument at the induction step fully repeats the proof of Proposition 6.11.12, cf. the proof of Lemma 6.11.11, which is actually a particular case of this construction for $L_0 = \mathbf{LP}^+_{h+1}$. ∎

So Proposition 6.11.12 yields an infinite axiomatisation for any nonconstant $\mathbf{t}$, since any set of the form $(Br^\Theta_{t_i})^\forall$ is infinite, while Proposition 6.11.13 allows us to find a finite axiom system for the case of increasing $\mathbf{t}$.

Viz., consider an increasing $h^\dagger$-sequence $\mathbf{t} = (t_i \mid i < h)$, i.e. $t_0 \leq t_1 \leq \ldots \leq t_{h-1}$, with the critical levels $0 = i_0 < i_1 < \ldots < i_k < h$ and put $i_{k+1} := h + 1$, so all these levels are $+$-critical except for 0. Let $\mathbf{t}'$ be the corresponding subsequence: $t'_j := t_{i_j}$ for $0 \leq j \leq k + 1$, so

$$t_i = \begin{cases} t'_j & \text{if } i \in [i_j, i_{j+1}[,\ j < k, \\ t'_k & \text{if } i_k \leq i < h. \end{cases}$$

Obviously,

$$i^+ = i^* = \min\{j > i \mid t_j > t_i\} = i_{j+1}$$

for $i = i_j$, $0 \leq j \leq k$.

Then we define the finitely axiomatisable logics

$$L(\mathbf{t}) := \mathbf{LP}^+_{h+1} + Br_{t'_k} + \{Br_{t'_j}[AP_{h+1-i_{j+1}}] \mid j < k\}.$$

Note that if $t'_k = \omega$, the axiom $Br_{t'_k} = \top$ is redundant.

Theorem 6.11.14 $\mathbf{IL}(\mathcal{K}(T_\mathbf{t})) = L(\mathbf{t})$ *for any increasing* $h^\dagger$*-sequence* $\mathbf{t}$.

Proof We apply Proposition 6.11.13 to $L_0 = L(\mathbf{t})$ and $\Theta_i = \{AP_{h+1-i}\}$ for each $i \in C(\mathbf{t}) = \{i_0, \ldots, i_k, i_{k+1}\}$ (where $i_0 = 0$, $i_{k+1} = h + 1$). Obviously $L(\mathbf{t}) \subseteq \mathbf{IL}(\mathcal{K}(T_\mathbf{t}))$ by Lemma 6.11.6, since the branching is $\leq t'_j$ at all points of levels $< i_{j+1}$, and $AP_{h+1-i_{j+1}} \in \mathbf{IL}(\mathcal{K}(T^i_\mathbf{t}))$ for $i \geq i_{j+1}$. The sets Θ_i are $\forall$-perfect in L_0 by 2.8.13(2).

For $i = i_j \in C(\mathbf{t})$, $j < k$ we have

$$\begin{aligned} L_0 + \Delta^{i^*-i}(L_0 + \Theta_{i^*}) &= L_0 + \Delta^{i_{j+1}-i_j}(L_0 + AP_{h+1-i_{j+1}}) \\ &= L_0 + \delta^0_{i_{j+1}-i_j} AP_{h+1-i_{j+1}} = L_0 + AP_{h+1-i_j} \end{aligned}$$

and $Br_{t_i}[\Theta_{i^+}] = Br_{t'_j}(AP_{h+1-i_{j+1}}) \in L_0$. Thus

$$L_0 + \Theta_i = L_0 + \Delta^{i^*-i}(L_0 + \Theta_{i^*}) + Br_{t_i}[\Theta_{i^+}].$$

Finally by 6.11.13, $\mathbf{IL}(\mathcal{K}(T_{\mathbf{t}})) = L_0 + \Theta_0 = L_0 + AP_{h+1} = L(\mathbf{t})$, since $AP_{h+1} \in L(\mathbf{t})$. ■

On the other hand, for all non-increasing $h^\dagger$-sequences $\mathbf{t}$ the corresponding logics $\mathbf{IL}(\mathcal{K}(T_{\mathbf{t}}))$ are not finitely axiomatisable [...]; a proof will be given in Volume 2.

Let us also present a simplified infinite axiomatisation for $\mathbf{IL}(\mathcal{K}(T_{\mathbf{t}}))$ for a decreasing $\mathbf{t}$.

Proposition 6.11.15 *Let $\mathbf{t} = (t_i \mid i < h)$ be a decreasing $h^\dagger$-sequence: $t_0 \geq t_1 \geq \ldots \geq t_{h-1}$; then*

$$\mathbf{IL}(\mathcal{K}(T_{\mathbf{t}})) = \mathbf{LP}^+_{h+1} + Br_{t_0} + \bigcup\{\delta^0_i(\{Br_{t_i}\}^\forall) \mid 0 < i < h,\ t_i \neq t_{i-1}\}.$$

Proof Readily follows from Proposition 6.11.13. Note that here $C^+(\mathbf{t}) = \{h+1\}$, $i^+ = h+1$ for any $i < h$, $Br_{t_i}[\Theta_{h+1}] = \{Br^\perp_{t_i}\}$ and $Br^\perp_{t_i}$ is $\mathbf{H}$-equivalent to Br_{t_i}. ■

Finally we obtain axiomatisations for uncountable levelwise uniform trees.

Theorem 6.11.16 *Let $\mathbf{t}$ be an h-sequence with some $t_i > \omega$ (i.e. $T_{\mathbf{t}}$ is uncountable). Then $\mathbf{IL}(\mathcal{K}(T_{\mathbf{t}})) = \mathbf{IL}(\mathcal{K}(T_{\mathbf{t}^\dagger}))$.*

Proof Obviously $T_{\mathbf{t}} \twoheadrightarrow T_{\mathbf{t}\leq\omega}$, so by Proposition 3.3.14, $\mathbf{IL}(K(T_{\mathbf{t}})) \subseteq \mathbf{IL}(\mathcal{K}(T_{\mathbf{t}^\dagger}))$.

On the other hand, $\mathbf{IL}(\mathcal{K}(T_{\mathbf{t}})) \supseteq \mathbf{IL}(\mathcal{K}(T_{\mathbf{t}\leq\omega}))$, since all the axioms used in the axiomatisation Θ_0 for $T_{\mathbf{t}\leq\omega}$ (Proposition 6.11.12), are also valid in $T_{\mathbf{t}}$.

More precisely, we can show (by induction on $h+1-i$) that $\Theta_i \subseteq \mathbf{IL}(\mathcal{K}(T_{\mathbf{t}^\dagger} \uparrow u))$ for points u of level i, for any $i \in C(\mathbf{t}^\dagger)$. The argument is almost obvious, cf. soundness proof for Lemma 6.11.11. In fact, recall that $Br^\Theta_\omega = \varnothing$, so the levels i with $t_i = \omega$ do not bring anything new to the axiomatisation Θ_i, and thus to Θ_j for $j \leq i$. ■

Therefore we really have an explicit recursive axiomatisation for an arbitrary levelwise uniform tree of any cardinality.

For an h-sequence of cardinals $\mathbf{t} = (t_i \mid i < h)$ let us also consider a poset $T_{\mathbf{t}} + 1$ obtained by adding the top element. Then the following holds.

Proposition 6.11.17 *Let $\mathbf{t} = (t_i \mid i < h)$ be an $h^\dagger$-sequence. Then $\mathbf{IL}(\mathcal{K}(T_{\mathbf{t}} + 1)) = \mathbf{QHP}^+_{h+2} + J + \hat{\Theta}_0$, where the sets $\hat{\Theta}_i$ for $i \in C[\mathbf{t}]$ are constructed just as in Proposition 6.11.12, but with a difference at the beginning*

$$\hat{\Theta}_{h+1} := \{AP_1\},^4$$

$$\hat{\Theta}_i := \delta^0_{i^*-i}(\hat{\Theta}_{i^*}) \cup (Br_{t_i}[\hat{\Theta}_{i+}])^{\forall} \text{ for } i < h.$$

All of them are $\forall$-perfect in $\mathbf{QHP}^+_{h+1}$.

This logic is strongly complete (w.r.t. $\mathcal{K}(T_{\mathbf{t}} + 1)$).

We can also obtain a finite axiomatisation of $\mathbf{IL}(\mathcal{K}(T_{\mathbf{t}}+1))$ for an increasing $\mathbf{t}$ (cf. Theorem 6.11.14) and a simplified infinite axiomatisation for a decreasing $\mathbf{t}$ (cf. Proposition 6.11.15). We also have

$$\mathbf{IL}(\mathcal{K}(T_{\mathbf{t}} + 1)) = \mathbf{IL}(\mathcal{K}(T_{\mathbf{t}^\dagger} + 1))$$

for an uncountable $\mathbf{t}$ (cf. Theorem 6.11.16).

One can also consider logics with equality and obtain the following consequence from the results of Section 3.9:

Proposition 6.11.18 *Let $\mathbf{t}$ be an h-sequence ($h \geq 0$). Then*

$$\mathbf{IL}^=(\mathcal{KE}(T_{\mathbf{t}})) = (\mathbf{IL}(\mathcal{K}(T_{\mathbf{t}})))^=,$$
$$\mathbf{IL}^=(\mathcal{K}(T_{\mathbf{t}})) = (\mathbf{IL}(\mathcal{K}(T_{\mathbf{t}})))^{=d},$$

and similarly for $T_{\mathbf{t}} + 1$. The corresponding logics are strongly complete w.r.t. these classes of Kripke frames or sheaves.

[4]So to say, the depths increase by 1.

Chapter 7

Kripke completeness for constant domains

In this chapter we prove completeness results in Kripke semantics with constant domains for modal logics containing Barcan axioms and superintuitionistic logics containing CD.

7.1 Modal canonical models with constant domains

Let us first describe canonical models with constant domains for modal logics. Since the Barcan formula is not valid in frames with varying domains, it is clear that every m.p.l. containing Ba is not V-canonical. Thus canonical models should be restricted to L-places having a fixed denumerable domain S_0.

Here is a precise definition.

Definition 7.1.1 *Let L be an N-modal logic containing Ba_i for $1 \leq i \leq N$. Let CP_L be the set of all L-places (from VP_L) with the set of constants S_0. The* canonical frame *and the* canonical model *with a constant domain for L are defined as $CF_L := VF_L|CP_L$, $CM_L := VM_L|CP_L$.*

For the sake of simplicity, the notation of relations and the domain function in CF_L is the same as in VF_L. So in particular, D_L is constant: $D_L(\Gamma) = S_0$.

It is clear that mentioning the superset S^* is not necessary. Thus we can consider CP_L as consisting of (L, S_0)-Henkin theories rather than L-places.

Now we have to show that the restricted relations R_{Li} are selective on CP_L. Unfortunately, a direct analogue of Lemma 6.1.9 does not hold for the case of constant domains, because one cannot extend an L-consistent theory $\Gamma \subseteq MF_{S_0}^{(=)}$ to an (L, S_0)-Henkin theory Γ without adding new constants.[1] But the following lemma suggests the way out.

[1]The corresponding counterexample for the intuitionistic case was constructed by Ghilardi.

Lemma 7.1.2 *Let L be an N-modal predicate logic, $L \vdash Ba_i$ for every $i \leq N$. Let $\Gamma \in CP_L, A \in MF_N^{(=)}(S_0)$ and $\Diamond_i A \in \Gamma$ for some i. Then there exists $\Gamma' \in CP_L$ such that $\Box_i^- \Gamma \subseteq \Gamma'$ and $A \in \Gamma'$.*

This lemma means that the relations R_{Li} on CP_L are selective, cf. Definition 6.1.31 (recall that $\Gamma R_{Li} \Gamma' \Leftrightarrow \Box_i^- \Gamma \subseteq \Gamma' \Leftrightarrow \Diamond_i \Gamma' \subseteq \Gamma$, cf. Definition 6.1.14).

Proof Let us enumerate the set of modal S_0-sentences of the form $\exists x A(x)$, viz. $\{\exists x A_k(x) \mid k > 0\}$. Then we construct a sequence of finite subsets $(\Gamma_k | k \in \omega)$ of $MF_{S_0}^{(=)}$ such that for any $k \in \omega$

$$\Diamond_i \left(\bigwedge \Gamma_k\right) \in \Gamma. \tag{$\Diamond$}$$

We put $\Gamma_0 := \{A\}$ and $\Gamma_k := \Gamma_{k-1} \cup \{\exists x A_k(x) \supset A_k(c)\}$ for *some* $c \in S_0$ (say, for the first one in an enumeration of S_0), such that ($\Diamond$) holds. Let us show the existence of this c.

In fact, by the inductive hypothesis, $\Diamond_i(\bigwedge \Gamma_{k-1}) \in \Gamma$. We also have

$$L \vdash \Box_i \exists y (\exists x A_k(x) \supset A_k(y))$$

(say, for a new variable y). Thus

$$\Diamond_i \exists y \left(\bigwedge \Gamma_{k-1} \wedge (\exists x A_k(x) \supset A_k(y))\right) \in \Gamma.$$

Now by the Barcan axiom,

$$\exists y \, \Diamond_i \left(\bigwedge \Gamma_{k-1} \wedge (\exists x A_k(x) \supset A_k(y))\right) \in \Gamma,$$

and thus by property (vi) of Henkin theories

$$\Diamond_i \left(\bigwedge \Gamma_{k-1} \wedge (\exists x A_k(x) \supset A_k(c))\right) \in \Gamma \text{ for some } c \in S_0.$$

Let $\Gamma_\omega := \bigcup\limits_{k \in \omega} \Gamma_k$. Then $\Box^- \Gamma \cup \Gamma_\omega$ is L-consistent. In fact, suppose $\Box_i^- \Gamma \vdash_L \neg(\bigwedge \Gamma_k)$ for some k. Then $\Gamma \vdash_L \Box_i \neg(\bigwedge \Gamma_k)$, i.e. $\neg \Diamond_i(\bigwedge \Gamma_k) \in \Gamma$. This contradicts ($\Diamond$).

Finally by Lemma 6.1.5, $\Box_i^- \Gamma \cup \Gamma_\omega$ can be extended to an (L, S_0)-complete theory Γ'. Clearly Γ' is an (L, S_0)-Henkin theory, and $\Box_i^- \Gamma \cup \{A\} \subseteq \Gamma'$. ∎

Remark 7.1.3 The above lemma implies that every L-consistent theory of the form $\Box^- \Gamma \cup \Gamma_0$ with finite Γ_0, can be extended to an (L, S_0)-Henkin theory. In fact, 7.1.2 is applicable, since in this case $\Diamond_i(\bigwedge \Gamma_0) \in \Gamma$ (otherwise $\neg \Diamond_i(\bigwedge \Gamma_0) \in \Gamma$, which implies $\Box_i \neg(\bigwedge \Gamma_0) \in \Gamma$, i.e. $\neg(\bigwedge \Gamma_0) \in \Box_i^- \Gamma$).

This argument fails if Γ_0 is infinite. In fact, if $\Diamond_i(\bigwedge \Gamma_0') \in \Gamma$ for any finite subset Γ_0' of Γ_0 and $\exists y A(y)$ has Henkin property (w.r.t. Γ) for any Γ_0', then there exist constants $c' \in S_0$ satisfying $\bigwedge \Gamma_0' \wedge A(c')$ for all finite $\Gamma_0' \subseteq \Gamma_0$, but we cannot state the existence of a *single* constant c for *all* these Γ_0', and thus for the whole infinite set Γ_0.

Now since the relations R_{Li} are selective on CP_L, we readily obtain the main property of the canonical model with a constant domain (cf. Lemma 6.1.21):

Theorem 7.1.4 *For any* $\Gamma \in CP_L$ *and* $A \in MF_N^{(=)}(S_0)$,

$$CM_L, \Gamma \vDash A \text{ iff } A \in \Gamma.$$

Corollary 7.1.5 *For a logic L with Barcan axioms, for any formula A,*

$$CM_L \vDash A \text{ iff } L \vdash A.$$

Proof (Cf. 6.1.22). We can consider only the case when A is a sentence. 'Only if' follows from 6.1.3 and Theorem 7.1.4. To check 'if', suppose $A \notin L$; then by 6.1.10, we have $A \notin \Gamma$ for some (L, S_0)-Henkin theory Γ; hence $CM_L, \Gamma \nvDash A$, by Theorem 7.1.4. ∎

Definition 7.1.6 *A m.p.l.(=) is called C*-canonical *if* $CF_L \vDash L$.

Corollary 7.1.7 *Every C-canonical m.p.l. is strongly* $\mathcal{CK}$*-complete. Every C-canonical m.p.l.= is strongly* $\mathcal{CKE}$*-complete.*

Proof Similar to Corollary 6.1.24. ∎

For an N-m.p.l.(=) $\mathbf{\Lambda}$ put $\mathbf{\Lambda C} := \mathbf{\Lambda} + \bigwedge_{i=1}^{N} Ba_i$.

The next result is an analogue of 6.1.29 for constant domains; the proof is very similar.

Theorem 7.1.8 *Let* $\mathbf{\Lambda}$ *be a propositional PTC-logic. Then the logics* $\mathbf{Q\Lambda C}$, $\mathbf{Q\Lambda C}^=$ *are C-canonical.*

Proof The argument for closed axioms is trivial, cf. 6.1.29.

If $A = \Diamond_\beta \Box_k p \supset \Box_\gamma p \in L = \mathbf{Q\Lambda}^{(=)} + Ba$, then $R_{L\beta}^{-1} \circ R_{L\gamma} \subseteq R_{Lk}$. In fact, suppose $\Gamma R_{L\beta} \Delta_1$, $\Gamma R_{L\alpha} \Delta_2$, $\Box_k B \in \Gamma$. Then $\Diamond_\beta \Box_k B \in \Gamma$ (note that $B \in \mathcal{L}(\Delta_1) = \mathcal{L}(\Gamma)$, since the domain is constant), and thus from $\Diamond_\beta \Box_k B \supset \Box_\gamma B \in L \subseteq \Gamma$ it follows that $\Box_\gamma B \in \Gamma$. Therefore $B \in \Delta_2$, since $\Gamma R_{L\gamma} \Delta_2$. So we obtain $\Delta_1 R_k \Delta_2$, and eventually $CF_L \vDash A$. ∎

This theorem is actually a particular case of a more general completeness result by Tanaka–Ono, which will be proved in Section 7.4.

7.2 Intuitionistic canonical models with constant domains

Let us now consider intuitionistic models.

Let L be a superintuitionistic predicate logic (with or without equality) containing the formula CD. Then L is definitely not V-canonical, and we have to extract a subframe of VF_L with a fixed denumerable domain S_0.

Definition 7.2.1 *Let CP_L be the set of all $(L\exists\forall, S_0)$-complete intuitionistic theories; they are called* CDL-places.

Recall that (Section 7.2) for L containing CD, $L\exists\forall$-completeness is equivalent to the property

$$(Ac) \qquad \forall x A(x) \in \Gamma \text{ iff for all } c \in S_0 \; A(c) \in \Gamma.$$

This reflects the property of forcing in intuitionistic Kripke models with constant domains:

$$u \Vdash \forall x A(x) \text{ iff for all } c \in D(u) \; u \Vdash A(c).$$

As in the modal case, we omit S_0 from our notation.

Definition 7.2.2 *The canonical frame and the canonical model with a constant domain for a superintuitionistic logic L containing CD are (respectively) $CF_L := VF_L|CP_L$ and $CM_L := VM_L|CF_L$.*

Again we use the same notation of relations and domains as in VF_L; so $(D_L(\Gamma) = S_0)$ and $\Gamma R_L \Gamma'$ iff $\Gamma \subseteq \Gamma'$.

Note that R_L coincides with $\leq$ on CP_L, since every proper extension of Γ contains additional S_0-formulas (see Definition 6.2.11).

Now let us show that the relation R_L is selective on CP_L. The property 6.2.12($\forall$) follows readily from (Ac); so it remains to check 6.2.12($\supset$).

Lemma 7.2.3 *Let L be a superintuitionistic predicate logic containing CD, $\Gamma \in CP_L$, and let $(\Gamma \cup \Gamma_0, \Delta_0)$ be an L-consistent theory with* finite *Γ_0, Δ_0. Then there exists an $L\exists\forall$-complete theory $\Gamma' \succeq (\Gamma \cup \Gamma_0, \Delta_0)$ (i.e. $\Gamma \cup \Gamma_0 \subseteq \Gamma', \Delta_0 \cap \Gamma' = \varnothing$).*

Proof Similar to Lemma 6.2.6. We enumerate $IF^{(=)}(S_0)$ as $\{B_k \mid k \in \omega\}$ and construct an increasing sequence of finite theories $(\Gamma_0, \Delta_0) \preceq \ldots \preceq (\Gamma_k, \Delta_k) \preceq \ldots$ such that $(\Gamma \cup \Gamma_k, \Delta_k)$ is consistent for any $k \in \omega$. The construction is as follows.

(1) If B_k is not of the form $\exists x A(x)$ or $\forall x A(x)$, we define

$$(\Gamma_{k+1}, \Delta_{k+1}) := \begin{cases} (\Gamma_k \cup \{B_k\}, \Delta_k) \text{ if } (\Gamma \cup \Gamma_k \cup \{B_k\}, \Delta_k) \text{ is } L\text{-consistent,} \\ (\Gamma_k, \Delta_k \cup \{B_k\}) \text{ otherwise.} \end{cases}$$

Then the consistency of $(\Gamma \cup \Gamma_k, \Delta_k)$ implies the consistency of $(\Gamma \cup \Gamma_{k+1}, \Delta_{k+1})$; this is proved as in Lemma 6.2.6.

(2) If $B_k = \exists x A(x)$ and $(\Gamma \cup \Gamma_k \cup \{B_k\}, \Delta_k)$ is L-consistent, then

$$\Gamma \nvdash_L \bigwedge \Gamma_k \wedge \exists x A(x) \supset \bigvee \Delta_k,$$

and thus

$$\forall x (\bigwedge \Gamma_k \wedge A(x) \supset \bigvee \Delta_k) \notin \Gamma.$$

Hence by (Ac),

$$(\bigwedge \Gamma_k \wedge A(c) \supset \bigvee \Delta_k) \notin \Gamma \text{ for some } c \in S_0,$$

and so

$$\Gamma \nvdash \bigwedge \Gamma_k \wedge A(c) \supset \bigvee \Delta_k,$$

by Lemma 6.2.5(v). Thus

$$\Gamma \nvdash \bigwedge \Gamma_k \wedge \exists x A(x) \wedge A(c) \supset \bigvee \Delta_k,$$

i.e. the theory $(\Gamma \cup \Gamma_{k+1}, \Delta_{k+1})$, where

$$\Gamma_{k+1} := \Gamma_k \cup \{\exists x A(x), A(c)\}, \ \Delta_{k+1} := \Delta_k,$$

is L-consistent.

Otherwise, if $(\Gamma \cup \Gamma_k \cup \{B_k\}, \Delta_k)$ is L-inconsistent, we put

$$(\Gamma_{k+1}, \Delta_{k+1}) := (\Gamma_k, \Delta_k \cup \{B_k\});$$

then as in (1), we obtain that $(\Gamma \cup \Gamma_{k+1}, \Delta_{k+1})$ is consistent.

(3) If $B_k = \forall x A(x)$ and

$$\Gamma \nvdash_L \bigwedge \Gamma_k \supset \bigvee \Delta_k \vee \forall x A(x),$$

then by axiom CD,

$$\forall x(\bigwedge \Gamma_k \supset \bigvee \Delta_k \vee A(x)) \notin \Gamma.$$

Hence

$$(\bigwedge \Gamma_k \supset \bigvee \Delta_k \vee A(c)) \notin \Gamma$$

for some $c \in S_0$. Take this c and define

$$(\Gamma_{k+1}, \Delta_{k+1}) := (\Gamma_k, \Delta_k \cup \{\forall x A(x), A(c)\});$$

then $(\Gamma \cup \Gamma_{k+1}, \Delta_{k+1})$ is L-consistent.

Otherwise, if

$$\Gamma \vdash_L \bigwedge \Gamma_k \supset \bigvee \Delta_k \vee \forall x A(x),$$

we define

$$(\Gamma_{k+1}, \Delta_{k+1}) := (\Gamma_k \cup \forall x A(x), \Delta_k);$$

then $(\Gamma \cup \Gamma_{k+1}, \Delta_{k+1})$ is again L-consistent.

Eventually, the theory $\Gamma' := \Gamma \cup \bigcup_{k \in \omega} \Gamma_k$ (or $\bigcup_{k \in \omega} \Gamma_k$ itself) is $L\exists\forall$-complete, and $(\Gamma \cup \Gamma_0, \Delta_0) \preceq \Gamma'$ as required. ■

Note that only property (Ac) of the original theory Γ is essential in the above proof, and the existential property is not used.

Lemma 7.2.4 *Suppose $L \vdash CD$, $\Gamma \in CP_L$, $(A_1 \supset A_2) \notin \Gamma$. Then there exists $\Gamma' \in CP_L$ such that $\Gamma \subseteq \Gamma'$ (i.e. $\Gamma R_L \Gamma'$), $A_1 \in \Gamma'$, $A_2 \notin \Gamma'$. In other words, the relation R_L on CP_L satisfies 6.2.12 ($\supset$) and is selective.*

Proof If $(A_1 \supset A_2) \notin \Gamma$, then the theory $(\Gamma \cup \{A_1\}, \{A_2\})$ is L-consistent, and so we can apply Lemma 7.2.3. ∎

Remark 7.2.5 Note that in Lemma 7.2.3 the sets Γ_0, Δ_0 are finite, and thus it is actually equivalent to Corollary 7.2.4. If $\Gamma_0 \cup \Delta_0$ is infinite, one cannot always find an $L\exists\forall$-complete extension for an L-consistent theory $(\Gamma \cup \Gamma_0, \Delta_0)$ (as we have mentioned, this was shown by Ghilardi).

From Corollary 7.2.4 and the conditions 6.2.12($\supset$), ($\forall$) we obtain the main property of the canonical model:

Theorem 7.2.6 *For any $\Gamma \in CP_L$ and $A \in IF^{(=)}(S_0)$*

$$CM_L, \Gamma \Vdash A \text{ iff } A \in \Gamma.$$

Corollary 7.2.7 *For a logic L containing CD, for any formula A*

$$CM_L \Vdash A \text{ iff } L \vdash A.$$

Proof By Lemma 6.2.6, for an L-unprovable S_0-sentence A there exists an $L\exists\forall$-complete theory $\Gamma \in CP_L$ such that $A \notin \Gamma$. Thus $CM_L, \Gamma \nVdash A$, by Theorem 7.2.6. ∎

Now we can repeat Definition 7.1.6 for the intuitionistic case.

Definition 7.2.8 *An s.p.l (=) L is called C-*canonical *if $CF_L \models L$.*

From Corollary 7.2.7 we obtain

Proposition 7.2.9 *Every C-canonical s.p.l. containing CD is Kripke-complete.*

7.3 Some examples of C-canonical logics

Unlike the case of varying domains, there are quite a few natural examples of C-canonicity; some of them were first found by H. Ono.

Let **QHC** := **QH** + CD, **QHCK** := **QHC** + KF. Sometimes we also use the notation **QΛC**, **QΛCK** for an arbitrary intermediate propositional logic Λ.

Theorem 7.3.1 *The logics* **QHC**$^=$, **QHC**, **QHC**$^{=d}$, **QHC**$^{=s}$ *are C-canonical.*

Proof Similar to Proposition 7.2.15. ∎

Proposition 7.3.2 *Classical logic* **QCL** *is C-canonical.*

Proof The frame $CF_{\mathbf{QCL}}$ is discrete, i.e. $\Gamma R_{\mathbf{QCL}}\Gamma'$ iff $\Gamma = \Gamma'$, cf. the proof of Lemma 6.3.4(i) (recall that $R_L = \subseteq$ in CF_L). ■

C-canonicity also holds for many logics of finite depth.

Lemma 7.3.3 *Let $k > 0$, and let L be a superintuitionistic logic.*

(1) If $CD \wedge AP_k \in L$, then CF_L is of depth $\leq k$.

(2) If $L = \mathbf{QHC} + AP_k$, then CF_L is of depth k.

Proof

(1) Suppose $\Gamma_k \subset \Gamma_{k-1} \subset \ldots \subset \Gamma_1 \subset \Gamma_0$ in CF_L, and let $A_i \in (\Gamma_{i-1} - \Gamma_i)$ for $1 \leq i \leq k$. Then by induction on i it follows that

$$B_i := A_i \vee (A_i \supset A_{i-1} \vee (A_{i-1} \supset \cdots \supset A_2 \vee (A_2 \supset A_1 \vee \neg A_1)\ldots)) \notin \Gamma_i$$

(for $1 \leq i \leq k$). In fact, $B_{i+1} = A_{i+1} \vee (A_{i+1} \supset B_i)$ for $0 \leq i \leq k-1$ (where $B_0 = \bot$). Then $B_0 \notin \Gamma_0$, by consistency. Suppose $B_i \notin \Gamma_i$, but $B_{i+1} \in \Gamma_{i+1}$. Then $A_{i+1} \in \Gamma_{i+1}$ or $(A_{i+1} \supset B_i) \in \Gamma_{i+1}$. The first option contradicts the choice of A_{i+1}. In the second case we have $(A_{i+1} \supset B_i) \in \Gamma_{i+1} \subseteq \Gamma_i$. Since $A_{i+1} \in \Gamma_i$, we obtain $B_i \in \Gamma_i$, which contradicts the inductive hypothesis.

Therefore $B_k \notin \Gamma_k$, while B_k is a substitution instance of AP_k. This contradicts Corollary 7.2.7.

(2) Recall that $L \nvdash AP_{k-1} = q_{k-1} \vee (q_{k-1} \supset \cdots \supset (q_2 \vee (q_2 \supset q_1 \vee \neg q_1)))$ (since AP_{k-1} is refuted in a k-element chain with a constant domain). By Corollary 7.2.7 there exists $\Gamma_{k-1} \in CP_L$ such that $\Gamma_{k-1} \nVdash AP_{k-1}$. Then $\Gamma_{k-1} \nVdash q_{k-1}$, $\Gamma_{k-1} \nVdash q_{k-1} \supset AP_{k-1}$, and so there exists $\Gamma_{k-2} \supseteq \Gamma_{k-1}$ such that $\Gamma_{k-2} \Vdash q_{k-1}$, $\Gamma_{k-2} \nVdash AP_{k-2}$. By induction we obtain a chain $\Gamma_{k-1} \subset \Gamma_{k-2} \subset \cdots \subset \Gamma_1 \subset \Gamma_0$ such that $\Gamma_i \nVdash AP_i$, $\Gamma_i \nVdash q_i$, $\Gamma_{i-1} \Vdash q_i$ for $i > 0$.

■

Remark 7.3.4 Note that the argument used in the above proof of (1) fails for the case of varying domains, because the formula A_i can contain additional constants not occurring in Γ_i, thus we may not find an appropriate substitution instance of AP_k refuted in Γ_k. This observation somewhat explains why propositional axioms cannot describe the finite depth in this case and we need predicate axioms taking new individuals into account.

Lemma 7.3.5 *Let L be a superintuitionistic logic containing Z. Then every cone in CF_L is linearly ordered, i.e. for any $\Gamma, \Gamma', \Gamma'' \in CP_L$, $\Gamma \subseteq \Gamma'$ and $\Gamma \subseteq \Gamma''$ imply that Γ' and Γ'' are $\subseteq$-comparable.*

Proof Suppose $A_1 \in (\Gamma' - \Gamma'')$ and $A_2 \in (\Gamma'' - \Gamma')$. Then $\Gamma \nVdash A_1 \supset A_2$, $A_2 \supset A_1$, and thus $\Gamma \nVdash (A_1 \supset A_2) \vee (A_2 \supset A_1)$. This contradicts $CM_L \Vdash L$. ■

Proposition 7.3.6 *The following superintuitionistic predicate logics are C-canonical:*

(1) $\mathbf{QLC} + CD = \mathbf{QH} + CD \wedge AZ$*;*

(2) $\mathbf{QH} + CD \wedge AP_k$*;*

(3) $\mathbf{QH} + CD \wedge AZ \wedge AP_k$ *for* $k > 0$.

They are respectively determined by the following classes of Kripke frames with constant domains:

(1) linear Kripke frames;

(2) Kripke frames of depth k*;*

(3) k*-element chains.*

These logics are also determined by countable frames of the corresponding kind.

(1) was proved in Ono [Ono, 1983], [Minari, 1983].

Remark 7.3.7 The logic $\mathbf{QLC} + CD$ was also characterised algebraically in [Horn, 1969]. Note that it is also equivalent to the logic IF introduced in [Takeuti and Titani, 1984].

Remark 7.3.8 Note that canonical and quasi-canonical frames with varying domains (for any logic) obviously do not validate AZ. In fact, two extensions of an L-place with incomparable sets of new constants are always R_L-incomparable. Thus the logic $\mathbf{QLC}$ and all its extensions are not V-, U-, or $U^{\leq}$-canonical. Still $\mathbf{QLC}$ is Kripke complete, as we showed in Section 6.7.

Now consider the logics

$$\mathbf{QHCK} := \mathbf{QHC} + KF,\ \mathbf{QS4CK}^{+} := \mathbf{QS4C} + MK^{+},$$

where

$$MK^{+} := \Diamond\forall x\Box(P(x) \supset \Box P(x)).$$

Note that $\mathbf{QS4CK}^{+}$ contains $\mathbf{S4.1}$.

Lemma 7.3.9 *(1) Let* L *be a superintuitionistic logic containing Kuroda formula* KF*. Then its C-canonical frame* CF_L *is coatomic.*

Moreover, for any CDL-place $\Gamma \in CP_L$ *there exists a CD*$\mathbf{QCL}$*-place* $\Gamma' \supseteq \Gamma$.

(2) Let L *be an m.p.l. containing* $\mathbf{QS4CK}^{+}$*. Then* CF_L *is coatomic.*

Note that $\mathbf{QCL}$-complete theories are maximal in CF_L (cf. Lemma 6.3.4(i)).

Proof (1) For $\Gamma \in CP_L$, we construct a sequence $\Gamma = \Gamma_0 \subseteq \Gamma_1 \subseteq \dots$ in CP_L as follows

Let $A_0, A_1 \dots$ be an enumeration of IF. For a CDL-place Θ a formula A is said to be *critical* if $CM_L, \Theta \not\Vdash \overline{\forall}(A \vee \neg A)$.

Since $L \vdash KF$, we have $\vdash_L \neg\neg\overline{\forall}(A \vee \neg A)$, which follows from $\vdash_{\mathbf{QCL}} \overline{\forall}(A \vee \neg A)$ by the Glivenko theorem. Hence $\Theta \not\Vdash \neg\overline{\forall}(A \vee \neg A)$ for any Θ and A.

Now assume that Γ_n is constructed and A_n is critical for Γ_n. Since $\Gamma_n \not\Vdash \neg\overline{\forall}(A_n \vee \neg A_n)$, there exists $\Gamma_{n+1} \supset \Gamma_n$ such that $\Gamma_{n+1} \Vdash \overline{\forall}(A_n \vee \neg A_n)$. If A_n is not critical for Γ_n, we put $\Gamma_{n+1} := \Gamma_n$.

This procedure makes all formulas noncritical — A_n is always noncritical for Γ_{n+1}.

Eventually we obtain a theory $\Gamma' := \bigcup\limits_n \Gamma_n$. It is clear that $(\Gamma', -\Gamma')$ is L-consistent, otherwise $\Gamma_n \vdash_L \bigvee \Delta$ for some n and finite $\Delta \subseteq -\Gamma' \subseteq -\Gamma_n$ — but then $(\Gamma_n, -\Gamma_n)$ is L-inconsistent. Since every Γ_n has the (EP), so does Γ'.

So Γ' is an L-place, and moreover, $\overline{\mathbf{QCL}} \subseteq \Gamma'$. In fact, by the deduction theorem, $B \in \overline{\mathbf{QCL}}$ implies $\overline{Sub}(p \vee \neg p) \vdash_{\mathbf{QH}} B$; hence there exist formulas $A_0, \dots, A_n$ such that $\overline{\forall}(A_0 \vee \neg A_0), \dots, \overline{\forall}(A_n \vee \neg A_n) \vdash_{\mathbf{QH}} B$; therefore $\Gamma' \vdash_{\mathbf{QH}} B$, whence $B \in \Gamma'$

By 6.3.3, Γ' is a **QCL**-place, and therefore an $L\forall$-place (by 6.3.2), and it is maximal in CF_L.

(2) Similar to (1). Again starting from $\Gamma \in CP_L$, we construct a sequence $\Gamma = \Gamma_0 R_L \Gamma_1 \dots$ in CF_L.

Let MF_1 be the set of all S_0-formulas with at most one parameter, and let $A_0, A_1 \dots$ be an enumeration of MF_1. Now we call a formula A *critical* for $\Theta \in CP_L$ if $CM_L, \Theta \not\Vdash \overline{\forall}\Box(A \supset \Box A)$.

Suppose Γ_n is constructed and A_n is critical for Γ_n. Since $L \supseteq \mathbf{QS4CK}^+$, we have $\Gamma_n \vDash \Diamond\overline{\forall}\Box(A_n \supset \Box A_n)$, so there exists $\Gamma_{n+1} \in R_L(\Gamma_n)$ such that $\Gamma_{n+1} \vDash \overline{\forall}\Box(A_n \supset \Box A_n)$. If A_n is not critical for Γ_n, we put $\Gamma_{n+1} := \Gamma_n$.

Then we claim that the theory

$$\Gamma' := \bigcup_n \Box^- \Gamma_n$$

is L-consistent. In fact, otherwise $\bigcup\limits_{n=1}^{k} \Box^- \Gamma_n$ is L-inconsistent for some finite k. But this is impossible, since $\Gamma_n R_L \Gamma_{k+1}$ for $n \leq k+1$, and thus $\bigcup\limits_{n=1}^{k} \Box^- \Gamma_n \subseteq \Gamma_{k+1}$.

Now by the Lindenbaum lemma, there exists an L-complete $\Delta \supseteq \Gamma'$. It follows that Δ is a Henkin theory. In fact, consider an arbitrary formula $A_n(y)$ of the form $\exists x B(x) \supset B(y)$, where $FV(B(x)) \subseteq \{x\}$ as usual. By our construction and reflexivity, $\Gamma_{n+1} \vDash \forall y(A_n(y) \supset \Box A_n(y))$. Since Γ_{n+1} is a Henkin theory, there exists $c \in S_0$ such that $\Gamma_{n+1} \vDash A_n(c)$. We also have $\Gamma_{n+1} \vDash A_n(c) \supset \Box A_n(c)$, so $\Gamma_{n+1} \vDash \Box A_n(c)$. Hence $A_n(c) \in \Box^- \Gamma_{n+1} \subseteq \Delta$. Therefore $\Delta \in CP_L$.

It remains to note that Δ is maximal — assuming $\Delta R_L \Delta'$, let us show that $\Delta' = \Delta$ (or $\Delta \subseteq \Delta'$). Suppose $A = A_n \in \Delta$ (i.e. $\Delta \vDash A_n$) for an S_0-sentence

A_n. By construction $\Gamma_n R_L \Delta$, so $\Delta \models \bar{\forall}(A_n \supset \Box A_n)$, which is the same as $\Delta \models A_n \supset \Box A_n$, and thus $\Delta \models \Box A_n$, which implies $A_n \in \Delta'$. ∎

Remark 7.3.10 Note that the proof in case (1) almost repeats the proof of Lemma 7.2.3, when we construct a **QCL**$\exists\forall$-complete extension Γ' of Γ. But in our case we cannot directly apply Lemma 7.2.3 to the logic $\mathbf{QCL}^{(=)}$ and empty Γ_0, Δ_0, because the basic theory Γ is not in general $\mathbf{QCL}^{(=)}\exists$-complete. Nor we can apply Lemma 7.2.3 to an $L\exists$-complete Γ and $\Delta_0 = \varnothing$, $\Gamma_0 = \mathbf{QCL}^{(=)}$ (or $\Gamma_0 = \{\bar{\forall}(A \vee \neg A) \mid A \in IF^{(=)}\}$), since Γ is infinite.

Proposition 7.3.11 *The logics* **QHCK** *and* **QLCCK** *are C-canonical. The former is determined by Kripke frames with constant domains and the McKinsey property, and the latter by chains with the greatest element and constant domains and also by countable chains with the greatest element and constant domains.*

Recall that $\mathbf{QH} + CD \wedge AP_k \vdash KF$ for any k, since a poset of a finite depth is clearly coatomic.

Remark 7.3.12 Note that the logic of the weak excluded middle with constant domains $\mathbf{QH}+CD\wedge AJ$ is not C-canonical. Recall that a Kripke frame validates AJ iff it is directed, i.e. $\forall v_1 \forall v_2 \exists w(v_1 Rw \;\&\; v_2 Rw)$. This property fails for the canonical frame, and our 'extension lemma' 8.2.3 does not allow us to extend the L-consistent theory $\Gamma' \cup \Gamma''$ (where $\Gamma \subset \Gamma'$ and $\Gamma \subset \Gamma''$) to an $L\exists\forall$-complete theory with the same set of constants S_0. The logic $\mathbf{QH} + CD \wedge AJ$ is actually Kripke-incomplete [Shehtman and Skvortsov, 1990; Ghilardi, 1989], see also Volume 2. But there is no difficulties with the logic $\mathbf{QH} + CD \wedge AJ \wedge KF$, as we shall see now.

Lemma 7.3.13 *Let L be a superintuitionistic logic containing $CD \wedge AJ \wedge KF$. Then for any $\Gamma \in CP_L$ the set $\{\Delta \in CP_L \mid \Gamma \subseteq \Delta\}$ has the greatest element.*

Proof By 7.3.9, there exists a CD**QCL**-place $\Delta \supseteq \Gamma$. Suppose Δ is not the greatest in $CF_L {\uparrow} \Gamma$, then there exists another

CD**QCL**-place $\Delta_1 \supseteq \Gamma$, and thus $\Delta \not\subseteq \Delta_1$, since Δ is maximal. So there exists $A \in IF(S_0)$ such that $\Delta \Vdash A$, $\Delta_1 \not\Vdash A$; hence $\Delta_1 \Vdash \neg A$ by maximality of Δ_1. Thus $\Gamma \not\Vdash \neg A \vee \neg\neg A$, which contradicts $\vdash_L \neg A \vee \neg\neg A$. ∎

Proposition 7.3.14 *The logics* $\mathbf{QHC} + AJ \wedge KF$ *and* $\mathbf{QHC} + CD \wedge AJ \wedge AP_k$ *for $k > 0$ are C-canonical. The former is determined by Kripke frames with constant domains and top elements, and the latter, by the same kind of frames of depth k.*

Definition 7.3.15 *A* subordination L-map with constant domains *is a natural L-map from a subordination frame to CF_L.*

Lemma 7.3.16 *Let $L \supseteq \mathbf{QH} + CD$ and let Γ be an $L\forall$-place. Then there exists a subordination L-map with constant domains $g : IT_\omega \longrightarrow CF_L$ such that $g(\curlywedge) = \Gamma$.*

Proof From Lemma 7.2.4 and an analogue to Lemma 6.4.16. ■

Lemma 7.3.17 *Let $L \supseteq \mathbf{QH} + CD + KF + J$ and let Γ be an $L\forall$-place. Then there exists a subordination L-map with constant domains $h : IT_\omega + 1 \longrightarrow CF_L$ such that $h(\curlywedge) = \Gamma$.*

Proof Let g be the subordination map given by Lemma 7.3.16. Let $\Gamma' = \{A \mid \neg\neg A \in \Gamma\}$.

Suppose $A \in g(\alpha)$ for some $\alpha \in IT^\omega$. Then $\vdash_L \neg A \vee \neg\neg A$ by AJ, $\Gamma \nvdash_L \neg A$, therefore $\Gamma \vdash_L \neg\neg A$ and $A \in \Gamma'$.

Moreover, Γ' is a $\mathbf{QCL}\forall$-place. In fact, it is clearly L-consistent and closed under deduction in L. Since for every D_Γ-sentence A, $\vdash_L \neg\neg A \vee \neg\neg\neg A$, Γ contains either $\neg\neg A$ or $\neg\neg\neg A$, so Γ' contains either A or $\neg A$.

To check the existence property, suppose $A(c) \notin \Gamma'$ for all c. Then by AJ we deduce that $\neg A(c) \in \Gamma$ for all c and so $\forall x \neg A(x) \in \Gamma \subseteq \Gamma'$. Since $\vdash_{\mathbf{QH}} \forall x \neg A(x) \supset \neg\exists x A(x)$, we have $\neg\exists x A(x) \in \Gamma'$ and so $\exists x A(x) \notin \Gamma'$.

To show the coexistence property, suppose $\forall x A(x) \notin \Gamma'$. Then $\neg\forall x A(x) \in \Gamma'$. But $\mathbf{QH}+KF \vdash \neg\neg(\neg\forall x A(x) \supset \exists x \neg A(x))$, hence $(\neg\forall x A(x) \supset \exists x \neg A(x)) \in \Gamma'$ and $\exists x \neg A(x) \in \Gamma'$. By existence property, $\neg A(c) \in \Gamma'$ for some c and so $A(c) \notin \Gamma'$.

Since Γ' is maximal, we see that $A \in \Gamma'$ and $B \notin \Gamma'$ whenever $(A \supset B) \notin \Gamma'$. Therefore h extending g by $h(\infty) := \Gamma'$ is natural. ■

From Lemmas 7.2.3 and 7.3.17, we obtain

Theorem 7.3.18 *The logic $\mathbf{QH} + CD + KF + AJ$ is determined by the subordination frame with the greatest element and constant domains.*

7.4 Predicate versions of subframe and tabular logics

The results on canonicity of some families of predicate logics presented in this and the next section are due to H. Ono, T. Shimura and Y. Tanaka. But the original proofs are essentially simplified, thanks to canonical models and some results on propositional logics.

Definition 7.4.1 *Let L be a modal or superintuitionistic predicate logic with constant domains. The* canonical general frame of L *is*

$$C\Phi_L := ((CF_L)_\pi, \mathcal{W}_L),$$

where $\mathcal{W}_L$ is $\{\theta_L(A) \mid A \in MF_N(S_0)\}$ in the modal case or $\{\theta_L^I(A) \mid A \in IF(S_0)\}$ in the intuitionistic case.

Lemma 7.4.2 *$C\Phi_L$ is well-defined.*

Proof Similar to 1.6.7. Note that $\mathcal{W}_L$ is a subalgebra of $MA((CF_L)_\pi)$, since by definition

$$\theta_L(\neg A) = -\theta_L(A),\ \theta_L(A \wedge B) = \theta_L(A) \cap \theta_L(B),\ \theta_L(\Box_i A) = \Box_i \theta_L(A).$$

■

Lemma 7.4.3 *$C\Phi_L$ is refined.*

Proof It is again similar to the propositional logic.

$C\Phi_L$ is differentiated, since $\Gamma \neq \Delta$ implies $\Gamma \not\subseteq \Delta$ (or vice versa), and thus by Theorem 7.1.4 (or 7.2.6 in the intuitionistic case), $\Gamma \in \theta_L(A),\ \Delta \notin \theta_L(A)$ for some A.

Tightness in the modal case

$$\forall A\, (\Gamma \models \Box_i A \Rightarrow \Delta \models A) \Rightarrow \Gamma R_{L,i} \Delta$$

follows from the definition of $R_{L,i}$ and 7.1.4. In the intuitionistic case this is written as

$$\forall A\, (\Gamma \models A \Rightarrow \Delta \models A) \Rightarrow \Gamma \subseteq \Delta$$

and follows from 7.2.6. ■

Remark 7.4.4 $C\Phi_L$ is not descriptive, as S. Ghilardi noticed.

Proposition 7.4.5 *For any m.p.l. or s.p.l. L, $C\Phi_L$ validates L_π.*

Proof Again we consider only the modal case. We can argue as in the proof of 1.7.5. Viz., consider a formula $A \in L_\pi\lceil k$ and a valuation ψ in $C\Phi_L$. Choose B_i such that $\psi(p_i) = \theta_L(B_i)$, and let $S := [B_1, \ldots, B_k/p_1, \ldots, p_k]$. Then $SA \in L$, so $\psi(A) = \theta_L(SA) = CP_L$ by 1.2.9 and the canonical model theorem 7.1.4. ■

Proposition 7.4.6 *If a propositional modal or intermediate logic $\mathbf{\Lambda}$ is r-persistent, then the predicate logic $\mathbf{Q\Lambda C}$ is C-canonical.*

Proof Let $L = \mathbf{Q\Lambda C}$. By 7.4.5, $C\Phi_L \models \mathbf{\Lambda}$. By 7.4.3, this frame is refined, so by r-persistence, we obtain $CF_L \models \mathbf{\Lambda}$. Since CF_L has constant domains, it also validates the Barcan formula, and thus the whole L. ■

Hence we obtain the main result from [Tanaka and Ono, 1999]:

Theorem 7.4.7 (Tanaka–Ono) *For any universal propositional modal logic $\mathbf{\Lambda}$, the logic $\mathbf{Q\Lambda C}$ is C-canonical. [and so, strongly Kripke complete]*

Proof By Theorem 1.12.8, $\mathbf{\Lambda}$ is r-persistent, so $\mathbf{Q\Lambda C}$ is C-canonical by Proposition 7.4.6. ■

This allows us to construct many examples of canonical predicate logics. In almost the same way we prove the result from [Shimura, 1993]:

Theorem 7.4.8 (Shimura) *If* $\mathbf{\Lambda}$ *is a subframe intermediate logic, then* $\mathbf{Q\Lambda C}$ *is C-canonical.*

Proof By Theorem 1.12.27, $\mathbf{\Lambda}$ is r-persistent. So we can apply Proposition 7.4.6. ∎

The next result for the intuitionistic case was proved in [Ono, 1983]; the modal version is probably new.

Theorem 7.4.9 *If a propositional modal or intermediate logic* $\mathbf{\Lambda}$ *is tabular, then* $\mathbf{Q\Lambda C}$ *is C-canonical.*

Proof By Proposition 1.14.7, $\mathbf{\Lambda}$ is r-persistent. So we can apply 7.4.6. ∎

7.5 Predicate versions of cofinal subframe logics

In this section we prove completeness results from [Shimura, 2001].

Now we use an auxiliary 'pre-canonical' model dual to the Lindenbaum algebra of S_0-sentences.

Definition 7.5.1 *Let* L *be an* N*-m.p.l. with constant domains,* C^+P_L *the set of all* L*-complete theories with the set of constants* S_0*. We define the following propositional frames and models:*

- *the* pre-canonical frame of L *is* $C^+F_L := (C^+P_L, R_1, \dots, R_N)$*, where* R_i *is defined as in the canonical model:*

$$\Gamma R_i \Delta \iff \forall A \in MF_N(S_0)\,(\Box_i A \in \Gamma \Rightarrow A \in \Delta);$$

- *the* pre-canonical model of L *is* $C^+M_L := (C^+F_L, \theta_L^+)$*, where*

$$\theta_L^+(p_i) := \{\Gamma \in C^+P_L \mid p_i \in \Gamma\};$$

- *the* pre-canonical general frame of L *is*

$$C^+\Phi_L := (C^+F_L, \{|A| \mid A \in MF_N(S_0)\}),$$

where $|A| := \{\Gamma \in C^+P_L \mid A \in \Gamma\}$.

Lemma 7.5.2 $C^+\Phi_L$ *is well-defined.*

Proof We have to show that the sets $|A|$ constitute a subalgebra of $MA(C^+F_L)$. This follows from the equalities:

$$|\neg A| = -|A|,\ |A \wedge B| = |A| \cap |B|,\ |\Box_i A| = \Box_i |A|.$$

The first two are just reformulations of 6.1.2(iii) and 6.1.3(i). For the third, note that the inclusion $|\Box_i A| \subseteq \Box_i |A|$ readily follows from the definition of R_i.

The converse $|\Box_i A| \supseteq \Box_i|A|$ is proved as in propositional logic and similarly to 6.1.16. In fact, suppose $\Box_i A \notin \Gamma$. Then the theory

$$\Delta_0 := \Box_i^- \Gamma \cup \{\neg A\}$$

is L-consistent. For otherwise we have $\Box_i^- \Gamma \vdash_L A$, hence $\Gamma \vdash_L \Box_i A$ by Lemma 2.., which contradicts $\Box_i A \notin \Gamma$.

Now by the Lindenbaum lemma 6.1.5 there exists an L-complete $\Delta \supseteq \Delta_0$. So $\Delta \in |\neg A| = -|A|$, and also $\Gamma R_i \Delta$. Thus $\Gamma \notin \Box_i |A|$. ∎

As complete theories may be not Henkin, the canonical model theorem for this model holds only for propositional formulas; this explains the name 'pre-canonical'.

Proposition 7.5.3 *Let L be an N-m.p.l. with constant domains.*

(1) For any N-modal propositional formula A, for any L-complete Γ

$$C^+M_L, \Gamma \vDash A \text{ iff } A \in \Gamma.$$

(2) For any N-modal propositional formula A,

$$C^+M_L \vDash A \text{ iff } A \in L.$$

(3) $\mathbf{ML}(C^+\Phi_L) = L_\pi$.

Proof

(1) Standard, by induction on the length of A, using properties of complete theories 6.1.3. We also need an analogue of 6.1.16 for complete theories; its proof uses 6.1.5.

(2) 'If' follows from (1), since L-complete theories contain $\overline{L}$. For 'only if' use the Lindenbaum lemma 6.1.5.

(3) ($\subseteq$) If $A \notin L_\pi$, then by (2), $C^+M_L \nvDash A$, and thus $C^+\Phi_L \nvDash A$.

($\supseteq$) The claim (1) can be written as $\theta_L^+(A) = |A|$. So we can argue as in the proof of 1.7.5. Viz., consider a formula $A \in L_\pi \lceil k$ and a valuation ψ in $C^+\Phi_L$. Choose B_i such that $\psi(p_i) = |B_i| = \theta_L^+(B_i)$, and consider a substitution $S := [B_1, \dots, B_k/p_1, \dots, p_k]$. Then $SA \in L$, so $\psi(A) = \theta_L^+(SA) = C^+P_L$ by 1.2.9 and (2). ∎

We also have an analogue of Theorem 1.7.13:

Theorem 7.5.4 *Let L be an m.p.l. with constant domains. Consider the modal algebra $Lind(\overline{L(S_0)})$ of all S_0-sentences up to the equivalence $\vdash_L A \equiv B$. Then*

$$Lind(\overline{L(S_0)})_+ \cong C^+\Phi_L.$$

Lemma 7.5.5 *$C^+\Phi_L$ is descriptive.*

Proof It is again similar to the propositional case.

$C^+\Phi_L$ is differentiated, since $\Gamma \neq \Delta$ implies $\Gamma \not\subseteq \Delta$, and thus $\Gamma \in |A|$, $\Delta \notin |A|$ for some A.

Tightness

$$\forall A\, (\Gamma \in \Box_i |A| \Rightarrow \Delta \in |A|) \Rightarrow \Gamma R_i \Delta$$

follows from the definition of R_i and the equality $|\Box_i A| \subseteq \Box_i |A|$ we have already proved.

For compactness, note that if $\mathcal{V}$ is a set of interior sets with non-empty finite intersections, then the theory $\Gamma_0 := \{A \mid |A| \in \mathcal{V}\}$ is L-consistent. In fact, if $\Gamma_0 \vdash_L \bot$, then $\vdash_L \neg \bigwedge\limits_{i=1}^{k} A_i$ for some $|A_1|, \ldots, |A_k| \in \mathcal{V}$, so $\bigcap\limits_{i=1}^{k} |A_i| = |\bigwedge\limits_{i=1}^{k} A_i| = \varnothing$ — a contradiction. By the Lindenbaum lemma, there exists an L-complete $\Gamma \subseteq \Gamma_0$, so $\bigcap \mathcal{V} \ni \Gamma$.

Alternatively, one can apply Theorem 7.5.4 and Proposition 1.7.14. ■

Quite similar constructions and proofs can be made in the intuitionistic case.

Definition 7.5.6 *For an s.p.l. L with constant domains, let C^+P_L the set of all L-complete intuitionistic theories with the set of constants S_0. Then we define the* pre-canonical model, *the* pre-canonical frame *and the* pre-canonical general frame *of L:*

$$\begin{array}{ll} C^+F_L & := (C^+P_L, \subseteq), \\ C^+M_L & := (C^+F_L, \theta_L^+), \\ C^+\Phi_L & := (C^+F_L, \{|A| \mid A \in IF(S_0)\}), \end{array}$$

where

$$\begin{array}{ll} \theta_L^+(p_i) & := \{\Gamma \in C^+P_L \mid p_i \in \Gamma\}, \\ |A| & := \{\Gamma \in C^+P_L \mid A \in \Gamma\}. \end{array}$$

Lemma 7.5.7 *In the intuitionistic case $C^+\Phi_L$ is well-defined.*

Proof Let us show that $\{|A| \mid A \in IF(S_0)\}$ is a subalgebra of $HA(C^+F_L)$. It is sufficient to check the equalities:

$$|\bot| = \varnothing,\ |A \wedge B| = |A| \cap |B|,\ |A \vee B| = |A| \cup |B|,\ |A \supset B| = |A| \to |B|.$$

The first one is trivial; the second and the third follow from Lemma 6.2.5.

The equality $|A \supset B| = |A| \to |B|$ is checked as in the propositional case. In fact, $|A \supset B| \subseteq |A| \to |B|$, since $|A \supset B| \cap |A| \subseteq |B|$. The latter follows from the implications

$$(A \supset B) \in \Gamma \;\&\; A \in \Gamma \Rightarrow \Gamma \vdash_L B \Rightarrow B \in \Gamma.$$

To show the converse, suppose $(A \supset B) \notin \Gamma$. Then the theory $(\Gamma \cup \{A\}, \{B\})$ is L-consistent. For otherwise we have $\Gamma, A \vdash_L B$, which implies $\Gamma \vdash_L A \supset B$, by the Deduction theorem, and next $(A \supset B) \in \Gamma$ by 6.2.5.

Thus by Lemma 7.2.3 there exists an L-complete $\Delta \succeq (\Gamma \cup \{A\}, \{B\})$, hence $\Gamma \subseteq \Delta \in (|A| - |B|)$, and therefore $\Gamma \notin (|A| \to |B|)$. ■

The next proposition is an analogue of 7.5.3.

Proposition 7.5.8 *Let L be an s.p.l. with constant domains. Then*

(1) for any intuitionistic propositional formula A, for any L-complete Γ

$$C^+M_L, \Gamma \Vdash A \text{ iff } A \in \Gamma;$$

(2) for any intuitionistic propositional formula A,

$$C^+M_L \Vdash A \text{ iff } A \in L;$$

(3) $\mathbf{IL}(C^+\Phi_L) = L_\pi$.

Proof Repeats the proof of 7.5.3, with minor changes; for example, use 1.2.13 instead of 1.2.9. We leave the details to the reader. ∎

Hence we obtain

Theorem 7.5.9 *For any s.p.l. L with constant domains, $C^+\Phi_L \cong Lind(\overline{L(S_0)})_+$, where $Lind(\overline{L(S_0)})$ is the Lindenbaum algebra of S_0-sentences in L.*

Lemma 7.5.10 *For any s.p.l. L with constant domains, $C^+\Phi_L$ is descriptive.*

Proof Similar to 7.5.5. ∎

Lemma 7.5.11 *In each of the following cases the pre-canonical frame $C^+\Phi_L$ is coatomic:*

(1) L is an s.p.l. containing **QHCK**;

(2) L is an m.p.l. containing $\mathbf{QS4CK}^+$.

Proof (1) Follows the lines of the proof of 7.3.9.
(2) Since $L \supseteq$ **S4.1**, we can apply the same argument as in the propositional case for **S4.1** — its canonical frame is coatomic. ∎

Now we can prove the main completeness results of this section. We begin with a theorem from [Shimura, 1993].

Definition 7.5.12 *A world t in a transitive frame is called* terminal *if it is accessible from every nonmaximal cluster and its cluster $t^\sim$ is maximal and either trivial or degenerate. A finite transitive frame is called* weakly directed *if it contains a terminal world.*

Theorem 7.5.13 *Let Λ be an intermediate propositional logic axiomatisable by cofinal subframe formulas of weakly directed finite rooted posets. Then* **QΛCK** *is C-canonical.*

Proof If $\mathbf{\Lambda} = \mathbf{H} + \{CSI(F) \mid F \in \mathcal{F}\}$ for a set $\mathcal{F}$ of weakly directed posets, $L = \mathbf{Q\Lambda CK}$, let us show that $CF_L \Vdash CSI(F)$ for $F \in \mathcal{F}$. Suppose the contrary, then CF_L is cofinally subreducible to some $F \in \mathcal{F}$.

By Theorems 1.12.22, 1.12.30, there exists $f : G \twoheadrightarrow F$ for a cofinal $G \subseteq CF_L \uparrow u_0$ (for some u_0) such that $f^{-1}(x)$ consists only of maximal points, whenever x is maximal in F.

Then let us prolong f to $f' : G' \twoheadrightarrow F$ for a cofinal $G' \subseteq C^+F_L \uparrow u_0$.

Since $KF \in L$, C^+F_L is coatomic by Lemma 7.5.11, so it suffices to extend f only to maximal points of C^+F_L. We can do this as follows.

Let $y \in C^+P_L - CP_L$ be a 'new' maximal point (so y is a classical non-Henkin theory). We put $f'(y) := t$, where t is a (fixed) terminal world of F.

Thanks to the McKinsey property, the subframe $G' := G \cup$ (maximal points of C^+F_L) is cofinal in C^+F_L. Since f' sends maximal points to maximal points, the lift property is preserved. The monotonicity of f' follows from the definition: if $uR^G y$, and $u \in G$, $y \in G' - G$, then u is nonmaximal, so $f(u)$ is nonmaximal by the choice of f, while $f'(y)$ is terminal; hence $f'(u) = f(u)R^F f'(y)$. (Here F^G, R^F denote the accessibility relations in G, F respectively.)

Therefore, C^+F_L is subreducible to F, so $C^+F_L \not\Vdash CSI(F)$, and thus $C^+F_L \not\Vdash \mathbf{\Lambda}$.

On the other hand, $C^+\Phi_L \Vdash \mathbf{\Lambda}$ by 7.5.8 and $C^+\Phi_L$ is descriptive (7.5.10), hence $C^+F_L \Vdash \mathbf{\Lambda}$ by d-persistence (Theorem 1.12.28).

This is a contradiction. ∎

The next lemma and theorem are also taken from [Shimura, 1993].

Lemma 7.5.14 *Let L be an s.p.l. containing* **QHK**, *and assume that a cone $CM_L \uparrow u$ has finitely many maximal worlds. Then $C^+M_L \uparrow u$ has the same maximal worlds.*

Proof Let $x_1, \dots, x_n$ be all maximal worlds in $CM_L \uparrow u$, and suppose that y is another maximal world in $C^+M_L \uparrow u$.

By distinguishability, there exist $A_i \in x_i - y$; then $\neg A_i \in y$, since y is classical. Hence $\neg A_1 \wedge \ldots \wedge \neg A_n \in y$, and thus $\neg(\neg A_1 \wedge \ldots \wedge \neg A_n) \notin u$, which is equivalent to $\left(\neg\neg \bigvee\limits_{i=1}^{n} A_i\right) \notin u$, and to $CM_L, u \not\Vdash \neg\neg \bigvee\limits_{i=1}^{n} A_i$. But then $CM_L, v \Vdash \neg \bigvee\limits_{i=1}^{n} A_i$ for some $v \in R_L(u)$, while by the McKinsey property, vR_Lx_i for some i. Then $x_i \Vdash A_i$ contradicts $x_i \Vdash \neg \bigvee\limits_{i=1}^{n} A_i$. ∎

Theorem 7.5.15 *Let $\mathbf{\Lambda}$ be a cofinal subframe intermediate propositional logic containing a formula $CSI(V_n)$ for some finite n, where $V_n = IT_n^2$. Then* **QΛCK** *is C-canonical.*

Proof By Theorem 7.5.13, $\mathbf{QHCK} + CSI(V_n)$ is C-canonical, so $CF_L \Vdash CSI(V_n)$ (where $L := \mathbf{Q\Lambda CK}$). This means that CF_L is not cofinally subreducible to V_n, and thus every cone $CF_L \uparrow u$ contains at most $(n-1)$ maximal points, otherwise, due to the McKinsey property 7.3.9, there exists a cofinal subreduction to V_n.

Now we can show that $CF_L \Vdash CSI(F)$ whenever $CSI(F) \in \mathbf{\Lambda}$. In fact, otherwise CF_L is cofinally subreducible to F and by 1.12.27, there exists $f : G \twoheadrightarrow F$ for a cofinal $G \subseteq CF_L \uparrow u$ such that for every maximal $x \in F$, $f^{-1}(x)$ contains only maximal points of CF_L.

Then G is cofinal in $C^+F_L \uparrow u$. In fact, by Lemma 7.5.14, $C^+F_L \uparrow u$ has the same maximal points as $CF_L \uparrow u$, and they are all in G, as G is cofinal in $CF_L \uparrow u$. By 7.5.11, the set of maximal points is cofinal in C^+F_L.

Therefore f is a cofinal subreduction from C^+F_L to F. Now we can use the same argument as in the proof of Theorem 7.5.13 to show that $C^+F_L \Vdash \mathbf{\Lambda}$, which leads to a contradiction. ∎

Hence we obtain another proof of 7.3.14.

Corollary 7.5.16 **QΛCK** *is C-canonical for* $\mathbf{\Lambda} = \mathbf{H} + AJ$.

Proof In fact, $\mathbf{\Lambda} = \mathbf{H} + CSI(V_2)$ — a cofinal subreduction to V_2 does not exist iff a frame is confluent. ∎

Let us now prove the main result from [Shimura, 2001] with slight additions.

Theorem 7.5.17 *Let* $\mathbf{\Lambda}$ *be a modal propositional logic axiomatisable above* **S4** *by cofinal subframe formulas of weakly directed posets. Then* $\mathbf{QΛCK}^+$ *is C-canonical.*

Proof Similar to the proof of 7.5.13.

It is sufficient to show that $CF_L \vDash CSM(F)$, provided $CSM(F) \in \mathbf{\Lambda}$ and F is a weakly directed poset. Suppose the contrary, then CF_L is cofinally subreducible to F.

By Theorems 1.12.22, 1.12.30, there exists $f : G \twoheadrightarrow F$ for a cofinal $G \subseteq CF_L \uparrow u_0$ such that for any maximal $x \in F$, $f^{-1}(x)$ consists only of maximal points.

Then we prolong f to $f' : G' \twoheadrightarrow F$ for a cofinal $G' \subseteq C^+F_L \uparrow u_0$.

Since by 7.5.11 C^+F_L is coatomic, we extend f to maximal points of C^+F_L by sending every $y \in C^+P_L - CP_L$ to a certain terminal world of F. Hence $C^+F_L \nvDash CSM(F)$.

On the other hand, $C^+\Phi_L \vDash CSM(F)$ by 7.5.3 and $C^+\Phi_L$ is descriptive (7.5.5), hence $C^+F_L \vDash CSM(F)$ by d-persistence (1.12.30(4), 1.12.22), which leads to a contradiction. ∎

Corollary 7.5.18

(1) Let $\mathbf{\Lambda}$ *be a cofinal subframe intermediate logic containing the formula* $SI(FS_3)$. *Then* $L = \mathbf{QΛCK}$ *is C-canonical.*

(2) The same holds for a Δ*-elementary cofinal subframe* $\mathbf{\Lambda} \supseteq \mathbf{S4.1}$ *containing* $SM(FS_3)$ *and* $L = \mathbf{QΛCK}^+$.

Proof

(1) As we know from 7.5.13, if $CSI(F) \in L$ and F is a weakly directed finite rooted poset, then $CF_L \Vdash CSI(F)$. So assuming that $CF_L \nVdash CSI(F)$, and F is not weakly directed we can show that $CSI(F) \notin \mathbf{\Lambda}$. By assumption, CF_L is cofinally subreducible to F.

Let $F = (W, R)$ and for any $a \in W$ put $\hat{a} := R(a) \cap \max(F)$. We claim that $\{\hat{a} \mid a \in W - \max(F)\}$ is a chain in $(2^{\max(F)}, \subseteq)$. In fact, if $\hat{a}, \hat{b}$ are $\subseteq$-incomparable and $m_1 \in \hat{a} - \hat{b}$, $\hat{m}_2 \in \hat{b} - \hat{a}$, then F contains a subframe (where 0 is the root):

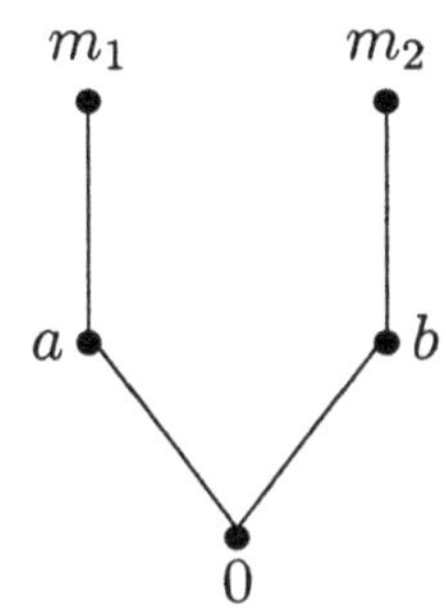

So F, and thus CF_L, is subreducible to FS_3. But then $CF_L \nVdash SI(FS_3)$, which contradicts the canonicity of $\mathbf{QHCK} + SI(FS_3) \subseteq L$.

Therefore the sets $\hat{a}$ constitute a chain, so there must be the least of them, say $\hat{a}$. But then every non-maximal a sees every $m \in \hat{a}_0$ which means that F is weakly directed. This is a contradiction.

(2) If $CSM(F) \in \mathbf{\Lambda}$, then we may assume that F is a blossom frame (by 1.12.30), and hence a poset (since $\mathbf{S4.1} \subseteq \mathbf{\Lambda}$). Now we can repeat the argument from the proof of (1) replacing CSI with CSM. ∎

Lemma 7.5.19

(1) A rooted **S4.1**-*frame is not cofinally subreducible to FS_1 iff the set of its inner points is a quasi-chain.*

(2) A rooted **S4.1**-*frame is not cofinally subreducible to FS_2 iff every maximal point is accessible from every inner point.*

Proof

(1) (If.) If f is a subreduction of G to $FS_1, f(a_1) = 1, f(a_2) = 2$, then both a_1, a_2 see points in $f^{-1}(3)$, so they are inner in G. a_1, a_2 are incomparable, since 1, 2 are. Thus inner points are not in a quasi-chain.

(Only if.) If G is coatomic, has a root a_0 and two incomparable inner points a_1, a_2, then the partial map $f : G \to FS_1$ sending a_i to i and every maximal point to 3 is a cofinal subreduction.

(2) (If.) If f is a cofinal subreduction of G to FS_2, $f(a_1) = 1$ and a_1 sees a maximal point a_2, then $f(a_2) = 2$. Similarly there exists a maximal $a_3 \in f^{-1}(3)$. So a_1 is an inner point and a_1 does not see a_3.

(Only if.) Suppose G is coatomic with root a_0 and has an inner point a_1, which does not see a certain maximal point a_3. Then the map defined on a_0, a_1 and all maximal points of G is a cofinal subreduction if it sends a_i to i and all other maximal points to 2.

■

Proposition 7.5.20

(1) Let $\mathbf{\Lambda}$ be an intermediate propositional cofinal subframe logic containing $CSI(FS_1)$ or $CSI(FS_2)$. Then $L = \mathbf{Q\Lambda CK}$ is C-canonical.

(2) The same holds for a Δ-elementary cofinal subframe $\mathbf{\Lambda} \supseteq \mathbf{S4.1}$ containing $CSM(FS_1)$ or $CSM(FS_2)$ and $L = \mathbf{Q\Lambda CK}^+$.

Proof (1) By Theorem 7.5.15, the logics $\mathbf{QHCK} + CSI(FS_1)$, $\mathbf{QHCK} + CSI(FS_2)$ are C-canonical, so $CF_L \Vdash CSI(FS_1)$ or $CF_L \Vdash CSI(FS_2)$.

Now assuming $CSI(F) \in \mathbf{\Lambda}$ let us show $CF_L \Vdash CSI(F)$. In fact, if $CF_L \nVdash CSI(F)$, i.e. CF_L is cofinally subreducible to F, then F is weakly directed. For, the validity of $CSI(FS_i)$ is preserved by cofinal subreductions, thus $F \Vdash CSI(FS_1)$ or $F \Vdash CSI(F_2)$. Therefore F is weakly directed by Lemma 7.5.19. Hence $CF_L \Vdash CSI(F)$ by Theorem 7.5.15.

To prove (2), use the same argument applying 7.5.17. ■

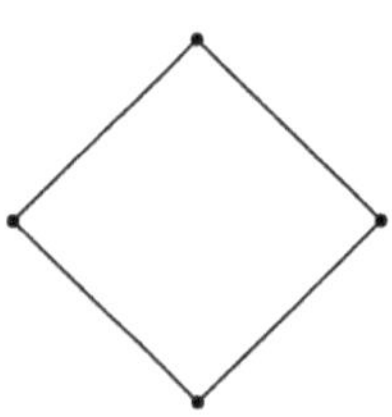

Figure 7.1. FS_1.

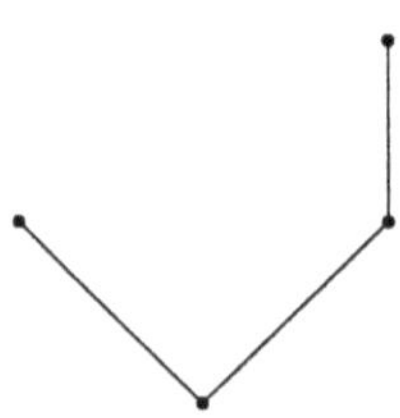

Figure 7.2. FS_2.

Figure 7.3. FS_3.

Since incompatible sets are always strongly incompatible in the canonical models of the intermediate *propositional* logics, we can prove that all cofinal subframe propositional logics are canonical.

Question 7.5.21 *Let L be an s.p.l. with constant domains. Is every incompatible subset of CP_L strongly incompatible?*

Question 7.5.22 [2] *Is* $\mathbf{Q}\mathbf{\Lambda}+CD+KF$ *strongly complete for every cofinal subframe logic* $\mathbf{\Lambda}$*? What happens in the case* $\mathbf{\Lambda}=\mathbf{H}+CSI(FS_3)$ *where* FS_3 *is the frame in Fig. 7.3?*

7.6 Natural models with constant domains

In this section we transfer the method described in section 6.4 to constant domains.

Lemma 7.6.1 *Let* M *be a Kripke model for a modal logic* L *with the constant domain* S_0*. Then for any world* $u \in M$*, for any* S_0*-sentence* A

$$M, u \vDash A \text{ iff } CM_L, \nu_M(u) \vDash A.$$

The same holds in the intuitionistic case, with obvious changes.

Lemma 7.6.2 *Let* M *be the same as in the previous lemma,* R_i *the accessibility relations in* M*. Then for any* $u, v \in M$

$$uR_iv \Rightarrow \nu_M(u)R_{Li}\nu_M(v).$$

Definition 7.6.3 *Let* L *be an m.p.l.(=) containing Barcan axioms or an s.p.l. (=) containing* CD*,* $F=(W, R_1, \ldots, R_N)$ *a propositional frame of the corresponding kind. A* CDL-map based on F *is a monotonic map from* F *to (the propositional base of)* $CF_L^{(=)}$.

Definition 7.6.4 *Let* h *be a* CDL*-map based on a propositional frame* F*. The* predicate Kripke frame associated with h *is* $\mathbf{F}(h) := F \odot S_0$.

If L *is a logic with equality, the* Kripke frame with equality associated with h *is* $\mathbf{F}^{=}(h) := (F, D, \asymp)$*, where*

$$c \asymp_u d \ \text{ iff } \ (c = d) \in h(u).$$

The CDL-model associated with h *is* $M^{(=)}(h) := (\mathbf{F}^{(=)}(h), \xi(h))$*, where*

$$\xi(h)_u(P_k^m) := \{\mathbf{c} \in S_0^m \mid P_k^m(\mathbf{c}) \in h(u)\}$$

for $u \in F$.

Definition 7.6.5 *A* CDL*-map* $h : F \longrightarrow CF_L$ *and the associated* CDL*-model* $M(h)$ *are called* natural *if* $h = \nu_{M(h),L}$*, i.e. for any* $u \in F$*,* $A \in \mathcal{L}(u)$

$$M(h) \vDash (\Vdash)A \text{ iff } CM_L, h(u) \vDash (\Vdash)A \ (\Leftrightarrow A \in h(u)).$$

Lemma 7.6.6 *For a modal logic* L*, a* CDL*-map* $h : F \longrightarrow CF_L$ *is natural iff it is selective (in the sense of Definition 6.4.7).*

---

[2] [Shimura, 1993]

Proof The same as for 6.4.8. ■

Definition 7.6.7 *An intuitionistic CDL-map* $h : F \longrightarrow CF_L$ *based on* $F = (W, R)$ *is called* selective *if it satisfies the condition*

($\supset$) *if* $(A_1 \supset A_2) \in (-h(u))$, *then* $A_1 \in h(v)$, $A_2 \notin h(v)$ *for some* $v \in R(u)$.

Note that now we do not postulate the analogue of condition ($\forall$) from Definition 6.4.9; it obviously holds for $v = u$, since $h(u)$ satisfies (Ac).

Lemma 7.6.8 *An intuitionistic CDL-map* $h : F \longrightarrow CF_L$ *is natural iff it is selective.*

Proof Similar to 6.4.10. The only difference is the case $A = \forall x B$ in the 'if' part:

If $A \notin h(u)$, then by (Ac), there exists $c \in D_u$ such that $B(c) \notin h(u)$. By induction hypothesis, $u \not\Vdash B(c)$. Hence $u \not\Vdash A$.

The other way round, if $u \not\Vdash \forall x B$, then since the domain is constant, there exists $c \in D_u$ such that $u \not\Vdash B(c)$. Thus $h(u) \not\Vdash B(c)$, which implies $h(u) \not\Vdash A$. ■

Lemma 7.6.9 *Let* L *be a modal (resp., superintuitionistic) logic containing Barcan formulas (resp. CD),* $M_1 \subseteq CM_L$ *a selective submodel,* F_1 *the propositional frame of* M_1, $h : F \twoheadrightarrow F_1$. *Then* h *is a natural CDL-map.*

Proof Cf. Lemma 6.4.11. ■

Lemma 7.6.10

(1) Let L *be an* N*-modal logic containing Barcan formulas for all* $\Box_i$, $\Gamma \in CP_L$. *Then there exists a CD-Kripke model* M *based on a standard greedy tree* F *such that* $\nu_M(\curlywedge) = \Gamma$ *and* $M \models L$.

(2) If L *contains* $\Diamond_i \top$ *for* $1 \leq i \leq N$, *then one can take* $F = F_N T_\omega$.

(3) If L *is 1-modal and* $L \supseteq$ **QS4**, *then the claim holds for* $F = IT_\omega$.

Proof Cf. Lemma 6.4.12. ■

Proposition 7.6.11

(1) $\mathbf{QK}_N\mathbf{C} = \mathbf{ML}(\mathcal{CK}\mathcal{F}_N\mathcal{T}_\omega)$, $\mathbf{QD}_N\mathbf{C} = \mathbf{ML}(\mathcal{CK}F_N\mathcal{T}_\omega)$,

(2) $\mathbf{QK}_N\mathbf{C}^= = \mathbf{ML}^=(\mathcal{CKE}(\mathcal{F}_N\mathcal{T}_\omega))$, $\mathbf{QD}_N C^= = \mathbf{ML}^=(\mathcal{CKE}(F_N\mathcal{T}_\omega))$.

Proof Follows from the previous lemma, cf. 6.4.13. ■

Proposition 7.6.12 *Let* $\mathbf{\Lambda}$ *be an* N*-modal propositional PTC-logic. Then*

(1) $\mathbf{Q\Lambda C} = \mathbf{ML}(\mathcal{CK}\mathcal{GT}(\mathbf{\Lambda}))$.

(2) $\mathbf{Q\Lambda C}^= = \mathbf{ML}^=(\mathcal{KE}(\mathcal{GT}(\mathbf{\Lambda})))$.

Proof Similar to Proposition 6.4.14.

(1) By 7.1.8, $\mathbf{Q\Lambda C}$ is $\mathcal{CK}$-complete. Then by 1.11.5 and 3.12.8 we obtain $L = \mathbf{ML}(\mathcal{CK}\mathbf{V}_0(\mathbf{\Lambda}))$, and hence $L = \mathbf{ML}(\mathcal{CK}\mathbf{V}_1(\mathbf{\Lambda}))$, by 3.3.21. (We use the same notation $\mathbf{V}_0$, $\mathbf{V}_1$ as in the proof of 6.4.14.) Now by 1.11.11 and 3.3.14 it follows that $\mathbf{ML}(\mathcal{CKGT}(\mathbf{\Lambda})) \subseteq L$.

(2) The proof is similar. ■

In the same way we obtain an analogue of Proposition 6.4.15

Proposition 7.6.13

(1) $\mathbf{QK}_N\mathbf{C}^= + CE = \mathbf{ML}^=(\mathcal{CKF}_N\mathcal{T}_\omega)$, $\mathbf{QD}_N\mathbf{C}^= + CE = \mathbf{ML}^=(\mathcal{CK}F_N T_\omega)$.

(2) Let $\mathbf{\Lambda}$ *be a propositional PTC-logic. Then* $\mathbf{Q\Lambda C}^= + CE = \mathbf{ML}^=(\mathcal{CKGT}(\mathbf{\Lambda}))$.

Natural models with constant domains are a particular case of natural models described in section 6.4; only the notation is slightly simplified in this case.

Definition 7.6.14 *Let L be a predicate logic, F a propositional Kripke frame, and let R'_i $(1 \leq i \leq N)$ or R' in the intuitionistic case be relations on CP_L.*

An $(L, R'_1, \dots, R'_N)$-model on F is a mapping from F to CP_L, i.e. $N = (\Gamma_u \mid u \in F)$ where Γ_u are (L, S_0)-places such that

$$uR_iv \Rightarrow \Gamma_u R'_i \Gamma_v.$$

An L-model N is called natural *if the conditions from Definitions 6.4.7, 6.4.9 (for the modal or the intuitionistic case respectively) holds (with S_0 replacing S_u).*

Recall that in the intuitionistic case the condition (∀) from Definition 6.4.9 holds obviously already with $v = u$, since theories in CP_L are $L\forall$-complete, i.e they satisfy Ac from Definition 7.2.1.

As in Section 6.4, a natural (L, R_L)-model is called a *natural L-model.* (Recall that $\bar{R}_L^{\subseteq} = R_L$ in the intuitionistic case with constant domains.)

All main properties of natural models and crucial lemmas from Section 6.4 (together with their proofs) can be directly transferred to the case of constant domains: namely 6.4.8, 6.4.10, 6.4.11, etc. (now one has to use Γ or Γ_u instead of L-places (S, Γ) or (S_u, Γ_u) respectively). Therefore the logics with Ba_i or with CD (in the N-modal or in the intuitionistic case respectively) satisfy the main completeness results about the universal tree T_ω (or its subtrees). Let us formulate these completeness theorems for the intuitionistic case.

Proposition 7.6.15

(1) $\mathbf{QH} + CD = \mathbf{IL}(\mathcal{CK}(IT_\omega))$.

(2) $\mathbf{QH}^= + CD = \mathbf{IL}^=(\mathcal{CKE}(IT_\omega))$.

(3) $\mathbf{QH}^{=d} + CD = \mathbf{IL}^{=}(\mathcal{CK}(IT_\omega))$.

Proposition 7.6.16 $\mathbf{QH} + CD \wedge AP_h = \mathbf{IL}(\mathcal{CK}(IT_\omega^h))$ *for* $h > 0$ *and similarly for the corresponding logics with equality.*

Note that in the case of constant domains a natural analogue of 7.4.12 holds for any logic L containing AP_h.

Now let us prove completeness results (w.r.t. coatomic trees) for the logic $\mathbf{QH}+CD\wedge KF$. Here the situation is subtler than for varying domains. Namely, for an arbitrary path $w = (u_0, u_1, \ldots)$ in IT_ω we cannot always extend an L-consistent theory $\Gamma = \bigcup\limits_{i\in\omega} \Gamma_{u_i}$ to an $L\forall$-place Γ_ω in the same language (i.e. with constants from S_0). For example, if $S_0 = \{c_i \mid i \in \omega\}$, and for each i, $P(c_i) \in \Gamma_{u_i}$, while $\neg\forall x P(x) \in \Gamma_{u_0}$, then Γ_ω satisfying (Ac) does not exist. That is why we cannot prove completeness of $\mathbf{QH}+CD\wedge KF$ w.r.t the coatomic tree $\overline{IT}_\omega$ with a constant domain. Moreover $\mathbf{QH}+CD\wedge KF$ is not determined by $\overline{IT}_\omega \odot \omega$, cf. Lemma 3.12.17. We do not know if this logic is complete w.r.t. $\overline{IT}_\omega \odot V$ with uncountable V, but anyway natural models with denumerable domains are insufficient.

Instead we can show completeness w.r.t. the class of (denumerable) coatomic trees of the form $\overline{IT}_w^G$ (for $G \subseteq IT_w^\sharp$) (cf. [Gabbay, 1972]).

Lemma 7.6.17 *Let* L *be a superintuitionistic predicate logic containing* $CD \wedge KF$. *Then for any* (L, S_0)*-place* Γ *there exists a subset* $G \subseteq IT_\omega^\sharp$ *such that* $\overline{IT}_\omega^G$ *is coatomic and for any* $G' \subseteq G$ *there exists an* L*-natural model* $N = (\Gamma_u \mid u \in \overline{IT}_\omega^{G'})$ *(with the constant domain* S_0*) such that* $\Gamma_\lambda = \Gamma$.

Proof First we construct an L-natural model $(\Gamma_u \mid u \in IT_\omega)$ on IT_ω. Recall that a subset X of IT_ω is called *dense* if $\forall u\ R(u) \cap X \neq \varnothing$. We put $X_B := \{u \in IT_\omega \mid B \in \Gamma_u\}$ for $\mathbf{QCL}$-theorems $B \in IF^{(=)}$ (without constants from S_0); actually it is sufficient to consider only theorems of the form $B = \bar\forall(A \vee \neg A)$, cf. Remark 7.3.1). These sets are dense since $L \vdash \neg\neg B$ (by the Glivenko theorem), thus $\neg B \notin \Gamma_u$ for any u.

Call a path *generic* if it intersects every X_B; let G_Γ be the set of generic paths. Obviously, $\forall u \in IT_\omega \exists w \in G(u \in w)$, i.e. the tree $\overline{IT}_\omega^G$ is coatomic. Now for any generic path $w \in G_s$ we take the set $\Gamma_w := \bigcup\limits_{v\in w} \Gamma_v$. Then $\Gamma_w \in CP_L$. In fact, if $\vdash_L \bigwedge \Gamma_1 \supset \bigvee \Delta_1$ for finite $\Gamma_1 \subseteq \Gamma_2, \Delta_1 \subseteq -\Gamma_2$, then $\Gamma_1 \subseteq \Gamma_v, \Delta_1 \subseteq -\Gamma_v$ for some $v \in w$; also, if $\exists x A(x) \in \Gamma_\omega$, then $\exists x A(x) \in \Gamma_v$ and $A(c) \in \Gamma_v \subseteq \Gamma_w$ for some $v \in w, c \in S_0$. Also by genericity, $\overline{\mathbf{QCL}} \subseteq \Gamma_w$, thus Γ_w is $(\mathbf{QCL}, S_0)$-place and condition (Ac) for Γ_w (in a natural model on $\overline{IT}_\omega^{G'}$ for $G' \subseteq G_s$) holds. ∎

Recall that although G_Γ may be uncountable, there always exists a denumerable subset $G \subseteq G_\Gamma$ such that $\overline{IT}_\omega^G$ is coatomic. Therefore we obtain the following completeness result.

Proposition 7.6.18 *The logic* $\mathbf{QHCK} := \mathbf{QH} + CD \wedge KF$ *is complete (in semantics* $\mathcal{CK}$*) w.r.t. the class of all coatomic trees* $\overline{IT}^G_\omega$*; moreover one can take only denumerable coatomic trees.*

However we do not know if **QHCK** is determined by a single coatomic tree of the form IT^G_ω. But still it is determined by a single denumerable coatomic tree, e.g. by Smorynski's sum $\uplus \overline{IT}^{G_i}_\omega$ of all coatomic trees refuting nontheorems of the logic. Note that the sum over *all* denumerable atomic trees of the form $\overline{IT}^G_\omega$ is not itself denumerable.

Another alternative is to add maximal elements not above paths, but above points of IT_ω as explained below.

Definition 7.6.19 *A subset* $X \subseteq IT_\omega$ *gives rise to a denumerable tree* $IT^X_\omega := IT_\omega \cup X^*$ *where* $X^* = \{u^* \mid u \in X\}$ *is the set of maximal elements* $u^* := (u, -1)$ *such that* u^* *is an immediate successor of* u.

Obviously the tree IT^X_ω is coatomic iff X is a dense subset of IT_ω.

Lemma 7.6.20 *Let* L *be a superintuitionistic predicate logic containing* $CD \wedge K$. *Then for any* $\Gamma \in CP_L$ *and for any* $X \subseteq IT_\omega$ *there exists an* L*-natural model* N *(with constant domain) on* IT^X_ω *such that* $\Gamma_\lambda = \Gamma$.

Proof We simply extend an L-place Γ_u (for $u \in X$) to an L- place Γ_{u^*}, by Lemma 8.3.6. ■

Proposition 7.6.21 $\mathbf{QHCK} = \mathbf{IL}(\mathcal{CK}(IT^X_\omega))$ *for any dense* $X \subseteq IT_\omega$ *(in particular, for* $X = IT_\omega$*).*

A similar completeness result for logics with equality (w.r.t. $\mathcal{CKE}$ or w.r.t. $\mathcal{K}$) is quite obvious.

7.7 Remarks on Kripke bundles with constant domains

Now let us make some observations on completeness in more general semantics than $\mathcal{K}$ or $\mathcal{KE}$.

Proposition 7.6.16 clearly shows that $\mathbf{QH} + CD \wedge AP_h$ is determined by the tree IT^h_w in the semantics of Kripke sheaves or even quasi-sheaves (recall that AP_h is strongly valid in any Kripke quasi-sheaf based on a frame of depth $\leq h$). In particular, $\mathbf{QCL} = \mathbf{IL}(\mathcal{CK}(IT^1)) = \mathbf{IL}(\mathcal{CKE}(IT^1)) = \mathbf{IL}(\mathcal{KQ}(IT^1))$, where $IT^1 \cong Z_1$ is a one-element poset. On the other hand, for Kripke bundles the situation changes drastically:

Proposition 7.7.1 $\mathbf{IL}(\mathcal{CKB}(IT^1)) = \mathbf{QH} + CD$.

Proof Let M be a Kripke model based on IT_ω with constant domain ω refuting a nontheorem A, i.e. $M, \curlywedge \not\Vdash A$. Take a Kripke bundle F based on $\{u_0\}$ in which $D_{u_0} = \{a_u \mid u \in IT_\omega\} \cup \{b_i, c_i \mid i \in \omega\}$ is quasi-ordered by the following relation R_1:

$$a_u R_1 a_v \text{ iff } u \leqslant v \in IT_\omega, \text{ for } u, v \in IT_\omega,$$
$$b_i R_1 a_v \text{ for all } i \in \omega, v \in IT_\omega, \text{ and } b_i R_1 b_j \text{ for all } i, j \in \omega.$$

Obviously, F has a constant domain — every c_i is incomparable with all elements of D_{u_0}; every a_u has infinitely many predecessors b_i ($i \in \omega$). Consider the following map χ from D_{v_0} onto ω:

$$\chi(c_i) = i \text{ and } \chi(b_i) = \chi(a_u) = 0 \text{ for any } i \in \omega,\ u \in IT_\omega,$$

and consider the model M' over F for the formula A^1 (with an additional parameter z) such that:

$$M', u_0 \Vdash P'(a_2, d_1, \ldots, d_n) \Leftrightarrow M, u \Vdash P(\chi(d_1), \ldots, \chi(d_n))$$

where the predicate letter P' is substituted for P in A^1.

One can easily check that

$$M', u_0 \Vdash B^1(a_u, d_1, \ldots, d_n) \text{ iff } M, u \models B(\chi(d_1), \ldots, \chi(d_n))$$

for any formula $B(x_1, \ldots, x_n)$ and $u \in IT_\omega$ (by induction on B). Thus, $M', v_0 \not\Vdash A^1(\curlywedge)$, and so $A \notin \mathbf{IL}(F)$. ■

In the same way we can simulate every Kripke model with a constant domain by a Kripke bundle model based on one-element poset, so that individuals replacing additional parameters of formula A^1 correspond to worlds of the original Kripke frame.

Corollary 7.7.2 $\mathbf{IL}(\mathcal{CKB}(F)) = \mathbf{QH} + CD$ *for any propositional base* F.

Therefore in the semantics of Kripke bundles the propositional base is not so important as the structure of individual domains.

Remark 7.7.3 Note that the above corollary does not transfer to logics with equality — e.g. the formula $\forall xy(x = y) \supset p \vee \neg p$ is strongly valid in every Kripke bundle over a one-element base.

Remark 7.7.4 One can also try to describe the logic $\mathbf{IL}(\mathcal{KB}(IT^1))$ determined by a one-element base with varying domains. We may conjecture that this logic is $\mathbf{QH}$, but it is still unclear how to simulate arbitrary Kripke frames with varying domains within a single domain in a single world.

Remark 7.7.5 Let us also mention the semantics of Kripke quasi-sheaves. One can easily show that $\mathbf{IL}(\mathcal{CKQ}(IT_1)) = \mathbf{QH} + CD$ and $\mathbf{IL}(\mathcal{KQ}(IT_1)) = \mathbf{QH}$ for the ω-chain IT_1; in fact, every Kripke model based on IT_ω can be simulated in a Kripke quasi-sheaf over IT_1 similarly to Proposition 7.7.1. Branching of individuals in a quasi-sheaf replaces the branching of worlds. For the case of varying domains, note that for constructing a natural model over IT_ω we may assume that individual domains of all worlds at the same level coincide.

Similarly, $\mathbf{IL}(\mathcal{CKQ}(F)) = \mathbf{QHCK}$ and $\mathbf{IL}(\mathcal{KQ}(F)) = \mathbf{QH} + KF$ for an $(\omega+1)$-chain F; atomic trees $\overline{IT}_\omega^G$ or $\overline{IT}_\omega^X$ can be used here. On the other hand, note that $\mathbf{IL}(\mathcal{CKQ}(F)) = \mathbf{QH} + CD \wedge AP_h$ and $\mathbf{IL}(\mathcal{KQ}(F)) = \mathbf{LP}_h^+$ for any poset F of finite height $h > 0$ (in particular, for an h-element chain).

7.8 Kripke frames over the reals and the rationals

In this section we consider specific extensions of the intermediate logic of linear ordered sets with constant domains **QLCC**, viz. the logic of frames over the rational and the real line. Both logics happen to be finitely axiomatisable. The first result is not a big surprise, but the second is a 'happy chance' making a contrast with classical logic. These results were first proved in [Takano, 1987] who also observed that strong completeness theorem holds in both cases.

We use the following two axioms:

Ta. $(\forall x P(x) \supset \exists x Q(x)) \supset \exists x(P(x) \supset r) \vee \exists x(r \supset Q(x))$.

Ta'. $(\forall x P(x) \supset \exists x Q(x)) \supset \exists x(P(x) \supset \exists y Q(y)) \vee \exists x(\exists y Q(y) \supset Q(x))$.

Lemma 7.8.1 $\mathbf{QHC} + Ta = \mathbf{QLCC} + Ta'$.

Proof ($\subseteq$). It suffices to show that $\mathbf{QLCC} + Ta' \vdash Ta$. So we argue in $\mathbf{QLCC} + Ta'$.

We have

$$(Q(x) \supset r) \vee (r \supset Q(x))$$

by AZ, hence $(Q(x) \supset r) \vee \exists x(r \supset Q(x))$, $\forall x((Q(x) \supset r) \vee \exists x(r \supset Q(x)))$, and thus

(1) $$(\forall x(Q(x) \supset r) \vee \exists x(r \supset Q(x))$$

by CD. Similarly we have

(2) $$\forall x(r \supset P(x)) \vee \exists x(P(x) \supset r).$$

From (1) and (2) we obtain either the conclusion of Ta:

$$\exists x(P(x) \supset r) \vee \exists x(r \supset Q(x))$$

or

(3) $$\forall x(Q(x) \supset r) \wedge \forall x(r \supset P(x)).$$

Now assume the premise of Ta (congruent to the premise of Ta'):

$$(4) \qquad \forall x P(x) \supset \exists x Q(x).$$

Hence by Ta' we have

$$\exists x(P(x) \supset \exists y Q(y)) \vee \exists x(\exists y Q(y) \supset Q(x)).$$

Consider the first option and assume

$$(5) \qquad \exists x(P(x) \supset \exists y Q(y)).$$

By 2.6.15 we also have

$$\forall x(Q(x) \supset r) \vdash \exists y Q(y) \supset r.$$

So

$$(3), P(x) \supset \exists y Q(y) \vdash P(x) \supset r,$$

and thus

$$(3), (5) \vdash \exists x(P(x) \supset r).$$

Now assume

$$(6) \qquad \exists x(\exists y Q(y) \supset Q(x)).$$

Since by 2.6.15

$$\forall x(r \supset P(x)) \vdash r \supset \forall x P(x),$$

we have

$$(3) \vdash r \supset \forall x P(x),$$

hence

$$(3), (4) \vdash r \supset \exists x Q(x),$$

and thus

$$(3), (4) \vdash r \supset \exists y Q(y).$$

Now obviously

$$r \supset \exists y Q(y), \exists y Q(y) \supset Q(x) \vdash r \supset Q(x),$$

hence

$$(3), (4), \exists y Q(y) \supset Q(x) \vdash r \supset Q(x),$$

and thus

$$(3), (4) \vdash \exists y Q(y) \supset Q(x) \bullet \supset \bullet\, r \supset Q(x).$$

Therefore by monotonicity

$$(3), (4) \vdash \exists x(\exists y Q(y) \supset Q(x)) \supset \exists x(r \supset Q(x)),$$

and eventually

$$(3), (4), (6) \vdash \exists x(r \supset Q(x)).$$

So we have proved the conclusion of Ta.

($\supseteq$). We have $\mathbf{QH} + Ta \vdash Ta'$, since obviously $[\exists x Q(x)/r]\, Ta \doteq Ta'$. To show $\mathbf{QH} + Ta \vdash AZ$, note that

$$[p, p/P(x), Q(x)]\, Ta = (\forall x p \supset \exists x p) \supset \exists x(p \supset r) \vee \exists x(r \supset p)$$

is clearly equivalent to $(p \supset r) \vee (r \supset p)$. ■

In this section we consider subsets of $\mathbb{R}$, in particular, $\mathbb{Q}$, $\mathbb{R}_+ = [0, +\infty)$, $\mathbb{Q}_+ = \mathbb{R}_+ \cap \mathbb{Q}$ as propositional Kripke frames with the accessibility relation $\leq$.

Theorem 7.8.2

(1) $\mathbf{QLCC} = \mathbf{IL}(\mathcal{CK}\mathbb{Q}) = \mathbf{IL}(\mathbb{Q} \odot \omega)$.

(2) **QLCC** *is strongly complete w.r.t.* $\mathcal{CK}\mathbb{Q}$ *and* $\mathbb{Q} \odot \omega$.

Proof

(1) The inclusion $\subseteq$ follows readily from Proposition 7.3.6. The proof of $\supseteq$ is quite similar to 6.7.3. By Lemma 6.7.2, $\mathbb{Q}_+ \twoheadrightarrow F$ for any rooted countable chain F. Then we can apply Propositions 7.3.6, 3.3.14.

(2) By the Shimura theorem, **QLCC** is strongly Kripke complete, hence it is also strongly complete w.r.t. the class of countable chains with domain ω. By 6.7.2 we know that $\mathbb{Q}_+ \twoheadrightarrow F$ for any countable chain F. Hence $\mathbb{Q}_+ \odot \omega \twoheadrightarrow F \odot \omega$ by 5.1 and eventually strong completeness w.r.t. $\mathbb{Q}_+ \odot \omega$ follows by 5.1.

■

Definition 7.8.3 *Let* $\mathbf{F}_0 \subseteq \mathbf{F}$ *be linear Kripke frames with a constant domain* V, R *the accessibility relation in* $\mathbf{F}$. *Let* $M = (\mathbf{F}, \theta)$, $M_0 = (\mathbf{F}_0, \theta_0)$ *be intuitionistic Kripke models.* M *is called a* right extension *of* M_0 *if for any* V*-sentence* A

$$\theta^+(A) = R(\theta_0^+(A)). \qquad (*r)$$

So we have

$$M, u \vDash A \text{ iff } \exists v \in \mathbf{F}_0\, (vRu \;\&\; M_0, v \Vdash A)$$

Note that $(*r)$ readily implies $\theta_0^+(A) = \theta^+(A) \cap \mathbf{F}_0$, i.e. $M_0 \subseteq M$.
A right extension is obviously unique (if it exists).

Definition 7.8.4 *Let* $\mathbf{F}_0 \subseteq \mathbf{F}$ *be linear Kripke frames with a constant domain* V, R *the accessibility relation in* $\mathbf{F}$. *Let* $M = (\mathbf{F}, \theta)$, $M_0 = (\mathbf{F}_0, \theta_0)$ *be intuitionistic Kripke models.* M *is called a* left extension *of* M_0 *if for any* V*-sentence* A*, for any* $u \in M$

$$u \in \theta^+(A) \text{ iff } R(u) \cap \mathbf{F}_0 \subseteq \theta_0^+(A).$$

The latter condition can also be written as follows

$$\theta^+(A) = \Box_R \theta_0^+(A), \tag{$*l$}$$

where

$$\Box_R U := \{u \mid R(u) \cap \mathbf{F}_0 \subseteq U\}.$$

($*l$) is also equivalent to

$$M, u \not\Vdash A \text{ iff } \exists v \in \mathbf{F}_0\, (uRv \;\&\; M_0, v \not\Vdash A).$$

Lemma 7.8.5 *Let $F \odot V$ be a linear Kripke frame over $F = (W, R)$, $W_0 \subseteq W$ and let $M_0 = (F|W_0, \theta_0)$ be a Kripke model such that*

(1r) $R(W_0) = W$,

(2r) for any V-sentence $\forall x B(x)$

$$\bigcap_{a \in V} R(\theta_0^+(B(a))) \subseteq R(\theta_0^+(\forall x B(x))).$$

Then there exists a right extension of M_0 over F.

Note that the converse to ($2r$) obviously holds, so ($2r$) is equivalent to

$$R(\theta_0^+(\forall x B(x))) = \bigcap_{a \in V} R(\theta_0^+(B(a))).$$

Proof We define the valuation θ by the equality from 7.8.3

$$\theta^+(A) = R(\theta_0^+(A)) \tag{$*r$}$$

for all atomic V-sentences A. θ is obviously intuitionistic, since R is transitive. Then we check ($*r$) for any V-sentence A by induction.

(I) Consider the case $A = B \supset C$. By the induction hypothesis, it suffices to show that for any u

$$u \in R(\theta_0^+(A)) \text{ iff } \forall v \in R(u)\ (v \in R(\theta_0^+(B)) \Rightarrow v \in R(\theta_0^+(C)).$$

('Only if'). Assuming $u \in R(\theta_0^+(A))$, uRv and $v \in R(\theta_0^+(B))$, let us check $v \in R(\theta_0^+(C))$.

By assumption there exists $u_1 R u$ such that $u_1 \Vdash B \supset C$ and $v_1 R v$ such that $v_1 \Vdash B$ (where $\Vdash$ refers to M_0). Let $w_1 = \max(u_1, v_1)$; then $w_1 R v$. Now $u_1 \Vdash B \supset C$ implies $w_1 \Vdash B \supset C$; $v_1 \Vdash B$ implies $w_1 \Vdash B$. Thus $w_1 \Vdash C$, and so $v \in R(\theta_0^+(C))$.

('If'). We suppose $u \notin R(\theta_0^+(A))$ and show

$$\exists v \in R(u)\ v \in R(\theta_0^+(B)) - R(\theta_0^+(C)).$$

By assumption ($1r$), there exists $v_1 \in R^{-1}(u) \cap W_0$. Since $u \notin R(\theta_0^+(A))$, we have $v_1 \not\Vdash A$. Then for some $v_2 \in R(v_1), v_2 \Vdash B$, but $v_2 \not\Vdash C$.

Now put $v = \max(u, v_2)$. Then $v_2 \Vdash B$ implies $v \Vdash B$, and thus $v \in R(\theta_0^+(B))$. It remains to check that $v \notin R(\theta_0^+(C))$.

If $v = v_2$, then $v \notin R(\theta_0^+(C))$ follows from $v_2 \nVdash C$.

If $v = u$, then $u \notin R(\theta_0^+(B \supset C))$ implies $u \notin R(\theta_0^+(C))$. This is because $C \supset (B \supset C)$ is an intuitionistic tautology, so $M_0 \Vdash C \supset (B \supset C)$, i.e. $\theta_0^+(C) \subseteq \theta_0^+(B \supset C)$.

(II) If $A = B \wedge C$, by the induction hypothesis it suffices to show

$$R(\theta_0^+(B \wedge C)) = R(\theta_0^+(B)) \cap R(\theta_0^+(C)). \qquad (\#)$$

The inclusion '$\subseteq$' follows from $\theta_0^+(B \wedge C) \subseteq \theta_0^+(B)$ and $\theta_0^+(B \wedge C) \subseteq \theta_0^+(C)$.

The other way round, suppose $u \in R(\theta_0^+(B)) \cap R(\theta_0^+(C))$. Then $v_1 \in \theta_0^+(B)$, $v_2 \in \theta_0^+(C)$ for some $v_1, v_2 \in R^{-1}(u)$. If $v = \max(v_1, v_2)$, then obviously $v \in \theta_0^+(B \wedge C)$, and thus $u \in R(\theta_0^+(B \wedge C))$. This proves '$\supseteq$' in (#).

The remaining cases are left to the reader; note that for the case $A = \forall x B(x)$ one can use the assumption $(2r)$. ■

Remark 7.8.6 $(r1), (r2)$ are not only sufficient, but necessary for the existence of a right extension. $(r2)$ follows from $(*r)$ for $A = \forall x B(x)$, $(r1)$ from $(*r)$ for $A = \top$.

Lemma 7.8.7 *Let $F \odot V$ be a linear Kripke frame over $F = (W, R)$, $W_2 \subseteq W$ and let $M_0 = (F|W_0, \theta_0)$ be a Kripke model such that*

(1l) $R^{-1}(W_0) = W$,

(2l) for any V-sentence $\exists x A(x)$

$$\bigcup_{a \in V} \Box_R \theta_0^+(A(a)) \supseteq \Box_R \theta_0^+(\exists x A(x)).$$

Then there exists a left extension of M_0 over F.

Proof Similar to Lemma 7.8.5. Now we define

$$\theta^+(A) := \Box_R \theta_0^+(A),$$

for atomic V-sentences A and prove $(*l)$ by induction for arbitrary A.

Suppose $A = B \vee C$. By the induction hypothesis, $(*l)$ for A is equivalent to

$$\Box_R \theta_0^+(B) \cup \Box_R \theta_0^+(C) = \Box_R \theta_0^+(B \vee C).$$

The inclusion '$\subseteq$' is obvious, so let us check '$\supseteq$'. Suppose $u \notin \Box_R \theta_0^+(B) \cup \Box_R \theta_0^+(C)$. Then

$$\exists v, w \in R(u)\, (v \notin \theta_0^+(B) \;\&\; w \notin \theta_0^+(C)).$$

Since R is linear, vRw or wRv. Then respectively, $v \notin \theta_0^+(B \vee C)$ or $w \notin \theta_0^+(B \vee C)$, so anyway $u \notin \Box_R \theta_0^+(B \vee C)$.

Suppose $A = B \supset C$. By the induction hypothesis we can rewrite $(*l)$ as

$$u \notin \Box_R\theta_0^+(A) \Leftrightarrow \exists v \in R(u)\, v \in (\Box_R\theta_0^+(B) - \Box_R\theta_0^+(C)).$$

($\Rightarrow$) Suppose $u \notin \Box_R\theta_0^+(A)$. Then $w \notin \theta_0^+(A)$ for some $w \in R(u)$, and thus $v \in \theta_0^+(B) - \Theta_0^+(C)$ for some $v \in R(w) \subseteq R(u)$.
Obviously, $v \notin \Box_R\theta_0^+(C)$. Since M_0 is intuitionistic, $\theta_0^+(B) \subseteq \Box_R\theta_0^+(B)$.
This proves ($\Rightarrow$).
($\Leftarrow$) Suppose uRv, $v \in (\Box_R\theta_0^+(B) - \Box_R\theta_0^+(C))$. Then $w \notin \theta_0^+(C)$ for some $w \in R(v) \subseteq R(u)$ and $v \in \Box_R\theta_0^+(B)$ implies $w \in \theta_0^+(B)$. Hence

$$w \notin \theta_0^+(A), \text{ so } u \notin \Box_R\theta_0^+(A).$$

The case $A = B \wedge C$ is obvious. The case $A = \exists x B(x)$ follows from $(2l)$, since its converse holds trivially.
If $A = \forall x B(x)$, we have

$$u \notin \theta^+(A) \Leftrightarrow \exists a \in V\; u \notin \theta^+(B(a)) = \Box_R\theta_0^+(B(a)), \tag{**}$$

since the domain is constant and by the induction hypothesis. But $\theta_0^+(A) \subseteq \theta_0^+(B(a))$, so $u \notin \theta^+(A)$ implies $u \notin \Box_R\theta_0^+(A)$.
The other way round, if $u \notin \Box_R\theta_0^+(A)$, then $v \notin \theta_0^+(\forall x B(x))$ for some $v \in R(u) \cap \mathbf{F}_0$. So $v \notin \theta_0^+(B(a))$ for some $a \in V$, and thus $u \notin \theta^+(A)$ by (**). ■

Lemma 7.8.8 *Let $M = (\mathbb{Q}_+ \odot V, \theta)$ be an intuitionistic Kripke model, in which every instance of Ta' is true. Then there exists an extension θ^* of θ such that $(\mathbb{R}_+ \odot V, \theta^*)$ is an intuitionistic model.*

Proof The proof consists of two parts.
(I) First we define a model over the set

$$S := \mathbb{Q}_+ \cup \{\sigma_{B(y)} \mid \exists y B(y) \text{ is a } V\text{-sentence}, M, 0 \nVdash \exists y(\exists z B(z) \supset B(y))\},$$

where

$$\sigma_{B(y)} := \inf_{\mathbb{R}} \theta(\exists y B(y)).$$

We would like to define a model M_0 with a valuation θ_0 over S as a right extension of θ:

$$\theta_0(A) := \ \leq(\theta^+(A)).$$

By Lemma 7.8.5, it suffices to prove that for the relation $\leq$ on S, for every V-sentence $\forall x A(x)$,

$$\bigcap_{a \in V} \leq(\theta^+(A(a)) \subseteq \leq(\theta^+(\forall x A(x))) \tag{7.1}$$

To prove this, we assume that for a certain $\sigma \in S$

$$\forall a \in V\; \exists q \leq \sigma\; M, q \Vdash A(a). \tag{7.2}$$

and show that for some $q \in \mathbb{Q}_+$

$$q \leq \sigma \ \& \ M, q \Vdash \forall x A(x). \tag{7.3}$$

Case 1. $\sigma \in \mathbb{Q}_+$.

Then $M, \sigma \Vdash \forall x A(x)$, so $\sigma \in \theta^+(\forall x A(x)) \subseteq \leq(\theta^+(\forall x A(x)))$, so $q = \sigma$ satisfies (7.3).

Case 2. $\sigma = \sigma_{B(y)} \notin \mathbb{Q}_+$.

By definition, $M, 0 \nVdash \exists y(\exists z B(z) \supset B(y))$. Now let us show

$$(\flat) \qquad M, 0 \nVdash \exists x(A(x) \supset \exists z B(z)).$$

Suppose the contrary. Then $M, 0 \Vdash A(a) \supset \exists z B(z)$ for some $a \in V$. By the assumption ($\sharp$), $M, q_a \Vdash A(a)$ for some $q_a \leq \sigma$. So $M, q_a \Vdash \exists z B(z)$, and hence $\sigma \leq q_a$, by the definition of $\sigma_{B(y)}$. Therefore $\sigma = q_a \in \mathbb{Q}$, and we have a contradiction. Hence ($\flat$) holds.

On the other hand, by the assumption of the lemma,

$$M, 0 \Vdash (\forall x A(x) \supset \exists y B(y)) \supset \exists x(A(x) \supset \exists z B(z)) \vee \exists y(\exists z B(z) \supset B(y)).$$

Hence $M, 0 \nVdash \forall x A(x) \supset \exists y B(y)$, so there is $q \in \mathbb{Q}_+$ such that $M, q \Vdash \forall x A(x)$ but $M, q \nVdash \exists y B(y)$. Hence $q \leq \sigma$, by the definition of σ. Thus q satisfies (7.3). This proves (7.1).

(II) Next, we construct M^* with a valuation θ^* over $\mathbb{R}^+$ as a left extension of M_0:

$$M^*, \alpha \Vdash A \Leftrightarrow \forall \sigma \in S\, (\alpha \leq \sigma \Rightarrow M_0, \sigma \Vdash A).$$

We use Lemma 7.8.7 to show the existence of M^*. So we have to check that for every V-sentence $\exists x B(x)$

$$\bigcup_{a \in V} \Box_{\leq} \theta_0^+(B(a)) \supseteq \Box_{\leq} \theta_0^+(\exists x B(x)). \tag{7.4}$$

To prove this, suppose $\alpha \in \Box_{\leq} \theta_0^+(\exists x B(x))$, that is

$$\forall \sigma \in S(\alpha \leq \sigma \Rightarrow M_0, \sigma \Vdash \exists y B(y)). \tag{7.5}$$

Let us show that $\alpha \in \bigcup_{a \in V} \Box_{\leq} \theta_0^+(B(a))$, i.e. for some $a \in V$

$$\forall \sigma \in S(\alpha \leq \sigma \Rightarrow M_0, \sigma \Vdash B(a)). \tag{7.6}$$

Now we argue in M_0.

Case 1. $0 \Vdash \exists y(\exists z B(z) \supset B(y))$.

Then there is $a \in V$ such that $0 \Vdash \exists z B(z) \supset B(a)$, and this a is what we need.

Case 2. $0 \nVdash \exists y(\exists z B(z) \supset B(y))$.

So $0 \nVdash \exists y(\exists z) \supset B(y))$, and hence $\sigma_{B(y)} \in S$.

Let us show that $\sigma_{B(y)} \nVdash \exists y B(y)$. Suppose the contrary. Then $\sigma_{B(y)} \Vdash \exists y B(y)$, so $\sigma_{B(y)} \Vdash B(a_1)$ for some $a_1 \in V$. Then by the assumption of Case 2,

$$0 \nVdash \exists z B(z) \supset B(a_1),$$

so there exists $\sigma \in S$ such that $\sigma \Vdash \exists z B(z)$, and $\sigma \nVdash B(a)$. Thus $\sigma < \sigma_{B(y)}$, since $\sigma_{B(y)} \Vdash B(a)$.

Hence

$$\exists q \in \mathbb{Q}_+ \ \sigma < q < \sigma_{B(y)},$$

and then

$$M_0, q \Vdash \exists z B(z),$$

which implies

$$M, q \Vdash \exists z B(z).$$

But this contradicts $q < \sigma_{B(y)}$.

Now, by (7.5), $M_0, \sigma \nVdash \exists y B(y)$ implies $\sigma_{B(y)} < \alpha$.

But then there is $q \in \mathbb{Q}$ such that $\sigma_{B(y)} < q < \alpha$. Since $\sigma_{B(y)} < q$, it follows that $M_0, q \Vdash \exists y B(y)$. Hence $M_0, q \Vdash B(a)$ for some $a \in V$, and this a satisfies (7.6). Thus (7.4) is proved. ∎

Theorem 7.8.9 $\mathbf{QHC} + Ta = \mathbf{IL}(\mathcal{CK}\mathbb{R})$.

Proof ($\subseteq$). Let us show that $\mathbb{R} \odot V \Vdash Ta$. So consider an arbitrary model over $\mathbb{R} \odot V$, $\alpha \in \mathbb{R}$ and show

$$\alpha \Vdash (\forall x P(x) \supset \exists x Q(x)) \supset \exists x(P(x) \supset r) \vee \exists x(r \supset Q(x)). \tag{7.7}$$

Suppose the contrary. Then there exists $\beta \leq \alpha$ such that

$$\beta \Vdash \forall x P(x) \supset \exists x Q(x), \text{ but } \beta \nVdash \exists x(P(x) \supset r), \ \beta \nVdash \exists x(r \supset Q(x)).$$

Thus for any $a \in V$, there is $\gamma_a \geq \beta$ such that $\gamma_a \Vdash P(a)$, $\gamma_a \nVdash r$. Similarly, for any $b \in V$ there is $\delta_a \geq \beta$ such that $\delta_b \Vdash r$, $\delta_b \nVdash Q(a)$. From $\gamma_a \nVdash r$ and $\delta_b \Vdash r$ it follows that $\gamma_a < \delta_b$, for every $a, b \in V$.

Now put $\varepsilon := \sup_{a \in V} \gamma_a$. Then $\beta \leq \varepsilon$ and $\varepsilon \Vdash \forall x P(x)$. Also $\beta \Vdash \forall x P(x) \supset \exists x Q(x)$, hence $\varepsilon \Vdash \exists x Q(x)$.

On the other hand, $\varepsilon \leq \delta_b$ for any $b \in V$, so $\varepsilon \nVdash \exists x Q(x)$. This contradiction proves (7.7)

($\supseteq$). We assume $\mathbf{QHC} + Ta \nvdash A$ for a sentence A and construct a countermodel for A over $\mathbb{R}_+$. Since $\mathbb{R}_+$ is a generated subframe, this implies $A \notin \mathbf{IL}(\mathcal{CK}\mathbb{R})$.

By Lemma 7.8.1, $\mathbf{QLCC} + Ta' \nvdash A$. By Theorem 7.8.2(2), we obtain a model $M = (\mathbb{Q}_+ \odot V, \theta)$ and $q_0 \in \mathbb{Q}$ such that every instance of Ta' is true at M, q_0 and $M, q_0 \nVdash A$. We may assume that $q_0 = 0$. By Lemma 3.3.18, $M' := M \restriction \mathbb{Q}_+$ is a model, in which every instance of Ta' is true. Hence by Lemma 7.8.8 there is an extension M^* of M' over $\mathbb{R}_+$. ∎

Proposition 7.8.10 *(1)* $\mathbf{QLCC} + KF = \mathbf{IL}(\mathcal{CK}([0,1] \cap \mathbb{Q}))$.

(2) $\mathbf{QHC} + Ta + KF = \mathbf{IL}(\mathcal{CK}[0,1])$.

We skip the proof, since it is very similar to Theorems 7.8.2 and 7.8.9.

Proposition 7.8.11 *For any linearly ordered set F, the following conditions are equivalent:*

(1) F is Dedekind complete (i.e. every non-empty subset with an upper bound has a supremum);

(2) $\mathbf{IL}(\mathcal{CKF}) \vdash Ta'$.

Proof

The proof of (1) $\Rightarrow$ (2) is the same as for ($\subseteq$) in 7.8.9.

Let us show (2)$\Rightarrow$(1). Suppose (2), but not (1). Then there exists a non-empty $X \subseteq W$ with an upper bound, but without a supremum. Let Y be the (non-empty) set of upper bounds of X. We index both X and Y by elements of V of cardinality $|W|$ so that $X = \{\xi_a \mid a \in V\}$ and $Y = \{\eta_a \mid a \in V\}$. Consider a model M over $W \odot V$, in which

(i) $w \Vdash P(a) \Leftrightarrow \xi_a \leq w$,

(ii) $w \Vdash Q(a) \Leftrightarrow \eta_a < w$.

Let us show that for any $w \in W$

(3) $\quad w \in Y \Leftrightarrow w \Vdash \forall x P(x) \Leftrightarrow w \Vdash \exists x Q(x)$.

The first equivalence is obvious:

$$w \Vdash \forall x P(x) \Leftrightarrow \forall a\ \xi_a \leq w \Leftrightarrow w \in Y$$

by (i) and the definition of Y.

Next, by (ii)

$$W \Vdash Q(a) \Leftrightarrow w > \eta_0,$$

which implies $w \in Y$.

Conversely, suppose $w \in Y$. Since w is not a supremum of X, there is $a \in V$ such that $\eta_a < w$. Hence $w \Vdash Q(a)$, and thus $w \Vdash \exists x Q(x)$.

Therefore

(4) $\quad M \Vdash \forall x P(x) \supset \exists y Q(y)$.

Now consider any $\xi \in X$. Let us show that

(5) $\quad \xi \nVdash \exists x(P(x) \supset \exists z Q(z))$.

In fact, for any a

$$\xi \leq \eta_a,\ \eta_a \nVdash Q(a),\ \eta_a \Vdash \exists z Q(z)$$

by (3), hence

$$\xi \nVdash \exists z Q(z) \supset Q(a).$$

(6) $\quad \xi \Vdash \exists y(\exists z Q(z) \supset Q(y))$.

In fact, let us show

(7) $\xi \not\Vdash P(a) \supset \exists y Q(y)$

for any $a \in V$. Consider $\xi' = \max(\xi, \xi_a)$. Then $\xi' \in X$, $\xi \leq \xi'$, $\xi' \Vdash P(a)$, and $\xi' \not\Vdash \exists y Q(y)$ by (3). This implies (7) and (6).

But (2) implies

$$\xi \Vdash (\forall x P(x) \supset \exists y Q(y)) \supset \exists x(P(x) \supset \exists z Q(z)) \vee \exists y(\exists z Q(z) \supset Q(y)),$$

which contradicts (4), (5), (6). ■

Proposition 7.8.12

(1) $\mathbf{IL}(\mathcal{CK}\mathbb{Q}) \subset \mathbf{IL}(\mathcal{CK}\mathbb{R})$.

(2) If a formula A does not contain $\forall$ *or* $\exists$, *then* $\mathbf{IL}(\mathcal{CK}\mathbb{R}) \vdash A$ *iff* $\mathbf{IL}(\mathcal{CK}\mathbb{Q}) \vdash A$.

Proof

(1) $\mathbf{IL}(\mathcal{CK}\mathbb{Q}) \subseteq \mathbf{IL}(\mathcal{CK}\mathbb{R})$ readily follows from Theorems 7.8.2 and 7.8.9, while $\mathbf{IL}(\mathcal{CK}\mathbb{Q}) \neq \mathbf{IL}(\mathcal{CK}\mathbb{R})$ by Proposition 7.8.11.

(2) Let us consider only formulas without occurrences of $\forall$.

Let $A(x_1, \ldots, x_n)$ be a $\forall$-free formula, and suppose $\mathbf{IL}(\mathcal{CK}\mathbb{Q}) \not\vdash A$. Consider a Kripke model $M_0 = (\mathbb{Q} \odot V, \theta)$, $q \in \mathbb{Q}$, $a_1, \ldots, a_n \in V$, such that $M_0, q \not\Vdash A(a_1, \ldots, a_n)$. We define M over $\mathbb{R}$ similarly to the proof of 7.8.4:

($\sharp$) $M, \alpha \Vdash B \Leftrightarrow \exists q \in \mathbb{Q}\, (q \leq \alpha \;\&\; M_0, q \Vdash B$

for every $\alpha \in \mathbb{R}$ and atomic V-sentence B. Now we prove ($\sharp$) for any $\forall$=free B. but unlike Lemma 7.8.5, we do not need the assumption (7.1). Hence $M, q \not\Vdash A(a_1, \ldots, a_n)$.

$\exists$-free formulas are considered similarly using Lemma 7.8.7 instead of Lemma 7.8.5.

■

Takano [Takano, 1987] also proved that $\mathbf{IL}(\mathcal{K}(\mathbb{R} - \mathbb{Q})) = \mathbf{QLC} + CD$. Actually his method proves the following statement.

Theorem 7.8.13 *If* $W \subseteq \mathbb{R}$ *and both* W *and* $F - W$ *are dense in* $\mathbb{R}$, $F = (W, <)$, *then* $\mathbf{IL}(\mathcal{CK}F) = \mathbf{IL}(F \odot \omega) = \mathbf{QLC} + CD$.

Proof It is sufficient to show that $\mathbf{IL}(F \odot \omega) \subseteq \mathbf{IL}(\mathbb{Q} \odot \omega)$. So given a sentence B and a countermodel $M = (\mathbb{Q} \odot \omega, \xi)$ for B, we construct a model M_1 over $F \odot \omega$ refuting B. The construction consists of two stages.

(I) At the first stage we take a denumerable subset W_0 of W that is dense in $\mathbb{R}$, $F_0 = (W_0, <)$, and construct a model $M_0 = (F_0 \odot \omega, \xi_0)$ refuting B and satisfying the following condition:

$$\inf_{\mathbb{R}} \xi_0^+(A) \notin (W - W_0) \text{ for every } \omega\text{-sentence } A. \qquad (+)$$

Put

$$S := \{\inf \xi^+(A) \mid A \text{ is an } \omega\text{-sentence}\} - \mathbb{Q}$$

and enumerate the sets $\mathbb{Q} \cup S$ and W_0, i.e. put $\mathbb{Q} \cup S = \{u_k \mid k \in \omega\}$ and $W_0 = \{v_n \mid n \in \omega\}$. Now since W_0 and $\mathbb{R} - W$ are dense in $\mathbb{R}$, there exists an order embedding $h : \mathbb{Q} \cup S \longrightarrow \mathbb{R}$ such that $h[\mathbb{Q}] = W_0$ and $h[S] \subseteq \mathbb{R} - W$.

To construct h, we can apply the back-and-forth method and define in turn $h(u_k)$ and $h^{-1}(v_n)$ for new elements of $\mathbb{Q} \cup S$ and W_0 respectively, preserving the order $<$. Or we can define $h(u_k)$ in the following direct way, by induction on k. Suppose $h(u_i)$ for $i < k$ are already defined, let

$$u_l := \max\{u_i \mid i < k, u_i < u_k\}, \ u_r := \min\{u_i \mid i < k, u_k < u_i\}.$$

If $u_k \in \mathbb{Q}$, we put $h(u_k) = v_n$ for the least n such that $v_n \in (h(a), h(u_r))$. And if $u \in S$, we choose $h(u_k)$ in $(h(u_l), h(u_r)) - W$. Then definitely every $v_n \in W_0$ is $h(u_k)$ for a suitable k, so to say, n becomes 'the least possible' at some stage of the construction.

We leave the routine technical details to the reader.

Let us show that

$$\forall X \subseteq \mathbb{Q} \ (\inf_{\mathbb{R}} X \in \mathbb{Q} \cup S \Rightarrow \inf_{\mathbb{R}} h[X] = h(\inf_{\mathbb{R}} X)). \qquad (*)$$

In fact, let $\inf_{\mathbb{R}} X = u_k$. Then $h(u_k) \leq h(u)$ for all $u \in X$. Suppose $\inf_{\mathbb{R}} h(X) > h(u_k)$. By density, there exists $v \in W_0$ such that $h(u_k) < v < \inf_{\mathbb{R}} h(X)$; let $v = h(u), u \in \mathbb{Q}$. Then u is a lower bound of X, so $u \leq u_k$, and hence $v = h(u) \leq h(u_k)$. This is a contradiction.

Obviously, the restriction $h_0 = h|\mathbb{Q}$ is an order isomorphism between $\mathbb{Q}$ and W_0, so $h_0^{-1} : W_0 \to \mathbb{Q}$ is a p-morphism. Now by Lemma 5.1 we obtain a model $M_0 = (W_0 \odot \omega, \xi_0)$ such that

$$\xi_0^+(A) = h[\xi^+(A)]$$

for every ω-sentence A. This model is a required one, because $(*)$ readily implies $(+)$. In fact,

$$\inf_{\mathbb{R}} \xi_0^+(A) = h(\inf_{\mathbb{R}} \xi^+(A)) \in h[\mathbb{Q} \cup S] \subseteq W_0 \cup (\mathbb{R} - W).$$

(II) Now we apply Lemma 7.8.4 and obtain a right extension M_1 of M_0. This model is a required one over W. It is sufficient to check the condition $(r2)$.

So we take an arbitrary $w_0 \in W$, assume that

$$\forall a \in V \; \exists u \in W_0 \; (u \leq w_0 \;\&\; M_0, u \Vdash A(a)), \tag{7a}$$

and find $u \in W_0$ such that

$$u \leq w_0 \text{ and } M_0, u \Vdash \forall x A(x). \tag{7b}$$

Put $u_0 := \inf_{\mathbb{R}}\{u \in W_0 \mid M_0, u \Vdash \forall x A(x)\}$.

Case 1. $u_0 \in W_0$ and $M_0, u_0 \Vdash \forall x A(x)$.

Let us show that $u_0 \leq w_0$. Suppose the contrary: $w_0 < u_0$. Then by density there exists $u_1 \in W_0$ such that $w_0 < u_1 < u_0$. On the other hand, for every $a \in V$ there exists $u \in W_0$ such that $u \leq w_0 < u_2$ and $M_0, u \Vdash A(a)$, so $M_0, u_1 \Vdash A(a)$ as well. Hence $M_0, u_1 \Vdash \forall x A(x)$ and $u_1 < u_0$, which contradicts the choice of u_0.

Thus we conclude that $u = u_0$ satisfies (7b).

Case 2. $u_0 \in W_0$ and $M_0, u_0 \not\Vdash \forall x A(x)$. Then $M_0, w_0 \not\Vdash A(a)$ for some $a \in V$. By (7a), there exists $u \in W_0$ such that $u \leq w_0$ and $M_0, u \Vdash A(a)$. Then $u_0 < u$, and so $M_0, u \Vdash \forall x A(x)$ by the definition of u_0. Thus u satisfies (7b).

Case 3. $u_0 \notin W_0$. Then $u_0 \notin W$ by (+). Let us show that $u_0 < w_0$. Suppose the contrary. Then $w_0 < u_0$, since $w_0 \in W$ and $u_0 \notin W$. So by density, there exists $w \in W_0$ such that $w_0 < w < u_0$. Then $M_0, w \not\Vdash \forall x A(x)$, by the definition of u_0. So $M_0, w \not\Vdash A(a)$ for some $a \in V$. On the other hand, $M_0, u \Vdash A(a)$ for some $u \leq w_0$, and hence $M_0, w \Vdash A(a)$ as well. Since $u \leq w_0 < w$, this is a contradiction.

Now since $u_0 < w_0$, there exists $u \in W_0$ such that $u \leq w_0$ and $M_0, u \Vdash \forall x A(x)$. This u satisfies (7b).

■

Corollary 7.8.14 $\mathbf{IL}(\mathcal{CK}(\mathbb{R} - \mathbf{Q})) = \mathbf{IL}((\mathbb{R} - \mathbb{Q}) \odot \omega) = \mathbf{QLC} + CD$.

Remark 7.8.15 The above argument can be readily transferred to the logics with equality. In fact, Lemmas 7.8.5 and 7.8.7 are extended easily.

The conditions $(r*)$ and $(l*)$ for equality:

$$M, u \Vdash a = b \Leftrightarrow \exists v \in W_0 (vRw \;\&\; M_0, u \Vdash a = b)$$

and

$$M, u \not\Vdash a = b \Leftrightarrow \exists v \in W_0 \, (wRv \;\&\; M_0, v \not\Vdash a = b)$$

respectively follow from (r1) and (l1).

Thus we obtain the following analogues of Theorems 7.8.2, 7.8.9 and 7.8.12:

(1) If W and $(\mathbb{R} - W)$ are dense in $\mathbb{R}$ then

$$\mathbf{IL}^{=}(\mathcal{CKE}(W)) = \mathbf{QLC}^{=} + CD, \quad \mathbf{IL}^{=}(\mathcal{CK}(W)) = \mathbf{QLC}^{=d} + CD;$$

in particular

$$\mathbf{IL}^{=}(\mathcal{CKE}\mathbb{Q}) = \mathbf{IL}^{=}(\mathcal{CKE}(\mathbf{R} - \mathbb{Q})) = \mathbf{QLC}^{=} + CD,$$
$$\mathbf{IL}^{=}(\mathcal{CK}\mathbb{Q} = \mathbf{IL}^{=}(\mathcal{CK}(\mathbf{R} - \mathbb{Q})) = \mathbf{QLC}^{=d} + CD.$$

(2)

$$\mathbf{IL}^{=}(\mathcal{KE}\mathbb{R}) = \mathbf{QLC}^{=} + CD + Ta,$$
$$\mathbf{IL}^{=}(\mathcal{K}(\mathbb{R})) = \mathbf{QLC}^{=d} + CD + Ta.$$

These results imply corollaries for the Kripke semantics with nested domains. Recall that $\mathbf{IL}(\mathcal{K}(\mathbb{Q})) = \mathbf{QLC}$, $\mathbf{IL}^{=}(\mathcal{KE}\mathbb{Q}) = \mathbf{QLC}^{=}$. $\mathbf{IL}^{=}(\mathcal{K}\mathbb{Q}) = \mathbf{QLC}^{=d}$. Hence by the method from section 3.9 we obtain

Proposition 7.8.16 *If subsets W and $\mathbb{R} - W$ are dense in $\mathbb{R}$, then*

$$\mathbf{IL}(\mathcal{K}W) = \mathbf{QLC},\ \mathbf{IL}^{=}(\mathcal{KE}W) = \mathbf{QLC}^{=},\ \mathbf{IL}^{=}(\mathcal{K}W) = \mathbf{QLC}^{=d}.$$

Let us finally mention some simple consequences of the above results.

Theorem 7.8.17 *The logics $\mathbf{IL}^{=}(\mathcal{K}\mathbb{R})$, $\mathbf{IL}^{=}(\mathcal{KE}\mathbb{R})$ are recursively axiomatisable.*

Proof Follows from 7.8.9 and 3.9.4. ■

Corollary 7.8.18 $\mathbf{IL}(\mathcal{K}\mathbb{R})$ *is recursively axiomatisable.*

Proof In fact, we already know that $\mathbf{IL}^{=}(\mathcal{K}\mathbb{R})$, $\mathbf{IL}^{=}(\mathcal{KE}\mathbb{R})$ are its conservative extensions (by 3.8.6 and 2.16.13). ■

As we know, the completeness proof for **QLC** can be extended to logics with equality, so

$$\mathbf{IL}^{=}(\mathcal{CKE}\mathbb{Q}) = \mathbf{QLC}^{=},\ \mathbf{IL}^{=}(\mathcal{CK}\mathbb{Q}) = \mathbf{QLC}^{=d}.$$

As we know from 7.8.2 and 7.8.14

$$\mathbf{IL}(\mathcal{CK}(\mathbb{R} - \mathbb{Q})) = \mathbf{IL}(\mathcal{CK}\mathbb{Q}) = \mathbf{QLC} + CD.$$

Thus 7.8.18 implies

Proposition 7.8.19

$$\mathbf{IL}(\mathcal{K}(\mathbb{R} - \mathbb{Q})) = \mathbf{IL}(\mathcal{K}\mathbb{Q}) = \mathbf{QLC};$$
$$\mathbf{IL}^{=}(\mathcal{KE}(\mathbb{R} - \mathbb{Q})) = \mathbf{IL}^{=}(\mathcal{KE}\mathbb{Q}) = \mathbf{QLC}^{=};$$
$$\mathbf{IL}^{=}(\mathcal{K}(\mathbb{R} - \mathbb{Q})) = \mathbf{IL}^{=}(\mathcal{K}\mathbb{Q}) = \mathbf{QLC}^{=d}.$$

Perhaps Takano's proof for $\mathbb{R} - \mathbb{Q}$ can also be transferred to the case with nested domains, but the embedding method provides the result rather easily.

As for an explicit axiomatisation for $\mathbf{IL}(\mathcal{KR})$, $\mathbf{IL}^{=}(\mathcal{KR})$, $\mathbf{IL}^{=}(\mathcal{KER})$, one can easily see that Ta is refuted in a frame over $\mathbb{R}$. In fact, informally speaking, Ta means the following.

Suppose $\forall x P(x) \supset \exists y Q(y)$ is true, say at world $0 \in \mathbb{R}$. Let v be the g.l.b. of the set $\{u \mid u \Vdash r\}$ (if this set is empty, then Ta holds trivially). Now if $v \not\Vdash \forall x P(x)$, i.e. $u \not\Vdash P(a)$ for some $u \geq v$, then $0 \Vdash P(a) \supset r$. Otherwise $v \Vdash \forall x P(x)$, and then $v \Vdash \exists y Q(y)$, i.e. $r \Vdash Q(b)$ for some b, so $0 \Vdash r \supset Q(b)$.

If the domain is constant, both individuals a, b are in D_0, but in the case of nested domains a, b can exist only in larger domains D_u or D_v, so the formula Ta is refutable. Constructing a counterexample is left as an exercise for the reader.

Conjecture. $\mathbf{IL}(\mathcal{KR}) = \mathbf{QLC}$

By Theorem 3.8.6, this conjecture implies the equality

$$\mathbf{IL}^{=}(\mathcal{KER}) = \mathbf{QLC}^{=},$$

and (as one can show),

$$\mathbf{IL}^{=}(\mathcal{KR}) = \mathbf{QLC}^{=d}.$$

Bibliography

[Alechina, van Lambalgen, 1994] N. Alechina and M. van Lambalgen. Correspondence and completeness for generalized quantifiers. *Bulletin of the IGPL*, 3 167–190, 1994

[Bell, 2001] J. Bell. *Logical options: An introduction to classical and alternative logics* Broadview Press Ltd, Canada (1 Jun 2001).

[Birkhoff, 1979] G. Birkhoff. *Lattice theory*, 3rd revised edition. American Mathematical Society Colloquium Publications, 25. American Mathematical Society, Providence, R.I, 1979.

[Blackburn, de Rijke and Venema, 2001] P. Blackburn, M. De Rijke, and Y. Venema. *Modal logic.* Cambridge University Press, 2001.

[Boolos, Burgess, and Jeffrey, 2002] G.S. Boolos, J.P. Burgess, and R.C. Jeffrey. *Computability and logic.* Cambridge University Press, 2002.

[Borceaux, 1994] F. Borceaux. *Handbook of categorical algebra, v. 1–3.* Cambridge University Press, 1994.

[Bourbaki, 1968] N. Bourbaki. *Elements of mathematics: theory of Sets.* Hermann, 1968.

[Chagrov and Zakharyaschev, 1997] A. V. Chagrov and M. V. Zakharyaschev. *Modal logic.* Oxford University Press, 1997.

[Church, 1996] A. Church. *Introduction to mathematical logic.* Princeton University Press, 1996.

[Corsi and Ghilardi, 1989] G. Corsi and S. Ghilardi. Directed frames. *Archive for math. logic*, 29:53–67, 1989.

[Corsi, 1989] G. Corsi. A logic characterized by the class of linear Kripke frames with nested domains. *Studia Logica*, 48:15–22, 1989.

[Corsi, 1993] G. Corsi. Quantified modal logics of positive rational numbers and some related systems. *Notre Dame Journal of Formal Logic*, 34:263–283,1993.

[Corsi, 1992] G. Corsi. Completeness theorem for Dummett's **LC** quantified and some of its extensions. *Studia Logica*, 51:317–335, 1992.

[Došen, 1993] K. Došen. Rudimentary Kripke models for the intuitionistic propositional calculus. *Annals of Pure and Applied Logic*, 62:21 – 49, 1993.

[Dragalin, 1973] A. G. Dragalin. On intuitionistic model theory (in Russian). In *Istoriya i metodologiya estestvennyx nauk, v. XIV: Matematika, mexanika*, pages 106–126. Moscow, MGU, 1973.

[Dragalin, 1988] A. G. Dragalin. *Mathematical intuitionism. Introduction to proof theory.*. American Mathematical Society, 1988. Translation of the Russian original from 1979.

[Dummett and Lemmon, 1959] M. Dummett and E. Lemmon. Modal logics between S4 and S5. *Zeitschrift für Math. Logik und Grundlagen der Math.*, 5:250–264, 1959.

[Esakia, 1974] L. L. Esakia. Topological Kripke models. *Soviet Mathematics Doklady,* 15:147-151, 1974.

[Esakia, 1979] L. L. Esakia. On the theory of modal and superintuitionistic systems (in Russian). In V. A. Smirnov, editor, *Logical inference (Moscow, 1974)*, pages 147–172. Nauka, Moscow, 1979.

[Fine, 1974] K. Fine. An ascending chain of S4 logics. *Theoria*, 40:110–116, 1974.

[Fine, 1974] K. Fine. Logics containing K4, part I. *Journal of Symbolic Logic*, 39:31–42, 1974.

[Fine, 1978] K. Fine. Model theory for modal logic. I. The de re/de dicto distinction. *Journal of Philosophical Logic*, 7:125–156, 1978.

[Fine, 1985] K. Fine. Logics containing K4, part II. *Journal of Symbolic Logic*, 50:619–651, 1985.

[Fine, 1988] K. Fine. Semantics for quantified relevance logic. *Journal of Philosophical Logic*, 17:27–59, 1988.

[Fine, 1989] K. Fine. Incompleteness for Quantified Relevance Logics. In J. Norman and R. Sylvan, editors, *Directions in Relevant Logic*, pages 205–225. Kluwer Academic Publishers, Dordrecht, 1989.

[Fitting and Mendelsohn, 1998] M. Fitting and R. L. Mendelsohn. *First-order modal logic.* Synthese Library, v. 277. Kluwer Academic Publishers, 1998.

[Fourman and Scott, 1979] M. Fourman and D. Scott. Sheaves and logic. In *Lecture Notes in Mathematics, v. 753*, pages 302–401. Springer, 1979.

[Gabbay, Reynolds and Finger, 2000] D. Gabbay, M. Reynolds and M. Finger. *Temporal logic*, Volume 2 Oxford University Press, 2000

[Gabbay and Pitts, 2002] M. J. Gabbay and A.M. Pitts. A new approach to abstract syntax with variable binding. *Formal Aspects of Computing*, 13:341–363, 2002.

[Gabbay *et al.*, 2002] D. M. Gabbay, A. Kurucz, F. Wolter, and M. Zakharyaschev. *Many-dimensional modal logics: theory and applications*. Elsevier, 2003

[Gabbay, 1972] D. M. Gabbay. Application of trees to intermediate logics. *Journal of Symbolic Logic*, 37: 135–138, 1972.

[Gabbay, 1975] D. M. Gabbay. A modal logic that is complete for neighbourhood frames but not in Kripke frames. *Theoria* , 41:148–153, 1975.

[Gabbay, 1976] D. M. Gabbay. Investigations in modal and tense logics with applications. *Synthese Library, 92. Reidel, 1976.*

[Gabbay, 1981] D. M. Gabbay. Semantical investigations in Heyting's intuitionistic logic. *Synthese Library, 148. Reidel, 1981.*

[Gabbay, 1996] D. M. Gabbay. *Labelled deductive systems.* Oxford University Press, 1996.

[Gabbay, 1998] D. M. Gabbay. *Fibring logics.* Oxford University Press, 1998.

[Gabbay and Shehtman, 1998] Products of modal logics, part I. *Logic Journal of the IGPL*, 6: 73–146, 1998.

[Gabriel and Zisman, 1967] P. Gabriel and M. Zisman. *Calculus of fractions and homotopy theory.* Ergebnisse der Mathematik und ihrer Grenzgebiete, Bd. 35. Springer-Verlag, 1967.

[Garson, 1978] J. Garson. Completeness of some quantified modal logics. *Logique et analyse*, 82–83:152–164, 1978.

[Gasquet, 1995] O. Gasquet. A new incompleteness result in Kripke semantics. *Fundamenta Informaticae*, Vol. 24 No. 4, 1995, 407–415.

[Gerson, 1975a] M. Gerson. An extension of S4 complete for the neighbourhood semantics but incomplete for the relational semantics. *Studia Logica*, 34:333–342, 1975.

[Gerson, 1975b] M. Gerson. The inadequacy of neighbourhood semantics for modal logics. *Journal of Symbolic Logic*, 40:141–147, 1975.

[Ghilardi and Meloni, 1988] S. Ghilardi and G. Meloni. Modal and tense predicate logic: models in presheaves and categorical conceptualization. In *Categorical algebra and its applications (Louvain-La-Neuve, 1987)*, Lecture Notes in Mathematics, 1348, 130–142. Springer, Berlin, 1988.

[Ghilardi, 1989] S. Ghilardi. Presheaf semantics and independence results for some non-classical first-order logics. *Archive for Mathematical Logic*, 29: 125–136, 1989.

[Ghilardi, 1992] S. Ghilardi. Quantified extensions of canonical propositional intermediate logics. *Studia Logica*, 51:195–214, 1992.

[Godement, 1958] R. Godement. *Topologie algebrique et theorie des faisceaux.* Hermann, Paris, 1958.

[Goldblatt and Thomason, 1975] R. Goldblatt and S.K. Thomason. Axiomatic classes in propositional modal logic. *Algebra and Logic, ed. by J.N. Crossley, Lecture Notes in Mathematics 450, Springer-Verlag, 1975, 163–173.*

[Goldblatt, 1976] R. Goldblatt. *Metamathematics of modal logic, Part I.* Reports on Mathematical Logic, 6: 41–78, 1976.

[Goldblatt, 1984] R. Goldblatt. *Topoi. The categorial analysis of logic.* Studies in Logic and the Foundations of Mathematics,v.98. North-Holland Publishing Co., 1984.

[Goldblatt, 1987] R. Goldblatt. *Logics of time and computation.* CSLI Lecture Notes, No.7, 1987.

[Goldblatt and Maynes, 2006] R. Goldblatt, E. Maynes. A general semantics for quantified modal logic. *Advances in Modal Logic, vol 6. College Publications, 2006, 227–246.*

[Harel and Tiuryn, 2000] D. Kozen D. Harel and J. Tiuryn. *Dynamic logic.* Foundations of Computing Series. MIT Press, Cambridge, MA, 2000.

[Harel, 1979] D. Harel. First-order dynamic logic. *Lecture Notes in Computer Science, 68. Springer-Verlag, Berlin-New York, 1979.*

[Higgs, 1984] D. Higgs. Injectivity in the topos of complete heyting algebra valued sets. *Canad. J. Math.*, 36:550–568, 1984.

[Hilbert and Bernays, 1934, 1939] D. Hilbert and P. Bernays. *Grundlagen der Mathematik I,II.* Springer-Verlag, 1934, 1939.

[Horn, 1969] A. Horn. Logic with truth values in a linearly ordered Heyting algebra. *Journal of Symbolic Logic*, 34: 395–408, 1969.

[Hosoi, 1967] T. Hosoi. On intermediate logics, I. *J. Fac. Sci. Univ. Tokyo, Sect. 1*, 14:2, 293–312, 1967.

[Hosoi, 1969] T. Hosoi. On intermediate logics, II. *J. Fac. Sci. Univ. Tokyo, Sect. 1*, 16:1, 1–12, 1969.

[Hosoi, 1976] T. Hosoi. Non-separable intermediate logics. *Journal of Tsuda College*, 8: 13–18, 1976.

[Hughes and Cresswell, 1996] G. Hughes and M. Cresswell. *A new introduction to modal logic.* Routledge, 1996.

[Jankov, 1969] V. A. Jankov. Conjunctively indecomposable formulas in propositional calculi. *Izvestiya Akademii Nauk SSSR*, 33:13–38, 1969. English translation: *Mathematics of USSR, Izvestiya*, 3:17–35, 1969.

[Kleene, 1952] S. C. Kleene. *Introduction to metamathematics.* D. Van Nostrand Co., 1952.

[Kleene, 1963] S. C. Kleene. Addendum to "Disjunction and existence under implication". *Journal of Symbolic Logic*, 27:11–18, 1962.

[Kleene, 1967] S. C. Kleene. *Mathematical logic.* John Wiley & Sons, 1967.

[Kolmogorov and Dragalin, 2004] A. N. Kolmogorov and A. G. Dragalin. *Mathematical logic*, 2nd edition, (in Russian). Editorial, Moscow, 2004.

[Komori, 1983] Y. Komori. Some results on the superintuitionistic predicate logics. *Reports on Mathematical Logic*, No. 15 13–31, 1983.

[Kracht and Kutz, 2005] M. Kracht and O. Kutz. Elementary models for modal predicate logics, Part 2: Modal individuals revisited. Intensionality, ed. by R. Kahle, Lecture Notes in Logic 22, Association for Symbolic Logic, 2005, 60–96.

[Kripke, 1963] S. Kripke. Semantical considerations on modal logic. *Acta Philosophica Fennica*, 16:83–94, 1963.

[Krongauz, 1998] M.A. Krongauz. *Prefixes and verbs in Russian language* (in Russian). Moscow, 1998.

[Kuznecov, 1974] A. V. Kuznecov. On superintuitionstic logics. In *Proc. ICM 1974, Vancouver*, 1974.

[Lewis, 1968] D. Lewis. Counterpart theory and quantified modal logic. *Journal of Philosophy*, 65:113–126, 1968.

[Makkai and Reyes, 1977] M. Makkai and G. E. Reyes. First-order categorical logic. model-theoretical methods in the theory of topoi and related categories., 1977.

[Makkai and Reyes, 1995] M. Makkai and G. E. Reyes. Completeness results for intuitionistic and modal logic in a categorical setting. *Ann. Pure Appl. Logic*, 72:25–101, 1995.

[Maksimova and Rybakov, 1974] L. L. Maksimova and V. V. Rybakov. The lattice of normal modal logics (in Russian). *Algebra i Logika*, 13:188–216, 235, 1974.

[Mares and Goldblatt, 2006] E. Mares and R. Goldblatt. An alternative semantics for quantified relevant logic. *The Journal of Symbolic Logic*, vol. 71, no. 1, March 2006, 163-187.

[McKay, 1968] C. G. McKay. The decidability of certain intermediate logics. *Journal of Symbolic Logic*, 33:258–262, 1968.

[McKinsey and Tarski, 1944] J. C. C. McKinsey and A. Tarski. The algebra of topology. *Annals of Mathematics*, 45:141–191, 1944.

[Mendelson, 1997] E. Mendelson. *Introduction to mathematical logic. Fourth edition.* Chapman & Hall, London, 1997.

[Minari, 1983] P. Minari. Completeness theorems for some intermediate predicate calculi. *Studia Logica*, 42: 431–441, 1983.

[Mints, 1967] G. E. Mints. Embedding operations connected with the 'semantics' of S. Kripke (in Russian). In *Zapiski nauch. seminarov LOMI, v.4*, Studies in constructive mathematics and mathematical logic I, pp. 152–159. 1967.

[Nagai, 1973] S. Nagai. On a semantics for non-classical logics. In *Proceedings of the Japan Academy*, 49:337–340, 1973.

[Naumov, 1991] P.G. Naumov. Modal logics that are conservative over intuitionistic predicate calculus. *Moscow Univ. Math. Bull.*, 46:58–61, 1991.

[Novikov, 1977] P. S. Novikov. *Constructive mathematical logic from the point of view of classical logic (in Russian).* Nauka, Moscow, 1977.

[O'Hearn and Pym, 1999] P. W. O'Hearn and D. J. Pym. The logic of bunched implications. *The Bulletin of Symbolic Logic*, Vol. 5, No. 2. (Jun., 1999), pp. 215-244.

[Ono and Nagai, 1974] H. Ono and S. Nagai. Application of general Kripke models to intermediate logics. *Journal of Tsuda College*, 6:9–21, 1974.

[Ono, 1972/73] H. Ono. A study of intermediate predicate logics. *Publications of the Research Institute for Mathematical Sciences*, 8:619–649, 1972/73.

[Ono, 1973] H. Ono. Incompleteness of semantics for intermediate predicate logics. I. Kripke's semantics. *Proc. Jap. Acad.*, 49:711–713, 1973.

[Ono, 1983] H. Ono. Model extension theorem and Craig's interpolation theorem for intermediate predicate logics. *Reports on Mathematical Logic*, 15: 41–58, 1983.

[Ono, 1988] H. Ono. On finite linear intermediate predicate logics. *Studia Logica*, 47: 391–399, 1988.

[Pankratyev, 1989] N.A. Pankratyev. Predicate calculus and arithmetic with Grzegorczyk logic. *Moscow Univ. Math. Bull.*, 44:80–82, 1989.

[Peters and Westerstahl, 2006] S. Peters and D. Westerstahl, *Quantifiers in language and logic,* Oxford University Press, 2006

[Rasiowa and Sikorski, 1963] H. Rasiowa and R. Sikorski. *The mathematics of metamathematics.* Warsaw, 1963.

[Rogers, 1987] H. Rogers. *Theory of recursive functions and effective computability,* MIT Press, 1987.

[Sahlqvist, 1975] H. Sahlqvist. Completeness and correspondence in the first and second order semantics for modal logic. *S. Kanger, ed., Proceedings of the 3d Scandinavian Logic Symposium,* pp. 110–143. North-Holland, 1975.

[Shoenfield, 1967] J. Shoenfield. *Mathematical logic.* Addison-Wesley, 1967.

[Schütte, 1968] K. Schütte. *Vollständige Systeme modaler und intuitionistischer Logik.* Ergebnisse der Mathematik und ihrer Grenzgebiete, 42. Springer-Verlag, Berlin-New York, 1968.

[Scott, 1970] D. Scott. Advice on modal logic. In *Philosophical Problems in Logic. Some Recent Developments,* pages 143–173. Reidel, Dordrecht, 1970.

[Shehtman and Skvortsov, 1990] V. Shehtman and D. Skvortsov. Semantics of non-classical first order predicate logics. In: P. Petkov, editor, *Proceedings of Summer School and Conference on Mathematical Logic, Sept. 13–23 in Chaika, Bulgaria,* pages 105–116. Plenum Press, 1990.

[Shehtman, 1980] V. Shehtman. Topological models of propositional logics (in Russian). *Semiotika i informatika,* 15:74–98, 1980.

[Shehtman, 2005] V. Shehtman. On neighbourhood semantics thirty years later. In: Sergei N. Artemov et al., eds. *We Will Show Them! Essays in Honour of Dov Gabbay, v.2. College Publications 2005,* pp.663-692.

[Shimura and Suzuki, 1993] T. Shimura and N.-Y. Suzuki. Some superintuitionistic logics as the logical fragments of equational theories. *Bulletin of the Section of Logic,* 22:106–112, 1993.

[Shimura, 1993] T. Shimura. Kripke completeness of some intermediate predicate logics with the axiom of constant domain and a variant of canonical formulas. *Studia Logica,* 52:23–40, 1993.

[Shimura, 2001] T. Shimura. Kripke completeness of predicate extensions of cofinal subframe logics. *Bulletin of the Section of Logic,* 30:107–114, 2001.

[Shirasu, 1997] H. Shirasu. *Hyperdoctrinal semantics for non-classical predicate logics.* PhD thesis, Japan Advanced Institute of Science and Technology, 1997.

[Skvortsov and Shehtman, 1993] D. P. Skvortsov and V. B. Shehtman. Maximal Kripke-type semantics for modal and superintuitionistic predicate logics. *Annals of Pure and Applied Logic*, 63:69–101, 1993.

[Skvortsov, 1989] D. Skvortsov. On axiomatizability of some intermediate predicate logics (summary). *Reports on Mathematical Logic*, 22:115–116, 1989.

[Skvortsov, 1991] D. Skvortsov. An incompleteness result for intermediate predicate logics. *Journal of Symbolic Logic*, 56:1145–1146, 1991.

[Skvortsov, 1995] D. Skvortsov. On the predicate logics of finite Kripke frames. *Studia Logica*, 54:79–88, 1995.

[Skvortsov, 1997] D. Skvortsov. Dmitrij P. Skvortsov: Not every 'tabular' predicate logic is finitely axiomatizable. Studia Logica 59(3): 387-396, 1997.

[Skvortsov, 2004] D. Skvortsov. On intermediate predicate logics of some finite Kripke frames, I. Levelwise uniform trees. Studia Logica 77(3): 295-323, 2004.

[Skvortsov, 2005] D. Skvortsov. On the predicate logic of linear Kripke frames and some of its extensions. Studia Logica 81(2): 261-282, 2005.

[Skvortsov, 2006] D. Skvortsov. On non-axiomatisability of superintuitionistic predicate logics of some classes of well-founded and dually well-founded Kripke frames. *Journal of Logic and Computation*, 16:685–695, 2006.

[Suzuki, 1993] N.-Y. Suzuki. Some results on the Kripke sheaf semantics for super-intuitionistic predicate logics. *Studia Logica*, 52:73–94, 1993.

[Suzuki, 1995] N.-Y. Suzuki. Constructing a continuum of predicate extensions of each intermediate propositional logic. *Studia Logica*, 54:173–198, 1995.

[Suzuki, 1999] N.-Y. Suzuki. Algebraic kripke sheaf semantics for non-classical predicate logics. *Studia Logica*, 63:387–416, 1999.

[Takano, 1987] M. Takano. Ordered sets R and Q as bases of Kripke models. *Studia Logica*, 46:137–148, 1987.

[Tanaka and Ono, 1999] Y. Tanaka and H. Ono. The Rasiowa – Sikorski Lemma and Kripke completeness of predicate and infinitary modal logics. In: *The Goldblatt variations*, Uppsala prints and preprints in philosophy, No.1, 108–127, 1999.

[Takeuti and Titani, 1984] G. Takeuti and S. Titani. Intuitionistic fuzzy logic and intuitionistic fuzzy set theory. *Journal of Symbolic Logic*, 49: 851–866, 1984.

[Thomason, 1972] S. Thomason. Semantic analysis of tense logic. *Journal of Symbolic Logic*, 37:150–158, 1972.

[Trachtenbrot, 1950] B. A. Trachtenbrot. Impossibility of an algorithm for the decision problem in finite classes (in Russian). *Doklady Akademii Nauk*, 70:569–572, 1950. English translation: *AMS Translations*, 23:1–5, 1963.

[Troelstra and van Dalen, 1988] A. S. Troelstra and D. van Dalen. *Constructivism in mathematics. Vol. I, II.* Studies in Logic and the Foundations of Mathematics, 121, 123. North-Holland Publishing Co., Amsterdam, 1988.

[Umezawa, 1956] T. Umezawa. On logics intermediate between intuitionistic and classical predicate logic. *Journal of Symbolic Logic*, 24:141–153, 1956.

[Umezawa, 1959] T. Umezawa. On logics intermediate between intuitionistic and classical predicate logic. *Journal of Symbolic Logic*, 24: 141–153, 1959.

[van Benthem, 1983] J. van Benthem. *Modal logic and classical logic.* Bibliopolis, 1983.

[van Benthem, 1991] J. van Benthem. *Language in action: categories, lambdas, and dynamic logic* MIT Press, Foundations of Mathematics, Volume 130, North-Holland, Amsterdam.

[van Dalen, 1973] D. van Dalen. Lectures on intuitionism. In *Cambridge Summer School in Mathematical Logic (Cambridge, 1971)*, Lecture Notes in Math., v. 337, pages 1–94. Springer, 1973.

[van Lambalgen, 1996] M. van Lambalgen. Natural deduction for generalized quantifiers. In Jaap van der Does and Jan van Eijk, editors, *Quantifiers, Logic, and Language.* CSLI (Center for the Study of Language and Information) Lecture Notes, No. 54. Stanford, California: CSLI Publications (distributed by Cambridge University Press). 1996.

[Viganò, 2000] Luca Viganò. *Labelled non-classical logics.* Kluwer Academic Publishers, 2000.

[Wojtylak, 1983] P. Wojtylak. Collapse of a class of infinite disjunctions in intuitionistic propositional logic. *Reports on Mathematical Logic*, 16: 37–49, 1983.

[Wolter, 1995] F. Wolter. Lattices of subframe logics: a survey. In *Proceedings of Non-classical Logics and their Kripke Semantics*, N. Y. Suzuki, ed., pp .187–208. RIMS Kokyuoku 927, Kyoto, 1985.

[Wolter, 1997] F. Wolter. *The structure of lattices of subframe logics*, Annals of Pure and Applied Logic 86, 1997, 47 - 100.

[Yokota, 1989] S. Yokota. Axiomatization of the first-order intermediate logics of bounded Kripkean heights, I. *Zeitschrift für Mathematische Logik und Grundlagen der Mathematik*, 35:415–421, 1989.

[Yokota, 1991] S. Yokota. Axiomatization of the first-order intermediate logics of bounded Kripkean heights, II. *Zeitschrift für Mathematische Logik und Grundlagen der Mathematik*, 37: 17–26, 1991.

[Zakharyaschev, 1989] M. V. Zakharyashchev. Syntax and semantics of superintuitionistic logics. *Algebra i Logika*, 28:402–429, 1989. English translation: *Algebra and Logic*, 28:262–282, 1989.

Index

FSC
www.fsc.org
MIX
Papier | Fördert gute Waldnutzung
FSC® C083411

Kolibri 360 GmbH
Ferdinand-Jühlke-Straße 7
99095 Erfurt, Deutschland
produktsicherheit@kolibri360.de